# AIR POLLUTION

## THIRD EDITION

## VOLUME IV

## Engineering Control of Air Pollution

ENVIRONMENTAL SCIENCES

An Interdisciplinary Monograph Series

Editors: Douglas H. K. Lee, E. Wendell Hewson, and Daniel Okun

A complete list of titles in this series appears at the end of this volume.

# AIR POLLUTION

## THIRD EDITION

## VOLUME IV

## Engineering Control of Air Pollution


Edited by

**Arthur C. Stern**
Department of Environmental Sciences and Engineering
School of Public Health
University of North Carolina at Chapel Hill
Chapel Hill, North Carolina

ACADEMIC PRESS New York San Francisco London 1977

A Subsidiary of Harcourt Brace Jovanovich, Publishers

ACADEMIC PRESS, INC.
111 Fifth Avenue, New York, New York 10003

*United Kingdom Edition published by*
ACADEMIC PRESS, INC. (LONDON) LTD.
24/28 Oval Road, London NW1

**Library of Congress Cataloging in Publication Data**

Stern, Arthur Cecil.
Engineering control of air polution.

(His Air pollution ; v. 4) (Environmental sciences)
Includes bibliographical references and index.
1. Air–Pollution. I. Title. II. Series: Environmental sciences.
TD883.S83 1976 vol. 4 628.5′3 76-8288
ISBN 0–12–666604–0

PRINTED IN THE UNITED STATES OF AMERICA

To Robert C. and Patricia, Mary and Ersi

# Contents

## 3. Source Control by Centrifugal Force and Gravity

Knowlton J. Caplan



## 4. Filtration

Koichi Iinoya and Clyde Orr, Jr.



## 5. Electrostatic Precipitation

Sabert Oglesby, Jr., and Grady B. Nichols



## 6. Scrubbing

Seymour Calvert

## 7. Mist Elimination

Werner Strauss



## 8. Adsorption

Amos Turk



## 9. Combustion

Chad F. Gottschlich



# PART C PROCESS EMISSIONS AND THEIR CONTROL

## 10. Fuels and Their Utilization

Richard B. Engdahl and Richard E. Barrett

## 14. The Control of Motor Vehicle Emissions

Donel R. Olson



## 15. Agriculture and Agricultural-Products Processing

W. L. Faith



## 16. The Forest Products Industry

E. R. Hendrickson



## 17. Mineral Product Industries

Victor H. Sussman

## 18. Chemical Industries

Stanley T. Cuffe, Robert T. Walsh, and Leslie B. Evans



## 19. Petroleum Refining

Harold F. Elkin



## 20. Nonferrous Metallurgical Operations

Kenneth W. Nelson, Michael O. Varner, and Thomas J. Smith

## 21. Ferrous Metallurgical Operations

Bruce A. Steiner

## List of Contributors

Numbers in parentheses indicate the pages on which the authors' contributions begin.

Richard E. Barrett (379, 425), Combustion Systems Section, Energy and Environmental Processes Department, Battelle-Columbus Laboratories, Columbus, Ohio

John O. Burckle (43), 3909 Middleton Court, Cincinnati, Ohio

Seymour Calvert (257), Air Pollution Technology, Inc., San Diego, California

Knowlton J. Caplan (97), School of Public Health, University of Minnesota, Minneapolis, Minnesota

Richard C. Corey (531), Liquefaction Directorate, Fossil Energy, U.S. Energy Research and Development Administration, Washington, D.C.

Stanley T. Cuffe (735), Industrial Studies Branch, Emission Standards and Engineering Division, Office of Air Quality Planning and Standards, U.S. Environmental Protection Agency, Research Triangle Park, North Carolina

Harold F. Elkin (813), Environmental Affairs, Sun Oil Company, Philadelphia, Pennsylvania

Richard B. Engdahl (379, 425), Combustion Systems Section, Energy and Environmental Processes Department, Battelle-Columbus Laboratories, Columbus, Ohio

Leslie B. Evans (735), Industrial Studies Branch, Applied Technology Division, Office of Air Programs, U.S. Environmental Protection Agency, Research Triangle Park, North Carolina

W. L. Faith (655), 2540 Huntington Drive, San Marino, California

Melvin W. First (3), Department of Environmental Health Sciences, School of Public Health, Kresge Center for Environmental Health, Harvard University, Boston, Massachusetts

F. E. Gartrell* (465), Environmental Research and Development Division, Tennessee Valley Authority, Chattanooga, Tennessee

Chad F. Gottschlich (365), Engineering Science Section, Research and Development, Selas Corporation of America, Dresher, Pennsylvania

E. R. Hendrickson (685), Environmental Science and Engineering, Inc., Gainesville, Florida

Koichi Iinoya (149), Department of Chemical Engineering, Kyoto University, Kyoto, Japan

David W. Locklin (425), Combustion Systems Section, Energy and Environmental Processes Department, Battelle-Columbus Laboratories, Columbus, Ohio

Kenneth W. Nelson (845), Environmental Affairs, ASARCO, Inc., New York, New York

Grady B. Nichols (189), Environmental Engineering Division, Southern Research Institute, Birmingham, Alabama

Sabert Oglesby, Jr. (189), Department of Engineering and Applied Sciences, Southern Research Institute, Birmingham, Alabama

Donel R. Olson (595), Olson Engineering, Inc., Huntington Beach, California

Clyde Orr, Jr. (149), School of Chemical Engineering, Georgia Institute of Technology, Atlanta, Georgia

Thomas J. Smith (845), Department of Family and Community Medicine, University of Utah Medical Center, Salt Lake City, Utah

Paul W. Spaite (43), 6315 Grand Vista Avenue, Cincinnati, Ohio

Bruce A. Steiner (889), Armco Environmental Engineering, Armco Steel Corporation, Middletown, Ohio

Werner Strauss (293), Department of Industrial Science, University of Melbourne, Parkville, Victoria, Australia

Victor H. Sussman (705), Office of Stationary Environmental Control, Ford Motor Company, Dearborn, Michigan

Amos Turk (329), Department of Chemistry, The City College of the City University of New York, New York, New York

Michael O. Varner (845), Department of Environmental Sciences, ASARCO, Inc., Salt Lake City, Utah

Robert T. Walsh (735), Performance Standards Branch, Standards Development and Implementation Division, Office of Air Programs, U.S. Environmental Protection Agency, Research Triangle Park, North Carolina

* Present address: 3114 Colyar Drive, Chattanooga, Tennessee.

# Preface

This third edition is addressed to the same audience as the previous ones: engineers, chemists, physicists, physicians, meteorologists, lawyers, economists, sociologists, agronomists, and toxicologists. It is concerned, as were the first two editions, with the cause, effect, transport, measurement, and control of air pollution.

So much new material has become available since the completion of the three-volume second edition that it has been necessary to use five volumes for this one. Volumes I through V were prepared simultaneously, and the total work was divided into five volumes to make it easier for the reader to use. Individual volumes can be used independently of the other volumes as a text or reference on the aspects of the subject covered therein.

Volume I covers two major areas: the nature of air pollution and the mechanism of its dispersal by meteorological factors and from stacks. Volume II covers the effect of air pollution on plants, animals, humans, materials, and the atmosphere. Volume III covers the sampling, analysis, measurement, and monitoring of air pollution. Volume IV covers two major areas: the emissions to the atmosphere from the principal air pollution sources and the control techniques and equipment used to minimize these emissions. Volume V covers the applicable laws, regulations, and standards; the administrative and organizational strategies and procedures used to administer them; and the energy and economic ramifications of air pollution control. The concluding chapter of Volume II discusses air pollution literature sources and gives guidance in locating information not to be found in these volumes.

To improve subject area coverage, the number of chapters was increased from 54 of the second edition (and 42 of the first edition) to 71. The scope of some of the chapters, whose subject areas were carried over from the second edition, has been changed. Every contributor to the second edition was offered the opportunity to prepare for this edition either a revision of his chapter in the second edition or a new chapter if the scope of his work had changed. Since 8 authors declined this offer and one was deceased, this edition includes 53 of the contributors to the second edition and 46 new ones.

The new chapters in this edition are concerned chiefly with aspects of air quality management such as data handling, emission inventory, mathematical modeling, and control strategy analysis; global pollution and its monitoring; and more detailed attention to pollution from automobiles and incinerators. The second edition chapter on Air Pollution Standards has been split into separate chapters on Air Quality Standards, Emission Standards for Stationary Sources, and Emission Standards for Mobile Sources. Even with the inclusion in this edition of the air pollution problems of additional industrial processes, many are still not covered in detail. It is hoped that the general principles discussed in Volume IV will help the reader faced with problems in industries not specifically covered.

Because I planned and edited these volumes, the gap areas and instances of repetition are my responsibility and not the authors'. As in the two previous editions, the contributors were asked to write for a scientifically advanced reader, and all were given the opportunity of last minute updating of their material.

As editor of this multiauthor treatise, I thank each author for both his contribution and his patience, and each author's family, including my own, for their forbearance and help. Special thanks are due my secretary, Patsy Garris, and her predecessors, who carried ninety-nine times the burden of the other authors' secretaries combined, and Eleanor G. Rollins for preparing the Subject Index for this volume. I should also like to thank the University of North Carolina for permitting my participation.

Arthur C. Stern

## Contents of Other Volumes

# Part A

# CONTROL CONCEPTS

# 1

## Control of Systems, Processes, and Operations

Melvin W. First

## I. Introduction

Hatch has wisely pointed out (*1*) that "Prevention of community air pollution from industrial operations starts within the factory or mill." Resort to air and gas cleaning devices to eliminate or reduce emissions of air polluting substances and to tall discharge stacks to disperse and dilute offensive substances to acceptable ground-level concentrations becomes unnecessary when process, operational, and system control is effective in preventing the formation and release of air pollutants.

Even when gas cleaning and atmospheric dispersion must be used as final steps, process, operational, and system control is a means of minimizing the quantities of substances entering cleanup systems and, ultimately, being discharged to the atmosphere. Process, operational, and system control will concentrate contaminants in the smallest possible volume of air. This is important as the cost of control equipment is based principally on the volume of gas that must be handled and not on the amount or concentration of the substances that must be removed. Also, most air and gas removal equipment is more efficient when handling higher concentrations of contaminants, all else being equal. Control of emissions with highly efficient industrial air and gas cleaning devices often costs in excess of \$2.5/m$^3$/second of installed capacity and entails high and continuing operational and maintenance expenses.

Emission of untreated gases to the atmosphere can also be costly. The 1974 construction cost of a 330-m reinforced concrete chimney, complete with free standing steel liner, was estimated at \$15,000/m of height, or \$5 million per chimney (*2*). Shorter stacks were less expensive; a 120-m stack costing approximately \$1 million. Much innovative process, operational, and system control can be engineered into production activities for sums of this magnitude. Although seldom the subject of technical publications, the chief air pollution control officer of Pittsburgh, Pennsylvania, commenting on the improvements that had occurred in the atmosphere of that city, stated that "A major portion of our success has been attained by process changes, as opposed to installing control devices" (*3*).

Reduction of air polluting emissions by process, operational, and system control is not only an important adjunct to air and gas cleaning technology and to atmospheric dispersion but is a definitive response to the concept of zero emissions when it can be employed for total control.

Many innovative methods of exercising operational control of processes that emit undesirable substances to the atmosphere have been proposed. They include modulating the scale of operations or the concentration of emitted pollutants to match the variable dispersive powers of the atmosphere and, thereby, to maintain ground-level concentrations within prescribed air quality standards. When modulating operational scale, the concentration of pollutants in the off-gases remains unchanged but the total volume of emissions is variable; whereas, when modulating concentration, the volume of emissions remains constant but the concentration of pollutants is altered by such methods as changing the sulfur content of fossil fuels utilized for generating electricity.

The United States Environmental Protection Agency (USEPA) (*4*) permits individual states to establish zones in which air polluting emissions may be increased over existing levels so long as applicable air quality standards are not exceeded. This makes it possible to utilize presently untapped air resources to promote the commercial and industrial developments that are considered essential for a region's economic well-being without risking detrimental human health effects. Decentralization of the decision-making process to the local level will permit balancing the effects of increased air contamination levels on agriculture, tourism, industrial production, etc., so as to optimize community well-being.

Widespread adoption of restrictive emission limitations and rigorous air quality standards makes it essential that all methods of air pollution control be utilized to the maximum possible degree in areas where the air environment is threatened with excessive pollution.

## II. Elimination of Air Pollution Emissions

### *A. Substituting Products*

Offensive substances often may be eliminated entirely from processes that cause air pollution by substituting materials that perform equally well in the process but discharge innocuous products to the atmosphere, or none at all. When this method of air pollution control can be applied, it usually produces very satisfactory results at trivial cost. The numerous new chemical products that enter the commercial market annually make it worthwhile to maintain a continuing search for substitute materials of low pollution potential, even when past searches have failed.

An example from industry of a product change controlling an air pollution problem is substitution of a cold-setting synthetic resin for rubber in the manufacture of paint brushes. Before this change, it was necessary to vulcanize the rubber bond at the base of the bristles for a period of many hours, causing severe odor nuisances in the vicinity of the factory from the emission of sulfur-containing volatile products. The cold-setting resins selected as rubber substitutes produce no odors and completely eliminate air polluting emissions from this operation. Another example involves the application of finishing agents to knitted and felted textiles to give them a number of desirable surface characteristics. Acrylic latex, commonly used for this purpose, contains a few hundred parts per million (ppm) of unreacted ethyl acrylate monomer that is volatilized into the curing oven off-gases and often creates a severe odor nuisance in the vicinity of the textile plant. The cost of stripping ethyl acrylate monomer from hot curing oven gases would be excessively costly. However, an investigation of the latex manufacturing process showed that ethyl acrylate monomer, which has a distinctive unpleasant smell and an odor threshold of 0.002 ppm, can be vacuum stripped to a few parts per million in the latex during manufacture and the residual monomer totally destroyed by the addition of a suitable redox catalyst prior to application, thereby totally eliminating the odor nuisance from this operation at a trivial increase in the cost of the latex product (*5*).

In the field of transportation there has been considerable interest in electric battery power as a substitute for the internal combustion automobile engine for commuting and intraurban family use. A battery-powered automobile would be virtually pollution free, in contrast to emissions from conventional automobiles of carbon monoxide, nitrogen and sulfur oxides, and a long list of organic compounds. However, the great weight, limited driving range, and high cost of lead-acid battery-powered vehicles have restricted their use to light urban delivery service (e.g., home milk routes) up to the present.

Concentrations of lead in urban air originate principally from exhaust emissions of automobiles using leaded fuel. It has proved possible to substitute unleaded gasoline (i.e., less than 0.0225 gm of lead per liter) for use in 1975 model United States automobiles by reducing engine compression ratios and increasing the octane rating of unleaded gasolines with the addition of larger amounts of ring compounds (e.g., benzol). Future cars will be engineered to burn unleaded fuels and the manufacture of leaded gasoline may be discontinued some time in the future (USEPA has proposed 1979). When this occurs, lead levels in urban air will decrease in proportion. Hydrogen has been proposed (*6*) as a noncarbonaceous fuel for motor vehicles that burns without forming carbon monoxide, hydrocarbons, or, when pure oxygen is used as the oxidizing agent, nitrogen oxides, thereby totally freeing the motor vehicle of air polluting engine exhaust emissions.

In power generation, little technological effort is required to substitute low-sulfur for high-sulfur fuels but these desirable fuels have limited availability. Complete freedom from sulfur oxides and other gaseous emissions associated with the burning of fossil fuels for heat and power may be obtained by changing to water power and nuclear fuels, but there appears to be little prospect that these sources of electricity will be able to satisfy the entire demand for the foreseeable future.

There are prospects for nonpolluting electrical energy from solar, geothermal, and tidal sources. At this time, 88,000 kg-Cal are required per kilowatt-hour from a dry steam field whereas only 33,000 kcal are needed for 1 kW-hour from a modern fossil fuel-burning plant. Geothermal hot water systems contain even less energy for conversion. Solar power is still three times as expensive as nuclear, and tidal power can be harnessed in only a few places. Until one or more of these new electricity sources becomes practical, major reliance must be placed on water and nuclear power as the principal new sources for electricity that do not pollute the air.

### *B. Changing Processes*

Process changes can be as effective as product substitutions in eliminating air pollution emissions. The chemical and petroleum refining industries have undergone radical changes in processing methods which emphasize continuous automatic operations, often computer controlled, and completely enclosed systems that minimize release of materials to the atmosphere. It has been found possible, and often profitable, to control loss of volatile materials by condensation and reuse of vapors (e.g., condensation units on volatile petroleum product storage tanks) and by recycling noncondensible gases for additional reactions (e.g., polymerization and alkylation of gaseous hydrocarbons to produce gasoline).

Examples of process changes from power production are (a) the development of ultrahigh voltage systems for long-distance transmission of electricity which has made it economically feasible to substitute distant water-powered turbines for urban fuel-burning steam stations; and (b) a substantial reduction in oxidation of $SO_2$ to $SO_3$ in coal- and oil-burning boilers which can be obtained by reducing excess air from 15 to 20% to less than 1% when burning fossil fuels. This helps to eliminate sulfuric acid formation but the absence of excess air tends to result in greater soot production. Firing with low excess air is also effective for reducing the formation of nitrogen oxides in the flame when burning gas and oil. Further reductions in nitrogen oxides can be obtained by supplying substoichiometric quantities of primary air to burners and accomplishing complete burnout of fuel "by injecting secondary air at lower temperatures, where NO formation is limited by kinetics" (7).

Examples of process changes from agriculture include: (a) the use of liquid and gaseous fertilizer chemicals, e.g., anhydrous ammonia, applied by injection into the earth instead of being spread across the surface as finely divided powders subject to wind entrainment, which reduces fugitive dust pollution; and (b) use of natural biological enemies and synthetic sex lures to disrupt mating for control of insect pests instead of widespread spraying of persistent pesticides such as DDT.

Many innovative methods for handling municipal waste disposal begin with mechanical methods for separating recyclable materials for reprocessing and reuse, thereby reducing the quantities of materials that must be handled by conventional waste disposal methods that are often severe local air pollution sources (*8*).

## III. Minimizing Emissions of Gaseous and Gas-Borne Wastes

### *A. Ground-Level Wide-Area Pollution Sources*

Although primary emphasis is usually placed on emissions from stacks, polluting substances may enter the air in other ways, usually at or near ground level and often over extended areas. Both situations tend to restrict natural atmospheric dilution and create severe local pollution problems. For example, sulfite process pulp mills often discharge odorous liquid wastes into streams that slowly release offensive substances to the air by degassing or volatilization for many miles downstream. Areas subjected to air pollution by this means may be larger than from a tall stack, with the added disadvantage of not being relieved, even intermittently, by changes in wind direction. Treatment of this waste liquor prior to discharge to a river, by removal of volatile and putrescible fractions, will reduce its air pollution potential to a level that can be met by the natural cleansing capabilities of the stream.

Similar considerations apply to air pollution from open dump burning; now outlawed nationwide in the United States. This can be an extensive source that releases intensely offensive airborne substances at ground level that sweep across large land areas without substantial dilution. Substitution of sanitary landfilling for open dump burning has largely eliminated air pollution from this operation. When sanitary landfill sites are unavailable, modern high-temperature central-station incinerators equipped with dust-removal equipment minimize air polluting emissions from the disposal of solid wastes.

Other wide-area sources of fugitive dusts that originate at ground level are land clearance and site preparation for road building and erection of structures. Although these sources are almost invariably of limited duration, they often produce intense local air pollution levels for a year or longer. Measures for minimizing emissions from site preparation are wetting the ground with water or solutions containing wetting agents, applications of asphalt and oil, covering vehicles carrying dusty cargos, and prompt removal of spillage (*9*). Open storage piles of chemicals, such as sulfur and bauxite, and minerals, such as sand and clay, are often local pollution sources during dry, windy weather. The application of persistent surface coatings such as plastics, combined with supplementary spraying during dry weather, is effective for suppression of dust from inactive storage piles of coal and similar granular materials. For active piles of crushed stone, sand, etc., the continuous application of water, with or without wetting agents, is an important control measure, especially for elevated conveyor belts handling pulverized products. Enclosed storage buildings, widely used for the curing of phosphate fertilizers, eliminate loss of particulate matter by wind erosion.

Commercial lumbering operations produce large volumes of slash (bark, branches, etc., deposited on the forest floor from logging) that dry and, after a time, represent a serious forest fire threat. Considerable controversy exists at the present time as to whether it is preferable to burn the slash immediately to reduce the later forest fire danger or to allow the slash to remain and avoid the immediate air pollution effects of widespread ground burning. Immediate burning under controlled conditions seems to have won out for the moment as a lesser threat to air quality and forest resources. Control measures for fugitive dust from all types of open industrial operations, storage piles, and quarrying have only recently been accorded serious study.

### B. Good Building Construction, Plant Layout, and Housekeeping

Industrial operations may be controlled to minimize the production and release of air pollutants with the aid of careful building construction, plant layout, and housekeeping. These precautions are of especial importance for all processes that handle food or inedible putrescible material. The traditional "offensive trades" (slaughtering, rendering, leather tanning, and pig farming) earned this designation because they frequently cause severe odor nuisances. The application of modern sanitary science can prevent this from occurring in most of these industries, although the virtual impossibility of deodorizing garbage-fed pig colonies has resulted in their complete disappearance from the outskirts of most cities.

Food processing plants are usually under strict sanitary surveillance by health and agricultural authorities but processing of animal products for other than human consumption is not similarly supervised. When these operations are conducted under primitive sanitary conditions, they result in the release to the air of offensive substances. The production from trash fish and gurry (fish remains after filleting) of fish meal (for animal feed supplements) and fish oils (for soap, paint, and varnish) has been associated with air pollution wherever a plant is located near a population center. Using ice to halt the deterioration of fish during storage is not an industry practice so that fish are often in an advanced state of decomposition when oil and meal processing begins.

The principal processing steps (see also Chapter 15, this volume) are grinding the raw fish, cooking to separate oil, drying the nonoil nonsoluble residues, and then grinding them to obtain a storable oil-free whole-fish product. All these processes, including raw fish storage, give off airborne products that cause air pollution. These emissions may be minimized by proper plant design and maintenance: First, all floors, walls, and fixtures in the plant should be smooth, hard, and impervious to water so that they may be hosed down to prevent fish scraps from lodging in cracks and crevices to putrefy. Second, scrupulous cleanliness is required to prevent accumulation of wastes. Cleaning water should contain a few parts per million of residual chlorine to discourage microbiological activity and all surfaces should be pitched to drain to sumps for treatment of the cleaning water. Third, seawater condensers—most of these plants are located on the shore for easy transfer of fish from boats to plant—should be used in the steam vents of the fish cookers to minimize the escape of condensible gases and vapors from the kettles. Fourth, steam-heated shell driers should be used in preference to direct-fired units to minimize local overheating and scorching of the meal and the discharge of malodorous burnt protein decomposition products with the moist flue gases. Attempts have been made to reduce fuel consumption in direct-fired fish meal driers by recycling a fraction of the discharged gases through the heating flame, but this results in burning and charring of the airborne dust and produces extremely unpleasant odor problems.

All of these recommended practices minimize the amounts of gases, vapors, and particles generated by, and emitted from, the processing vessels and make it possible to treat the remaining air pollutants by direct flame incineration or hypochlorite scrubbing in an economical and technically satisfactory manner. Proper plant construction, careful equipment maintenance to prevent breakdown of essential control equipment, and sanitary operation give good air pollution control results when combined with residual gas cleaning.

Rendering plants that produce fats and meal by similar processes offer the same opportunities for minimizing air polluting emissions by correct plant design, selection of suitable processing equipment, and sanitary operation. Tanning of hides, production of glue and gelatin, and many other necessary manufacturing operations are dependent on similar techniques to reduce pollution levels.

### C. Process and Operational Modifications and Material Substitutions

#### 1. Industry

Substantial benefits and economies may be realized even when process modifications and material substitutions merely reduce air pollution emissions, rather than completely eliminating them. For example, a change in the steel industry from raw ore to briquetted or pelletted sintered ore has greatly reduced dust production during ore handling and helped reduce blast furnace "slips" which result in emission to the atmosphere of enormous amounts of uncleaned blast furnace gas when the safety dampers open and allow the gases to bypass the dust collectors.

Abnormally large emissions to the atmosphere result from operating certain kinds of production equipment at excessive rates. For example, the output of the rotary sand and stone drier controls the production rate of hot mix asphalt plants (see also Chapter 17, this volume). When hot gas velocity through the drier is increased above the design rate to improve drying capacity, the quantity of dust carried out increases in greater proportion than the increase in gas volume. This relationship, unfavorable from the air pollution standpoint, occurs because in the range of viscous particle flow, i.e., particles less than 100 $\mu$m diameter, the size of particles entrained is related to the square of the diameter whereas the weight of dust entrained is related to the cube of particle diameter. Because this industry uses scrubbers having a collection efficiency that is relatively insensitive to changes in dust loading, the weight discharged to the atmosphere increases in proportion to the dust load carried out of the rotary drier.

Many air pollution problems caused by the hot mix asphalt paving industry in cold climates stem from the traditional practice of postponing maintenance procedures until winter shutdown. This means that the machinery progressively declines in effectiveness from startup time in early spring until seasonal shutdown in late autumn. By the time the dry, windy autumn months arrive, considerable unrepaired damage has occurred to dust collectors, exhaust systems, and equipment enclosures which results in excessive and unnecessary air pollution.

Additional examples of industrial process and component changes to reduce air pollution have been noted by Rose *et al.* (*10*). These include substitution of bauxite flux for fluorine-containing fluorspar in open hearth practice and the use of borate salts as a substitute for elemental sulfur, used as an antioxidant and flux when casting molten magnesium metal. Both changes result in decreased emission of air pollutants.

## 2. Transportation

The amount of carbon monoxide and hydrocarbons produced by well-maintained well-tuned engines is only a fraction of the quantity emitted from those that have not been given proper maintenance. In only a very few communities are automobiles that trail dense clouds of blue smoke ordered off the highways until suitable maintenance procedures have been taken to eliminate excessive emissions.

Injection of air into the hot exhaust manifold of a gasoline engine to continue the combustion of carbon monoxide and hydrocarbons that was halted in the cylinders by the depletion of oxygen represents an equipment modification designed to minimize the emission of polluting substances by the automobile (see also Chapter 14, this volume). Modifications to carburetion and ignition can produce similar results.

Changes in automotive use patterns by the entire urban community could play a decisive role in controlling air pollution. These include: (a) enlarging and improving urban mass transportation systems to reduce the need for bringing private vehicles into the city, (b) speeding traffic flow to decrease average trip time and to take advantage of decreased per kilometer contaminant emission rates at higher speeds, and (c) encouraging, by taxation policy, the use of small, low-horsepower vehicles that emit less contaminants per kilometer of travel. These kinds of solutions for vehicle pollution obviously require public support and long-range urban planning. Although difficult to achieve, pollution control plans of this nature hold promise of substantial relief because they produce fundamental alterations of contaminant generation patterns.

## 3. Agriculture

Air pollution arising from agricultural operations may be minimized by a number of techniques (see also Chapter 4, Vol. I, and Chapter 15, this volume). Spraying for insect and weed control (by terrestrial and aerial methods) should be conducted during windless weather to confine the spread of insecticides and weed control chemicals to the intended areas. The damage that may be caused by imprudent spraying is demonstrated by the fact that 1 gm of 2,4-dichlorophenoxyacetic acid (2,4-D) can mark all cotton on 100,000 $m^2$ and 15 gm can injure this amount of cotton permanently. In one aerial spraying incident, 40 $km^2$ of rice were damaged by 2,4-D, 25–32 km downwind of the spraying site.

Dust storms originating in the midwestern prairies of the United States during the 1930's were a severe local problem and ultimately obscured the sun thousands of kilometers distant. These dust storms were the result of poor farming techniques. Present-day agricultural practices in these same regions are designed to prevent land erosion and have been effective in reducing airborne dust from this source to an acceptable level.

In recent years, the pollen content of urban atmospheres has been reduced by greater weed control on nearby farms and by routine weed eradication activities of highway and public works departments. Although pollen travels long distances by wind action, often the greatest annoyance is produced by pollen emitted locally.

## 4. Fuel Burning for Space Heating, Steam, and Power

Particulate emissions from burning coal, especially from pulverized coal, have been significantly reduced as a result of coal washing that reduces the ash content of run-of-mine coal. In a modern pulverized coal-burning furnace, approximately 80% of the coal ash finds its way into the flue gas. Reductions in ash content by washing are reflected in corresponding decreases in the amount of fly ash produced. Electrostatic precipitators are now used routinely in modern coal-burning stations to collect fly ash. As they are purchased for large, new stations with a guarantee that they will collect 99.9% of the fly ash in the flue gases, emissions to the atmosphere are related directly to the ash content of the coal, and reductions in incombustible residue by coal washing result in lower particulate emissions.

Coal washing also results in a reduction in the sulfur content of coal and an additional reduction occurs in the pulverizer. For some coals, sulfur reduction by washing and pulverizing may reach as high as 40%; whereas for others, the decrease in sulfur may be insignificant. The physical state and chemical nature of the sulfur compounds that are present determine the result. When sulfur is present principally in the form of organic compounds that are intimately mixed with coal, little or no sulfur reduction may be obtained by coal cleaning techniques. Satisfactory sulfur removal processes that are applicable to most of the available coal deposits are still being sought but it has been found practical to remove sulfur from oil at the refinery (see also Chapter 19, this volume). Prior to fall 1973, when crude oil was selling for less than $3 per barrel (42 U.S. gallons), the added cost of oil desulfurization (approximately $1 per barrel) was considered excessive. When crude prices increased to more than $10 per barrel, the added cost of desulfurization became a much smaller percentage of the overall cost but it still has a substantial effect on the overall cost of oil for power generation. Methods of stretching limited supplies of low-sulfur fuel are (a) blending high- and low-sulfur grades, (b) storing low-sulfur fuel and issuing it for use as a substitute for a high-sulfur fuel when unfavorable weather conditions reduce the natural processes of atmospheric dilution and dispersion, and (c) requiring high efficiency flue gas $SO_2$ cleaning systems for large fuel users so they may burn high sulfur fuels safely, and then allocating the limited supply of low-sulfur fuel to the large numbers of small users who discharge flue gases close to ground level and cannot operate flue gas $SO_2$ cleaning systems economically.

It is well to recall that similar problems were faced and resolved when limitations on the volatile content of coals used for commercial and industrial purposes were instituted in the United States many years ago to assist in reducing smoke and soot emissions. Following the 1952 London air pollution disaster, Britain adopted similar prohibitions for domestic fuels in areas designated as "smokeless zones" and gradually extended the zones to encompass all of Britain.

Technological improvements in heat transfer equipment, turbine generators, and furnaces have resulted in a reduction in the amount of coal needed to generate a kilowatt of electricity from 0.64 kg (6660 kcal/kg) in 1927 to only slightly more than 0.32 kg today. This extraordinary reduction in fuel requirements results in an equal reduction in the emission of sulfur gases and other contaminants to the atmosphere per unit of electricity produced.

The technology to reduce pollution effects from emissions associated with space heating includes (a) increasing the efficiency of space heating by better insulation of buildings so that less fuel is required to produce a given indoor temperature, (b) providing heat to groups of buildings from a single source in the form of steam, hot water, or electricity, (c) gasifying solid fuels and using the resulting gas for space heating, and (d) reducing emissions from space heaters by improved combustion efficiency, tighter fuel specifications with respect to ash, sulfur, viscosity, etc., and limitation of allowable fuels. Current objectives directed toward fuel conservation have given powerful reinforcement to the rapid employment of all of these air pollution reduction techniques applicable to fuel usage.

## IV. Concentration of Air Pollutants at the Source for Effective Treatment prior to Release to the Atmosphere

### A. Sources of Contaminants

Off-gases result from diverse manufacturing steps. Many are generated in totally enclosed processes. Frequently, these off-gases contain volatile or entrained waste products that the process is designed to remove as a manufacturing step. Processes for drying a wide variety of products, from sand to fish meal, place the moist product in intimate contact with a stream of warm, dry air so that excess water may be extracted and carried off in the effluent gases. Were water the only product carried out of a drier, no air pollution problems would result. However, other less innocuous volatile substances and entrained particles are also carried out of the processing vessel. The minimum volume of drying air and maximum drying temperature are determined by the amount of moisture to be removed and the heat-resistant properties of the materials being dried. It is therefore generally impossible to reduce the volume of gases needing cleaning or atmospheric dilution. Many process tail gases contain valuable products in concentrations too low to make recovery economically worthwhile. Nonetheless, these lean tail gases often represent concentrated air pollution sources. In the manufacture of nitric acid the tail gases contain approximately 0.2–0.3% $NO_x$ when operating under ideal conditions. The price of nitric acid does not justify additional $NO_x$ removal so that these tail gases are often discharged to the atmosphere through tall stacks. To eliminate this source of air pollution, special gas cleaning devices must be installed to reduce the nitrogen oxides to elemental nitrogen and oxygen before release of the waste gases. Similar considerations apply to the recovery of useful products from tail gases from sulfuric acid manufacturing and many other processes.

### *B. General versus Local Exhaust Ventilation*

Pollution also originates from the discharge of local exhaust ventilation systems installed to protect workers and prevent in-plant damage. Included are systems that ventilate open surface tanks, crushing and grinding operations, high temperature melting of metals and minerals, curing of elastomers, and use of products containing volatile solvents. As pollution controls may add as much as 40% to the capital cost of electric arc furnace equipment, it is not surprising that efforts are made to minimize the size and cost of collection systems. This is accomplished by enclosing the furnace in a closely fitting exhaust hood to prevent the escape of fumes and the use of water-cooled hood walls and ducts to prevent damage to the apparatus from the high temperatures that are generated. However, some electric furnace operators have elected, for easy access to their furnaces, to leave them unhooded, and permit dust, fumes, and heat to rise to the roof of the factory building before collection and cleaning. Crane operators are protected in air-conditioned cabs supplied with air from a clean source (see also Chapter 21, this volume).

To keep dusts, fumes, and temperatures in the working zone at levels tolerable for employees, the number of factory air changes per hour must be maintained at high rates, and the volumes of air that must be handled are an order of magnitude greater than with hooded furnaces utilizing local exhaust ventilation. The air, however, is at lower temperature, so that ordinary ductwork may be used to conduct it from the roof monitors to the equipment which cleans it before discharge to the atmosphere.

These two systems of controlling the escape of dust and fumes from electric arc furnaces illustrate the advantages and disadvantages of each method. The trend toward automation to increase production rates and reduce labor costs makes manual access to the processing equipment less important than formerly. The very considerable cost of high performance dust collection equipment for submicrometer-sized fume particles, now required to meet federal emission standards, is an important force tending to favor the use of local exhaust ventilation systems designed to concentrate the materials evolved from the process into the least possible volume of air.

### *C. Effects of Contaminant Concentration on Atmospheric Dispersion*

After contaminants have been discharged to the atmosphere for dilution and dispersion, it makes relatively little difference at some distance downwind whether the polluting substances were emitted in concentrated or dilute form as downwind ground-level concentration will be proportional to mass rate of emission rather than to concentration in the effluent gas. (See also Chapter 9, Vol. I.) When hot gases are discharged with little or no dilution by cool air, benefits may be obtained from increased effective stack height. This favors the use of process enclosures that conserve gas temperature and minimize discharge volume. On the other hand, when gases are already cool, greater effective stack height may be obtained by increasing the volume and velocity of the stack discharge, as by adding dilution air. Generally, however, greater increases in effective stack height may be obtained by modest temperature elevations of the stack discharge than by large increases in volume rate and velocity of discharge. Only when those subjected to pollution are very close to the emission source may substantial benefits be obtained from diluting stack discharges with clean air to reduce the concentration of contaminants. For this situation, the laws governing the behavior of air jets are more applicable than accepted atmospheric diffusion equations that become fully effective only when the stack emission has assumed the speed and direction of the wind and has reached reasonable density equilibrium with the atmosphere. In most cases, local exhaust ventilation of processes that produce contaminants is preferable to general factory ventilation, and concentrated exhaust streams are more easily cleaned than dilute ones.

Therefore, properly designed and installed local exhaust systems are an important aspect of process control to minimize the emission of contaminants to the atmosphere.

### *D. Principles of Local Exhaust Hood Design*

### 1. Dispersive Forces and Control Velocities

Local exhaust hoods are designed to create a controlled air velocity that will prevent the escape of contaminants from an exhaust-ventilated enclosure to the general workroom air, or to draw inward, or capture, contaminants generated at a distance from the hood face. The air velocity that will just overcome dispersive forces, plus a suitable safety factor, is known as the "control velocity." Control velocity is adjusted to the least airflow rate that gives satisfactory results to keep gas volumes to a minimum and contaminant loadings to a maximum in order to reduce cost and increase the effectiveness of air and gas cleaning operations. Optimum control velocities are determined by a number of factors that include the nature of the process and the dispersive properties of the immediate environment. When partial enclosure of a process is possible, as shown in Figure 1 (*11*), the control velocity is the average velocity through the open hood face.

The control velocity must be great enough to overcome disruptive air turbulence generated by (a) movement of work pieces into and out of ventilated enclosures, e.g., lowering and raising workbaskets during solvent degreasing, (b) man and machine traffic in the immediate vicinity of the exhaust hood, and (c) air currents from open windows, loading platforms, and mechanical heating and ventilating equipment. Air cur-

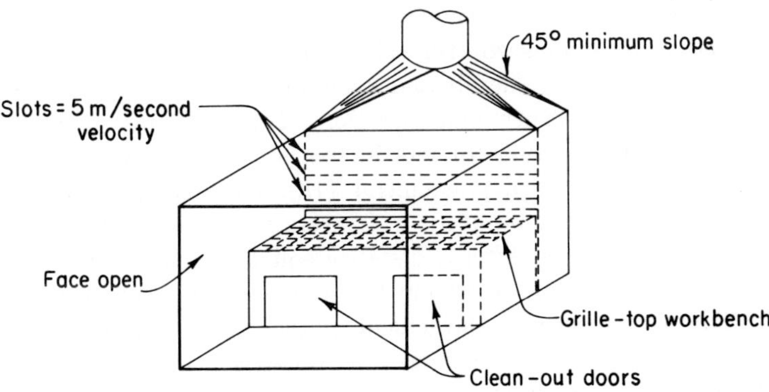


Figure 1. Metallizing booth (*11*).

rents generated by work activities seldom exceed 0.5 m/second near hoods, but air currents that sweep into the workroom from nearby open doors and windows may reach several times this value. Therefore, operations requiring local exhaust ventilation should be placed in areas that are protected from drafts if minimum control velocities are to be effective.

Control velocities of 0.5–0.75 m/second usually are adequate to prevent loss of contaminant to the general room atmosphere but there are exceptions. Hot processes are capable of generating strong thermal currents that may become reflected from hood and machinery surfaces and directed out of the hood openings at speeds that exceed the inward velocity. This malfunction will become much worse if adequate account has not been taken of the expansion of the induced air after it enters the hot exhaust enclosure. The required exhaust flow rate varies directly with temperature and surface area of the process (*12*).

a. Rotary Machines. Sources of disruptive air currents may be found inside the exhaust ventilation enclosure itself. Grinding wheels and high speed lathes behave as low efficiency centrifugal fans, i.e., the rotating parts act as fan blades. Depending on peripheral speed and roughness, the fan effect may generate high speed air currents that are difficult to confine and require high control velocities. The movements of bucket conveyor scoops inside conveyor housings act as paddle wheels to impart velocity to the air in contact with them.

b. Percussive Machines. Drop forge hammers compress hot air between the hammer and anvil and "squirt" air out horizontally at high velocity just before impact. These intermittent high velocity jets entrain gases, die lubricants, and fragments of metal scale and must be counteracted by a continuous inward flow of air at a velocity that exceeds the dispersive properties of the machine. There is little or no theory that can be used to determine control velocities for these and similar situations. Although rules of thumb and experience are the principal guides for estimating the velocity and direction of air currents generated by such machines, the behavior of jets in free air (Fig. 2) is of assistance in explaining the observed phenomena (*13*). At first (zone 1), the jet travels outward as a column of air which retains its initial cross-sectional dimensions and velocity. After a distance equal to about four or five jet diameters, the jet begins to entrain the quiescent atmosphere in contact with it. This results in an enlargement of the jet cross section (zone 2) and a reduction of velocity as energy is transferred to the entrained air. During this phase, which covers a distance of about eight jet diameters, centerline velocity decreases as the square root of the distance of travel.

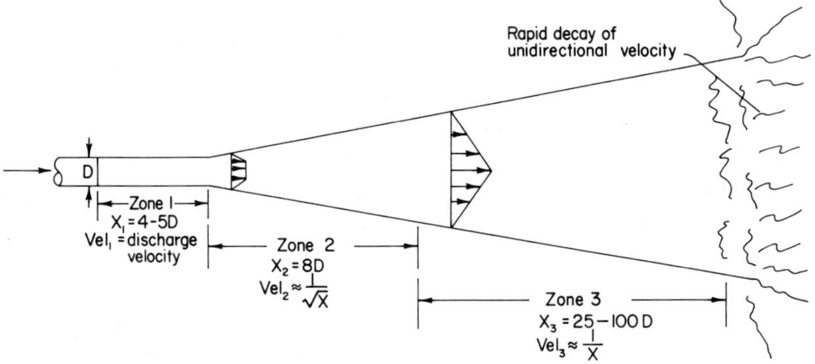


Figure 2. Jet in free air. $D$ = diameter of jet opening; $X$ = centerline distance of flow zone.

In the last phase (zone 3), velocity decreases more rapidly with distance and the cross section spreads more rapidly until, after 25–100 jet diameters, it degenerates into generalized turbulence. To apply jet theory to evaluate the magnitude of the air currents generated by a drop forge hammer, all that is needed is knowledge of the cross section and speed of the hammer. If a 0.6-$m^2$ hammer descends at the rate of 5 m/second, it displaces air at the rate of 1.8 $m^3$/second. As the hammer nears the anvil and the height of the cross section through which displaced air can escape sideways becomes very small, high horizontal air velocities result. For example, when the distance between hammer and anvil decreases to 2.5 cm, the instantaneous ejection velocity of the displaced air in all four directions becomes $1.8/4 \times 0.6 \times 0.025 = 30$ m/second. Horizontal jet velocities become increasingly larger as the hammer approaches the anvil, but theory indicates that persistence of the jet and transference of momentum to the surrounding air are influenced by the absolute size of the jet. Therefore, velocity increases during the final instant are less effective in creating disruptive air currents than are the somewhat lower velocities associated with jets of larger cross section. From Figure 2, it is possible to estimate jet velocity at various distances from the bed of the forge and to determine the control velocity that will be needed to confine the air contaminants generated by the forge. Although the airborne products generated during the forging of iron and steel have little air pollution significance, the processing of powdered beryllium and other highly toxic metals by this fabrication technique requires careful design and construction of local exhaust systems.

c. PARTICLE PROJECTION. Not only are particles entrained by dispersive air currents generated in machines, but in many instances they are given substantial acceleration. This is the case with grinding wheels. The peripheral velocities of large, high speed grinding wheels exceed 100 m per second and particles are discharged tangentially from the surface of the wheel at speeds that approximate those of the wheel surface. Fine particles, i.e., those in the Stokes' law range, move through air in laminar flow and rapidly attain the speed and direction of the entraining air currents. This is especially true for particles in the respirable size range, i.e., less than 10 μm diameter. For these particles, control velocities scarcely greater than those needed to contain the air currents generated by the fan effect of the wheel are all that will be needed to capture them. Large particles generated by coarse-grained grinding wheels may be projected many meters because of the momentum imparted by the wheel, but they settle from the air under gravitation force so rapidly that they represent only a plant housekeeping problem and have little significance for air pollution. According to Drinker and Hatch (*14*), a 2-mm particle thrown into the air from a grinding wheel with an initial velocity of 50 m/second travels 20 m before its speed is reduced to the point where the residual travel is in the streamlined flow range, whereas a 10-μm particle having the same initial velocity comes to rest in still air within a distance of less than 5 cm and a 1-μm particle travels only about 400 μm in still air before coming to rest. Therefore, it may be concluded that control of turbulent air currents generated by grinding wheels and similar contaminant-producing machines is of greater importance in the design of local exhaust hoods and the selection of control velocities than the dynamic projection velocity or distance of travel of particles independent of entraining air movement.

d. Falling Granular Materials. During free fall, materials such as crushed stone, sand, and finely divided powders of diverse origin entrain air that is expelled at the bottom of the drop. This air separates explosively and at high velocity from the granular material with which it was mixed and, in turn, entrains large fractions of the finest dust. This dust-laden air must be released from the storage bin, transfer station, or pile into or onto which the granular material has been deposited. If this air and its entrained dust is released directly to the atmosphere, pollution results. For control, the dust-laden air must be conducted to dust collecting devices with a minimum amount of induced air. Required local exhaust velocities and air volumes are determined from a knowledge of the amount of air entrained during the free fall of the granular materials.

Air displacement is related to the rate of flow of material, the height of fall, and the size and shape of the granular material. Measurements by Pring *et al.* (*15*) show that the induced air volume is roughly propor-

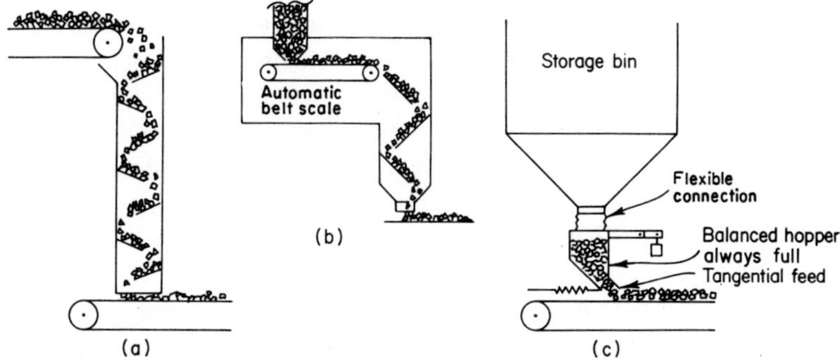


Figure 3. Illustrating ways in which height of free fall of solids can be minimized. (a) Stone ladder in vertical chute. (b) Sloping surfaces taking material from continuous weighing machine. (c) Balanced secondary hopper below storage bin with tangential feed of material onto belt.

tional to the cube root of the weight rate of granular material flow and inversely proportional to the diameter of the granules. Drinker and Hatch (*14*) using observations of Chirico (*16*) for crushed rock falling in chutes, showed that the ratio

$$Q/(Wh^2)^{1/3}$$

has a value between 0.025 and 0.053 for fall distances ($h$) from 3 to 9 m, material flow rates ($W$) from 9 to 50 kg/second, air displacements ($Q$) from 0.15 to 1.2 $m^3$/second, and the material sizes ranging from fine to coarse. Though far from precise, this empirical relationship provides a basis for estimating the order of magnitude of the induced airflows that must be controlled during conveying of granular materials.

Free-fall air entrainment is modified by the degree of enclosure of the dropping stream and the ease with which air can enter and leave the enclosure. Modification of materials handling equipment (Fig. 3) is capable of reducing the height of fall of materials to very low levels, thereby minimizing air displacement and entrainment of fine dust.

e. Gas Diffusion. Gaseous contaminants that must be confined for air pollution control purposes possess diffusive properties which cause them to migrate in all directions from the point of release. The rate at which one gas, i.e., a contaminant, diffuses through another, i.e., the air, is related to temperature, pressure, and the molecular weights and volumes of the gases. Viles (*17*) examined the relative importance of diffusion as a dispersive force in laboratory fume hoods and concluded that when control velocities are adequate to prevent contaminant losses from drafts and traffic-induced turbulence, no escape of fumes occurs because of the diffusive properties of even highly volatile materials.

f. Hot Processes. Heat sources within a local exhaust enclosure generate convective air currents whose upward velocity is proportional to the rate of heat transferred to the surrounding air and the height of rise of the heated air. When hot gases rise through a confining enclosure which acts as a chimney, the vertical velocity of rise is proportional to the square root of the difference in weights of the heated air column and the weight of an equal column of the surrounding ambient air—the so-called "stack" or "chimney" effect. However, when a column of hot air rises in a free atmosphere, it entrains the surrounding cooler air in the manner of a low velocity jet—cooling and spreading laterally as it rises. This is a much more complex system to define and analyze than the hot chimney, and only approximate solutions are available. Hemeon (*18*) has suggested the following approximate formula for estimating the quantity of airflow ($Q$) induced by the presence of a hot body inside a ventilated enclosure:

$$Q = 0.61H^{1/3}A^{2/3}X^{1/3} \tag{1}$$

where

$Q$ = $m^3$/second
$H$ = heat loss from the hot body, kcal/second
$A$ = cross-sectional area of hot body, $m^2$
$X$ = height of rise of hot air column, m

Control velocities for hot processes must take into account the expansion of the air heated by the hot body, and adequate duct and blower sizes must be provided to handle the larger air volume or else part of the induced air will spill out of the enclosure. Careful system design, in addition to making provision for air expansion from heating, will shape the hood structure in such manner that the rapidly rising thermal currents will flow naturally into the hood exit duct (*12*). This is important for torch cutting, welding, and scarfing hoods.

## 2. Hood Design

Although complete enclosure of a process that emits dusts, fumes, vapors, and gases results in the least possible exhaust volume and minimizes the difficulties of preventing air pollution, many processes cannot be enclosed. For these, local exhaust velocities must be generated at a distance from the source to entrain and remove contaminants. Unlike air jets, that retain a substantial fraction of their initial discharge velocity at distances as great as 100 jet diameters from the point of release, exhaust velocities decrease rapidly with distance from the source of suction [Fig. 4, from Alden (*19*)]. The reason for this difference is that

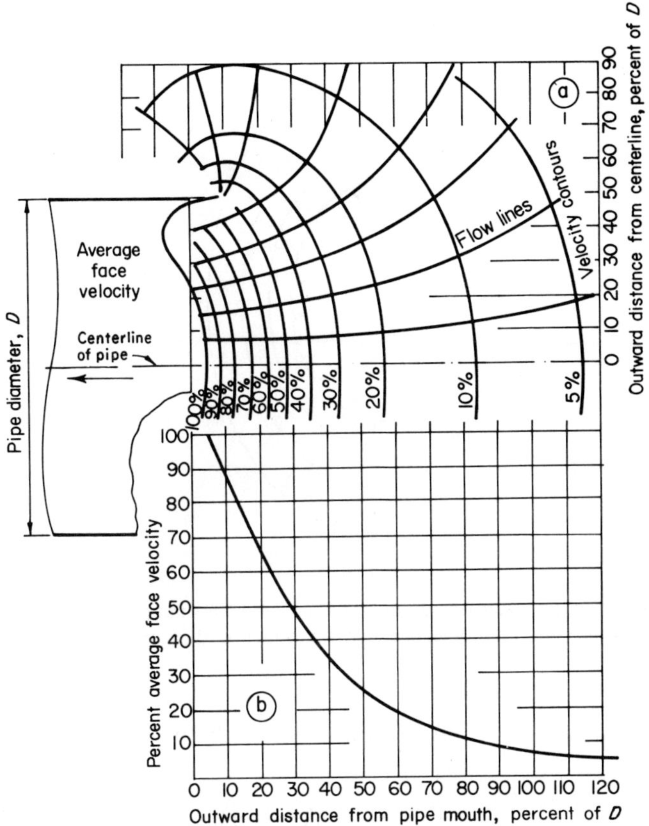


Figure 4. (a) Velocity contours and (b) flow directional lines in radial plane of circular suction pipe (*19*).

whereas a jet retains its cross section almost unchanged for many jet diameters downstream, a suction opening draws its air from all directions simultaneously and creates quasispherical or cylindrical surfaces of equal flow that rapidly decrease in velocity as the distance from the suction opening increases. In Figure 4, lines of equal velocity at graduated distances from a round suction opening (expressed in multiples of the pipe diameter) are shown as percent of suction opening face velocity. The stream lines are shown perpendicular to the spherical envelopes of equal velocity. As may be seen, control velocity falls off very rapidly in free air.

The relationship among suction opening face area ($A$), total exhaust volume rate ($Q$), and velocity ($V$), at a distance ($X$) out along the centerline from the suction opening (for round, square, and rectangular cross

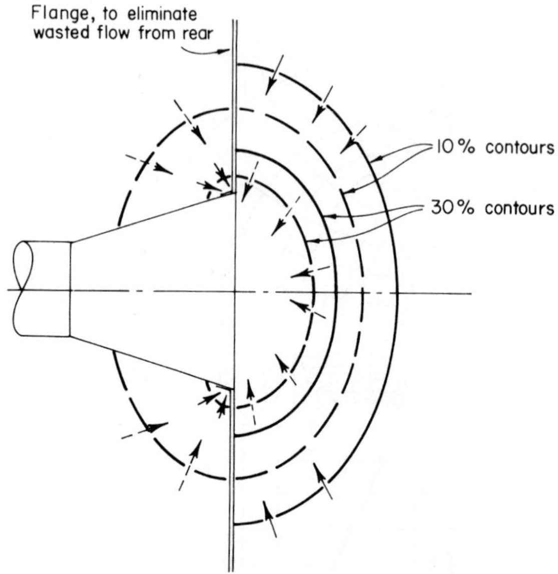


Figure 5. Relative position of contours for flanged and simple hoods. Dashed line: plain hood; solid line: flanged hood. Note outward displacement of contours for flanged hood.

sections with side ratios up to 1:3) was determined by DallaValle (*20*) to be

$$V = Q/(10X^2 + A) \tag{2}$$

Velocity at a distance is somewhat higher than would be calculated from a theoretical point source of suction; and the larger the suction opening, the less the velocity falls off with distance. This is because real exhaust hoods, both simple and flanged, have finite dimensions that impede the flow of air from areas to the rear of the opening (Fig. 5).

The velocity distribution outward from a long exhaust slot—the usual exhaust ventilation arrangement for controlling losses from downdraft tables (Fig. 6a) and open surface tanks (Fig. 6b) (characteristic of electroplating, solvent degreasing, and similar operations)—corresponds only approximately to a theoretical cylinder of equal velocity enveloping a line source of suction in space. This is because the surface of the table or the liquid and the finite dimensions of the exhaust hood profoundly modify the stream lines—in effect, drawing air from an area having the approximate shape of a half or quarter cylinder. This suggests that flanged hoods that restrict airflow from all areas except those immediately in front of the suction opening may be used effectively to reduce the exhaust volume needed to generate satisfactory control velocities.

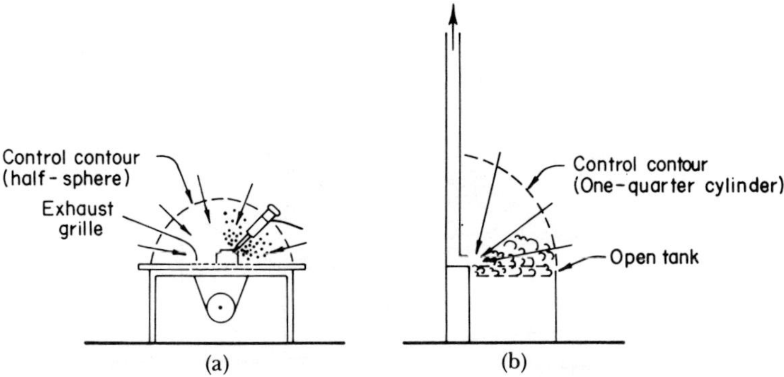


**Figure 6.** An illustration of the confining effect of adjacent plane surfaces on the pattern of airflow into exhaust hoods. (a) A downdraft hood on bench, with work surface acting as a confining plane. (b) Exhaust slot along back of tank draws air from only a quarter cylinder because of confining vertical wall and horizontal tank surface.

Overhead canopy-type exhaust hoods have been applied to many potential contaminant emitting operations including paint dipping, metal pickling, and a variety of hot processing steps that emit gases, vapors, and aerosols. Figure 7 shows the velocity contour around a typical canopy hood over a dip tank. The entire open perimeter of the hood must be used when calculating the air volume needed to produce appropriate control velocities. This is seldom less than 0.5 $m^3$/second/$m^2$ of opening. When the nature of the work permits, erection of side enclosures, shown in Figure 8, results in savings from two sources: (a) the open face area across which control velocity must be maintained is substantially reduced and (b) lower control velocities are possible as the emitting source is well shielded from drafts and turbulence. The laboratory fume hood is basically a canopy hood enclosed on three sides. Its effectiveness for re-

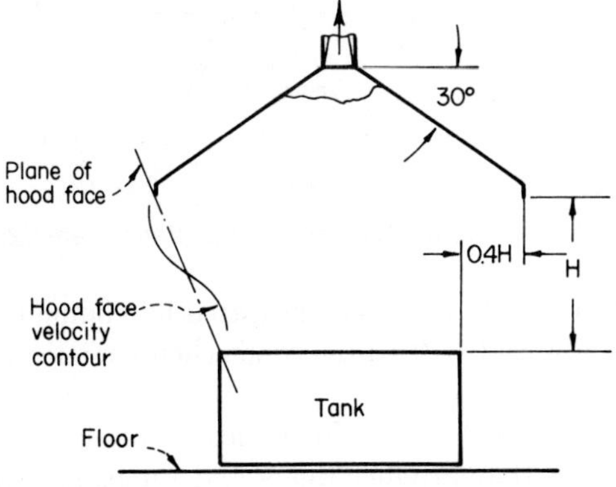


**Figure 7.** Plain canopy hood (*19*).

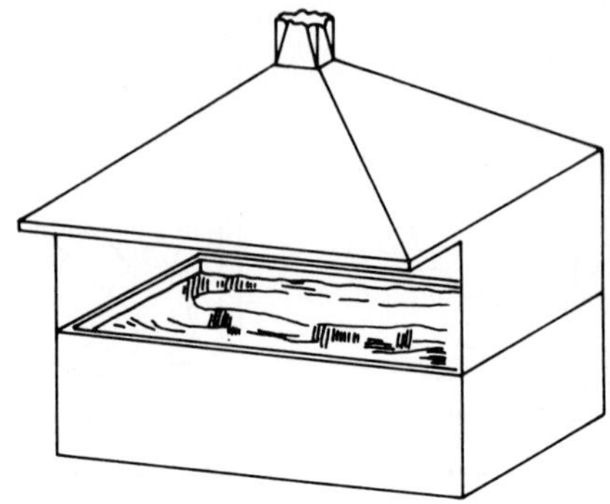

Figure 8. Canopy hood with side enclosures (*19*).

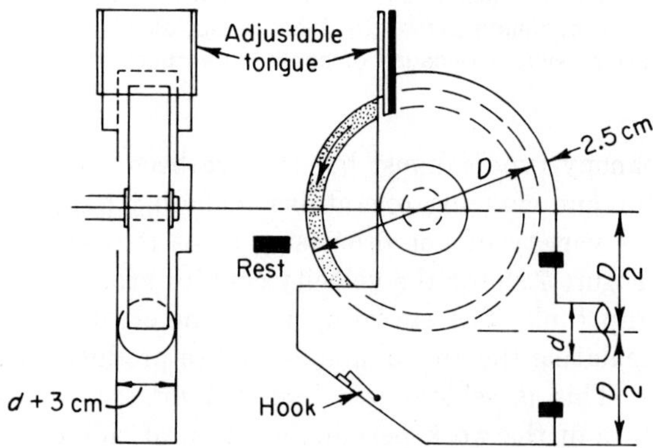


Figure 9. Hood for stationary grinder (*20*).

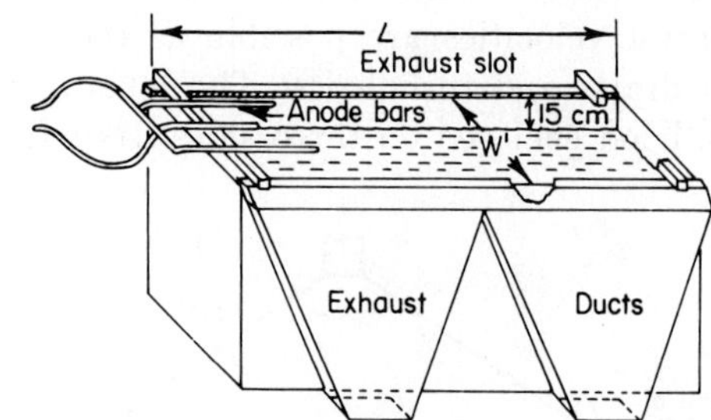


Figure 10. Lateral slots for chrome plating tank (*20*).

taining fumes at low airflow rates is indicated by commonly recommended face velocities as low as 0.3–0.4 m/second when chemicals of low toxicity are used.

Additional exhaust hood shapes for a variety of commonly encountered industrial operations that produce dusts and fumes are shown in Figures 9 through 12.

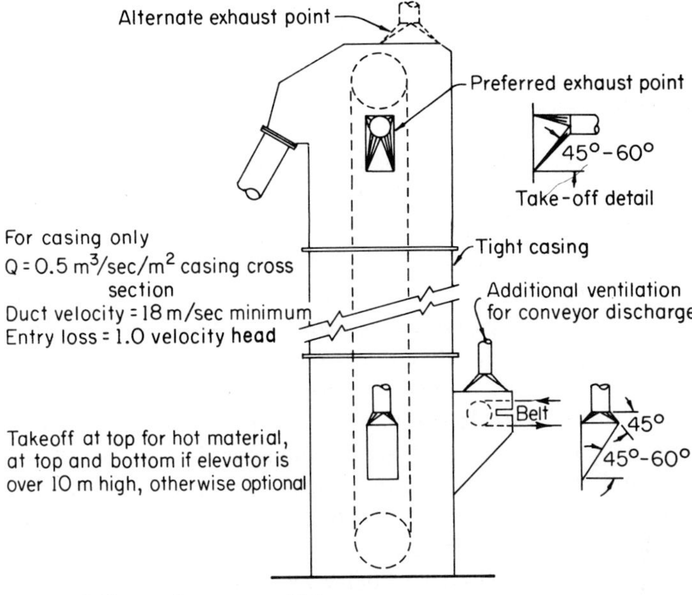


| Belt speed | Volume |
|---|---|
| Less than 1 m/sec | -0.0055 $m^3$/sec/cm of belt width. Not less than 0.0025 $m^3$/sec/cm of opening |
| Over 1 m/sec | -0.008 $m^3$/sec/cm of belt width. Not less than 0.003 $m^3$/sec/cm of opening |

Figure 11. Bucket elevator ventilation (*11*).

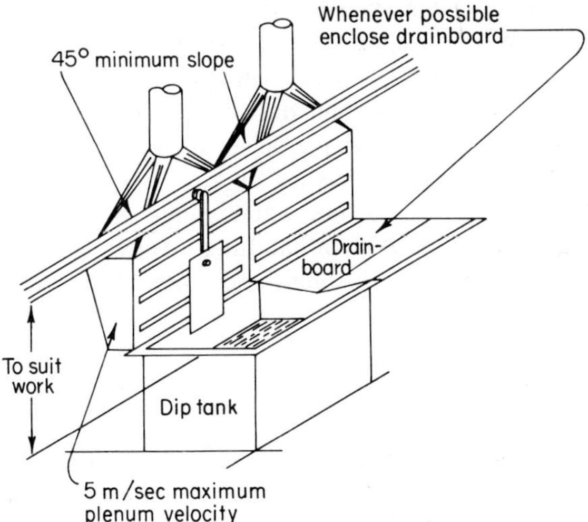


Figure 12. Dip tank (*11*). Slot velocity = 10 m/second. Entry loss = 1.78 slot velocity pressure + 0.25 duct velocity pressure.

### *E. Principles of Exhaust System Piping*

#### 1. Conveying Velocity

Pipes that carry particulate contaminants from exhaust hoods must be sized to prevent settlement. Settlement is most likely to occur in long horizontal pipe runs and particle conveying velocities suitable for horizontal runs will be satisfactory for other, less critical, parts of the system. Although it is possible to calculate theoretical minimum conveying, or transport, velocities based on a knowledge of particle size, shape, specific gravity, and dust loading, calculated values have been shown to be inadequate. Therefore, higher velocities, based on experience, are employed. Conveying velocities of 18–23 m/second are recommended for mineral dusts (the higher velocities for small diameter ducts); 13–18 m/second for plastic and rubber dust and for many substances of vegetable origin such as coffee, beans, and grain; and 8–13 m/second for cotton lint, wood sander dust, and similar fluffy materials.

Considerations of conveying velocity are absent when the exhaust system carries only gases and vapors. For this service, duct diameters may be selected to minimize the total cost of the system over its predictable lifetime. Savings on installation cost by selecting small diameter ducts and high pipe velocities may be offset within a few years by the higher annual power costs associated with the greater pressure drop that is characteristic of high velocity systems. Duct velocities of 10–15 m/second have been found economical for systems of moderate size that handle only gases and vapors. When designing large systems, it is generally advisable to account for all expense items and then optimize the size of the system for minimum total cost. Maximum duct velocities are usually based on the erosive properties of aerosol particles, such as silica and rock dust, and on the high noise characteristics of air movers when total system pressure drop exceeds 250–300 mm water gauge. Alden (*19*) notes that "The elbows of a conveying system handling a heavy tonnage of abrasive materials will cut through in 12 months or so while the piping will last about twice as long."

#### 2. Materials of Construction

Galvanized steel sheet, in thicknesses from 16 to 24 gauge, is the most common construction material for exhaust piping—up to diameters of 100 cm for light service and up to 50 cm for the transport of abrasive particles. Longitudinal crimped and soldered pipe seams are commonly used. The interior of the seam is hammered flat to give smooth inner surfaces that will not interfere with the flow of air or collect materials. Long pipe runs are assembled from smaller sections with telescoping joints that have the small end of each section oriented to face downstream, thereby offering least resistance to flow. Heavier piping is constructed of welded iron sheet and sections are joined with bolted flanges or by circumferential welding of the ends. Provision must be made for the expansion and contraction of long straight runs of all-welded piping that occur when temperature changes.

Rigid polyvinylchloride (PVC) and glass-reinforced polyester are used extensively for hood and pipe construction whenever corrosive gases and vapors are to be handled. Generally these materials are assembled into gastight, seamless systems by chemical bonding techniques. Centrifugal blowers, gas cleaning devices of many types, and discharge stacks made from these two materials are also available for severe corrosive conditions. For example, rigid PVC and glass-reinforced polyester structures have almost entirely replaced metal and wood in exhaust ventilation systems for electroplating service. When a temperature greater than about 150°C is associated with corrosive contaminants, stainless steels and metals such as titanium and tantalum must be used as construction materials.

### 3. Pipe Fittings

Pipe fittings should be constructed with smooth, unobstructed internal surfaces throughout to reduce air flow resistance and the tendency of abrupt changes in direction and internal roughness to collect dusts, fibers, etc., with eventual plugging of the system. Recommended fittings include (a) long sweep elbows, e.g., a 90° elbow having a centerline radius of two pipe diameters has less than one-quarter the airflow resistance of a miter elbow, (b) gradual merging of branch ducts and mains, e.g., a branch joining a main at an included angle of 30° has one-seventh the energy loss of a tee joining at 90°, and (c) gradual duct expansions and contractions, e.g., an abrupt expansion has more than three times the energy loss of a gradual, 20° expansion of the same cross-sectional change.

The high negative pressures developed in most local exhaust ventilation systems make it essential that in-leakage be eliminated if adequate control velocities are to be maintained at the hood faces. Careful attention must be paid to sealing longitudinal pipe seams, the joints between duct sections, and duct fittings. The flimsy metal gauges and S-type joints that are commonly employed in low pressure supply air systems are certain to fail when employed for exhaust ventilation service. There is no practical way of sealing the multiple openings which characterize this latter method of construction or the additional leaks which occur as a result of flexure of thin-walled ducts when subjected to high negative pressure. Certain fittings, such as slide dampers, are likely to leak excessively when constructed in the customary manner. For exhaust ventilation service, special attention to airtight construction is required for every component of the system.

## 4. Blowers

Good practice in the design of local exhaust ventilation systems calls for blowers to be the last component before the discharge stack. This arrangement makes it possible to keep most of the system (and usually all of it located within workrooms) under negative pressure so that leakage, if any, will be inward. If maintained under positive pressure, contaminant-laden air would spurt out into the workroom atmosphere through cracks and leaks in the system. When an air cleaner is used to remove particulate matter or acid gases, it should be placed immediately upstream of the blower to protect the blower from the erosive and corrosive effects of the contaminants. Recommended methods for balancing the various branches of complex exhaust systems and for calculating the system volumetric airflow rate and resistance are illustrated and explained in detail in a number of accepted texts on this subject. These include Drinker and Hatch (*14*), Alden (*19*), the ACGIH publication "Industrial Ventilation" (*11*), and the ASHRAE Guide (*21*). With this information, it is possible to select a blower having the required delivery rate and static pressure characteristics.

## 5. United States Occupational Safety and Health Administration Standards

The United States Occupational Safety and Health Administration (OSHA) has published ventilation standards (*22*) that have the force of law for certain industrial processes including buffing, grinding, polishing, spray painting, open surface tanks, and degreasing. Reference is made in thse OSHA "Rules and Regulations" to American National Standards Institute (ANSI) concensus standards ANSI Z9.2 and Z33.1 (*23*) covering "Fundamentals Governing the Design and Operation of Exhaust Systems."

## 6. Discharge Stacks

Discharge stacks from exhaust ventilation systems should extend high enough above the roof of the building in which they are installed, as well as above nearby structures, to be out of the zone of influence of turbulence wakes that tend to turn the stack effluent groundward. Similar unfavorable downward flow effects are created when stacks are equipped with devices such as rain caps and goose necks for preventing the entry of rain. Not only do such devices direct the discharge downward, and thereby eliminate all beneficial effects that might be gained from the increased effective stack height associated with upward discharge velocity and the buoyancy of elevated gas temperature, but they are, in fact, unnecessary for protection against rain because when the system is in operation, no precipitation can penetrate the upward air current emerging from a vertical, uncapped stack. When the system is shut down, the small amount of precipitation that collects in a stack (for example, a 2.5-cm rain will deposit less than 15 liters of water in a 90-cm diameter stack)

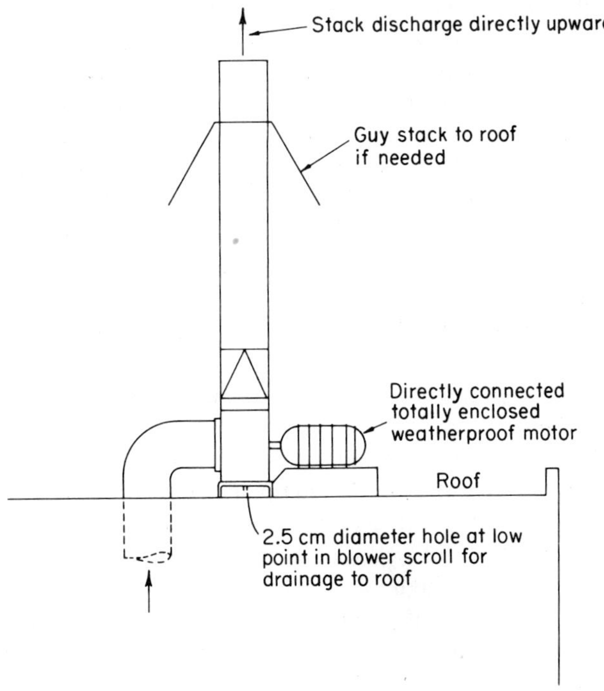


Figure 13. Exhaust stack and blower.

can be drained away by any one of several arrangements, such as the one illustrated in Figure 13, which shows a free-standing stack supported on a centrifugal blower casing that has been rotated to the correct upward discharge position. Drainage is accomplished through a 2.5-cm diameter hole drilled in the bottom of the scroll casing of the blower. The motor shown is a direct-connected, totally enclosed, weatherproof unit, thereby eliminating the need for a V-belt drive and a motor-drive weatherproof housing.

## V. Managing the Air Resource

### A. Objectives of Air Use Management

Air pollution is primarily an urban phenomenon and results from the crowding together of large numbers of people who produce and consume the numberless products and services that contribute to the contamination of the atmosphere. It has been estimated (*24*) that two-thirds of the population in the United States now live on less than 9% of the land area and that by the end of this century more than three-fourths of a much larger population will be living on this same space. This means that the number of cubic meters of air available to each city dweller for operating his automobile, heating his home, producing his electricity, manufacturing the products he consumes, burning the refuse from these products, and breathing, is already limited and will continue to shrink. However, the atmosphere over the other 91% of the United States is being utilized at far less than its natural capacity and the 70% of the world covered by oceans is being used hardly at all.

The techniques of rational air use management are based on the principle that air quality standards have been set at levels that protect the entire population from harm with an adequate margin of safety. Acceptance of this concept is not unanimous at this time and those who disagree advocate a serious effort to achieve "zero emissions" at the earliest possible date. It will be obvious that those who press for "zero emissions" will find the concept of "air use management" anathema. Nonetheless, the idea of air use management has received official sanction from the USEPA through its rules (*4*) that permit individual states to designate regions where existing high air quality may be lowered to levels that approach current air quality standards. Therefore, it is useful to examine possible applications of air use management to define its strengths and pitfalls.

The objective of air use management is to minimize the cost and inconvenience to the public of achieving and maintaining prescribed air quality standards. It is intended to cope with predicted population increases and the provision of the many additional services, goods, and jobs this added population will require.

### B. Assimilative Power of the Atmosphere

Management of the air resource is based on full utilization of the considerable self-purifying properties of the atmosphere. (See also Chapter 8, Vol. I.) On a bright, breezy day, atmospheric dilution and dispersion may be 1000 or more times as effective as during periods of atmospheric stagnation that have been responsible for fatal air pollution episodes. Therefore, more stack discharges are permissible during periods of good atmospheric mixing. In many parts of the world, good mixing conditions are normal and periods of atmospheric stagnation, abnormal. In these areas, it may be feasible to increase and decrease emissions depending upon atmospheric conditions. This is the basis for plans to conserve scarce low-sulfur-containing fuels by using readily available high-sulfur fuels during periods of excellent atmospheric ventilation and switching to low-sulfur fuels when stagnation periods are predicted by the weather services. Thus, the dispersive properties of the atmosphere can be utilized to produce prompt dilution of emissions for the protection of close-in areas whereas the self-purifying properties of the atmosphere may be relied upon to avoid long-distance transport of air contaminants. That the self-purifying properties of the atmosphere are finite has been demonstrated by recent high acid content of rain in Scandinavia, the appearance of much higher concentrations of very fine particle clouds over the North Atlantic regions (*25*) (both originate from sulfur oxide and particulate emissions from stacks located in the industrial heartland of Western Europe), and the presence of high ozone levels in unpopulated areas of the United States remote from large urban centers. These events, along with increased concern about the toxic effects of sulfates in the atmosphere, suggest that current air quality and emission standards require continuing critical review to take account of new knowledge about toxic effects, gas-phase reactions, and long-distance atmospheric transport of air contaminants. Nevertheless, it is known that substances emitted to the atmosphere have different residence times and variable clearance mechanisms. These range from (a) gravitational settling to the earth of large primary particles within minutes to (b) conversion of carbon monoxide to carbon dioxide over a number of years. Within these constraints, some degree of low-cost air pollution control can be achieved by intelligent application of air use management. This requires recognition of regional differences in the magnitude of the air resource, e.g., the much greater natural ventilation rate of coastal northeastern United States cities compared to those in southern California, and, hence, a variable ability to dilute atmospheric emissions to levels below air quality standards.

### *C. Utilization of Untapped Air Resources*

Untapped air resources can be utilized effectively when mine-mouth coal-burning generating stations produce electricity in remote, lightly populated areas and send it many hundreds of miles to crowded urban centers for consumption without adding to the existing pollution load from vehicles and other activities that cannot be relocated. This, and similar practices, should not subject a small, local population in a remote area to excessive air pollution as a substitute for exposing a large urban population. Rather, the largely unused atmospheric resource at the mine-mouth location can be made available to disperse and dilute emissions from a single power station without exceeding recommended air quality criteria designed to protect even the most sensitive element in the population. This plan worked well in northwestern Pennsylvania because excellent emission controls were included in the plant design but was highly unsatisfactory when applied to the very large coal-burning electricity-generating Four Corners Station constructed without adequate emission controls in the southwestern United States.

Some measure of air pollution control may be obtained by staggering work hours. This would avoid overburdening the assimilative capabilities of the atmosphere during peak traffic hours and under-utilizing the full air resource during off-peak hours. A project is under study to utilize the untapped air resources of the oceans. It calls for waste collections to be transported to docks and transferred to an ocean-going vessel (*26*). After loading, the vessel puts to sea, burns the wastes, and discharges the residue into deep holes in the ocean floor. Studies have shown that proper selection of offshore burning sites will avoid the movement of stack discharges to the shore under all weather conditions without significant degradation of the ocean air, decreasing the recreational use of the water, or interfering with commercial and sport fishing.

### *D. Operational and System Control to Maximize Beneficial Use of Urban Atmospheres*

#### 1. Intermittent and Interruptible Operations

There are a number of serious air polluting operations of widespread usage and importance for which no substitute or reasonable control method has been devised to date. These operations include open burning of brush from land clearing and agricultural waste burning for parasite control. "Such use of open burning need not be unregulated" (*27*) and specific recommendations for controlled burning of brush and other bulky objects have been formulated (*28*). These control measures are oriented around proper preparation and stacking of the material to be disposed of plus a "meteorologically scheduled open burn" (*29*). Satisfactory application of a plan to maximize air usage for this purpose depends on an ability to store debris until favorable meteorological conditions occur and on very close supervision by a competent professional meteorologist who designates periods suitable for burning. Although serious attempts have been made in many areas to outlaw open burning for any purpose, experience has demonstrated that exemptions and variances have become common after a few years of prohibition. This is obviously an area that calls for greater efforts to generate acceptable alternatives, although rigorously controlled open burning seems to have generated a minimum of reported air pollution complaints.

## 2. Land Use Management

Land use management has become an important tool for the control of new sources of air polluting emissions from stationary and mobile sources. This is reflected in a requirement of the USEPA (*30*) for the preparation of environmental impact statements as a preliminary to the construction or enlargement of power-generating stations, airports, shopping malls, industrial parks, and similar important large area sources of stationary and mobile emissions. Land usage as a basis for air resource management planning can be related directly to the nature and quantity of anticipated emissions originating from the activities scheduled to occur, and this information can be employed for the development of regional air resource management plans.

a. Areal Emission Flux. It has been suggested (*31*) that the term "air resource" takes on real meaning only when each parcel of land is associated with "emission rights" that may be used by the occupant or leased to an abutter. According to this plan, it becomes possible to maintain a predetermined areal flux of emissions based on topographic, meteorological, and other factors in a free-market setting.

The current use of air pollution control imperatives as the only guiding principle in land use planning is placing undue emphasis on just one of the many important factors that must be considered. For example, airport master planning is being hampered unnecessarily by an insistence on maintaining ambient air quality standards everywhere inside the airport grounds when, in fact, the exposure of passengers and those who accompany them is infrequent and of brief duration and the exposure of employees should be guided by 40 hour per week occupational standards rather than by ambient air standards that are based on 168 hour per week exposures.

b. Airport Planning. A principal protective measure within the control of the airport planner is the distance between potentially disturbing airport operations and the general community. Although airports servicing large civilian and military jet aircraft require very large land areas to establish adequate buffer zones, the total airport area is less significant than its shape and the way the various facilities are located relative to the community. This is because aircraft land and takeoff into the wind and the areas lateral to the main runway may be occupied by automobile parking areas, commercial and warehousing facilities, passenger terminals, and other auxiliary functions and services that produce lesser amounts of air polluting emissions and serve to shield adjacent communities from the full impact of aircraft operations.

It is not always possible to put these principles into practice at older airports. For example, the Los Angeles County (California) Air Pollution Control District stated that in 1971 motor vehicles entering and leaving airport property accounted for 65% of carbon monoxide and 25% of total pollutants emitted within the boundaries of the airport and recommended that vehicle access routes be designed to minimize motor vehicle congestion (*32*). In some communities, opportunities exist to locate airports on, or immediately adjacent to, water bodies, thereby tapping an enormous unused air resource relatively remote from human activity.

## 3. Load and Fuel Switching by Power Stations

The unfavorable impact of fossil fuel combustion on adjacent land areas can be greatly moderated by decreasing electrical output and burning low-sulfur high-quality fuel during periods of poor atmospheric dispersion; and reversing this policy during periods of good atmospheric conditions. (See also Chapter 8, Vol. I, and Chapter 12, this volume.) Load switching depends on the existence of integrated interregional power networks with excess capacity so that during periods of severe air stagnation in one region a major part of the electrical requirements for power can be supplied by plants operating outside the affected region.

Fuel-switching programs may be seasonal, e.g., permitting high-sulfur-fuel use during warm weather when emissions associated with space heating are absent and the atmospheric burden is reduced proportionately (therefore, automatically regulated by the calendar) or they may be weather-based, e.g., permitting high-sulfur-fuel use whenever atmospheric conditions make it possible to avoid exceeding air quality standards, regardless of season.

Fuel-switching plans of the second type are based on an extensive array of field sensors that continuously telemeter air measurements to a computer facility programmed to analyze air concentrations, meteorological conditions, and, with the aid of a weather forecast, predict the need for fuel switching well in advance of unfavorable weather. A highly developed fuel-switching plan for power stations (*33*) proposes to use 2.2% sulfur oil whenever predicted ground concentrations are below 0.06 ppm $SO_2$ (57% of the daily standard). Whenever predicted concentrations would exceed 0.06 ppm while burning 2.2% sulfur fuel, 0.5% sulfur fuel (one-half the sulfur content required by law) would be substituted, thereby effecting an improvement in air quality above that required by law during periods of unfavorable atmospheric dispersion. Ability to predict ground-level concentrations of $SO_2$ within $\pm 0.01$ ppm was demonstrated to be close to 100% over an extensive trial period.

## 4. Transportation Regulation

Transportation strategies have been proposed for reducing emissions to urban air from privately operated motor vehicles (*34*). Many involve severe restrictions on the use of private automobiles by barring half the cars from the city center on an odd–even day sticker plan or by reducing parking spaces during the entire work day or for only the 7–10 a.m. period. Other plans seek to apply economic incentives by such strategies as graduated tolls on access roads into city centers that are inversely proportional to the number of passengers per vehicle and by reducing fares for public transportation.

Until travel restrictions are put into practice and/or emission controls for individual vehicles become totally effective, better management of the available air resource can be achieved by staggering work hours to eliminate the morning and late afternoon rush hour peaks that often severely overburden the assimilative capacities of the atmosphere for brief periods (*35*). Spreading automobile emissions evenly throughout the day make it possible to avoid exceeding air quality standards even though the same number of vehicle trips are made daily. Similar benefits may be attainable by resorting to 4-day work weeks, provided the 4-day periods are distributed reasonably evenly throughout the week. By these and similar means, the air resource can be used more effectively to avoid undesirable effects from excessive automotive emissions to the atmosphere.

## VI. Summary

It is clear that all available methods that are designed to minimize the production and emission of contaminants must be employed to the maximum possible extent if pollution of the environment of an overpopulated, overcongested world is to be avoided in the future. An important element of environmental pollution control is the wise use of as large a fraction of the total environment as may be accomplished.

## ACKNOWLEDGMENT

The author wishes to thank Professor T. Hatch for his kind permission to reuse material he presented in the first edition of this book.

## REFERENCES

1. T. Hatch, *in* "Air Pollution" (A. C. Stern, ed.), 1st ed., Vol. 2, p. 211. Academic Press, New York, New York, 1962.
2. Tennessee Valley Authority, Office of Engineering Design and Construction, Chattanooga, Tennessee, 1974 (private communication).
3. E. L. Stockton, *U.S. Publ. Health Serv.* **1649,** 228 (1966).
4. U.S. Environmental Protection Agency, *Fed. Regist.* **39,** 42515 (1974).
5. M. W. First, *in* "Proceedings of the Third International Clean Air Congress," p. E112. Ver. Deut. Ing., Dusseldorf, Federal Republic of Germany, 1973.
6. W. E. Winsche, K. C. Hoffman, and F. J. Salzano, *Science* **180,** 1325 (1973).
7. W. Bartok, A. R. Crawford, A. R. Cunningham, H. J. Hall, E. H. Manny, and A. Skopp, *in* "Proceedings of the Second International Clean Air Congress" (H. M. England and W. T. Beery, eds.), p. 801. Academic Press, New York, New York, 1971.
8. Office of Solid Waste Management Programs, Second Report to Congress, "Resource Recovery and Source Reduction," Publ. SW-122. U.S. Environmental Protection Agency, Supt. of Documents, Washington, D.C., 1974.
9. U.S. Environmental Protection Agency, *Fed. Regist.* **36,** 15495 (1971).
10. A. H. Rose, Jr., D. G. Stephan, and R. L. Stenburg, "Prevention and Control of Air Pollution by Process Changes or Equipment," SEC TR A58-11. U.S. Public Health Service, U.S. Department of Health, Education, and Welfare, Cincinnati, Ohio, 1958.
11. American Conference of Governmental Industrial Hygienists, Committee on Industrial Ventilation, "Industrial Ventilation, A Manual of Recommended Practice," 13th ed. A.C.G.I.H., Lansing, Michigan, 1974.

12. G. W. Siebert and D. A. Fraser, *Amer. Ind. Hyg. Ass., J.* **34,** 481 (1973).
13. A. Koestel, *Trans. ASHVE* (*Amer. Soc. Heat. Vent. Eng.*) **60,** 385 (1954).
14. P. Drinker and T. Hatch, "Industrial Dust," 2nd ed. McGraw-Hill, New York, New York, 1954.
15. R. T. Pring, J. F. Knudsen, and R. Dennis, *Ind. Eng. Chem.* **41,** 2442 (1949).
16. F. N. Chirico, unpublished data referred to by Drinker and Hatch (*14*).
17. F. J. Viles, Jr., *Coll. Univ. Business* **22,** No. 6, 41 (1957).
18. W. C. L. Hemeon, "Plant and Process Ventilation." Industrial Press, New York, New York, 1954.
19. J. L. Alden, "Design of Industrial Exhaust Systems for Dust and Fume Removal," 3rd ed. Industrial Press, New York, New York, 1959.
20. J. M. DallaValle, "Exhaust Hoods." Industrial Press, New York, New York, 1952.
21. American Society of Heating, Refrigerating, and Air Conditioning Engineers, "Systems," Handbook 1973, Chapter 22, p. 22.1. A.S.R.A.E., New York, New York, 1973.
22. U.S. Occupational Safety and Health Administration, Occupational Safety and Health Standards, *Fed. Regist.* **37,** 22144 (1974).
23. American National Standards Institute, "Fundamentals Governing the Design and Operation of Local Exhaust Systems," Z9.2-1960, Z33.1-1961, and ANSI Z9.1-1971. Amer. Nat. Stand. Inst., New York, New York.
24. Environmental Pollution Panel, President's Science Advisory Committee, "Restoring the Quality of Our Environment." The White House, Washington, D.C., 1965.
25. Environmental Pollution Panel, "Inadvertent Climate Modification: Report of the Study of Man's Impact on Climate." MIT Press, Cambridge, Massachusetts, 1971.
26. M. W. First, ed., "Municipal Waste Disposal by Shipborne Incineration and Sea Disposal of Residues," p. 1–14. Harvard Univ. School of Public Health, Boston, Massachusetts, 1972.
27. C. S. Brandt, *J. Air Pollut. Contr. Ass.* **16,** 85 (1968).
28. TS 2.3 Committee, *J. Air Pollut. Contr. Ass.* **22,** 858 (1972).
29. S. Duckworth, *J. Air Pollut. Contr. Ass.* **15,** 274 (1965).
30. U.S. Environmental Protection Agency, *Fed. Regist.* **39,** 42510 (1974).
31. E. J. Croke and J. J. Roberts, *in* "Proceedings of the Second International Clean Air Congress" (H. M. Englund and W. T. Beery, eds.), p. 1247. Academic Press, New York, New York, 1971.
32. Los Angeles County (California) Air Pollution Control District, "Study of Jet Aircraft Emissions and Air Quality in the Vicinity of the Los Angeles International Airport." Los Angeles, California, 1971.
33. Environmental Research and Technology, Inc., "The Greater Boston Air Monitoring, Analysis and Prediction System (Airmap) and its Role in the G.E. [General Electric Co.] Fuel Management Plan." Environ. Res. Technol., Inc., Lexington, Massachusetts, 1972.
34. Institute of Public Administration and Teknekron, Inc., "Evaluating Transportation Controls to Reduce Motor Vehicle Emissions in Major Metropolitan Areas," APTD-1364. U.S. Environmental Protection Agency, Research Triangle Park, North Carolina, 1972.
35. A. D. Cortese, "Ability of Fixed Monitoring Stations to Represent Personal Carbon Monoxide Exposure." Sc. D. Thesis, Harvard University, 1976.

# Part B

# CONTROL DEVICES

# 2

## Selection, Evaluation, and Application of Control Devices

Paul W. Spaite and John O. Burckle

## Nomenclature

| | |
|---|---|
| $C$ | Quantity of pollutant prevented from entering the atmosphere by the control device, or the quantity collected and retained |
| $C_j$ | The total quantity of particulate matter, having a particle size within the $j$th particle-size increment, which is collected and retained |
| $DF$ | Overall decontamination factor |
| $E_j$ | The efficiency of collection for particles having a size within the $j$th particle size increment (decimal percent) |
| $E_o$ | Overall collection efficiency (decimal percent) |
| $I$ | Quantity of pollutant entering the control device from the process, or inlet quantity from the upstream side |
| $I_j$ | The total quantity of particulate matter, having a particle size within the $j$th particle-size increment, which enters the control device from the process, or inlet quantity from the upstream side |
| $\theta$ | Quantity of pollutant leaving the control device and entering the atmosphere, or outlet quantity from the downstream side |
| $\theta_j$ | The total quantity of particulate matter, having a particle size within the $j$th particle-size increment, which leaves the control device and enters the atmosphere, or outlet quantity from the downstream side |
| $P_j$ | Penetration for the $j$th particle-size interval |
| $P_o$ | Overall penetration (decimal percent) |

## I. Selection of a Control System

### A. General Considerations

The selection of a control system is based upon the relationship of the process emission (as uncontrolled) to the quality of the emission needed to meet control objectives. Therefore, careful characterization of contaminants in an effluent stream and establishment of requirements for removal efficiency must precede selection of equipment for gas cleaning.

Minimum removal efficiencies of one or more components may be dictated by applicable local, state, provincial, or national regulation. Regulations will vary widely for different locations. The standards which are applicable are discussed in Chapter 12, Vol. V.

After identifying the requirements which must be met, it is important to fully and accurately characterize the emission stream to be controlled from the standpoint of both the components to be collected and the stream characteristics which affect the operation of collectors. Finally, after the problem has been defined in terms of what is being emitted and what must be removed, it is necessary to choose among collector alternatives. This process for the selection of a complete control system is shown in Figure 1 (*1*).

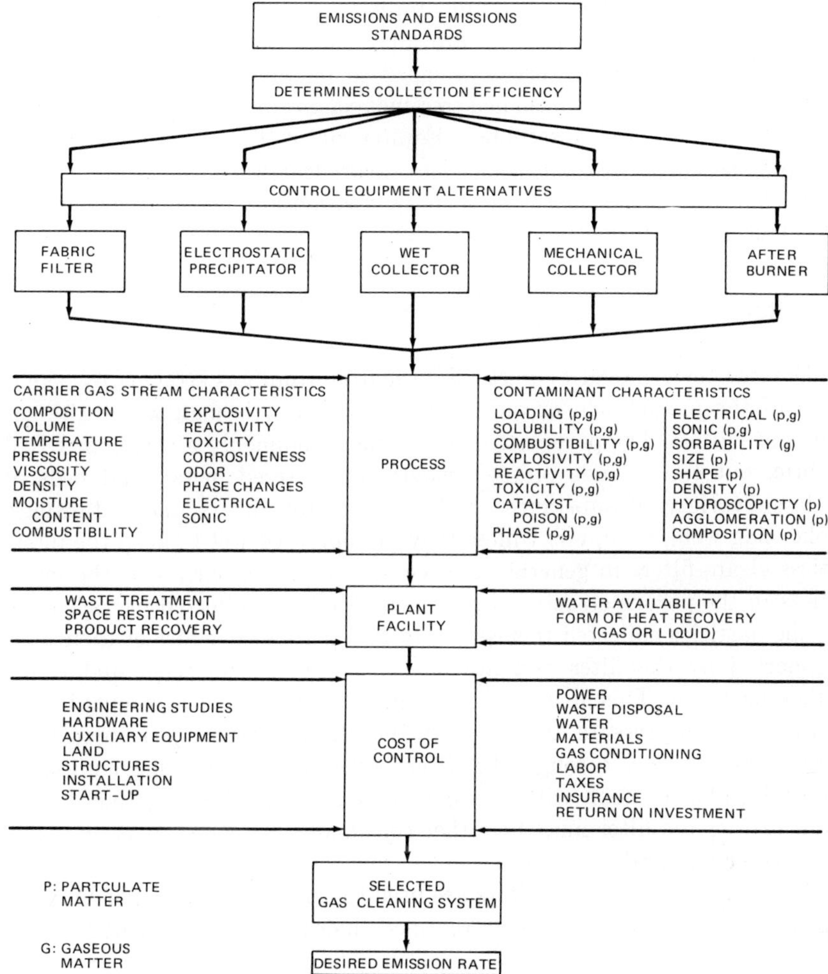


**Figure 1. Process for selection of gas-cleaning equipment (*1*).**

## *B. Types of Control Equipment*

Control equipment may be classified into several general types: filters, electrical precipitators, cyclones, mechanical collectors (other than cyclones), scrubbers, adsorbers, and equipment in which the contaminant is burned as the means for its control. This latter category includes afterburners, catalytic combustion, and similar apparatus. Some equipment combines elements of more than one type. Thus there are cyclones in which a liquid is sprayed, and there are scrubbers in which cyclonic action is employed to remove the liquid droplets. Packed bed filters, operated wet, and packed bed scrubbers are alike in construction, the difference being that when the device is designed to remove particulate matter it is a filter and when it is designed to remove a gas or vapor phase contaminant it is considered a scrubber. Equipment of different types are frequently used in series and sometimes incorporated into the same equipment housing. Thus filters commonly incorporate an integral settling chamber, a form of mechanical collector.

### 1. Filters

Devices for removal of particulate matter from gas streams by retention of the particles in or on a porous structure through which the gas flows are filters. The porous structure is most commonly a woven or felted fabric, but can include pierced, woven, or sintered metal, and beds of a large variety of substances such as vegetable fibers, metal turnings, coke, slag wool, sand, etc. Unless they are operated wet to keep the interstices clean, filters in general improve in retention efficiency as the interstices in the porous structure begin to be filled by collected particles and as the particles collected themselves form a porous structure of their own, supported by the filter and having the ability to intercept and retain other particles. This increase in retention efficiency is accompanied by an increase in pressure drop through the filter. Therefore, to prevent decrease in gas flow through the filter, either the gas-moving equipment must be able to cope with the increased pressure drop without loss of flow rate or the filter must be either continuously cleaned or periodically cleaned or replaced.

In special applications, filters are used to remove gas or vapor by reaction with the particulate matter retained on or in the porous structures.

### 2. Electrostatic Precipitators

Devices in which one or more high intensity electrical fields are maintained to cause particles to acquire an electrical charge and to cause the charged particles to be forced to a collecting surface are electrical precipitators. The collecting surface may be either dry or wet. Since the collecting force is applied to only the particles, not to the gas, the pressure drop of the gas is only that of flow through a duct having the configuration of the collector. Hence pressure drop is both very low and does not tend to increase with time. In general, collection efficiency increases with length of passage through an electrical precipitator. Therefore, additional precipitator sections are employed in series to obtain higher collection efficiency.

### 3. Cyclones

Devices in which organized vortex motion created within the collector provides the force to cause particles to be propelled to locations from which they may be removed from the collector are called cyclones. They may be operated either wet or dry. They may either deposit the collected particulate matter in a hopper, or concentrate it into a stream of carrier gas that flows to another separator, usually of a different type for ultimate collection. As long as the interior of the cyclone remains clean, pressure drop does not increase with time. Up to a certain limit, both collection efficiency and pressure drop increase with flow rate through a cyclone; beyond that limit only pressure drop continues to increase with flow rate increase. Cyclones are frequently used in parallel, seldom in series.

### 4. Mechanical Collectors (Other than Cyclones)

This category includes devices which collect particulate matter by gravity or centrifugal force but which do not depend upon a vortex, as in the case of cyclones. These devices include settling chambers, baffled chambers, louvered chambers, and devices in which the carrier gas–particulate matter mixture passes through a fan in which separation occurs. In general, collectors of this class are of relatively low collection efficiency. They are frequently used as precleaners preceding other types of collectors.

### 5. Scrubbers

Devices, in which contact with a liquid introduced into the collector for the purpose of such contact is the prime means of collection, are scrubbers. Scrubbers are primarily employed to remove gases and vapor phase contaminants from the carrier gas, but are sometimes used to remove particulate matter. The liquid may either dissolve or chemically react with the contaminant collected. Methods of effecting contact between scrubbing liquid and carrier gas include: spraying the liquid into open chambers, or chambers containing various forms of baffles, grillage, or packing; flowing the liquid into these structures over weirs; bubbling the gas through tanks or troughs of liquid; and utilizing gas flow to create droplets from liquid introduced at a location of high gas velocity. The liquid can frequently be recirculated to the scrubber after partial or complete removal of the collected contaminant from the liquid. In other cases all or a part of the liquid must be discarded to waste. In general, as long as the interior elements of the scrubber remain clean, pressure drop does not increase with time. Usually presure drop increases with increasing gas flow rate. The relationship between collection efficiency and gas flow rate depends upon design, generally tending to increased efficiency, provided the liquid feed keeps pace with gas flow and that carry-out of liquid with the effluent gas is effectively prevented.

### 6. Adsorbers

Devices in which contaminant gases or vapors are retained on the surface of the porous media through which the carrier gas flows are adsorbers. The media most commonly used is activated carbon. The design of an adsorber is similar to that of a filter for particulate matter in that the gas flows through a porous bed. However, in the case of an adsorber, the porous bed is frequently protected from plugging by particulate matter by preceding it with a filter so that the gases passing through the adsorption bed are free of particles. In true adsorption, there is no irreversible chemical reaction between the adsorbent and the adsorbed gas or vapor. The adsorbed gas or vapor can therefore be driven off the adsorbent by heat, vacuum, steam, or other means. In some adsorbers, the adsorbent is regenerated in this manner for reuse. In other applications the spent adsorbent is discarded and replaced with fresh adsorbent. Pressure drop through an adsorber which does not handle gas contaminated by particulate matter, should not increase with time, but should increase with gas flow rate. The relationship between retention efficiency and gas flow rate depends upon design.

### 7. Contaminant Combustors

In devices of this class, combustible organic contaminants are burned by the oxygen in the carrier gas to products of as complete combustion as possible. In some cases, combustion takes place on the surface of a catalyst; in others no catalyst is necessary. Combustion is rarely used for particulate contaminants, but mostly for contaminant organic gases and vapors.

## *C. Approach to Selection*

Collector selection involves two basic steps: first, the choice from among all collector types of those which will meet the technical requirements of the process; and second, the choice from among those meeting the requirements, of the type which will do the job at lowest overall cost. The technical requirements are set forth in terms of the carrier gas and the contaminants it carries. With respect to the carrier gas, one must know its physical and chemical properties, its rate of flow and variations of these properties with changes in both rate of flow and time. With respect to the contaminants, one must also know their physical and chemical properties, their concentration or loading in the carrier gas, and variations in both loading and properties with flow rate and with time.

The principal physical properties of the carrier gas and the contaminant are related to their chemical composition. It is not necessary that the chemical composition be precisely known if the physical characteristics are known. However, when collectors selected on the basis of assumed physical properties do not work, it is sometimes necessary to precisely determine chemical composition in order to know what changes to try so as to make the collector work properly.

The prime physical properties of the carrier gas, temperature and pressure, are usually independent of chemical composition unless constituents are reacting exothermally or endothermally. The major properties dependent upon chemical composition of the carrier gas are density, viscosity, moisture content, reactivity, combustibility, toxicity, and electrical and sonic properties. Many of these latter depend upon temperature and pressure as well as chemical composition. If reactivity is construed as including solubility and adsorbability, the above list of properties applies as well to the contaminant as to the carrier gas.

In the special case where the contaminant is a particulate solid, there is an additional group of physical properties related to the size, shape, density, and surface properties of the particulate. These properties are of concern both when the particulate is suspended in the gas stream, and after its collection when they affect the flow and packing properties of the collected material.

When the process is in operation, but without a collector, and the need is to select a collector to add to the process, many of these properties can be measured using accepted techniques. For some of the properties mentioned, there are no accepted techniques. Basic properties are not always translatable into a measure of the ease with which the pollutant may be collected. In these instances, tests must be devised for the specific collection technique being considered. Examples of these latter are those for electrical resistivity of dust and for the retentivity of gases in adsorbents. These tests are designed to provide a direct measure of "collectibility" by a given technique, and do not rely on measurement and interpretation of more basic physical properties. When the process is on the drawing board and the collector is to be specified as part of the plant design, the prime recourse is to experience on similar plants and processes. Where such experience does not exist, one must compute those properties such as carrier gas viscosity, density, humidity, reactivity, and combustibility from basic physical principles. However, methods are not yet sufficiently refined for the advance computation of many of the properties, such as those of the particulate phase, which are necessary for equipment design and selection.

When experience does not exist on a particular process, it is frequently possible to base design upon a closely parallel process for which design information exists. To provide a margin for error in the assumptions made, it is wise in such cases to incorporate a safety factor in the design so that changes required to improve collection efficiency after the plant goes into operation will not necessitate major reconstruction. Such safety factors include the initial provision of motors larger than the minimum horsepower computed, of ducts, pipes, and casings of ample dimensions, etc.

## D. Properties of the Carrier Gas

### 1. Composition

As was previously noted, gas composition is important only as it affects its physical and chemical properties. The chemical properties are important to the extent that there may be chemical reaction between the gas, the contaminant, and the collector—its structure or its contents. One common example of reaction between gas components and equipment is where gases containing sulfur oxides and water vapor corrode metallic parts of collectors.

### 2. Temperature

The two principal influences of temperature are on the volume of the carrier gas and on the materials of construction of the collector. The former influences the size and cost of the collector and the concentration of the contaminant per unit of volume, which, in turn, is a factor where concentration is itself the driving force for removal. Viscosity, density, and other gas properties are temperature dependent.

Adsorption processes are generally exothermic and are impracticable at higher temperatures, the adsorbability being inversely proportional to the temperature (when the reaction is primarily physical and is not influenced by accompanying chemical reaction). Similarly in absorption (where gas solubility depends on the temperature of the solvent) temperature effects may be of significance if the concentration of the soluble material is such that appreciable temperature rise results. In combustion as a means for contaminant removal, the gas temperature affects the heat balance, which is the vital factor in the process. In electrostatic precipitation, both dust resistivity and the dielectric strength of the gas are temperature dependent.

Wet processes cannot be used at temperatures where the liquid would either freeze, boil, or evaporate too rapidly. Filter media can only be used in the temperature range within which they are stable.

### 3. Pressure

In general, carrier gas pressure much higher or lower than atmospheric pressure requires that the control equipment be designed as a pressure vessel. Some types of equipment are much more amenable to being designed into pressure vessels than others.

Pressure of the carrier gas is not of prime importance in particulate collection except for its influence on gas density, viscosity, and electrical properties. It may, however, be of importance in certain special situations such as where the choice is between high efficiency scrubbers and other devices for collection of particulate. The available source pressure can be used to overcome the high pressure drop across the scrubber and thereby reduce the high power requirement which often limits the utilization of scrubbers. In absorption, high pressure favors removal and may be required in some situations.

### 4. Viscosity

Viscosity is of importance to collection techniques in two respects. First, it is important to the removal mechanisms in many situations (inertial collection, gravity collection, and electrostatic precipitation). Particulate removal techniques often involve migration of the particles through the gas stream under the influence of some removal force. Ease of migration decreases with increasing viscosity of the gas stream. Second, viscosity influences the pressure drop across the removal equipment and thereby becomes a power consideration.

### 5. Density

Density appears to have no significant effect in most real gas cleaning processes although the difference between particle density and gas density appears as a factor in the theoretical analysis of all gravitational and centrifugal collection devices. Particle density is so much greater than gas density that the usual changes in gas density have negligible effect.

### 6. Humidity

Humidity of the carrier gas stream may be important to the selection or performance of control equipment in any of several basically different ways. High humidity may lead to caking and blocking of inertial collectors, caking on filter media, or corrosion. In addition, the presence of water vapor may influence the basic removal mechanism in electrostatic precipitation and greatly influence resistivity. In catalytic combustion it may be an important consideration in the heat balance which must be maintained. In adsorption it may tend to limit the capacity of the bed if water is preferentially or concurrently adsorbed with the contaminant. Even in filtration it may influence agglomeration and produce subtle effects. Carrier gas humidity limits the utilization of evaporative cooling to cool gases prior to collection. In situations where humidity is a serious problem for one of the above reasons, scrubbers or absorption towers may be particularly appropriate devices.

### 7. Combustibility

The handling of a carrier gas which is flammable or explosive will require certain precautions. The most important of these precautions is to be sure that the carrier gas is either above the upper explosive limit or below the lower explosive limit with respect to any air admixture that may exist or occur. The use of water scrubbing or absorption may be an effective means of minimizing the hazards in some instances. Electrostatic precipitators are impractical where they tend to spark and may thereby ignite the gas.

### 8. Reactivity

A reactive carrier gas presents special problems. In filtration, for example, the presence of gaseous fluorides may eliminate the possibility of high temperature filtration using glass fiber fabrics. In adsorption, carrier gas must not react preferentially with the adsorbents. For example, silica gel is not appropriate for adsorption of contaminants when water vapor is present as a component of the carrier gas stream. Also, the magnitude of this problem may be greater when one is dealing with a high temperature process. On the one hand, devices involving the use of water may be eliminated from consideration if the carrier gas reacts with water. On the other hand, scrubbers may be especially appropriate in that they tend to be relatively small and require small amounts of material of construction so that corrosion-resistant components may be used with lower relative increase in cost.

### 9. Toxicity

When the carrier gas is toxic or irritant, special precautions are needed in the construction of both the collector, the ductwork, and the means of discharge to the atmosphere. The entire system up to the stack must be under negative pressure and the stack must be of tight construction. Since the collector is under suction, special means such as "airlocks" must be provided for removing the contaminant from the hoppers, if collection is by a dry technique. Special precautions may be required for service and maintenance operations on the equipment.

### 10. Electrical and Sonic Properties of the Carrier Gas

Electrical properties will be important to electrostatic precipitation in that the rate or ease of ionization will influence removal mechanisms.

Generally speaking, intensity of Brownian motion and gas viscosity both increase with gas temperature. These factors are important gas stream characteristics which relate to the "sonic properties" of the stream. Increases in either property will tend to increase the effectiveness with which sonic energy can be used to produce particle agglomeration.

## *E. Flow Characteristics of the Carrier Gas*

### 1. Flow Rate

The rate at which a carrier gas must be treated depends on its rate of evolution from the process, its initial temperature, and the means by which it is cooled, if cooling is used. These factors fix the rate at which gases must be treated and therefore the size of removal equipment and the rate at which gas passes through it. For economic reasons it is desirable to minimize the size of the equipment. Optimizing the size and velocity relationship involves consideration of two effects: (a) reduction in size results in increased power requirements for handling a given amount of gas because of increased pressure loss within the control device and (b) the effect of velocity on the removal mechanisms must be considered. For example, higher velocities favor removal in inertial equipment up to the point of turbulence but beyond this, increased velocity results in decreased efficiency. In gravity settling chambers, flow velocity determines the smallest size that will be removed. In venturi scrubbers, efficiency is directly proportional to velocity through the system. In absorption, velocity varies film resistance to mass transfer. In filtration, the resistance of the medium will often vary with velocity because of changes in dust cake permeability with flow. In adsorption, velocity across the bed should not exceed the maximum which permits effective removal. Optimum velocities have not generally been established with certainty for any of the control processes because they are highly influenced by the properties of the contaminant and carrier gas as well as the design of the equipment. Certain generalizations are possible, however. In gravity settling chambers, flows are generally limited to 10 ft/second to minimize reentrainment. For adsorption on activated carbon, velocities through the beds range from 20 to 120 ft/minute. For conventional fabric filters, filtration velocities range from $1\frac{1}{2}$ to 3 ft/minute.

## 2. Variations in Flow Rate

Rate variations result in velocity changes and thereby influence equipment efficiency and pressure drop. Various control techniques have differing ability to adjust to flow changes. In situations where rate variations are inescapable, it is necessary to: (a) design for extreme conditions, (b) employ devices that will correct for flow changes, or (c) use a collector which is inherently positive in its operation. Filtration is most adapted to extreme rate variations because it presents a positive barrer for particulate removal. This process is, however, subject to pressure drop variations and generally the air moving equipment will not deliver at a constant rate when pressure drop increases. In most other control techniques variations in flow will result in change in the effectiveness of removal unless the equipment has been designed for the least desirable flow condition.

One means for coping with rate variation is the use of two collectors in series, one which improves performance with increasing flow (e.g., multicyclone) and one whose performance decreases with increasing flow (e.g., electrostatic precipitator).

## 3. Changes in Carrier Gas Properties with Variations in Flow Rate

Variations in flow rate are of two main types: those where merely more or less of the same carrier gas flows and those where variations in flow are caused by process changes which also cause variation in the composition or temperature of the carrier gas. Since many carrier gas properties change when composition and temperature changes, equipment selection must give these changes recognition.

### 4. Changes in Properties with Time

This is the more general case of that just discussed. It recognizes that there are processes where flow rate remains reasonably uniform over a process cycle but where gas composition goes through a cyclic variation. The problems are essentially the same as when there is variation of both composition and rate.

### 5. Relationship to Air Mover Characteristics

Control techniques that result in progressively increasing collector pressure loss with time will require that consideration be given to the effect of such changes on air mover selection. Fabric filters are perhaps the best illustration of this effect. Accumulation of dust cake during the filtering cycle results in increased resistance to flow. The increase in resistance generally reduces centrifugal fan output. Where the resultant flow variation cannot be tolerated by the process, positive displacement blowers or other special precautions must be employed.

## *F. General Properties of the Contaminant*

### 1. Composition

In general, the composition of the contaminant is important only as it affects physical and chemical properties and the chemical properties are, in turn, important mainly as they affect physical properties. As a separate consideration, composition directly affects the use or value of the collected material which in turn frequently dictates the kind of collection device required. Thus, if the collected material is to be used in process or shipped dry, a dry collector is indicated, and if the collected material is of high intrinsic value, a very efficient collector is indicated.

a. Change in Composition with Time. Just as the carrier gas composition can change throughout a cyclic process, so can the composition of the contaminant. In the secondary smelting of aluminum, the period of evolution of extremely fine $AlCl_3$ fume lasts for only a few minutes of the 8–16 hours of the total cycle. Since chemical and physical properties vary with composition, a collector must be able to cope with cyclic composition changes.

### 2. Loading

Loading influences different types of collectors in different ways. Thus cyclone efficiency increases markedly at high dust loadings. Conversely, extremely high loading may overtax hopper, rapper, or shaker capacity. Processes such as sonic agglomeration are quite sensitive to changes in loading.

a. Changes in Loading with Time. Contaminant loading from many processes varies over a wide range for the operating cycle. Ten to one variation in loading is not uncommon. One example of such a process is the open hearth furnace; another is soot blowing in a steam boiler.

b. Changes in Contaminant Loading with Carrier Gas Flow Rate. A prime example of a process in which loading increases rapidly with flow rate is fly ash in flue gas from a stoker-fired coal furnace. An increase in flue gas flow rate is the result of an increase in upward velocity of air through the coal bed and increased gas velocity in the furnace, both of which increase the carry-over from the fuel bed into the gas stream.

### 3. Contaminant Phase

In most air cleaning operations, the contaminant to be removed will not undergo change of phase at temperatures near those normally existing in conventional collection equipment (unless such change of phase is related to the actual removal mechanism as it is in absorption where gaseous pollutants are put into solution). However, in some situations, determination of the temperature at which the gas should be cleaned may depend on the relationship between temperature and phase of the contaminant. For example, aluminum chloride is evolved from operations concerned with removal of magnesium from reprocessed aluminum. This aluminum chloride exists as a vapor at temperatures in excess of about 360°F. Since this change of phase takes place at temperatures which commonly exist in many types of control equipment, selection and control of the operating temperature is obviously critical. Change of phase with change in carrier gas temperature has been used as a means for the selective removal of the fumes of different metals in different banks of collectors assembled in series. In this application, the gas in the first collector bank is at a high enough temperature that only one metal fume is condensed to the particulate phase and collected, the fumes of all other metals present in the gas being still in the vapor phase. A drop of temperature between the first and second collector bank condenses another metal fume which is then collected in the bank, and so forth until all metals present have been removed as a fume in an appropriate collector bank.

### *G. Specific Properties of the Contaminant*

#### 1. Solubility

Solubility of contaminant is important to absorption, adsorption, and scrubbing. In absorption, the degree of solubility is one indicator of the ease of removal of the contaminant. In adsorption, solubility may be important to the ease with which the adsorbent may be regenerated. In scrubbing to remove particulate, solubility will provide a secondary removal mechanism to aid the basic separating forces.

#### 2. Sorbability

The sorbability, or ease with which a contaminant can be removed by absorption or adsorption techniques, is a function of a number of more basic properties. Generally adsorption is defined as the process whereby gases, vapors, or liquids are exposed to and concentrated on the surface or in the pores of a solid. Absorption is a similar process where the sorbent may be either liquid or solid and the combination is more permanent because it is accompanied by chemical reaction (which may be reversible) between the contaminant and the sorbent. The more basic properties of temperature, pressure of the system, chemical composition of contaminant and sorbent, solubility, as well as undefined properties such as the nature of the surface forces on adsorbing solids, are of controlling importance in any given situation.

#### 3. Combustibility

Generally, it is not desirable to use a collection system which permits accumulation of "pockets" of contaminant when the contaminant collected is explosive. Systems handling such materials must be protected against accumulation of static charges. Electrostatic precipitators are not suitable because of their tendency to spark. Wet collection by scrubbing or absorption methods may be especially appropriate. However, some dusts such as magnesium are pyrophoric in the presence of small amounts of water. In combustion (with or without a catalyst), explosability must be considered.

## 4. Reactivity

Certain obvious precautions must be taken in the selection of equipment for the collection of reactive contaminants. In filtration, selection of the filtering media may present a special problem. In adsorption, certain situations require that the adsorbed contaminant react with the adsorbent so that the degree of reactivity will be important. For uses where scrubbers are considered, aggravation of corrosive conditions must be balanced against the savings possible because of the fact that corrosive-resistant construction requires relatively small amounts of material.

## 5. Electrical and Sonic Properties

The electrical properties of the contaminant may influence the performance of any of several collector types. Electrical properties are considered to be a contributing factor influencing the buildup of solids in inertial collectors. In electrostatic precipitators, the electrical properties of the contaminant are of paramount importance in determining collection efficiency and influence the ease with which it is removed by periodic cleaning. In fabric filtration, electrostatic phenomena may have direct and observable influence upon the process of cake formation and the subsequent ease of cake removal. In spray towers or other forms of scrubber in which liquid droplets are formed and contact between these droplets and contaminant particles are required for particle collection, the electrical charge on both particles and droplets is an important process variable. The process is most efficient when the charges on the droplet attract rather than repel those on the particle. Sonic properties are significant where sonic agglomeration is employed.

## 6. Toxicity

The degree of contaminant toxicity will influence collector efficiency requirements and may necessitate the use of equipment which will provide ultrahigh efficiency. Toxicity will also affect the means for removal of collected contaminant from the collector and the means of servicing and maintaining the collector. However, the toxicity of the contaminant does not influence the removal mechanisms of any collection technique.

## 7. Contaminant Size, Shape, and Density

a. GENERAL. Size, shape, and density are the three factors that determine the magnitude of forces resisting movement of a particle through a fluid. These forces are a major factor in determining the effectiveness of removal by means of inertial collectors, gravity collectors, venturi scrubbers, and electrostatic precipitators. In fact, this force is a prime consideration for any device, except fabric filters, in which particulate is collected. In every instance, this force is balanced against some removal force which is applied in the control device and the magnitude of the net force tending to remove the particle will determine the effectiveness of the equipment. Even in the case of filters, size and shape of particle influence both collection efficiency and pressure drop.

The only analogous situation in gaseous removal is found in adsorption where minimum molecular weight relates to the adsorbability of the gas. Generally, molecular weights must be in excess of 40–45 before the contaminant can be effectively adsorbed.

b. SIZE DISTRIBUTION. Since size has the previously discussed importance to the ease with which individual particles are removed from the gas stream, it is apparent that size distribution will largely determine the overall efficiency of a particular piece of control equipment. Generally, the smaller the size to be removed, the greater the expenditure that will be required for power or equipment or both. To increase the efficiencies obtainable with scrubbers it is necessary to expend additional power either to produce high gas stream velocities as in the venturi scrubber or to produce finely divided spray water. Cyclones will require that a larger number of small units be used for higher efficiency in a given situation. Both the power cost (because of the increased pressure drop) and equipment cost (for a multiple unit installation) will be increased. Higher efficiencies for electrostatic precipitators will require that a number of units be used in series. Generally there is an approximately inverse logarithmic relationship between outlet concentration and the size of collection equipment. A precipitator giving 90% efficiency must be doubled in size to give 99% efficiency and tripled in size to give 99.9% efficiency.

## 8. Hygroscopicity

Hygroscopicity is not specifically related to any removal mechanism. However, it may be a measure of how readily particulate will cake or tend to accumulate in equipment if moisture is present. If such accumulation occurs on a fabric filter, it may completely blind it and prevent gas flow.

### 9. Agglomerating Characteristics

Collectors are sometimes used in series with the first collector acting as an agglomerator, the second as the collector of the particles agglomerated in the first one. Carbon black collection is an example of a process where extremely fine particles are first agglomerated so that they may be made practicably collectable.

### 10. Flow Properties

These properties are mainly related to the ease with which the collected dust may be discharged from the collector. Extreme stickiness may eliminate the possibility of using equipment such as fabric filters. Hopper size depends in part on the packing characteristics or bulk density of the collected material.

### 11. Catalyst Poisoning

The presence of traces of metals such as mercury, lead, and zinc will make catalytic combustion impractical even though effluent stream characteristics are such that it would otherwise be a suitable technique. Other than by mechanical attrition, catalysts are deteriorated by four other phenomena associated with stream content or condition: (a) surface coating of the granular structure by particulate contaminants within the gas stream, (b) coating by particulate products of oxidation, (c) chemical reaction with gaseous components of the stream, and (d) bed temperature levels which will cause sintering of the catalyst.

## *H. Discussion*

In general, collectors handle as dust between 0.25 and 3.0% by weight of the solid material being processed.

In selecting a collector, one of the principal choices is between wet and dry collectors. Wet systems range between those using 100% makeup water and those with elaborate works for separating the particulate matter from the water so the water may be reused with only minimum addition of makeup water. This same range is involved when dry collectors are used with hydraulic transport of particulate matter from the hoppers to the point of disposal. The factors to be weighed in considering a wet system include solubility of the aerosol, ultimate pH of the scrubbing liquid, its corrosion and erosion potential, special metals or protective coatings necessary to cope with these problems, availability of makeup water, disposal and treatment of waste water, and space required for liquid-handling equipment. Some types of wet systems can also separate aerosols and scrub gaseous pollutants simultaneously from a waste gas stream.

Among the other factors to be considered are:

(a) Operating life of collector components. Fabric filters require periodic replacement. Electrostatic precipitator elements need protection against condensation if corrosive-gas water-vapor mixtures near the dew point are processed. Cyclones can have holes worn in them by abrasive erosion.

(b) Leakage between "dirty" and "clean" compartments. In fabric filters and multiple cyclones a pressure differential exists between the dirty and clean sides of the separator. Any leakage between the two, which bypasses the normal flow path through the fabric or the cyclone, will carry aerosol into the clean compartment, thereby reducing overall efficiency. In fabric filters the most important sources of bypass flow are rips and tears through the fabric itself or its sewn seams and the fabric-to-metal connections required to allow easy fabric replacement. In multiple cyclones, in addition to gasket leakage, there is the possibility of backflow of gas and dust from the dust hopper into the dust discharge opening if there is partial or complete stoppages of the gas inlet to a unit.

(c) Uniformity of aerosol distribution across the inlet duct. Electrostatic precipitators require elaborate measures to ensure such uniformity; inertial collectors are less dependent, and fabric filters least dependent on such uniformity to maintain design efficiency.

## II. Gas Pretreatment

Three general approaches are sometimes used to modify the characteristics of an emission stream prior to its introduction into the primary collection equipment in a gas cleaning system. The first and most common is gas cooling where inlet temperature is reduced to permit the use of equipment with modest thermal resistance instead of more expensive high temperature resistant devices. The second involves the use of additives for any of several purposes, e.g., to neutralize objectionable components which might corrode, foul, or otherwise interfere with equipment operation, to change dust resistivity and enhance electrostatic precipitator performance, or to alter particle-size distribution and improve the filtration characteristics of a dust. The third involves the use of control devices in series so that some component is precleaned to simplify final cleaning of the outlet gas.

### A. Gas Cooling

The importance of carrier gas temperature as an economic factor must be considered in the design of any control system. Because cleaning a gas at high temperature is costly, gas cooling is frequently practiced. Three basic types of cooling can be used: radiation-convection, evaporation, and air dilution. Radiation-convection cooling has the advantage of reducing temperature without introduction of water, which can later condense, or of air, which increases the amount of gas to be processed. Cooling by this method has the disadvantage that it requires large equipment and considerable space. Evaporative cooling can be accomplished using more compact equipment, and increases in the amount of gas to be processed are slight. However the requirement of large amounts of water can complicate the operation of gas cleaning systems. Air dilution has the disadvantage of increased power requirements to move the diluted gas when cooling to low temperatures.

Cooling by any method affects viscosity and volume simultaneously with gas temperature. The effect of cooling by the three methods on power requirements can be important. This relationship is illustrated in Figure 2 which shows the change in power requirement for a filtration system when the stream to be treated is cooled (*2*). The curves show the effect of cooling from either 1800° or 1000°F to the temperatures shown along the abscissa. The three curves originating from each point illustrate the differing effects on power resulting from cooling by each of the three mechanisms independently.

It can be seen that filtering power requirements for an effluent stream will increase only gradually as the gas is cooled by simple air dilution down to as far as 750°F while cooling by either evaporation or radiation-convection over the same range will result in rapid decrease in power required. Even though dilution cooling is the most extravagant method from a power standpoint (cooling by either radiation-convection or evaporation will not add significantly to the total mass being handled), the increase in power which would result from adding the required mass of diluent air is nearly canceled by reductions in both viscosity and volume of the effluent when it is cooled from elevated temperatures to 750°F. While power requirements are not reduced as is the case with evaporative

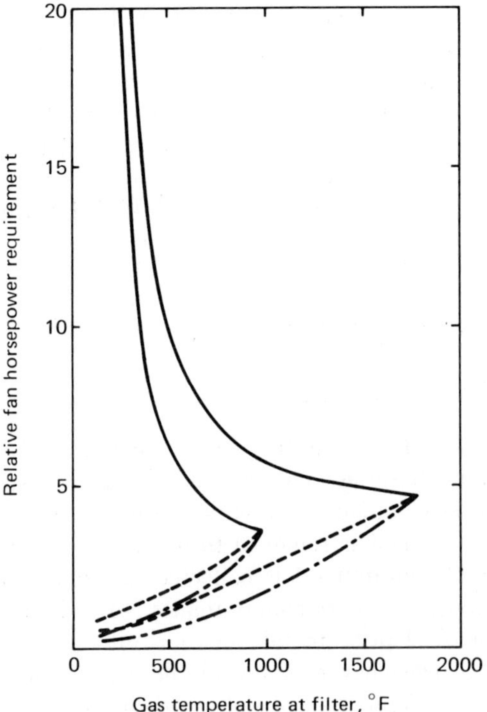


Figure 2. The relation of power requirements for gas cleaning to temperature conditioning (*2*). Solid line: dilution cooling; dashed line: evaporation cooling; dotted-dashed line: radiation and convection cooling.

or radiant cooling they do not begin to increase significantly until dilution to lower temperatures is practiced.

### *B. Additives for Gas Conditioning*

Additives have been used to alter gas stream characteristics prior to cleaning with fabric filters and electrostatic precipitators. Generally additives are used to deal with problems more or less unique to a given installation, for example dolomite has been added to protect fabric filters from acid attack when they were applied to control of an oil burning power plant. Also pretreatment of filter bags has been used on occasion to increase filtration efficiency where low loadings of contaminant are cleaned to very low levels. Because filter additives are used only in special situations, there is little in the way of generally available information on how filtration performance can be enhanced by their use.

Work with additives for electrostatic precipitators has been more extensive and better documented than that with filters. Water, sulfur trioxide, ammonia, and other chemical additives have been used to alter resistivity (*3, 4*). Despite past work, the use of conditioners to enhance precipitator performance is still somewhat of an art. It appears however that increasing interest in ways to maximize precipitator performance will lead to development of a more extensive data base for design of improved additive systems.

### C. Use of Collectors in Series*

#### 1. Purpose of Collectors in Series

Air pollution control systems frequently incorporate two or more contaminant collectors in series. This approach may be used for a variety of reasons but most commonly it occurs in three situations, as where an effluent stream component is removed to maximize the effectiveness with which the principal contaminant is collected, where simultaneous collection of gaseous and particulate contaminants is required, or where coarse particulate is removed prior to introduction into a dust collector being used for final high efficiency cleaning of the gas. Examples which illustrate these most common uses of collectors in series include: systems for precleaning of fly ash and sulfur trioxide from gases being introduced into $SO_2$ recovery equipment, lime and limestone scrubbing systems which incorporate two scrubbers in series to collect particulate and $SO_2$ from power plant emissions, and mechanical collectors in series with baghouses to control emissions from cement kilns.

#### 2. Specific Series Collector Arrangements Used in Dust Control

There are several series arrangements of multiple collectors used in common practice for control of particulate emissions.

(a) The primary collector is used to scalp the larger particles from the gas stream to enhance collection of the finer particles in the secondary collector. Scalping also reduces the load on secondary collector.

(b) The secondary collector(s) may act as a reserve in the event of malfunction in the upstream collector. This approach is used in fluidized catalytic cracking processes in petroleum refineries, where several identical cyclones may be used in series, because plugging of the dip legs

* See Stern (*5*).

(which discharge the collected particulate matter by gravity) is a common occurence.

(c) A primary collector may act as a fire protection device by preventing a source of ignition such as incandescent particles, or readily combustible material such as cotton lint, from entering the secondary collector.

(d) Series collectors may be used to classify materials in a mixture, i.e., one separator may remove a reusable or soluble material, the other a waste material.

(e) A primary collector may act as a plenum chamber from which several secondary collectors may be operated in parallel.

(f) A primary collector may act as a concentrator so that there are two exit gas streams, one with a small gas flow and a high particle loading, the other with a large gas flow and a low particle loading. Either or both of the exit gas streams may go to secondary collectors. A collector acting as concentrator may also discharge separated particles in addition to the two exit gas streams. Common variations of this use include a hopper purged to a secondary collector and the "skimmed" collector where the "cleaned" effluent from a centrifugal collector is split by skimming the gas stream at the outer radius of the exhaust and feeding it to a secondary collector in order to capture the larger particles in the exhaust.

## III. Control Efficiency

### *A. Objectives*

The concept of control efficiency is central to the limitation of emissions into the atmosphere by the use of air pollution control equipment. The efficiency of control is defined as the ratio of the quantity of emissions prevented from entering atmosphere by the control device to the quantity of emissions that would have entered the atmosphere (quantity input to the control device) if there had been no control.

Control equipment is designed on the basis of mathematical relationships, derived in part from the theoretical principles underlying the control mechanisms and in part from empirical data accumulated from tests of pilot and full-scale installations. While theoretical relationships are useful, they are not accurate in all cases, especially where there has been no previous experience with inlet stream conditions. Therefore it becomes necessary to test the performance of the control equipment to ensure that it satisfies in practice the objectives for which it was designed and installed.

The overall evaluation process may, depending on the type of equipment involved, include consideration of the control efficiency, capacity, size, pressure drop, power consumption, utility requirements, operability, operating reliability, and component life in addition to other factors. But from the viewpoint of environmental protection, control efficiency is the most important performance parameter to be considered and, for this reason, is discussed exclusively in the following sections.

There are two basic purposes for field testing the performance of equipment, whether for gaseous or particulate control: acceptance testing to show that newly installed equipment meets specifications and compliance testing to show that a system meets legal requirements.

## 1. Acceptance Tests

Because of the uncertainty inherent in the design of control equipment, it is normal practice to obtain from the vendor, warranties that the installed system will meet the performance objectives under the conditions accepted by the vendor. Acceptance tests are made to ensure that the installation meets the minimum levels of performance for which it was designed and for which warranties had been given. The extent of warranties and the nature of the tests can have a profound effect on the final costs of the control system. Warranty requirements that are too stringent add extra costs, while requirements that are too lenient can result in equipment which falls short of control objectives.

Therefore the objectives of the acceptance tests should be decided in the early planning stages—before the design of the installation has begun. Performance parameters for which warranties will be sought should be defined, and the test methods and procedures to be used should be decided upon and included in the bidding specifications.

## 2. Compliance Tests

Compliance tests are conducted to ensure that emissions are maintained within the regulatory limitations, i.e., "are in compliance." The applicable regulations may require periodic retesting upon demand to ensure continued compliance. Compliance testing should be an integral part of the acceptance test for those control objectives which are dictated by regulations or law. For this reason, it is of paramount importance that all parties to the installation of new equipment understand the significance of the test methods specified by applicable governmental regulations in relation to the design methods used to derive the equipment design efficiency. During the equipment design phase, a review of the engineering layout drawings should be made to determine that adequate space and acceptable sampling stations are available (*6*) to meet the provisions of the tests methods and procedures required by regulatory authorities.

### B. Measures of Control Efficiency

#### 1. Overall Efficiency

The efficiency of gas cleaning devices is expressed in a variety of ways, including control efficiency, penetration, and decontamination factor. The most common means for expressing the efficiency of performance is in terms of the control efficiency, which is defined as the ratio of the quantity of pollutant prevented from entering the atmosphere by the control device to the quantity that would have been emitted (inlet quantity to the gas cleaning device) to the atmosphere had there been no control device.

$$E = \text{Collected}/\text{Inlet} = C/I \quad (1)$$

$$E = (\text{Inlet} - \text{Outlet})/\text{Inlet} = (I - \theta)/I \quad (2)$$

Penetration is defined as the ratio of the amount of pollutant escaping (penetrating without control) the gas cleaning device to the amount entering.

$$P = \theta/I \quad (3)$$

Hence penetration focuses attention upon the quality of the emission stream, but in reality it is actually another way of looking at control efficiency. Since

$$E = 1 - \theta/I \quad (4)$$

then

$$E = 1 - P \quad (5)$$

For a control efficiency of 99.999% (i.e., $E = 0.99999$), the penetration is 0.00001, or $10^{-5}$. Alternatively, efficiency can be expressed as the decontamination factor *DF* which is defined as the ratio of the inlet amount to the outlet amount.

$$DF = I/\theta = 1/(1 - E) \quad (6)$$

For a percentage control efficiency of 99.999%, *DF* is $10^5$. The logarithm to the base 10 of the decontamination factor is the decontamination index. In the numerical example above, this index is 5.0. These efficiency terms are used to represent the overall efficiency of control of a single device or any combination of control devices.

## 2. Efficiency of Collectors in Series

Where several collectors are arranged in series (Fig. 3), their overall efficiency is given, for $k$ stages, by

$$E_0 = 1 - (1 - E_1)(1 - E_2)(1 - E_3) \ldots (1 - E_i) \ldots (1 - E_k) \quad (7)$$

Where $E_1$ is the efficiency of the first stage, $E_2$ of the second, etc. The overall efficiency is then seen to be given by the quantity:

"1.0 minus the product of the penetration values for each stage"

There are several common series arrangements of collectors used in practice (*5*). Where the primary collector acts only as a concentrator to split the gas into two outlet streams, one dirtier and one cleaner than the influent stream, and no dust is removed from the primary collector, removal being effected only from the dirtier gas stream by the secondary collector, $E_1$ is considered "efficiency as a concentrator," and the computation becomes

$$E_0 = E_1 E_2 \quad (8)$$

In this case $E_1$ is the ratio of the amount of particulate matter in the concentrated stream to that in the inlet.

$$E_o = \frac{I_1 - \theta_k}{I_1} = 1 - \frac{\theta_k}{I_1}$$

but since $\theta_i = I_i(1-E_i)$ and $I_i = \theta_{i-1}$

then $\theta_i = \theta_{i-1}(1-E_i)$

and $E_o = 1-(1-E_1)(1-E_2)(1-E_3) \ldots (1-E_i) \ldots (1-E_k)$ follows

$$C_o = \sum_{i=1}^{k} C_i$$

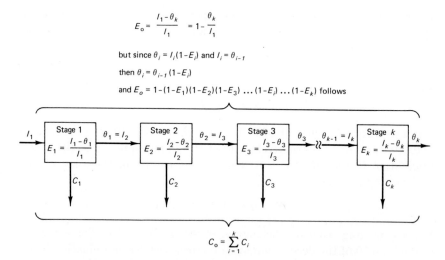

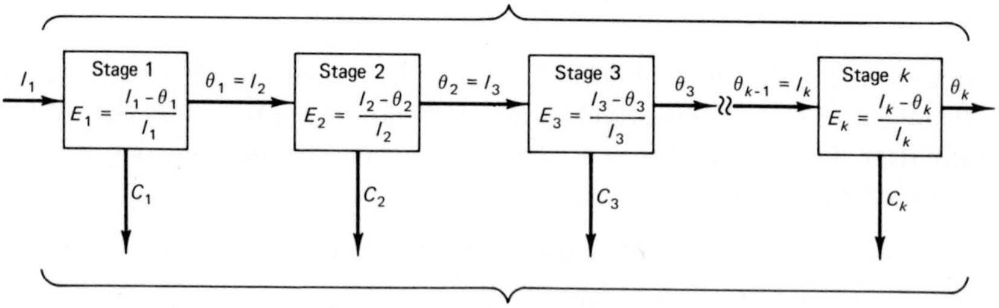

**Figure 3.** Material balance diagram for series collectors.

For the collector with a purged dust hopper, i.e., a hopper which is placed under suction by a bleed line going to a secondary collector, the equation is

$$E_0 = E_1 + ZE_1(E_2 - 1) \tag{9}$$

where $Z$ = fraction of dust collected by the primary collector which leaves the hopper in the purge to the secondary collector.

Another case is that for the skimmed collector, i.e., one which has a normal dust hopper but in which the cleaned gas stream is split into two streams, one dirtier and one cleaner than the average, for which the equation is

$$E_0 = E_1 + XE_2(1 - E_1) \tag{10}$$

where $X$ = percent of dust leaving the primary collector which goes to the secondary collector.

## 3. Fractional Efficiency

In the case of particulate collectors, the efficiency of collection is affected by the size of the particles. This aspect of particulate control introduces the concept of fractional efficiency, i.e., the efficiency over a given particle-size range with which the particles within that range are collected.

The fractional efficiency is defined as follows: for the $j$th particle-size increment, the fractional efficiency $E_j$ is the ratio of the amount of particulate matter in the $j$th increment collected $C_j$ to the total amount in the $j$th increment entering the control device $I_j$.

$$E_j = C_j/I_j \tag{11}$$

As explained before, it will normally be necessary to calculate the collection efficiency from the inlet and outlet quantities of particulate

$$E_j = 1 - \theta_j/I_j \tag{12}$$

where $\theta_j$ is the amount of particulate matter in the $j$th size interval in the outlet stream of the control device. The overall efficiency for $k$ intervals is

$$E_0 = (1/I)(E_1I_1 + E_2I_2 + \cdots + E_jI_j + \cdots E_kI_k) \tag{13}$$

By appropriate substitution and rearrangements of terms this equation for $E_0$ can be expressed in terms of $\theta$ and $C$ as follows:

$$E_0 = \frac{\Sigma[E_j\theta_j/(1 - E_j)\theta_k]}{1 + \Sigma[E_j\theta_j/(1 - E_j)\theta_k]} \tag{14}$$

where $\theta_k$ = the amount from the last stage

$$E_0 = \left(\sum \frac{C_j}{E_j C}\right)^{-1} \tag{15}$$

where $C$ = the total amount collected in all stages.

Just as control efficiency can be expressed for particle-size ranges, fractional efficiency can be expressed in terms of penetration. Penetration of a given particle size or size interval through a control device is defined

$$P_j = 1 - E_j \tag{16}$$

The concept of fractional efficiency for the evaluation of particulate collectors is most useful in considering their selection and performance in the fine particulate range, about 0.01 to 5 $\mu$m.

### C. Testing for Control Efficiency

This section points out several special aspects of sampling and analysis which must be taken into consideration when testing to determine control efficiency, including fractional efficiency. The actual source sampling and analytical procedures are discussed in Chapters 1, 13, 14 in Vol. III.

#### 1. Measurement Basis

The first decision that must be dealt with in the setting up of a test for control efficiency is that of the selection of the response variable upon which the efficiency is to be based. In the case of a gas cleaning system for control of gaseous pollutants, such as sulfur oxides or nitrogen oxides, it is usual practice to measure concentration by volume (as this relates to molar concentration). When dealing with particulate, a number of techniques have been used in the past, the most important being mass loading, reduction in particle count, and soiling measurement (relative volume of gas, before and after the control device, necessary to produce equal soiling of a standard filter paper).

Regulatory requirements play a governing role in this decision, for the basis selected should satisfy those regulations applicable to the equipment to be installed and tested. Once again those situations dealing with particulate measurements are more complex than those for gaseous pollutant control. While the various properties for gas measurement can be related through the gas laws, the properties used to quantitate particulate loading cannot in practice be easily converted from one basis to another as they are comparable only for an aerosol in which all particles are of only one size and have identical characteristics of density, shape, and optical properties (*7*).

As an example consider the following hypothetical case (*5*). Under conditions of constant particle density and spherical shape, mass is proportional to the cube, and soiling to the square of the particle diameter. Therefore, efficiency based on mass is invariably higher than that based upon soiling, which in turn should be higher than that based upon particle count. To illustrate quantitatively, assume a dust which for every thousand 1-μm particles has a hundred 10-μm particles and one 100-μm particle. Next assume two collectors, one which separates all particles larger than 10 μm; the other all particles larger than 1 μm. Applying the second and third power proportionalities noted above, the collection efficiencies of these two collectors by mass, soiling, and particle count are shown in the tabulation below.

| | *Efficiency* (%) | | |
|---|---|---|---|
| *Collector* | *Mass basis* | *Soiling basis* | *Count basis* |
| Collects all above 10 μm | 91 | 48 | 0.09 |
| Collects all above 1 μm | 99.9 | 95 | 9 |

## 2. Test Conditions

Control equipment is designed and built on the basis of estimates or measurements made at some time past for the important variables which influence control efficiency, such as gas flow rate, temperature, composition, density, and viscosity as well as the aerosol concentration, size distribution, composition, shape, and density. Warranties for control efficiency are normally based upon specified conditions or ranges of inlet stream conditions, with the emission limitations specified in regulations often related to some parameter of process capacity.

Once the plant is in operation, the effects of the real values, which may differ from the design values for these variables, become manifest. For acceptance testing, the process should be operated so that the tests can be performed under the conditions specified as the design conditions, if at all possible. Where this is not possible, computational procedures can be applied, to a limited extent, to obtain a mathematical correction. But, to be useful, these procedures should be agreed upon in the negotiation stage of the procurement process and should be included in the warranty agreement.

The equations for converting from test to warranty conditions are based upon the physical laws governing the performance of the control device. Corrections for gas temperature, pressure, density, and viscosity are fairly straightforward. As an example, the correction from test condition a, to warranty condition b, for pressure drop $h$ in cyclone collectors when there are differences in gas flow rate $Q$, gas density $\rho_g$, and gas temperature T (absolute), is (*5*)

$$\frac{h_a}{h_b} = \frac{Q_a{}^2\rho_{ga}}{T_a}\frac{T_b}{Q_b{}^2\rho_{gb}} \tag{17}$$

The corrections to pressure drop for differences in gas viscosity $\mu$, particle density $\rho_p$, and particle size are negligible for cyclones, and that for dust loading is empirical.

Pressure drop and fan horsepower will be different for a bare collector and the same collector plus its housing, ductwork, and straightening vanes. The changes in housing, ductwork, or vanes needed to meet efficiency guarantees may cause pressure drop and power guarantees to be exceeded. The correction to efficiency of cyclone collectors for changes in gas conditions is (*5*)

$$\frac{1 - E_a}{1 - E_b} = \left(\frac{Q_b}{Q_a}\right)^{0.5}\left(\frac{\rho_p - \rho_{gb}}{\rho_p - \rho_{ga}}\right)^{0.5}\left(\frac{\mu_a}{\mu_b}\right)^{0.5} \tag{18}$$

The correction for change in gas temperature is made by appropriately correcting density and viscosity for the temperature change.

Efficiency correction factors based upon substantial differences in aerosol characteristics are much more difficult to develop. As a result, corrections for dust loading are empirical and are usually treated as proprietary information by the equipment suppliers. Among the most difficult factors to correct are those for characteristics derivative to aerosol composition, such as electrical resistivity, wettability, and agglomeration properties.

When testing a particulate control system, whether pilot scale or a plant installation, measured efficiencies higher than true efficiencies, including tests showing over 100% efficiency, can also result from dust, built up within the system prior to test, augmenting hopper collection during the test. The reverse can occur if the buildup occurs during tests where hopper collection is one of the factors used in efficiency calculation. A factor of vital importance in selection of a particulate control device based on pilot scale tests is the representativeness of the test aerosol used to establish the control efficiency upon which the full-scale equipment design has been based. Performance estimates based upon samples brought to a test collector in a drum introduces possibilities for error, the most important of which is the use of material collected by a method which allowed an appreciable proportion of the original fines to escape. The worst case of this type of error is to use a test dust taken from a collector with a fractional efficiency curve similar to, or poorer than, that of the collector being evaluated.

In any case, the inlet conditions and operation of the control device should not be permitted to deviate outside of the ranges applicable to obtaining a given control efficiency during testing or the result will be in doubt since the warranties are based upon the inlet conditions. If acceptance testing is performed under conditions outside the warranty range, then the appropriate correction factors must be applied to adjust the test results to the corresponding values within the warranty range.

In the case of compliance testing, operating conditions should be adjusted to those normally used in day-to-day plant operations, as these conditions could be markedly different from those used for nominal design purposes or as otherwise specified by applicable regulations. Because it is known that the conditions of the inlet gas stream to the control system may have a profound influence upon the control efficiency and because, in many cases, this influence cannot be reliably predicted, any significant change in operating conditions, especially where feed raw materials are concerned, should be accompanied by an emissions test to ensure that the modified operation is in compliance.

Should some process upset occur which causes a shift of inlet conditions or control device operation outside allowable ranges, then the portion of the test sample taken during the upset period would not be valid. Where pollutant measurements are made with continuous monitoring instrumentation, corrections can be made for the values obtained during the upset period. However where measurements are obtained with a time-integrated sampling technique, as is the usual case for particulate matter and noncontinuous wet chemical methods, the test must be interrupted during the upset period or it will be invalid and a repeat test will likely be necessary. For these reasons, it is important to closely monitor the inlet conditions and the operating parameters of the control device in addition to the emitting process equipment so that such test errors can be detected and prevented from influencing the test results.

## 3. Test Location

For acceptance tests, it is usual practice to test both the inlet and outlet gas streams of the control device to determine the quality of control performance. In small scale lab or pilot operations used for equipment development it may be feasible to measure the collected portion, but in large scale equipment this approach usually is not feasible, because of the difficulty, expense, and large errors involved. For such installations it becomes necessary to measure the amount of pollutant in the inlet and outlet streams to determine the efficiency of control. However, when compliance testing, it is the usual practice to sample only the outlet gas stream for the quantity of pollutant carried into the atmosphere.

The process of selecting the exact test locations for both gaseous and particulate pollutant measurements should include special attention to leaks, flow conditions, and traversing requirements.

a. LEAKS. Where leaks occur into the equipment, dilution occurs; therefore, concentration measurements alone may not be used in place of total quantities to determine efficiency or compliance. A study of inleakage (*8*) has shown that gradients in gas concentration are formed because of dilution. These results are a strong indication that particulate concentration gradients should be expected where inleakage occurs. On the other hand, where leaks occur out of the equipment, concentration measurements must be used in place of quantities for calculation, since the total effluent measured would be reduced resulting in an apparent and artificial increase in measured efficiency. Testing should be avoided, if possible, in the area of leaks, especially inleakage.

b. FLOW CONDITIONS. Under conditions of turbulent flow, fully developed turbulent flow free of mechanically induced turbulence is preferred. The sampling location should be selected where such turbulence is at a minimum for particulate sampling and at a maximum for gas sampling; refer to Chapter 14, Vol. III on source sampling for guidelines and detailed discussion.

c. SAMPLING STATIONS. Sampling stations must be located to provide sufficient space to permit traversing where large duct sizes are encountered.

### 4. Sampling Sequence

Unlike the case of compliance testing where only the outlet loadings are of interest, a high degree of coordination between measurement of inlet and outlet loading is required to accurately test for acceptance. Inlet loadings will normally vary somewhat due to variabilities in process raw materials and operations.

Outlet conditions are dependent upon inlet conditions and are related through the control efficiency equation. Therefore the best practice is to measure the inlet and outlet loadings simultaneously. While it is possible to test the inlet and outlet at different times under presumably identical operating conditions, this implicitly assumes that the inlet and outlet loadings are truely at steady state over the total test period. Simultaneous testing removes the necessity of making this assumption and will reduce the overall testing errors.

## 5. Traversing

Traversing should always be employed when sampling particulate matter, when sampling gases in very large ducts, or when sampling for particulate matter or gases where leaks occur into the system between the inlet and outlet test points. Pretest gas traversing should be employed in negative pressure systems to determine if leakage is occuring into the selected test zone. Furthermore, where flow stratification is expected (as in the case of a large electrostatic precipitator) the traverse patterns for simultaneous inlet/outlet sampling should be arranged so that the corresponding inlet and outlet flow segments across the collector are simultaneously sampled. Where testing is accomplished at steady-state operation (which is normally the test condition sought after), pollutant concentration is proportional to quantity and may be used in place of the amounts of pollutant entering and leaving the gas cleaning device (i.e., for inlet volumetric flow rate equal to outlet volumetric flow rate at standard conditions) for efficiency calculations. In the case of non-steady-state flow conditions, the concentration is not proportional to the amount of the pollutant and therefore the amount must be quantified for use in calculating efficiency.

## 6. Test-Method Selection

In the case of acceptance testing, sampling of the inlet and outlet pollutant loadings for a highly efficient control device introduces special problems which must be carefully addressed in the selection and application of a test method and a test procedure. In this case, the outlet loadings are of the order of 5 to 0.001% of the inlet loading (corresponds to 95 and 99.999% control efficiency, respectively). Since test validity is enhanced by the use of identical test methods and procedures, including an identical sampling period, at both the inlet and outlet, the test method must have a considerable rangeability (i.e., trade-off among the factors of sample rate, sample size, nozzle size, and filter tare weight). In compliance testing, this aspect is not important, because only the outlet is tested, and no consideration need be given to the requirements for inlet testing. In the process of selecting equipment to conduct a composite (i.e., integrated in space and time) test, the relationship of pollutant loading to the collection medium is highly important. In the case of gas sampling, the quantity and concentration of the sorbing medium must be carefully selected so that the resulting sample solution will be within the effective range of the analytical technique. Where particulate samples are to be obtained, the relation of the filter tare weight to the total sample weight is the critical factor. As a rule of thumb, the catch weight should be at least 10 to 20% of the filter tare weight.

### 7. Sampling Rate

The sampling rate, i.e., the rate at which the sample is aspirated from the main gas stream, is important to the accuracy with which the sample represents the true pollutant emission rate. Particulate samples must be taken at the isokinetic rate. Composite samples for measurement of gaseous pollutants must be acquired under conditions of proportional rate sampling, while concentration measurements made with real-time analyzers and continuous recording instrumentation may be taken at any convenient constant sampling rate providing an acceptable time lag with stack gas flow rate measurement. The importance of isokinetic rate in sampling for particulate concentration is illustrated by consideration of the aerosol conditions used previously in Section II,C,1 above. For this same aerosol, the apparent results for 50% over- or undersampling in the inlet gases of particles larger than 10 μm would be as shown in the tabulation below (*5*).

| | *Mass efficiency (%)* | | |
|---|---|---|---|
| | *Oversampling* | *True* | *Undersampling* |
| Collects all above 10 μm | 48 | 91 | 160 |
| Collects all above 1 μm | 52 | 99.9 | 183 |

This example stresses the importance of accurate sampling of particles over 10 μm in the inlet gas stream if erroneous mass efficiency measurements are to be avoided. The possibility of this type of error is much less when sampling the outlet gases from a collector, since the outlet gases are generally free of large particles. Because of this, in installations where the catch for a given time interval can be weighed accurately (i.e., to within 1%) there is less chance for error in calculating efficiency on catch weight and outlet sampling than on other bases.

### *D. Testing for Fractional Efficiency*

#### 1. General Considerations

In order to determine the fractional efficiency of a control device, the investigator must obtain data for the particle concentrations as a function of particle size at the inlet and outlet of the control device. The accuracy of any measurement of particle-size distribution for an aerosol is ultimately affected by the representativeness of the overall sample obtained (i.e., sampling), the degree to which the particles are accurately classed into their appropriate size ranges (i.e., classification), and the accuracy of the method used to determine the amount of particles in each size range (i.e., assay). The most significant factor influencing accurate measurements for fractional efficiency determinations are the physical properties of the particles, their flow behavior in the conveying gas stream, and spatial and temporal variations in the particle-size distribution of the source, as coupled with the method and actual equipment selected for size discrimination of the particles and assay of the respective fractions.

#### 2. Sampling

Those factors which influence sampling accuracy for size distribution measurements used to determine the fractional efficiency of particulate control devices include those discussed previously for testing overall efficiency, and, in addition, the problem of probe deposition. Any time a sample is withdrawn from the main gas stream and transported through a probe, the particles comprising the aerosol sample are subject to a variety of phenomena which can alter the size distribution of the aerosol. This transport process can cause fracture of agglomerates or fragile particles to form smaller particles, selective deposition of particles in the probe, or deposition followed by agglomeration to form larger particles, some of which may become reentrained. The mechanisms of deposition have been investigated for certain narrow size ranges, but a sampling technique to prevent such deposition mechanisms from affecting the size distribution of an aerosol over a large size range has not yet been devised. For this reason the investigator must check his equipment carefully after each test for evidence of deposition of particles, and should question the validity of any test where deposition is in evidence. It is recommended that sampling equipment which does not require the sample to be withdrawn from the stack be used.

## 3. Analysis

Another constraint in the choice of a method is the useful particle-size range of the analytical technique. Particle concentrations are commonly expressed in terms of particle number, area, or mass. Additionally, there are a number of different measures of particle size, such as inertia, terminal settling velocity, a characteristic physical dimension, surface area, or volume, each of which is associated with particular measurement techniques. The various methods of analysis measure a "particle size" on the basis of one of these size discriminating techniques relating to these measures of size (*9*). This means that the same sample subjected to several methods of analysis will yield a number of different size distributions, some of which will be significantly different. Therefore, when it is desired to compare the results of the analyses to those of other studies, the method of analysis becomes a significant factor which can prevent useful comparisons of data. For these reasons, when making determinations of fractional efficiency, the investigator must decide first what type of concentration measurement is desired and then select the size parameter, appropriate sampling-analytical technique, and equipment which most nearly satifies the measurement problem he has identified. The aerodynamic size (as measured by inertial classifiers) is considered to be the most important measure of particle size in air pollution work.

## 4. Measurement Systems

There are a great number of techniques for particle-size measurements which have been reduced to practice (*9–11*) and a few newer approaches under development. They are too numerous to discuss individually here, however they may be conveniently catorgorized into classes (Table I) on the basis of the sampling and analysis procedures utilized in their employment.

a. Gross Sampling Laboratory Size Classification and Assay. This method consists of obtaining a gross sample from the source by filtration (either in-stack or out-of-stack), with subsequent redispersion and analysis by any of the laboratory techniques such as the Bahco (inertial), Coulter Counter (electrical resistance), light microscopy (characteristic dimension), and equipment based upon sedimentation or terminal velocity (Stokes diameter). This method provides a time-average sample over a long sampling period (an hour or more), thus permitting traversing while minimizing the number of samples taken. However use of this method introduces a prime problem affecting the accuracy of the mea-

**Table I Methods for Particle-Size Distribution Analysis for Fractional Efficiency Determinations**[a]

| *Method* | *Sampling* | *Classification* | *Assay* | *Example* |
|---|---|---|---|---|
| Gross sampling | C | L | L | Optical microscopy<br>Sedimentation<br>Bahco<br>Coulter Counter<br>Aerosol spectrometer |
| Classified sampling | C | R | L | Impactor<br>Cyclone<br>Diffusion battery |
| Continuous sampling | C | R | R | Single particle optical analyzer |
| *In situ* intermittent sampling | I | R | L | Holography |
| *In situ* continuous sampling | I | R | R | Laser Doppler velocimeter |

[a] Key to abbreviations: (C), Collection, i.e., the sample is obtained by aspiration, removed from the main gas stream and is collected in-stack or out-of-stack for analysis. (I), *In situ,* i.e., the sample is not collected and continues to move with the main gas stream. (L), Denotes that operation is performed off-line, usually in laboratory. (R), Denotes that operation is performed in real-time concurrent with sampling.

surement, namely, modification of the particle-size distribution by probe deposition mechanisms, agglomeration on the filter and within the bulk sample during sampling and transport to the laboratory, and preparation and redispersion for analysis, in addition to those problems associated with the many techniques in use to analyze samples so obtained. Because the actual relation of the state of redispersion of the sample to the aerosol as it exists in the source emission cannot be defined, techniques which require redispersion for size classification (i.e., such as elutriation, sedimentation, particle counting by microscope) must be considered suspect until this relationship can be demonstrated to have minimal effect. Also it is nearly impossible to quantitatively recover small particles embedded within filter media for redispersion. This need for a method to circumvent the gross sampling and provide real-time classification of the aerosol as it is sampled from the source has led to the development of the techniques discussed below.

b. Real-Time Classification/Laboratory Assay. This method employs equipment which causes the sample to be separated into its constituent size fractions during the sampling operation (i.e., real-time classification). The sample fractions are then assayed in the laboratory. The only equipment commercially available for this type of measurement are impactors. The various types of commercially available impactors have been evaluated for application to control devices under field test conditions. The impactor technique can be used to provide useful information on the fractional efficiency of control devices in the size range from about 10 to 0.20 μm, if the following points are observed (*12*):

(a) Collection stage tare weight, ranging from 4 to 60 gm is much too high for accurate weighing in view of the very small sample weights that can be collected before reentrainment occurs. For this reason it is necessary to use a light weight collection substrate, ranging in weight from 10 to 500 mg.

(b) For reliable determinations of mass loading as a function of particle size for sizes below 5 μm, a weighing precision of at least 30 μg is required and 10 μg is desirable.

(c) Particle bounce and reentrainment of noncohesive particles in the smaller size ranges may occur; this condition can be prevented by application of a suitable silicon grease to the collection substrates and/or by operation at much reduced jet velocities (however such reductions limit the lower level of particle size which can be measured).

(d) Total sample size for each stage is limited by the critical clearance (i.e., the clearance at which reentrainment occurs) between the accumulated sample and the jets above. As a result of this construction feature, it is necessary to control the rate of accumulation of sample upon the limiting stage by operating at a sample rate over a sampling time period appropriate to the particle concentrations existing at the sampling location.

(e) Sampling time can be extended by the use of a scalping cyclone in those applications where the aerosol contains high concentrations of particles larger than the nominal cut size of the first stages.

(f) Low flow rate impactors operated externally to the stack or duct cannot provide useful particle-size distribution information because of losses of particles to the probe walls.

Impactors have been suitable only for application where particulate loadings are very low, as in the case of very high efficiency control equipment. Where particulate loadings were other than very low, sampling times at reasonable flow rates become quite short (Fig. 4) (*13*). This problem becomes even more aggrevated when dealing with the inlet to control equipment. Under such conditions, one must employ very low sample flow rates or limit sampling times to less than just a few minutes. Such short sampling times would require that individual samples be taken for each traverse point. Obviously a fairly large number of samplers (suitable for the combinations of flow rates and sample times for each

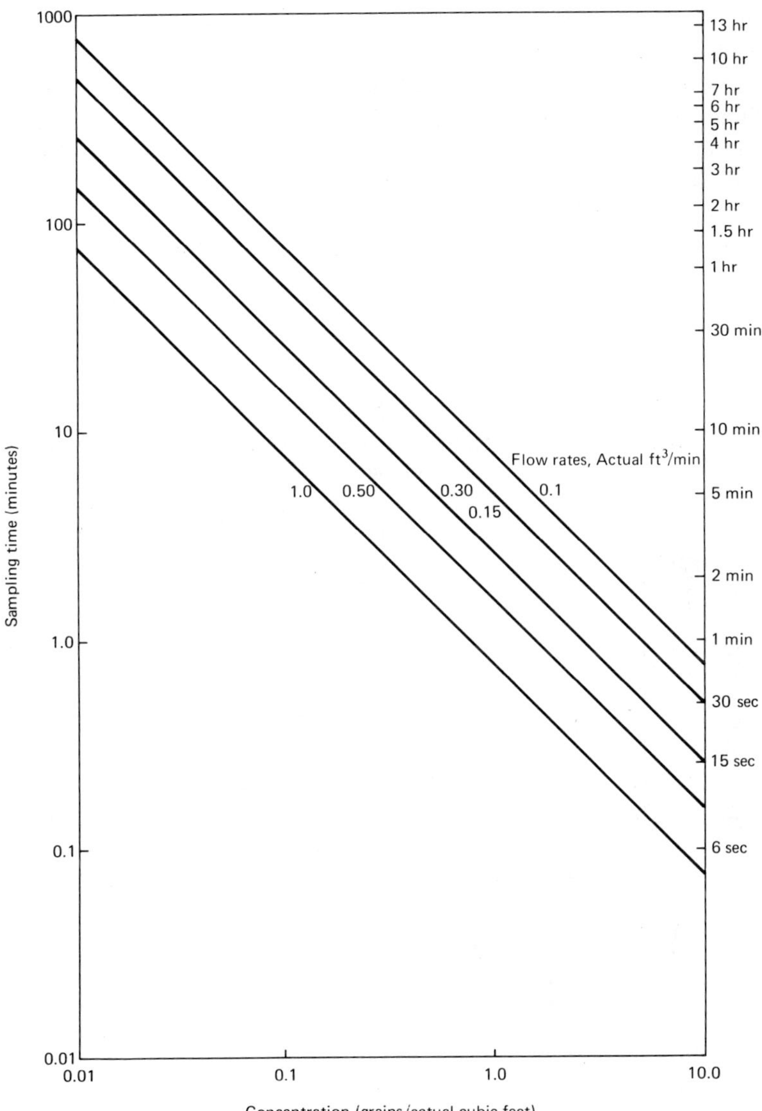


**Figure 4.** Relation of applicable sampling times and rates for particulate loadings for impactors (*13*).

sample point in the traverse pattern) and analyses would be required in performing a fractional efficiency test on a large-scale collector.

Possible future alternatives to the cascade impactor for obtaining particle-size distribution data based on inertial characteristics of particles include miniature cyclones, diffusion batteries, and the apparent surface impactor (impaction into cavities rather than onto solid surfaces).

A prototype parallel cyclone (*14*) assembly, consisting of large scalping cyclone, three smaller cyclones, and a filter, sampling from a manifold at the outlet of the scalping cyclone, has been demonstrated to produce size distribution measurements which agree reasonably well with those obtained with the in-stack classifiers (*12*). Differential size distributions for representative runs are shown in Figure 5. Evidently the high sample

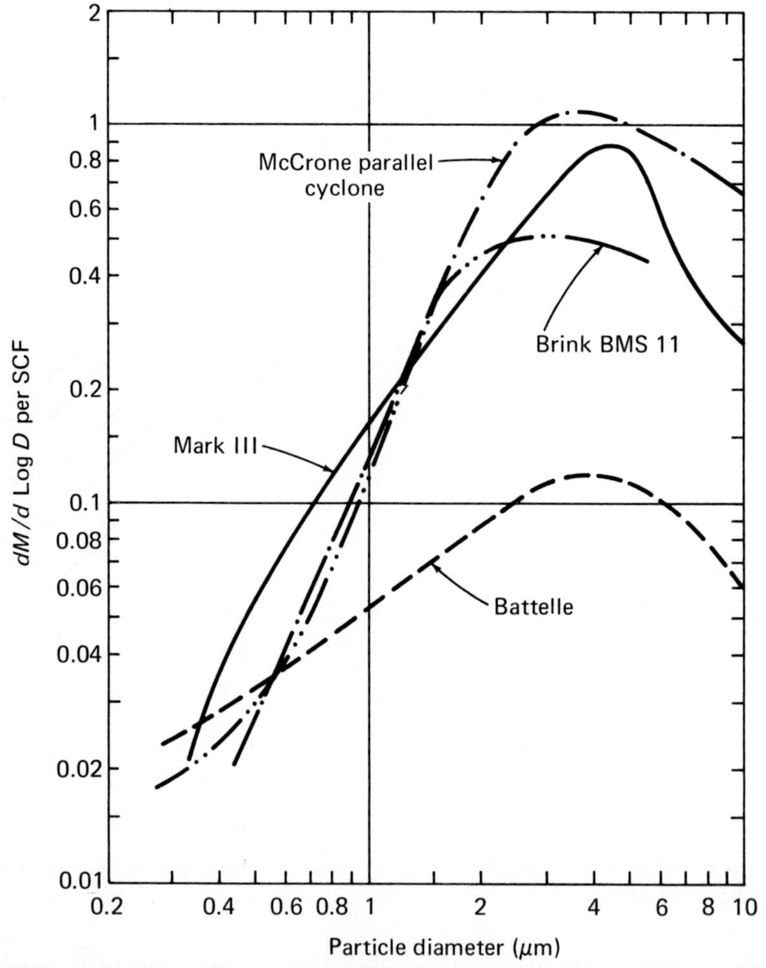


Figure 5. Performance of selected aerodynamic sizing devices (*12*). Inlet size distribution as measured with three out-of-stack samplers (Battelle, Brink BMS 11, McCrone parallel cyclone) compared with the distribution obtained with the Mark III in-stack sampler.

flow rate of the cyclone classifier prevented errors from probe deposition experienced in tests with the low flow rate impactors. Although the larger size and weight of the prototype parallel cyclone make this unit cumbersome to transport and operate in comparison to in-stack impactors, the concept does have particular advantages over existing types of impactors. Sampling times at the control device inlet from 9 to 15 minutes, which resulted in sample volumes between 0.6 and 1.3 $m^3$ (22 to 45 $ft^3$), have been achieved in practice. In comparison, the total volume sampled with impactors is typically about 0.03 $m^3$ (1 $ft^3$). Thus, the parallel cyclone was able to sample for about two to six times as long as impactors with 20 to 50 times the total gas sample volume. The sampling time of the parallel cyclone is limited by the pressure drop across the back-up filters. Also the device can be operated isokinetically while impactors are limited to a constant flow rate.

The diffusion battery technique (*15*) (Fig. 6) is capable of extending the measurement range below the 0.2-$\mu$m point to as low as perhaps 0.01 $\mu$m. Application of the difusion battery technique has provided results (Fig. 7) for the fractional collection efficiency of an electrostatic precipitator below that possible with impactors.

The apparent surface cascade impactor (*16*) accomplishes classification by inertial separation of particles from the sample stream by impingement into a series of cavities, through which a very slow moving flow is maintained to provide for collection of the aerosol particles on a filter substrate. Such a device is shown in the diagram of Figure 8 (*9*). This device is essentially a cascade impactor in which the impaction occurs onto an apparent surface formed by the cavity rather than a solid surface as in a conventional design.

An apparent surface impactor possesses a number of important advantages over the conventional impactor designs, namely, (a) use of high sampling flow rates; (b) long sampling times; (c) elimination of problems of particle pile-up, reentrainment, and particle bounce; (d) collection in a form convenient for other analyses. Provision of sensing instruments in place of the static filter results in automation of the device. Such an automated device has been developed for use for ambient air measurement (*17*).

c. Continuous Sampling and Analysis. In this method, a sample is continuously withdrawn from the stack through a probe and is presented to a monitor in which its size distribution is continuously measured. Only those types of instrumentation which rely upon single particle analysis by light scattering are now available for application to this technique (*9*). Development of the apparent surface cascade impactor (discussed

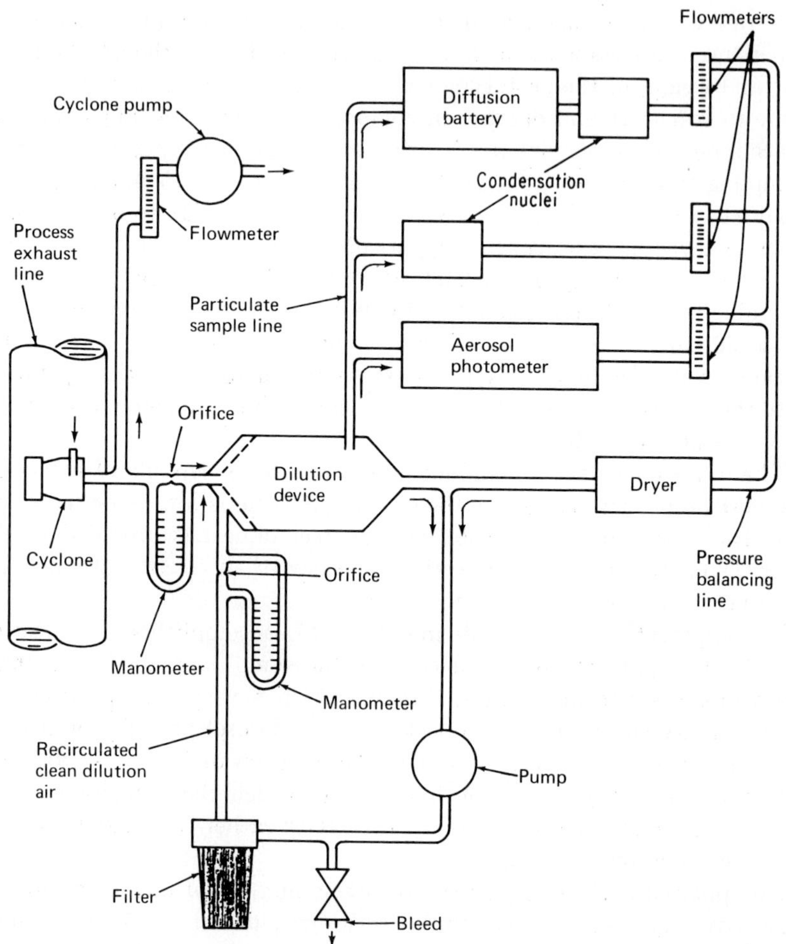


**Figure 6. Diagram of diffusion battery test system of the Southern Research Institute (*15*).**

above) into an automatic instrument would provide another instrument in this category. The problem of modification of size distribution in probes also pertains to out-of-stack continuous monitors. Additionally, available single-particle analyzers are subject to several important limitations in application to stack gases, namely, low sample rate, higher concentrations require sample dilution, and particle-size resolution limited by interaction of electromagnetic radiation with the particle.

d. *In Situ* MEASUREMENT. Advanced electrooptical techniques are under development for routine application to stack measurements. The

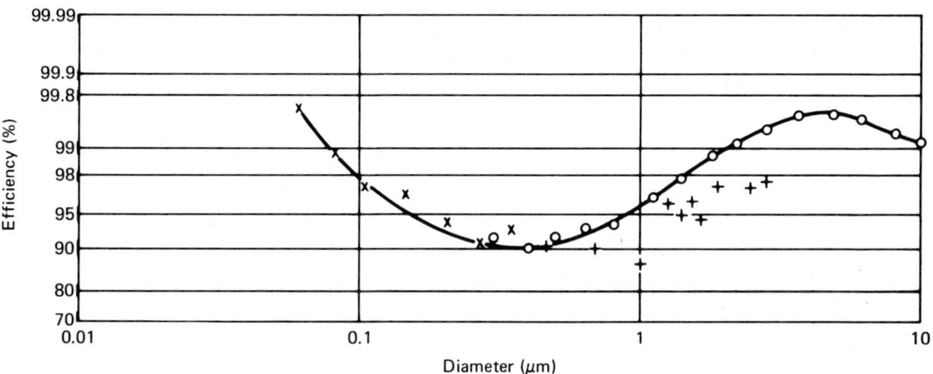


**Figure 7.** Typical fractional efficiency curve for a pilot electrostatic precipitator (*15*). +: optical; ○: impactor; X: diffusional and optical.

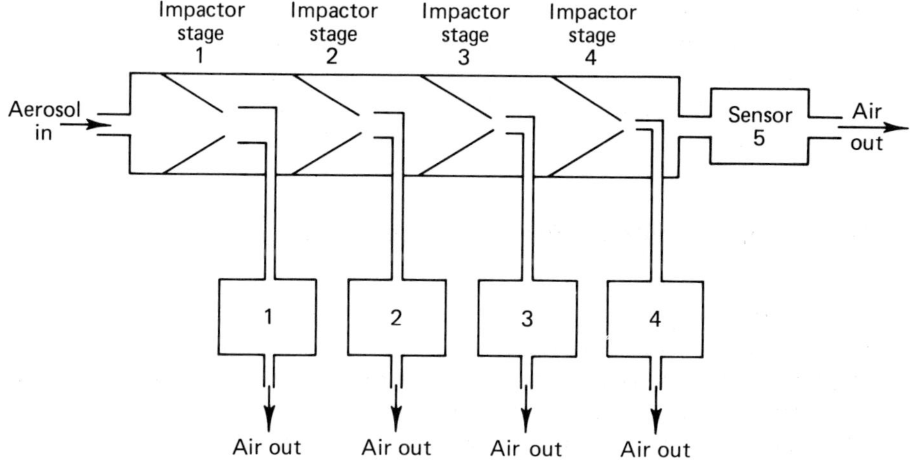


Figure 8. Apparent surface cascade impactor (*9*).

"sampling" is performed *"in situ,"* that is the particle or particle field is measured without physical removal from the main gas stream and the "sample" continues to flow undisturbed with the main gas stream. The discrimination of the size of the particles is made on the basis of the interaction of electromagnetic radiation with the particles. This approach obviates the sampling problems attendant to obtaining and physically removing a sample from the main gas stream. However, because the analysis relies upon the interaction of electromagnetic radiation with the individual particles, the analytical capability is subject to the same limitations as inherent in other optical techniques. The applicable technologies are based on holography and laser doppler velocimetry.

In the holographic process, a recording, which in essence constitutes a sample upon which the aerosol is frozen in its *in situ* condition, is made on film, and later the assay operation is performed on the reconstructed holograph in the laboratory to provide quantitative data for the sizes of the particles recorded as the sample. The size determined is a physical dimension (diameter); and the distribution resulting is a number distribution. Among the more important limitations of this method are: the lowest absolute level of size discrimination is limited by the physics of particle interaction with light and at present this level is about 0.2 $\mu$m; a given minimum particle size which may be detected is limited by range, i.e., as the range (depth-of-field) increases, the minimum detectable size increases (*7, 18, 19*).

The technology of the laser dopler velocimeter (*20, 21*) is similar to that employed by the optical single particle analyzers, i.e., to extract information regarding particle size from the interaction of electromagnetic radiation incident upon a single particle passing through a view volume, and hence is subject to the same limitations in absolute minimum particle-size detection and the type of information furnished.

## IV. Cost, Applicability, and Efficiency of Collectors

It is difficult to generalize relative to the cost, applicability, and efficiency of control devices. Site-related factors can control decisions with respect to which device is preferred and can greatly affect both capital and operating costs. Further, cost will be strongly influenced by required efficiency and degree of reliability which is required by control objectives. A procedure for making a cost estimate for installation and operation of a control system is shown in Figure 9 (*1*). Several factors influencing the cost estimate are given in Table II. Pressures for more effective control of emissions have resulted in changes which make "rule of thumb" generalities increasingly unreliable especially where cost is concerned. With these qualifications in mind, the following data are presented to give a feeling for cost, efficiency, and applicability relationships.

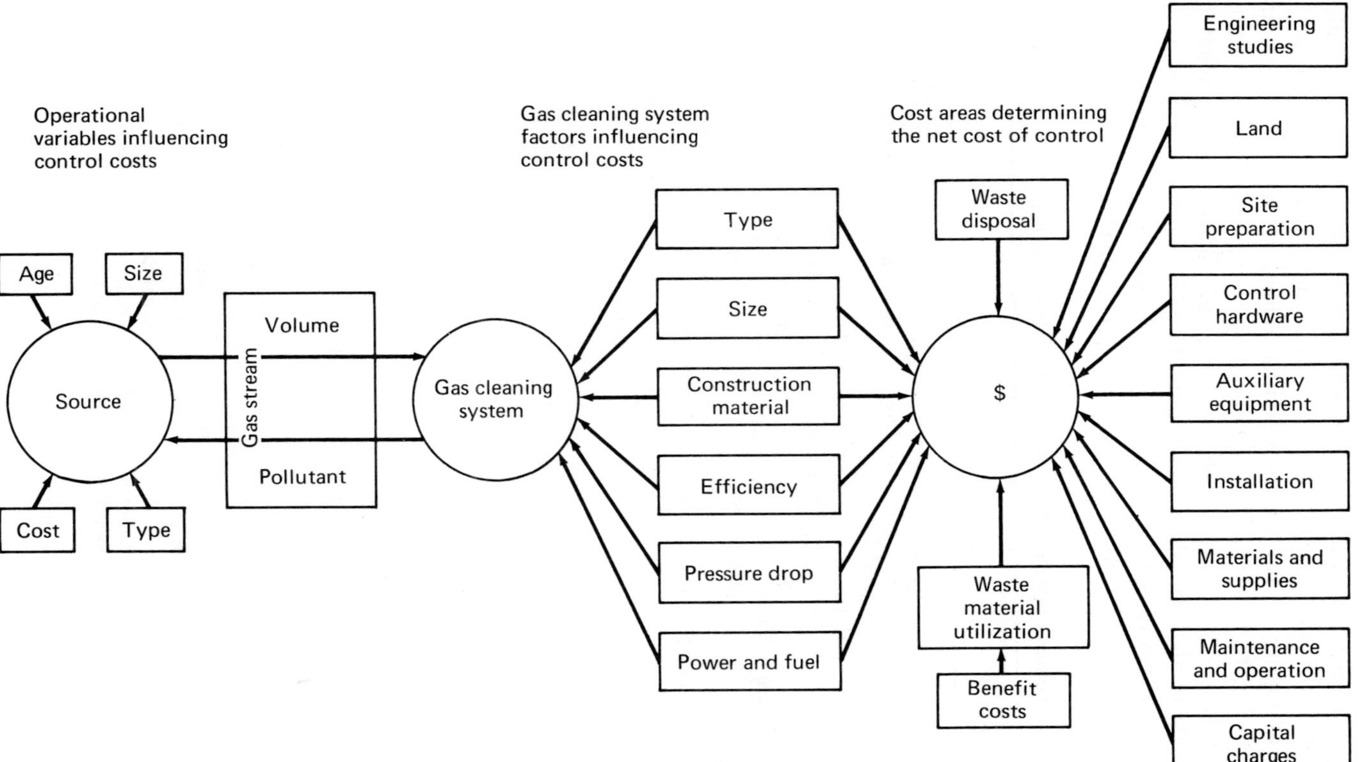


**Figure 9. Diagram of cost evaluation scheme for a pollutant control system (*1*).**

The general relationship between efficiency of collection and installed cost for fabric filters, electrostatic precipitators, and wet scrubbers is shown in Figure 10 (*22*). Corresponding annualized operating costs are shown in Figure 11 (*22*). The annualized operating costs for high voltage precipitators as they are affected by level of efficiency and amount of gas to be treated are shown in Figure 12 (*23*). Cost of equipment for collection of power plant fly ash as it is influenced by efficiency and ash characteristics is shown in Figure 13 (*24*). The effect of power plant size

**Table II Conditions Affecting Cost of Control Devices Installed (*1*)**

| *Cost category* | *Low cost* | *High cost* |
|---|---|---|
| Equipment transportation | Minimum distance; simple loading and unloading procedures | Long distance; complex procedure for loading and unloading |
| Plant age | Hardware designed as an integral part of new plant | Hardware installed into confines of old plant requiring structural or process modification or alteration |
| Available space | Vacant area for location of control system | Little vacant space requires extensive steel support construction and site preparation |
| Corrosiveness of gas | Noncorrosive gas | Acidic emissions requiring high alloy accessory equipment using special handling and construction techniques |
| Complexity of start-up | Simple start-up, no extensive adjustment required | Requires extensive adjustments; testing; considerable downtime |
| Instrumentation | Little required | Complex instrumentation required to assure reliability of control or constant monitoring of gas stream |
| Guarantee on performance | None needed | Required to assure designed control efficiency |
| Degree of assembly | Control hardware shipped completely assembled | Control hardware to be assembled and erected in the field |
| Degree of engineering design | Autonomous "package" control system | Control system requiring extensive integration into process, insulation to correct temperature problem, noise abatement |
| Utilities | Electricity, water, waste disposal facilities readily available | Electrical and waste treatment facilities must be expanded, water supply must be developed or expanded |
| Collected waste material handling | No special treatment facilities or handling required | Special treatment facilities and/or handling required |
| Labor | Low wages in geographical area | Overtime and/or high wages in geographical area |

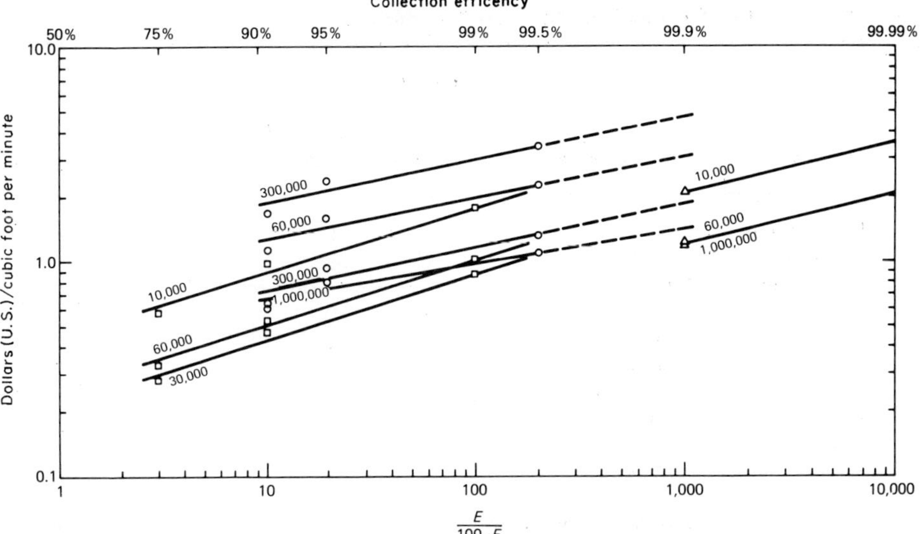


Figure 10. Installed costs for control equipment (*22*). Parameter on curves: equipment capacity in CFM; △: fabric filter; ○: electrostatic precipitator; □: wet scrubber.

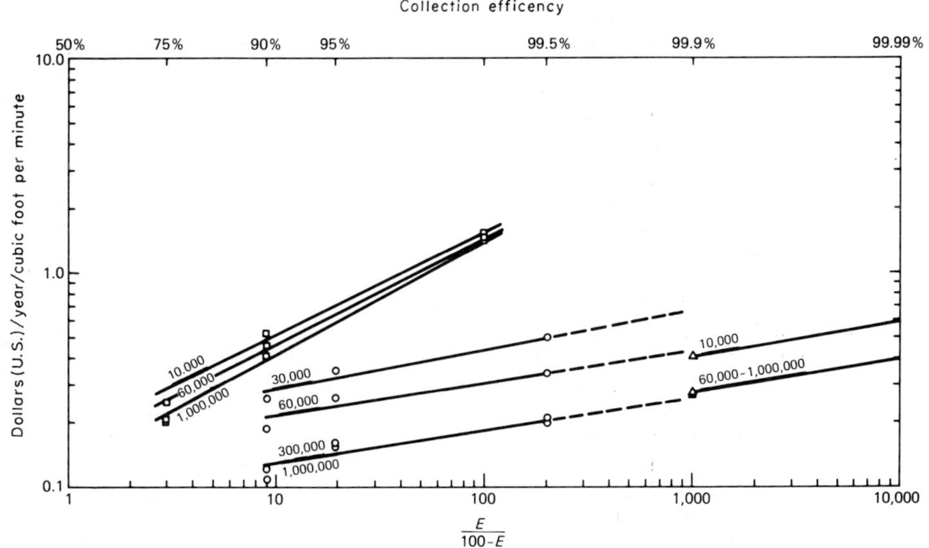


Figure 11. Annualized costs for control equipment (*22*). Parameter on curves: equipment capacity CFM; △: fabric filter; ○: electrostatic precipitator; □: wet scrubber.

**Table III Use of Particulate Collectors by Industry[a,b]**

| *Industrial classification* | *Process* | *Electrostatic precipitator* | *Mechanical collector* | *Fabric filter* | *Wet scrubber* | *Other[c]* |
|---|---|---|---|---|---|---|
| Utilities and industrial power plants | Coal | ⊕ | ⊕ | — | — | — |
| | Oil | ⊕ | ⊕ | — | — | — |
| | Natural gas | Not required | | | | |
| | Lignite | ⊕ | ⊕ | — | — | — |
| | Wood and bark | + | ⊕ | — | + | — |
| | Bagasse | — | ⊕ | — | — | — |
| | Fluid coke | ⊕ | + | — | — | + |
| Pulp and paper | Kraft | ⊕ | — | — | ⊕ | — |
| | Soda | ⊕ | — | — | ⊕ | — |
| | Lime kiln | — | — | — | ⊕ | — |
| | Chemical | — | — | — | ⊕ | — |
| | Dissolver tank vents | — | ⊕ | — | — | + |
| Rock products | Cement | ⊕ | ⊕ | ⊕ | + | — |
| | Phosphate | ⊕ | ⊕ | ⊕ | ⊕ | — |
| | Gypsum | ⊕ | ⊕ | ⊕ | ⊕ | — |
| | Alumina | ⊕ | ⊕ | ⊕ | + | — |
| | Lime | ⊕ | ⊕ | + | — | — |
| | Bauxite | ⊕ | ⊕ | — | — | — |
| | Magnesium oxide | + | + | — | — | — |
| Steel | Blast furnace | ⊕ | — | — | ⊕ | + |
| | Open hearth | ⊕ | — | — | + | + |
| | Basic oxygen furnace | ⊕ | — | — | ⊕ | — |
| | Electric furnace | + | — | ⊕ | ⊕ | — |
| | Sintering | ⊕ | ⊕ | — | — | — |
| | Coke ovens | ⊕ | — | — | — | + |
| | Ore roasters | ⊕ | ⊕ | — | + | — |
| | Cupola | + | — | + | ⊕ | — |
| | Pyrites roaster | ⊕ | ⊕ | — | ⊕ | — |
| | Taconite | + | ⊕ | — | — | — |
| | Hot scarfing | ⊕ | — | — | + | — |
| Mining and metallurgical | Zinc roaster | ⊕ | ⊕ | — | — | — |
| | Zinc smelter | ⊕ | — | — | — | — |
| | Copper roaster | ⊕ | ⊕ | — | — | — |
| | Copper reverberatory | ⊕ | — | — | — | — |
| | Copper converter | ⊕ | — | — | — | — |
| | Lead furnace | — | — | ⊕ | ⊕ | — |
| | Aluminum | ⊕ | — | — | ⊕ | + |
| | Elemental phosphorus | ⊕ | — | — | — | — |
| | Ilmenite | ⊕ | ⊕ | — | — | — |
| | Titanium dioxide | + | — | ⊕ | — | — |
| | Molybdenum | + | — | — | — | — |

**Table III** *(Continued)*

| *Industrial classification* | *Process* | *Electrostatic precipitator* | *Mechanical collector* | *Fabric filter* | *Wet scrubber* | *Other*[c] |
|---|---|---|---|---|---|---|
| Mining and metallurgical | Sulfuric acid | ⊕ | — | — | ⊕ | ⊕ |
| | Phosphoric acid | — | — | — | ⊕ | ⊕ |
| | Nitric acid | — | — | — | ⊕ | ⊕ |
| | Ore beneficiation | + | + | + | + | + |
| Miscellaneous | Refinery catalyst | ⊕ | ⊕ | — | — | — |
| | Coal drying | — | ⊕ | — | — | — |
| | Coal mill vents | — | + | ⊕ | — | — |
| | Municipal incinerators | + | ⊕ | — | ⊕ | + |
| | Carbon black | + | + | + | — | — |
| | Apartment incinerators | — | — | — | ⊕ | — |
| | Spray drying | — | ⊕ | ⊕ | + | — |
| | Machining operation | — | ⊕ | ⊕ | + | + |
| | Hot coating | — | — | — | ⊕ | ⊕ |
| | Precious metal | ⊕ | — | ⊕ | — | — |
| | Feed and flour milling | — | ⊕ | ⊕ | — | — |
| | Lumber mills | — | ⊕ | — | — | — |
| | Wood working | — | ⊕ | ⊕ | — | — |

[a] Reference (*24*), pp. 2637–2638 (E. L. Wilson).
[b] ⊕ = most common: + = also used.
[c] Other = Packed towers, mist pads, slag filter, centrifugal exhausters, flame incineration, settling chamber.

on fly ash precipitator cost is shown in Figure 14 (*24*). Cost data on control systems (wet scrubbers or dry dry mechanical) for small foundry cupola emissions are shown in Figure 15 (*24*).

Table III shows general industrial use patterns for different types of particulate collection equipment. It serves to illustrate some of the important applications for each type of equipment and illustrates where different devices are competitive.

As indicated earlier there is growing interest in defining the fractional efficiency of control equipment. There is no comprehensive body of empirical data upon which to predict equipment capabilities, but estimates of performance to be expected from various types of equipment have been made (Fig. 16) (*25*).

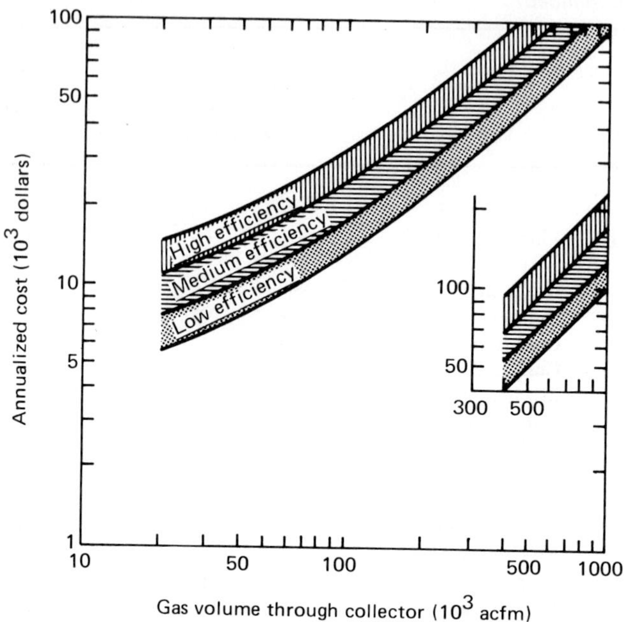


**Figure 12. Annualized cost for operation of high-voltage electrostatic precipitators (*23*).**

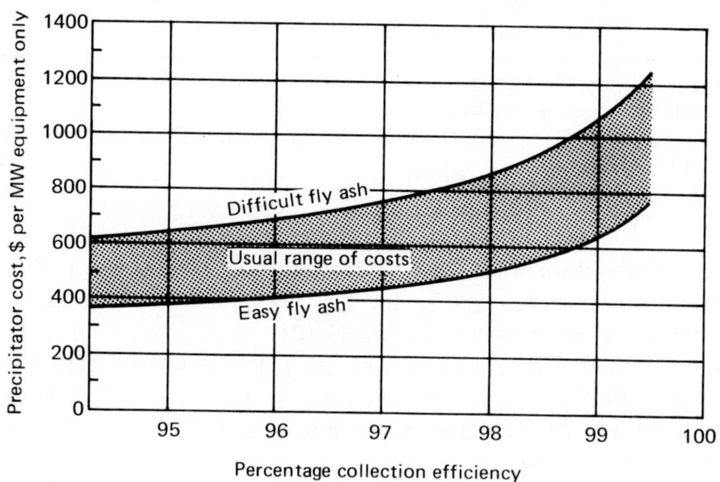


**Figure 13. Relation of precipitator cost to collection efficiency (*24*).**

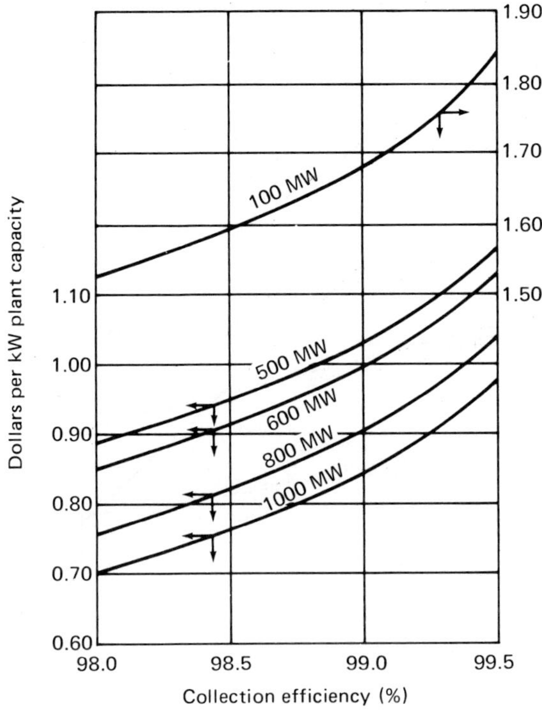


Figure 14. Relation of precipitator cost to capacity (*24*). Approximate average equipment cost per kilowatt for fly ash precipitators. Erected price = 1.68 × FOB. Ash removed system = 0.13 per kW.

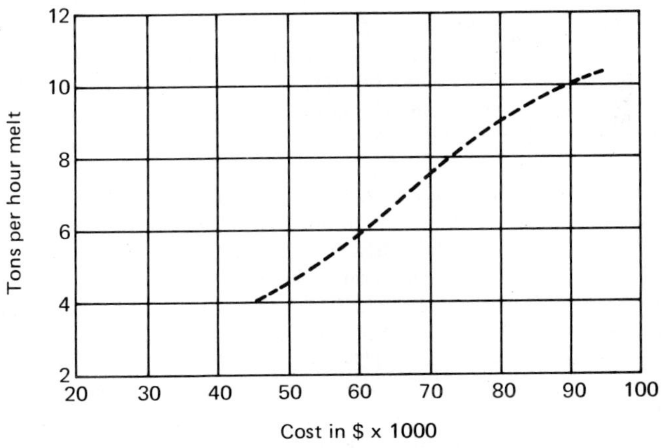


Figure 15. Cupola emission control system costs (*24*). Total installation cost of small foundry cupola emission control system. Wet collectors or dry mechanical collectors.

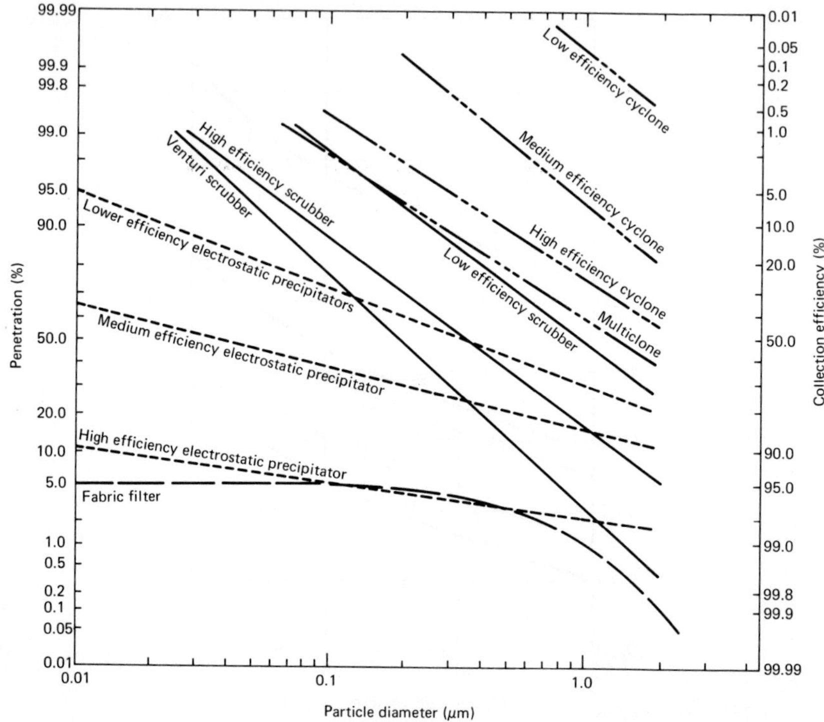


Figure 16. Extrapolated fractional efficiency of control equipment (*25*).

## REFERENCES


1. A. E. Vandergrift, L. J. Shannon, E. W. Lawless, P. G. Gorman, E. E. Sallee, and M. Reichel, "Particulate Pollutant Systems Study, Vol. III, Handbook of Emission Properties," USEPA APTD-0745, p. 548, 1971 (NTIS:PB 203–522, U.S. Dept. of Commerce, Springfield, Virginia).
2. P. W. Spaite, D. G. Stephan, and A. H. Rose, *J. Air Pollut. Contr. Ass.* **11,** 243 (1961).
3. H. J. White, *J. Air Pollut. Contr. Ass.* **24,** No. 4, 313–338 (1974).
4. S. Oglesby and G. B. Nichols, "A Manual of Electrostatic Precipitator Technology: Part I—Fundamentals," USEPA APTD-0610, pp. 166–186, 1970 (NTIS:PB 196-380, U.S. Dept. of Commerce, Springfield, Virginia).
5. A. C. Stern, "Air Pollution" (A. C. Stern, ed.), 2nd ed., Vol. III, pp. 319–358. Academic Press, New York, New York, 1968.
6. S. S. Ross, *Chem. Eng.* **79,** No. 13, 112–118 (1972).
7. G. J. Sem, J. A. Borgos, J. G. Olin, J. P. Pilney, B. Y. H. Liu, N. Barsic, K. T. Whitby, and F. D. Dorman, *in* "State-of-the-Art 1971—Instrumentation for Measurement of Particulate Emissions from Combustion Sources, Vol. 2, Particulate Mass—Detail Report," USEPA APTD-0734, pp. 1–223, 1972 (NTIS: PB 202-666, U.S. Dept. of Commerce, Springfield, Virginia).

8. A. Zakak, R. Siegel, J. McCoy, S. Arab-Ismali, J. Porter, L. Harris, L. Forney, and R. Lisk, "Procedures for Measurement in Stratified Gases," Vol. 1, USEPA EPA-650/2-74-086a, pp. 246–251, 1974.
9. G. J. Sem, J. A. Borgos, K. T. Whitby, and B. Y. H. Liu, *in* "State-of-the-Art 1971—Instrumentation for Measurement of Particulate Emissions from Combustion Sources, Vol. 3, Particle Size," USEPA APTD-1524, pp. 1–81, 1972 (NTIS:PB 233-393/AS, U.S. Dept. of Commerce, Springfield, Virginia).
10. C. E. Lapple, *Chem. Eng.* **75,** No. 11, 149–156 (1968).
11. L. D. Carver. *Ind. Res.* **13,** No. 8, 40–43 (1971).
12. J. D. McCain, K. M. Cushing, and A. N. Bird, Jr., "Field Measurements of Particle Size Distribution with Inertial Sizing Devices," USEPA EPA 650/2-73-035, pp. 1973 (NTIS:PB 226-292AS, U.S. Dept. of Commerce, Springfield, Virginia).
13. D. B. Harris, "Guidelines to Conduct Fractional Efficiency Evaluations of Particulate Control Systems: Part I—Impactors," USEPA, 1975 (in press).
14. "Prototype Construction and Field Demonstration of the Parallel Cyclone Sampling Train," USEPA R2-73-220, pp. 1–28, 1972 (NTIS:PB 221-291, U.S. Dept. of Commerce, Springfield, Virginia).
15. G. B. Nichols, "Proceedings of the Symposium on Control of Fine Particulate Emissions from Industrial Sources," Joint US–USSR Working Group, San Francisco, USEPA 600/2-74-008, pp. 137–168, 1974 (NTIS:PB 235-829, U.S. Dept. of Commerce, Springfield, Virginia).
16. W. D. Conner, *Air Pollut. Contr. Ass.* **16,** No. 1, 35 (1966).
17. J. Wagman and C. M. Peterson, "Proceedings of the 3rd International Clean Air Congress." VDI Verlag, GMbH, Dusseldorf, Federal Republic of Germany, 1973.
18. J. B. Allen, L. M. Boggs, D. M. Meadows, and R. F. Tanner, "Velocity of Particulate in Laminar and Turbulent Gas Flow by Holographic Techniques," USEPA APTD-0918, pp. 1–115, 1971 (NTIS:PB 206-950, U.S. Dept. of Commerce, Springfield, Virginia).
19. J. H. Matkin, "Determination of Aerosol Size and Velocity by Holography and Steam-Water Critical Flow." PhD. Dissertation, Univ. of Washington, Seattle, Washington, 1968 (University Microfilms, Ann Arbor, Michigan, and High Wycomb, England).
20. D. G. Andrews and H. S. Seifert, "Investigation of Particle Size Determination from the Optical Response of a Laser-Doppler Velocimeter," Project Squid Technical Rep. SU-1-PH. Project Squid Headquarters, Purdue University, Lafayette, Indiana, 1972.
21. W. M. Farmer, "On the Measurement of Particle Size, Number Density, and Velocity Using a Laser Interferometer," Contract No. F40600-72-c-0003. Arnold Engineering Development Center, Arnold Air Force Station, Tennessee, 1971.
22. L. J. Shannon, P. G. Gorman, and W. Park, "Feasibility of Emission Standards Based on Particle Size," USEPA EPA-600/5-74-007, pp. 96, 97, 1974.
23. Anonymous, "Control Techniques for Particulate Air Pollutants," USEPA AP-51, pp. 173, 1969 (NTIS:PB 190-253, U.S. Dept. of Commerce, Springfield, Virginia).
24. Hearings before the Subcommittee on Air and Water Pollution of the Committee on Public Works, U.S. Senate, 90th Congress, 1st Session, S.780-Air Pollution-1967 (Air Quality Act) Part 4, March 15–18, 1967. U.S. GPO, Washington, D.C., 1967.
25. L. J. Shannon, P. G. Gorman, and M. Reichel, "Particulate Pollutant Systems Study, Vol. 2—Fine Particle Emissions," USEPA APTD-0744, p. 59, 1971 (NTIS:PB 203-522, U.S. Dept. of Commerce, Springfield, Virginia).

# 3

# Source Control by Centrifugal Force and Gravity

Knowlton J. Caplan

## I. Introduction

Particles suspended in a gas possess inertia and momentum and are acted upon by gravity. These properties may be used to create centrifugal forces acting on the particle if the gas stream is forced to change direction. Centrifugal force is the primary mechanism of particle collection in cyclone separators and in most types of dust collection equipment loosely classed as "inertial separators." Inertia of particles and centrifugal forces also play a part in filtration, scrubbing, and other methods of gas cleaning, but other mechanisms are also important in such equipment.

The cyclone collector is widely used and cyclone theory is the basis for many inertial separator designs. The cyclone is simple, inexpensive, has no moving parts, and can be built of any reasonable material of construction. If the gas cleaning problems do not require high efficiency—viz, nontoxic, coarse, low-value dusts—the collection efficiency of properly designed cyclones is frequently adequate. Disappointing cyclone installations are usually due to overoptomistic estimates of efficiency. Also, because cyclones can be constructed of so many materials, they are frequently used under severe service conditions where erosion, corrosion, or plugging would create problems with any type of equipment. The basic purpose of this chapter is to further the proper application of cyclones and related equipment.

Solid or liquid particulates settle slowly through gas under the influence of gravity. Settling chambers utilizing this mechanism are practical only for relatively coarse material, and have high space requirements.

## II. Cyclone Collectors

### *A. Definition*

A cyclone collector is a structure without moving parts in which the velocity of an inlet gas stream is transformed into a confined vortex from which centrifugal forces tend to drive the suspended particles to the wall of the cyclone body.

### *B. Types of Cyclones*

The necessary elements of a cyclone consist of a gas inlet which produces the vortex; an axial outlet for cleaned gas; and a dust discharge

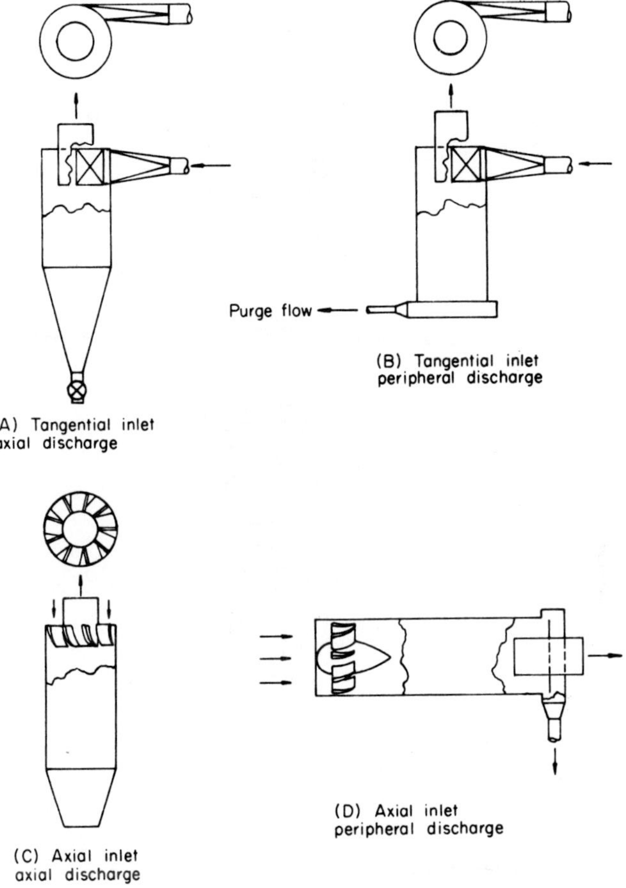


Figure 1. Types of cyclones in common use.

opening. The various arrangements of these elements lead to a classification system of types (Fig. 1A–D):

A. the common cyclone, tangential inlet with axial dust discharge
B. tangential inlet with peripheral dust discharge
C. axial inlet through swirl vanes, with axial dust discharge
D. axial inlet through swirl vanes, with peripheral dust discharge

### C. Mechanism of Cyclone Operation

#### 1. Properties of the Vortex

The common cyclone will be used as the basis for describing the cyclone vortex. The gas entering the tangential inlet near the top of the cylin-

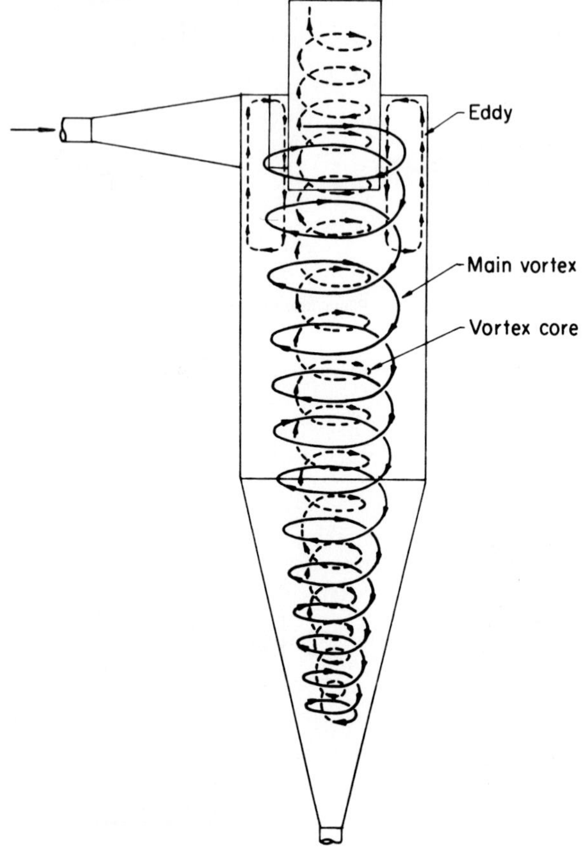


Figure 2. Vortex and eddy flows in a cyclone.

drical body creates a vortex or spiral flow downward between the walls of the gas discharge outlet and the body of the cyclone. This vortex, called the "main vortex," continues downward even below the walls of the gas outlet, and at some region near the bottom of the cone, the vortex reverses its direction of axial flow but maintains its direction of rotation, so that a secondary or inner vortex core is formed traveling upward to the gas outlet (Fig. 2).

The tangential velocity of gas in the vortex ($V_t$) increases as the radius decreases from the radius of the cylindrical body ($R_p$) to a maximum at some intermediate radius; and from this intermediate point inward to the axis of the cylinder the tangential velocity decreases.

In the main vortex (*1*)

$$V_t = V_{tp}(R_p/R)^n \tag{1}$$

where

$V_t$ = tangential velocity at radius $R$, ft/second
$V_{tp}$ = tangential velocity at perimeter (body wall), ft/second
$R_p$ = radius of cyclone body, ft
$R$ = radius, ft
$n$ = exponent, dimensionless

For an ideal gas $n$ would equal 1. Real values are between 0.5 and 1, depending upon the radius of the cyclone body and gas temperature.

The vortex core is generally smaller in diameter than the gas outlet. The radius of the core is between 0.2 and 0.4 times the radius of the gas outlet; and the radius of maximum tangential velocity is from 0.4 to 0.8 times the radius of the gas outlet (*2*).

In the annular space between the cyclone body and the gas outlet, near the top of the body where the tangential inlet enters, the tangential gas velocity increases uniformly from the body wall to the outlet wall, and there is generally downward vortex flow. Because the radius of the outlet is greater than the radius of maximum velocity, the gas does not attain the maximum velocity in the annulus that it can attain later in the main body of the cyclone. In addition, there is upward gas flow along the body wall surface near the top of the cylinder. This upward flow, known as an eddy, carries gas (and dust particles) up along the body wall, inward across the top, and downward along the gas outlet wall. From this point dust particles are lost into the gas outlet. The longer the gas outlet projection into the body, the more pronounced the eddy; elimination of the gas outlet protrusion, however, does not eliminate the eddy. Proprietary designs have been furnished to combat the effects of this eddy current as shown in Figure 3. The axial inlet cyclone (Fig. 1) has no such eddy.

## 2. Separation of Dust Particles in the Vortex

Particulate matter is separated from the gas by centrifugal force which drives the particles toward the cyclone wall across the stream lines of the gas flow. The radial force imparted to the particle is

$$F_s = M_p V_p^2 / gR \tag{2}$$

where

$F_s$ = separating force, pounds
$M_p$ = particle mass, pounds
$V_p$ = particle tangential velocity, ft/second
$g$ = gravitational constant, 32.2 ft/second/second
$R$ = radius of rotation, ft

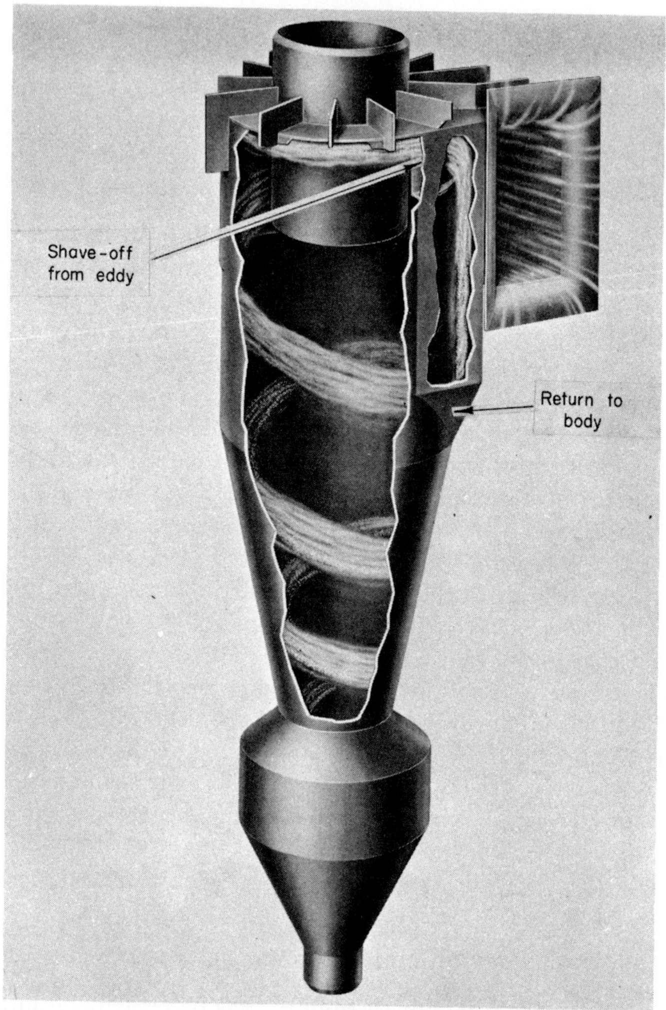


Figure 3. Shave-off to reduce dust loss from eddy current. (Courtesy Buell Engineering Co.)

By assuming that the particle velocity is the same as the gas velocity in the tangential direction, and introducing the parameters for particle mass, the separating force becomes

$$F_s = \beta \rho_p D_p{}^3 V_{tp}{}^2 R_p{}^{2n} / g R^{(2n+1)} \tag{3}$$

where

$\rho_p$ = particle density, lb/ft³
$D_p$ = particle diameter, ft
$\beta$ = volume shape factor, dimensionless

It can be shown that the Stokes' law force resisting the motion of a particle through the gas in the particle-size range 3–100 $\mu$m is

$$F_r = K\mu D_p u \tag{4}$$

where

$F_r$ = frictional resistance to flow, pounds
$K$ = proportionality constant, dimensionless
$\mu$ = gas viscosity, lb/ft-second
$u$ = particle velocity with respect to gas, ft/second

Thus the separating force increases with the cube of the particle diameter but the resistance to particle flow toward the cyclone wall increases only linearly with the particle diameter. In any practical cyclone, the particulate matter is spread over the width of the dirty gas inlet, so that the radius of rotation, tangential velocity, and distance from the cyclone wall vary from the layer entering next to the outer wall of the cyclone to that nearest the wall of gas outlet. Therefore, the cyclone separator is a poor classifier and does not make a sharp size cut between particles separated and passed. Lapple (*3*) has presented an equation typical of those purported to permit calculation of the size particle which will be collected in a given cyclone. Many such equations have been proposed (*1*), and this is presented merely as typical, and to demonstrate the parameters involved.

$$D_{cp} = \sqrt{\frac{9\mu W_i}{2\pi N_e V_i(\rho_p - \rho)}} \tag{5}$$

where

$D_{cp}$ = "cut size," that size collected at 50% efficiency
$\mu$ = gas viscosity, lb/ft-second
$W_i$ = inlet width, ft
$N_e$ = effective number of turns in cyclone (5 to 10 for typical cyclone)
$V_i$ = gas inlet velocity, ft/second
$\rho_p$ = particle density, lb/ft$^3$
$\rho$ = gas density, lb/ft$^3$

Batel (*4*) more recently has given a formula for computation of cut size which is somewhat more direct but which involves a constant, presumably experimentally determined for a given design. His formula also ignores the geometry of the inlet (unless it is intended to be included in the experimental constant), which most investigators (*1*, *5*, *6*) regard as being of prime importance.

### 3. Discharge of Separated Dust

The application of centrifugal force to drive particles out of the gas stream toward the walls of a cyclone collector results in a concentrated dust layer swirling slowly down the walls of the cyclone body. The dust is yet to be finally separated from the gas stream. The purpose of the discharge is to retain the dust or liquid in a container and to prevent its reentrainment into the gas stream at the base of the vortex. The length and dimension ratios of the cyclone body and cone, of course, affect such reentrainment. Smoothness of the inner walls of the cyclone is essential to prevent small eddy currents which would bounce the dust layer out into the active zone of the vortex. Recirculation of gas or in-leakage of gas into the dust outlet will be harmful to attempts to discharge the dust without reentrainment, and conversely, a small purge flow of gas outward from the dust outlet will be helpful; similarly an air lock material discharge valve or dip leg is an aid to discharging dust without reentrainment.

Because the outlet duct for the gas discharge consists essentially of a cylinder to confine and conduct the vortex core out of the cyclone, any dust which escapes into this gas stream is still subjected to centrifugal forces and tends to be concentrated near the walls of the duct. Devices for skimming off the outside layer of this vortex can be employed to improve the overall efficiency of dust separation.

## *D. Efficiency and Pressure Drop of Cyclones*

### 1. General Conditions

Cyclones are frequently divided into two classes, conventional and "high efficiency." High efficiency cyclones merely have a smaller body diameter to achieve greater separating forces, and there is no sharp dividing line between the two groups. High efficiency cyclones are generally considered to be those with body diameters up to about 9 in.

Cyclones as a class of equipment provide the lowest collection efficiency as well as the lowest initial cost for devices in general commercial use to control particulate air pollution sources. Ranges of efficiency to be expected from cyclone collector installations are shown in Table I.

In general, efficiency will increase with increase in dust particle size or density, gas inlet velocity, cyclone body or cone length, and ratio of body diameter to gas outlet diameter. Conversely, efficiency will decrease with increase in gas viscosity or density, cyclone diameter, gas outlet diameter, and inlet width or inlet area.

**Table I Efficiency Range of Cyclones**

| Particle size range, μm | Efficiency range, wt % collected | |
|---|---|---|
| | *Conventional* | *"High efficiency"* |
| Less than 5 | Less than 50 | 50–80 |
| 5–20 | 50–80 | 80–95 |
| 15–40 | 80–95 | 95–99 |
| Greater than 40 | 95–99 | 95–99 |

For a given cyclone arrangement, the resistance varies with the square of the air volume and therefore with the square of the inlet velocity. Since the velocity pressure or velocity head of a flowing gas also varies with the square of the velocity, it is convenient to express cyclone pressure drop in terms of the number of inlet velocity heads.

Unfortunately, changes in parameters which tend to increase collection efficiency also tend to increase pressure drop. The loss of pressure of the gas stream through the cyclone is of the order of one to four inlet velocity heads, with a resulting range of resistance from $\frac{1}{4}$ to 8 in. water gauge (wg). Most actual installations will show a resistance between 2 and 7 in. wg. Efficiency increases with increasing inlet velocity, but this increase is at lower rate than the increase in pressure drop. In addition, for any practical installation there is a limiting value of inlet velocity above which turbulence causes a decrease in collection efficiency with further increase in inlet velocity.

The only practical upper limit to dust loadings occurs with the smaller body diameters of high efficiency cyclones, where as much as 10 to 15 grains/ft$^3$ of a fine or sticky dust may cause plugging of the narrow passages. For ordinary cyclones, there is no practical upper limit to the dust loadings and the efficiency increases with an increasing dust load. Even though the percent collected increases with increasing dust load, the total weight rate of contaminant discharged will be higher for higher inlet loadings because the efficiency increase is not nearly as rapid as is the increase in loading.

An extensive survey (*1*) was made of the various proposed equations for calculating cyclone pressure drops, and the most promising of these methods were tested against three different sets of field data which included 29 different cyclone installations. It was concluded that there was no known satisfactory method for predicting cyclone resistance from cyclone dimensions which was accurate over a wide range of different types of construction. For commercial units, pressure drop should be determined experimentally on a geometrically similar prototype. Where this is not possible, the following method developed by Alexander (7) is presented as being equivalent to other published methods and yielding results which are as good as can be obtained in the present state of the art.

According to Alexander, cyclone resistance is assumed to be a function of gas inlet area and outlet area in the form

$$\Delta P_c = CH_i W_i / d_o^2 \tag{6}$$

where

$\Delta P_c$ = cyclone resistance, number of inlet velocity heads
$C$ = proportionality constant
$H_i$ = inlet height, ft
$W_i$ = inlet width, ft
$d_o$ = gas outlet diameter, ft

The proportionality constant $C$ can be determined from

$$C = 4.62 \frac{R_o}{R_p} \left\{ \left[ \left( \frac{R_p}{R_o} \right)^{2n} - 1 \right] \left( \frac{1 - n}{n} \right) + f \left( \frac{R_p}{R_o} \right)^{2n} \right\} \tag{7}$$

where

$R_o$ = gas outlet radius, ft
$R_p$ = body radius, ft
$f$ = varies with $n$

| $n$ | 0 | 0.2 | 0.4 | 0.6 | 0.8 |
|---|---|---|---|---|---|
| $f$ | 1.90 | 1.94 | 2.04 | 2.21 | 2.40 |

The value of the exponent $n$ can be determined from Figure 4, and Equation (7) has been reduced to graphical form in Figure 5.

## 2. Design Factors Affecting Efficiency and Pressure Drop

a. Body Diameter and Dimension Ratios. In the study of design parameters which affect efficiency and pressure drop, a typical cyclone has been selected as the starting point. The dimensions of the various body elements (Fig. 6) are based on the proportion of the dimension to the gas outlet diameter.

Starting with this design, a cyclone of higher efficiency and higher pressure drop could be designed by increasing the length of the cyclone, decreasing the inlet width, or increasing the ratio of body diameter to outlet diameter while at the same time providing a smaller body diameter.

The length of the cyclone body is of importance. An increase in length provides for a longer residence time of gas in the vortex and therefore

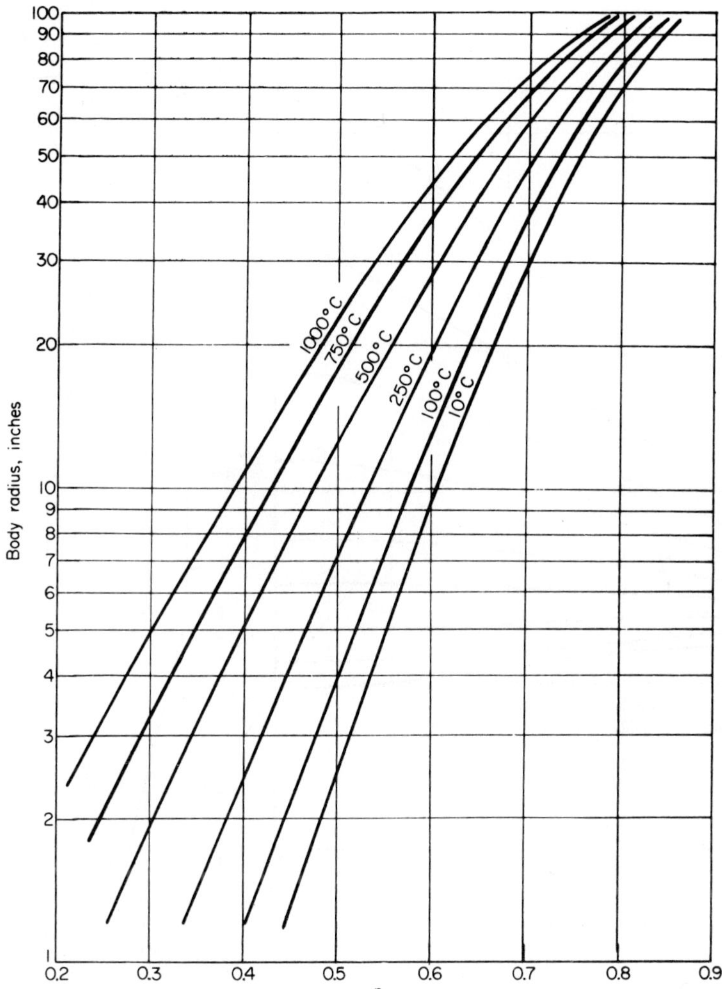


Figure 4. Values of *n* in Equation (7).

for more revolutions or turns in the vortex. A very pratical consideration involving the total length of the cyclone is that one of the most common defects leading to lower efficiency is reentrainment of dust into the vortex core from the region of the dust discharge port. The longer the cyclone below the gas outlet, the greater the opportunity for reentrained dust to be precipitated out of the vortex core before it enters the gas outlet.

A number of investigators (*7–10*) have presented data or conclusions which are in substantial agreement that the height of the main vortex zone should be at least 5.5 times the gas outlet diameter, preferably more

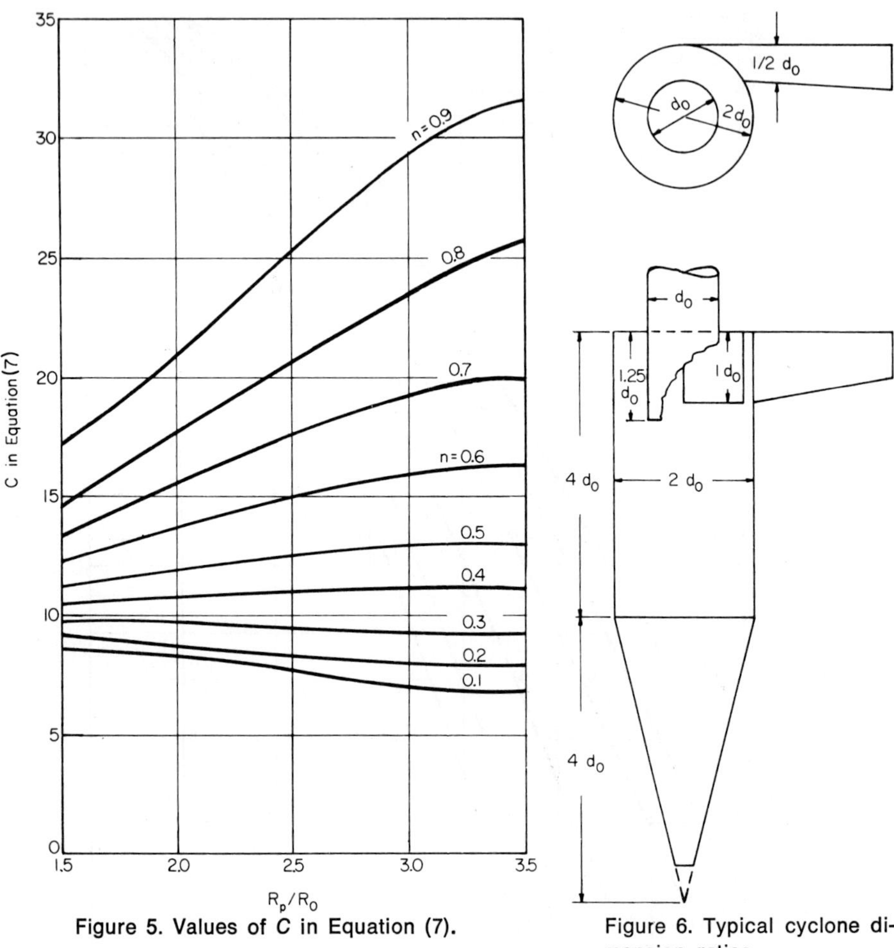


Figure 5. Values of $C$ in Equation (7).

Figure 6. Typical cyclone dimension ratios.

perhaps up to 12 times the outlet diameter. The total length for the typical cyclone (Fig. 6) of 8 times the outlet diameter seems to meet these criteria. Contrary to general statements previously made, increasing the length of the cyclone without changing any other dimension ratios will achieve an improvement in efficiency with no penalty in terms of increased pressure loss.

Increasing the ratio of the body diameter to the gas outlet diameter does show an increase in efficiency up to a ratio of about 3 with relatively small gain above that. On the other hand, there is a corresponding increase in pressure drop as this ratio increases so that the optimum ratio would appear to be between 2 and 3.

Theoretically, efficiency should continue to increase with a decrease in cyclone diameter, but this has not been proved in practice. In a very small cyclone, the gas outlet is dimensionally very close to the region where the dust is concentrated along the cyclone wall. Therefore any bouncing of large particles or local eddies caused by turbulence are more likely to result in accidental loss of dust to the gas outlet merely because the dimensions are so small. Furthermore, the corresponding dimensions of gas inlet and dust discharge are small, leading to practical difficulties with many industrial dust dispersions of a sticky nature.

b. Cone Design. The definition of a cyclone makes no mention of a cone. If a cone is present its design is important, but it is not necessary for a cyclone to have a cone. Neither is a cone essential to cyclone theory, since the main vortex will transform to the upflowing vortex core in a long cylinder without a cone. The various types of cyclones without a cone will be discussed later under the subject of dust discharge. However, a cone does serve the practical function of delivering the dust to a central point for ease in disposal, and forces the main vortex to transform to the vortex core in a shorter total length than would occur in a straight cylinder. In nature, the axis of a free vortex is frequently curved. A similar curvature or eccentricity of the vortex core has been observed in cyclone operation *(11)* and may amount to as much as one-fourth the gas outlet diameter. Theoretically, the diameter at the apex of the cone should be greater than one-fourth the gas outlet diameter to prevent the vortex core from touching the wall of the cone and reentraining collected dust. For cyclones of larger sizes, a cone apex of such dimensions may be unreasonable. This merely reemphasizes the need for adequate total cyclone length so that any dust reentrained at the cone apex may be separated again before it reaches the gas outlet.

c. Inlet Design. The design of the cyclone inlet is of critical importance to both cyclone efficiency and pressure drop. Unfortunately, little design data are available for the axial inlet type, since most cyclones of this design are proprietary. However, there has been much effort to experiment with the design of the tangential inlet to improve cyclone performance. The different types of common tangential inlets are shown in Figure 7.

The helical inlet design is provided to impart a downward velocity to the gas to avoid interference between the incoming gas and the mass of gas already rotating in the annulus. Existing test data are conflicting as to whether or not this design actually does provide a lower pressure drop, and there is some indication that a lower efficiency is obtained. Most commercial cyclones do not have a helical inlet.

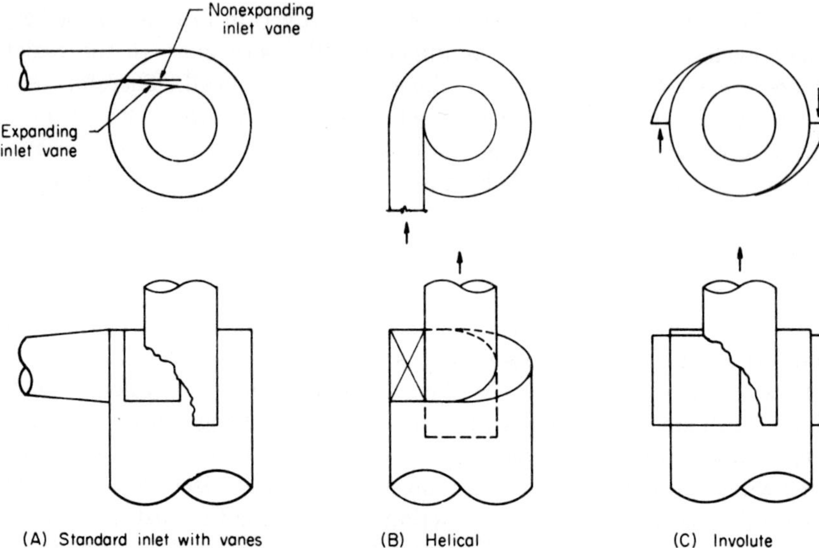


Figure 7. Types of tangential inlets.

As the inlet gas enters the annular space between the cyclone body wall and the wall of the gas outlet duct, it undergoes a squeeze between the body wall and the rotating air mass already in the annulus. The involute inlet design has been developed to minimize the interference between these gas streams. Use of multiple involute inlets has the further advantage that for the same inlet area and height, the inlet width is reduced.

No quantitative test data are available on the effects of multiple inlets on efficiency or pressure drop. However, in most practical cases, the multiple inlet designs involve small "high efficiency" cyclones where the inlet gas is taken from a plenum serving all inlets simultaneously. If practical, a bell-mouth inlet from the plenum to the cyclone inlets will reduce pressure loss and possibly improve efficiency.

The acceleration of gas associated with the previously described squeeze of gas entering the annulus is an important part of the total pressure loss of the cyclone. Inlet vanes (Fig. 7A) have been tried as a method of reducing this pressure loss. Nonexpanding inlet vanes result in one-half the pressure drop of the same cyclone without vanes; expanding vanes correspondingly in one-fourth the pressure drop. However, both of these types of inlet vanes decreased dust collection efficiency by preventing the formation of a vortex in the upper part of the annulus. Here again we are confronted with the dilemma of designs to decrease the pressure drop resulting in a corresponding decrease in efficiency. A cyclone is a device for the creation of a vortex, and it is the vortex that does the work in separating the particulate matter from the gas. Therefore it seems unwise to make modifications which are intended to, or which result in, suppressing the vortex.

The approach duct to a common cyclone is usually round, and if a proper inlet height-to-width ratio is to be obtained, the round duct must be transformed to a rectangular inlet. Such transformation should be gradual, if possible with a maximum included angle of 15° or less, in order to minimize shock losses if inlet velocity is increased from duct velocity. Similarly, if the inlet velocity is lower than the duct velocity, a gradual transformation will result in maximum static pressure regain. To conserve space an involute inlet can be used as shown in Figure 7.

The orientation of the inlet duct is also important. If the inlet is downwardly inclined from the horizontal to a vertical axis cyclone, the pressure drop is increased and the efficiency is decreased (*12*). Although no supporting data are found, it is commonly assumed that an approach inclined upward will improve performance since it is the converse of the arrangement which decreased performance. It is also assumed that an elbow in the horizontal plane leading to a vertical axis cyclone will improve the cyclone performance if it is the same hand as the cyclone vortex, and will decrease performance if it is opposite hand.

d. Dust Discharge. It is essential in any cyclone design to remove the separated dust from the cyclone cone or body as immediately, completely, and continuously as possible. Many different schemes have been developed for accomplishing these results. The most simple system is a hopper or dust bin closed at the bottom and open at the top to the cyclone discharge. There will be a vortex in the bin as well as in the core. If this upflow is excessive, the normal discharge of dust is prevented. The vortex in the connection between the cyclone cone and dust bin can be suppressed by straightening vanes in the dust discharge pipe; or by baffles (disks or cones) installed about two dust discharge diameters above the apex of the cone so that there is approximately a 3-in. annular space between the edge of the baffle and the cone wall. An axial disk near the dust discharge is a trend toward the peripheral dust discharge cyclone design discussed later.

Another common method of minimizing upflow through the dust discharge pipe is to use some type of valve to prevent such flow. For most dust collection systems, where the negative pressure at the bottom of the cyclone is in the range of inches water gauge, ordinary rotary valves are sufficiently gas-tight. If the negative pressure is higher, however, much better valving is required—usually a double set of valves which are capable of providing airtight closure in the presence of solids. Choke discharge screw conveyors may also be used at the bottom of a cyclone as a gas seal.

Various commercial designs of automatic flap valves are available, whereby the dust accumulates in the cyclone or in an intermediate position in the valve until its weight forces a counterweighted flap valve open. Since flap valves do not provide a positive closure, various mechanical improvements have been offered to upgrade their performance.

Cyclones inside of pressure vessels, for example in fluid catalytic crackers, may rely solely upon a full leg of dust, called a dip leg, to prevent reverse flow. Such dip legs are of course subject to plugging, and the location of the dip leg terminus so that the proper pressure relationships are obtained is frequently difficult. Dip legs are useful only where valves or other mechanisms would be inaccessible for service.

Since an inward flow of gas at the dust discharge is harmful, one method of correcting this situation is to install the cyclone at such a location in the system that it is under positive pressure with respect to the atmosphere or to the dust retention bin. Although this solves the problem of gas inflow at the dust discharge, it causes two other problems which may be of equal importance. First, the dust retention bin and disposal system may itself become a source of air pollution; second, such an arrangement requires the entire dust load to be handled by the fan, in many cases with excessive erosion or fan unbalance.

Inflow of gas at the dust discharge can also be prevented by installing an auxiliary fan to maintain a positive gas flow outward—commonly called "purge" flow. A purge flow of 10% of gas throughput may decrease dust emission from a relatively efficient cyclone by as much as 20–28% (*9*).

So many combinations of devices have been proposed and used at the dust discharge, with such varying degrees of success according to particular circumstances, that a summary evaluation of specific designs is impractical. However, the design of the dust discharge merits as much or more engineering attention as any other aspect of cyclone design. For successful operation, it is necessary to meet the criteria of immediate, complete, and continuous discharge of dust, with prevention of inflow, and if possible, inducement of outflow of gas through the dust discharge.

Peripheral discharge cyclones offer some advantages if dust discharge purge is to be used. With this design, the dust discharge area can frequently be made smaller than the dust discharge of a conventional cone bottom cyclone, with the result that the necessary volume of purge flow is reduced. A secondary advantage of purge flow is that the dust retention bin may be remotely located from the cyclone itself. In the Mark III

Dunlab cyclone (Fig. 1B) the purge rate is normally 1% of the throughput *(13)*. Another type of peripheral discharge, sometimes called "uniflow," has the gas and dust discharge at the same end. This has advantages in that there is no inlet eddy current and no reversal of gas flow in the axial direction. On the other hand, it requires higher purge rates, on the order of 25%, to attain high efficiency *(8)*.

A method for skimming dust out of the eddy current at the inlet annulus has previously been described. The disposition of this small flow is of importance in connection with dust discharge. Some designs conduct the flow from the skimming slots to the dust bin under the bottom of the cyclone. The static pressure at the outer wall of the cyclone in the region of the annulus is higher than at any other zone, so the induced flow of gas to the dust bin would return upward from the dust bin through the main dust discharge, thus inducing an inflow of gas at a point where it is highly undesirable. It is preferable to make any connection from the slots in the annulus back into the body of the cyclone in the upper or central part of the main vortex zone. The pressure at the outer wall of the annulus is somewhat higher than at the wall lower in the body or cone. There is not much static pressure difference available to induce flow through such a skimming arrangement, but whatever flow is induced may be helpful and will not be harmful.

e. Gas Outlet Design. The eddy currents in the annulus of the cyclone require that the gas outlet have an extension into the body of the cyclone in order to minimize loss of dust through the gas outlet. The optimum length of the gas outlet extension has been determined *(8)* to be about one gas outlet diameter. It is also generally assumed that this extension should terminate slightly below the bottom of the gas inlet. The shorter the outlet extension into the cyclone, the lower the pressure drop attained. No outlet extension results in the lowest pressure loss, but dust collection efficiency under such circumstances will generally be unsatisfactory.

The gas flowing out of the gas outlet extension is in vortex flow, and therefore contains energy in excess of normal flow and also is continuing to separate dust from the stream by centrifugal force. Various devices have been developed for recovery of the pressure and energy in the outlet pipe. Those devices which consist of straightening vanes, baffles, etc., in or just below the gas outlet pipe achieve their energy recovery by suppressing the vortex; and as has been noted previously, this will also decrease the dust collection efficiency, and is undesirable. The most successful pressure recovery devices on the outlet are the involute scroll or the outlet drum as shown in Figure 8. The involute scroll is designed as a conventional duct expansion wrapped around a circle, receiving the

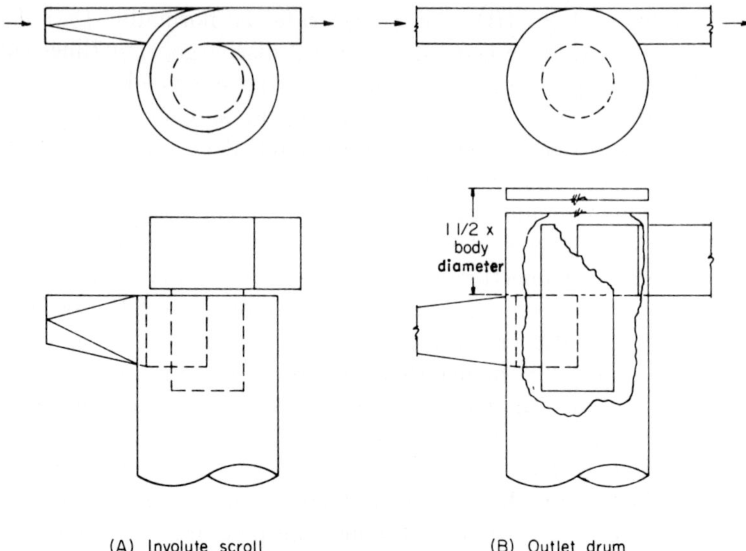


Figure 8. Gas outlet pressure recovery.

gases at high rotational velocity near the wall of the gas outlet pipe and converting the kinetic energy to static pressure by gradual expansion. The outlet drum operates somewhat as a cyclone in reverse tending to convert vortex flow back into linear flow. Outlet devices of this type can decrease the pressure drop of the cyclone from 5 to 10% without any detrimental effects on the dust collection efficiency.

The vortex in the gas outlet pipe also concentrates dust near the wall of the outlet (*14*), and the installation of skimmers to shave off the dust-rich layer can be used to further improve the collection efficiency. The practical problem is that there is no place in the cyclone where the pressure relationships are proper so as to induce a flow from a skimmer in this location to any place within the cyclone which will not result in reentrainment of dust. The purge flow through such skimmers must be induced by a separate fan. The only practical further treatment is to return it in a thin layer adjacent to the outer wall of the inlet, or to handle the purge flow separately in smaller diameter cyclones or some other type of high efficiency dust collector.

f. Effect of Internal Roughness. An extensive experimental and theoretical investigation of cyclone design and performance (*2*) resulted in the conclusion that the wall friction in the cyclone was a negligible portion of the pressure drop. The pressure drop is due almost entirely to the vortex, and to the design of the inlet and gas outlet. Increased roughness of the internal wall of the cyclone, probably by the inducement of local eddy currents and increased local wall friction, reduces vortex intensity, with the overall result that cyclone pressure drop is reduced. Dust collection efficiency is also reduced.

An experiment (*15*) wherein coatings of sand particles of definite sizes were applied to the wall of the cyclone showed that the cyclone pressure drop was reduced from 8 inlet velocity heads with a smooth wall to 4.1 velocity heads with a heavy coating of sand of 0.5- to 1.0-mm particle size. Another experiment where $\frac{1}{2}$-in. diamond mesh liner was applied to a cyclone collecting fly ash at 3500 ft/minute inlet velocity showed a decreasing collection efficiency with an increase in the amount of internal area lined with diamond mesh, as shown in the following tabulation.

| *Area lined* (%) | *Collection efficiency* (%) |
|---|---|
| 0 | 87.5 |
| 11.6 | 86.1 |
| 75.5 | 81.9 |
| 87.0 | 78.0 |

Roughness at the cyclone wall will cause local eddy currents to carry dust away from the wall and defeat any effort to concentrate it at the wall and separate it from the gas stream. All seams should be ground smooth on the inside or consist of carefully matched flanged joints. If access doors are provided in the cyclone body, they should be designed so that there is a minimum crack and so that the inner surface of the door is flush with the inner cyclone surface. A buildup of adhesive dust sticking to the internal wall surface, or erosive wear of the metallic surfaces, will also result in surface roughness. No data are available as to the quantitative effect of this degree of roughness. Certain types of cyclones are lined with erosion-resistant refractory linings, and if this lining were to fail, exposing the metal mesh or ties used to hold the lining in place, the effect on efficiency would almost certainly be quite large.

### *E. Effect of Operating Variables on Cyclone Performance*

#### 1. Flow Rate

The pressure drop theoretically varies with the square of the flow rate and therefore with the square of the inlet velocity. This is the reasoning behind the expression of cyclone pressure drop in terms of the number of inlet velocity heads. Experimentally, deviations have been found (*1*). For example, the exponent of flow rate in a test of pressure drop on 13 different cyclone designs resulted in values ranging from 1.5 to over 2.0. However, out of the 13 designs, 8 of them exhibited exponents of flow rate in the range of 1.75 to 2.0. Unless there are test data to show otherwise, it is customary to assume that pressure drop varies with the square of flow rate.

Pressure drop through a given cyclone also varies with the properties of the gas, as will be discussed later.

Variation in flow rate also has a marked effect on efficiency. Efficiency increases with increasing flow rate up to some limiting velocity, above which the internal turbulence increases more rapidly than the separation, thus causing a decrease in efficiency with further increase in flow rate. Hejma (*5*) investigated the dust concentration and particle-size distribution as influenced by turbulence as well as other design parameters. He concluded that the finer particles are much more strongly influenced by turbulence than are the larger sizes. Ebert (*16*) presents a method for measuring and calculating the characteristics of the boundary layer flow, which is the region of most significance in the relation of turbulence to efficiency. The necessity to limit velocity and therefore turbulence is not a troublesome problem in most practical designs, because it generally ocurs at velocities above 4000–5000 ft/minute, in a range where the pressure drop would be excessive anyway.

There is wide disagreement as to the quantitative relationship of efficiency with inlet velocity. In an analysis of all available data (*1*) the following conclusions were reached.

1. As inlet velocity is increased from zero, there is at first a substantially straight-line increase in efficiency. The slope of the line is steeper for larger particles and for small cyclones than it is for finer particles and larger cyclones. This straight line continues to about 70% of the maximum efficiency obtainable, after which the slope decreases rapidly. Cyclones are almost never used in this portion of the curve.

2. Following the initial portion of straight-line efficiency increase, the line gradually decreases in slope until it becomes essentially flat. The inlet velocity at which the curve becomes essentially flat is lower for small-diameter cyclones and large particles. In most cyclone applications, the flat portion of the curve is never reached.

3. In many practical installations, the variation of efficiency with inlet velocity is masked by an increase of particle size or an increase of dust loading with flow rate, both of which independently affect efficiency. Other variables, such as the presence or absence of agglomerates, also

affect efficiency and frequently are affected by flow rate in a given installation.

4. Plots of cyclone efficiency versus pressure drop are more useful for cyclone selection than are plots of efficiency versus flow. Typical curves of the various parameters discussed are shown in Figures 9 and 10.

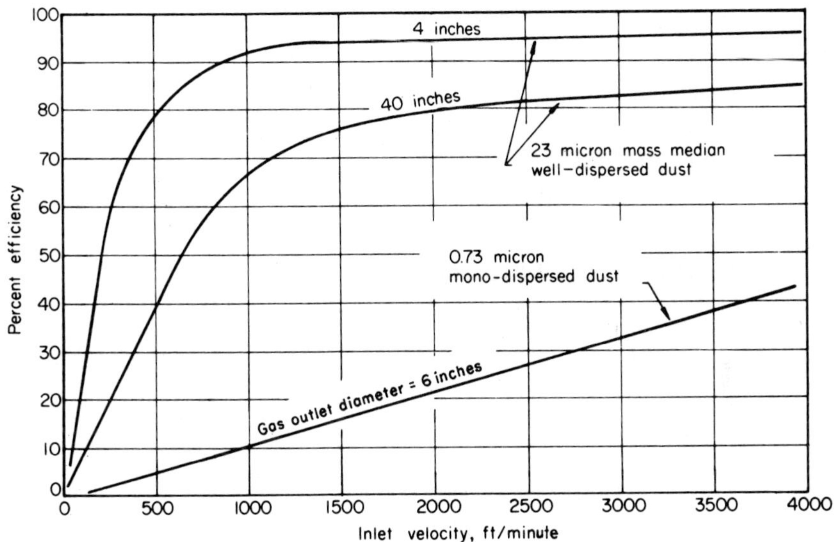


Figure 9. Variation of efficiency with inlet velocity.

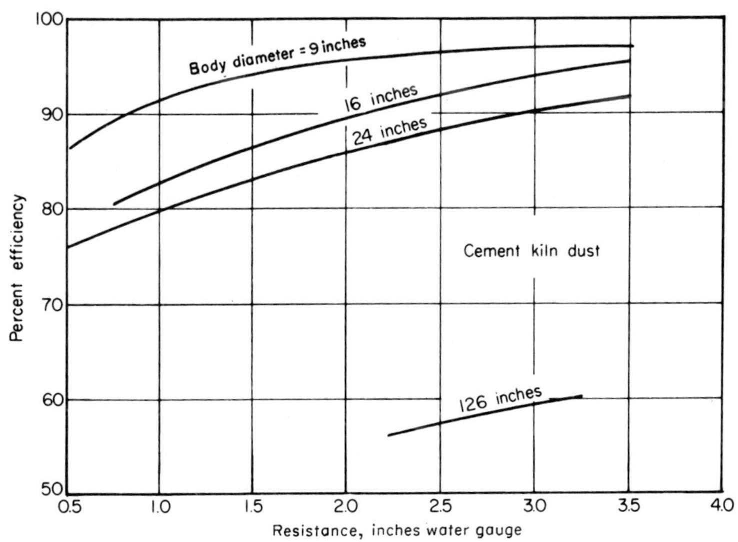


Figure 10. Variation of efficiency with pressure drop.

5. Specific test data for the cyclone and dust in question are to be preferred over any other method of estimating. However, in the absence of such data, variation of efficiency with flow rate over short ranges of flow may be estimated by

$$\frac{100 - \eta_a}{100 - \eta_b} = \left(\frac{Q_b}{Q_a}\right)^{0.5} \tag{8}$$

where

$\eta_a$ = collection efficiency, weight percent at condition a
$\eta_b$ = collection efficiency, weight percent at condition b
$Q_b$ = flow, $ft^3$/minute at condition b
$Q_a$ = flow, $ft^3$/minute at condition a

## 2. Physical Properties of the Gas

The pressure drop of a cyclone is affected by the temperature, density, and pressure of the gas as shown by

$$h = KQ^2 p \rho_g / T \tag{9}$$

where

$h$ = pressure drop, inches water gauge
$K$ = proportionality constant
$Q$ = flow rate, $ft^3$/minute
$p$ = absolute pressure, atmospheres
$\rho_g$ = gas density, lb/$ft^3$
$T$ = absolute temperature °R

The viscosity of the gas through which the dust particle must be driven to the cyclone wall obviously will have an effect on collection efficiency. The viscosity of gases increases with increasing temperature, and correspondingly the efficiency will decrease with increasing temperature, all other factors being constant. This has been verified by several experimenters.

On the basis of formulas for critical size of particles separated, the relation between efficiency and gas viscosity at constant flow rate may be estimated by

$$\frac{100 - \eta_a}{100 - \eta_b} = \left(\frac{\mu_a}{\mu_b}\right)^{0.5} \tag{10}$$

where $\mu$ = viscosity at conditions a and b in any consistent units

Although the density of the gas is theoretically a factor in separation of particles, its value is so small compared to particle density that for most practical cases any variation in gas density is negligible. Of course, at very high pressure or density it should be taken into consideration. The relationship between efficiency and gas density may be estimated by

$$\frac{100 - \eta_a}{100 - \eta_b} = \left(\frac{\rho_p - \rho_{gb}}{\rho_p - \rho_{ga}}\right)^{0.5} \tag{11}$$

where

$\rho_p$ = density of dust particle, lb/ft$^3$
$\rho_g$ = density of gas at condition a and b, lb/ft$^3$

## 3. Properties of the Dust

The properties of the dust to be collected represent the most important variable in cyclone efficiency, and are probably the most difficult to evaluate. Many physical properties of the dust will affect the efficiency, but the only ones that have been investigated quanitiatively are particle size and particle density. Other physical and chemical properties which make the dust hard or easy to handle will also affect the practical aspects of cyclone operation. These aspects are discussed later in terms of erosion and fouling in cyclones.

The particle-size distribution of a dust should be expressed at a mass distribution if cyclone efficiency is to be considered on a weight basis. Manipulation of various types of particle-size distribution data, as well as the determination of particle size, are involved and complicated subjects which are referred to detailed discussions (*17*, *18*).

A standard method for determining the properties of dust has been published by the American Society of Mechanical Engineers as Power Test Code 28. This method standardizes the determination of fractional settling velocity of a dust, as well as the determination of particle specific gravity. Settling velocity, taking into account particle density and shape as well as particle size, is more meaningful in relation to cyclone efficiency than is particle size alone, because these additional factors also affect the separation of particles in the cyclone. Manipulation of fractional efficiency data to obtain total efficiency is identical whether the fractional size units are in terms of microns or settling velocity.

The advantages of the settling velocity method are

1. It is more valid technically for cyclone collectors and other equipment dependent primarily on centrifugal or inertial forces.

2. The method of analysis is much more rapid, and requires less skill, than a microscopic determination of particle size.

This method has not yet gained widespread acceptance, and during the changeover period, conversion may be made between micron particle size and settling velocity by a method also described in Power Test Code 28, which also requires a knowledge of the particle density and certain assumptions concerning particle shape.

Any realistic consideration of the effect of particle size on cyclone efficiency requires a fractional efficiency curve for the type of cyclone in question which can be obtained only from test data (Fig. 11). The fractional efficiency curve should also specify all the other factors affecting the performance such as the gas, temperature, pressure, dust load, nature of dust, true density of dust particles, and any unusual characteristics of the dust.

If the fractional efficiency curve of the cyclone and the size distribution of the dust to be collected are known, the overall collection efficiency may be calculated. Using the particle-size data in Figure 12 and the fractional efficiency curve of Figure 11, the particle size corresponding to various intervals of the size distribution is tabulated (columns 1 and 2 of Table II). The corresponding efficiency for that particle size is taken from Figure 11 and tabulated as column 3. Column 1 is plotted against column 3 as shown in Figure 13. The area under the curve may be determined by graphical integration, and the mean ordinate determined. The mean ordinate represents the total dust collection efficiency of the cyclone for the size distribution of the dust considered.

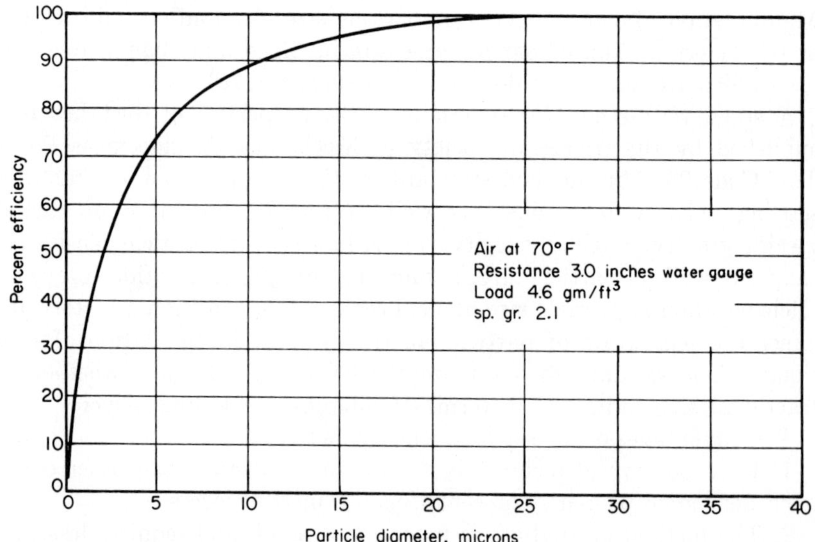


Figure 11. Typical fractional efficiency curve.

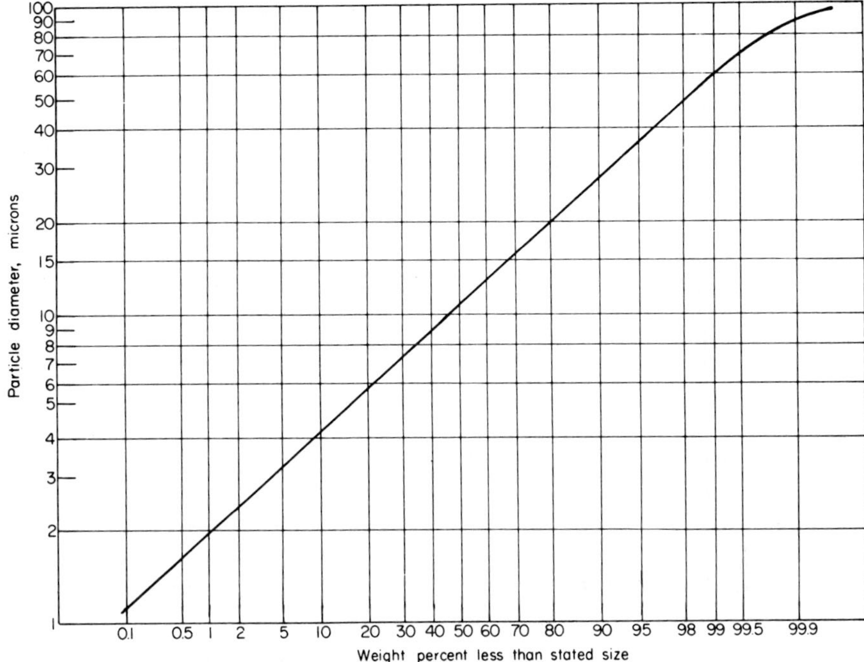


Figure 12. Typical particle-size distribution.

Test data which give the overall collection efficiency for a specified dust and do not give the fractional efficiency data are useful for extrapolation to other dusts only if the particle-size distributions and particle shapes of both dusts are closely similar. Corrections for other factors such as particle density, dust load, gas characteristics, etc., can be made. No correction can be made for differences in particle-size distribution if the fractional efficiency curve of the cyclone is not known.

In the absence of test data, cyclone efficiency may be only crudely estimated. A rough approximation of the efficiency may be obtained by determining the cut size [Eq. (5)] and assuming the overall collection efficiency will be equal to the cumulative percentage of the dust larger than the cut size. Procedures for "optimizing" the compromise between pressure loss and efficiency are discussed in Section II,F.

The efficiency of cyclone dust collection is greater for particles of high density than for low density. Although theoretically the efficiency might be expected to vary as the square root of particle density, experimental data (*19*) do not confirm this, as shown in Figure 14. In all considerations of efficiency, it is the true density of the particle, and not the bulk density of the dust, that is of importance.

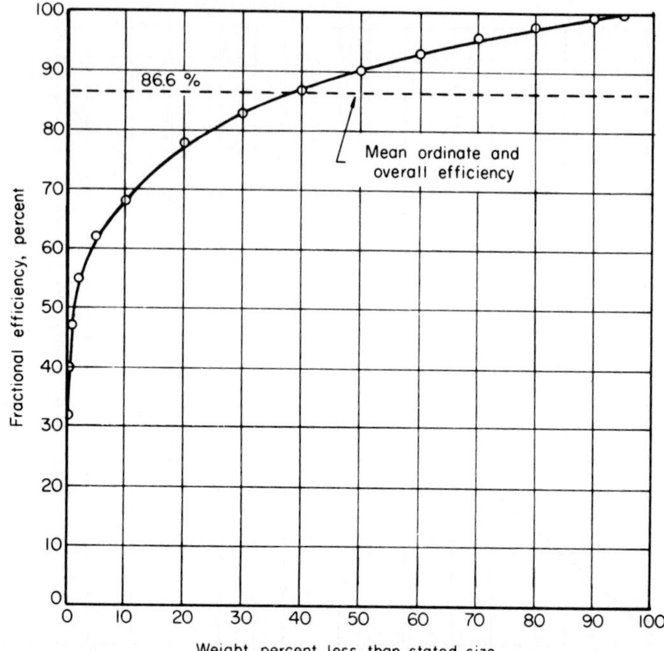


Figure 13. Total efficiency graph (from Table II).

**Table II Calculation of Overall Efficiency from Fractional Efficiency and Particle-Size Data**

| *1*<br>*Wt %*<br>*less than size* | *2*<br>*Size,*<br>*μm* | *3*<br>*Efficiency*<br>*for size, %* |
|---|---|---|
| 0.1 | 1.1 | 32 |
| 0.5 | 1.6 | 40 |
| 1.0 | 1.9 | 47 |
| 2.0 | 2.4 | 55 |
| 5.0 | 3.2 | 62 |
| 10.0 | 4.2 | 68 |
| 20.0 | 5.9 | 78 |
| 30.0 | 7.3 | 83 |
| 40.0 | 8.8 | 87 |
| 50.0 | 10.7 | 90 |
| 60.0 | 13.0 | 93 |
| 70.0 | 15.5 | 96 |
| 80.0 | 19.8 | 98 |
| 90.0 | 27.2 | 99.5 |
| 95.0 | 35.5 | 100 |
| 98.0 | 48.0 | 100 |
| 99.0 | 58.0 | 100 |
| 99.5 | 69.0 | 100 |
| 99.9 | 87.0 | 100 |

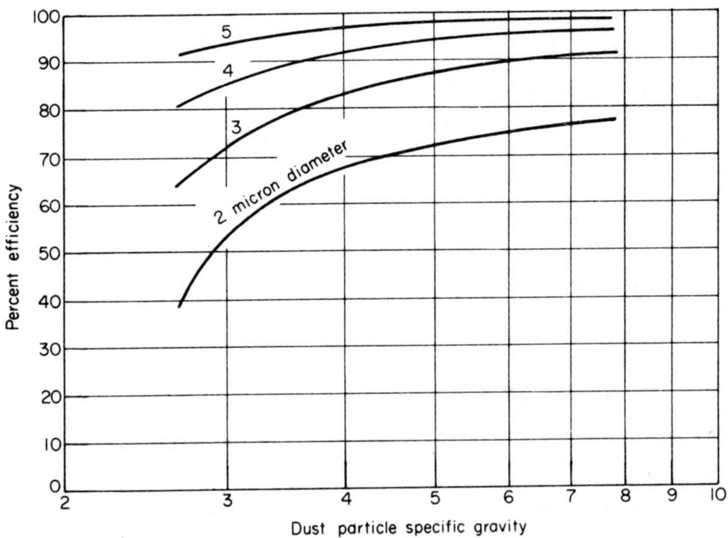


Figure 14. Variation of efficiency with particle density.

## 4. Dust Loading

Increased inlet dust loading to a cyclone causes the pressure drop to be lower and the efficiency to be increased. Apparently the pressure drop at a heavy dust load of 75–100 grains/ft³ will be in the range of 75–85% of that with essentially clean air. One author (*20*) gives an equation for this relationship as

$$\Delta P_{\mathrm{d}} = \Delta P_{\mathrm{c}}/(0.013\sqrt{C_{\mathrm{i}}} + 1) \tag{12}$$

where

$\Delta P_{\mathrm{d}}$ = pressure drop with dust load
$\Delta P_{\mathrm{c}}$ = pressure drop with clean air
$C_{\mathrm{i}}$ = inlet dust concentration, grains/ft³

Dust collection efficiency increases with dust loading. At very high loadings, all cyclones have relatively high percentage efficiencies although the gas discharge still contains progressively higher dust loadings. Apparently the details of design become less important to efficiency as loading increases, but because erosion and plugging become more important factors, cyclones are designed to overcome those problems. Therefore, small-diameter cyclones are seldom used for very heavy dust loads.

At very high loadings, the movement of the larger dust particles toward the cyclone wall creates an air drag which also sweeps some of the finer particles in the same direction. In addition, there is some impaction of large particles on small particles. A less important reason is the fact that a higher inlet velocity is obtainable at the same pressure drop because the higher dust loading results in a lower pressure drop than would be the case with clean air. W. A. Baxter has developed the following equation from an analysis of 15 different studies of dust loading versus collection efficiency:

$$\frac{100 - \eta_a}{100 - \eta_b} = \left(\frac{C_{bi}}{C_{ai}}\right)^{0.182} \tag{13}$$

where

$\eta$ = efficiency at conditions a and b
$C$ = inlet concentration at conditions a and b, grains/ft$^3$

## 5. Parallel Cyclone Operation

Where high efficiency cyclones are used, it is customary to operate a number of them in parallel in order to achieve practical gas volume. If the number of cyclones in parallel is small, each cyclone should have its own inlet and its own dust bin. The inlet should be a well-designed duct branch from the dust collection header, arranged so as to provide relatively uniform distribution of gas and dust to each of the cyclones. The gas outlets may be direct to the atmosphere or may be manifolded together into a common duct.

When the number of cyclones in parallel is large, the only practical arrangement is to use a common inlet plenum chamber, a common dust bin, and a common outlet plenum chamber. Under such circumstances, new operating problems arise. To obtain efficiencies near those obtainable with a single tube of the same size, it is necessary to equalize the gas and dust load distribution to the cyclones to prevent backflow through individual cyclones, plugging of cyclones, or reentrainment from the dust bin. Although in many practical installations it has been observed that the efficiency of a bank of parallel cyclones is noticeably less than for an individual tube, if all these problems are prevented, it is possible to obtain essentially the same efficiency.

The inlet and outlet plenums and the dust bin should be designed so that the pressure relationship between these three chambers is essentially the same at all portions of the housing. High plenum velocities, poor manifold layouts, or variations in tube sizes or shape will cause excessive flow through some tubes and backflow through others. Obviously if backflow occurs from the dust bin up through the gas discharge opening of a tube, the collection efficiency of that tube will be nil.

If the dust discharge of any of the individual tubes should plug, that particular tube will begin to pass 100% of the dust entering its inlet. The dust outlet of tubes may be plugged by collected dust in the hopper below building up to that point. Therefore dust bin design should take into account the fact that the dust may not be uniformly deposited but may be banked up against one wall.

One of the ways of minimizing dust outlet plugging in parallel cyclone operation is to provide a small purge out of the dust collection hopper. This is accomplished by using a separate purge fan which returns the flow either to the common inlet plenum or to a separate collector. Purge rates in this application are usually about 5% of the total gas flow.

If it is necessary to valve off individual cyclones in a parallel arrangement in order to maintain efficiency at reduced gas loading, it is necessary to close both the inlet and the outlet in order to prevent entrainment of dust from the bin through the dust outlet and gas outlet. In a large installation involving a number of plenums and hoppers, entire plenums should be valved off.

A typical arrangement of multiple, small-diameter high efficiency cyclones is shown in Figure 15.

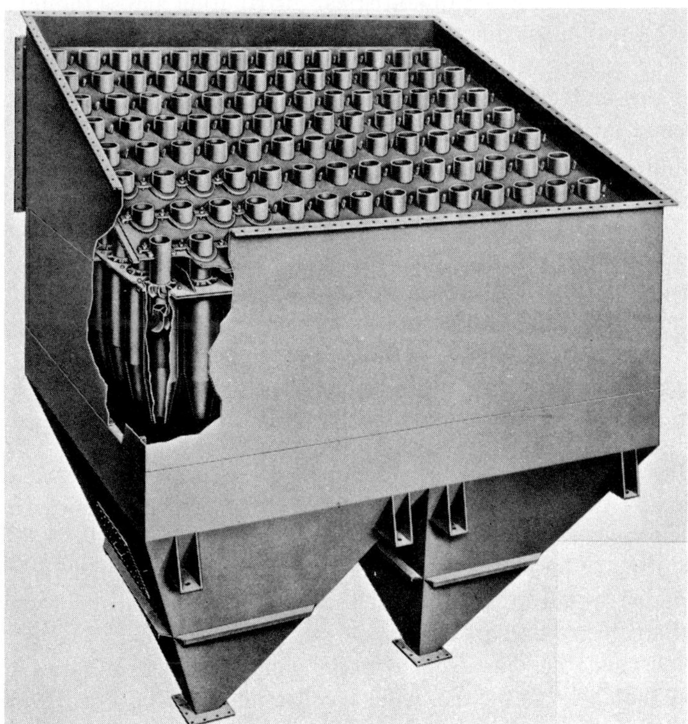

Figure 15. Typical bank of small-diameter high efficiency cyclones. (Courtesy Research-Cottrell, Inc.)

## 6. Series Cyclone Operation

The efficiency of two cyclone dust collectors operating in series on the gas flow is expressed by

$$\eta = \eta_p + \eta_s(100 - \eta_p) \quad (14)$$

where

$\eta$ = efficiency of the combination of both cyclones
$\eta_p$ = efficiency of the primary cyclone
$\eta_s$ = efficiency of the secondary cyclone (based on the inlet dust load to it)

Previous discussion has indicated the difficulty of calculating cyclone efficiency in the absence of test data. The calculation of the efficiency of the second cyclone in series is even more difficult. The second cyclone serves to separate particles which could have been but were not collected in the first cyclone owing to statistical distribution across the inlet, accidental reentrainment due to roughness or eddy currents, or reentrainment in the vortex core. The efficiency of the second cyclone will be less than that of the first cyclone. It is frequently assumed that the efficiency of the second cyclone is half that of the first in series. It has been shown *(21)* that the total collection efficiency of one cyclone, operated at an inlet velocity such that its pressure drop is equal to the total of two geometrically similar cyclones in series, is higher for the single cyclone than for the two in series. It is known that cyclones have a limit in ability to increase efficiency with increasing inlet velocity, so the range of velocities over which this relationship holds must also be subject to a similar limit.

Sometimes conditions prevail which make the use of series cyclones advantageous for practical reasons. Examples of such circumstances are

1. Dust agglomerates which are fragile and subject to degradation of particle size at high velocities may be collected more advantageously in two cyclones in series at low velocities.
2. A primary large-diameter cyclone may be used to collect coarse material which would otherwise foul the smaller passages of more efficient small-diameter secondary cyclones.
3. If dependable operation is paramount, cyclones in series may be used so that a large degree of dust collection is maintained even if the dust outlet of the primary cyclone plugs. If this were to happen, the secondary cyclone then acts as a primary cyclone as far as collection efficiency is concerned. Three cyclones in series is the maximum number

that has been used commercially, and this only under very severe service conditions as typified by fluid catalytic cracking.

Cyclone collectors are frequently successful when installed in series with other types of dust-collecting equipment. One of the most common applications is to install high efficiency cyclones ahead of electrostatic precipitators. In an installation of this kind, the cyclone exhibits an increased efficiency with an increase of gas load and/or dust loads; on the other hand, the precipitator shows an increase in efficiency with a reduced gas load or dust load. Thus the characteristics of the two types of equipment compensate for each other with a tendency to maintain a good efficiency over a wide range of gas flow and dust loading.

If the dust size distribution is such that most of the material is amenable to cyclone separation, very high overall efficiencies can be obtained by arranging cyclones in series, using a small purge flow on each stage so that it acts as a dust concentrator. Wilson and Miller (*22*) describe a fluid catalytic cracker system using two-stage series cyclones internal to the regenerator, followed by a third stage consisting of small diameter axial inlet peripheral discharge tubes made of ceramic. The purge flow from the third stage is 2% of the main gas flow and is passed to a fourth stage separator, the purge from which represents only 0.1% of the original gas flow. This stream is small enough to be economically treated in a fabric filter, high efficiency precipitator, or high energy scrubber. The overall efficiency of the system is 99.998%. In this instance, the gas is under high pressure in the regenerator so that pressure drop through cyclones represents very low cost energy.

### F. Optimizing Cyclone Design and Performance

In recent years a number of authors have presented procedures for "optimizing" cyclone design and performance. The question that is really attacked is the balance, or trade-off, between pressure loss and efficiency. The question of cost and value of these two parameters is left to the reader. Obviously, if cost of improved design (to obtain lower pressure loss) or of lower efficiency is omitted, the "optimized" design is that which gives the minimum pressure loss and maximum efficiency. Since the choices available to the designer almost universally effect the two desired results in opposite ways, what is available is a "compromise" rather than an "optimum."

A restatement of the objective, as usually pursued by most authors on optimizing, would be—cyclone design to achieve a required efficiency on a known dust dispersion at minimum pressure loss, cost factors of design excluded.

Of several recent and complete discussions of the subject (*6, 23, 24*), that of Leith and Mehta (*6*) is the most recent and possibly the most complete. They describe in detail a procedure applicable only to the "standard" cyclone, as shown in Fig. 1A.

The basis of the procedure is as follows: The term

$$CD^2/ab \tag{15}$$

where

$C$ = cyclone geometry coefficient
$D$ = cyclone diameter
$a$ = gas inlet height
$b$ = gas inlet width

is defined to include either explicitly or implicitly all seven cyclone dimension ratios $a/D$, $b/D$, $D_e/D$, $S/D$, $h/H$, $H/D$, and $B/D$ where $a$, $b$, and $D$ are as above, and $D_e$ is gas exit duct diameter, $S$ is gas exit duct length, $h$ is cylinder height, $H$ is overall height, and $B$ is dust outlet diameter.

For practical reasons, constraints are placed on the above ratios as follows:

$$a/D \leq S/D \tag{16}$$
$$S/D \leq h/D \tag{17}$$

Combining Equation (6) with (15) and the fundamental relationship between velocity head and velocity yields

$$P = g\rho_L D^4\, \Delta P/8\rho_g Q_g^2 \tag{18}$$

where

$P$ = cyclone pressure drop coefficient
$g$ = acceleration of gravity
$\rho_L$ = density of gauge liquid
$D$ = cyclone diameter
$\Delta P$ = pressure drop
$\rho_g$ = density of gas
$Q_g$ = gas throughput

After consideration of length of vortex and other practical geometric relations and constraints, an equation is obtained:

$$\frac{CD^2}{ab} = \pi P^2 \left(\frac{D_e}{D}\right)^4 \left\{2\left[1 - \left(\frac{D_e}{D}\right)^2\right]\left[\frac{H}{D} - 2.3P^{1/3}\left(\frac{D_e}{D}\right)^{5/3} - \frac{1}{P(D_e/D)^2(1 - D_e/D))}\right] + \frac{1}{3}\left(\frac{H-h}{D}\right)\left[1 + \frac{B}{D} + \left(\frac{B}{D}\right)^2\right] + \left[1 - \left(\frac{D_e}{D}\right)^2\right]\left[2.3P^{1/3}\left(\frac{D_e}{D}\right)^{5/3}\right] + \frac{h}{D} - \frac{H}{D}\right\} \tag{19}$$

Further constraints are imposed, to keep Equation (19) within the ranges that have been validated experimentally, as follows:

$$H/D = 5 \tag{20}$$

$$h/D = 3 \tag{21}$$

Then, upon assuming a value for $B/D$ (a most important ratio) and substituting Equations (20) and (21) into Equation (19), an expression for $CD^2/ab$ is obtained which depends only on $D_e/D$ and $P$. By use of a digital computer, values of $CD^2/ab$ may be maximized in terms of $D_e/D$ for a range of values of $P$, and all other dimension ratios calculated as a function of $D_e/D$. The value of "optimized" $C$ is then computed from

$$C = \frac{\pi D^2}{ab}\left\{2\left[1 - \left(\frac{D_e}{D}\right)^2\right]\left(\frac{S}{D} - \frac{a}{2D}\right) + \frac{1}{3}\left(\frac{S + l - h}{D}\right)\left[1 + \frac{d}{D} + \left(\frac{d}{D}\right)^2\right] + \frac{h}{D} - \left(\frac{D_e}{D}\right)^2\frac{l}{D} - \frac{S}{D}\right\} \tag{22}$$

where

$$\frac{l}{D} = 2.3\,\frac{D_e}{D}\left(\frac{D^2}{ab}\right)^{1/3} \tag{23}$$

and all other symbols as before.

For the assumed value $B/D = 0.375$, Leith and Mehta present two graphs from which efficiency and dimension ratios can be found for any specific case of gas and dust properties. The entire procedure must be repeated for each assumed value of $B/D$.

The results of such an optimized computation were compared to the performance of the "standard" cyclone proposed by Stairmand (*9*), at identical operating conditions, as shown in the tabulation following:

| | *Percent efficiency* | |
|---|---|---|
| | *Optimized* | *Stairmand* |
| Efficiency at 2 μm | 75 | 67 |
| Efficiency at 5 μm | 92 | 86 |
| Pressure loss, in. water gauge | 3.0 | 3.0 |

These results apparently were not verified experimentally.

### G. Erosion in Cyclones

Erosion in cyclones is caused by the impingement and rubbing of dust particles on the cyclone wall. Erosion is worse with high dust loadings, high inlet velocities, and large or hard dust particles. Any defect in cyclone design or operation which tends to concentrate dust moving at high velocity will accelerate erosion.

The areas most subject to erosive wear are those along welded seams or mismatched flange seams, near the bottom of the cone; and opposite the inlet. Surface irregularities at welded joints, and the annealing softening of metal adjacent to the weld will induce rapid wear in the weld region. Welded seams should be kept to a minimum, and heat treated if necessary to maintain the hardness of the metal adjacent to the weld.

The importance of proper dust discharge has been stressed previously in discussions of efficiency. It is similarly important in preventing erosion. If dust is not effectively and continuously discharged from the bottom of the cone, a high circulating dust load is maintained in that region, leading to excessive wear of the cone. If the dust outlet should plug, the entire circulating dust load is conducted through the cyclone, including the gas discharge pipe, and it may cause erosive wear at any point.

Excessive wear of the cyclone shell opposite the inlet may occur, particularly if large particles are handled. This can be cured by the provision of removable wear plates of abrasive-resistant metal or rubber designed so as to be flush with the inside surface of the shell.

Combinations of dust loading and velocity which will, if exceeded, induce erosion have been shown in the following tabulation (*1*).

| *Dust load* (*grains/ft*$^3$) | *Velocity* (*ft/min*) |
|---|---|
| 0.3 | 7000 |
| 3.0 | 4000 |
| 3000 | 400 |

It has also been determined that typical dust particles smaller than the 5- and 10-$\mu$m range do not cause appreciable erosion.

It is possible to design a cyclone to reduce erosion by increasing the diameter of the cyclone body without increasing the diameter of the gas outlet. This results in reduced velocity at the body wall without reducing maximum velocities and separating force of the vortex. It also results in increased pressure drop. Consequently, at high loadings of abrasive dust, large-diameter cyclones are required to control erosion. For more moderate conditions, small-diameter cyclones have an advantage since they usually do not have seams or welds.

Erosion-resistant linings consisting of a troweled or cast refractory may be used. Such linings are usually $\frac{5}{8}$ to 1 in. thick and must be supported by metal mesh or ties. It is necessary to maintain the integrity of such refractory linings to maintain efficiency and it is recommended that inlet velocities be kept below the range of 60–75 ft/second if such linings are used.

### *H. Fouling in Cyclones*

Fouling of cyclones results in decreased efficiency, increased erosion, and increased pressure drop. Fouling is generally found to occur either by plugging of the dust outlet or by the buildup of materials on the cyclone wall.

Dust outlets become plugged by large pieces of extraneous material in the system, by the overfilling of the dust bin, or by the spalling of material caked upon the walls of the cyclone. The valves used to discharge dust from the collection bin should not be smaller than 4 in. in most applications, and in all except large pneumatic conveying or fluidized bed applications need not be larger than 14 in. A vertical axis cyclone is somewhat less subject to plugging of the dust outlet because gravity helps to remove large objects through the discharge.

It is of utmost importance to prevent overfilling of dust hoppers, particularly under multiple banks of small-diameter cyclones. If a hopper has filled sufficiently to plug the outlets and later has been emptied, the dust plugs may remain in the outlets. In large-diameter cyclones, cleanout openings can be provided, but this is not practical for large banks of small-diameter tubes.

The buildup of sticky materials on the wall of the cyclone is primarily a function of the dust. In general, the finer and softer the dust, the greater the tendency to cake on the wall. Chemical and physical properties will also affect this behavior. Condensation of moisture on the walls of the cyclone will also contribute to the accumulation of material. In many cases, buildup of sticky material on the walls can be minimized by keeping the inlet velocity above 50 ft/second. Smoothness of the cyclone walls is also important, and some applications have even used electropolished walls to minimize the buildup of powdered milk or coffee dust.

The cure for excessive wall buildup frequently must be tailored to meet the particular circumstance. In one case, the removal of an inertial precleaner so as to permit the coarser material to also traverse the cyclone walls may present a cure; in another, periodically inducing a reverse gas flow from the dust bin by the introduction of compressed air to the bin may be successful. If wall condensation is the cause, it must be eliminated by insulation or other appropriate methods. In some cases, the use of water or other fluids is necessary to wash accumulations out of the tubes. As far as the design is concerned, the important features to minimize fouling are good removal of dust from the dust discharge, adequate size of dust bin discharge, prevention of dust bin overfilling, choice of proper inlet velocity, and prevention of wall condensation.

## *I. Wet Cyclones*

### 1. Problems of Definition

Most types of commercial gas cleaning equipment utilize, either accidentally or by design, more than one of the possible theoretical mechanisms for separating particulate matter. This leads to difficulty in classifying the numerous types of equipment. Cyclones and inertial separators, when operated wet, cannot avoid some droplet collision phenomena associated with wet scrubbers; and furthermore, various features of the construction or appurtenances may be modified to adapt the equipment to handling liquid or slurry. For the purposes of this discussion, however, wet cyclones are defined as equipment which meets the basic definition of a cyclone, and is handling liquid, whether in the form of droplet or mist contamination of the entering gas, or deliberately introduced sprays or flushing streams.

### 2. Droplet Collection

Cyclones are frequently used for removal of liquid droplet contamination of a gas stream. Practically 100% collection can be obtained for droplets of 100 μm and over, typical of entrainment from boiling liquids. Efficiency of well-designed wet cyclones range from 95 to 99% for reasonably light loadings of 5- to 50-μm droplets and 99% or more with heavy loadings of droplets over 50 μm.

Various proprietary designs of devices known as cyclone scrubbers consist of a primary device designed to wet the dust particles by impaction and a secondary cyclone serving to collect the droplet dispersion. These devices are usually considered scrubbers since the major objective in design is to wet the particles. Large amounts of power are required to achieve artificial droplet suspensions on the order of 100 μm in size, and the cyclone portion of such scrubbers is usually simple in design and effective in performance.

## 3. Wet Operation for Dry Dust

Cyclones operated wet for the collection of dry dust improve efficiency, prevent wall buildup and fouling, and reduce erosion. The efficiency is considerably higher because the dust particles are trapped in a liquid film and are not easily reentrained. The usual application of water to an otherwise dry cyclone consists of spraying countercurrent to the gas flow in the inlet duct, at water rates of from 5 to 15 gal/minute per 1000 $ft^3$/minute capacity. Additional sprays may be installed in the duct upstream of the cyclone inlet if desired.

Care must be exercised to direct spray nozzle patterns and flushing streams so that no portion of the cyclone or duct is merely moist, since this condition will lead to caking and plugging. Troublesome areas will be the inlet duct, the dust discharge, and under some conditions, the gas outlet. If the cyclone is cone bottom, swirl patterns in the lower part of the cone may concentrate the water steams, leaving unflushed areas between them, and auxiliary flushing may be needed.

## 4. Preventing Reentrainment from Wet Cyclones

Reentrainment from a wet cyclone is very deleterious to its efficiency, but is rather easily prevented. It is necessary to continuously and completely drain the liquid from the bottom of the cyclone in order to prevent the vortex core from shearing droplets off the axial peak of liquid, which would occur if it were retained in the bottom of the cyclone. If the tangential wall velocity is high enough, it can shear the film from the wall in droplets too small to be recollected. An inlet velocity of 150 ft/second maximum is recommended for atmospheric air and water, and corresponding maximum tangential velocities are recommended for other combinations.

Excessively high liquid loadings sometimes give difficulty, probably because the vortex of gas is suppressed. The recommended solution for a problem of this nature is to reduce the inlet gas velocity or to install cyclones in series.

One of the most common causes of reentrainment from wet cyclones is the shearing of droplets from a liquid film which tends to creep along the walls to the gas outlet lip. A cylindrical or conical baffle installed around the gas outlet extension (as shown in Fig. 16) provides a point for the liquid to drop, or to be sheared off while still in the separating zone of the vortex.

Care should be taken to prevent the impingement of the inlet gas stream on such a skirt.

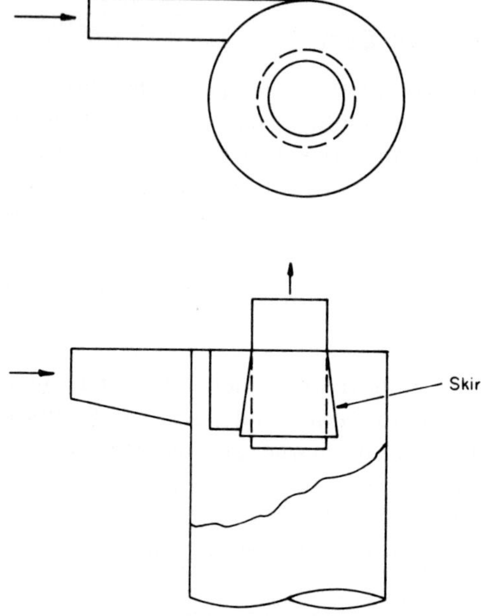


Figure 16. Gas outlet skirt for wet cyclones.

## 5. Advantages and Disadvantages of Wet Operation

The operation of cyclones as wet collectors for the removal of droplets, or even for the collection of dry dust, presents a number of advantages provided that the droplet size is sufficiently large and that reentrainment of liquid from the cyclone is prevented. The liquid droplets collect on the cyclone wall and form a continuous film which is less subject to reentrainment in the gas stream than is a dispersion of dry dust. In addition, liquid can be drained from any point in the cyclone bottom, thus permitting a location away from the axis of the vortex core. Gas flow into the liquid outlet can be completely prevented by a simple liquid leg seal. Under such conditions wet cyclones operate with a higher efficiency and less erosion and plugging difficulties than do dry cyclones.

The major disadvantage of wet cyclone operation is caused by the corrosion problem. If corrosive dusts or gases are handled, the presence of water usually makes the problem that much worse. Other fluids may be advantageous but are more expensive. The other disadvantage to the wet operation of cyclones is the additional cost of water, and the cost of recirculating or disposing of the contaminated water.

## *J. Scope of Cyclones in Field Use*

### 1. Range of Efficiency

The range of efficiency of cyclones in field use is as shown in Table I. If operated wet, a conventional cyclone will yield efficiencies comparable to that of the high efficiency cyclone. For practical reasons, banks of multiple high efficiency tubes are seldom operated wet. Cyclones are frequently of great practical value in minimizing air pollution sources if the dust dispersion is not too fine and if the contaminant is not highly valuable, or highly toxic.

Efficiencies can be improved by design changes which usually result in higher pressure drop. No designs are available which yield a very high efficiency on very fine contaminant dispersions.

### 2. Range of Pressure Drop

Pressure drop through cyclones under field conditions usually ranges between one and four inlet velocity heads, corresponding to 1–7 in. water gauge. Pressure drop increases with the square of the inlet velocity, and efficiency also increases, but not as rapidly as pressure drop. All devices intended to minimize pressure drop result in decreased efficiency except those which recover the energy in the vortex flow leaving the gas outlet.

### 3. Range of Loading and Operating Conditions

Cyclones can be designed to handle any dust loads and the percentage collection efficiency increases with increasing dust loads. However, at loads above a few hundred grains per cubic foot, erosive and plugging conditions become severe and only large-diameter cyclones are practical. Cyclone efficiency decreases with a decreasing dust load, and cyclones are seldom applied to applications below 1 grain/$ft^3$. The dust load entering a cyclone can be reduced by the installation of settling chambers or inertial traps upstream of the cyclone, but this is seldom found necessary in practice.

Cyclone dust collectors can be made of any reasonable material of construction and have no moving parts. Consequently, as a class, they can be designed to handle a wider range of chemical and physical conditions than most other types of collecting equipment. Sticky materials which tend to block the air passages or dust passages in cyclones present one of the main limiting factors in their application. Many such problems can be solved by operating the cyclone wet, and others may be solved by using large-diameter cyclones as a precleaner ahead of a bank of small-diameter high efficiency tubes.

## III. Rotary Stream Dust Separators

### A. Definition and Description

Although not conforming exactly to the definition of a cyclone collector, the recently developed rotary stream dust separator (drehströmungstauber) is basically akin to the cyclone in that it uses primarily centrifugal forces in a confined vortex, and is without moving parts. It may be defined as a cylindrical flow tube in which the main gas stream forms an inner vortex, opposed by a secondary outer vortex flowing in the same tangential direction but in the opposite axial direction, so that the combined centrifugal forces and radial gas flows concentrate the dust particles in an annular ring. A typical device is shown in Figure 17. The bulk of the description and performance data given below is from Klein (*25*) with additional material from Nickel (*26*), Schaufler (*27*), and Schmidt (*28*). Budinsky (*29*) describes the velocity field and motion of particles in the separator. Alt and Schmidt (*30*) compare the separator to ordinary cyclones and find its use advantageous below about 3-$\mu$m particle size.

### B. Mechanism of Operation

The entering dusty gas passes through the annular space between the dobbas and the inside walls of the inlet port. The inside diameter of the inlet port increases slightly toward the open end, and the presence of the dobbas stabilizes the flow pattern and aids in establishing the inner vortex at a lower point in the tube. The outer vortex caused by the secondary air jets creates a vortex pattern of flow in the entering gas.

The secondary air jets, entering at a tangential angle near the top of the tube, create an outer vortex ("potential flow") with a downward axial component and an inward radial component. This potential flow creates an inner vortex ("rotational flow") in the same direction of rotation but with an upward axial component. The result of these forces is the concentration of the dust in an annular space, called the "mixed flow" region. This dust-rich layer is conducted downward, outside the lips of the inlet port, and into the dust bunker below (or withdrawn separately) as part

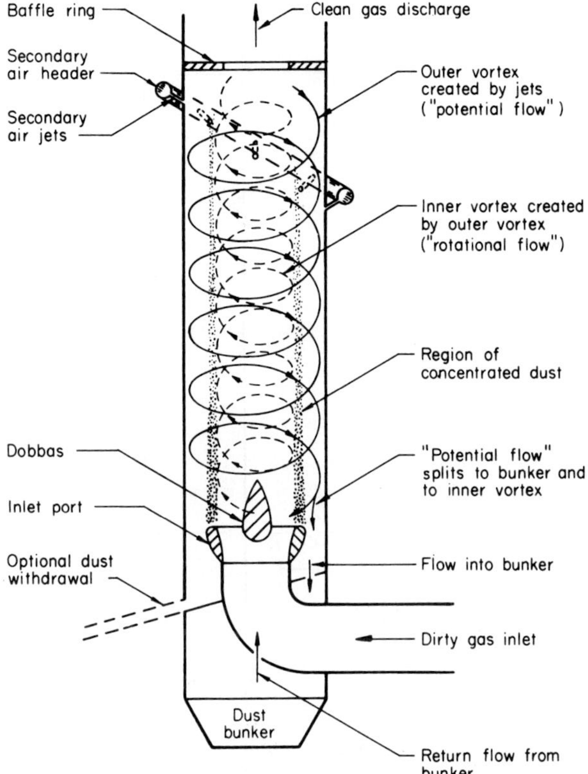


Figure 17. Typical rotary stream dust separator and major gas flows.

of a purge flow, with a compensating return flow from the bunker to the dirty gas inlet.

The baffle ring at the top serves to block the portion of the uppermost jet flow that would otherwise escape from the top of the tube, and markedly improves the efficiency.

Typical dimensions of a 200-mm-diameter experimental tube are shown in Figure 18.

### C. Efficiency and Pressure Drop

Two pressures are involved—the pressure loss in the main gas stream, and the pressure required for optimum operation of the secondary air jets. For a 200-mm-diameter unit handling 500 $m^3$/hour (300 cfm) of dusty gas, the jet pressure used was 480 mm water gauge, resulting (according to the investigators) in a total "equivalent" pressure drop of 200

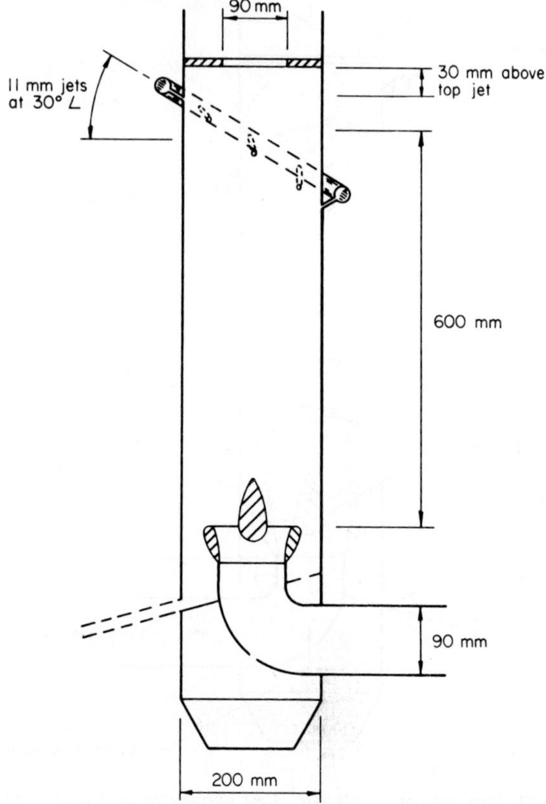


Figure 18. Major dimensions of experimental 200-mm-diameter tube.

mm (approximately 8 in.) water gauge. Under these conditions, using various test dusts, the following efficiencies were obtained.

| *Particle-size frequency maximum, μm* | *% Collection efficiency* |
|---|---|
| 2.5 | 92 |
| 5 | 95 |
| 10 | 98.5 |

Fractional efficiencies are stated to be 100% for 5-μm particles and 90% for 1-μm particles, of density 2.65 gm/cm³.

The efficiency variation with dust load is negligible over the range of 1 to 200 grains/m³ (0.4–88 grains/ft³). This is attributed to the fact that the dust is introduced near the zone of separation, and is separated in a space filled with well-defined and controllable flow patterns, without deleterious eddy currents.

The amount of bleed flow conducting the dusty concentrate out of the tube affects the efficiency. Again in the 200-mm-diameter tube treating 500 $m^3$/hour of gas, a bleed flow of 10 $m^3$/hour (2%) resulted in 88% efficiency, while a bleed flow or 50 $m^3$/hour (10%) yielded 92.5%. Further increases in bleed flow lowered efficiency, because the vortex patterns were upset.

The efficiency characteristics of a 200- and 500-mm-diameter tube are essentially equal, but larger sizes (viz, 1000 mm diameter) show a decreased efficiency. On "fine grain coal dust," 98% collection was obtained with 200- and 500-mm units, 97% with the 1000-mm unit. On a larger scale field trial, a 2000-mm-diameter unit treated the cyclone discharge from the dry drum of a tar–macadam plant. The gas was at 200°C, dust concentration 4–5 gm/$m^3$, with a particle size maximum frequency of 2 $\mu$m and a maximum particle size of 12–15 $\mu$m. Collection efficiency was 90%.

### D. Comparison with Conventional Cyclones

In terms of efficiency, the rotary stream dust separator appears to be definitely superior to conventional cyclones, and at comparable total power costs. The efficiency is in the range of the better wet scrubbers and ordinary applications of dry electrostatic precipitators. In spite of the fact that increasing body diameter is less harmful to efficiency than it is with cyclones, it is still deleterious; and the highest efficiency must be obtained with multiple-bank installations of small tubes. These, as with banks of small cyclones, must be designed for reasonably uniform distribution of pressure in the dirty gas plenum. Other problems of dust load distribution, etc., appear less difficult than with cyclones.

Another important advantage is the absence of erosion problems on the tube wall, and the absence of a need for smooth internal construction of the tube. Minor turbulence caused by welding beads, etc., has no detrimental effect.

On the other hand, the rotary stream separator does require a source of higher pressure secondary air, and presumably, therefore, a high pressure blower. Furthermore, for optimum performance the secondary air pressure and flow rate must be controlled in relation to the flow of primary gas; this could, on occasion, be an advantage, because by exercising such control, efficiency may be kept at a peak over a wide range of operating conditions.

## IV. Gravity Settling Chambers*

### A. Definition

A gravity settling chamber is a chamber large enough so that the gas velocity is reduced sufficiently to permit dust or droplets to settle out by the action of gravity. The chamber may contain horizontal plates to reduce the distance through which particles must settle. A settling chamber usually consists of a horizontal rectangular chamber with an inlet at one end and an outlet at the other, with or without horizontal plates.

### B. Theory

If the velocity in the settling chamber is low enough so that turbulence is minimized and the gas flow entering and leaving the chamber is well distributed, the performance is expressed by

$$\eta = 100 u_t L / HV \tag{24}$$

where

$\eta$ = efficiency, weight percent of particles of settling velocity $u_t$
$u_t$ = settling velocity of dust, ft/second
$L$ = chamber length, ft
$H$ = chamber height, ft
$V$ = gas velocity, ft/second

Under the above assumptions and combining Stokes' law with Equation (24) the minimum particle size that can be completely separated may be calculated by

$$D_p = \sqrt{\frac{18\mu HV}{gL(\rho_p - \rho)}} \tag{25}$$

where

$D_p$ = minimum size particle collected at 100% efficiency
$\mu$ = gas viscosity, lb/ft-second
$H$ = chamber height, ft
$V$ = gas velocity, ft/second
$g$ = gravitation constant, 32.2 ft/second/second
$L$ = chamber length, ft
$\rho_p$ = particle density, lb/ft$^3$
$\rho$ = gas density, lb/ft$^3$

* See Reference (*31*).

Horizontal plates in the settling chamber, arranged as shelves, reduce the vertical distance through which the particles must settle. The performance of such a chamber is given by Equations (26) and (27).

$$\eta = N u_t W L / q \tag{26}$$

$$D_p = \sqrt{\frac{18 \mu N q}{g W L (\rho_p - \rho)}} \tag{27}$$

where

$N$ = number of plates
$W$ = width of chamber, ft
$q$ = gas flow rate, ft$^3$/second
other units as in Equation (24)

### C. Applications

The gravity settling chamber has the advantage of utmost simplicity, and can be constructed of almost any material. However, the space required for such chambers is large, and they are seldom used to remove particles smaller than 40- to 50-$\mu$m diameter. Their most practical use is in removing very large particles as an aid or adjunct to more efficient subsequent gas cleaning equipment. The particles which can be removed by the settling chamber itself are seldom of air pollution significance.

Careful design of the settling chamber is necessary to provide good distribution of gas entering and leaving the settling chamber. The usual types of design include gradual transitions, splitters, or perforated distributing plates. The gas velocity in the chamber should generally be restricted to 600 ft/minute or less in order to prevent excessive reentrainment. Settling chambers with horizontal plates offer greater efficiency in a smaller space but present difficult cleaning problems; probably the most effective method of cleaning is to flush the plates with water sprays.

## V. Inertial Separators*

### A. Definition

Inertial separators are devices which, by causing sudden changes in direction of the gas stream, cause particles to be separated by a combination of the inertia of the dust particle, impaction on a target, and centrifugal forces. The separated particles may be retained on the impaction target, separated into a dust bin, or separated in a dust-enriched side stream which is conducted to a secondary high efficiency dust collector.

* See Reference (*31*).

The specific designs of collectors which employ such principles are legion. Most of the designs are proprietary, and they have corresponding advantages and disadvantages for particular applications.

### B. Theory

Impingement or impaction separation occurs when the gas undergoes a sudden sharp change in direction of flow, the principles of which are illustrated in Figure 19.

The target efficiency of impingement, defined as the fraction of particles in the field volume swept by the impinging target which will impinge on that target, has been determined for simple shapes as shown in Figure 20 (*32*).

The separation number in Figure 20 is defined as

$$N_s = D_p{}^2 V \rho_p / 18 \mu D_b \tag{28}$$

where

$N_s$ = separation number, dimensionless
$D_p$ = particle diameter, ft
$V$ = relative velocity gas to target, ft/second
$\rho_p$ = particle density, lb/ft$^3$
$\mu$ = gas viscosity, lb/ft-second
$D_b$ = target diameter, ft (ribbon width)

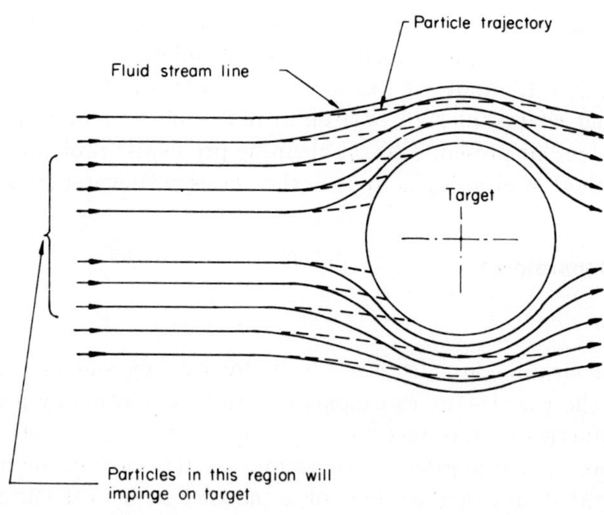


Figure 19. Mechanism of impingement.

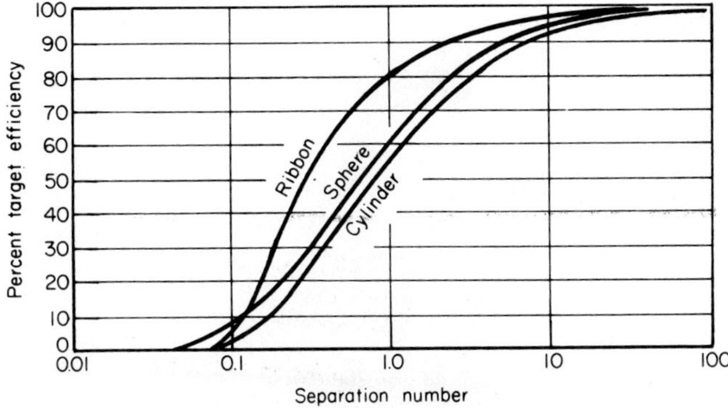


Figure 20. Target efficiency of impingement.

Figure 20 and Equation (28) serve to demonstrate the relationship of the parameters involved in impingement. Higher target efficiencies will be obtained for larger or denser particles, small impinging targets, and higher relative velocities between particles and target.

Although the theory of impingement separation has been well developed (*31–34*), such theoretical approaches afford little practical information in evaluating the commercial inertial separators on the market. The most practical approach to efficiency and pressure drop considerations is the use of experimental data. Fractional size efficiency data and overall efficiency calculations may be handled in the same manner as previously described for cyclone separators.

### *C. Some Commercial Types*

Commercial cyclone and inertial separators are furnished in a very wide variety of arrangements and combinations. Many of these have evolved in response to a particular need, and have advantages for such specific applications. Space does not permit a comprehensive review of all the commercially available equipment of this general type. A few selected examples, illustrating the wide range of combinations possible, are shown in Figures 21–25.

### *D. Range of Performance*

In general, the range of performance of inertial separators is similar to that of high efficiency cyclones. Particular designs may be more suited for particular applications because they may occupy less space, or may

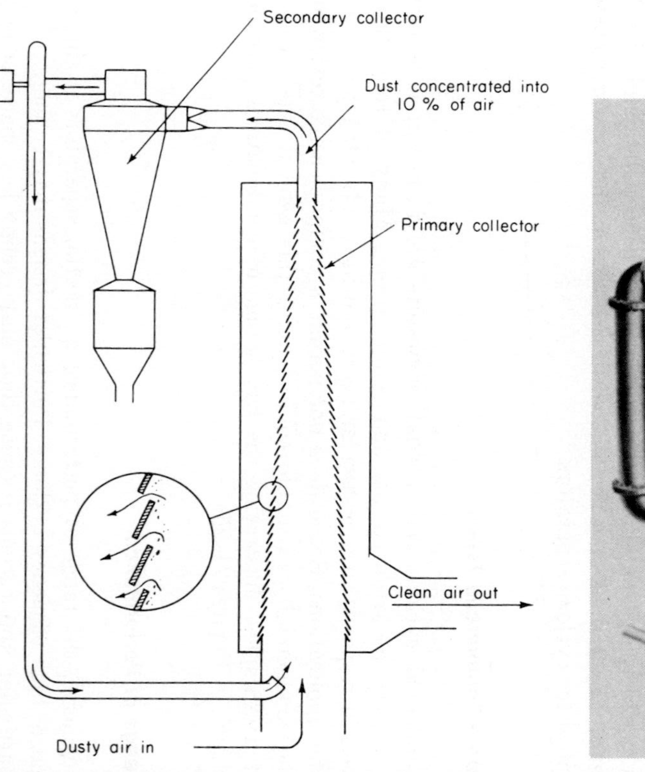


Figure 21. Aerodyne tube. (Courtesy Green Fuel Economizer Co., Inc.)

Figure 22. Aerodyne tube inertial collector arrangement. (Courtesy Green Fuel Economizer Co., Inc.)

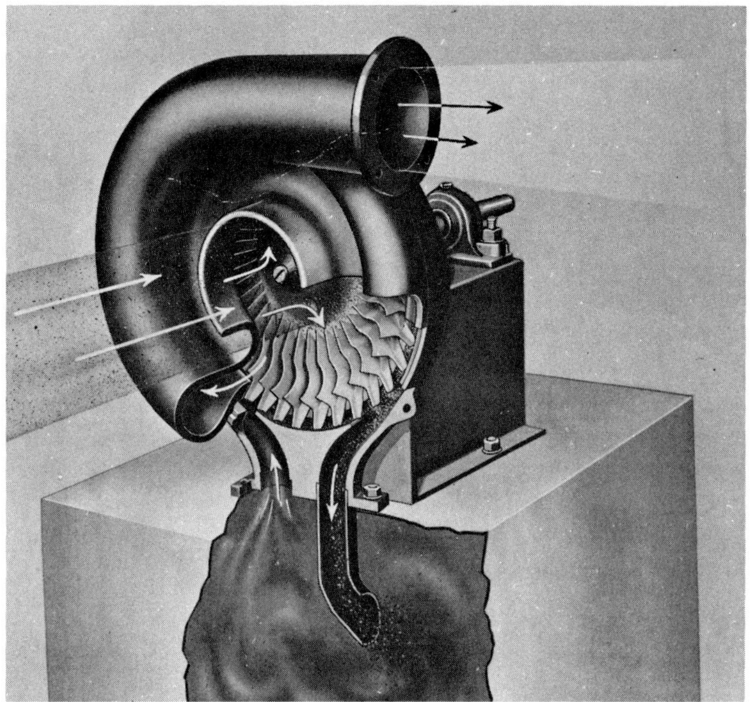

Figure 23. Type D Rotoclone, a mechanical inertial collector. (Courtesy American Air Filter Co.)

Figure 24. "Low draft loss" fly ash collector. (Courtesy Buell Engineering Co., Inc.)

**Table III Representative Performance of Cyclones and Inertial Separators**

| *Collector type* | *Process* | *Material* | *Air flow, ft³/min* | *Pressure drop, in. wg* | *Efficiency, wt %* | *Inlet load, grains/ft³* | *Inlet mass median size, μm* |
|---|---|---|---|---|---|---|---|
| Series cyclones | Fluid catalytic cracking | Catalyst | 40,000 | High | 99.98 | 2800 | 37.0 |
| Special cyclone | Laboratory test | Fly ash | 185 | 23.0 | 91.2 | 0.06 | 3.0 |
| Special cyclone | Laboratory test | Micronized talc | 185 | 23.0 | 83.9 | 5.6 | 2.3 |
| Cyclone | Abrasive cleaning | Talc | 2,300 | 0.33 | 93.0 | 2.2 | — |
| Cyclone | Drying | Sand and gravel | 12,300 | 1.9 | 86.9 | 38.0 | 8.2[a] |
| Cyclone | Grinding | Aluminum | 2,400 | 1.2 | 89.0 | 0.7 | — |
| Cyclone | Planning mill | Wood | 3,100 | 3.7 | 97.0 | 0.1 | — |
| Rotary stream separator | Test dust (2.65 sp. gr.) | — | 300 | 8 | 100 | 0.4 | 5 |
| | | | | | 90 | 88 | 1 |
| Rotary stream separator | Tar–macadam plant | — | — | — | 90 | 2 | 2[b] |
| Inertial | Cyclone outlet | Sand and gravel | 1,700 | 4.0 | 50.0 | 5.8 | 5.3[c] |
| Mechanical | Grinding | Iron scale | 11,800 | 4.7 | 56.3 | 0.15 | 3.2[d] |
| Mechanical | Rubber dusting | Zinc stearate | 3,300 | 9.0 | 88.0 | 0.6 | 1.7 |

[a] Outlet mass median size, μm = 3.2.
[b] Size frequency maximum.
[c] Outlet mass median size, μm = 1.8.
[d] Outlet mass median size, μm = 2.5.

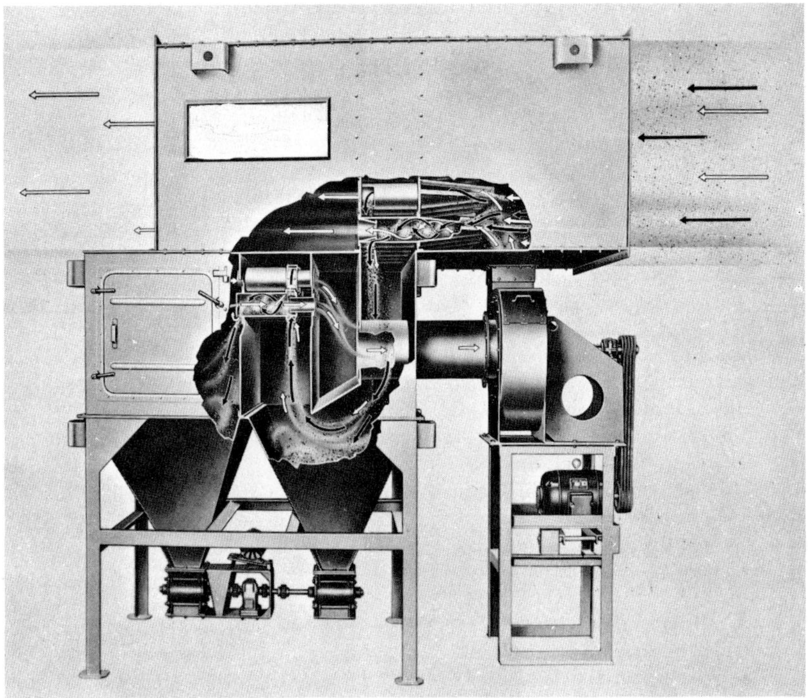

Figure 25. Horizontal, peripheral discharge cyclones arranged for partial series operation. (Courtesy American Air Filter Co.)

be specially designed to compensate for particular problems such as erosion, necessity to minimize pressure drop, etc. In general, those without moving parts may be constructed of any reasonable material of construction, while those with moving parts must include considerations of stress and accelerated corrosion and erosion of the moving parts. Most inertial separators can be operated wet, if desired, and under such conditions they exhibit a higher collection efficiency due to the same factors that operate in the case of cyclone dust collectors. A tabulation of representative data from various sources is shown in Table III.

Additional pertinent material can be found in Danielson (*35*) and in the American Industrial Hygiene Association "Air Pollution Manual" (*36*).

## REFERENCES

1. A. C. Stern, K. J. Caplan, and P. D. Bush, "Cyclone Dust Collectors." Amer. Petrol. Inst., New York, New York, 1955.

2. M. W. First, Ph.D. Thesis. Harvard University, Cambridge, Massachusetts, 1950.
3. C. E. Lapple, *Ind. Hyg. Quart.* **5**, No. 11, 40–48 (1950).
4. W. Batel, *Staub* **32** (9), 1–7 (1972).
5. J. Hejma, *Staub* **31** (7), 22–28 (1971).
6. D. Leith and D. Mehta, *Atmos. Environ.* **7**, **527** (1973).
7. R. McK. Alexander, *Australas. Inst. Mining Met., Proc.* [3] **152/153,** 202–228 (1949).
8. A. J. ter Linden, *Proc. Inst. Mech. Eng.* **160,** 233 (1949).
9. C. J. Stairmand, *Trans. Inst. Chem. Eng.* **29,** 356–383 (1951).
10. F. B. Schneider, *Gen. Elec. Rev.* **53,** 22–29 (1950).
11. F. Schulz, *Eng. Dig.* **5,** 49 (1948).
12. H. L. M. Larcombe, *Mining Mag.* **77,** 137–148, 208–217, 273–278, and 356–347 (1947).
13. *Bitum. Coal Res., Inc., Tech. Rep.* **14,** No. 1, 8–12 (1954).
14. L. Silverman *et al.,* Contract No. AT130-1, Gen-238, Rept. NYO 1527. U.S. At. Energy Comm., Washington, D.C., 1950.
15. K. Iinoya, *Mem. Fac. Eng., Nagoya Univ.* **5,** No. 2 (1953).
16. F. Ebert, *Staub* **29,** (7), 5–9 (1969).
17. P. Drinker and T. Hatch, "Industrial Dust," 2nd ed. McGraw-Hill, New York, New York, 1954.
18. K. T. Whitby, *Minn., Univ., Eng. Exp. Sta., Bull.* **32** (1950).
19. R. Dennis *et al.,* Contract No. **AT-30-1,** 841, Rep. NYO-1583. U.S. At. Energy Comm., Washington, D.C. 1952.
20. L. W. Briggs, *Trans. Amer. Inst. Chem. Eng.* **42,** No. 3, 511–526 (1946).
21. E. Fiefel, *Schweiz. Bauztg.* **68,** 247–251 (1950).
22. J. G. Wilson and D. W. Miller, *J. Air Pollut. Contr. Ass.* **17** (10), 682–685 (1967).
23. E. Muschelknautz, *Staub* **30** (5), (1970).
24. H. Rumpt, K. Broho, and H. Reichert, *Staub* **29** (7), (1969).
25. H. Klein, *Staub* **23** (11), 501–509 (1963).
26. W. Nickel, *Staub* **23** (11), 509–512 (1963).
27. E. Schaufler, *Staub* **23** (4), 228–230 (1963).
28. K. R. Schmidt, *Staub* **23** (11), 491–501 (1963).
29. K. Budinsky, *Staub* **32** (3), 1–6 (1972).
30. C. Alt and P. Schmidt, *Staub* **29** (7) (1969).
31. C. F. Gottschlich, "Gravity Inertial, Sonic and Thermal Collectors." Amer. Petrol. Inst., New York, New York, 1961.
32. S. K. Freidlander *et al.,* "Handbook on Air Cleaning." U.S. At. Energy Comm., Washington, D.C., 1952.
33. W. E. Ranz, *Pa. State Univ., Miner. Ind. Exp. Sta., Bull.* **66** (1956).
34. J. B. Wong *et al., J. Appl. Phys.* **26,** 244–249 (1955).
35. J. A. Danielson, ed., "Air Pollution Engineering Manual," 2nd ed., Chapter 4. U.S. Environmental Protection Agency, Washington, D.C., 1973.
36. American Industrial Hygiene Association, "Air Pollution Manual," Part II, Chapter 4. Amer. Ind. Hyg. Ass., Akron, Ohio, 1968.

# 4

# Filtration

Koichi Iinoya and Clyde Orr, Jr.



## Nomenclature

$A$ Constant
$C_{\mathrm{m}}$ Cunningham coefficient

| | |
|---|---|
| $D_{BM}$ | Diffusion coefficient of particle, $m^2$/second |
| $D_f$ | Fiber diameter, m or $\mu$m |
| $D_p$ | Particle diameter, m or $\mu$m |
| $D_{ps}$ | Specific area diameter of particle, m or $\mu$m |
| $E_0$ | Overall collection efficiency, dimensionless |
| $g$ | Gravitational acceleration, 9.8 m/$sec^2$ |
| $g_c$ | Gravitational conversion factor, 9.8 kg m/kg-force $sec^2$ |
| $G$ | Gravitational parameter, $C_m D_p^2 g/18\mu u$, dimensionless |
| $K$ | Constant, dimensionless |
| $L$ | Filter thickness |
| $m$ | Collected dust loading on the filter, kg/$m^2$ |
| $N_D$ | Diffusion parameter, $D_{BM}/D_f u$, dimensionless |
| $N_R$ | Interception parameter, $D_p/D_f$, dimensionless |
| $\Delta P$ | Pressure loss, kg-force/$m^2$, mm $H_2O$ |
| $\Delta P_0$ | Pressure loss due to the filter itself, kg-force/$m^2$ or mm $H_2O$ |
| $\Delta P_p$ | Pressure loss due to the deposited particle layer, kg-force/$m^2$ or mm $H_2O$ |
| Re | Reynolds number, $D_f u \rho_g/\mu$, dimensionless |
| $u$ | Average true filtering gas velocity, $u_s/\epsilon_f$ or m/second |
| $u_s$ | Filtering gas velocity (superficial), m/second |
| $\bar{\alpha}$ | Average specific resistance of collected particle layer, m/kg |
| $\beta$ | Coefficient, dimensionless |
| $\epsilon_f$ | Volumetric void fraction of fibrous bed, $<1$, dimensionless |
| $\epsilon_p$ | Apparent volumetric void fraction of particle layer, dimensionless |
| $\rho_g$ | Density of gas, kg/$m^3$ |
| $\rho_p$ | Density of particle, kg/$m^3$ |
| $\mu$ | Gas viscosity, kg/m second |
| $\eta_0$ | Initial collection efficiency of an isolated fiber, dimensionless or % |
| $\eta_\epsilon$ | Initial collection efficiency of a single fiber with the influence of neighboring fibers, dimensionless or % |
| $\zeta_0$ | Pressure loss coefficient for clean filter, 1/m |

### *Subscripts*

| | |
|---|---|
| D | Diffusion |
| f | Fiber |
| I | Inertia |
| 0 | Isolated fiber |
| p | Particle |
| R | Interception |
| $\epsilon$ | Single fiber with influence of neighboring fibers |

## I. Introduction

Filtration is among the most reliable, efficient, and economical methods by which particulate matter may be removed from gases. It also is one of the relatively few air pollution control methods capable of meeting most present particulate emission standards. Recently many practical studies, most sponsored by the United States Environmental Protection Agency (USEPA), have been published (*1—1c*).

As an industrial air pollution control means, gas filters may be broadly classified into one of two kinds—fabric, or cloth, filters and in-depth, or bed, filters. The former is represented by various fabric bag arrangements while the latter is most frequently encountered as a fibrous array, a paperlike mat, and occasionally, as a deep packed bed. Fabric filters are generally utilized with gas or airstreams having a dust loading of the order of 1 $gm/m^3$; fibrous packings, paper filters, and packed beds are applied when the particulate concentration is several orders of magnitude less, perhaps 1 $mg/m^3$.

There are three major performance criteria for a filter: pressure loss; collection efficiency; and lifetime, which is related to endurance and dust holding capacity. Pressure loss is usually expressed in terms of millimeters of water (water column) and is directly proportional to the required fan or blower horsepower, or energy. The pressure loss is a major index of the operating cost of a filtration system.

Collection efficiency is probably the single most important factor in the performance of a filter; it is usually evaluated in terms of the mass percentage of dust retained but sometimes an optical density measure is employed. There are two kinds of collection efficiency—an instantaneous one and a cumulative one—because the efficiency changes with the dust captured by the filter and with the filtering time.

The lifetime of a filter is very important from the economic standpoint because the cost of the filtering medium is a major portion of the initial expense as well as of long-term operating costs. It is difficult, however, to estimate filter life from desired operating conditions for any application unless a background of experience is first established.

Filter media for gas filtration are prepared from fibers, aggregates, or porous substances. Woven fabrics, felt, loosely packed beds, and paper filters are all compositions of fibers. A dense packed bed is an assemblage of a great many bits and pieces through which the particulate-laden gas is made to pass by confining the pieces in one chamber. And a porous filter is a rigid, or semirigid, structure within which are a multitude of passageways. Filters are produced from both natural and synthetic materials. Each filter material must, therefore, be selected for a particular use from the viewpoint of both economics and collection characteristics. Most popular are polyester or glass cloth for fabric filters, and synthetic or glass fibers for mat filters. The diameter of individual filaments should be as small as possible; less than 1 $\mu m$ diameter for individual fibers gives best collection. Deep packed beds are prepared from crushed stone or brick, wire screens, or fibers of many types arranged individually or in combination.

## II. Fabric (Bag) Filters

Fabric filters are extensively employed to control harmful or obnoxious emissions in operations involving abrasives, irritating chemical dusts, and the exhausts from electric furnaces, oil-fired boilers, and oxygen-fed converters for steel making (*1d–5*).

### A. Principle of Operation

With fabric bags, filtration is principally accomplished by the particle layer that accumulates on the fabric surface (*6*). Therefore, it is difficult to predict filter pressure loss and collection efficiency without making preliminary tests. The bag filter has not been successfully analyzed theoretically, because of the filtering action of the dust layer. When the pressure loss increases because of this accumulated load and the gas velocity decreases concomitantly to a prescribed lower limit, the filtering operation must be stopped and the filter cleaned by a dust-dislodging operation. Filtration is then begun again. Many fabric bag filters assembled in one unit form what is commonly called a bag house.

#### 1. Weave Characteristics

Table I presents the basic properties of fibers that are widely used at the present time (*7–11*), and Figure 1 shows weave types—twill and sateen being favored. The performance of any fabric is greatly influenced by thread density, fiber composition, and nap, i.e., the hairy or downy texture of the cloth surface. Cotton is the least expensive fiber. Silicone- or graphite-coated glass fiber cloth is commonly employed in applications

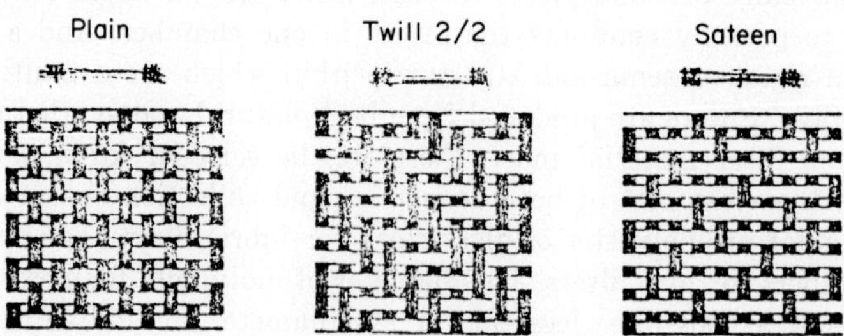


Figure 1. Filter cloth weaves.

**Table I Properties of Fiber Materials**

| Fiber | *Physical characteristics* | | | | *Relative resistance to attack by* | | | Other attribute |
|---|---|---|---|---|---|---|---|---|
| | *Relative strength* | *Specific gravity* | *Normal moisture content (%)* | *Maximum usable temperature (°F)* | *Acid* | *Base* | *Organic solvent* | |
| Cotton | Strong | 1.6 | 7 | 180 | Poor | Medium | Good | Low cost |
| Wool | Medium | 1.3 | 15 | 210 | Medium | Poor | Good | — |
| Paper | Weak | 1.5 | 10 | 180 | Poor | Medium | Good | Low cost |
| Polyamide (nylon) | Strong | 1.1 | 5 | 220 | Medium | Good | Good[a] | Easy to clean |
| Polyester (Dacron) | Strong | 1.4 | 0.4 | 280 | Good | Medium | Good[b] | — |
| Acrylonitrile (Orlon) | Medium | 1.2 | 1 | 250 | Good | Medium | Good[c] | — |
| Vinylidene chloride | Medium | 1.7 | 10 | 210 | Good | Medium | Good | — |
| Polyethylene | Strong | 1.0 | 0 | 250 | Medium | Medium | Medium | — |
| Tetrafluoroethylene | Medium | 2.3 | 0 | 500 | Good | Good | Good | Expensive |
| Polyvinyl acetate | Strong | 1.3 | 5 | 250 | Medium | Good | Poor | — |
| Glass | Strong | 2.5 | 0 | 550 | Medium | Medium | Good | Poor resistance to abrasion |
| Graphitized fiber | Weak | 2.0 | 10 | 500 | Medium | Good | Good | Expensive |
| Asbestos | Weak | 3.0 | 1 | 500 | Medium[d] | Medium | Good | — |
| "Nomex" nylon | Strong | 1.4 | 5 | 450 | Good | Medium | Good | Poor resistance to moisture |

[a] Except phenol and formic acid.
[b] Except phenol.
[c] Except heated acetone.
[d] Except $SO_2$.

up to 250°C (*12–15*). Acceptable cloth life is generally considered to be 1 year or more. Factors upon which fabric life depend are the operating temperature, the duration and manner of cleaning or dislodging the dust load from the cloth, the characteristics of the particulate matter, the nature of the gas, and the care with which the installation is designed for the particular application (*16, 17*).

The electrostatic properties of both the dust and the collecting fabric influence filtering and cleaning performance (*18–22*). Fibers and fabrics conform to a triboelectric series as indicated in Table II and dusts may be classified in a similar series. The charge intensity developed on either the fabric or the dust, or both, depends upon the processing conditions, as well as on the nature of the materials themselves. The charge dissipation rate is an especially important property in the dust cleaning process. A fabric bag into which is woven stainless-steel fibers or wire is most effective when dislodging dust that is particularly adhesive because of retained electrostatic charges and for protection from fire arising from electrostatic discharges. As yet, not enough is known about how best to utilize these factors in the design and operation of particular systems.

While in the past most filter media were woven fabrics, recent years have seen the introduction of nonwoven materials such as fleeces and felts (*23*). The term "fabric filters" is not applied to this type of medium; instead the more general term "filter cloth" is used.

**Table II Electrostatic Charging Order of Filter Fibers** (*20*)

| *Material* | *Relative charge generation* |
|---|---|
| Wool | +20 |
| Silicon-treated glass (filament and spun) | +15 |
| Woven wool felt | +11 |
| Nylon (spun) | +7 to +10 |
| Cotton (sateen) | 6 |
| Orlon (filament) | +4 |
| Dacron (filament) | 0 |
| Dynel (spun) | −4 |
| Orlon (spun) | −5 to −14 |
| Dacron (spun) | −10 |
| Steel | −10 |
| Polypropylene (filament) | −13 |
| Acetate | −14 |
| Saran | −17 |
| Polyethylene (filament and spun) | −20 |

## 2. Pressure Loss

In general, gas flow through fabric filters is laminar in character (*24–26*). The pressure loss, being the sum of the loss due to the filter itself and the deposited particle layer, may be written

$$\Delta P = \Delta P_0 + \Delta P_p = (\zeta_0 + \bar{\alpha} m) u_s \mu / g_c \tag{1}$$

The value of the pressure loss coefficient $\zeta_0$ for a clean filter depends upon the fabric; it usually is negligibly small in practical applications. However, its value should be larger than a certain minimum for initial collection performance. For clean fabric, the filtering air velocity $u_s$ at which the pressure loss $\Delta P_0$ equals $\frac{1}{2}$ in. of water (12.7 mm $H_2O$) is called the permeability.

The value of the average specific resistance $\bar{\alpha}$ of a collected particle bed depends upon the particle size, the volumetric voids in the bed, and the density of the particles. Values as indicated by Figure 2 (*27*) are typical. The terms of Equation (1) might have values as follows: $\zeta_0 = 7 \times 10^7$/m, $\bar{\alpha} = 3.0 \times 10^{10}$ m/kg, $m = 0.12$ kg/m$^2$, $u_s = 1.0$ m/min $=$ 0.017 m/sec ($=$ 0.67 in./second), $\mu = 1.81 \times 10^{-5}$ kg/m second (at 760 mm Hg and 20°C), and $g_c = 9.8$ m kg/sec$^2$ kg. Substituted into the equation, these result in a pressure loss of 115 kg/m$^2$ or mm $H_2O$, or 4.4 in. $H_2O$.

The change in pressure loss with particle collection may be obtained

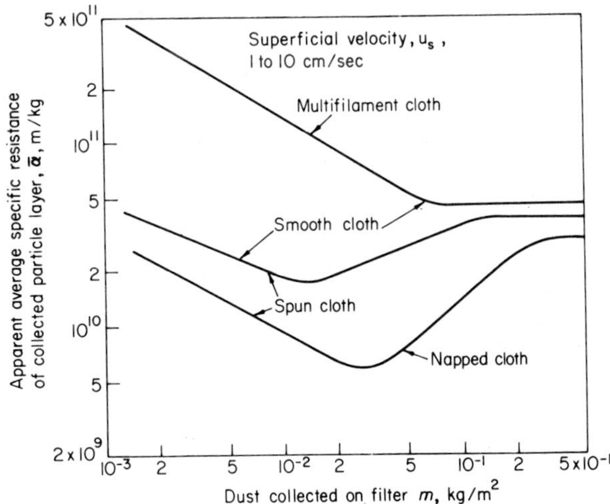


Figure 2. Effect of fabric on average specific resistance of dust layer (27).

from the Kozeny–Carman relationship (*27*) written

$$\frac{d\,(\Delta P_p)}{dm} = \frac{180\mu u_s(1 - \epsilon_p)}{g_c \rho_p D_{ps}{}^2 \epsilon_p{}^3} \tag{2}$$

where the apparent volumetric void $\epsilon_p$ is obtained from Figure 3. For example, using the values of $u_s$ and $\mu$ as above, $\rho_p = 3000$ kg/m³, $D_{ps} = 0.40 \times 10^{-6}$ m, and $\epsilon_p = 0.96$, Equation (2) gives a value for $d(\Delta P_p)/dm$ of 532 kg/kg or mm of $H_2O$/kg of dust/m² of cloth area. However, the permeability of the dust cake on the cloth varies with the operating conditions of the bag filter in a way that significantly affects the pressure drop (*28*).

Typical gas flow behavior to be expected within a multicompartment baghouse (*29–31*) is presented in Figures 4 and 5. The effect of the number of compartments on the cleaning cycle is shown in Figure 6, which indicates that many compartments prolong the cleaning period (*32*).

A nonwoven cloth is often the more favorable filter medium because the pressure loss after cleaning is lower and remains essentially constant if a strong cleaning method, for example, a pulsating reverse air jet, is employed (*34*).

The following values are common for conventional bag filters under normal operations: $\Delta P = 100$–$200$ mm $H_2O$; $m = 0.05$–$0.03$ kg/m; $\bar{\alpha} = 10^{10}$–$10^{11}$ m/kg; $u_s = 0.5$–$3$ m/minute.

## 3. Collection Efficiency

Particle separation by fabric is more than simple entrapping of the particles by single fibers, since the open spaces through such media are

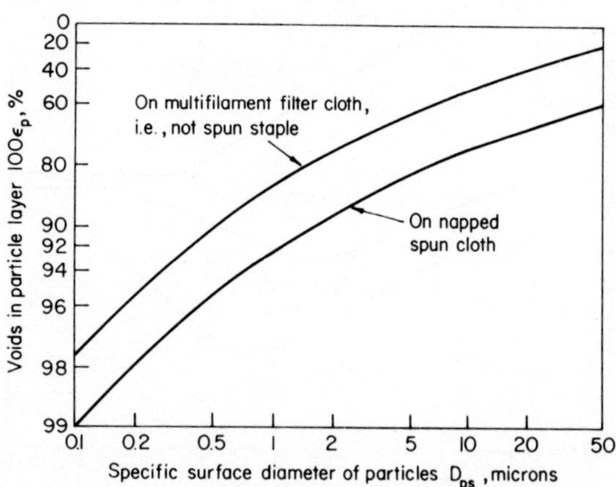


**Figure 3.** Relation between specific surface diameter of particles to be collected and voids of collected particle layer (*27*).

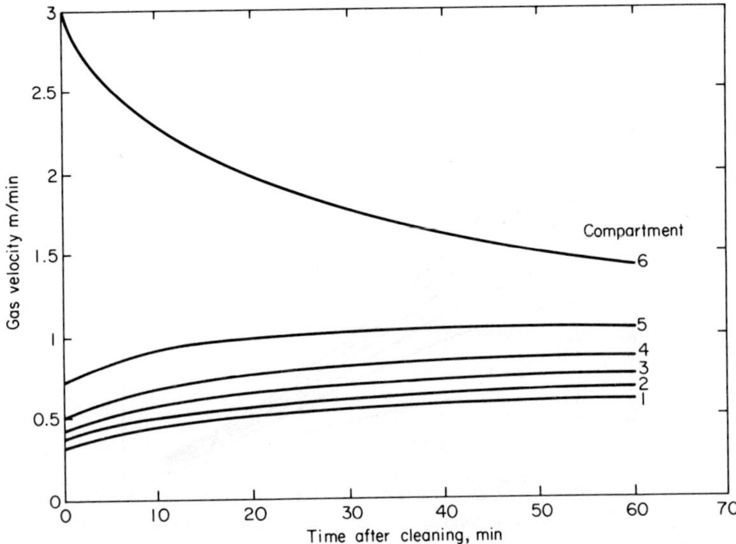


Figure 4. Velocity pattern in six-compartment bag house as a function of time (*29*). Average velocity, 1 m/min; initial compartment pressure loss, 17 mm $H_2O$/m/min; final compartment pressure loss, 170 mm $H_2O$/min.

usually many times the size of the individual particles which are collected. Actually, separation with filters is poor until enough particles have been captured to form an arching bed across the openings (*33*). Once the particle bed has been formed, separation efficiency will rise to values near 99% as indicated in Figure 7. Efficiencies of more than 99.9% are

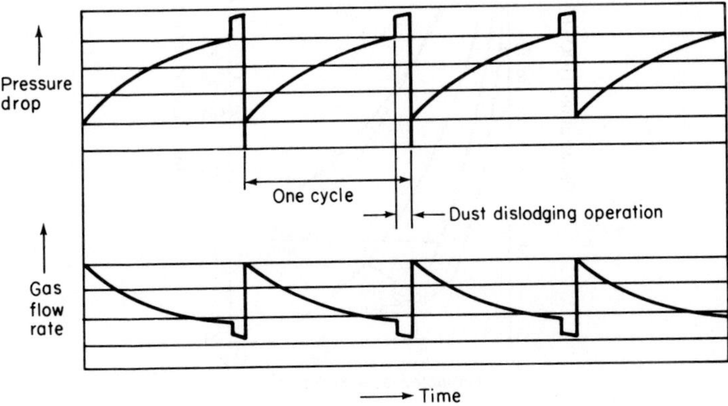


Figure 5. Flow and pressure loss variations as a function of time in a multicompartment baghouse. Average velocity, 1 m/min; initial compartment pressure loss, 17 mm $H_2O$/m/min; final compartment pressure loss, 170 mm $H_2O$/m/min.

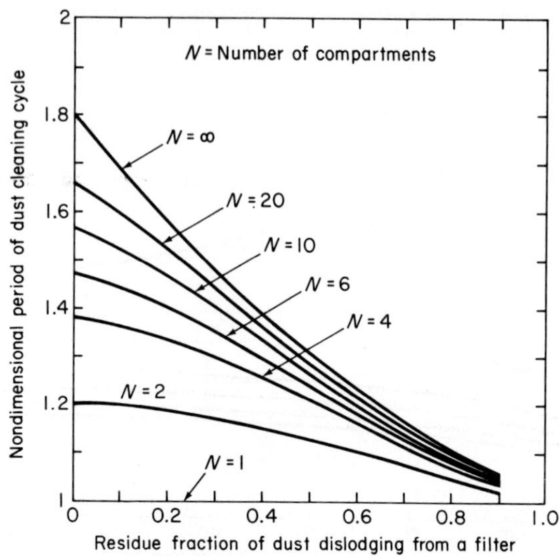


Figure 6. Effect of the number of compartments on cleaning cycle periods (*32*).

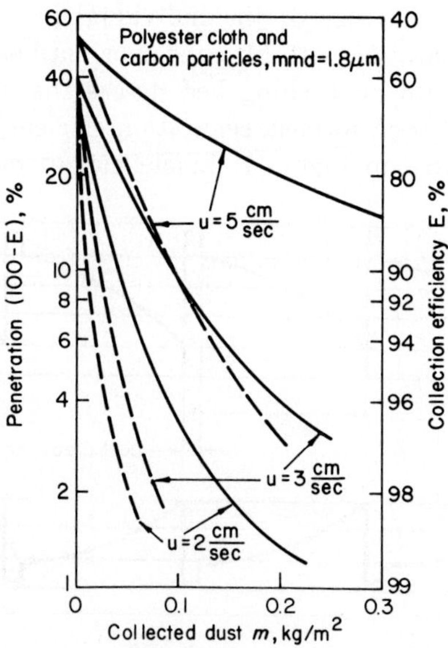


Figure 7. Collection efficiency of fabric filters. The solid lines show cumulative collection efficiencies and the broken lines show instantaneous ones. (mmd = Mass median diameter.)

seldom encountered (*34*). Failure to achieve high collection efficiency is almost always due to excessive cleaning, torn bags, bypass leakage, or an excessive gas flow rate which produces pinholes within the deposited particle bed. Instantaneous efficiency values usually are higher than cumulative collection efficiencies for an operating cycle, because the collection efficiency generally increases with the dust deposit (*35–37*).

Twill or broken twill fabrics usually give less residual pressure loss and better collection efficiency than other weaves (*38*).

### 4. Cleaning or Dust-Dislodging Methods

Methods of cleaning filter media are shaking, low pressure continuous fan scavenging, jet-pulsing, and reverse-jet dedusting. Generally, large-scale bag filter installations are cleaned by shaking or by countercurrent air scavenging (*39, 40*). In small-scale applications, jet-pulse filters utilizing short blasts of higher pressure air into the individual filter bags during the filtration process are widely used. Bag filters with traveling jet cleaning are being less used now than heretofore. A cleaning cycle usually requires only a very few minutes. The layer of dust deposited on the bags in successive compartments is periodically cleaned in a continuous cycle.

## *B. Equipment Types*

Bag filters are classified by the type of cleaning method and the shape of the filter bag. Filters are commonly made in the form of a tubular bag or an envelope or pocket slipped over a wire frame. The tubular bags range from 120 to 400 mm in diameter and may be up to 12 m long. Particle-laden gas is usually introduced inside the tube in such a manner as to allow the larger particles to settle, or to be projected, into the dust hopper before the gas actually enters the tubes (*41*).

### 1. Bag Filter with Mechanical Shaking

Bags with mechanical shaking constitute one of the oldest and most widely used type of fabric filter. An installation generally consists of several compartments, each compartment usually containing many bags, in order that one compartment may be removed from service while its bags are being cleaned. Periodically, either according to a predetermined program or by monitoring the draft loss, the collector goes into its cleaning cycle. During the cleaning cycle the normally open outlet butterfly valve is closed. The upper ends of the bags are usually shaken either by mechanical or pneumatic means with or without the assistance of low pressure reverse airflow (Fig. 8). In some designs, the middle sections

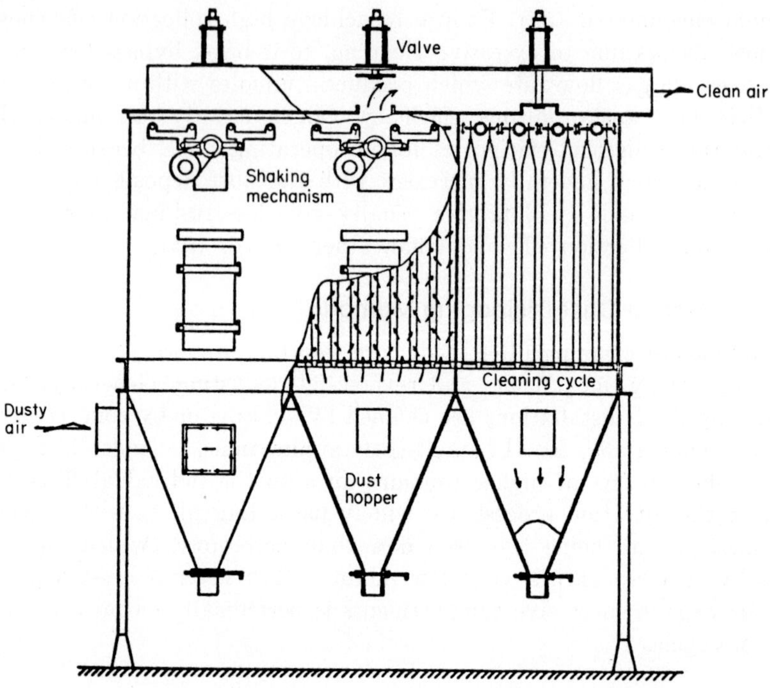


Figure 8. Typical bag filter with mechanical shaking of upper ends.

of bags are shaken with higher frequency and smaller amplitude than are the ends; the dislodged dust is removed from the hopper by a sealed screw conveyor.

## 2. Bag Filters with Reverse Flow

In reverse-flow filters, clean air from the outside, or other appropriate source is blown in a reverse direction into the compartment requiring cleaning. During the cleaning cycle the normally open outlet butterfly valve is closed. At the same time a small air vent valve is opened, allowing atmospheric air to rush into the casing and collapse the bags. This will break the filter deposit and allow it to drop into the hopper, aided by the reverse flow. This secondary airflow is maintained by the suction through the still-open common inlet manifold (Fig. 9).

## 3. Bag Filters with Reverse Pulse Jets

In reverse pulse-jet filters the casing is divided into an upper and lower portion by a tube sheet as shown in Figure 10. The upper portion serves

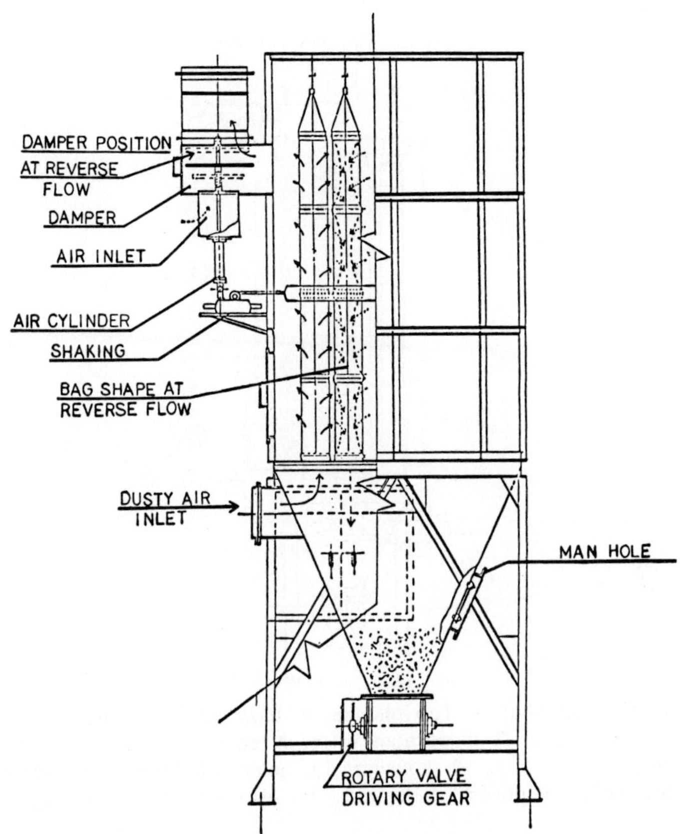


Figure 9. Bag filter with mechanical shaking of middle ring and reverse flow.

as a common cleaned gas discharge manifold. For each felted filter bag there is a venturi-shaped thimble. A compressed air inlet is mounted above each tube. The felt filter medium is slipped over an internal frame. The lower part of the casing serves as a settling chamber and collects coarse dust without the need for filtering. the filter cake is built up on the outside of the bag. Periodically solenoid valves introduce compressed air jets, thereby blowing air with a 0.015-sec pulse at high pressure (5 to 7 atm) through the venturi into individual bags. About 1 to 2 $m^3$ of compressed air are required per 1000 $m^3$ of collector capacity. The force of the pulse jet snaps the bag outward, thereby dislodging the accumulated filter cake. The cleaning cycle usually cleans several bags at a time through set of manifolds. Filters equipped with pulsing or reverse jet devices can handle gases at higher flow rates (1–5 m/minute) than previously mentioned types without the need of separate bypass units.

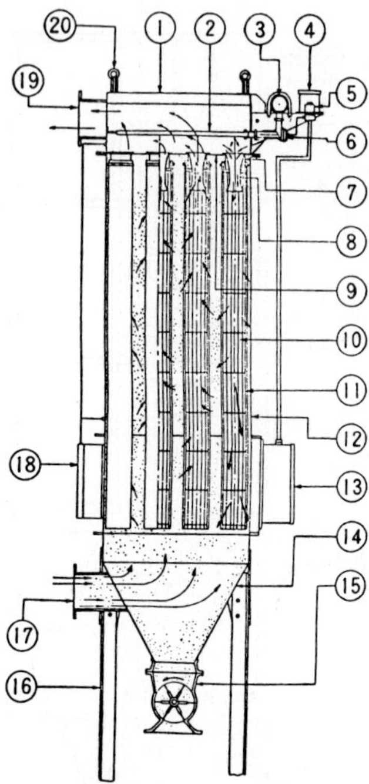


Figure 10. Pulse-jet filter. 1. Top plenum; 2. blow tube; 3. header-pipe; 4. pilot valve; 5. control valve; 6. diaphragm valve; 7. collar; 8. venturi; 9. tube sheet; 10. bag retainer; 11. bag; 12. housing; 13. timer; 14. hopper; 15. rotary valve; 16. support; 17. air inlet; 18. manometer; 19. air outlet; 20. lifting lug.

This kind of filter permit higher inlet dust loads, but it is more complex; hence it has a greater first cost.

## 4. Envelope

In an envelope filter, the individual filters range from 0.6 to 1 m wide and are only 3 to 5 cm thick. Gas passes from the outside inward in these units since the filters are prevented from collapsing by the inner framework. As in bag filters, the envelope panels are mounted in multiples (*42*).

Figure 11 gives an example of an envelope filter with cleaning provided by a pulsating reverse airflow. The high frequency vibrating flow is generated for felt cloths by a rotary valve in the reverse air line; it is effective for a humid cake. The filtering velocity is between 1 to 2.5 m/minute in such units.

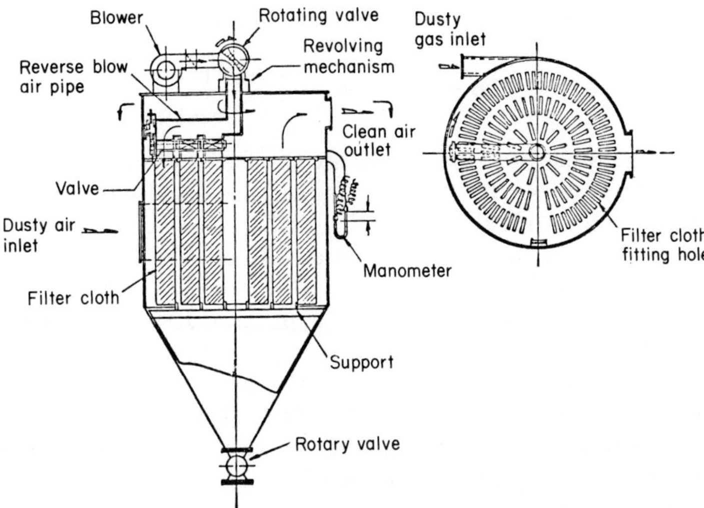


Figure 11. Envelope-type filter with vibrating reverse flow.

## III. Fibrous Mat Filters

Fibrous-mat filters find their most extensive use in air conditioning, heating, and ventilating systems, but high efficiency mat filters are also employed as aftercleaners and less efficient ones as roughing units to protect still more efficient equipment for pollution control purposes.

### *A. Theory of Operation*

Fibrous mat filters are the most completely understood class of filters in terms of their operating characteristics. Since 1950, they have been studied extensively, with and without electrostatic and gravitational effects being considered. The effect of fiber shape on performance has also been studied in terms of both fiber elongation and circularity (*43*). Despite this, it is still difficult to estimate their performance under dust loading conditions by theoretical calculation.

#### 1. Pressure Loss

Many relationships for the pressure loss of clean fibrous mats have been proposed (*44–49*). From a large amount of experimental data on glass and steel fibrous filters, an empirical equation

$$\Delta P = \left(0.6 + \frac{4.7}{\sqrt{\mathrm{Re}}} + \frac{11}{\mathrm{Re}}\right) \frac{2\rho_g u^2 L(1 - \epsilon_f)}{\pi g_c D_f} \tag{3}$$

has been suggested (*50*). It is generally satisfactory for practical conditions of $10^{-3} < \mathrm{Re} < 100$ (mostly laminar flow) and $0.92 < \epsilon_f < 0.98$. Here Re ($= D_f u \rho_g / \mu$) is the Reynolds number, $\epsilon_f$ the void fraction of the bed, $D_f$ the fiber diameter, $L$ the thickness of the bed, and all other quantities expressed in consistent units. If $\epsilon_f = 0.95$, $L = 0.025$ m (1 in.), $D_f = 2 \times 10^{-5}$ m (20 $\mu$m), $\rho_g = 1.20$ kg/m$^3$, $\mu = 1.81 \times 10^{-5}$ kg/m second, $u_s = 1.0$ m/second. With the surface gas velocity $u$ defined as $u_s/\epsilon_f$, the Reynolds number will be 1.4 and the pressure loss calculated by Equation (3) will be 61 mm $H_2O$ (61 kg/m$^2$). The pressure loss across a homogeneously mixed, multicomponent filter is always lower than the pressure loss across a filter composed of layers of the same components in the same weight fractions (*51*).

The pressure loss can only be estimated by means of theoretical analysis under dust loading conditions (*52, 53*). However, experimental results such as those given by Figure 12 are helpful for practical estimation.

## 2. Collection Mechanism and Efficiency

The removal by fibrous filters of fine, uncharged particulate matter suspended in a gas stream is mainly due to inertial effects, direct intercep-

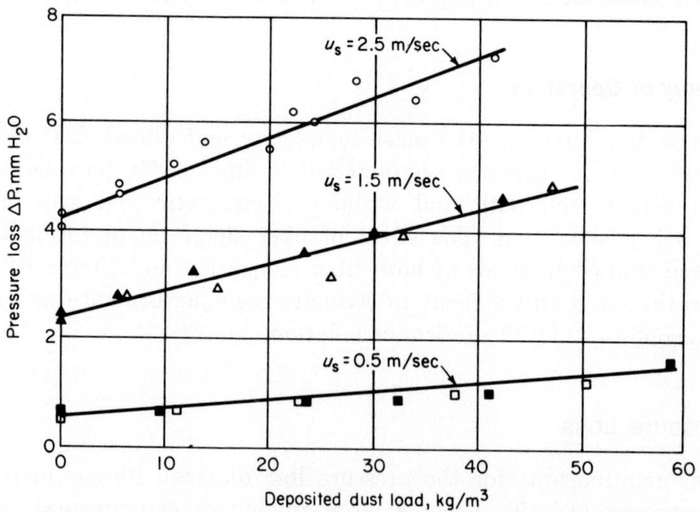


Figure 12. Pressure drop of fibrous filters with dust loading.

tion, and diffusional (Brownian) motion. Gravity influences the collection of relatively large particles only. Inertial effects are of significance when they cause particles to deviate from the initial path of gas flow as an obstruction is approached. A particle may or may not contact the obstruction even though it departs from the flow streamlines. This depends upon the size of the obstruction, the particle size and inertia, and the particle position relative to the obstruction when the approach began. Inertia is not considered significant in collecting particles below 1 $\mu$m in diameter, but it becomes increasingly important as size increases. Collection by direct interception is paramount for 1- to about 0.2-$\mu$m-diameter particles. While such particles are not inertialess, their inertia is relatively insignificant. They tend to follow the flow streamlines around an obstruction and to contact the obstruction only if the path of the streamlines carries them sufficiently close. Very small particles—less than about 0.2 $\mu$m in diameter—do not follow the gas streamlines because collisions of individual gas molecules knock then about, resulting in what is known as Brownian motion. This random movement results in enhanced chances for contact with a collecting surface and becomes the predominant collection mechanism for very small particles. Obviously, there are overlapping influences from the several mechanisms; this is considered further subsequently.

The collection mechanism of a fibrous bed may be arrived at by considering the removal of a spherical particle by a single cylindrical fiber *(54)* and summing over all particles and all fibers. The initial efficiency of the filter, i.e., the efficiency at the beginning of use, is

$$E_0 = 1 - \exp \frac{-4(1 - \epsilon_f)L}{\pi \epsilon_f D_f} \eta_c \tag{4}$$

on the basis of the so-called log-penetration law. The initial efficiency $\eta_\epsilon$ of a single fiber here takes into account an influence of neighboring fibers. The initial collection efficiency $\eta_0$ of an isolated fiber, i.e., the efficiency without the influence of neighboring fibers, is correlated with the above single fiber efficiency $\eta_\epsilon$ as follows *(44, 55)*:

$$\eta_\epsilon = \eta_0[1 + \beta(1 - \epsilon_f)] \tag{5}$$

where the parameter $\beta$ is between 5 and 20 when inertial effects predominate.

By no means do all the particles that contact a fiber adhere to it. The adherence or nonadherence of particles upon impact depends, among other factors, upon the angle of impaction *(56)*. Some of the unexplained filtration results may be due in part to particles escaping capture on contact. Reentrainment always begins to be a significant factor after a certain dust load is accumulated on a mat. The particular condition under which this occurs depends primarily on the filter medium itself, the particles, and the gas velocity (*57*). It is not presently predictable.

The leakage around filter media sometimes contributes to the deterioration of overall filter efficiency.

a. Diffusional Motion. The initial collection efficiency by diffusion for an isolated fiber may be written (*58*)

$$\eta_{0D} = 2.9K^{1/3}N_D^{2/3} + 0.62N_D \tag{6}$$

where

$$K = -0.5\ln(1 - \epsilon_f) + [1 - \epsilon_f - 0.25(1 - \epsilon_f)^2] - 0.75$$
$$N_D = D_{BM}/D_f u \qquad \text{(diffusion parameter)} \tag{7}$$

The diffusion parameter $N_D$ is the reciprocal of the Peclet number, and the diffusion coefficient $D_{BM}$ of a particle is shown in Figure 13 for air at

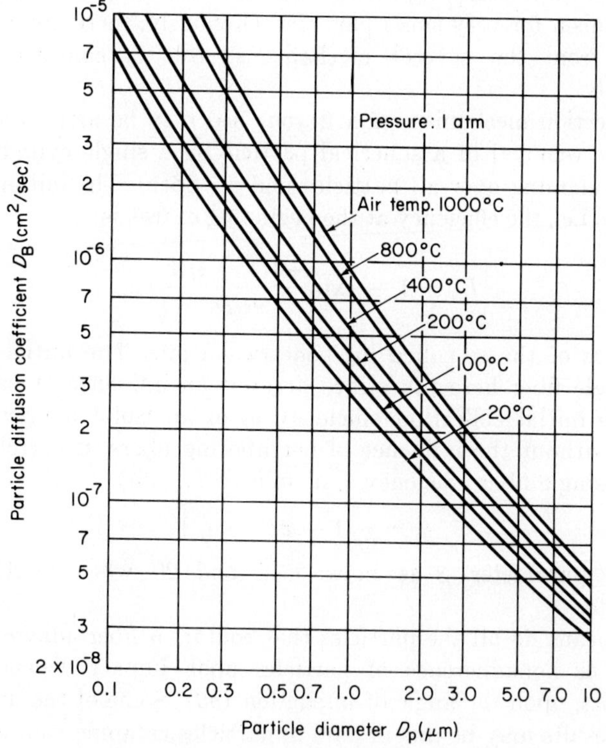


Figure 13. Diffusion coefficient of a particle.

various temperatures. Friedlander (*59*) gives the following relationship for diffusional collection:

$$\eta_{0D} = 6\ \mathrm{Re}^{1/6} N_D^{2/3} \tag{8}$$

b. Diffusion and Interception. A general formula for the initial collection efficiency of a single fiber considering both diffusional motion and interception effects may be written (*60*)

$$\eta_{\epsilon DI} = 8(A\, N_R N_D)^{1/3} + \tfrac{1}{3} A\, N_R^2 \tag{9}$$

where

$$A = \frac{\frac{1}{2}\epsilon_f}{-\ln(1-\epsilon_f) + 2(1-\epsilon_f) - \frac{1}{2}(1-\epsilon_f)^2 - \frac{3}{2}}$$

$$N_R = D_p/D_f \qquad \text{(interception parameter)} \tag{10}$$

c. Gravity and Interception. The initial collection efficiency of an isolated fiber due to gravity and interception effects is analytically obtained by solving the equations of particle motion given in Table III (*61, 62*) where the gravitational parameter $G = C_m D_p^2 g/18\mu$, $C_m$ being the

**Table III Collection Efficiency of an Isolated Cylinder Considering Gravity and Interception Parameters** (*61*)

| *Flow condition* | *Flow direction* | *Efficiency by gravity and interception* | *Gravitational efficiency* |
|---|---|---|---|
| Laminar | Horizontal | $\dfrac{1+R}{(1+G^2)\sqrt{G^2 + R^4/(2-\ln \mathrm{Re})^2}} \times \left(\dfrac{R^2[1/(1-R)^2 - 1 + \ln(1+R)^2]}{2(2-\ln \mathrm{Re})^2} + G^2\right)$ | $\dfrac{G}{\sqrt{1+G^2}}$ |
| Laminar | Vertical downward | $\dfrac{1+R}{1+G}\left[\dfrac{1}{2(2-\ln \mathrm{Re})} \times \left(\dfrac{1}{(1+R)^2} - 1 + \ln(1+R)^2\right) + G\right]$ | $\dfrac{G}{1+G}$ |
| Potential | Horizontal | $\dfrac{1}{(1+R)\sqrt{1+G^2}} \times \sqrt{(1+R)^4 G^2 + [(1+R)^2 - 1]^2}$ | $\dfrac{G}{\sqrt{1+G^2}}$ |
| Potential | Vertical downward | $\dfrac{1}{1+G}\left(G(1+R) - \dfrac{1}{1+R} + (1+R)\right)$ | $\dfrac{G}{1+G}$ |

Cunningham coefficient. Aerosol penetration for upflow is always equal to or greater than that for downflow (*63*).

d. INERTIA, GRAVITY, AND INTERCEPTION. The equations of particle motion for this case have to be numerically solved by a computer. The solution gives the theoretical initial collection efficiency of an isolated fiber (*64, 65*). Then, the total initial collection efficiency can be obtained using the above value by use of Equations (4) and (5). Figure 14 shows comparisons between calculated values and experimental data for the initial

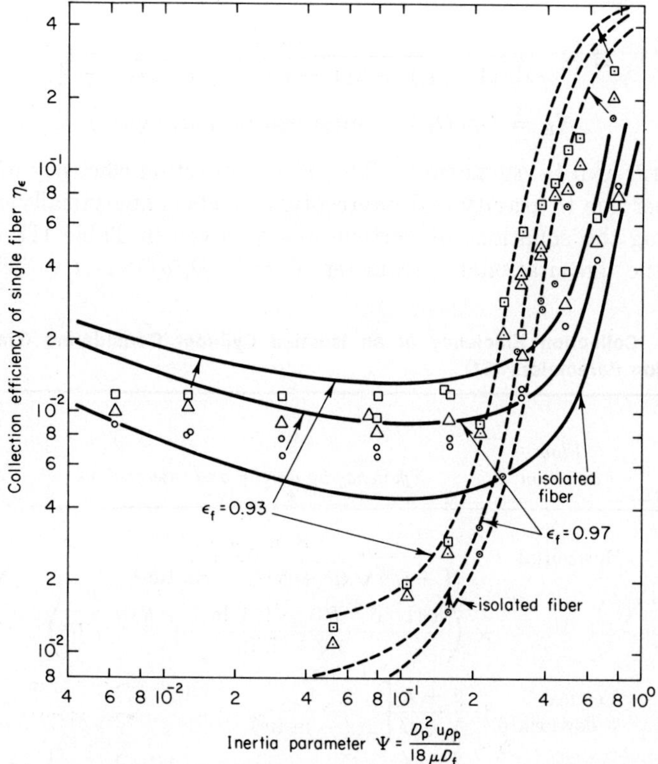


Figure 14. Theoretical initial collection efficiency of an isolated fiber compared with experimental results (*62a*). Values for $\eta_\epsilon$ and $\Psi$ are dimensionless; $D_p = 1.0$ $\mu$m; solid line: $D_f = 10.46$ $\mu$m; dashed line: $D_f = 31.9$ $\mu$m.

| $D_f = 10.46$ $\mu$m | | $D_f = 31.9$ $\mu$m | |
|---|---|---|---|
| *Symbol* | $1 - \epsilon_f$ | *Symbol* | $1 - \epsilon_f$ |
| ○ | 0.03 | ⊙ | 0.0252 |
| □ | 0.05 | ⊡ | 0.05 |
| △ | 0.07 | △ (dotted) | 0.07 |

collection efficiency over a range of operating conditions where particle inertia is significant. It is still difficult to estimate the collection efficiency for dust loading conditions (*66, 67*), these being of more practical importance than the initial efficiency. A probabilistic theory of aerosol penetration has been developed; it seems to give results in good agreement (*68*).

### 3. Holding Capacity (Filter Media Life)

Dust holding capacity is defined as the maximum dust load per unit area of the filter medium for a certain limit of pressure drop. Higher holding capacity is desirable in practice because of the longer life of the filter medium. Its value ranges usually between 0.3 to 2 kg of dust per square meter of filter surface.

### 4. Electrostatic Charging

Electrostatic attraction will draw particles from the gas stream to the fibers if the two are oppositely charged (*69*). Even if only one of the materials, i.e., the particles or the fibers, is charged, an induced charge will be created on the other, producing a polarization force that results in attraction and particle deposition (*70–72*). Electrostatic charge also influences particle agglomeration which increases the likelihood of particle entrapment. Charges may be developed through frictional effects during the filtering operation (*73*) even if not initially present.

Applying an electrostatic field to a fibrous filter of a dielectric material by means of embedded electrodes also enhances the collection efficiency (*74–80*). The dielectric effect can satisfactorily compensate for poor collection efficiency of a fibrous filter when the interception mechanism predominates as shown in Figure 15. In comparison with conventional fibrous filters, higher efficiencies can be achieved. The efficiency is affected by the dielectric constant of the fibers and depends upon the volumetric mixture ratio in case of two different packing media (*81*).

Another type of electrostatic fibrous filter, one containing so-called electroconductive fibers (e.g., graphite) to which a high dc voltage is applied, has been studied (*82, 83*). Figure 16 shows experimental results. In the case of a single stage of conductive fibers, the collection efficiency increases drastically with the absolute value of applied voltage, but the polarity makes little difference in performance. On the other hand, the efficiency of a filter having two separated conductive layers does not depend upon the absolute voltage, but upon the electric field strength between the two layers. It is also shown that the less the volume fraction of fibers, the filter thickness, the superficial filtering velocity, and the distance between the two layers, the better the collection.

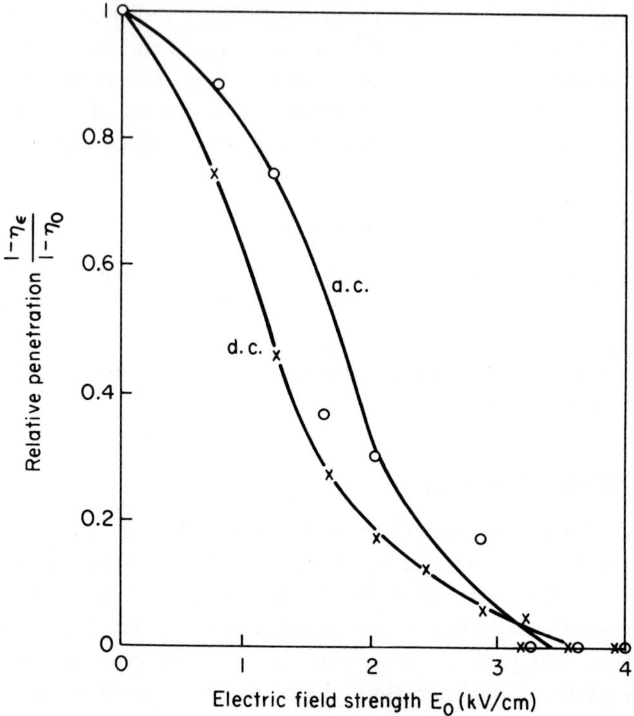


Figure 15. Effects of electric field on collection performance of dielectric fibrous filter (*60*). $D_p$: 1 $\mu$m (dioctyl phthalate); *L*: 2.5 cm/second; $D_f$: 12 $\mu$m (glass); $u_s$: 17.6 cm/second; $\epsilon_f$: 0.98.

### *B. Equipment Types*

The fibrous filter is one of the more commonly used air filters, being particularly suitable for low dust concentrations. Maintenance is easy because of its simple construction. There are two types of this filter. One is a throwaway panel type, in which the filter medium is packed in a frame and replaced together with the frame after a certain period of time. The other is one in which the medium only is renewed by a renewal operation.

#### 1. Throwaway Panel Filters

Throwaway filters are usually constructed with fibers of metal, glass, or plastic packed loosely into a frame, generally about 50 cm (20 in.) square and 2 to 5 cm (1 to 2 in.) thick. Superficial gas velocities employed with these mats range from 0.3 to 2 m/sec (60 to 600 ft/minute) and their pressure losses are from 5 to 50 mm $H_2O$ (0.2 to 2 in. $H_2O$). Figure

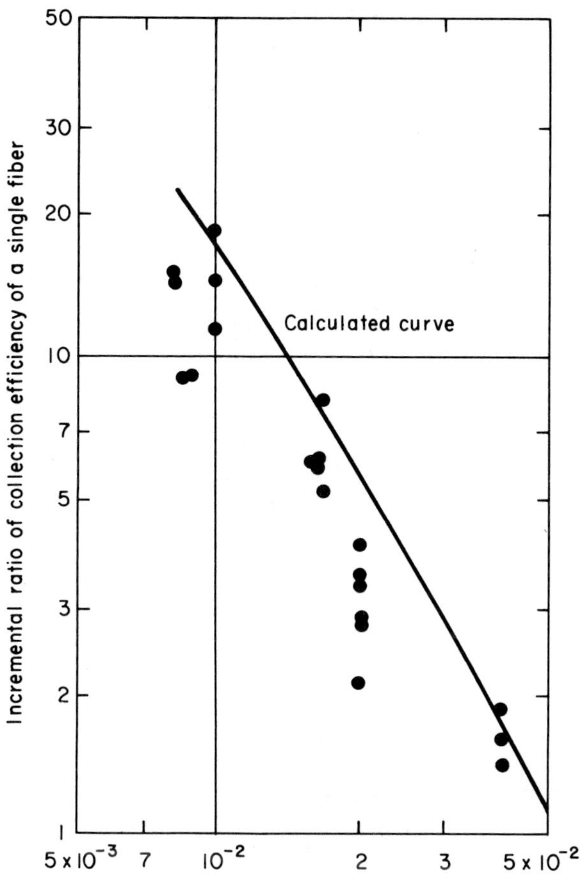


Figure 16. Collection performance of electroconductive fibrous filter (single stage of conductive fibers) (*60*). $u$: 2.7 cm/second; $D_f$: 18 $\mu$m; $D_p$: 0.9 $\mu$m; $L$: 2.24 cm; $V$: 10 kV.

17A shows an example of a throwaway fibrous filter. Throwaway panel filters are often installed in a zig-zag construction because of larger filtering surface area.

Filters of metallized, ceramic, or mineral fibers are available for high temperature applications. Their cost limits their use.

## 2. Replaceable-Medium Filters

Replaceable-medium filters employ a roll, or pack, of medium, and the medium is automatically renewed periodically. In the partial-renewal

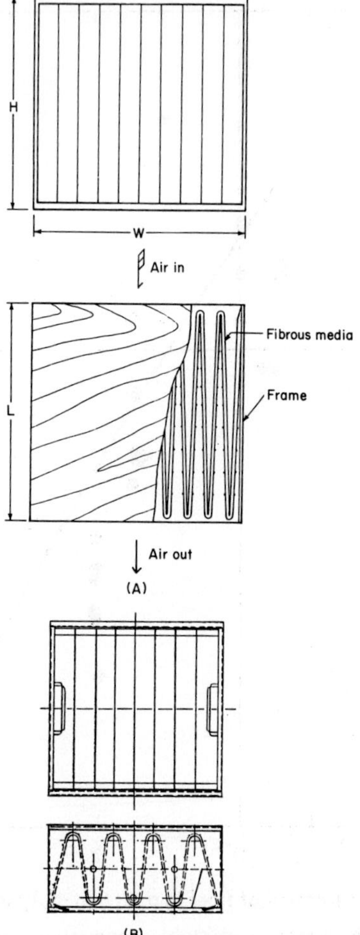


Figure 17. Renewable fibrous filters.

type the used portion of the medium in the filtering zone is intermittently replaced by the fresh one, and in the continuous-renewal type the medium is continuously displaced at a specified velocity (Fig. 17B). Furthermore, two methods are used to control the renewal operation, one utilizing a pressure switch and the other a timer. Figure 18 shows a typical partial-renewal system employing a dielectric fibrous mat with corona discharge. Both the partial and continuous renewal types are remarkably superior to the throwaway type from the economic standpoint of medium consumption and reduced fluctuation of pressure drop and flow rate. The

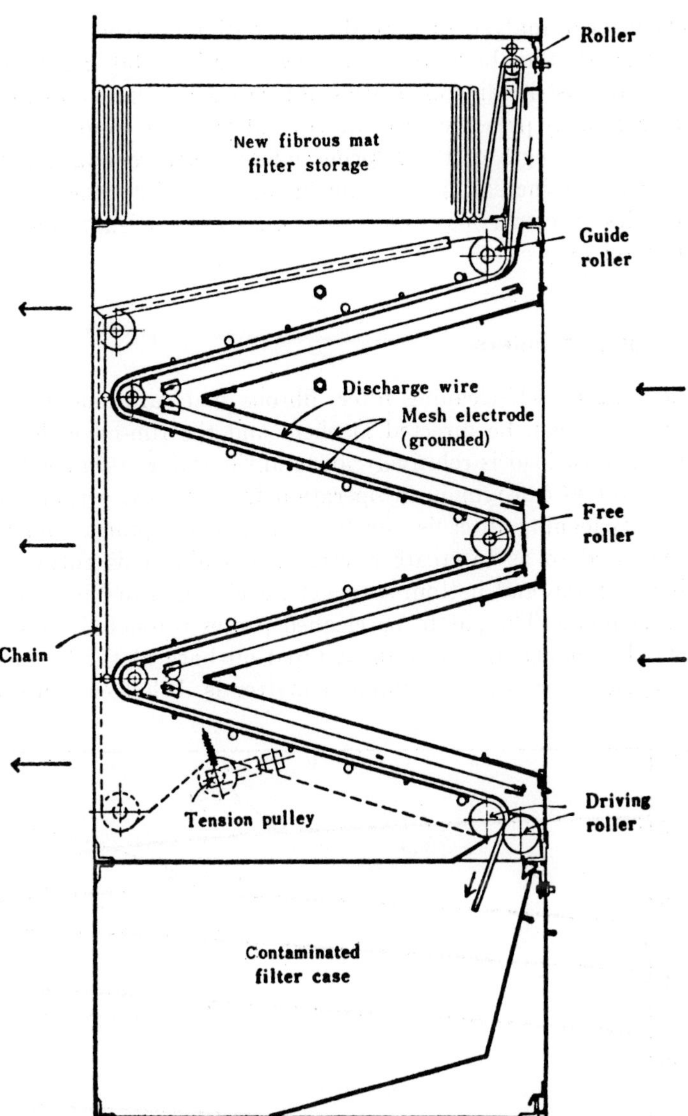


Figure 18. Partially renewable dielectric fibrous filter with corona discharge.

merits of partial renewal are little affected by changes in the properties of the medium and dust or by the maximum pressure drop.

The interesting experimental result has been reported that a partial-renewal filter controlled by a pressure switch can hold about twice the dust load of the replaceable-medium or the throwaway type of filter

(*84*). Theoretical analysis of partial- and continuous-renewal-type filters (*85, 86*) suggests the above mentioned result to be valid. Figure 19 gives an example of dust-holding capacities in terms of the ratio $\alpha$ of the initial and the maximum pressure drop at several renewal ratios $k$, which is the ratio of the length of the filtering zone to each renewal length. Another method for increasing filter media life is to introduce a prefilter, which has less pressure loss and lower collection efficiency than that of the main filter (*87*).

## 3. Self-Cleaning Filters

In one type of self-cleaning filter, fibrous mats are mounted so that sprays of water can be directed at them and the run-off collected in a basin. If the dust load is relatively low and the water is changed periodically, such filters can remain in operation for extended periods. Another type of self-cleaning filter is constructed of overlapping, hinged filter panels mounted so as to create a vertical curtain. This curtain travels on chains over top and bottom sprockets, each panel making one circuit in about 24 hours. The gas to be cleaned passes through a double layer of panels. The panels dip into an agitated oil bath at the bottom point of their cycle where the accumulated matter is dislodged. Cleaned and

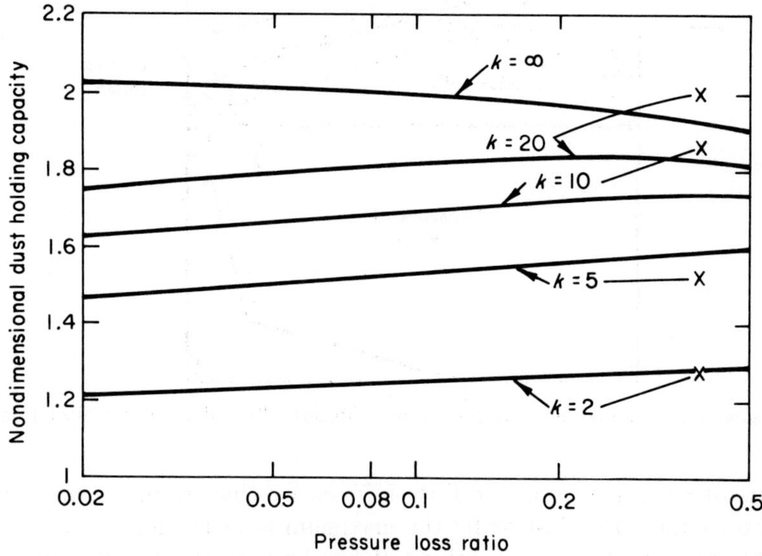


Figure 19. Dust-holding capacity of partially renewable fibrous filter for constant maximum pressure loss (*85*). Solid line: calculated data; ×: experimental data.

freshly oiled, the panels, upon rising, return to their overlapping position to form again the filtering curtain.

## IV. Miscellaneous Filters

### A. Paper

Paper filters come in a variety of grades. The ordinary paper filter is said to have conventional characteristics while those with superior filtering ability are called high efficiency paper (HEPA) filters. Figure 20 presents a correlation between pressure drop and dust load for various kinds of paper filters (*88*).

#### 1. Conventional

Conventional filter paper, such as that for laboratory analysis, has less then 90% collection efficiency for 0.3 $\mu$m particles (*89*). It is widely employed in engine air cleaners, households vacuum cleaners, and the like. Pressure drop is about 10 mm $H_2O$ at 1 cm/second of superficial filtering velocity at no dust load, and is proportional to the velocity either with or without a dust deposit. In these filters, the fiber diameter is about 20 $\mu$m and the volume fraction of fibers is about 0.25 to 0.30. The thickness of these filters is usually from 0.2 to 0.3 mm.

The collection efficiency of a single fiber in a paper filter can be calculated by the equation (*88*)

$$\eta_\epsilon = \frac{10}{2 - \ln \mathrm{Re}} [1 + 10^6(1 - \epsilon_f)^6] N_R{}^2 \tag{11}$$

where $\mathrm{Re} = D_f \rho_g u/\mu$ and $N_R = D_p/D_f$, from which the overall efficiency can then be estimated using Equations (4) and (5).

These filters are constructed upon frames of several sizes and are installed in airtight banks. Each filter element, or cell, ordinarily has 10 to 15 times its apparent filtering area because the paper medium is arranged in accordionlike pleats, folds, or pockets. Paper filters cannot be cleaned and reused. When it becomes necessary to change them, the paper is either replaced on the frame, or both the frame and the paper are discarded and new units are installed. The latter course is necessary, of course, when radioactive or pathogenic contaminants are collected. Gas velocities with paper filters are generally of the order of that for cloth filters (*90*).

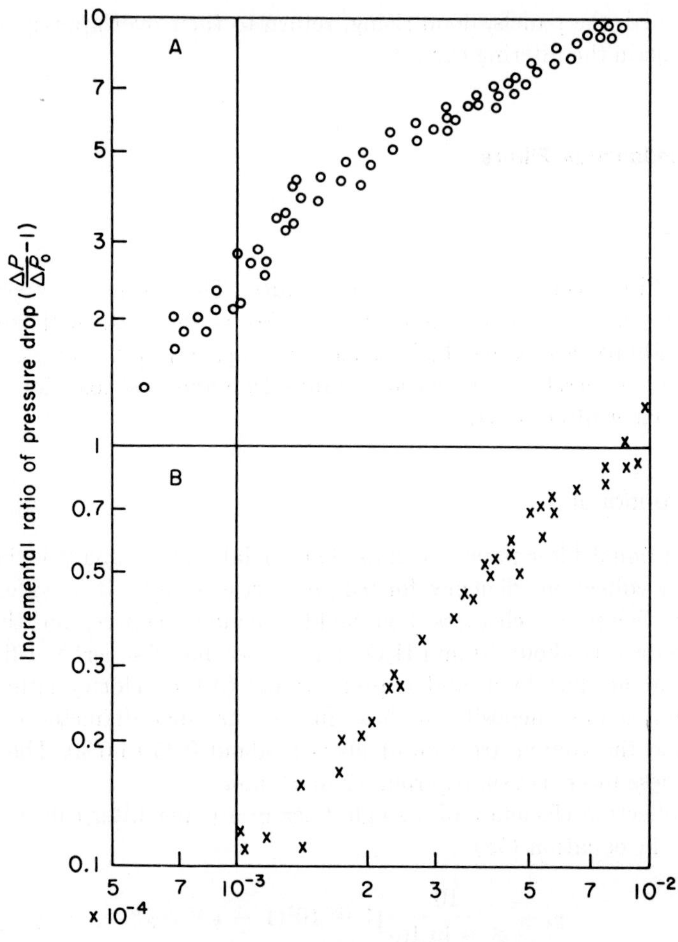


Figure 20. Pressure loss characteristics for most filter papers with dust load (*88*). A: Conventional paper filter; $D_f$ = 17 $\mu$m, $\epsilon$ = 0.75–0.70, $L$ = 220–280 $\mu$m, $D_p$ = 0.25–1 $\mu$m, $u_s$ = 3–10 cm/second. B: High performance paper filters; $D_f$ = 1 $\mu$m, $L$ = 310–360 $\mu$m, $\epsilon$ = 0.9.

Figure 21 shows a correlation between pressure drop and filtering velocity for various kinds of clean paper filters. Pressure loss is allowed to rise to about 20 to 100 mm $H_2O$ before the filter is replaced. The useful service life ranges from $10^3$ to $10^4$ hours when filtering ordinary city air, if protected by a less expensive prefilter.

The collection efficiency rating of paper filters (*91–94*) depends on the method of evaluation. A radioactive aerosol affords a convenient means

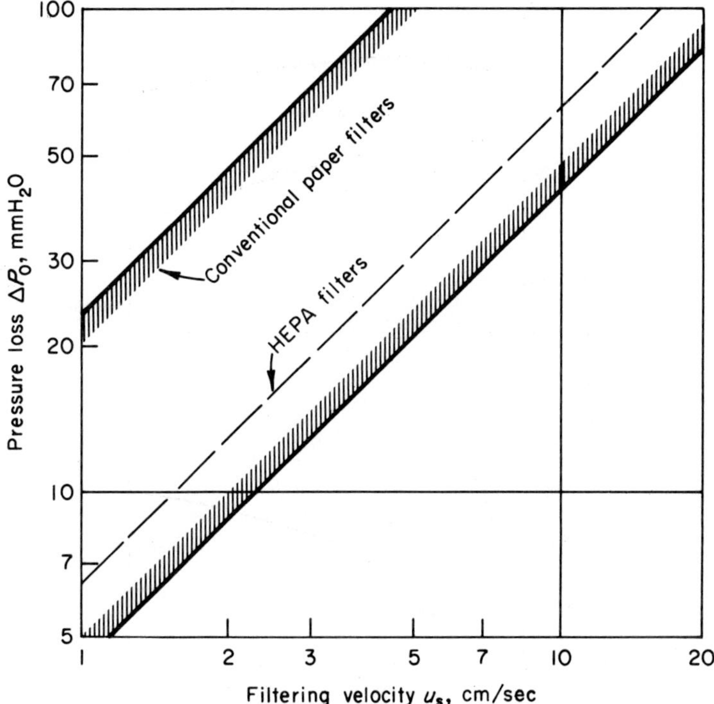


Figure 21. Pressure loss of filter papers.

for measuring the efficiency of dry filter media (*95*). Figure 22 shows a correlation between collection efficiency for a 0.25-$\mu$m-diameter aerosol and filtering velocity.

## 2. High Performance

High efficiency paper (HEPA) filters, in addition to being used for sampling or testing purposes, find practical applications where radioactive dusts or bacterial particulates need to be controlled (*96, 97*). Clean rooms, as used in pharmaceutical plants, in electronic instrument assembly areas, in operating rooms, etc., also utilize these high efficiency paper filters to obtain air essentially free of particulates. Their service life is often prolonged in clean-room operations by use of high efficiency prefilters and recirculation of the treated air.

Incremental ratios of pressure loss for high performance paper filters are given in Figure 20 as a function of dust load. They are approximately one-tenth those of conventional paper filters. The best commercial

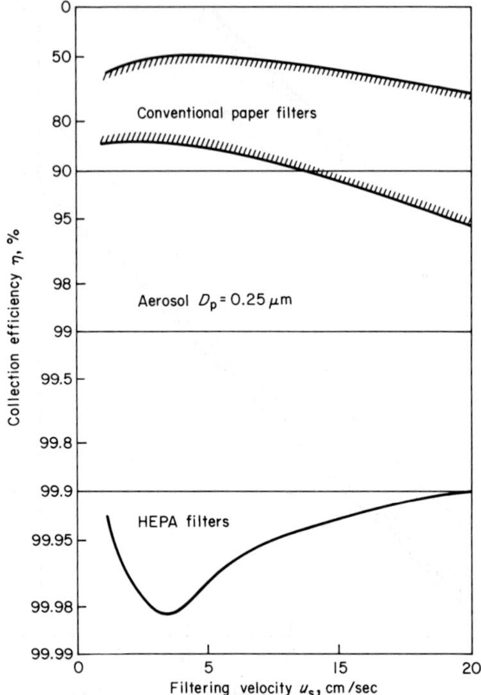


Figure 22. Collection efficiency of paper filters (*88*).

HEPA filters show more than 99.97 wt % collection efficiency for particles having less than 0.3 μm diameter (*98*).

Diffusion accounts for much of the collection in HEPA filters as is evident from their performance under reduced pressure (*99*). Electric charging also contributes, for charged aerosols with 0.25 to 0.55 μm-diameter particles are more efficiently collected (*10*) than are uncharged aerosols of otherwise similar characteristics.

## *B. Packed Bed*

Aggregate packed bed filters can operate either dry or wet.

### 1. Dry Aggregate

Applications of this type are restricted to special conditions, such as high temperature or corrosive gas treatment. They are operated both statically and dynamically. Static or fixed beds have to be cleaned by replacing, vibrating (*101*), agitating with reverse flow, or washing the filter media. Figure 23 shows a gravel bed with agitating rakes and re-

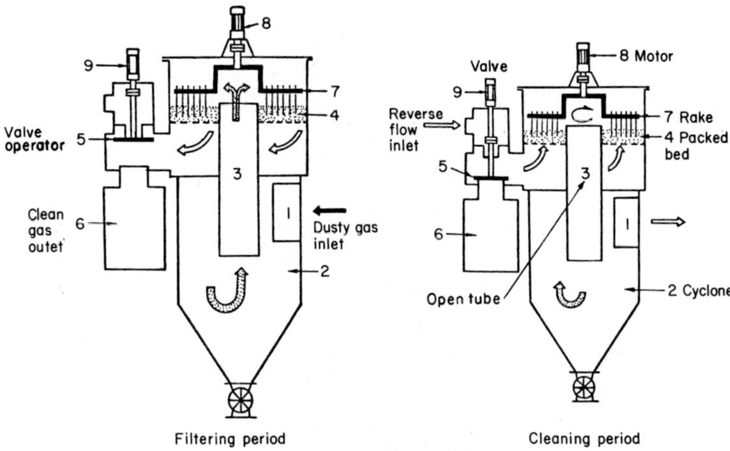


Figure 23. Gravel bed filter (*102*). GFE-Drallschlicht-filter.

verse flow dampers (*102*). Examples of applications are packed towers for acid mist recovery (*103, 104*) and sand beds for fly ash removal (*105, 106*). Dynamic beds are fluidized or continuously moved downward for cleaning. Fluidized bed filters are sometimes employed for the removal of mists and dusts from gas streams (*106–109*).

## 2. Wet Aggregate

A wetted packing provides an impingement surface that prevents re-entrainment. The liquid also provides a means for washing off dust and conveying it away as a slurry or solution. The packing may be fixed or it may be a fluidized bed of low density spheres. The advantages are low cost, simplicity, corrosion resistance, and no moving parts. Dust collecting in some cases is secondary to direct-contact cooling and gas absorption. The usual fixed packed tower can be cleaned by washing the face of the packing with spray nozzles in parallel flow, while the body of the packing is irrigated from the top as shown in Figure 24. In a fluidized bed of plastic spheres, the movement helps to free the solids.

### *C. Porous Media*

There are two kinds of porous filters, a soft type, like foam rubber, and a rigid type, like ceramic or sintered metal. The former is employed as an air filter just as are fibrous filters, but it has better dust holding capacity (*60, 110, 111*). Open-pore foam filters show little pressure increase during dust loading because of their pore shape. Their collection efficiency is calculated in a manner similar to that of a glass fibrous mat

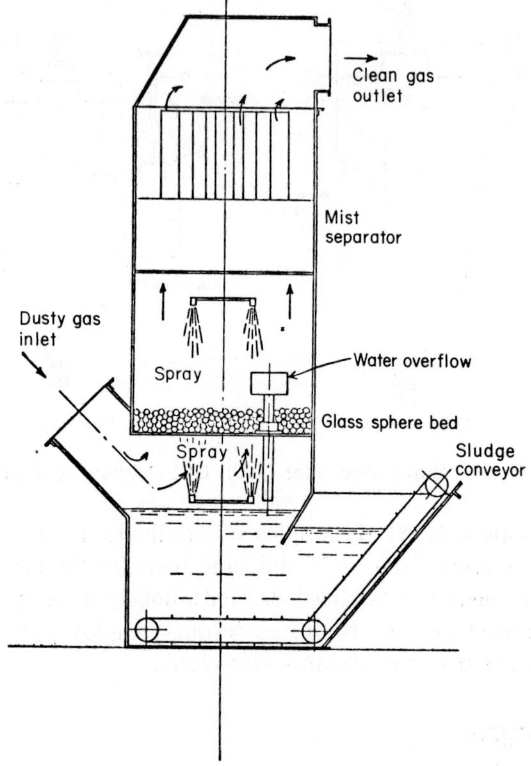


Figure 24. Fixed bed wet aggregate filter.

with an apparent fiber diameter derived from initial pressure loss data. Figure 25 shows experimental results of pressure drop and collection efficiency.

Another kind of soft media filter is the membrane filter, which is a thin plasticlike sheet having many tiny holes; they give high collection efficiency. Their filtration properties have been studied under various conditions, for example, electrostatic charge (*112*), extreme conditions (*113*), penetration depth (*114*), and other factors (*115–118*).

Rigid filters are often employed for process gas sampling because they resist high temperature, moisture, and chemical attack.

## D. Mist Collection*

There are two kinds of mist separators. The one is a fibrous or wire mesh mat for mists less than 10 μm in diameter; commercially it is called

* See also Chapter 7, this volume.

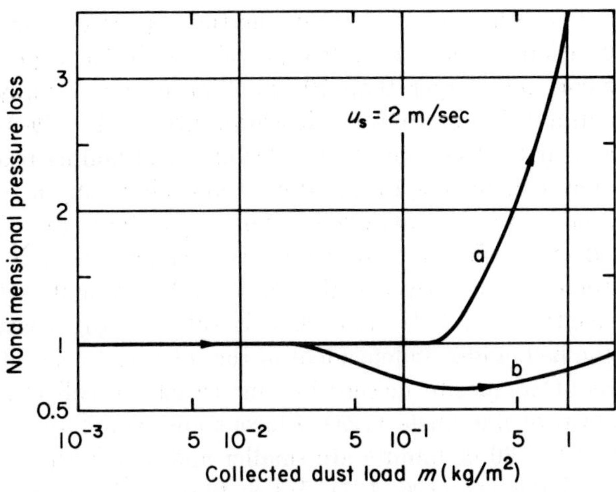


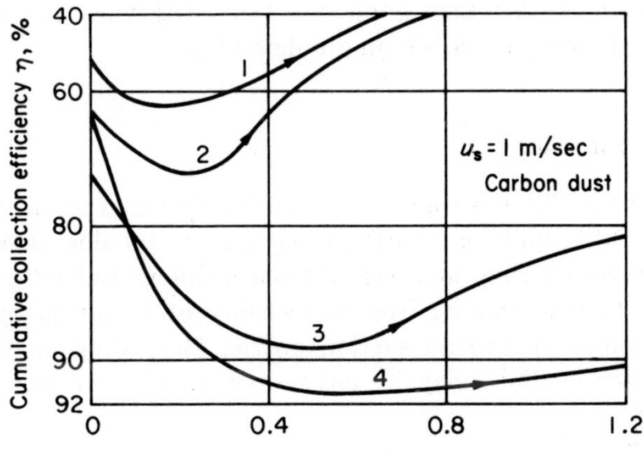


Figure 25. Pressure drop and collection efficiency of foam-rubber-type porous filters (*110*).

| Label | Thickness, mm | Void | Equivalent diameter of cell wall, $\mu$m | Number of cells per inch |
|---|---|---|---|---|
| a | 19 | 0.92 | 41 | 40 |
| b | 20 | 0.97 | 184 | 9 |
| 1 | 10 | 0.97 | 91 | 18 |
| 2 | 10 | 0.96 | 73 | 22 |
| 3 | 10 | 0.96 | 69 | 28 |
| 4 | 10 | 0.97 | 40 | 31 |

a demister. The other is more of an inertial separator, for example, a packed bed, a louver, or a zig-zag-type collector (*103, 104*); it is applicable to coarse mist larger than 10 μm. Fine mist is usually generated in acid manufacturing or from condensation processes (*119–122*). Coarse mist generally arises from spraying, splashing, and boiling processes such as encountered in scrubbers, evaporators, and cooling towers.

Fluidized beds are also employed as mist separators (*108*). For a given bed, the collection efficiency for acid mists, composed of 2- to 14-μm-diameter droplets, is improved with increasing bed weight and superficial filtering velocity (*123, 124*). Efficiency is substantially constant during the life of the bed, and is independent of the inlet mist concentration. The development of composite meshes has increased the collection efficiency for the removal of fine mists (*125*). Fibrous mist eliminators for filtering velocities of 10 to 30 m/minute are smaller and less expensive than inertial types of demisters (*126, 127*). When fibrous filters are loaded with mist particles, there is typically an increase in the amount of liquid passing through (*128*). A rotary impaction filter with wire filaments is also effective in removing 10- to 40-μm droplets (*129*).

### E. Rotary Drum Filter

A rotary drum filter is sometimes used for filtration of dust from room air, e.g., in textile mills or return air systems. As the dust deposit builds, the drum starts rotating at a signal from a differential pressure switch. The dust is then evacuated from the surface of the drum by means of a simultaneously traversing suction nozzle. Depending on the type of dust, the filter medium may be either a washable fine mesh screen or a filter mat.

## V. General Considerations

### A. Cost

When evaluating the use of a filter versus other types of dust removal systems, cost factors to be considered include the initial purchase price and installation, operation, maintenance, and dust disposal costs.

All collector prices per cubic meter per minute of gas handled will decrease as gas flow rates increase. There is a break point, however, at about 1000 to 2000 $m^3$/minute, depending on design, where cost per cubic meter per minute will level off (*130*). If the filter can be installed for pressure operation, i.e., with the blower located ahead of the filter, the initial cost may be 20 to 30% less than if a suction-type installation is required, because the outside walls of the bag house need not be air tight and can be open to the atmosphere. Installation costs may equal or even exceed initial equipment purchase cost. This depends on the method of shipment, equipment location, electrical supply, dust disposal facilities, and labor charges. Cost of components such as blowers, motors, drives, conveyors, etc., also must be included and may amount to a considerable part of the total. Annual maintenance costs for a filter system are usually higher than those of other particulate collectors. They are less than 10% of the initial costs in fabric filters, not including power costs, but may reach three times the initial costs for throwaway filters.

The purchase cost of a pulse-jet filter is approximately $30/m^3$/minute of capacity, which is comparable to that of an electrostatic precipitator. A fabric filter provided with intermittent mechanical cleaning costs about $20/m^3$/minute, which is approximately three times that of cyclone collectors. Figure 26 shows a typical relationship between purchase cost and gas flow rate (*131*).

The purchase costs of fibrous mat filters are generally about one-tenth those of cloth filters, but their maintenance costs, including replacement charges, can easily reverse the cost situation, especially when the dust concentration is high.

From the standpoint of installation and operating costs, the optimum filtering velocity for a fabric filter will usually be between 1 and 3 m/minute. This is so because lower velocities require larger filtering areas and greater initial costs, while higher velocities are accompanied by greater pressure losses which give rise to higher operating costs. The actual filtering velocity is often set lower than the above figure in order to obtain longer fabric life and greater collection efficiency (*132*, *133*).

### B. Humidity

A bag filter has to be operated above the dew point of the gas because of the choking of the filter medium which condensed moisture would cause. Dryers and kilns sometimes discharge gas with a high water-vapor content. In these cases, the bag house should be thermally insulated from the atmosphere. Troubles are most likely to occur during the winter, especially if filter units are located out of doors. The possible effects of condensation should always be considered in start-up and shutdown of a filter operation. Hot air, from the heat exchanger of a gas precooler, is sometimes introduced into the exhaust in order to diminish the white plume.

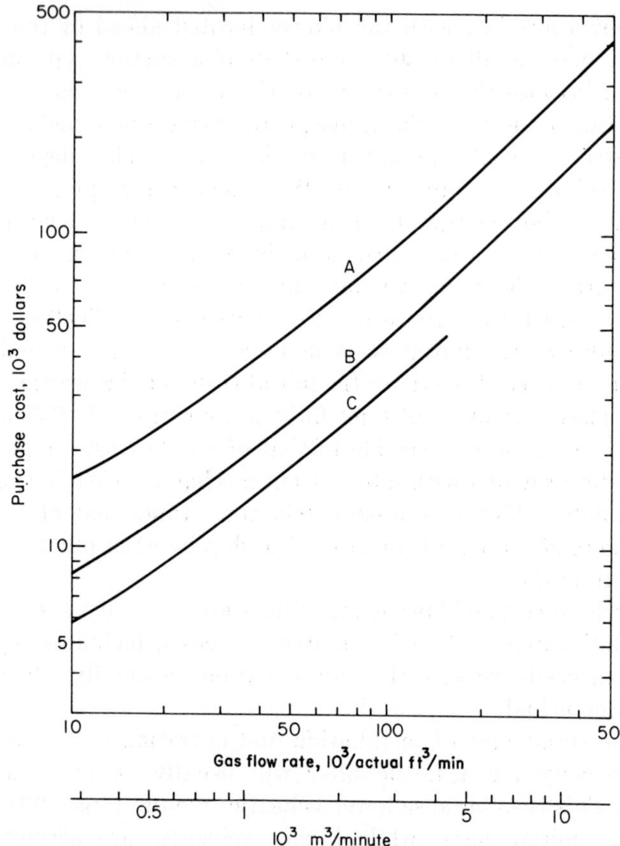


Figure 26. Purchase cost of fabric filters (*131*). A: High temperature synthetic, woven and felt; continuously automatically cleaned. B: Medium temperature synthetic, woven and felt; continuously automatically cleaned. C: Woven natural fibers; intermittently cleaned, single compartments. Costs for equipment of indicated construction may vary by 20% of reported figure.

### C. Fire Hazard

Filtration of oxidizable substances can constitute a serious fire hazard. Sparks or accidental sources of ignition such as hot cinders can be dangerous. Fires in filtration units may quickly attain high temperatures unless the airflow is cut off and sprinkler systems or other fire control mechanisms are available to control the conflagration. Because of this, all large filter installations having this characteristic should be equipped with automatic devices for closing off the flow of air in case of fire, and should be provided with adequate sprinklers or chemical fire control apparatus. The use of spark arresters in ducts leading to filter units is recommended.

### *D. Maintenance and Service*

As with other collectors, bag filters should be set up on a routine maintenance cycle. At least once a month, bag tension should be checked and bags should be inspected visually for failures. Faulty bags should be replaced immediately because adjacent bags may be damaged by the bag that has failed. A leak is easy to find since it will often discolor adjacent bags or structures and cause buildup of dust at the bottom of the casing.

The drive assembly should be checked once a month for V-belt tension and wear. At least once every 6 months, ducts leading to and from the filter should be inspected for dust buildup. Fans should be checked at the same time to be certain that no buildup or unusual wear has occurred on the blades or housing. With good preventive maintenance, it is not unusual for a bag to last 10 years or more.

## REFERENCES

1. U.S. Environmental Protection Agency, *J. Air. Pollut. Contr. Ass.* **24,** No. 12 (1974).

1a. C. E. Billings *et al.,* "Handbook of Fabric Filter Technology," Vols. 1–4. Nat. Tech Inform. Center, Springfield, Virginia, 1970.

1b. Air Pollution Control Association, "The User and Fabric Filtration Equipment, Buffalo, New York," Specialty Conf. Proc., Air Pollut. Contr. Ass., Pittsburgh, Pennsylvania, 1974.

1c. Air Pollution Control Association, "Design, Operation, and Maintenance of High Efficiency Particulate Control Equipment, St. Louis, Missouri," Specialty Conf. Proc., Air Pollut. Contr. Ass., Pittsburgh, Pennsylvania, 1973.

1d. W. E. Ballard, *Rock Prod.* **65** (10), 61 (1962).

2. J. E. L'Anson, T. L. Harsell, Jr., and R. T. Pring, *Amer. Foundryman* **23** (1), 61 (1953).

3. R. L. Adams, *J. Air Pollut. Contr. Ass.* **14,** 299 (1964).

4. R. A. Herrick, J. W. Olsen, and F. A. Ray, *J. Air Pollut. Contr. Ass.* **16,** 7 (1966).

5. F. A. Bagwell, L. F. Cox, and E. A. Pirsh, *J. Air Pollut. Contr. Ass.* **19,** 149 (1969).

6. J. L. Venturini, *J. Air Pollut. Contr. Ass.* **20,** 808 (1970).

7. R. C. French, *Chem. Eng.* **70,** 177 (1963).

8. P. A. F. White and S. E. Smith, "High Efficiency Air Filtration," p. 100. Butterworth, London, England, 1964

9. G. W. Edwald, *Air Eng.* **7,** 22 (1965).

10. L. L. Dollinger, *Hydrocarbon Process.* **48,** 88 (1969).

11. J. Dyment, *Filtr. Separ.* **7/8,** 441 (1970).

12. J. J. Gussman and R. C. Horton, *J. Air Pollut. Contr. Ass.* **13,** 266 (1963).

13. P. W. Spaite, J. E. Hagan, and W. F. Todd, *Chem. Eng. Progr.* **59** (4), 54 (1963).

14. R. L. Adams, *Chem. Eng. Progr.* **62** (4), 66 (1966).
15. P. W. Spaite and R. E. Harrington, *J. Air Pollut. Contr. Ass.* **17**, 310 (1967).
16. C. A. Snyder and R. T. Pring, *Ind. Eng. Chem.* **47**, 960 (1955).
17. W. Strauss, "Industrial Gas Cleaning," p. 257. Pergamon, Oxford, England, 1966.
18. L. Silverman, E. W. Conners, and D. M. Anderson, *Ind. Eng. Chem.* **47**, 952 (1955).
19. K. T. Whitby, D. A. Lundgren, A. R. McFarland, and R. C. Jordan, *J. Air Pollut. Contr. Ass.* **11**, 503 (1961).
20. E. R. Frederick, *Chem. Eng.* **68** (13), 107 (1961).
21. E. Butterworth, *Mfg. Chem.* **35** (1), 66; (2) 65 (1964).
22. D. A. Lundgren and K. T. Whitby, *Ind. Eng. Chem., Process Des. Develop.* **4**, 345 (1965).
23. S. Rüdiger, *Staub* **31**, 439 (1971).
24. C. N. Davies, *Proc. Inst. Mech. Eng.* **1B**, 185 (1952).
25. G. E. Cunningham, G. Broughton, and R. R. Kraybill, *Ind. Eng. Chem.* **46**, 1196 (1954).
26. K. Iinoya and N. Yamamura, *Kagaku Kogaku* **20**, 163 (1956).
27. N. Kimura and K. Iinoya, *Kagaku Kogaku* **29**, 166 (1965); *Kagaku Kogaku (Abr. Ed. Engl.)* **3**, 193 (1965).
28. R. H. Borgwardt, R. E. Harrington, and P. W. Spaite, *J. Air Pollut. Contr. Ass.* **18**, 387 (1968).
29. G. W. Walsh and P. W. Spaite, presented at *Annu. Meet. ASME, New York, December, 1960.*
30. J. W. Robinson, R. E. Harrington, and P. W. Spaite, *Atmos. Environ.* **1**, 499 (1967).
31. W. Solvach, *Staub* **29**, 24 (1969).
32. N. Tanaka, K. Makino, and K. Iinoya, *Kagaku Kogaku* **37**, 718 (1973).
33 F. Loeffler, *Staub* **30**, 518 (1970); *Staub-Reinhalt. Luft* **30** (12), 27 (1970).
34. R. Dennis, G. A. Johnson, and L. Silverman, *Chem. Eng.* **59**, 196 (1952).
35. N. Kimura and M. Shirato, *Kagaku Kogaku* **34**, 984 (1970).
36. N. Kimura and K. Iinoya, *J. Res. Ass. Powder Tech.* **7**, 124 (1970) (in Japanese).
37. K. Iinoya, "Manual of Dust Collection Techniques," p. 49. Nikkan Kogyo Shinbun, Japan (in Japanese).
38. A. Nakai and K. Iinoya, *J. Res. Ass. Powder Tech.* **9**, 399 (1972).
39. R. E. Kunkle, *J. Air Pollut. Contr. Ass.* **13**, 274 (1963).
40. H. Karsten, *Staub* **24**, 205 (1964).
41. F. R. Wiesbaden, *Keram. Z.* **23**, 324 (1971).
42. K. J. Caplan, *Amer. Ind. Hyg. Ass., J.* **28**, 567 (1967).
43. N. Kimura and K. Iinoya, *Kagaku Kogaku* **33**, 1008 (1969).
44. C. Y. Chen, *Chem. Rev.* **55**, 595 (1955).
45. J. B. Wong, W. E. Ranz, and H. F. Johnstone, *J. Appl. Phys.* **27**, 161 (1956).
46. L. A. Clarenburg and H. W. Piekaar, *Chem. Eng. Sci.* **23**, 765 (1968).
47. L. A. Clarenburg and F. C. Schiereck, *Chem. Eng. Sci.* **23**, 773 (1968).
48. C. P. Kyan, D. T. Wasan, and R. C. Kintner, *Ind. Eng. Chem., Fundam.* **9**, 596 (1970).
49. R. W. Wetner and L. A. Clarenburg, *Ind. Eng. Chem., Process Des. Develop.* **4**, 288 (1965).
50. N. Kimura and K. Iinoya, *Kagaku Kogaku* **23**, 792 (1959).
51. L. A. Clarenburg and R. M. Werner, *Ind. Eng. Chem., Process Des. Develop.* **4**, 293 (1965).

52. J. Juda and S. Chrosciel, *Staub* **30,** 196 (1970).
53. N. Kimura and K. Iinoya, *Kagaku Kogaku* **33,** 1255 (1969).
54. J. Pich, *Staub* **25,** 186 (1965).
55. N. Kimura and K. Iinoya, *Kagaku Kogaku* **29,** 538 (1965); *Kagaku Kogaku* (*Abr. Ed. Engl.*) **4,** 27 (1965).
56. T. Gillespie, *J. Colloid Sci.* **10,** 266, 289, and 299 (1955).
57. D. C. Freshwater and J. I. T. Stenhouse, *AIChE J.* **18,** 786 (1972).
58. A. A. Kirsch and N. A. Fuchs. *Ann. Occup. Hyg.* **11,** 299 (1968).
59. S. K. Friedlander, *Ind. Eng. Chem.* **50,** 1161 (1958).
60. K. Makino and K. Iinoya, *Kagaku Kogaku* **33,** 1261 (1969).
61. N. Yoshioka, H. Emi, C. Kanaoka, and M. Yasunami, *Kagaku Kogaku* **36,** 313 (1972).
62a. H. Emis, K. Okuyama, and T. Aoki, presented at *Annu. Meet. Soc. Chem. Eng. (Japan), Tokyo, 1973.*
63. J. W. Thomas, D. Rimberg, and T. J. Miller, *J. Aerosol Sci.* **2,** 31 (1971).
64. N. Yoshioka, H. Emi, H. Matsumura, and M. Yasunami, *Kagaku Kogaku* **33,** 381 (1969).
65. J. A. Harrop and J. I. T. Stenhouse, *Chem. Eng. Sci.* **24,** 1475 (1969).
66. N. Kimura and K. Iinoya, *Kagaku Kogaku* **28,** 39 (1964); *Kagku Kogaku* (*Abr. Ed. Engl.*) **2,** 136 (1964).
67. N. Yoshioka, H. Emi, M. Yasunami, and H. Sato, *Kagaku Kogaku* **33,** 1013 (1969).
68. L. A. Clarenburg, *Aerosol Sci.* **3,** 461 (1972).
69. G. Zebel, *Staub* **29,** 62 (1969).
70. V. Havlicék, *Int. J. Air Water Pollut.* **4,** 225 (1961).
71. N. Yoshioka, H. Emi, M. Hattori, and Y. Tamori, *Kagaku Kogaku* **32,** 815 (1968).
72. D. Hochrainer, *Staub* **29,** 67 (1969).
73. A. T. Rossano and L. Silverman, *Heat Vent.* **51,** 102 (1954).
74. J. W. Thomas and E. J. Woodfin, *Trans. Amer. Inst. Elec. Eng.* **78,** II, 276 (1959).
75. R. D. Rivers, *J. Amer. Soc. Heat. Refrig. Air Cond. Eng.* **4,** 37 (1962).
76. R. Flossman and A. Schuetz, *Staub* **23,** 443 (1963).
77. W. Walkenhorst and G. Zebel, *Staub* **24,** 444 (1964).
78. G. Zebel, *Staub* **26,** 281 (1966).
79. K. Makino and K. Iinoya, *Kagaku Kogaku* **33,** 684 (1969).
80. A. A. Kirsch, *Aerosol Sci.* **3,** 25 (1972).
81. K. Makino and K. Iinoya, *Kagaku Kogaku* **32,** 99 (1968).
82. K. Iinoya, K. Makino, and N. Kimura, *Kagaku Kogaku* **29,** 574 (1966); *Kagaku Kozaku* (*Abr. Ed. Engl.*) **4,** 35 (1966).
83. K. Makino and K. Iinoya, *Kagaku Kogaku* **33,** 701 (1969).
84. T. Higuchi and A. Iba, *J. Res. Ass. Powder Tech.* **6,** 317 (1969) (in Japanese).
85. N. Tanaka, K. Makino, and K. Iinoya, *J. Chem. Eng. Jap.* **5,** 401 (1972).
86. N. Tanaka, K. Makino, and K. Iinoya, *J. Chem. Eng. Jap.* **6,** 102 (1973).
87. S. Przyborowski, *Staub* **25,** 291 (1965).
88. K. Iinoya, K. Makino, O. Inoue, and T. Imamura, *Kagaku Kogaku* **34,** 632 (1970).
89. D. Rimberg, *Amer. Ind. Hyg. Ass., J.* **30,** 394 (1969).
90. J. A. Wheat, *Can. J. Chem. Eng.* **67,** (April 1963).
91. R. F. Hounam, *Ann. Occup. Hyg.* **4,** 301 (1961).

92. R. H. Collingbourne and H. E. Painter, *Int. J. Air Water Pollut.* **8,** 159 (1964).
93. H. J. Ettinger, J. D. Defield, and D. A. Bavis, *Amer. Ind. Hyg. Ass., J.* **30,** 20 (1969).
94. R. G. Stafford and H. J. Ettinger, *Amer. Ind. Hyg. Ass., J.* **32,** 493 (1971).
95. J. K. Skrebowski and B. W. Sutton, *Brit. Chem. Eng.* **6,** 12 (1961).
96. E. Stafford and W. J. Smith, *Ind. Eng. Chem.* **43,** 1346 (1951).
97. P. A. F. White and S. E. Smith, *Research (London)* **13,** 228 (1960).
98. J. B. Harsted and M. E. Filler, *Amer. Ind. Hyg. Ass., J.* **30,** 280 (1969).
99. S. C. Stern, H. W. Zeller, and A. I. Schekman, *J. Colloid Sci.* **15,** 546 (1960).
100. G. G. Goyer, R. Gruen, and V. K. LaMer, *J. Phys. Chem.* **58,** 137 (1954).
101. H. L. Engelbrecht, *J. Air Pollut. Contr. Ass.* **15,** 43 (1965).
102. G. Funke, *Zem.-Kalk-Gips* **23,** 101 (1970).
103. S. Jackson and S. Calvert, *AIChE J.* **12,** 1075 (1966).
104. R. Germerdonk and H. Gunther, *Chem.-Ing.-Tech.* **41,** 649 (1969).
105. A. N. Squires and R. Pfeffer, *J. Air Pollut. Contr. Ass.* **20,** 534 (1970).
106. L. Paretsky, L. Theodore, R. Pfeffer, and A. M. Squires, *J. Air Pollut. Contr. Ass.* **21,** 204 (1971).
107. H. P. Meissner and H. S. Mickley, *Ind. Eng. Chem.* **41,** 1242 (1949).
108. D. S. Scott and D. A. Guthrie, *Can. J. Chem. Eng.* **200,** (1959).
109. J. P. Pilney and E. E. Erickson, *J. Air Pollut. Contr. Ass.* **18,** 684 (1968).
110. N. Kimura, F. Hayashi, and K. Iionya, *Kagaku Kogaku* **29,** 622 (1965); *Kagaku Kogaku (Abr. Ed. Engl.)* **4,** 65 (1965).
111. J. F. Roesler, *J. Air Pollut. Contr. Ass.* **16,** 30 (1966).
112. B. Binek and S. Przyborowski, *Staub* **25,** 533 (1965).
113. K. Spurny and J. Hrbek, *Staub* **29,** 70 (1969).
114. V. Lössner, *Staub* **24,** 217 (1964).
115. K. Spurny and J. Pich, *Staub* **24,** 250 (1964).
116. K. Spurny and J. P. Lodge, *Staub* **28,** 179 (1968).
117. K. Spurny, J. P. Lodge, E. R. Frna, and D. C. Cheesley, *Environ. Sci. Technol.* **3,** 453 (1969).
118. K. Spurny, G. Pfefferkorn, and R. Blaschke, *Staub* **31,** 317 (1971).
119. J. A. Brink, *Can. J. Chem. Eng.* **41,** 134 (1963).
120. J. H. Nichols and J. A. Brink, *Electrochem. Technol.* **2,** 233 (1964).
121. J. A. Brink, W. F. Burggrabe, and L. E. Greenwell, *Chem. Eng. Progr.* **62** (4), 60 (1966).
122. J. W. Coykendall, E. F. Spencer, and O. H. York, *J. Air Pollut. Contr. Ass.* **18,** 315 (1968).
123. A. Burkholz, *Chem.-Ing.-Tech.* **42,** 1314 (1970).
124. O. H. York and E. W. Poppele, *Chem. Eng. Progr.* **66** (11), 67 (1970).
125. O. H. York and E. W. Poppele, *Chem. Eng. Progr.* **59** (6), 45 (1963).
126. J. A. Brink, W. F. Burggrabe, and J. A. Rauscher, *Chem. Eng. Progr.* **60** (11), 68 (1964).
127. J. A. Brink, W. F. Burggrabe, and L. E. Greenwell, *Chem. Eng. Progr.* **64** (11), 82 (1968).
128. H. Mohrmann, *Staub* **30,** 317 (1970).
129. B. W. Soole and H. C. W. Meyer, *Aerosol Sci.* **1,** 147 (1970).
130. J. S. Munsen, *Chem. Eng. (New York)* 147 (Oct. 14, 1968).
131. N. G. Edmisten and F. L. Bunyard, *J. Air Pollut. Contr. Ass.* **20,** 446 (1970).
132. K. Iinoya, K. Makino, and N. Tanaka, *Kagaku Kogaku* **38,** 453 (1974).
133. N. Tanaka, K. Makino, and K. Iinoya, *Kagaku Kogaku* **37,** 1226 (1973).

# 5

## Electrostatic Precipitation

Sabert Oglesby, Jr., and Grady B. Nichols

## Nomenclature

| | |
|---|---|
| $A$ | Collecting surface area, $m^2$ |
| $A_c$ | Cross-sectional area, $m^2$ |
| $A_d$ | Area of resistivity probe, $m^2$ |
| $a$ | Radius of corona wire, m |
| $a_p$ | Particle radius, m |
| $B$ | Constant in Cunningham correction factor, dimensionless |
| $b$ | Electrode spacing, m |
| $C$ | Effective velocity of ion cluster, m/second |
| $d$ | Relative air density, defined as $(T_0/T)(P/P_0)$, where $T_0$ is 298°K and $P_0$ is 1.0 atm |
| $E$ | Electric field, V/m |
| $E_c$ | Corona starting field, V/m |
| $E_d$ | Electric field in dust deposit, V/m |
| $E_0$ | Average electric field in interelectrode space, V/m |
| $E_p$ | Electric field at the collecting electrode, V/m |
| $E(r)$ | Electric field at radius $r$, V/m |
| $e$ | Electronic charge, C |
| $F$ | Force, N |
| $F_a$ | Drag force, N |
| $F_e$ | Force due to electric field, N |
| $i$ | Current per unit length, A/m |
| $j$ | Current density conducted through dust layer, $A/m^2$ |
| $K_a$ | Constant in Anderson equation, dimensionless |
| $k$ | Boltzmann constant, J/°K |
| $L$ | Loss fraction, dimensionless |
| $m$ | Wire roughness factor, dimensionless |
| $N_0$ | Free ion density, number/$m^3$ |
| $Q$ | Gas volume flow rate, $m^3$/second |
| $q$ | Particle charge, C |
| $q_s$ | Saturation charge, C |
| $q(t)$ | Charge at time $t$, C |
| $R$ | Resistance, Ω |
| $r$ | Radius, m |
| $r_0$ | Radius of corona glow region, m |
| $S$ | Collection length, m |
| $T$ | Temperature, °K |
| $t$ | Time, seconds |

| | |
|---|---|
| $V$ | Potential, V |
| $V_a$ | Potential at particle surface, V |
| $V_c$ | Corona initiation potential, V |
| $V_d$ | Voltage drop across dust layer, V |
| $V_0$ | Potential minimum in vicinity of particle due to opposition of self field and applied field, V |
| $v$ | Gas velocity, m/second |
| $v_r$ | Resultant particle velocity, m/second |
| $v_0$ | Gas velocity, average, m/second |
| $w$ | Migration velocity, m/second |
| $w_p$ | Precipitation rate parameter, m/second |
| $x_d$ | Thickness of dust layer, m |
| $\gamma(v)$ | Velocity distribution function |
| $\epsilon$ | Dielectric constant, dimensionless |
| $\epsilon_d$ | Permittivity of dust, $C^2/J\text{-}m$ |
| $\epsilon_0$ | Permittivity of free space, $C^2/J\text{-}m$ |
| $\eta$ | Collection efficiency |
| $\lambda$ | Molecular mean free path, m |
| $\mu$ | Mobility, $m^2/V$-second |
| $\mu_i$ | Mobility of ion, $m^2/V$-second |
| $\nu$ | Gas viscosity, kg/m-second |
| P | Resistivity, Ω-cm |
| $P_d$ | Resistivity of dust, Ω-cm |
| $\rho$ | Total space charge, $C/m^3$ |
| $\rho_i$ | Space charge due to ions, $C/m^3$ |
| $\rho_p$ | Space charge due to particulate, $C/m^3$ |
| $\tau$ | Time constant, seconds |

## I. Introduction

The process of electrostatic precipitation involves the removal of dust or liquid aerosol from a gas stream by utilizing the force resulting from an electrical charge in the presence of an electric field. The electrostatic forces of attraction or repulsion have been recognized since the time of the early Greek philosophers. Quantitative electrostatic force relationships involving the inverse square law were developed by Coulomb in the late 1700's, and these form the fundamental basis of electrostatic precipitator theory today.

Industrial interest in electrostatic precipitators in the United States can be traced to the work of Cottrell (*1*). His work was concerned principally with the removal of air pollutants, mainly sulfuric acid mists from copper smelters. As a result of his pioneering work, the term "Cottrell precipitator" has for many years been used almost interchangeably with electrostatic precipitator. Experimental work by Cottrell in the elimination of acid mists led to the use of precipitators for the collection of metal oxides from the effluents, and this added further impetus to the growth of precipitators in the primary smelter industry. This work was followed by the application of precipitators to other nonferrous metal processes, such as lead blast furnaces, ore roasters, and reverberatory furnaces. Success in the nonferrous metals industry was in turn followed by application of precipitators to the collection of dust from cement kilns. From these beginnings, the use of precipitators has expanded to include a wide variety of uses ranging from unique problems of recovering precious metals to the removal of fly ash from coal-fired boilers for electric power generation. The principal uses of precipitators today are in gas-cleaning applications in which high collection efficiencies of small particles are required for processes that emit large gas volumes. Since the separation force in a precipitator is applied to the particle itself, the energy required for gas cleaning is less than that for equipment in which energy is applied to the entire gas stream. This unique characteristic of precipitators results in lower gas pressure drops and usually lower operating costs than other methods of gas cleaning.

The precipitation process requires: (a) a method of providing an electrical charge on a particle, (b) a means of establishing an electric field, and (c) a method for removing the particle from the precipitator.

The process of electrically charging a particle involves the addition to or removal of electrons from the material or the attachment of ionized gas molecules. Almost all small particles in nature acquire some charge as a result of naturally occurring radiation, triboelectric effects due to transport through a duct, flame ionization, or other processes. These charges are generally too small to provide effective precipitation, and in all industrial precipitators, charging is accomplished by the attachment of electrical charges produced by an electrical corona.

The electrical field in an industrial precipitator is provided by the application of a high dc voltage to a dual electrode system. Electrical corona is established in this electrode system, one of which is a small diameter wire or other configuration which gives the small radius of curvature required to produce a highly nonuniform electric field. The other electrode can be a cylinder concentric with the corona electrode or a plate parallel to the plane of the corona wires. The corona generated in the high field region provides the charges necessary for electrical collection. The charged particles entering the region of the electric field are urged toward one of the electrodes where they are collected and held by electrical, mechanical, and molecular forces. Liquid particles, such as acid mists or tars, coalesce on the collection plate and drain into a sump at the bottom of the precipitator. Solid particles are usually removed by periodic rapping of the collection electrode, which permits the dust to fall into hoppers at the base of the precipitator. Solid material can also be removed by irrigation of the collection electrode with water or other fluid in what is termed a "wet" precipitator.

The mechanical arrangement of equipment to carry out the precipitation functions varies with the application as well as other considerations. Functionally, precipitators can be classified as single stage or two stage, depending on whether the collecting electric field is an extension of the corona field or is separated from it. In the latter case, a charging section consisting of conventional wire and plate electrodes is followed by parallel plates alternately connected to a high voltage source and to the ground, to provide the field for particle collection.

Almost all industrial precipitators are of the single-stage design, although the mechanical details may vary considerably between manufacturers and types of service in which the precipitator is used.

Figure 1 is a schematic diagram of the precipitation process illustrated with wire and concentric tube electrodes. The high voltage applied to

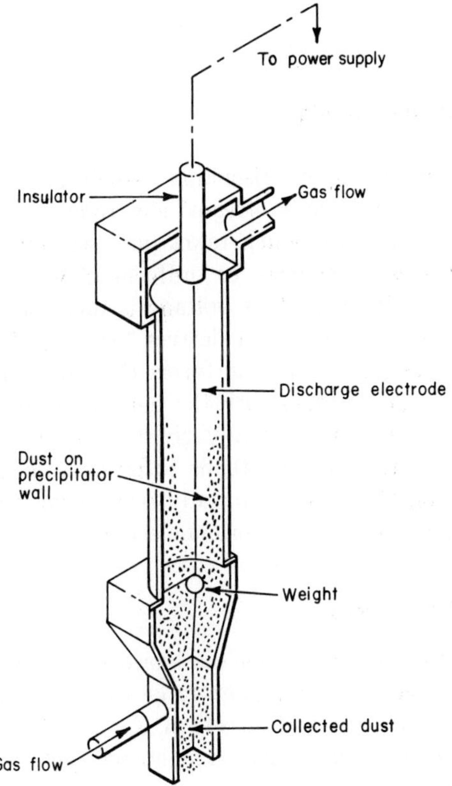


Figure 1. Schematic of a wire and pipe precipitator.

the wire or corona electrode results in a high electric field near the wire and an associated corona, not unlike that experienced in high voltage electrical transmission. The corona, either directly or indirectly, produces gas ions in the interelectrode region. These ions collide with and are held by the dust particles and provide them with an electric charge. The electric field in the interelectrode region, acting in conjunction with the charged particle, produces a coulomb force which propels the dust particles to the grounded or collection electrode. The collected dust is then removed by imparting a rap to the plates, causing the dust to be dislodged and fall into the hopper below.

In both theory and practice, the operation of precipitators is governed largely by the magnitude of the charge, the electric field, and the extent of reentrainment of collected dust. Factors that determine precipitator performance are those which establish limits on each of these parameters. The method by which each of them influences precipitator behavior can best be understood by a review of each of the major steps in the precipitation process.

## II. Generation of the Corona

An electrical corona is the discharge associated with gaseous breakdown near a highly electrically stressed electrode. A manifestation of the corona is a luminous glow which takes on a variety of shapes such as bright spots, brushes, streamers, or uniform glows and is accompanied by a rapid increase in current as voltage is increased beyond the point of corona onset. The term corona derives from the French *couronne* or crown, which represents one of the forms that have been observed. The appearance of the glow differs with the polarity of the corona electrode. A positive electrode corona generally gives a uniform glow over the surface, whereas a negative corona tends to produce localized tufts, bright spots, or streamers. Coronas are usually termed positive or negative depending upon the polarity of the corona electrode.

### A. *The Avalanche Process*

The phenomenon of the corona discharge can best be understood by considering the process of conduction in gases. Neutral gas molecules are not influenced by the application of an electric field, other than by a slight polarization, and hence there is no current flow. However, naturally occurring radiation does produce electrons and positive ions in an otherwise insulating gas. These electrons and positive ions move under the influence of an electric field and constitute charge carriers, although the current resulting from the flow of these carriers alone is too low to be of practical significance.

If, however, a sufficiently high voltage is applied to a pair of electrodes of which one is in the form of a wire or any shape with a small radius of curvature, the electric field near the surface will be high. Free electrons within this high field will be accelerated to velocities sufficient to strip an electron from the outer shell of a gas molecule on impact, creating a positive ion and another free electron. This additional electron is, in turn, accelerated to a velocity sufficient to cause further impact ionization. This process, called an avalanche, is repeated many times so that large quantities of electrons and positive ions are produced within the corona region.

### *B. Electron Capture and Formation of Space Charge*

If the corona electrode is negative, electrons produced by the avalanche process move rapidly away from the wire toward the grounded or positive electrode. The positive ions move toward the corona or negative electrode. As the electrons move away from the corona wire, the electric field rapidly diminishes. Since electron velocity is determined largely by the field strength, it rapidly diminishes to the point below that required for impact ionization. If electronegative gases, such as oxygen, water vapor, and sulfur dioxide, are present, the free electrons produced by the corona are captured and produce negative ions which also move toward the grounded electrode under the influence of the electric field. It is these electrons and negative ions which provide the source of charge for the particles that are to be collected.

A positive corona acts in much the same fashion as a negative corona, except that the direction of the electric field is such that the positive ions created as a result of the avalanche process move toward the grounded plate, whereas the electrons move toward the corona electrode. One significant difference between the positive and negative coronas is that the negative corona requires the presence of an electronegative gas to form ions necessary to produce the space charge required for a stable corona. In a positive corona, the process inherently produces the ions required for the stabilizing space charge. In almost all industrial processes, sufficient quantities of electronegative gases are present so that this requirement does not pose a serious problem. In fact, almost all industrial precipitators are of the negative corona type due to the more favorable voltage–current characteristics at the temperatures and pressures encountered in the majority of processes.

The use of positive-corona precipitators is generally limited to cleaning of air for inhabited space, since they produce less ozone. Positive coronas are also used on precipitators that operate at temperatures in the range of 1500°F or above.

### C. Corona Onset Voltage

In terms of electrostatic precipitation, the factors influencing the potential at which the corona is initiated and the voltage–current relationship of the corona discharge are of most practical significance. The current–voltage relationships in a precipitator depend upon the geometry of the electrodes, the composition and conditions of the gas, the thickness and properties of the collected dust layer, and the concentration and particle size of the suspended dust. From a mathematical standpoint, analysis of precipitator voltage–current relationships has been developed on the basis of concentric wire and cylinder configurations, since the symmetry yields equations that can easily be handled. Cooperman (*2*) has extended the analysis to wire and plate geometries; however, the mathematical relationships become somewhat unwieldy.

In a wire and cylinder electrode system the electric field at any point can be calculated by

$$E(r) = \frac{V}{r \ln(b/a)} \tag{1}$$

for the condition of no current flow.

As the voltage between the wire and cylinder is increased, the field near the wire increases until corona is initiated. The field strength required for the onset of corona depends upon geometrical factors as well as the properties of the gas. The most extensive investigation of corona initiation was the work of Peek (*3*), who developed an empirical relationship for corona initiation in air of the form

$$E_c = 3 \times 10^6 \, md \, (1 + 0.3\sqrt{ad}) \tag{2}$$

Since the applied voltage is the integral of the field strength, the voltage required for corona initiation can be found by

$$V_c = \int_a^b E(r)\, dr \tag{3}$$

which yields

$$V_c = 3 \times 10^4 \, amd \, (1 + 0.3\sqrt{d/a}) \ln(b/a) \tag{4}$$

The critical voltage required for the onset of corona can be altered by the geometry of the electrodes; the smaller the wire the lower the voltage required to initiate the corona.

As the voltage between the electrodes is increased beyond that required for corona initiation, there is a corresponding increase in current resulting from an increase in the number of ions produced in the corona process. The mathematical relationships between voltage and current for concentric cylindrical electrodes have been developed on the basis of Poisson's equation, which yields

$$V = V_c + \frac{V_c}{\ln(b/a)}\left(\left[1 - \left(\frac{b}{aE_c}\right)^2 \frac{i}{2\pi\epsilon_0\mu}\right]^{1/2} - 1 - \ln\frac{1}{2}\left\{1 + \left[1 + \left(\frac{b}{aE_c}\right)^2 \frac{i}{2\pi\epsilon_0\mu}\right]^{1/2}\right\}\right) \tag{5}$$

## *D. Effect of Gas Composition*

From the above equation, it is apparent that the voltage–current relationship can be altered by a number of factors. One of the principal sources of variation is the composition of the gas, which determines the molecular species of the charge carrier. The process of electron attachment to form negative ions is different for different gases. Hydrogen, nitrogen, and argon have no electron affinity, and these gases are incapable of electron attachment to form negative ions. However, oxygen and sulfur dioxide, which are present in most industrial effluent gases, readily capture electrons to form stable ions. Other gases present in industrial effluents, such as carbon dioxide or water vapor, have no electron affinity and attachment involves dissociation of the molecule, followed by the capture of the electron by the oxygen atom.

Thus, one would expect the current–voltage characteristics to vary with gas composition, as indeed is the case. Figure 2 shows the voltage–current characteristics for mixtures of nitrogen, oxygen, and sulfur dioxide as reported by White (*4*). The effect of gas composition on the current–voltage relation is due to the difference in probability of electron capture by each of the components of the mixture and the mobility of each. The mobility $\mu_i$ of a gas ion is defined by the equation

$$v_0 = \mu_i E \tag{6}$$

which relates the average velocity of the ion $v_0$ (m/second) to the electric field $E$ (V/m).

Table I shows the mobilities of both positive and negative ions for various pure gases (*5*). In industrial processes, effluent gases are primarily mixtures (as opposed to a single component). In such cases, electron attachment will depend upon the probability of attachment per collision

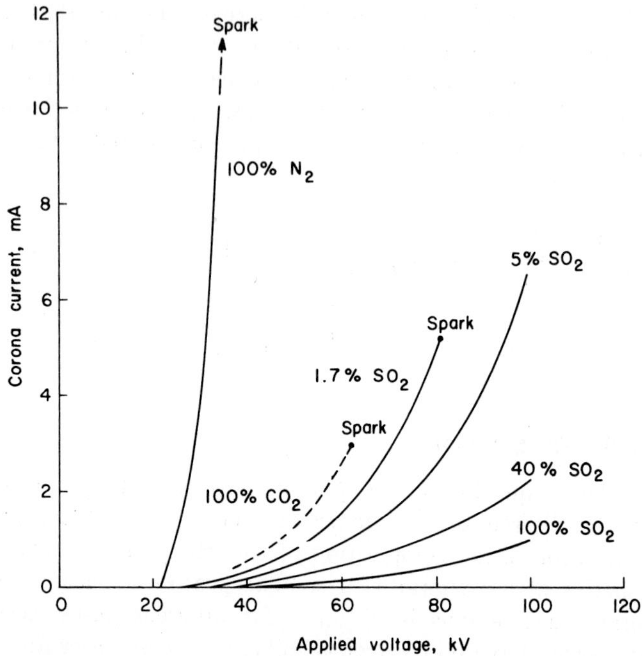


Figure 2. Negative corona curves for nitrogen–sulfur dioxide mixtures (6-in. tube, 0.109-in. wire) (*4*).

**Table I Mobilities of Singly Charged Gaseous Ions at 0°C and 1 atm** (*5*)

| | *Mobility* $[m^2/(second\text{-}V)] \times 10^{-4}$ | | | *Mobility* $[m^2/(second\text{-}V)] \times 10^{-4}$ | |
|---|---|---|---|---|---|
| *Gas* | $\mu_i$ (−) | $\mu_i$ (+) | *Gas* | $\mu_i$ (−) | $\mu_i$ (+) |
| He | —[a] | 10.4 | $C_2H_2$ | 0.83 | 0.78 |
| Ne | —[a] | 4.2 | $C_2H_5Cl$ | 0.38 | 0.36 |
| A | —[a] | 1.6 | $C_2H_5OH$ | 0.37 | 0.36 |
| Kr | —[a] | 0.9 | CO | 1.14 | 1.10 |
| Xe | —[a] | 0.6 | $CO_2$ (dry) | 0.98 | 0.84 |
| Air (dry) | 2.1 | 1.36 | HCl | 0.62 | 0.53 |
| Air (very dry) | 2.5 | 1.8 | $H_2O$ (100°C) | 0.95 | 1.1 |
| $N_2$ | —[a] | 1.8 | $H_2S$ | 0.56 | 0.62 |
| $O_2$ | 2.6 | 2.2 | $NH_3$ | 0.66 | 0.56 |
| $H_2$ | —[a] | 12.3 ($H_3^+$) | $N_2O$ | 0.90 | 0.82 |
| $Cl_2$ | 0.74 | 0.74 | $SO_2$ | 0.41 | 0.41 |
| $CCl_4$ | 0.31 | 0.30 | $SF_6$ | 0.57 | — |

[a] No electron attachment in the pure gas.

**Table II Impacts Required for Electron Attachment** (*6*)

| *Gas* | *β, average number of collisions* | *Gas* | *β, average number of collisions* |
|---|---|---|---|
| Inert gases | ∞ | $C_2H_6$ | $2.5 \times 10^6$ |
| $N_2$, $H_2$ | ∞ | $N_2O$ | $6.1 \times 10^5$ |
| CO | $1.6 \times 10^8$ | $C_2H_5Cl$ | $3.7 \times 10^5$ |
| $NH_3$ | $9.9 \times 10^7$ | Air | $4.3 \times 10^4$ |
| $C_2H_4$ | $4.7 \times 10^7$ | $H_2O$ | $4.0 \times 10^4$ |
| $C_2H_2$ | $7.8 \times 10^6$ | $O_2$ | $8.7 \times 10^3$ |
| | | $Cl_2$ | $<2.1 \times 10^3$ |

and on the concentrations of the various components. Table II shows the number of collisions required for electron attachment for various gases (*6*). Thus, for gas mixtures, the combination of the electron affinity, as indicated by the number of collisions required, and the mobility of each species determine the apparent mobility of the gas mixture.

In general, improved precipitator performance would be favored by gases with high electron affinity and low mobility, which would produce higher voltages and hence more intense electric fields. However, there is very little option of altering the gas composition in most processes, and changes in the current–voltage curves would normally be made by control of the size of the discharge electrode or other geometrical change.

### E. Temperature and Pressure Effects

Temperature and pressure of the gas modify both corona starting potential and the voltage–current relationship. One effect of pressure and temperature is to change the gas density and hence the electron mean free path. This, in turn, alters the electric field required to accelerate an electron to the velocity required for ionization.

Figure 3 shows the variation in corona starting potential with temperature and pressure as reported by Shale *et al.* (*7*). The curves indicate the general effect that increasing pressure and decreasing temperature serve to increase gas density and hence the corona starting potential.

A second effect of temperature and pressure is to alter the voltage–current curves by changing the effective mobility of the charge carriers. Shale *et al.* (*7*) have shown that effective ion mobilities increase with (a) decreasing gas density at constant temperature and field strength, (b) increasing temperature at constant density and field strength, and (c) increasing field strength at constant temperature and gas density.

The effect of the first factor is that at lower gas densities, electrons

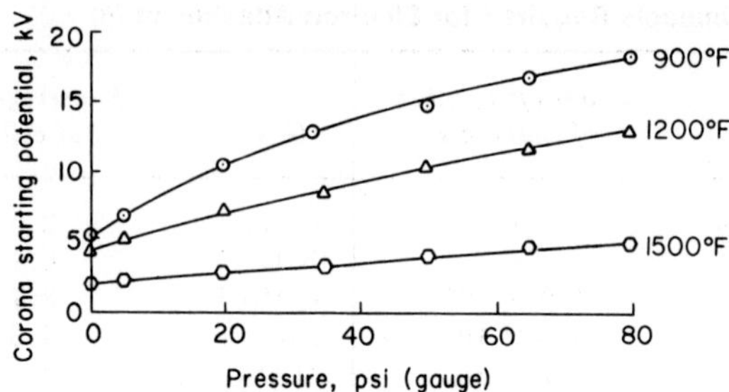


Figure 3. Variation in corona starting potential in air with pressure and temperature (negative corona, clean electrode) (7).

travel further before attachment; thus, the higher mobility of the electron appears as an increased velocity of the current carriers. The other factors are reportedly due to the increased number of electrons in the vicinity of the corona electrode caused by increased thermionic emission and increased secondary emission from positive ion impact on the corona electrode.

In a negative corona, the presence of negative ions in the interelectrode space provides a stabilizing space charge. With lower gas densities, the highly mobile electrons move rapidly away from the corona electrode and travel a longer distance before capture. The overall effect is a less favorable voltage–current relationship, wherein sparkover occurs at a voltage closer to the corona initiation voltage. The effect is quite pronounced in the case of very high temperatures (about 1500°F), although it also influences precipitator operation within the normal range of temperatures.

In the case of positive corona, the effects of temperature and pressure are not so pronounced, since the current carriers are positive ions. Consequently, for extreme temperature applications, positive corona may give more favorable precipitation conditions than negative corona, the reverse of that for the more normal temperature range.

## III. Electric Field

The electric field plays a central role in the precipitation process since it influences the charge on the particles to be collected, as well as the force acting on the charged particles. From a theoretical basis, the product of the field in the region where charging takes place and the field near the collection electrode is instrumental in determining the collection efficiency for a particle of a given size.

As previously indicated, the electric field in a precipitator is the result of the application of a high voltage to a pair of electrodes and the space charge effect due to the presence of ions and charged particles within the interelectrode region. If there is no current flow, the field consists only of the component resulting from the applied voltage and geometry. For the coaxial wire and cylinder geometry, the field at any point can be determined from the relationships given in Equation (1). Figure 4 shows the variation in field with distance from the corona wire for a wire and cylinder precipitator. Curve A is for the condition of no current flow.

Above the corona threshold voltage, the field is altered by the presence of ions in the interelectrode region. Since the mobility of a gas ion is much less than that of an electron, the comparatively immobile ions would give rise to an effective space charge. The influence of the ionic space charge can be visualized by assuming it to be a concentrated charge in the interelectrode region. Negative charges in the vicinity of the collection electrode would be accelerated by the repulsive forces due to the space charge in addition to that due to the electrostatic field. The effect would be to enhance the electric field near the collection electrode. On the other hand, negative ions or electrons near the corona electrode would be repelled by the space charge, thus reducing the effect of the electric field in the vicinity of the corona electrode. The overall effect of current

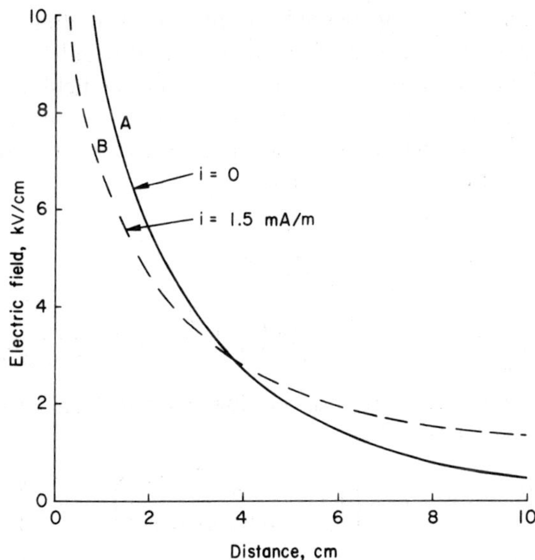


Figure 4. Change in electric field with distance from corona wire.

flow would be to alter the electric field as indicated by curve B of Figure 4.

The magnitude of the space charge is related to the number of charge carriers present and their mobility. Electrons with their very high mobility would be rapidly swept from the region and, hence, would not form an appreciable space charge. Charged dust particles, on the other hand, have much lower mobilities than ions and hence would develop a higher space charge than would ions alone.

Mathematically, the magnitude of the electric field can be determined from relationships based upon derivations of Poisson's equation by taking into account the charge carriers present and their mobilities. The field at any radius $r$ in a wire and cylinder electrode configuration is approximately given by

$$E(r) = -\left[\left(\frac{r_0 E_c}{r}\right)^2 - \left(\frac{r_0^2}{r^2} - 1\right)\frac{i\rho}{2\pi\epsilon_0\mu_i\rho_i}\right]^{1/2} \tag{7}$$

where $\rho = \rho_i + \rho_p$.

The approximation is valid when the amount of current carried by the charged particulate matter is negligible in comparison to that carried by the ions.

The voltage corresponding to any given field configuration can be determined by the integration of the electric field from the wire to the cylinder. This relationship emphasizes that alterations in one portion of the electric field must be compensated for by corresponding alterations in another, such that the integral of the field should equal the applied voltage. Thus, mathematically, an increase in electric field near the plate due to space charge enhancement is compensated for by a reduction in the field near the corona wire if the applied voltage remains constant. For the low current case, the voltage is

$$V = V_c + aE_c\left[\sqrt{1 + \frac{i}{2\pi\epsilon_0\mu_i}\frac{\rho}{\rho_i}\frac{b^2}{a^2E_c^2}} - 1 - \ln\left(\frac{1}{2} + \frac{1}{2}\sqrt{1 + \frac{i}{2\pi\epsilon_0\mu}\frac{\rho}{\rho_i}\frac{b^2}{a^2E_c^2}}\right)\right] \tag{8}$$

For moderate and high currents, the logarithmic term is small, so the expression can be reduced to

$$V = V_c + aE_c\left[\left(1 + \frac{i\rho}{2\pi\epsilon_0\mu_i\rho_i}\frac{b^2}{a^2E_c^2}\right)^{1/2} - 1\right] \tag{9}$$

In the field and voltage equations above, the effects of both ionic and particulate space charge are included in the term $\rho$, which is the sum of

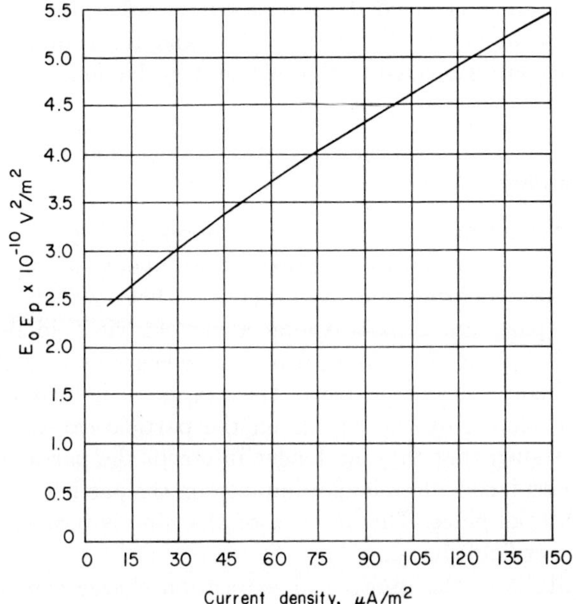


Figure 5. Variation of the product of space average field and collection field with corona current density.

the ionic space charge $\rho_i$ and the particulate space charge $\rho_p$. If there is no particulate charge, the term $\rho/\rho_i$ is unity.

From a practical standpoint, the field that is important in the particle charging process is the space average field in the interelectrode region, as will be discussed in the following section. The collection field which influences the effectiveness of collection for a particle of a given charge is that near the plate. A good measure of precipitator operation would, therefore, be the product of the space average field and the collection field $E_0E_p$. For a given corona wire size and collection electrode diameter, the product of the fields varies with corona current as indicated in Figure 5. In general, therefore, precipitator performance is increased by the corona current primarily because of its influence on electric field. Additional factors related to particle charging will be discussed subsequently.

## IV. Particle Charging

It is a fundamental requirement of the precipitation process that the particles be electrically charged to the maximum value consistent with other operating conditions.

Particle charging is normally considered to take place in the region between the boundary of the corona glow and the collection electrode, where particles are subjected to the rain of negative ions from the corona process.

### A. Field Charging

The predominant mechanism by which charging takes place varies with the particle size. Large particles, greater than about 0.5 $\mu$m in diameter, cause a localized deformation of the electric field such that field lines intercept the particles. Ions traveling along electric field lines impinge on the particles where they are held by image charge forces. As ions continue to impinge on a dust particle, the charge on it increases until the local field developed by the charge on the particle causes distortion of the field lines such that they no longer intercept the particle. When this condition exists, ions will no longer impinge on the particle and no further charging will take place. This method of charging is termed field-dependent charging or field charging.

The magnitude of the limiting or saturation charge can be computed with the assumptions that the particles are spherical, that there is no interaction between the fields of adjacent particles, and that the electric field is constant. In MKS units, the value of the saturation charge is

$$q_s = 12[\epsilon/(\epsilon + 2)]a_p{}^2\pi\epsilon_0 E_0 \tag{10}$$

Thus the magnitude of the charge depends primarily upon the particle size and the magnitude of the electric field modified by the dielectric constant of the material being collected. Since particle size enters the relationship as a squared term, it is the predominant factor influencing particle charge. Figure 6 shows the saturation charge for various particle sizes of interest in electrostatic precipitator applications for various values of electric field.

The time required for a particle to reach its saturation charge varies with the ion density in the region where charging takes place. The charging rate of a particle in a field of ion density $N_0$ can be shown to vary according to

$$q(t) = q_s \frac{1}{1 + \tau/t} \qquad \text{where} \qquad \tau = \frac{4\epsilon_0}{N_0 e \mu_i} \tag{11}$$

For normal conditions where relatively high currents can be maintained, charging times are short, often of the order of a few milliseconds. However, if currents are limited because of high dust resistivity or other factors, particle charging times can be relatively long, such that the mate-

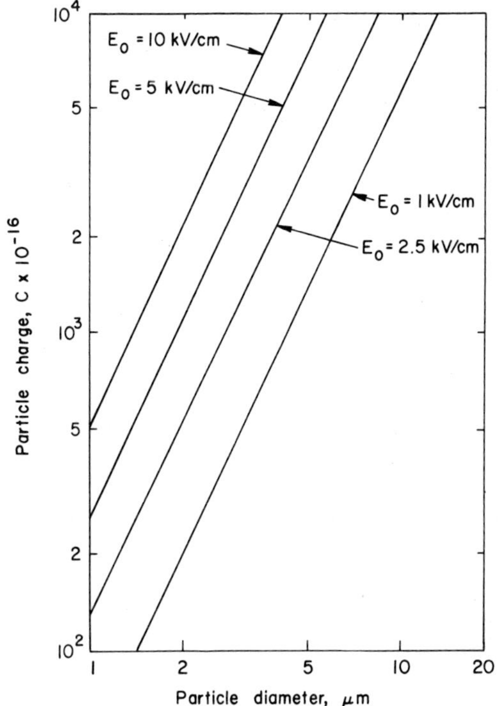


**Figure 6.** Saturation particle charge as a function of particle size (*4*).

rial to be collected can travel several feet through the precipitator before approaching its saturation charge. This effect is illustrated in Figure 7, which shows the charge for four values of current density typically encountered in precipitators when collecting normal and high resistivity dusts.

A second factor influencing particle charging time is the variation in the electric field with time. Despite the implication of the term electrostatic, the voltage applied to a precipitator is typically rectified, unfilered alternating voltage of 50–60 hertz. Consequently, the voltage waveform is a time-varying function. A particle in the interelectrode region will, therefore, be subjected to varying fields, and hence, the instantaneous saturation value of the charge will vary. The principal effect is that charging will be interrupted for that portion of the cycle during which the charge on the particle exceeds that corresponding to the saturation charge for the electric field existing at the time. This further adds to particle charging times and, in the case of high resistivity dust, degrades precipitator performance.

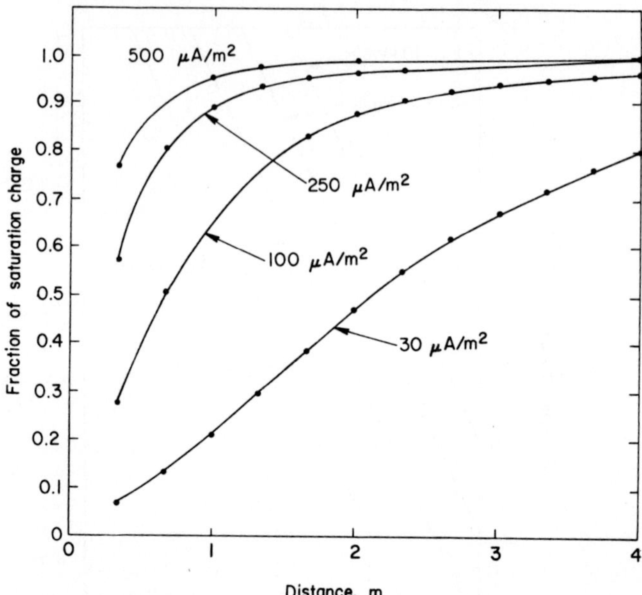


Figure 7. Computed fraction of saturation charge versus distance at 1.5 m/second for various current densities.

### *B. Diffusion Charging*

In addition to the process of charging by ion impact from the electric field, particles are also charged by ion attachment resulting from random thermal motion. This mode of charging is referred to as diffusion charging and is the dominant mode of particle charging for particles below about 0.2 $\mu$m diameter. In the strictest sense, diffusion charging is also influenced by the magnitude of the electric field, since ion movement is governed by electrical as well as diffusional forces.

Neglecting the electrical forces yields a simplified relationship of diffusion charging based upon the kinetic behavior of gases. The thermal motion of molecules causes them to diffuse through a gas and contact the particles. The rate of contact between a particle and gas ion is related to the density of ions in the vicinity of the particles and the root mean square velocity of the ions, the latter being related to the temperature and gas properties. As the particle becomes charged, it will repel additional gas ions. However, due to the distribution of thermal energy, there is always an ion with sufficient velocity to overcome the repulsion force; hence there is no theoretical saturation or limiting charge in the diffusion charging mode other than that limit imposed by field emission of electrons. However, the charging rate decreases as a particle acquires a charge. White (*4*) developed an equation for diffusion charging neglecting the influence of the electric field. This equation (MKS units)

$$q = (akT/e) \ln(1 + a_p v N_0 e^2 t / 4\epsilon_0 kT) \tag{12}$$

applies in the particle-size range below about 0.2 $\mu$m diameter, where field charging contributions are negligible.

### C. Combined Field and Diffusion Charging

The majority of industrial dusts contain size fractions in the range where both diffusion and field charging are significant. This is especially true of high efficiency precipitators where the smallest particles must be collected.

Because of the interdependence of the particle charge and charging rate, the method of computing the particle charge in the region where both charging mechanisms are significant becomes more complex. The technique for computing charge is to determine charging rates for each mechanism and compute an overall charging rate. This procedure leads to a nonlinear differential equation with no analytic solution.

Several investigators have studied particle charging in the particle-size region of 0.2 to 1 $\mu$m and several approaches to determining particle charge have been suggested.

The work of Liu and Yeh (*8*) has resulted in a method of calculating charge due to both diffusion and field charging in which the effect of electric field has been included. The calculations agree reasonably well with experimental data developed by Hewitt (*9*).

The Liu and Yeh approximation divides the charging process into two conditions: (a) in which the particle charge $q$ is less than the saturation charge by field charging and (b) in which the charge is greater. The charge applicable to condition (a) is

$$\frac{dq}{dt} = \frac{N_0 e \mu_i q_s}{4\epsilon_0}\left(1 - \frac{q}{q_s}\right)^2 + \frac{C\pi N_0 a^2 e}{2}\left(1 + \frac{q}{q_s}\right) \tag{13}$$

The first term in the above equation is the expression for field charging and the second term is a correction factor which approximates the additional charge imparted by the diffusion process.

For condition (b), the equation is

$$\frac{dq}{dt} = (\pi a^2 N_0 C e) \exp\left(-\frac{e\,\Delta V}{kT}\right) \tag{14}$$

in which $\Delta V = V_a - V_0$, $V_a$ being the electric potential at the particle

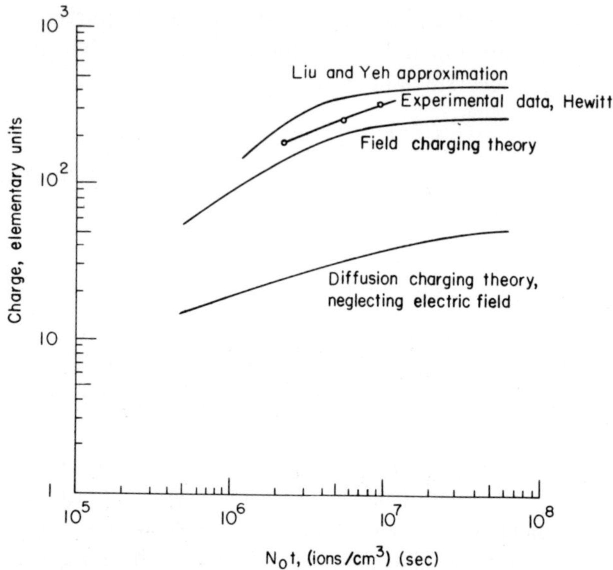


Figure 8. Comparison of theoretical and experimental data for 0.46 $\mu$m radius particle and high electric field intensity. 1 atm pressure, $a_p$ = 0.46 $\mu$m, $E_0$ = 9000 V/cm. Adapted from Liu and Yeh (*8*).

surface, and $V_0$ being the potential minimum in the vicinity of the particle due to the opposition of the self field and the applied field.

Figure 8 shows the charge as a function of ion density–time product as computed by Liu and Yeh and as compared with the experimental data of Hewitt. The values obtained by the Liu and Yeh approximation agree reasonably well with the experimental values. As shown in Figure 8, the charge predicted by Liu and Yeh for a 0.46 $\mu$m radius particle would be about a decade higher than that predicted by diffusion alone and considerably higher than that predicted by field charging alone. Such differences can significantly alter the predicted collection efficiencies of particles in the submicron region, and it is exactly these particles that are of greatest importance in high efficiency collectors.

## V. Particle Collection

### *A. Migration Velocity*

The motion of a particle in an electric field is governed primarily by electrostatic and aerodynamic forces. Electric forces are readily calculated from the charge on the particle and the electric field as

$$F_e = qE \tag{15}$$

Aerodynamic forces are primarily drag forces resulting from relative motion between the particle and gas, and can be computed from the equation

$$F_a = 6\pi\nu a_p w \tag{16}$$

The limiting or terminal velocity of a particle due to these forces occurs when the two forces are equal. Equating these and solving for the velocity gives

$$w = qE/6\pi a\nu \tag{17}$$

The term $w$ is generally referred to as the migration velocity. If the particle size is large, the charge is primarily the field charge contribution, and the migration velocity can be expressed in terms of the saturation charge on the particle due to the field charging process as

$$w = 2a_p\epsilon\epsilon_0 E_0 E_p/(\epsilon + 2)\nu \tag{18}$$

For particles less than about 0.5 $\mu$m diameter, the above expression is not valid because of the contribution of diffusion charging. Also in the case of small particles, a correction in the aerodynamic drag forces must be applied to account for the variation in velocity caused by the reduction in collisions with gas molecules. The correction factor $[1 + B(\lambda/a)]$ is referred to as the Cunningham correction to Stokes' law, and should be applied to migration velocity equations for particles less than about 0.5 $\mu$m.

Applying the correction gives the more complete equation for migration velocity:

$$w = (qE_p/6\pi a\nu)[1 + B(\lambda/a)] \tag{19}$$

The collection process in an electrostatic precipitator can be demonstrated by first considering the conditions existing in laminar gas flow. Under these conditions, a dust particle will have two components of velocity, as indicated in Figure 9. The component of velocity urging the

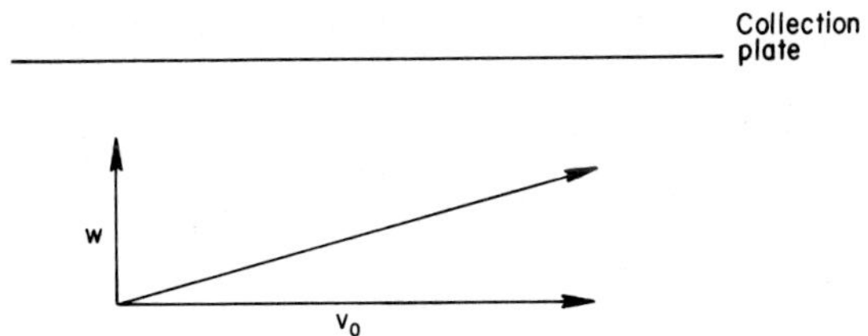


Figure 9. Velocity components for a charged particle in laminar gas flow.

particle toward the collection plate results in the migration velocity determined by the electrical force and the aerodynamic drag. The velocity along the precipitator axis is essentially the gas velocity less the slip. The resultant velocity is the vector sum of these components.

Referring to Figure 9, it is apparent that particles entering the precipitator near the corona electrode will be collected in a distance determined by the ratio of the gas velocity to the migration velocity and the precipitator dimensions. A shorter precipitator length will obviously yield a collection efficiency of less than 100%.

The laminar flow situation is primarily of academic interest only, since commercial precipitators operate with Reynolds numbers indicating gas flows that are well in the turbulent flow region. Under turbulent gas flow conditions, the motion of the smaller size fraction of the particles is determined primarily by aerodynamic forces and hence the trajectory of an individual particle cannot be deterministically predicted. Under turbulent flow conditions, collection efficiency is an exponential function of the collecting area, gas volume, electrical charge, electric field, and gas viscosity.

### B. The Deutsch Equation

The most extensive investigations relating precipitator efficiency to operating parameters have been the work of Anderson (*10*) in 1919 and Deutsch (*11*) in 1922. Anderson showed experimentally that the efficiency of a precipitator was described by the empirical equation

$$\eta = 1 - \exp(-K_a t) \tag{20}$$

The value of the empirical factor $K_a$ depends on the specific process and set of operating conditions.

The Deutsch equation was derived on the basis of theoretical considerations, the principal assumptions being that: (a) the particle concentration is uniform through the cross section, (b) the particles are fully charged immediately on entering the precipitator, and (c) there is no loss or reentrainment of the collected particles.

Within the boundary layer near the collection plate, each particle has a velocity component $w$ in the direction normal to the collection surface. Within the time interval $t$, particles within a distance $wt$ will be precipitated onto the collecting surface within a length $S$ (Fig. 10).

From these relationships and the concentration of particles in the boundary layer, it can be shown that the efficiency of particle removal is an exponential relationship of the form

$$\eta = 1 - \exp[-(A/Q)w] \tag{21}$$

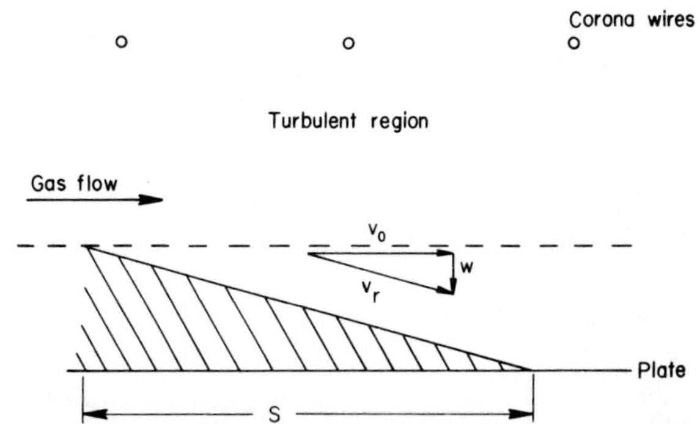


Figure 10. Collection of particles within the boundary layer in turbulent gas flow.

This equation is commonly referred to as the Deutsch equation and is widely used in precipitator design and analysis.

### *C. Theoretical Factors Affecting Precipitation Rate*

There are several relationships of interest in the use of the Deutsch equation. As derived, the equation predicts only the amount of material reaching the precipitator collection surface and does not consider effects of reentrainment by any of the various mechanisms by which collected dust can be reintroduced into the gas stream.

Because it generally describes the exponential relationships between efficiency, collecting area, and gas volume, the Deutsch equation has been used to describe precipitator performance by calculating the factor $w$ from measured efficiencies. Migration velocities so calculated have been used by precipitator manufacturers as a basis for precipitator sizing, where experience factors have been used in establishing these parameters for various industrial applications. Used in this sense, the migration velocity is more properly an empirical proportionality constant and should be referred to as the precipitation rate parameter to distinguish it from the more theoretically based migration velocity of the Deutsch equation. The equation used in this manner is often referred to as the Deutsch–Anderson equation since the exponential relationship was first reported by Anderson.

The theoretical migration velocity as used in the Deutsch equation indicates a considerable particle-size dependence. Therefore, a larger precipitator is required to reach a given efficiency when small particles are collected. Figure 11 shows the migration velocities for particles of various sizes for various operating conditions. The range of particle sizes in which

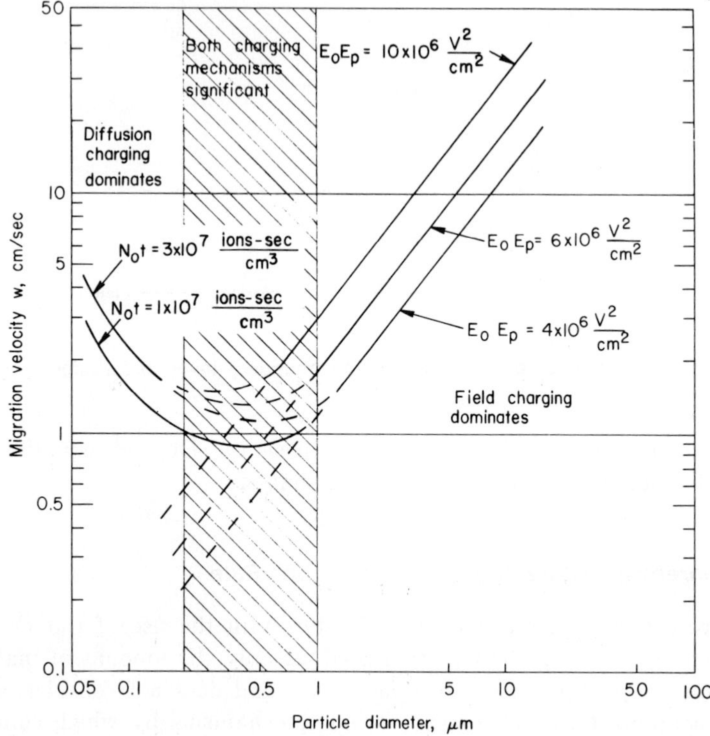


**Figure 11.** Theoretical migration velocity as a function of particle diameter for field and diffusion charging mechanisms.

field charging and diffusion charging predominate is indicated, together with the size range in which both mechanisms are significant. The migration velocity is adjusted by the Cunningham correction factor for particle sizes below 1 μm. This correction largely accounts for the increase in migration velocity in the very small particle-size range.

One consequence of the variation in migration velocity with particle size is that there will be a corresponding change in collection efficiency. Thus, for a polydisperse dust, the larger particles will tend to be collected in the inlet sections of a precipitator, resulting in a change in the median diameter of the dust as it progresses through the precipitator. Figure 12 shows the theoretical mass median diameters of the inlet dust—the dust collected in each hopper of a precipitator with four independently powered sections in series—and that leaving the precipitator. Figure 13 shows the theoretical collection efficiency of various particle sizes for typical operating conditions.

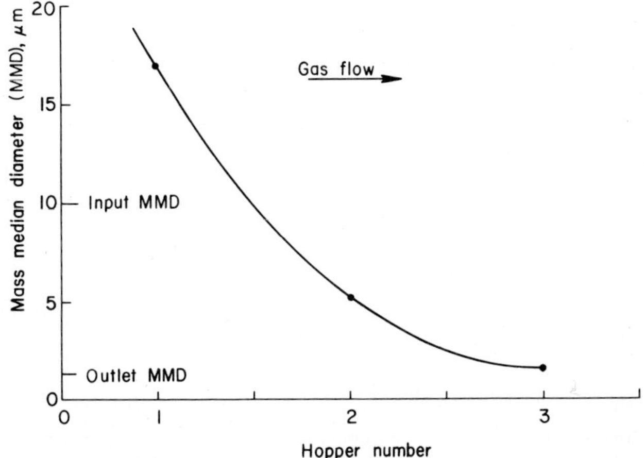


Figure 12. Expected particle-size distribution for three hoppers in normally operating precipitator.

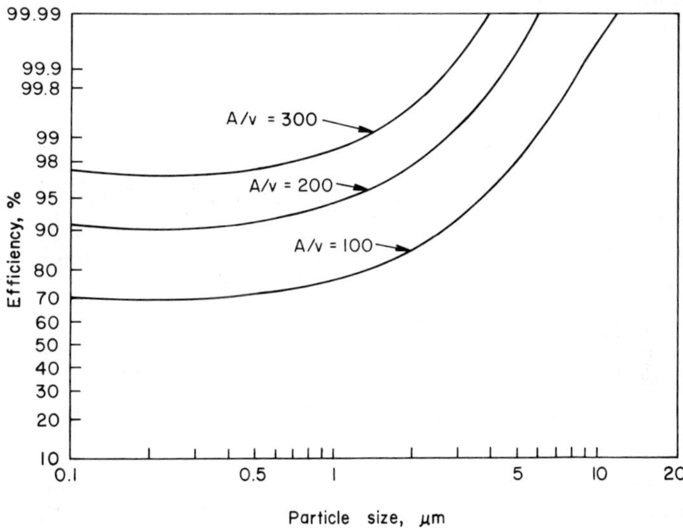


Figure 13. Collection efficiency for various particle-size fractions.

## VI. Removal of Collected Particles

Once collected, particles can be removed from the collection electrode by draining in the case of liquid aerosol, by flushing the plates with a liquid, or by periodically rapping or vibrating the plates in the case of solid particles. Draining of coalesced liquid on the plates constitutes a more or less straightforward process and is common in precipitators used in the removal of acid mist, tar, and similar materials. Removal of solid particles by irrigation of the collection plates has been used extensively in some metallurgical processes, notably blast furnaces, for a number of years. There is at present considerable renewed interest in so-called "wet" precipitators for a number of applications, including final mist elimination from gas scrubbers, collection of fine particles from aluminum production, and collection of high resistivity dust. Irrigation of the collection plates is accomplished by the use of a weir at the top of the plate which provides a flow of fluid down the plate, or by sprays which discharge horizontally or vertically into the interelectrode region.

By far the majority of precipitators in use are of the dry type, in which dust is removed by rapping. Successful removal of dust by rapping involves the formation of a coherent dust which, when dislodged, falls as a sheet or agglomerate. This requires that a dust layer of some appreciable thickness be accumulated between rapping cycles which vary from a few minutes to hours, depending upon dust loadings, dust properties, etc.

### A. Rapping

The mechanics of rapping vary among manufacturers, but they involve either a periodic vibration or impact which dislodges the dust and permits it to fall toward the hopper. The primary requirement for successful rapping is that it accomplishes removal of the dust without excessive reentrainment. This is physically accomplished by (a) adjusting the rapping intensity to prevent powdering or excessive breaking up of the dust layer, (b) adjusting the rapping frequency to give optimum dust layer thickness for the most effective removal, (c) maintaining proper airflow and baffling, and (d) rapping only a small portion of the precipitator at a time.

There has been surprisingly little investigation of the mechanism of dust removal in view of its importance to the precipitation process.

Sproull (*12*) reports that optimum rapping conditions occur if the dust layer slides vertically down the collection plate a distance of perhaps several feet following each rap. Under these conditions, the dust would proceed down the collection plate in discrete steps until it finally falls into the hopper.

It is apparent that a dust layer $\frac{1}{4}$ to $\frac{1}{2}$ in. thick falling from a 30-ft-high plate would reach a rather high terminal velocity in free fall. Thus, localized scouring could cause a large percentage of the dust to be reentrained.

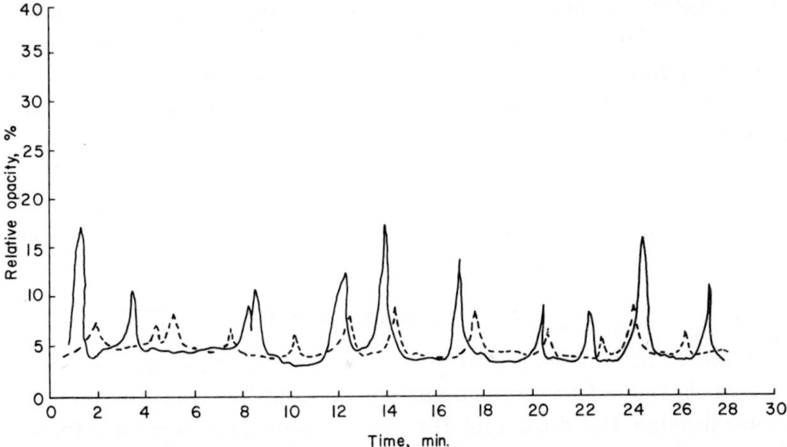


**Figure 14. Rapping puffs as shown by an obscuration meter in a power plant. Solid line: heavy rapping, dashed line: light rapping.**

Further, the fall of so large a quantity of dust into the hopper would generate a large dust cloud, which would be picked up by the gas stream and carried out of the section being rapped.

Successful rapping must therefore avoid the condition in which the dust layer is allowed to fall freely off the plates. This requires that the rapping intensity and rapping frequency be adjusted for optimum conditions which are related to the forces holding the dust layer to the collection surface.

At the opposite extreme, the optimum rapping conditions must prevent powdering of the dust by too severe a rap or by attempting to remove too thin a dust layer.

Losses of dust due to rapping are often visible at the discharge end of a precipitator as discrete puffs that coincide with the rapping cycle. These puffs can be used as a basis for optimizing rapping conditions. Figure 14 shows a series of rapping puffs for two levels of rapping intensity. The reduction in emissions by optimizing rapping intensity is obvious.

### *B. Factors Affecting Rapping and Reentrainment*

The factors influencing optimum rapping include the forces that hold the dust layer to the collection surface. These forces are electrical, molecular, and mechanical in nature. Molecular forces include van der Waals' forces at the surface of the particle, modified by molecules adsorbed on the surface. These forces can be influenced by the surface condition due to the presence of molecular surface layers. Mechanical forces are due primarily to the interlocking of particles and to interparticle friction.

Electrical forces result from the flow of current through the dust layer. The magnitude of the electrical force has been determined by Penney (*13*) to be

$$F = \tfrac{1}{2}\epsilon_0[E^2 - (j\mathrm{P}_\mathrm{d}\epsilon_\mathrm{d}/\epsilon_0)^2]^{1/2} \tag{22}$$

where $E$, the potential gradient in the gas adjacent to the dust surface, is given by

$$(V - j\mathrm{P}_\mathrm{d}x_\mathrm{d})/(b - x_\mathrm{d})$$

The force $F$ in this relationship is the result of two components. The first term is related to the coulomb forces resulting from the flow of ion current through the dust, and the second term is a repelling force.

In a practical sense, the electrical force influences the rapping requirements to a considerable degree. For high resistivity dusts, the force is so large that it becomes difficult to remove the dust from the collection plate by conventional rapping. In some extreme cases, the current must be turned off during rapping to release the dust. Such "power off" rapping generally gives rise to large rapping puffs and is to be avoided where other alternatives are available.

At the opposite extreme, very low resistivity dust results in an extremely low holding force and excessive rapping losses can occur even with a very light rap.

It appears that the phenomenon of recollection of agglomerates required to give optimum rapping conditions may be associated with dust resistivity. In the case of high-to-medium resistivity dust, a residual charge remains after the agglomerate is dislodged from the plate. In the case of very low resistivity dust, the residual charge can be low or even of the opposite polarity. In such cases reentrainment losses can be high.

The effect of reentrainment on overall precipitator performance depends upon the number of precipitator rapping sections, the composite precipitator efficiency, and the rapping loss in each section. The material lost from the inlet section as a result of rapping adds to the inlet dust burden of subsequent sections and is at least partially recollected. Rapping losses from the last section are not recollected and appear as direct dust losses. However, the amount of material collected in the last section is small for high efficiency collectors because of the exponential removal rates. Consequently, even though the percentage dust loss due to rapping may be high, its overall effect on performance is not as great as it might appear at first thought.

Figure 15 shows the influence on precipitator performance of various

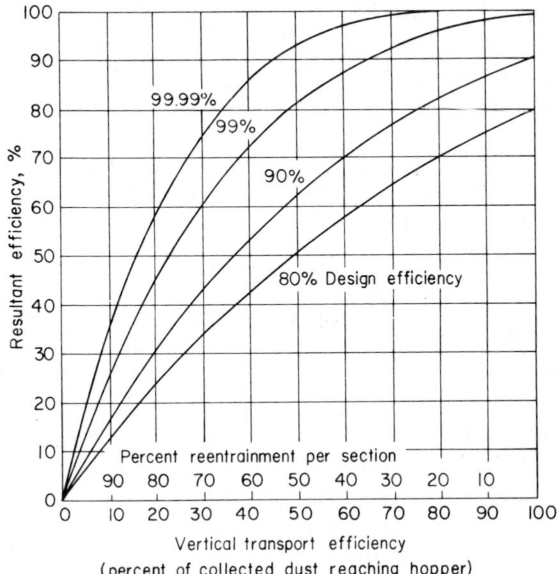


Figure 15. Effect of reentrainment on efficiency.

percentage rapping losses for a precipitator with four independently powered sections in series designed for different efficiencies. With rapping losses of 20% per section in a precipitator designed for 99% collection efficiency, efficiency due to rapping loss drops to about 95%. This constitutes a fivefold increase in exit dust loading.

## VII. Dust Resistivity

Dust resistivity provides a fundamental limitation to the operation of a precipitator. The configuration of a precipitator is such that the corona current must flow through the collected dust layer to reach the grounded collection electrode. In the case of dry precipitators, this current flow can result in large voltage drops across the dust layer if the electrical resistivity of the dust is high. In many applications, the resistivity of the dust is sufficiently high to impair precipitator performance.

Electrical resistivities of the dusts encountered in industrial gas cleaning applications can differ considerably. Some materials, such as carbon black, have very low resistivity, so that on contact with a grounded metal surface, the particles lose their charge and are easily reentrained into the gas stream. At the opposite extreme, dusts of insulating materials, such as alumina, can have a sufficiently high resistivity that the charge leaks off very slowly. In such cases, the electrical force holding the dust to the collection plate can be very high and the voltage drop across the dust layer can be sufficient to cause breakdown of the interstitial gases within the dust layer.

### A. Mechanism of Current Conduction in Dust Layer

The mechanism of current conduction in a dust layer has been studied by Bickelhaupt (*14*), McLean (*15*), and others. Two modes of conduction are possible in industrial dusts, depending upon the temperature and composition of the dust and flue gases. At elevated temperatures (above about 400°F) conduction takes place primarily through the bulk of the material. The resistivity of the dust in the temperature region where bulk conduction predominates is referred to as volume resistivity. At lower temperatures, moisture or other substances present in the flue gases are adsorbed and conduction occurs principally along the surface of the dust particles. In the temperature region where conduction takes place along the surface of the dust particle, the resistivity is referred to as surface resistivity.

Figure 16 illustrates a typical temperature–resistivity curve showing the regions where volume and surface resistivity predominate and the region where both are significant. The overall resistivity is analogous to that of parallel resistors, each of which is temperature dependent. The absolute value of resistivity varies depending upon the method and conditions of measurement, as discussed subsequently. However, the shape of the curve is similar for the various measurement techniques.

In the high temperature region, the resistivity depends on the chemical composition of the material. Bickelhaupt found that, for fly ashes of generally similar composition, the resistivity decreases with increasing sodium and lithium content, indicating that these ions are the primary charge carriers. The presence of iron causes a further decrease in resistivity, apparently by increasing the percentage of lithium and sodium ions that are capable of participating in the conduction process.

In the low temperature region, resistivity is thought to be an ion-transport phenomenon related to the adsorption of water vapor or other conditioning agents present in the flue gas. In the case of fly ash from coal-fired boilers, resistivity is primarily related in an inverse manner to the amount of sulfur trioxide ($SO_3$) and moisture present in the flue gas. The burning of coal containing sulfur produces sulfur dioxide ($SO_2$) in quantities dependent on the sulfur content in the coal. Under normal conditions, about 0.5 to 1% of the $SO_2$ present is oxidized to $SO_3$, which serves to reduce the resistivity of the fly ash, if the temperature is low

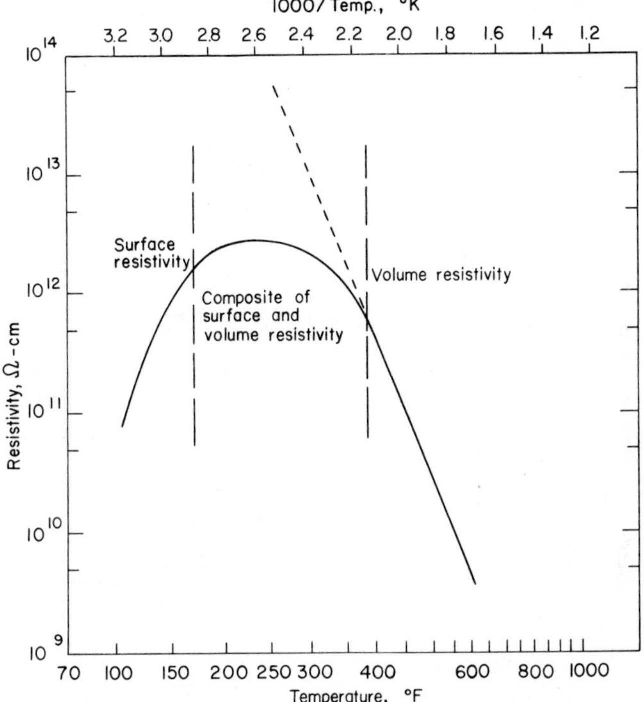


**Figure 16.** Typical temperature–resistivity relationship.

enough for the $SO_3$ to be adsorbed on the ash. Thus, high-sulfur coals tend to produce ash with lower resistivities than coals with lower sulfur content. In general, lowering the flue-gas temperature increases the rate of $SO_3$ adsorption, so that the resistivity of fly ash can be controlled to some extent by changes in flue-gas temperature. The presence of large quantities of CaO in the ash apparently has an adverse effect on conductivity. An ash high in lime tends to react initially with the available $SO_3$ to produce a sulfate which decreases the $SO_3$ available for conditioning.

For other processes, notably cement kilns and metallurgical furnaces, the principal conditioning agent is adsorbed moisture. Higher moisture content in the flue gases and lower temperature give lower resistivities. Resistivity–temperature relationships can, therefore, be represented as a family of curves with both the volume and surface resistivities changing with particulate and flue-gas composition. Figure 17 illustrates a range of resistivity values for a group of fly ash samples. The volume conductivity portions of the curves differ according to the concentration of alkali

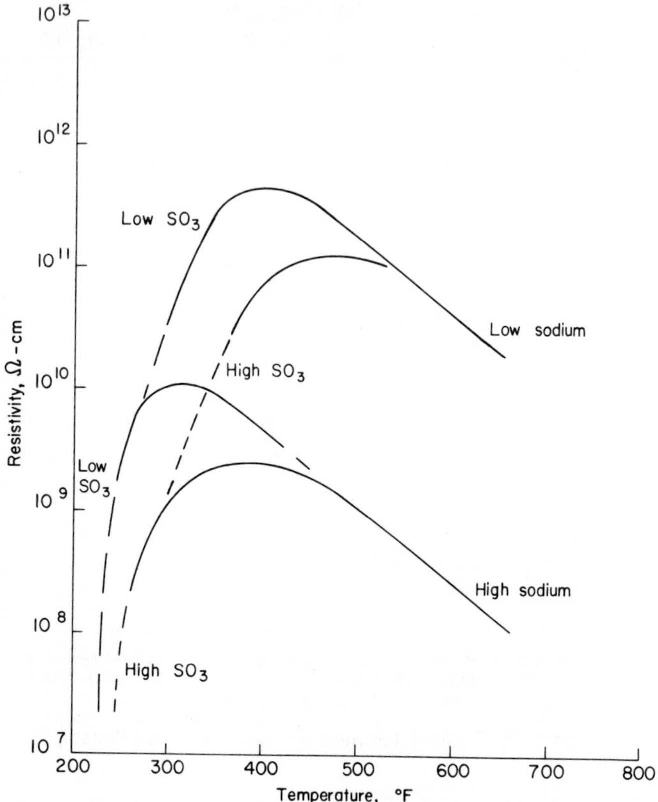


Figure 17. Resistivity versus temperature for high and low sulfur trioxide and high and low sodium.

present in the ash and the surface conductivity portions vary with the amount of sulfur in the coal. Curves such as these are generally used in estimating the range of resistivities expected for fly ash. Similar curves with flue-gas water content as a parameter are used for other types of dust.

### B. Effect of Resistivity on Precipitator Operation

The flow of corona current through the collected dust layer causes a voltage drop which is proportional to the current density, dust resistivity, and thickness. The electric field in the dust deposit is

$$E = j\mathrm{P}_{\mathrm{d}} \tag{23}$$

Electrical breakdown of the interstitial gases in the dust layer occurs when the electric field exceeds the breakdown strength of the gases. For most gases encountered in industrial precipitators, breakdown of the precipitated layer occurs when the electric field exceeds about 20 kV/cm. The exact value of the breakdown strength depends on the particle size and extent of packing of the dust and on the gas composition. The effect of breakdown of the dust layer can be illustrated by Figure 18.

If there is no dust layer present, the voltage in a precipitator can be increased with an accompanying increase in current until the gases in the interelectrode region break down. This breakdown takes the form of a sparkover or arc originating at the anode or collecting surface in a negative corona precipitator and propagating to the corona electrode. This condition, indicated by curve A, establishes the maximum current and voltage condition that can be obtained with the particular electrode geometry, spacing, and gas composition.

Curve B of Figure 18 illustrates the condition that exists with a dust deposit of intermediate resistivity. The displacement of the curve to the right of the clean plate curve is due to the voltage drop across the dust layer. As the voltage is increased, the current increases until the electric field in the dust deposit exceeds the breakdown voltage.

When breakdown occurs, the resistance of the localized area around the breakdown is reduced, so that the voltage that was across the dust layer is now applied to the space between the dust surface and the corona

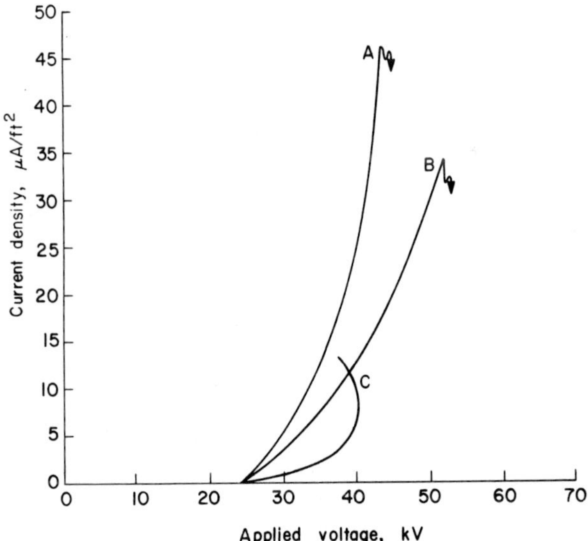


Figure 18. Voltage–current characteristics for various dust resistivities (plate spacing 10 in., wire diameter 0.019 in.). A: clean plate, B: 0.5-cm layer, $\rho_d = 5 \times 10^{11}$ Ω-cm, C: 0.5-cm layer, $\rho_d = 5 \times 10^{12}$ Ω-cm.

wire. This additional voltage can cause breakdown of the gases in the interelectrode region, resulting in a sparkover, provided the voltage across the electrode is sufficient to propagate a spark.

Curve C represents a condition of very high dust resistivity. With very high resistivity, electrical breakdown occurs at a voltage that is insufficient to propagate a spark across the interelectrode region. The result of this condition is a continuous breakdown of the dust layer without sparkover between electrodes. This breakdown is analogous to that occurring at the discharge electrode and similarly produces ion–electron pairs. The positive ions flow across the interelectrode region toward the discharge electrode. The net effect is a reduction of the charge on the particles and poor precipitation. This phenomenon is called back corona or reverse ionization and can be observed as a diffuse glow on the dust surface of the collection electrode under dark conditions.

Normally precipitators are set to operate at maximum voltage. The average high tension voltage applied to the corona electrode increases with increased spark rate up to an optimum at a rate of around 100 sparks/minute, as illustrated in Figure 19. Consequently, most precipitators are operated in the sparking mode for normal resistivity dusts al-

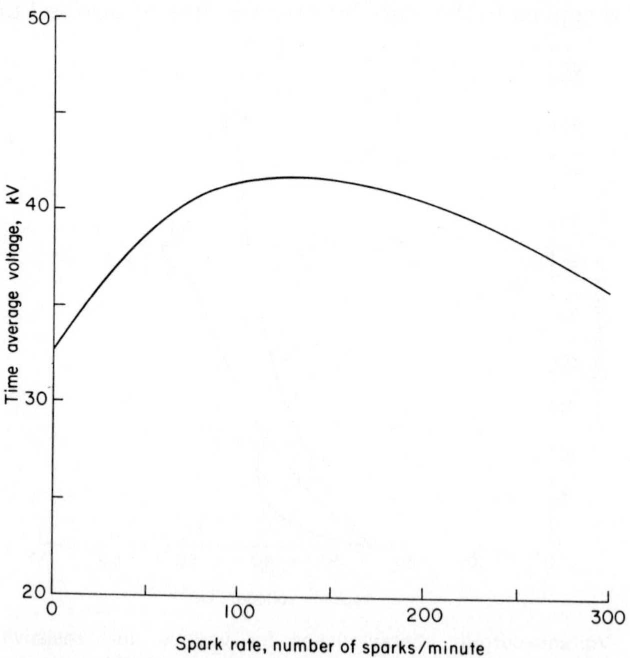


Figure 19. Change in operating voltage with spark rate (*4*).

though the spark rate is generally lower than 100/minute if large power supplies are used. When the rate of about 100 sparks/minute is exceeded, the average high tension voltage decreases and precipitator performance is reduced. When operating in a sparking mode, the electric energization equipment is generally set to a preselected spark rate, which determines the value of corona current at which the precipitator will operate. Either a current or voltage limit can set the operating point.

When dust resistivity is very high, back corona will cause deterioration of precipitator operation at voltages below that for sparking and hence the electrical energization equipment cannot be set for spark rate control. There is some uncertainty as to the point of most effective operation under a back corona condition. Lowe *et al.* (*16*) point out that certain types of back corona occur in localized areas, so that the positive ions streaming from the collecting plate to the corona electrode can be confined to a small portion of the interelectrode space. In such cases, improved precipitator operation can be achieved at voltages and currents above those required for the onset of back corona. When back corona is more generalized over the dust surface, the positive ions discharge the dust previously charged by negative ions and effective precipitation is not possible under these conditions. As indicated in curve C, Figure 18, the sharp increase in corona current at low voltage is characteristic of a severe back corona. Under these conditions, precipitators are generally operated with a current limit control at a point of optimum precipitator performance, which is usually experimentally determined.

### C. Methods of Altering Resistivity

When a dust with high electrical resistivity is to be collected, the options are (a) design a larger than normal precipitator to accomodate the lower precipitation rate or (b) alter the resistivity so that it is in a more favorable range for precipitation.

The electrical resistivity of a dust can be altered by a change in the operating temperature or by addition of moisture or other conditioning agents.

The preceding section discussed the characteristic change in dust resistivity in the temperature region where surface conduction predominates. Many industrial process precipitators operate with flue-gas temperatures in the range of 600–900°F. Two reasons for this practice are (a) less cooling is required and a greater plume buoyancy is achieved with the higher temperature, and (b) the electrical resistivity of the dust is generally low enough that it does not limit precipitator performance.

However, when heat recovery systems are used, the exit flue-gas temperature may be low enough that the volume resistivity is above that for good precipitator performance. Thus, in the collection of fly ash from electric power boilers, if the usual practice of passing the flue gas through a heat exchanger to heat the combustion air is followed, the temperature of the flue gas is reduced to about 300°F, at which temperature the resistivity of the fly ash may become too high for efficient collection. To overcome this problem, a substantial number of installations have been made in which the precipitator is located ahead of the air heater, where temperatures are generally in the range of 600–700°F. The increased gas volume and the complexity of the duct work required add to the cost of such installations; however, if the volume resistivity of the ash is sufficiently low at these temperatures, the technique is an effective method of combating the high resistivity problem.

At the opposite extreme, flue-gas temperatures can often be reduced with an accompanying reduction in resistivity associated with surface conduction. When low sulfur coals are being burned, the amount of sulfur trioxide present is low, so that corrosion of the air heater and precipitator may not constitute a problem, unless other corrosive materials, such as chlorine, are present. In such instances, low temperature operation may prove to be an effective and economical method of overcoming high resistivity. Berube (*17*) describes a precipitator installation collecting ash from a low sulfur coal in which the flue-gas temperature is reduced to about 220°F by increasing the amount of air passing through the air heater. The air in excess of that required for combustion is either dumped or used to reheat the exit gases to the stack.

A third method of controlling resistivity is through the addition of chemical conditioning agents to enhance surface conductivity. Historically, additives to flue gas for improving performance have been used in precipitators for the collection of catalyst dust in the petroleum industry. Water spray chambers have been used for the combined purpose of reducing temperature and adding moisture to condition the dust in precipitators used on municipal incinerators, metallurgical furnaces, and cement kilns.

Conditioning of flue gases from electric power boilers has been tried on both pilot and full scale by means of various additives. The most common additive is $SO_3$, in the stabilized anhydrous form, as vaporized $H_2SO_4$, or as $SO_3$ from the catalytic oxidation of $SO_2$. The quantity of $SO_3$ required for a given change in resistivity depends upon the composition of the ash. Effectively conditioned ash contains small quantities of free $H_2SO_4$ on the surface of the particles. As stated previously, a basic ash high in lime (CaO) tends to condition less well than a substantially neutral ash. Although the mechanism has not been definitely established, it has been suggested that the $SO_3$ deposited on a particle of basic ash probably reacts with water and with calcium ions to form a shell of calcium sulfate, which itself does not materially reduce resistivity. Once this layer of sulfate is formed, adsorption of additional $SO_3$ forms free $H_2SO_4$ on the surface, which reduces the dust resistivity.

The effectiveness of $SO_3$ as a conditioning agent has primarily been in its ability to reduce dust resistivity. However, Dalmon and Tidy (*18*) suggest that an added benefit of conditioning agents is in the increased adhesive and cohesive properties of the ash, which tends to reduce reentrainment, principally during rapping.

Ammonia ($NH_3$) has been used successfully as a conditioning agent for improving fly ash precipitator performance. Reese and Greco (*19*) report substantial improvements in performance of a precipitator used on a boiler burning high sulfur coal (3–4%) with flue-gas temperatures in the vicinity of 260°F. These conditions would normally be expected to produce an ash with very low resistivity. The improvement due to ammonia conditioning appears to be associated with factors other than resistivity changes. Studies of ammonia injection indicate that the ammonia probably reacts with the $SO_3$ in the flue gases to produce fine particles, probably ammonium sulfate, which alter the space charge electric field, primarily in the inlet section of the precipitator.

The use of ammonia to condition fly ash from low sulfur coals has also been attempted with varying degrees of success. Again, the beneficial effects of ammonia, when they have been noted, appear to be due to causes other than resistivity changes. Watson and Blecher (*20*), as well as others, have noted changes in electrical characteristics, especially in the inlet sections of precipitators, following $NH_3$ injections. These effects appear to be too rapid to be the result of dust resistivity changes, and again suggest the possibility of a space charge alteration. Furthermore, direct measurements of resistivity following ammonia injection have not indicated a resistivity change.

Flue-gas conditioning to control dust resistivity or to otherwise improve precipitator performance appears to be most useful when improvements are being made in existing installations to conform to changes in the type of fuel being burned or changes in emission requirements. The use of conditioning as a means for combating a high resistivity problem in a new installation is less readily acceptable than alternative methods.

### D. Measurement of Resistivity

Methods of measuring dust resistivity vary with respect to the method of collecting the dust sample and the manner in which current–voltage relationships are determined.

Dust for resistivity determinations is collected by electrostatic precipitation in some types of resistivity probes and by mechanical means, such as cyclones, in others. Resistivities are determined by measuring the current resulting from the application of a known voltage to a cell in which the geometry of the collected dust is known, or by measuring the change in voltage–current relationships in a precipitation-type resistivity apparatus.

Figure 20 shows a high-voltage resistivity cell of the parallel-disk type recommended as a standard by the American Society of Mechanical Engineers (ASME) (*21*). The apparatus consists of a shallow cup which contains the dust, a disk and annular guard ring which rest on the dust samples, and a weight for applying a constant load on the measuring disk. The ash cup is supported by ceramic insulators and is connected to the negative terminal of a high voltage dc power supply. The annular guard ring is grounded and the center disk is grounded through a current

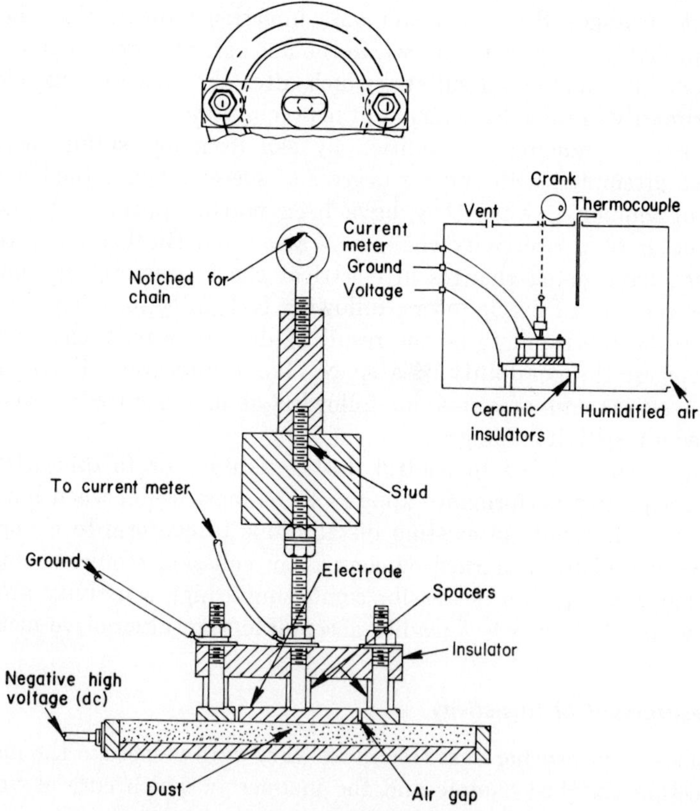


Figure 20. Bulk electrical resistivity apparatus.

meter. The positive terminal of the power supply is connected to ground, and the entire test assembly is housed in a chamber suitable for temperature and humidity control.

To measure resistivity of a dust sample, it is placed in the sample cup and screeded to give the proper thickness. The temperature and humidity in the chamber are brought to the desired conditions and the upper electrode assembly lowered to rest on the dust surface. Voltage is then applied to the electrodes and gradually increased while the current and voltage are recorded up to the point of electrical breakdown of the dust layer. The resistance of the dust layer is calculated as the ratio of voltage to current just prior to breakdown. Resistivity can then be computed as

$$\mathrm{P_d} = RA_{\mathrm{d}}/x_{\mathrm{d}} \tag{24}$$

An apparatus of this type is normally utilized for what are termed laboratory resistivity measurements, since there is no provision for depositing the dust layer other than by manually placing a previously collected sample in the apparatus.

An alternative apparatus for measuring resistivity is called a point-plane probe (Fig. 21). The point-plane probe differs from the ASME apparatus in that the dust is deposited onto one electrode surface by precipitation. During collection of the sample, an electrical corona is established between the point and the grounded electrode, and dust is precipitated onto the electrode surface in much the same manner it is in a precipitator. When a sufficiently thick deposit has been precipitated, the upper disk electrode is lowered to contact the dust surface. A voltage is applied across the electrodes and gradually increased while the corresponding current is measured. As with the ASME apparatus, the resistance of the dust layer is determined by the ratio of the current and voltage just prior to electrical breakdown. Resistivity is computed in the same manner as in the ASME apparatus; however, the thickness of the

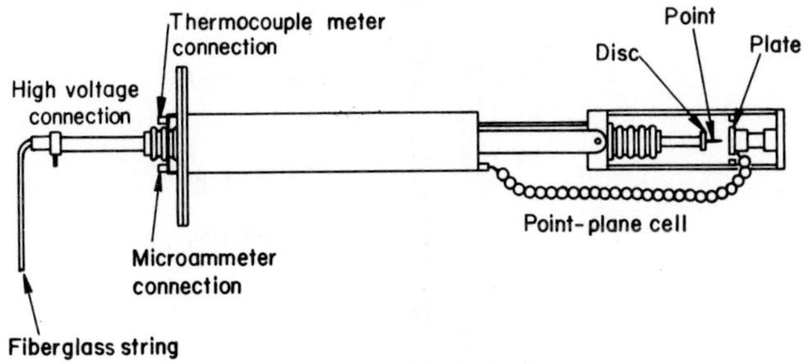


Figure 21. Point-plane resistivity probe.

dust layer in the point-plane apparatus varies with the time and conditions during precipitation and must be measured. This requirement presents some difficulties when measurement is being made *in situ* by insertion of the point-plane probe in the gas duct. To determine thickness, the probe must be withdrawn from the duct and the thickness measured as the separation between disk electrodes. When the duct is under heavy negative pressure, the rush of air into the duct as the probe is being withdrawn can aspirate the sample, so that it is not possible to determine thickness under these conditions.

Figure 22 shows a third type of probe, which permits the thickness of the sample to be measured while the probe is still within the duct. Thickness of the dust sample is measured by determining the separation of the plates due to the dust layer by means of a dial indicator, as shown in Figure 22. Such a probe overcomes the difficulty of losing the sample at the expense of added mechanical complexity.

Point-plane types of resistivity probes lend themselves to an alternative method of resistivity measurement which serves as a check and in-

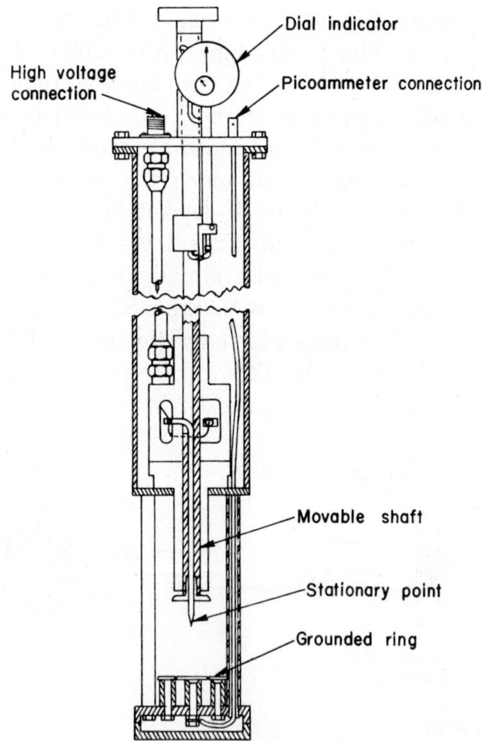


Figure 22. Point-plane resistivity probe equipped for thickness measurement.

creases confidence in the measurements made with the parallel disks. If voltage–current relationships are established for the specific gas and cell geometry both prior to precipitating a dust layer and following the collection of a layer of suitable thickness, the shift in the curves due to the voltage drop across the dust layer can be used to calculate resistivity, provided the dust layer thickness is known. Figure 23 shows typical voltage–current curves for a moderately high resistivity dust. The voltage drop at a given current density can be used to determine the average electric field in the deposit

$$E_{\mathrm{d}} = V_{\mathrm{d}}/x_{\mathrm{d}} \tag{25}$$

The resistivity can be calculated from

$$\mathrm{P}_{\mathrm{d}} = E_{\mathrm{d}}/j \tag{26}$$

This method of measuring resistivity only applies to moderate to high resistivity dusts, where there is sufficient voltage drop across the dust layer for accurate measurement. In such cases, good agreement is usually obtained between parallel disk and voltage–current methods of measuring resistivity.

Figure 24 shows an alternative type of resistivity apparatus, which

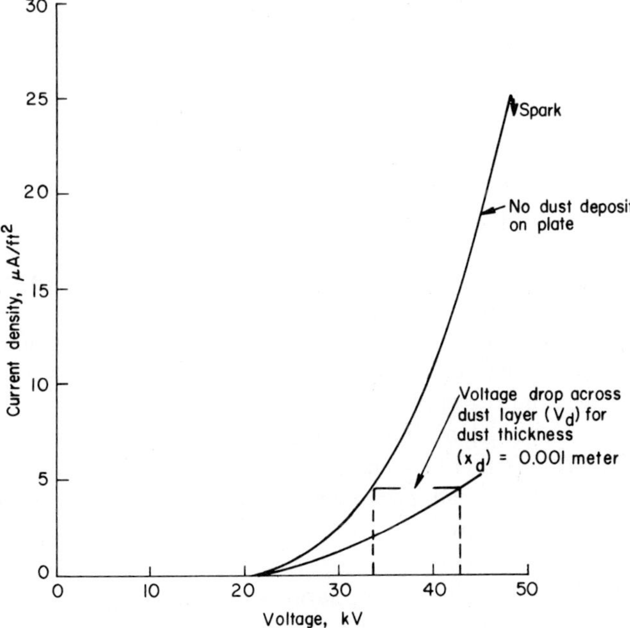


Figure 23. Typical voltage–current relationships for point-plane resistivity probe.

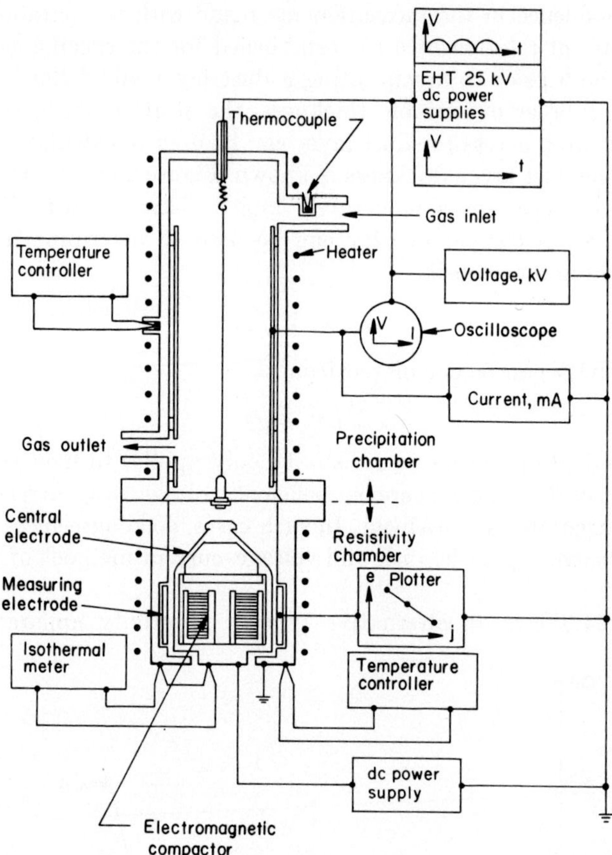


Figure 24. Kevatron resistivity probe (22).

utilizes a wire and cylinder design of electrostatic precipitator for obtaining a dust sample. When a sample is collected, it is rapped from the precipitator section and falls into a concentric cylinder type of resistivity cell. Resistivity is determined by measuring the current through the dust layer resulting from the application of a known voltage.

A commercial model of this apparatus, the Kevatron, was developed in Australia (*22*). It includes circuits for computing the resistivity which take into account the cell factor or geometry of the measurement cell. Resistivities are measured at a predetermined electric field in the dust layer of 0.15 or 1.5 kV/cm.

Figure 25 shows a resistivity apparatus described by Cohen and Dickinson (*23*) which utilizes a mechanical cyclone dust collector. A sample of the dust-laden gas stream is withdrawn through a probe. The cyclone

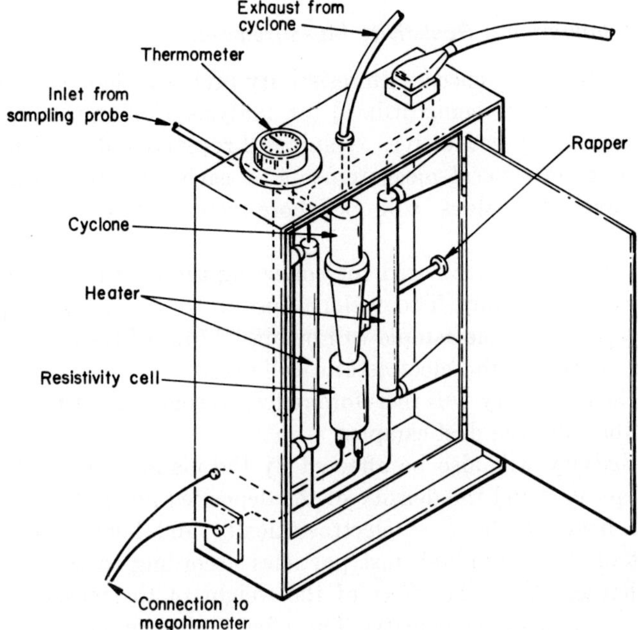


Figure 25. Resistivity apparatus using mechanical cyclone dust collector (*23*).

collects the dust from the gas stream; the collected dust is removed by vibrating the cyclone and allowing the dust to fall into a concentric cylinder resistivity cell. Resistivity is determined by measuring the current through the dust resulting from application of a voltage across the electrodes.

The cyclone and resistivity cell are located in a heated chamber external to the duct. Operating procedure calls for the chamber to be maintained at the same temperature as the gas stream in the duct.

Except for the ASME apparatus, each of the probes described can be used to measure resistivity of the dust in the flue-gas environment. In the temperature region where surface conduction predominates, measurements made *in situ* are preferable to laboratory measurements of previously collected dust because it is very difficult to accurately reproduce flue-gas conditions in the laboratory, and resistivities measured under laboratory conditions often bear little resemblance to those measured *in situ*. At higher temperatures, where volume conduction is the predominant mode, laboratory resistivity measurements agree quite closely with those measured *in situ* since, under these conditions, flue-gas composition has little or no influence on resistivity.

### *E. Factors Influencing Resistivity Measurements*

One difficulty experienced with resistivity probes is that relatively small quantities of dust are being utilized for analysis. Consequently, the samples may not be representative. Meaningful resistivity data should therefore be based on several measurements at each operating condition, so that the average values will be more representative of true dust resistivity.

The types of apparatus used for collecting the dust sample are by no means efficient collectors. The cyclone dust collector will not capture the smallest particles, so the sample will be biased toward the larger particles. The same is true of the electrostatic collectors, since the residence time is short. The resistivity will therefore reflect to some extent the size distribution of the collected dust samples.

The resistivity will also be affected by the manner in which the dust layer is deposited and the density of the deposited layer. The point-plane apparatus deposits the dust electrostatically and hence there is some alignment of the individual dust particles according to their shape and surface charges. Also, the effect of the weight of the measuring disk is to increase the packing density. The effects of alignment and packing density are not quantitatively known, but it is probable that there are significant differences in the results obtained with the various cells.

The resistivity of fly ash is known to vary significantly with changes in the electric field. The effects are different for various types of ash and flue-gas conditions. Figure 26 shows a typical relationship between electric field and resistivity for a medium-resistivity dust. It is apparent that differences of an order of magnitude or greater can be measured for the same ash if resistivity is measured in different electric fields.

Another phenomenon encountered in measuring dust resistivity with present techniques is the rapid change of current with time immediately following the application of voltage across the cell. The initial current surge is due to absorption current, a phenomenon well established in measuring properties of insulators.

It is apparent therefore that considerable care must be exercised in measuring dust resistivity so that the results are reproducible and can be correlated with precipitator operation. At present, there is no universally established procedure for resistivity measurement, and values reported by different investigators can vary by two orders of magnitude or more. Correction to the same electric field can reduce this variation significantly. The remaining discrepancies are apparently due to differences in density of the particles, to variations in the dust layer, and to differences in the extent of collection of particles of various sizes.

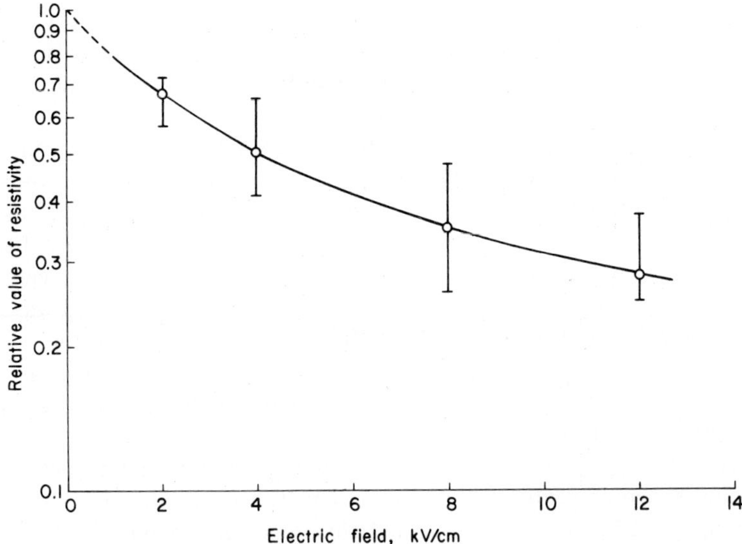

Figure 26. Typical variation of resistivity of fly ash with electric field.

## VIII. Gas Flow

Uniformity of gas flow has a profound influence on precipitator performance, especially when low-density dust is being collected. Nonuniformity of gas flow can cause severe reentrainment of dust and variable treatment times, the combination of which can cause marked decreases in precipitator efficiency. Other effects of poor gas flow are fallout or buildup of dust around turning vanes, elbows, and distribution plates. These conditions further alter the gas flow pattern, causing even less uniformity.

The best operating conditions for electrostatic precipitators would be provided by uniform gas flow. Such a condition is, however, never achievable in practice.

White (*4*) computes the effect of nonuniform flow on performance as

$$L = \int_0^{v_{\max}} [\exp - (Aw/A_c v)]\gamma(v)\, dv \tag{27}$$

The loss in efficiency calculated from this relationship is a result of the change in treatment time alone. It illustrates that the performance can be degraded significantly from that predicted from average gas flow conditions.

The effect of poor distribution of gas flow can be magnified by excessive reentrainment due to direct scouring of the collected dust or by increased rapping losses in the higher velocity regions.

In precipitators used for collecting dust from pulp mill recovery boilers, incinerators, or similar operations, where light, bulky, or fluffy dust may be involved, losses from reentrainment can be much more significant than those due to varying treatment time.

Quality of gas flow is generally expressed as a series of graphs showing velocity profiles, isopleths (contour plots) of lines of constant velocity. These are helpful in visualizing the gas flow pattern and in establishing corrective procedures to improve gas flow distribution. The quality of gas flow can also be observed from histograms showing the occurrence of velocity readings falling within a series of intervals. Good gas flow quality should approximate a Gaussian or normal distribution with a standard deviation on the order of 1.25 and a tendency to skew somewhat toward zero because of wall friction. The narrower the spread in gas flow, the greater the uniformity.

Acceptable gas flow quality is somewhat difficult to define because of the wide range of problems encountered in the field. The Industrial Gas Cleaning Institute has recommended that gas flow be such that 85% of the local velocities be within ±25% of the mean and that no single reading vary more than ±40%. This set of conditions corresponds to a standard deviation of about 23% of the average gas velocity. In practice, it is not uncommon to find examples of exceedingly poor gas flow distribution, and poor performance can often be traced to this cause.

Evidence of poor gas flow distribution can be found by several means. First, gas velocity can be measured by a velocity traverse at the precipitator inlet by hot-wire or propeller-type anemometers to quantitatively indicate the gas flow pattern. Second, poor gas flow distribution can often be observed in the pattern of dust buildup on precipitator plates, on distribution plates, and in the ductwork.

Particle-size distribution of the inlet and outlet dusts can also be used to indicate poor gas flow quality. Figure 27 shows the inlet and outlet dust particle sizes for a pulp mill recovery boiler with extremely poor gas flow quality. The outlet dust is distinctly bimodal, with a large percentage of the dust having been agglomerated and reentrained.

Design of precipitators should include gas flow model studies of the inlet ducting to ensure proper gas flow quality, especially on large precipitator installations and under conditions of unusual inlet ductwork design. Models are generally made $\frac{1}{4}$, $\frac{1}{8}$, or $\frac{1}{16}$ scale and may be constructed of pressed hardboard, Plexiglas, or other suitable materials. Plexiglas is often used since it permits direct observation of gas flow patterns by introduction of smoke or a similar material. Granulated cork is used to study dust fallout patterns and erosion.

Model studies require careful interpretation to ensure that the data

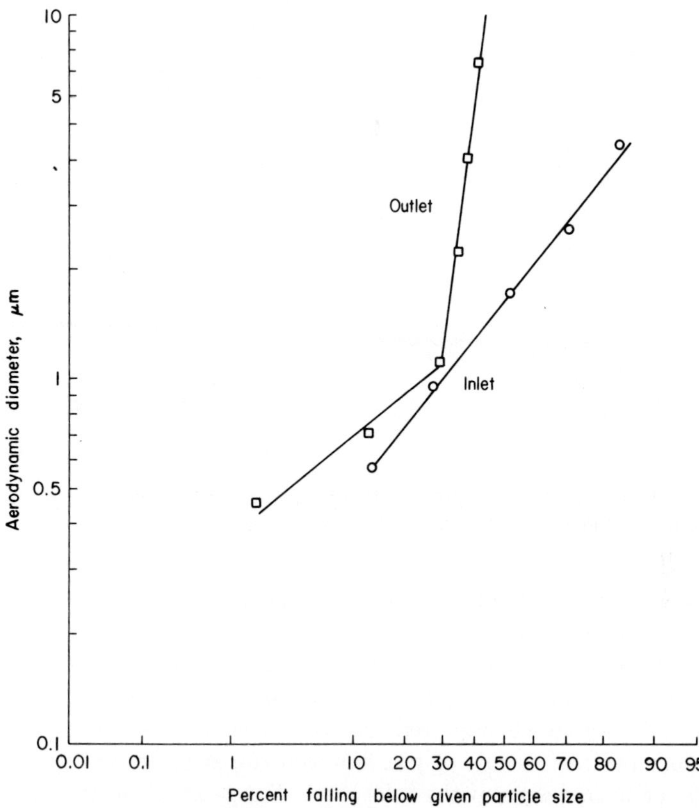


Figure 27. Particle-size distribution of inlet and outlet dusts for an electrostatic precipitator with poor gas flow quality.

are applicable to the full-scale system. It is, of course, impractical to get all the factors in the same scale relationship in the model as in the full-scale unit. Hence some experience factors are required. As an example, it is impractical to try to scale a gas diffusion plate to $\frac{1}{16}$ scale since it would be too thin. Hence, hole sizes should be altered to more nearly represent the full-scale condition. It is also usually impractical to achieve the same Reynolds number in a model as in the full size unit, since gas temperatures and hence viscosities are widely different. These do not invalidate the method, however.

Control of gas flow quality is achieved by the use of straighteners, splitters, turning vanes, and diffusion plates, either separately or in various combinations. Diffusion plates or screens are simply perforated plates or wire screens which improve gas flow quality by reducing the scale of turbulence to the order of magnitude of the holes and by providing a

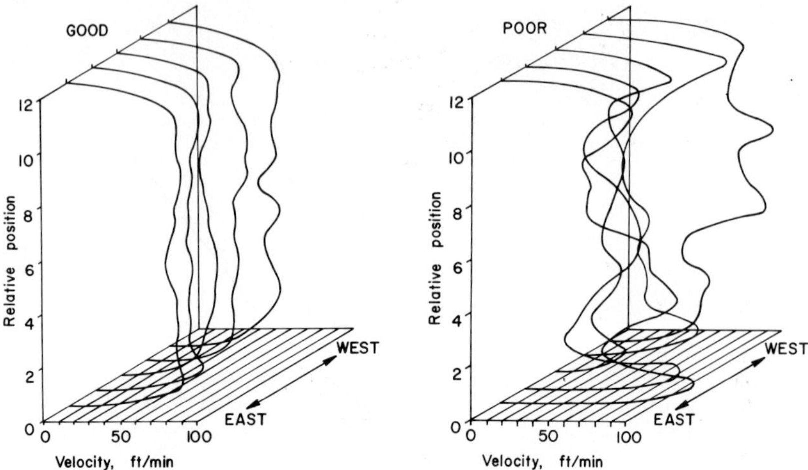


Figure 28. Example of the improvement of gas flow distribution in the inlet to a precipitator by the use of diffusion plates and flow straighteners.

pressure drop which tends to even out gas flow. One or more diffusion plates are often used on precipitator inlets to provide a final improvement in gas flow quality. However, if the gas flow distribution is extremely poor, it cannot usually be corrected by diffusion plates alone. When very poor distribution is encountered, correction should be made by use of other more positive means of gas flow control, such as turning vanes or splitters, or by redesign of the inlet duct. Figure 28 illustrates correction of a poor gas flow distribution to an acceptable one by the use of diffusion plates and flow straighteners.

Gas flow in precipitators is extremely critical in the region near the hoppers. Dust rapped from the plates and falling into the hoppers creates a suspension of dust particles which would be carried out of the precipitator if high velocity gas flow were permitted in the vicinity of the hoppers. For this reason, baffles and other means are generally provided to prevent gas flow through the hoppers and to minimize gas flow in the region between the lower edge of the collection plates and hoppers.

## IX. Design and Installation

### *A. Mechanical Design*

Mechanical design of precipitators varies with the type of dust control application and process to which they are applied. The majority of precipitators are of the dry collection type with plate-type collection electrodes and pyramidal-shaped hoppers. Gas flow is usually horizontal through the precipitator. Figure 29 shows an example of this kind of precipitator, which is typical for installations on electrical power boilers, cement kilns, and metallurgical furnaces.

Variations of this design are found in precipitator installations in pulp and paper mills for collecting salt cake dust from recovery boilers. These precipitators may be of the wet-bottom type, in which black liquor from the pulping operations is circulated through the precipitator bottoms. The salt cake particles rapped from the precipitator plates fall into the liquor, in which they are dissolved. Such installations were common until 1970, when emphasis on odor-free pulp mills caused a shift from wet-bottom construction to dry-bottom precipitators with scraper chain removal of

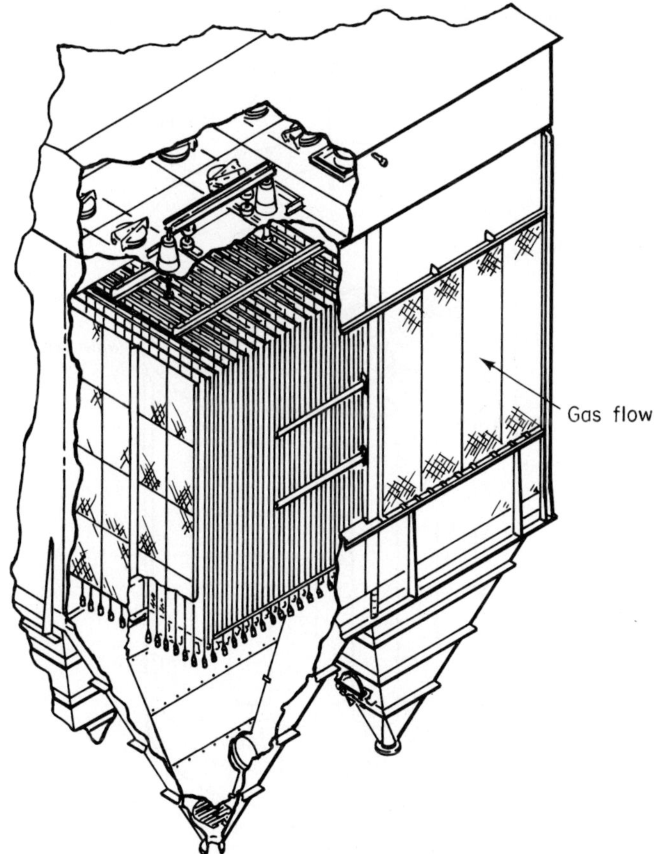


Figure 29. Parallel-plate electrostatic precipitator with pyramidal hoppers.

the collected salt cake. Earlier recovery boiler precipitators were of tile shell construction. However in the 1960's, there was a shift to conventional steel shell construction.

A second variation of precipitator construction is the wet-wall type. Blast furnace precipitators, which were first installed around 1930, were of the wet-wall type with tubular collection electrodes. The upper ends of the tubes form weirs, and water flows over the tube ends to irrigate the collection surface (Fig. 30).

Figure 31 shows an alternative design of wet precipitator with plate-type electrodes. In this design, sprays located in the ducts formed by adjacent collection electrodes serve to irrigate the plates. These are often supplemented by overhead sprays to ensure that the entire plate surface is irrigated. Design of such precipitators is similar to conventional construction except for the means for keeping insulators dry, measures to minimize corrosion, and provisions for removing the slurry.

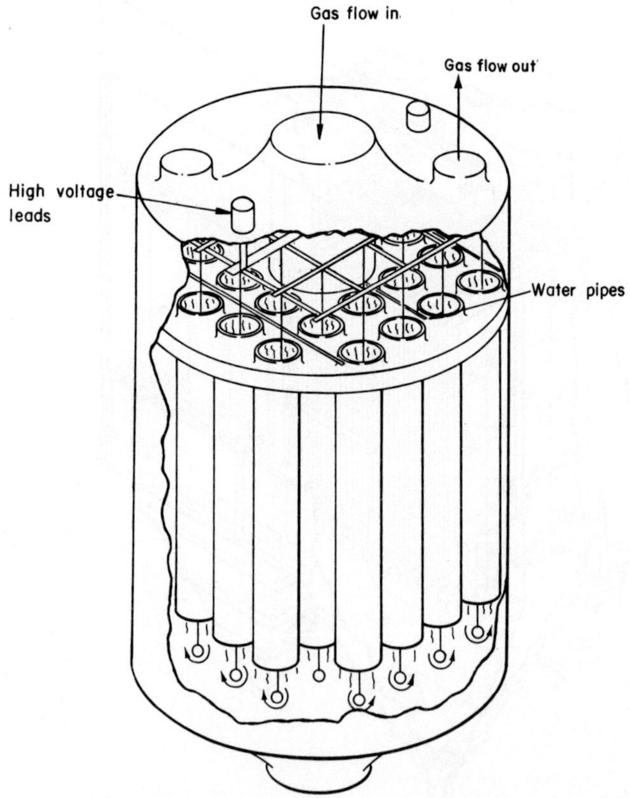


Figure 30. Wet-wall electrostatic precipitator with tubular collection electrodes.

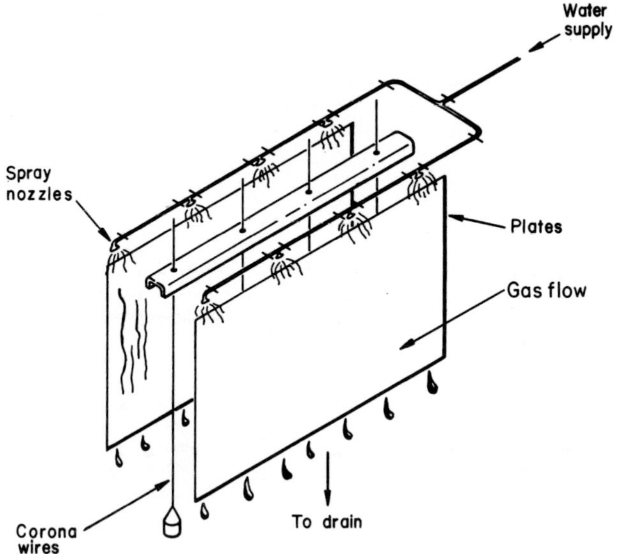


Figure 31. Wet electrostatic precipitator with plate collection electrodes.

Other types of shell construction have been used in specialized applications, such as removing tar from coke oven gas and high pressure gas cleaning.

In addition to differences in overall precipitator configuration, there are rather substantial differences in the electrodes and rappers used. Collection electrodes are generally constructed of 18 to 20 gauge mild sheet steel. Plates are formed or fabricated to provide stiffness for the sheet and to shield the collected dust against reentrainment, especially during rapping. Means of providing this stiffening and shielding vary with the manufacturer and range from roll-formed sheets with offset pockets to various shapes that protrude into the gas stream. Figure 32 illustrates some of the more common designs of collection plates.

Discharge electrode design is basically one of two types. The most prevalent type used by suppliers in the United States is the weighted-wire design illustrated in Figure 33. Figure 34 shows a supported electrode structure in which the active discharge electrodes are rigidly supported in frames or mastlike structures. The length of the active electrode generally is about 2 to 8 ft. Variations in these basic designs include a spiral wire electrode, in which the spiral acts as a spring to maintain electrode tension.

The shape of the active electrode differs with manufacturers and applications. It may be made of round, square or barbed wire, stamped sheets

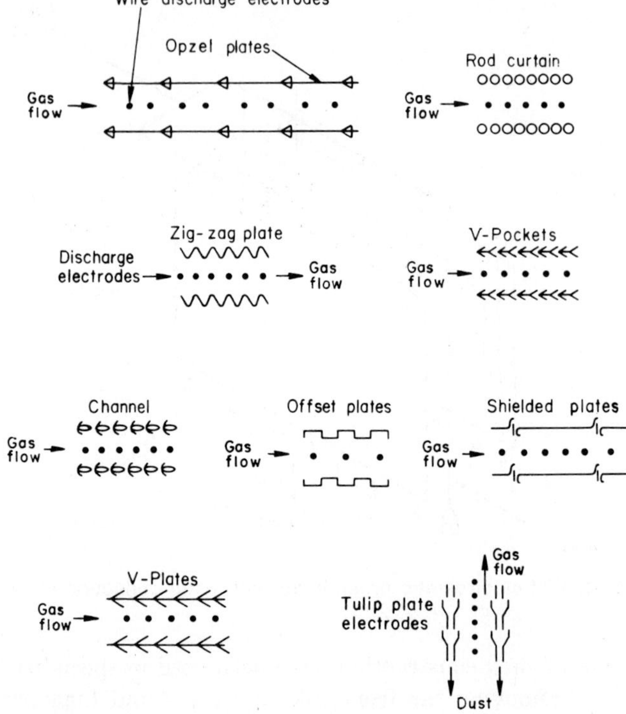


Figure 32. Various designs of collection electrodes.

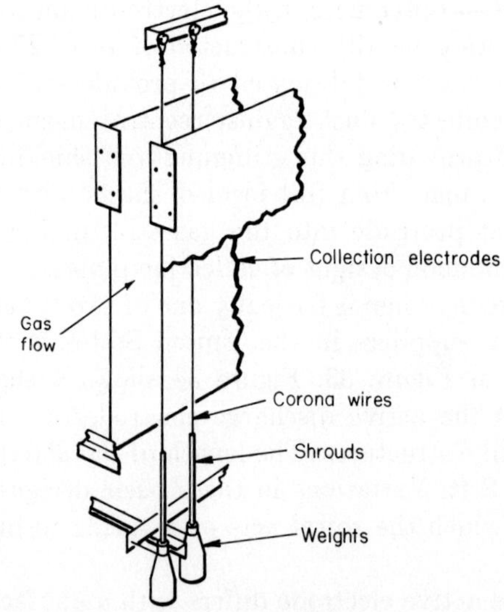


Figure 33. Weighted-wire discharge electrodes.

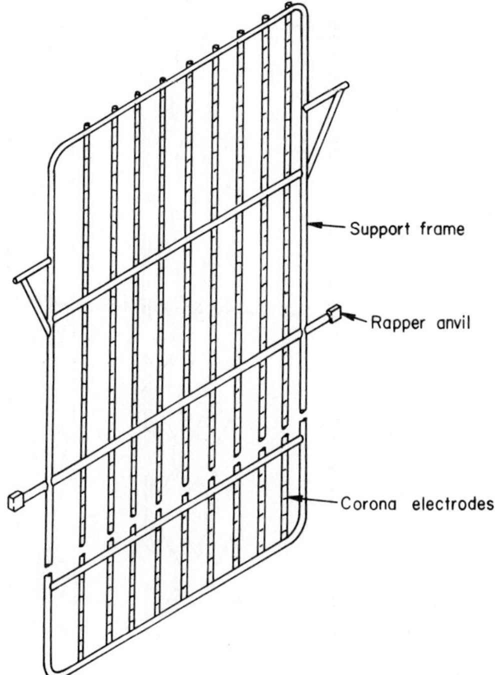


Figure 34. Supported electrode structure.

with various protrusions, or other forms. Figure 35 shows some typical electrode configurations. Discharge electrode design can be altered to give a desired voltage–current relationship by changing the sharpness of the emitting surface or radius of curvature of the wire. This permits some variations in the current to compensate for conditions of corona suppression due to the presence of fine particles.

Advantages are claimed for both the frame-supported and the weighted-wire electrode designs. It is argued that the frame-supported electrode structure gives more positive alignment, heavier sections to combat spark erosion, and less probability of a failure shorting out an entire field. Wire- and weight-type electrodes, on the other hand, are considered to be more easily removed without the necessity for shutdown, easier to locate once a failure occurs, and more economical to construct. Both types of electrodes have been used successfully in a variety of applications, although electrode failure still constitutes a major maintenance problem in precipitators.

Another variation in the design of the mechanical components concerns the method of rapping. Rappers can be of the tumbling hammer type,

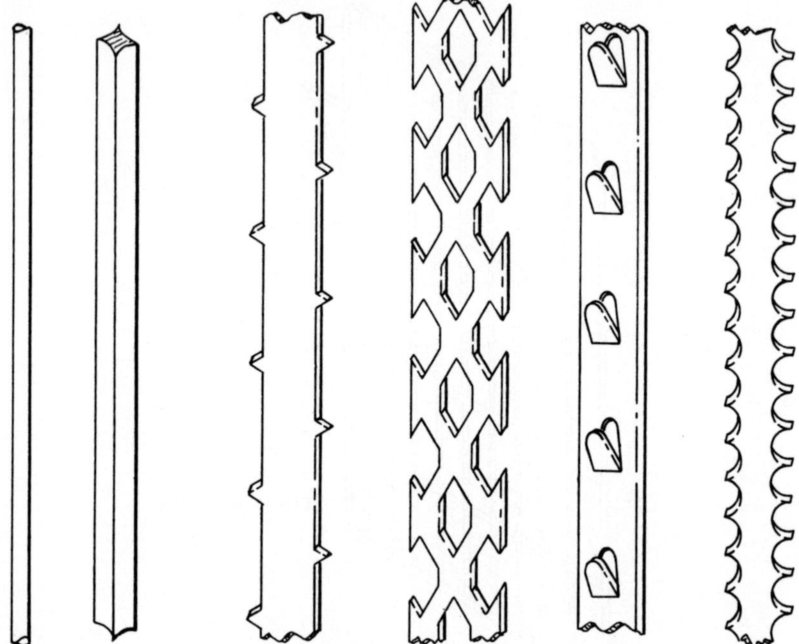

Figure 35. Typical forms of discharge electrodes.

in which a single shaft activates all the rappers in a given field. The hammers strike an anvil located on each collection plate or discharge electrode frame. The hammers can be physically located within the precipitator shell, or they can be located external to the shell with the impact being transmitted to the plates through rods extending through the shell wall, or through support insulators in the case of the discharge electrodes.

Rappers of the electromagnetic or pneumatic type are also used. These rappers are generally located on the precipitator roof and are coupled to the wire or plate frames through rods extending through the shell. In the case of discharge electrodes, rapping is coupled through insulators. Rapping can be accomplished by vibrating or by impact, the latter involving lifting a weight to a predetermined height and releasing it periodically so that it falls onto an anvil attached to the electrodes. Both vibrating and impact rappers of the pneumatic or electromagnetic type can usually be adjusted to change the intensity of each rap by changing the height to which the weights are lifted. Tumbing hammer rappers are usually not field adjustable.

Rapping frequency or interval is generally adjustable in all types of rapping equipment. Cycling times range from a few minutes to several hours, depending upon the type of dust and the rapping technique. Various manufacturers employ different techniques of rapping. Some believe a less frequent but more intense rap is better, whereas others attempt to optimize rapping intensity to minimize visible rapping puffs and to reduce losses during rapping to a minimum.

### *B. Erection*

Proper erection of a precipitator is of fundamental importance if high efficiencies are to be achieved and maintained. The primary concern is to keep good electrode alignment and constant spacing between collection and discharge electrodes. Variation in electrode spacing causes localized sparking at the point of closest spacing and thus limits the operating current and voltage of an entire field to that corresponding to the most closely spaced electrode, potentially causing an excessive rate of discharge electrode failure.

Poor control of electrode spacing can result from inadequate foundations which permit distortion of the entire shell, or from poor alignment of the electrodes during erection. Good practice calls for alignment to be held to within about $\frac{1}{4}$ in. or less, for 9- or 10-in. ducts.

The effects of poor electrode alignment can be seen from Figure 36, which shows the voltage that can be maintained with various degrees of electrode misalignment (*2*). Electrode misalignment is most often indicated by localized heavy sparking and wire failure or by visual inspection of the electrodes.

Mechanical difficulties other than alignment can also adversely influ-

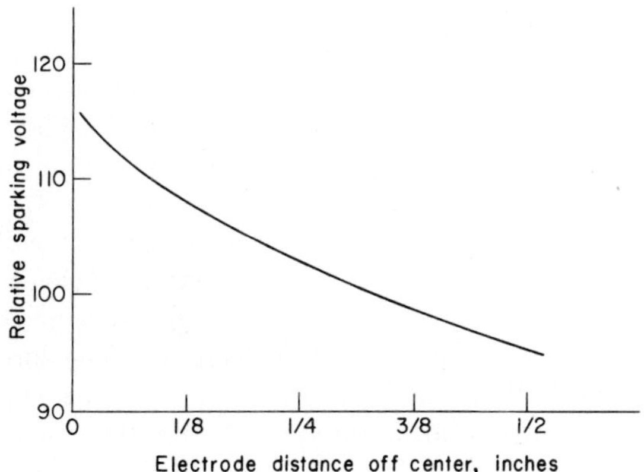


Figure 36. Effect of electrode misalignment on sparking voltage for pilot precipitator with 6-in. plate-to-plate spacing (2).

ence precipitator performance. Wire- and weight-type electrodes can be subject to resonant oscillations which can cause variations in electrode spacing (*24*). Such conditions are abnormal and can be corrected by changing weight or wire, or by other means, once the condition is identified.

### C. Sizing

Precipitators are generally purchased on the basis of a specified collection efficiency, exit dust loading, stack plume opacity, or all three. Specifications generally include the gas volume to be treated, gas composition, temperature, and information relative to the dust properties or type of fuel burned. To meet these conditions, precipitators must be sized to provide sufficient collecting surface area.

Precipitators have historically been sized by analogy with installations of a similar type. The basis for design is the exponential relationship among gas volume, plate area, and efficiency as given by the Deutsch–Anderson equation:

$$\eta = 1 - \exp[-(A/Q)w_p] \tag{28}$$

The factor $w_p$ used in this manner is an empirical parameter and includes effects or rapping losses, gas flow distribution, particle-size distribution, and dust resistivity and is properly termed a precipitation rate parameter to distinguish it from the theoretically developed $w$ in the Deutsch equation. The exponential relationship has been used in various forms for precipitator sizing. In spite of its various shortcomings, the equation has served as a useful tool in sizing so long as conditions remain about the same for comparable installations.

The method used in sizing of precipitators is to select a value of the effective migration velocity ($w_p$) based on experience of the manufacturer or user. For a specified efficiency and gas volume, the plate area can be computed. The values of $w_p$ depend upon many factors and can vary over a considerable range for a given application. Figure 37 shows typical values of specific collecting surface area (the ratio of collecting plate area to gas volume) required for 99% collection efficiency for various applications. The general trend is toward a higher specific collecting surface area as the particle size of the dust decreases. The spread in values shows the variations that can be expected for a given application and reflects the range of precipitation rate parameters. Because of this spread, other relationships have been developed which help to narrow the uncertainty in design.

The principal factors influencing the value of $w_p$ are particle size, exit

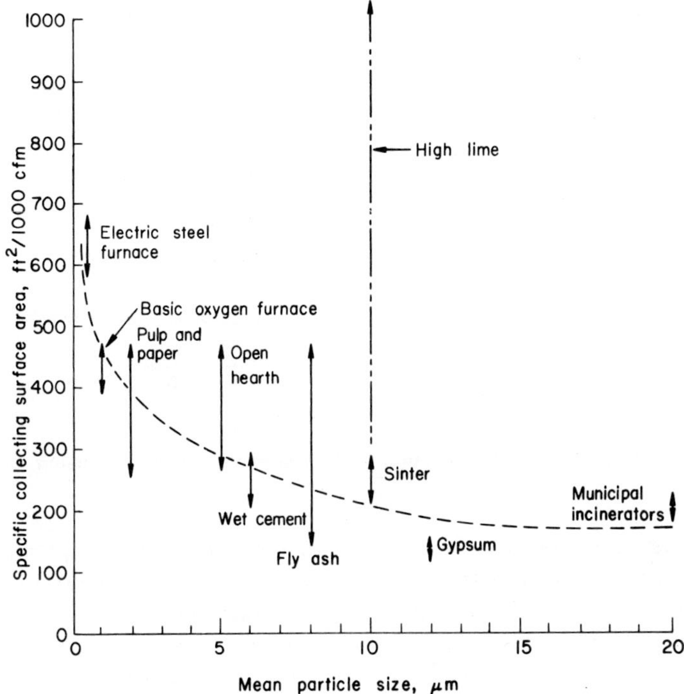


Figure 37. Variation in specific collecting surface area with dust particle size.

dust loading or efficiency, electrical resistivity of the dust, and tendency toward dust reentrainment. Efficiency plays an important role in determining the value of $w_p$ and, hence, the precipitator size. A high value of the precipitation rate parameter can be used if a relatively low efficiency will suffice, since precipitators are more effective in removing large particles. If a higher efficiency is required, a larger percentage of the small particles must be captured, and, hence, a proportionately larger plate area is needed. This requires the selection of a lower value of $w_p$ than if design were based on lower efficiency.

A second major factor in selecting the precipitation rate parameter is the electrical resistivity of the dust. If the resistivity is high, the allowable current density will be reduced, resulting in lower electric fields, lower charge on the particle, and longer particle charging times.

Figure 38 shows how the precipitation rate parameter might change with efficiency for a typical fly ash from a pulverized fuel electric power boiler. The figure shows the variation for two values of the product of charging and collecting fields ($E_c E_p$), which in turn are functions of the current density.

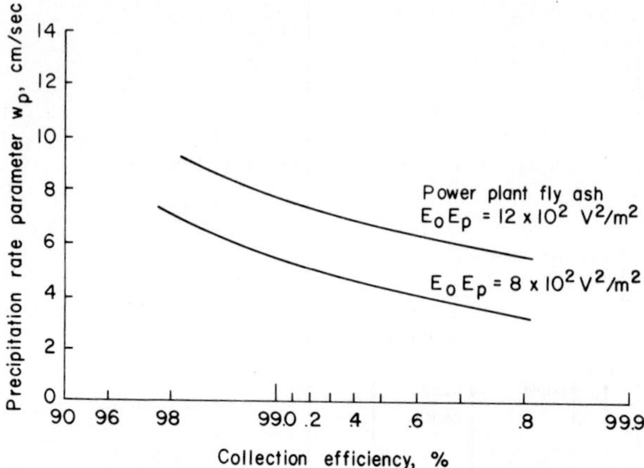


Figure 38. Decrease in precipitation rate parameter with increasing collection efficiency.

Various attempts have been made to empirically relate precipitation rate parameter to properties of the dust or fuel in an effort to narrow design requirements. Figure 39 shows the relationship between the design precipitation rate parameter $w_p$ and electrical resistivity of fly ash. These relationships give a reasonable basis for selecting $w_p$ for a given efficiency range. Ramsdell (*25*) has developed a series of curves based on precipi-

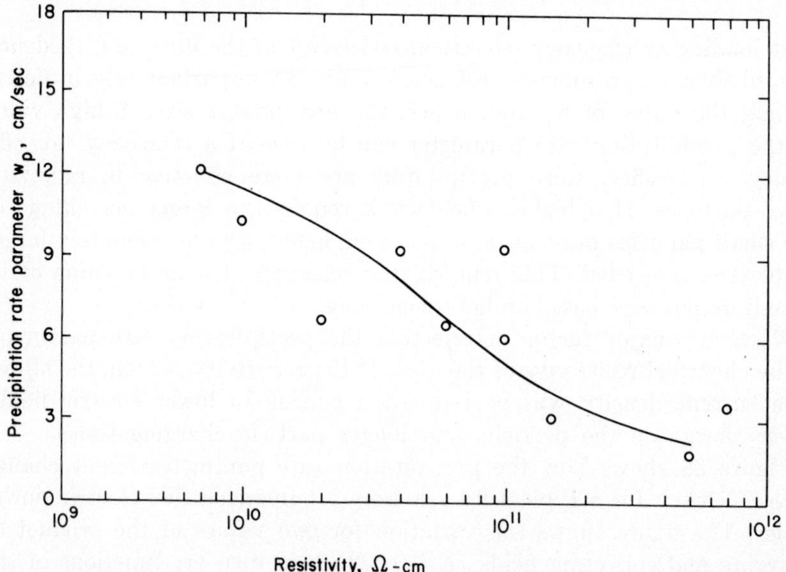


Figure 39. Variation of precipitation rate parameter with fly ash resistivity.

tator experience at the Consolidated Edison Company of New York. These curves (Fig. 40) show the specific collecting surface area required to reach a given efficiency with sulfur content of the coal as a parameter. This relationship has provided a useful tool for sizing precipitators for the types of coals covered in the study. However, the western coals, which are typically low in sulfur and high in lime, do not follow these general relationships. Also, these relationships do not apply to very high efficiency precipitators because of the change in $w_p$ due to particle size. In the process of precipitator size selection, each manufacturer selects a value of $w_p$ that is believed to be attainable under the conditions of the specific application. The data from previous installations form the general basis for the selection of the specific value of $w_p$, and these data must be modified to meet unusual or different circumstances such as abnormally high resistivity or higher efficiency. These procedures have been successful in many types of applications; however, there have been many installations which have failed to meet design specifications because of inadequate knowledge of these factors that influence performance. With the requirement to meet increasingly severe air pollution regulations, designs have tended to be more conservative, with greater attention being given to process variables.

Another technique that has been used in precipitator sizing for new applications is to use a small-scale pilot precipitator. There are several

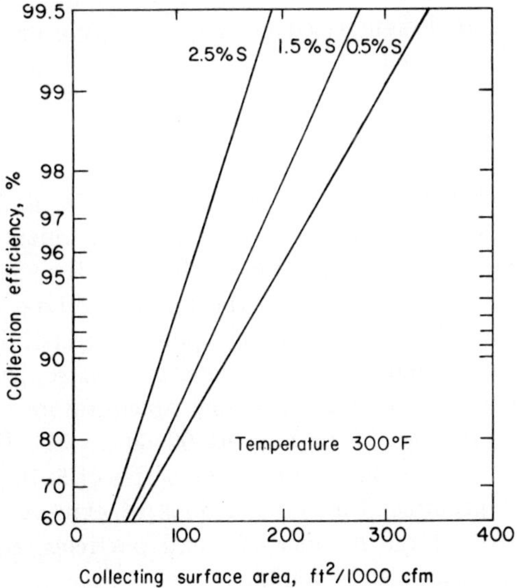


Figure 40. Collecting surface area versus collection efficiency for coals with various sulfur contents (*25*).

approaches to the use of pilot precipitators to derive design data. However, there are also several problems with the use of pilot precipitators, since they almost always perform better than full-size units. The primary reasons for this are that, in the pilot precipitator: (a) the gas flow distribution is almost always better, (b) the degree of sectionalization is better, and (c) better electrode alignment can be maintained. Because of better sectionalization and alignment, pilot precipitators can generally operate at much higher current densities and voltages than full-scale plants. Thus, if the pilot precipitator is operated spark limited, a scale factor must be applied to arrive at the full-scale precipitator size. This factor can often be quite large and can lead to uncertainties in the sizing of the full-scale plant.

An alternate approach is to set the current density on the pilot precipitator to that expected for the full-scale unit. This technique presupposes that the permissible operating current density is known. Since it is related to dust properties (resistivity and breakdown voltage of the dust layer), these properties must be known in order to have a basis for selecting the operating current density.

A computer model for determining precipitator performance based upon fundamental dust and gas properties and precipitator geometry has been developed (*26*). The model computes the performance based on known theoretical relationships and provides a basis for precipitator design and analysis. Further studies to upgrade and refine this model should provide a method of design based on a quantitative engineering rather than an empirical approach.

### D. Electrification

In addition to selecting precipitation rate parameter and required plate area, precipitator design includes selecting the energization equipment. Within the present state of technology, this selection is also based on empirical relationships developed by the manufacturers. Power requirements vary with application depending upon the resistivity of the dust, precipitator geometry, and extent of corona current suppression. Current densities for wet-wall precipitators or dry precipitators collecting material of low resistivity can be as high as 50–60 $\mu A/ft^2$. However, if the collected dust resistivity is high, current densities of 5–10 $\mu A/ft^2$ or lower may be found in practice. Under conditions of extreme corona current suppression due to a large number of small particles, current densities of about 1–2 $\mu A/ft^2$ may also be found.

The requirements for sectionalization are dependent upon a number of factors. The sparking conditions in a precipitator are related to the resistivity of the dust for high resistivity dusts or to the mechanical conditions within the precipitator for intermediate values of resistivity. The larger the collection electrode area served by a given power supply, the greater the probability of a sparkover. Thus, the correct choice of power-supply sectionalization is dependent upon economic factors; the trade-offs to consider are the additional costs associated with the large number of small power supplies in comparison with the higher probability of sparking with the large power supplies. The higher probability of sparking at a given voltage leads to a lower average value of voltage on each set for an installation with a small number of large power supplies. This lower average voltage will in turn require an increase in the total collection electrode area to attain a given collection efficiency. Manufacturers depend upon their past experience modified by their estimate of the specific conditions that are expected to prevail in the new installation to determine the proper sectionalization for a given application.

Another factor pertinent to the electrification of electrostatic precipitators is the selection of the voltage waveform to be used. The more common types of waveform are either full-wave or half-wave rectified sine wave, and recently there has been a renewed interest in pulse energization. Prior to the advent of fast response control systems with silicon controlled rectifiers (SCR), double half-wave energization was popular. The control elements for these systems were saturable reactors with reasonably slow time response. Double half-wave energization could provide increased sectionalization in that two electrical fields could be powered from a single supply.

The newer SCR controls operate better with full-wave electrification. The full-wave energization with the filtering action of the corona wire and collection electrode assembly works well with modern electronic equipment. The present trend in electrification follows this approach.

Pulse energization is a rather sharp departure from past practices in electrostatic precipitator energization. The philosophy of pulse energization was to provide a waveform that was conducive to spark suppression. It was expected that short-duration (0.1 msec) pulses with a repetition rate of 100/second would operate with a spark suppression time of 99% of the pulse period. Thus, higher average voltages would be expected to prevail.

These expectations must be modified in the light of the filtering action of the precipitator electrode system and the dust layer. Sharp rise times do occur, but the filter action of the precipitator tends to maintain the voltage at a sufficiently high value to preclude sparkover suppression.

White (*4*) made a study of the relationship between current density and precipitation rate parameter for a group of fly ash precipitators and

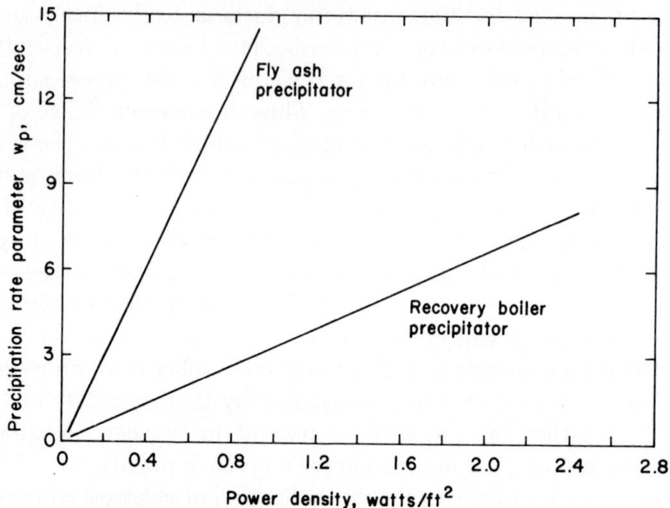


Figure 41. Variation of precipitation rate parameter with power density.

found reasonably good correlation, as indicated in Figure 41. This relationship generally holds for efficiencies in the 90–99% range. Curves such as these form the basis for selection of power supplies for a given application. Different curves apply to different processes because of variations due to particle size of the dust, reentrainment, etc.

Another design selection is the number of transformer-rectifier sets that will be used to supply the power. The trend both in the United States and in Europe has been toward larger power supplies with a lower degree of sectionalization. The trend is primarily dictated by the proportionately lower costs of the larger power supplies. However, there is considerable evidence to support the contention that larger power supplies with their inherently lower impedances cause poorer precipitator performance than that attainable with a larger number of smaller sets.

### E. Maintenance

Properly designed and installed precipitators should require a minimum of maintenance and upkeep. Except for rapping gear located within the precipitator housing, there are no moving parts to wear, and failures of a mechanical nature are generally not experienced. One of the goals in the design of precipitators is to ensure maintenance-free operation between major plant maintenance outages, which occur at 1- to 2-year intervals. Because of the expense associated with loss of production, excessive outages caused by malfunctioning of the dust control equipment cannot usually be tolerated.

Causes of failure of precipitator equipment can generally be traced to one of the following conditions: (a) wire breakage causing shorting of an electrical section, (b) excessive ash accumulation on discharge electrodes or collection plates or in flues, (c) failure of ash removal equipment to maintain acceptable ash level in the hoppers, (d) failure of the electric control circuits, and (e) rapper malfunctions.

Wire breakage or failure can result from several conditions. Causes of wire failure have been studied extensively by Detroit Edison Co. (*24*). Failure may occur due to excessive localized sparking, which causes erosion, overheating, and subsequent failure. If there is an area of high localized electric field, sparking will be concentrated at that location. Successive sparking over a period of time can erode the electrode leading to ultimate wire failure. This type of failure is often associated with inadequate shielding of the collection plate edge. Under these conditions, wire failures occur at distances along the wire corresponding to the distance between the top of the wire and the edge of the collection plate. Wires failed in this manner generally are eroded on one side, with the length of the eroded area extending only a few inches. The cross section of the wire on the side away from that where sparking occurs is generally round with the same radius as the original wire.

A second type of electrical failure results from wire swinging. In such cases, wires move in a circular motion and sparking occurs when the wires move near the collection plate. The appearance of wires that fail in this manner differs from those from other electrical failure modes in that the eroded area covers a longer length of wire and the cross section of the wire is circular, with a gradual decrease in area from the unaffected zone to the failed area.

Mechanical failure generally occurs near the wire supports and is due to excessive vibrations, improper design of the attachment fixture, or both. Mechanical failures are caused by fatigue of the wire in areas of high stress that is of a cyclic nature due to wire swinging. Such failures are easily identifiable by conventional fractographic analysis and are usually plain fractures across the wire with no evidence of necking or erosion.

Corrosion of electrodes can be due to deposits of acid in areas where the temperature falls below the acid dew point or galvanic corrosion due to the presence of dissimilar materials in a corrosive environment. Corrosion failures can be identified metallographically by the appearance of the grain boundaries and by the presence of corrosion products.

Electrode failures of the types described can be avoided by proper attention to detail in the design, erection, and operation of precipitators. Localized electrical failures can generally be traced to the presence of welds, sharp edges on collection electrodes, or similar factors that give rise to localized high electric fields. These conditions can be avoided by proper design, although corrections often can be difficult and expensive.

Wire swinging problems which are responsible for some types of electrical and mechanical failures can be avoided by maintenance of uniform gas flow and by design of the wire and weight system to avoid the possibility of resonant vibrations. If this condition occurs, it can be corrected by changing the weights, changing the type of electrode wire material to change the elastic modulus, providing additional restraints on the weight movement, or a combination of these. Again, experience together with adequate engineering analysis can prevent this condition from developing into a problem.

Corrosion problems are usually associated with the presence of acids in the flue gases and can be avoided by proper insulation of the portions exposed to the atmosphere and by elimination of air inleakage. Corrosion can occur if precipitator internals are cycled over a considerable temperature range for an extended period of time.

Ash accumulations on electrodes can be due to several factors. Very high resistivity dust can adhere tenaciously to collection surfaces resulting in the requirement of severe rapping acceleration. Boiler tube leaks, presence of high concentration of $SO_3$ coupled with low temperature operation, or other conditions that result in a sticky ash can usually be found to be the cause of excessive dust buildup. Deposits on discharge wires can take several forms, the most common being rings or doughnuts around the wires. These can generally be traced to inadequate transmission of the rapping impact or vibration to the wires, to inadequate gas distribution, or to an especially sticky or adherent ash.

Deposits of ash on the collection electrodes of $\frac{1}{4}$ to $\frac{1}{2}$ in. are not uncommon and are required for proper rapping conditions. Clean plates are neither desirable nor attainable in practice.

Discharge electrodes, on the other hand, should be kept relatively clean and large doughnuts or other formations on the electrode surfaces should be removed by adequate rapping or vibration of the discharge electrodes.

Failure of the ash removal equipment to maintain a proper level of ash in the hoppers is a common cause of difficulty in precipitators. When ash is not removed at the rate it is collected, it can build up to such an extent that it bridges between discharge and collection electrodes at the bottom of the plates and shorts out one or more electrical sections. Under some conditions, the ash can fuse to hard glassy deposits that require shutdown of the process to remove.

Buildup of ash in the hoppers can result from bridging of the dust in the hoppers as well as from inadequate ash removal capacity. To flow properly, ash must be kept hot and dry. Hoppers should be designed with proper hopper angle to facilitate flow, and they should be insulated and heated or vibrated when necessary to prevent this type of difficulty.

Electric energization equipment on precipitators consists of transformer-rectifier sets with the necessary sensing and control equipment to keep the current and voltage at the optimum conditions for most efficient operation. Transformers are of the oil-filled type and the major consideration from a maintenance standpoint is to prevent moisture from entering the transformer case. Transformers are equipped with dryers to remove moisture from the air, and these must be properly serviced and maintained.

Control elements vary depending upon manufacturer and can be of the saturable reactor or silicon types. These are normally trouble-free but must be kept reasonably clean and cool. Servicing should be done by properly trained personnel familiar with high voltage equipment.

Alignment of discharge and collection electrodes is an essential requirement for good precipitator performance. Poor alignment can be caused by faulty erection practice or by a shift in the precipitator foundations. Electrode misalignment can best be identified during inspection of the precipitator internals during an outage and corrected at that time. Correction of poor alignment requires some judgment as to the cause and can take the form of shims, braces, or other means to ensure proper spacing.

Also during inspection, conditions of plate hangers or supports, conditions of the electrodes, insulators, and other internal parts should be determined and repairs made at that time.

Rapper failure can occur in many forms. Rappers located within the shell are of the tumbling hammer type. Failures can occur in the form of bearing wear or seizure, breakage of the anvils, breakage of insulators on the drive shaft, or similar problems. Bearings in this type of service are generally cast-iron pillow block types and with time will wear to the point of permitting the shaft to be out of line. This condition can change the position of impact of the hammer on the anvil and alter the rapping intensity or cause the anvil support to fail.

Inspection of the rappers should be a part of routine maintenance when the precipitator is out of service, and repairs and replacements should be made to extend the life of the rappers until the next scheduled outage.

Rappers located external to the shell can be of the tumbling hammer type, electromagnetic or pneumatic impact type, or vibrating type. In these types of rappers, the shaft or rapper rods extend through the shell, and the area near the shell must be kept clear of ash deposits which would interfere with rapper function. Potential sources of failure of these types of rappers are the seals through the precipitator shell, electrical or pneumatic controls, bearings, and insulators. Inspection of these should form a part of the routine precipitator maintenance program.

Since each precipitator design varies, no general maintenance program can be written to cover all types. However, with the assistance of the precipitator supplier, regular inspection and maintenance procedures should be established to ensure proper operation over extended periods of time.

### F. Troubleshooting

When precipitators fail to achieve their design efficiencies, the difficulties can be traced to one or more causes. If the precipitator fails to meet specifications on start-up, the difficulties can be due to inability to achieve the desired electrical conditions, inadequate plate area, or poor gas distribution. Inadequate electrical energization can be due to high dust resistivity, poor electrode alignment, or current suppression due to large quantities of small particles. Resistivity measurements should show whether this condition is limiting performance. For moderately high resistivity, low currents accompany low voltages and limitations of both current and voltage are due to excessive sparking. For very high resistivities, back corona can occur prior to sparking and, under this condition, currents can be relatively high at low voltages. Secondary current–voltage curves are generally useful in identifying this condition.

Current suppression due to small particles is generally very pronounced in the first precipitation field. This condition is not abnormal, since the current is being used to charge the fine particulate and the low current is due to the decreased mobility of the charge carriers. However, charging times are longer under these conditions and additional precipitator length is required for proper collection.

Electrode misalignment can cause localized sparking at low current and voltage and can limit operating conditions. This condition can best be detected by physical inspection.

Poor gas distribution can cause excessive reentrainment of the dust in addition to variations in treatment time. Measurements of gas flow distribution are the most direct indication of this type of difficulty. However, observation of the hopper area and the outlet field can usually detect the presence of heavy dust reentrainment either by hopper sweepage or by scouring of the plates.

Since larger particles are preferentially reentrained, abnormal size distribution of the exit dust can be an indication of reentrainment due to poor gas flow quality.

If gas flow distribution is adequate and electrical conditions are normal, poor performance is perhaps due to inadequate precipitator size. If the inlet dust size is smaller than that considered in the design, the precipitator will be undersized for a given application. If it is determined that this is the case, the corrective solution is addition of more precipitator collecting surface.

When precipitators which have previously functioned satisfactorily fail to achieve performance levels, problems can generally be traced to changes in dust character due to fuel or process changes, deterioration of the mechanical condition of the precipitator such as electrode alignment or rapper wear, or ash accumulation in ducts and hoppers or on gas flow control devices. The same procedures for isolating the causes of the malfunction apply as for analyzing a new installation. These consist of determining whether the problem is electrical or mechanical in nature and identifying the specific factors responsible for the malfunction.

## REFERENCES

1. F. G. Cottrell, *J. Ind. Eng. Chem.* **3,** 542 (1911).
2. P. Cooperman, "The Dependence of the Electrical Characteristics of Duct Precipitators on their Geometry," Progr. Rep. No. 46. Research Corp., Bound Brook, New Jersey, 1952.
3. F. W. Peek, Jr., "Dielectric Phenomena in High Voltage Engineering." McGraw-Hill, New York, New York, 1929.
4. H. J. White, "Industrial Electrostatic Precipitation." Addison-Wesley, Reading, Massachusetts, 1963.
5. S. C. Brown and J. C. Ingraham, *in* "Handbook of Physics" (E. U. Condon and H. Odeshaw, eds.), pp. 4-180. McGraw-Hill, New York, New York, 1967.
6. L. B. Loeb, "Fundamental Processes of Electrical Discharge in Gases," p. 267. Wiley, New York, New York, 1939.
7. C. C. Shale, W. S. Bowie, J. H. Holden, and G. R. Strimbeck, *U.S., Bur. Mines, Rep. Invest.* **6325** (1963).
8. B. Y. H. Liu and H. C. Yeh, *J. Appl. Phys.* **39,** 1396 (1968).
9. G. W. Hewitt, *Trans. Amer. Inst. Elec. Eng. Part 1,* **76,** 300 (1957).
10. E. Anderson, Report, Western Precipitator Co., Los Angeles, California, 1919; see *Trans. Amer. Inst. Chem. Eng.* **16,** 69 (1924).
11. W. Deutsch, *Ann. Phys.* (*Leipzig*) [4] **68,** 335 (1922).
12. W. T. Sproull, *J. Air Pollut. Contr. Ass.* **22,** 181 (1972).
13. G. W. Penney and E. H. Klingler, *Trans. Amer. Inst. Elec. Eng. Part 1* **81,** 200 (1962).
14. R. E. Bickelhaupt, *J. Air Pollut. Contr. Ass.* **24,** 251 (1974).
15. K. J. McLean, "Electrical Conduction in High Resistivity Particulate Solids."

Ph.D. Thesis, University of New South Wales, Kensington, New South Wales, Australia, 1969.
16. H. J. Lowe, J. Dalmon, and E. T. Hignett, "Colloquium on Electrostatic Precipitators." Inst. Elec. Eng., London, England, 1965.
17. D. T. Berube, *Proc., Electrostat. Precipitator Symp., Birmingham, Alabama, 1971,* p. 223.
18. J. Dalmon and D. Tidy, *Atmos. Environ.* **6,** 81 (1972).
19. J. T. Reese and J. Greco, *J. Air Pollut. Contr. Ass.* **18,** 523 (1968).
20. K. S. Watson and K. J. Blecher, *Int. J. Air Water Pollut.* **10,** 573 (1966).
21. "Determining the Properties of Fine Particulate Matter," Power Test Code 28. American Society of Mechanical Engineers, New York, New York, 1965.
22. O. J. Tassicker, Z. Herceg, and K. J. McLean, *Trans. Inst. Eng., Aust.* **5** (2), 277 (1969).
23. L. Cohen and R. W. Dickinson, *J. Sci. Instrum.* **40,** 72 (1963).
24. J. H. Casiglia and H. R. Fletcher, "Unstable Vibration and Power Arc Induced Failures in Electrostatic Precipitator Discharge Electrode Wires," Report. Detroit Edison Co., Detroit, Michigan (n.d.).
25. R. G. Ramsdell, *Amer. Power Conf., Chicago, Illinois, 1968.*
26. S. Oglesby and G. B. Nichols, Rep. Nos. PB-196379, PB-196380, PB-196381. Nat. Tech. Inform. Serv. U.S. Dept. of Commerce, Springfield, Virginia, 1970.

# 6

## Scrubbing

Seymour Calvert



### Nomenclature

$a$ Transfer area per unit volume of scrubber ($cm^2/cm^3$)
$A$ Cross-section area ($ft^2$ or $cm^2$) or a constant in Equation (24)
$A_d$ Total collection surface area in scrubber ($cm^2$)

| | |
|---|---|
| $B$ | A constant in Equation (24) |
| $B_p$ | Packing factor ($cm^2/cm^3$) |
| $C'$ | Cunningham "slip" correction factor (dimensionless) |
| $C_{Ai}$ | Concentration of diffusing component A at the interface (same units as $q$) |
| $d_b$ | Bubble diameter (cm) |
| $d_c$ | Packing diameter (nominal) (cm) |
| $d_d$ | Drop diameter (cm) |
| $d_h$ | Sieve plate hole diameter (cm) |
| $d_p$ | Particle diameter ($\mu$m or cm) |
| $d_{p50}$ | Diameter of particle collected with 50% efficiency (aerodynamic) ($\mu$mA) |
| $d_{pa}$ | Aerodynamic particle diameter ($\mu$mA) |
| $d_{PC}$ | Performance cut diameter (aerodynamic) ($\mu$mA) |
| $d_{pg}$ | Geometric mean particle diameter (aerodynamic) ($\mu$mA) |
| $d_{RC}$ | Required separation cut diameter (aerodynamic) ($\mu$mA) |
| $D$ | Diffusivity of transferring compound ($cm^2/sec$): $D_l$, liquid; $D_g$, gas; $D_A$, component A; $D_B$, component B |
| $D_p$ | Particle diffusivity ($cm^2/sec$) |
| $E$ | Efficiency (fraction or %) |
| $f$ | Fractional approach to equilibrium in mass transfer: $f_p$, cocurrent; $f_e$, countercurrent |
| $f'$ | Empirical constant for sprays (dimensionless) |
| $F$ | Foam density ($g/cm^3$) |
| $F_c$ | Overall efficiency for countercurrent operation |
| $F_m$ | Overall efficiency of mass transfer |
| $F_p$ | Overall efficiency for cocurrent (parallel) operation |
| $g_c$ | Gravitational acceleration = 980 ($cm/sec^2$) |
| $G$ | Gas flow rate ($moles/sec\text{-}cm^2$) |
| $h$ | Height of scrubber (cm) |
| $H$ | Henry's law constant (atm/mole fraction) |
| $H'$ | Henry's law constant ($atm/mole/cm^3$) |
| $k$ | Mass transfer coefficient ($gm\ moles/sec\text{-}cm^2\text{-}atm$) |
| $k_g$ | Mass transfer coefficient based on driving force between the gas and the interface ($gm\ mole/sec\text{-}cm^2\text{-}atm$) |
| $k_l$ | Mass transfer coefficient based on driving force between the liquid and the interface ($gm\ mole/sec\text{-}cm^2\text{-}gm\ mole/cm^3$) |
| $k^*$ | Mass transfer coefficient with chemical reaction |
| $K_{og}$ | Mass transfer coefficient based on overall gas phase driving force ($gm\ moles/sec\text{-}cm^2\text{-}atm$) |
| $K_{ol}$ | Mass transfer coefficient based on overall liquid phase driving force ($gm\ mole/sec\text{-}cm^2\text{-}gm\ mole/cm^3$) |
| $K_p$ | Inertial impaction parameter (dimensionless) |
| $K_{pt}$ | In throat of venturi |
| $L$ | Liquid flow rate ($gm\ moles/sec\text{-}cm^2$) |
| $m$ | Phase equilibrium constant = $y/x$ at equilibrium |
| $M$ | Mass (gm) |
| $N$ | Mass transfer flux ($mole/cm^2\text{-}sec$) |
| $N_{OG}$ | Number of mass transfer units—overall gas phase driving force |
| $p$ | Partial pressure of compound (atm): $p_i$, inlet; $p_o$, outlet |
| $p^*$ | Partial pressure in equilibrium with liquid (atm) |
| $P$ | Total pressure (atm) |
| $P_A$ | Partial pressure of compound A (atm) |

| | |
|---|---|
| $Pt$ | Penetration = 1 − $E$ (fraction) |
| $\overline{Pt}$ | Average (integrated over particle-size distribution) penetration (fraction or %) |
| $\Delta P$ | Pressure drop (cm W.C., water column, or atm) |
| $\Delta P_a$ | Pressure difference due to acceleration (gm/cm²) |
| $\Delta P_a''$ | Pressure drop (cm W.C.) as defined by Equation (23) |
| $q$ | Concentation of reactive component in the bulk liquid |
| $q'$ | Mass of water condensed/mass dry gas (gm/gm) |
| $Q_G$ | Gas volumetric flow rate (m³/sec) |
| $Q_L$ | Liquid volume flow (m³/sec or liters/sec) |
| $r_d$ | Radius of drop (ft or cm) |
| $R$ | Gas constant = 82.057 (atm-cm²/mole-°K) |
| $t$ | Time (seconds) |
| $t_g$ | Contact time for gas (seconds) |
| $t_l$ | Contact time for liquid (seconds) |
| $T$ | Absolute temperature (°K) |
| $u_{BD}$ | Particle deposition velocity for Brownian diffusion (cm/sec) |
| $u_G$ | Gas velocity relative to duct (cm/sec) |
| $u_h$ | Gas velocity through hole (cm/sec) |
| $u_{PD}$ | Particle deposition velocity (cm/sec) |
| $u_r$ | Drop velocity relative to gas (cm/sec) |
| $u_t$ | Terminal settling velocity (cm/sec) |
| $v_G$ | Gas velocity relative to duct (cm/sec) |
| $v_L$ | Liquid velocity relative to duct (cm/sec) |
| $x$ | Mole fraction of compound in the liquid phase: $x_i$ inlet; $x_o$, outlet |
| $x_A$ | Mole fraction of compound A in the liquid |
| $x^*$ | Mole fraction of compound in the liquid phase at equilibrium with the gas contacting it |
| $y$ | Mole fraction of compound in the gas phase: $y_i$ inlet; $y_o$ outlet |
| $y^*$ | Mole fraction of compound in the gas phase at equilibrium with liquid contacting it |
| $Z$ | Height or length of scrubber (cm) |

### Greek

| | |
|---|---|
| $\epsilon$ | Fraction void volume space |
| $\rho$ | Density (gm/cm³): $\rho_g$, gas; $\rho_l$, liquid; $\rho_p$, particle |
| $\rho_M$ | Molar density (gm moles/cm³) |
| $\pi$ | 3.14159 |
| $\sigma$ | Surface tension (dyne/cm) |
| $\sigma_g$ | Geometric standard deviation of particle-size distribution |
| $\mu$ | Viscosity (poises) |
| μm | Micrometers (μm) = $10^{-4}$ cm |
| μmA | Aerodynamic size [μm(gm/cm³)$^{1/2}$] |
| $\eta$ | Collection efficiency |

## I. Introduction

Source control by liquid scrubbing involves the removal of contaminants in either vapor or particulate state from an effluent gas stream by means of a liquid. The transfer of contaminant requires the contacting of the gas and liquid and their subsequent separation into cleaned gas and contaminated liquid streams. For particulate matter, material transfer between the gas and liquid phases may be a variety of mechanisms. For gaseous molecules, it is basically by diffusion. The particulate collection mechanisms active in scrubbers involve inertial, gravitational, electrostatic, thermal, and diffusional phenomena. Gaseous mass transfer proceeds by diffusion, moving from a region of high concentration to one of low concentration.

While heat transfer may or may not be important in a given scrubber system, mass and momentum transfer always are. Mass transfer rate influences the efficiency of contaminant collection, while momentum transfer rate determines the amount of frictional pressure drop and, therefore, the power requirement. Efficiency is also dependent on the level of equilibrium between the gas and the liquid phases since this, by definition, describes the ultimate states of the two phases left in contact.

Capacity is generally determined by the flow rate at which the scrubber will become inoperable because of excessive or complete carry-over of liquid by gas, or at which pressure drop will become excessive. As a rule, the necessary height or length of the gas scrubbing path will increase with the efficiency required, and the necessary cross-sectional area will increase with the capacity required.

Background and detailed information on mass transfer is readily available *(1–7)* in the chemical engineering and related literature. The present volumes provide the necessary general background related to small particles and their separation from gases.

## II. Gas Absorption

### A. Equilibrium Considerations

The object of gas scrubbing for mass transfer is to get as much material as possible out of the gas and into the liquid at as low a cost as possible. There are, however, physical or chemical equilibrium limits to the solubility of material in liquid that cannot be exceeded regardless of how intimate the contacting of phases or how long the time of contact.

The solution of $SO_2$ in water is an example of physical equilibrium, even though there is chemical interaction with the water and some heat of solution. The chemical reactions involved are reversible and rapid, as compared to the rate of diffusion through the liquid. In contrast, the absorption of $SO_2$ in NaOH solution is a clear case of chemical reaction since the reaction is substantially irreversible and rapid at the temperature and pressure of the scrubber. Furthermore, the rate of absorption may be influenced by the chemical reaction rate or by the rate at which reactant (NaOH) can diffuse toward the interface to replace that which has been used up.

Gas solubilities in water can usually be described in terms of Henry's law

$$p_A = Hx_A \tag{1}$$

especially for the low partial pressures that are encountered in air pollution control equipment (Table I).

## B. Contacting Scheme

The amount of mass transfer depends not only on the equilibrium relationship but also on the contacting scheme. If liquid enters the top of a vertical column so that it runs down and gas enters the bottom so that it passes upward through the column, the contacting scheme is said to be countercurrent, and the effluent gas has its last contact with the entering liquid. Were the gas also to enter the top and pass downward through the column, the contact would be cocurrent and the effluent gas would contact last the effluent liquid.

A variant in the contacting scheme is when one stream flows in a direction perpendicular to the other stream. This is called cross, or crosscurrent, flow. Examples of this are spray chambers in which the gas travels horizontally while spray drops fall vertically, and trays across which liquid flows horizontally while gas flows vertically. These situations are complicated to analyze and are generally treated with the simplifying assumption that one phase is completely and instantaneously mixed along

**Table I Henry's Law Constants for Several Gases in Water**[a]

| | *Temperature (°C)* | | | | |
|---|---|---|---|---|---|
| *Gas* | *10* | *20* | *30* | *40* | *50* |
| CO | 44,000 | 53,600 | 62,000 | 69,000 | 75,000 |
| $O_2$ | 33,000 | 40,000 | 47,500 | 52,000 | 58,000 |
| NO | 22,000 | 26,400 | 31,000 | 35,000 | 39,000 |
| $CO_2$ | 1000 | 1450 | 1900 | 2300 | 2900 |
| $H_2S$ | 370 | 480 | 610 | 730 | 890 |
| $SO_2$ | 27 | 38 | 50 | 65 | 80 |

[a] $H$ = atm/mole fraction.

any plane parallel to the flow of the other phase. Thus in a cross-flow spray chamber one could assume that the gas is completely mixed along any vertical plane with the result that the gas composition is the same anywhere on that plane.

In a plate-type column the overall fluid motion is either co- or counter-current, but the motion on each plate is essentially crosscurrent. Combinations of these variations may occur in a single piece of equipment and any attempt to analyze carefully its performance requires that the liquid and gas flow paths be known and taken into account. Once the operating conditions are fixed, they, in conjunction with the equilibrium relationship, determine the ultimate performance possible.

For a cocurrent scrubber, no matter how long the phases are kept in contact or how efficient the device, the gas composition can never go below $mx_o$ nor the liquid composition go above $y_o/m$. In a countercurrent scrubber, more efficient contact is possible and the ultimate performance limitation is that the gas can never become leaner (or cleaner) than that in equilibrium with the inlet liquid for countercurrent flow, or with the outlet liquid for cocurrent flow. An alternative performance limitation for countercurrent contacting is when the outlet liquid is in equilibrium with the inlet gas.

### C. Approach to Equilibrium

The proper measure of scrubber performance is the degree to which the two streams approach the equilibrium limits noted above. Overall efficiency is the product of the fraction removed at equilibrium and the fractional approach to equilibrium. This equilibrium efficiency or fractional approach to equilibrium is defined in terms of the gas phase by

$$f = (y_i - y_o)/(y_i - mx_i) \qquad \text{for countercurrent operation} \qquad (2)$$

and

$$f = (y_i - y_o)/(y_i - mx_o) \qquad \text{for cocurrent operation} \qquad (3)$$

Overall efficiency is defined as

$$F_m = f\left(\frac{y_i - mx_i}{y_i}\right) = \frac{y_i - y_o}{y_i} \qquad \text{for countercurrent operation} \qquad (4)$$

and with $x_o$ replacing $x_i$ for cocurrent operation.

Mass transfer rate and contacting time determine equilibrium efficiency. Transfer rate may be described mathematically as

$$N = k_g(p - p^*) = k_gP(y - y^*) \qquad (5)$$

Equation (5) describes mass transfer in the gas phase in terms of the partial pressure driving force between the bulk of the gas phase and the liquid surface, and a coefficient which depends on how rapidly the transferring material diffuses through the gas phase. Mass transfer within the liquid phase is described by

$$N = k_l(c_l^* - c_l) \tag{6}$$

For low concentrations we may use the approximation that

$$c_l = x\rho_M \tag{7}$$

Thus

$$N = k_l\rho_M(x^* - x) \tag{8}$$

Equations (5) and (8) can be used to define mass transfer rates if $y^*$ or $x^*$ are known. These equilibrium values are generally assumed to be those which exist just at the gas–liquid interface.

In preference to using Equations (5) and (8), mass transfer relationships based on the overall driving force between the bulk phase compositions are used. The definitions for overall gas and liquid phase coefficients are

$$N = K_{og}(p - H'c_l) = K_{ol}(p/H' - c_l) \tag{9}$$

These overall coefficients are related to the individual phase coefficients by the following equations when the material balance and equilibrium relationships are linear.

$$1/K_{og} = 1/k_g + H'/k_l \tag{10}$$

and

$$1/K_{ol} = 1/k_l + 1/H'k_g \tag{11}$$

One or the other overall coefficient is used, depending on whether the absorption is gas phase or liquid phase resistance controlled.

Gas phase resistance controls when the gas is very soluble in the liquid. Liquid phase resistance controls when the gas is only slightly soluble in the liquid. It can be appreciated that when the capacity of the liquid is high, the burden is placed upon the gas phase to maintain the transfer rate because the liquid can carry material away from the interface with ease. When the solubility is low the gas phase can easily keep the interface nearly saturated, and the rate of transfer will depend on how rapidly the liquid can move material from the interface.

The significance of whether a gas or liquid phase controls stems from the fact that gas phase transfer is much more rapid than liquid phase transfer. The type of equipment used and its size will, therefore, depend upon this fact. A method for the estimation of which phase controls can be based on a rearrangement of Equation (10):

$$k_g/K_{og} = 1 + H'k_g/k_l \tag{12}$$

If $k_g$ is nearly equal to $K_{og}$, the transfer is gas phase controlled and a ratio of $k_g/K_{og} \leq 1.1$ indicates a clear-cut case of this condition. At the other extreme, a ratio of 10 or greater is the criterion for liquid phase control. The ratio of transfer coefficients ($k_g/k_l$) is relatively constant. Therefore most of the variation in ($H'k_g/k_l$) is due to variation of $H'$. If $H'$ is less than about 3.0, the system is gas phase controlled and, if it is greater than about 3000, the liquid phase controls. This range of values will compensate for the probable extremes of transfer coefficients and is not large in relation to the variation in gas solubilities. Thus to a first approximation, the efficiency of a gas absorber can be related to solubility alone. A better definition of efficiency, in terms of both solubility and the ratio of liquid to gas, is discussed later.

## 1. Prediction of Coefficients

Mass transfer coefficients have been predicted and correlated by a variety of approximate methods, among which the penetration theory (*8*) does as well as any other. It has the attractive virtues of permitting predictions to be made for new cases, and of providing a logical mechanistic model for conceptual purposes. This theory states that mass transfer from or to turbulent streams may be considered to be the consequence of many small elements of fluid transferring material to or from an interface for a short period of time. Before and after their time of contact with the interface, the fluid elements are thoroughly mixed with the main stream. Development of this idea with further assumptions leads to the following equations:

$$k_g' = (2/RT)(D_g/\pi t_g)^{1/2} \tag{13}$$

and

$$k_l' = 2\left(\frac{D_l}{\pi t_l}\right)^{1/2} \tag{14}$$

Many substances have diffusivities in water of about $1.5 \times 10^{-5}$ cm²/sec and in air of about 0.1 cm²/sec.

Taking a column packed with 2.5 cm (1 in.) Raschig rings as a typical device, the liquid and gas contact times are estimated as the time it takes either phase to move a distance of one packing diameter. Within the range of data given by Shulman (*9*) and at 50% of flooding, $t_l$ ranges from 0.525 to 0.18 second. The gas contact time might range from 0.01 to 0.05 second. Using these approximations, we estimate

$$k_g' = \left(\frac{2}{82.06 \times 298}\right)\left(\frac{0.1}{0.01\pi}\right)^{1/2} = 1.45 \times 10^{-4}$$

for a gas contact time of 0.01 second and $0.65 \times 10^{-4}$ for $t_g = 0.05$. Similarly, we can estimate that the liquid phase coefficient $k_l'$ would range from about $6.0 \times 10^{-3}$ to $1.0 \times 10^{-2}$. Most design calculations and performance reporting are not on the basis of coefficients, but rather in terms of "transfer units." Equation (15) defines an overall gas phase transfer unit for the dilute gas case:

$$N_{OG} = \int_{p_o}^{p} \frac{dp}{p - Hx} = \frac{K_{og}aZP}{G} \tag{15}$$

The integral equal to the number of transfer units is a measure of the difficulty of transfer and is approximately equal to the change in composition divided by the average driving force for transfer. More complex forms must be used when the concentration of the transferring gas is greater than 5 or 10% (*1*) to account for changes in total gas flow rate and the effect of diffusion of inert species toward the interface.

Integration of Equation (15) for constant $L/G$ and $H$, and cocurrent flow, gives the following expression for equilibrium efficiency:

$$f_p = 1 - \exp(-K_{og}azP/G)(1 + HG/PL) \tag{16}$$

Overall efficiency for cocurrent operation is

$$F_g = \frac{p_i - p_o}{p_i}\left(1 + \frac{HG}{PL}\right)^{-1} f_p \tag{17}$$

For countercurrent contact we have an implicit definition of equilibrium efficiency as

$$N_{OG} = \left(1 - \frac{HG}{PL}\right)^{-1} \ln\left[\left(1 - \frac{HG}{PL}\right)\left(\frac{1}{1 - f_c}\right) + \frac{HG}{PL}\right] \tag{18}$$

and the overall efficiency is

$$F_c = [(p_i - Hx_i)/p_i]f_c \tag{19}$$

The efficiencies which are defined in Equations (16)–(19) are based on changes in gas composition and are referred to as gas phase efficiencies. Figure 1 is a plot of $F_p$ and $f_c$ vs $HG/PL$ with $N_{OG}$ as the parameter. If we consider the case of countercurrent contact where the inlet liquid concentration is zero, then $F_c = f_c$, and Figure 1 shows the overall efficiencies for the two modes of contact. We can see that for low effi-

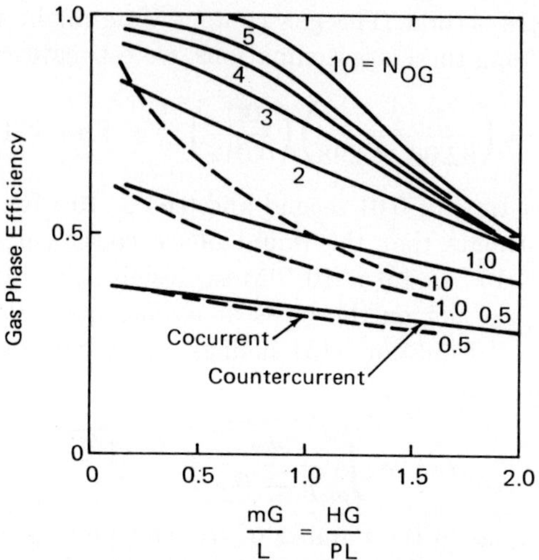


**Figure 1.** Relationship between mass transfer efficiency, transfer parameter, and number of transfer units for cocurrent and countercurrent flow.

ciency, such as 0.5 transfer units will provide, there is little advantage of counter over cocurrent. For more difficult separations, such as 10 transfer units can provide, countercurrent contact is superior.

An estimate of the column height required for a desired efficiency can quickly be made with the help of Figure 1. Economic operation is usually in the region of $(HG/PL)$ between 0.5 and 1.0 for countercurrent operation. Thus, if we need 95% efficiency it will take something like 5 to 10 transfer units to do the job, depending on the amount of liquid we use to scrub the required gas volume. Since the column height per transfer unit generally runs from 0.3 to 1.2 m (1 to 4 feet), with 0.6 m (2 feet) being a good average for approximation, we will need about 3 to 6 m (10 to 20 feet) of packed height. The exact choice will depend on the economic balance between cost of tower and higher gas pumping power for more transfer units and the higher liquid pumping and purification costs for fewer transfer units.

## 2. Chemical Reaction

Higher gas absorption efficiency and greater absorbing capacity of the liquid phase can result from the use of a chemically reactive liquid. Components such as the ethanol amines, dimethyl aniline, and other organic compounds capable of forming weak complexes with contaminants such as $H_2S$, $SO_2$, and $CO_2$ can be used. Slurries (suspensions) of reactive or adsorptive particles in liquids, such as lime slurry for acid gas (*10*) and activated carbon slurry for adsorbable gases (*11*) may also be used.

The ultimate performance for a reactive liquid occurs when all "back pressure" of the transferring gas above the liquid is eliminated by the gas reacting with it instantly. If the concentration of the reactant is high enough and the reaction rate fast enough, Equation (15) becomes

$$N_{OG} = \int_{p_o}^{p_i} \frac{dp}{p} = \ln \frac{p_i}{p_o} \tag{20}$$

In addition to decreasing the number of transfer units, chemical reaction will also reduce the height of a transfer unit because of the increase in liquid phase mass transfer rate. One can get an idea of the possible benefit due to a chemical reactant from the following equation relating the liquid phase coefficient with chemical reaction to that for physical absorption.

$$k_l^*/k_l = [(D_A/D_B)^{1/2} + (D_A/D_B)^{1/2} q/C_{Ai}] \tag{21}$$

Equation (21) is valid for a very rapid second-order reaction and has been successfully applied to a few systems such as the absorption of $H_2S$ in NaOH and KOH solutions. A coefficient 12 times that for physical absorption has been obtained with a hydroxyl ion concentration of about 1.0 (gm mole/liter) and the relationship was linear as indicated by Equation (21).

Unfortunately, it is not presently possible to do much more in the way of predicting absorption with chemical reaction. Some data and design equations are available for specific systems (*3, 4*) but there is no general method for predicting the complex combination of kinetic, transport, thermal, and concentration effects which are involved. If full-scale data are not available, small-scale or pilot plant data must be used for reliable design.

Sulfur oxides control scrubbing with chemical reactants and have been studied and used to a greatly increased extent in recent years; an extensive literature has developed (*10, 12*). In general, the slurry and dissolved reactant scrubbing systems have proved capable of $SO_2$ removal of 90% overall efficiency and better. However, there have been serious problems such as scaling caused by deposition of suspended and/or dissolved solids. These and other operating problems are being resolved by various design, operating, and chemical control modifications.

### D. Method of Contact

Gas–liquid contacting can be done in a variety of ways in scrubbers which may be classified according to their overall geometry or their "unit mechanisms" (7). The unit mechanisms are the essential gas absorption or particle collection elements which account for the scrubber's capability. Within any unit mechanism, the gas or particles may be separated from the gas phase by means of one or more of the basic processes such as molecular diffusion, inertial impaction, sedimentation, etc. An understanding and analysis of any scrubber can be reached by determining which combinations of unit mechanisms and deposition phenomena are involved. Once the basic elements of the scrubber are determined and their performance capabilities defined by mathematical equations or charts, the performance of the scrubber can be predicted.

Bubbles of gas in a tank of liquid is a simple example of a dispersed gas system. One can enhance contacting in such a system by mechanical stirring or by adding baffling elements in the form of packing or plates. Packing elements may serve to spread the liquid out in thin films, to promote the stirring of the liquid as well as the gas phase, and to induce centrifugal forces on particles due to changing the direction of gas flow.

Dispersed liquid in the form of drops moving through gas is the inverse analogy to the bubbles in a tank system. Here too, one can enhance contacting by mechanical agitation and by the use of packing or baffles. Generally, the liquid is a single phase but it is also possible to use a suspension of chemically reactive or absorptive particles. In this way one can utilize the properties of insoluble solids while retaining the advantages of fluid handling.

### E. Types of Equipment

Scrubber classifications in general terms of geometry and unit mechanisms are shown in Table II. Equipment for gas phase dispersed systems

**Table II Scrubber Classification**

| *Geometric type* | *Unit mechanism for particle collection* |
|---|---|
| Plate | Jet impingement, bubbles |
| Massive packing | Sheets (curved or plane), jet impingement |
| Fibrous packing | Cylinders |
| Preformed spray | Drops |
| Gas atomized spray | Drops, cylinders, sheets |
| Centrifugal | Sheets |
| Baffle and secondary flow | Sheets |
| Impingement and entrainment | Sheets, drops; cylinders, jets |
| Mechanically aided | Drops, cylinders, sheets |
| Moving bed | Bubbles, sheets |
| Combinations | |

is usually of the plate column type. Various forms of gas dispersing elements such as bubble caps, perforated plates, and variations are incorporated at one or more levels (stages or plates) in a vertical column. Usually, the scrubbing liquid is introduced at the top and runs down from stage to stage, either through downcomer pipes or by dripping through perforations. During operations the gas causes violent agitation of the liquid, forming a dense froth ranging from 10 to 30 cm deep. A pressure drop of 5 to 10 cm of water per plate is common. Packed columns are the most common apparatus used for liquid phase dispersal in gas absorption. The reason is that packing supports liquid contact area throughout the range of flow rates in countercurrent flow usually desired for gas absorption. Packing materials are continually being invented and range from dense packings such as rings, saddles, and spheres to low density fabrications of expanded metal, screen, and fibers. The choice of one packing material over another, or a packed tower instead of a plate column, will depend upon economics, the availability of suitable materials of construction, and design constraints such as limitation on pressure drop or tendency to become plugged. Larger diameter packed columns require special provisions for liquid distribution to counteract the channeling which develops as liquid flows down.

Spray contactors (Fig. 2) are not well suited to countercurrent gas absorption, but are superior for particle collection without danger of plugging with accumulated mud. Liquid drops may be generated by spray nozzles or by the atomization of liquid by a high velocity stream of gas. Some scrubbers, such as the Doyle scrubber, whose action seems at first glance to be based on the impingement of gas on liquid or solid phase, may actually depend on impingement upon atomized drops torn from the liquid mass. The high energy gas atomizing types such as flooded disk, wedge, orifice, and venturi scrubbers have been given more attention in recent years as the demand for high efficiency has grown.

## F. Capacity

Pressure drop for gas flow through scrubbers is caused by friction with stationary surfaces and by the acceleration of liquid. Frictional loss is very dependent upon the geometry of the scrubber and must generally be determined experimentally. Acceleration loss is fairly insensitive to scrubber geometry and is frequently the predominant cause of pressure drop. If the liquid is introduced at a velocity $v_l$ in the direction of gas flow, and finally attains the gas velocity $v_G$ then

$$\Delta P_a = \frac{(v_G - v_L)}{g_c A} \frac{dM}{dt} \tag{22}$$

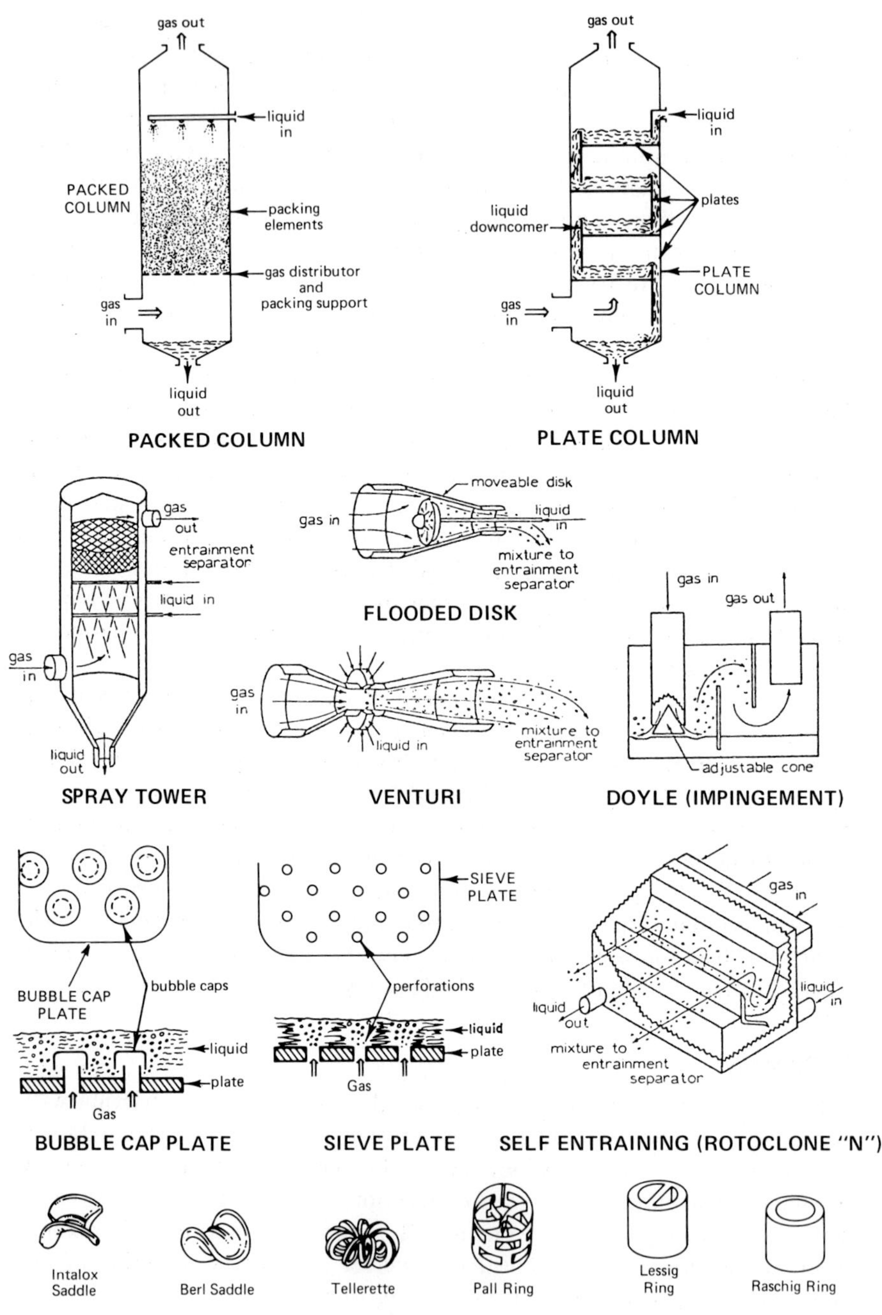


Figure 2. Schematic drawings of spray and atomizing collectors.

For pressure drop $\Delta P_a'$, in cm of water, Equation (22) reduces to

$$\Delta P_a' = 1.0 \times 10^{-3}(v_G - v_L)^2(Q_L/Q_G) \tag{23}$$

Equation (23) predicts roughly a 15% higher pressure drop than experimentally measured except at low liquid rates, where the frictional losses for the gas alone are significant. At higher liquid rates it appears that 80–90% of the momentum of the liquid is lost rather than serving to recompress the gas. This is in keeping with the observation that the pumping efficiency of an ejector venturi is generally less than 15%.

Frictional pressure drop through packed columns is dependent on many factors, and there is no simple and exact relationship to define it. For specific information, one must consult manufacturers' data or more extensive sources of information. Eckert (*13*) presented the generalized correlation shown in Figure 3, of pressure drop and fluid flow parameters. The packing factor $B_p$ is the ratio of packing surface per unit of tower volume to the cube of void fraction and ranges from 8 to 22 $cm^2/cm^3$ for 1.3 cm packings, 1.6 to 4.4 $cm^2/cm^3$ for 2.5 cm packings (Pall rings, saddles, and Rasching rings), 0.8 to 3.5 $cm^2/cm^3$ for 3.8 cm packings, and 0.6 to 2.2 $cm^2/cm^3$ for 5 cm packings.

### Flooding

Scrubber capacity is limited to the gas flow at which the amount of liquid carried over (entrained) by the gas becomes excessive, and at which flooding occurs, i.e., when not enough liquid will pass through the device so as to give proper operation. Flooding in countercurrent packed column operation will occur at a gas velocity about 1.5 times that which causes 1.3 cm $H_2O$ pressure drop, as shown by the flooding lines in Figure 3.

Entrainment of liquid requires that the drop size be small enough and the gas rate high enough for the drops to be carried out. If the drops are already formed, as in a spray scrubber, the conditions for entrainment can be predicted simply from drop dynamics. When drops are not initially present or when gas velocity is high enough to shatter large drops, it is necessary to consider the atomization process.

## III. Particle Collection

### *A. Mechanisms and General Approach*

Particle collection in scrubbers is due to one or more of the same phenomenon which may be operative in other types of collection equipment.

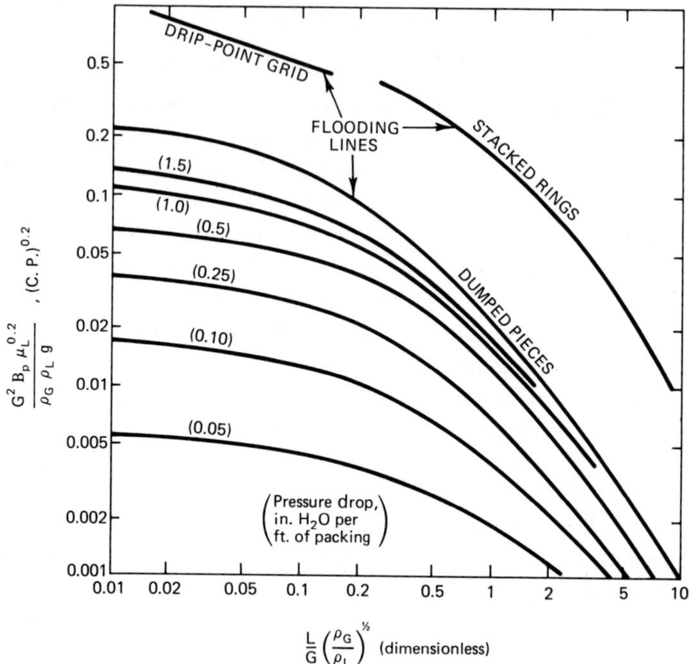


**Figure 3. Generalized flooding and pressure drop correlation.**

Deposition within the context of any unit mechanism may be due to inertial impaction, interception, Brownian diffusion, turbulent diffusion, gravitational force, electrophoresis, diffusiophoresis, thermophoresis, photophoresis, and magnetophoresis. The purpose of the liquid is to provide the collecting surface and/or to wash a solid collecting surface.

Relationships between particle collection performance and design parameters may be predicted from consideration of the unit mechanisms involved, or measured by experiment, or both. A specific performance relationship between collection efficiency and particle size (often called a "grade-efficiency curve") can be integrated over the particle-size distribution to yield the overall collection efficiency. Alternatively, a generalized and somewhat idealized method, which is described below, can be used for rapid prediction of scrubber performance.

## 1. Difficulty of Separation

The "cut diameter" method for scrubber performance prediction (7) is based on the idea that the most significant single parameter to define both the difficulty of separating particles from gas and the performance of a scrubber is the particle diameter for which collection efficiency is 50%.

For inertial impaction, the most common particle separation process in presently used scrubbers, aerodynamic diameter defincs the particle properties of importance:

$$d_{pa} = d_p(\rho_p C')^{1/2}, \qquad \mu m(gm/cm^3)^{1/2} \equiv \mu mA$$

When other separation mechanisms are important, other particle properties may be more significant but this will occur generally when "$d_p$" is less than a micron.

When a range of sizes is involved, the overall collection efficiency will depend on the amount of each size present and on the efficiency of collection for that size. We can take these into account if the difficulty of separation is defined as the aerodynamic diameter at which collection efficiency must be 50%, in order that the necessary overall efficiency for the entire size distribution be attained. This particle size is the required "separation cut diameter" $d_{RC}$, and it is related to the required overall penetration $Pt$ and the size distribution parameters. The number and weight size distribution data for most industrial particulate emissions follow the log-probability law. Hence, the geometric mean weight diameter $d_{pg}$ and the geometric standard deviation $\sigma_g$ adequately describe the size distribution.

Penetration for many types of inertial collection equipment can be expressed as

$$Pt = \exp(-Ad_{pa}{}^{B}) = 1 - E \tag{24}$$

In some cases one is concerned with particles larger than 1 $\mu$m diameter or where the particle-size distribution is log-normal in terms of physical rather than aerodynamic diameter. It may be convenient to use the simplifying assumption that penetration is related to physical diameter by

$$Pt = \exp(-A_p d_{pa}{}^{B}) \tag{25}$$

Packed towers, centrifugal scrubbers, and sieve plate columns follow the first relationship. For the packed tower and sieve plate column, $B$ has a value of 2. For centrifugal scrubbers, $B$ is about 0.67. Venturi scrubbers also follow the above relationship and $B \simeq 2$ when the throat impaction parameter [see Eq. (37)], is between 1 and 10.

The overall (integrated) penetration $\overline{Pt}$ of any device on a dust of any type of size distribution will be

$$\overline{Pt} = \int_0^w (dw/w)Pt \tag{26}$$

The right-hand side of the above equation is the integral of the product of each weight fraction of dust times the penetration on that fraction.

If Equation (26) is solved for a log normal size distribution and collection as given by Equation (24), the resulting equation can be solved to yield Figures 4 and 5.

Figure 4 is a plot of $\overline{Pt}$ vs $(d_{p50}/d_{pg})^B$ with $B \ln(\sigma_g)$ as a parameter. For a required $\overline{Pt}$ one can find the value of $d_{RC}$ when $d_{pg}$, $\sigma_g$, and $B$ are given. Figure 5 is presented as a plot of $\overline{Pt}$ vs $(d_{p50}/d_{pg})$ with $\sigma_g$ as the parameter when $B = 2$.

To illustrate the use of the separation cut diameter, assume that 2% penetration is needed for dust with $d_{pg} = 10$ μmA, $\rho_p = 3$ gm/cm³, and $\sigma_g = 3$. If a scrubber such as a packed bed, sieve plate, or venturi is to be used, Figure 5 shows the cut diameter $d_{p50}$ must be $0.09 \times (d_{pg}) = 0.9$ μmA. The corresponding physical diameter is 0.44 μm (C′ for air is about 1.35). Of course, if the scrubber is capable of a smaller cut diameter, that is good; so $d_{RC}$ is the maximum cut diameter acceptable. Some scrubbers, such as venturis, are only approximately fitted by relating penetration to $\exp(d_{pa}^2)$ and more accurate plots can be prepared by using more representative performance equations.

## 2. Scrubber Performance

Scrubber capability is defined by "performance cut diameter" $d_{PC}$ which is the aerodynamic particle diameter at which the scrubber gives

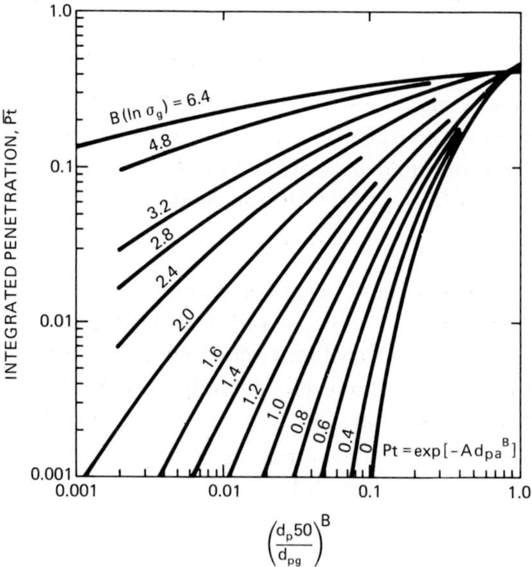


Figure 4. Integral (overall) penetration as a function of cut diameter, particle parameters, and collector characteristic.

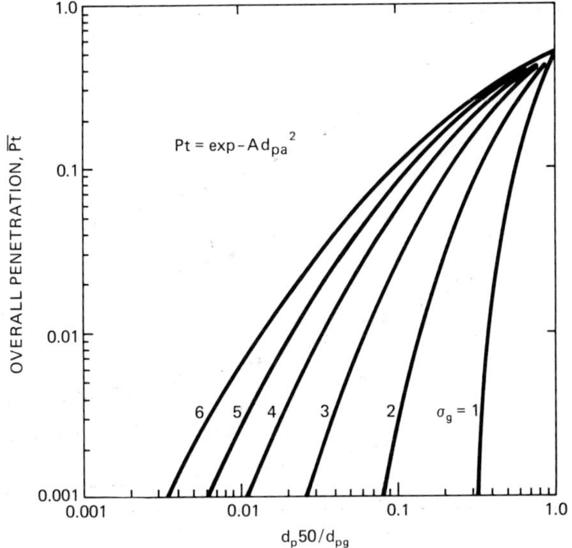


Figure 5. Overall penetration as a function of cut diameter and particle parameters for common scrubber characteristic.

50% collection efficiency. Once a scrubber type, size, and operating conditions are chosen by matching the "separation" and "performance" cut diameters (i.e., $d_{RC} = d_{PC}$), a more accurate efficiency diameter relationship can be developed and a more accurate computation of overall penetration can be made. The reason this step is necessary is that the relationship between overall penetration and separation cut diameter as shown in Figures 4 and 5 is correct only for packed beds and similar devices and is an approximation for others.

### *B. Plate Columns*

Particle separation in sieve (perforated) and impingement (Peabody-type) plates can be defined mathematically by starting from the basic mechanisms of particle collection in bubbles, on drops, and jet impaction.

Experimental data on the collection of hydrophilic particles by water on sieve plates (*14*) are correlated by

$$Pt = \exp(-40FK_p) \tag{27}$$

where

$$K_p = u_h d_{pa}^2 / 9\mu_G d_h \tag{28}$$

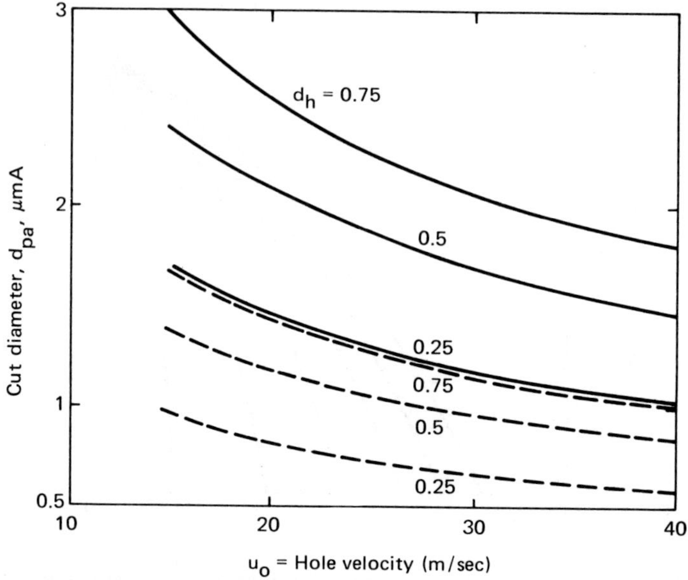


**Figure 6. Performance cut diameter prediction for typical sieve plate conditions. Solid line: $F$ = 0.4; dashed line: $F$ = 0.65.**

By setting $Pt = 0.5$ in Equation (27), the following relationship for cut diameter in a sieve plate results;

$$d_{pc} = 0.4(\mu_G d_h / u_h F)^{1/2} \tag{29}$$

For impingement plates there are no reliable experimental data available so the efficiency is predicted based on the impingement of round jets on plane surfaces. The cut diameter is given by

$$d_{pa50} = \left( \frac{1.37 \mu_G n_H d_h^3}{Q_G} \right)^{1/2} \tag{30}$$

Some examples of the performance predictions of Equations (29) and (30), for inertial collection, are given in Figures 6 and 7, respectively.

### *C. Packings*

Particle collection in packed columns (*15*) can be described in terms of gas flow through curved passages, and performance for a variety of packing shapes, such as saddles, rings, and spheres, can be correlated simply by the packing diameter, as shown in Equation (31):

$$Pt = \exp - (7.0 Z K_p / \epsilon d_c) \tag{31}$$

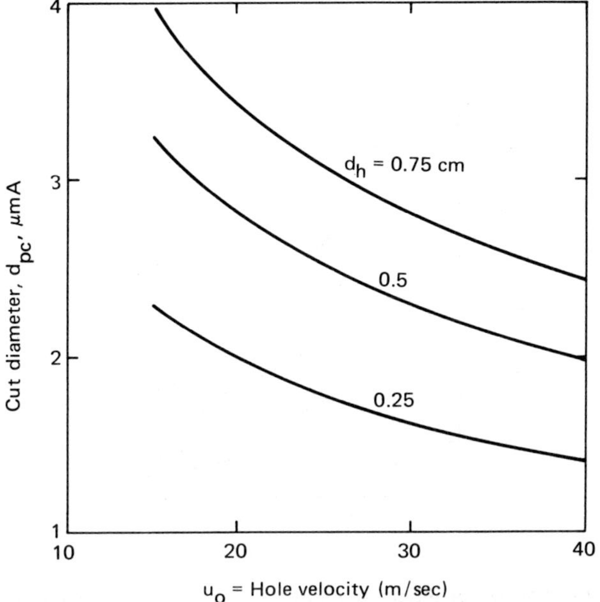


Figure 7. Performance cut diameter predictions for typical impingement plate conditions.

where $K_p$ is defined as in Equation (28), but with gas velocity equal to the superficial velocity through the total bed area and collector dimension ($d_h$) taken as the packing diameter. Aerodynamic cut diameter is given by

$$d_{pc} = \left(\frac{\epsilon d_c{}^2 \mu_G}{\mu_G Z}\right)^{1/2} \tag{32}$$

Figure 8 is a plot of cut diameter versus packed bed height, with packing size and air velocity as parameters when bed porosity is 0.75 and gas viscosity is the same as for air.

## D. Preformed Sprays

A spray chamber consists essentially of a round or rectangular chamber into which water is introduced through one or more sprays. Drop size depends upon liquid pressure drop and the type of nozzle used. Water pressure drop varies from 1.4 to 7.3 atm (20 to 100 psi), and water consumption is usually in the range of 0.67–2.68 liters/m³ (5–20 gal/MCF). In practice a gas velocity of 0.6–1.2 m/second (2–4 ft/second) is used and the gas pressure drop is about 2 cm of water.

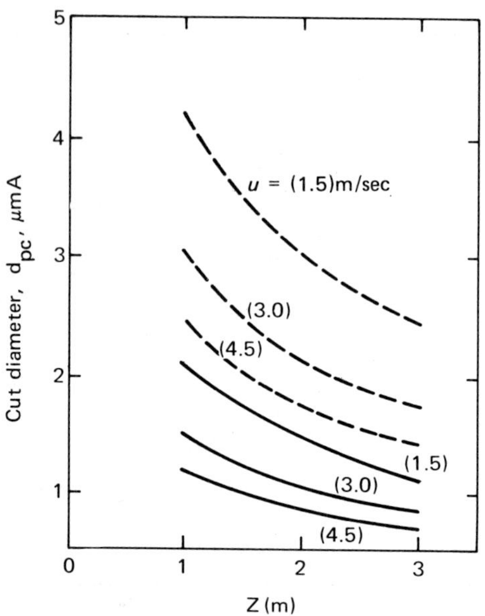


Figure 8. Performance cut diameter predictions for typical packed bed conditions. Solid line: $d_c$ = 2.5 cm; dashed line: $d_c$ = 5.0 cm; $\epsilon$ = 0.75.

## 1. Vertical Countercurrent Flow—Interial Impaction

Starting from a material balance over a small section of the tower, one gets after integration,

$$Pt = \exp - \left[ \frac{3Q_L u_t Z \eta}{4Q_G r_d (u_t - u_G)} \right] = \exp - 0.25 \left[ \left( \frac{A_d u_t \eta}{Q_G} \right) \right] \quad (33)$$

where $A_d = (3Q_L Z)/r_d(u_t - u_G)$ = total surface area of all drops in the scrubber, assuming no liquid reaches the scrubber wall.

Target efficiency is greatly influenced by the drop Reynolds number and is calculated from an interpolation between the target efficiencies for viscous and potential flow around the drop.

Some solutions of the equations for inertial collection in a countercurrent spray chamber are plotted in Figure 9 as $d_{pc}$ versus column height, with drop diameter, air velocity, and water-to-air ratio as parameters. Standard air and water properties have been used, and no liquid flow on the wall is assumed. Actually, only a small fraction of the drops remain in suspension, and as little as 20% of $Q_L$ may be effective, depending on scrubber size.

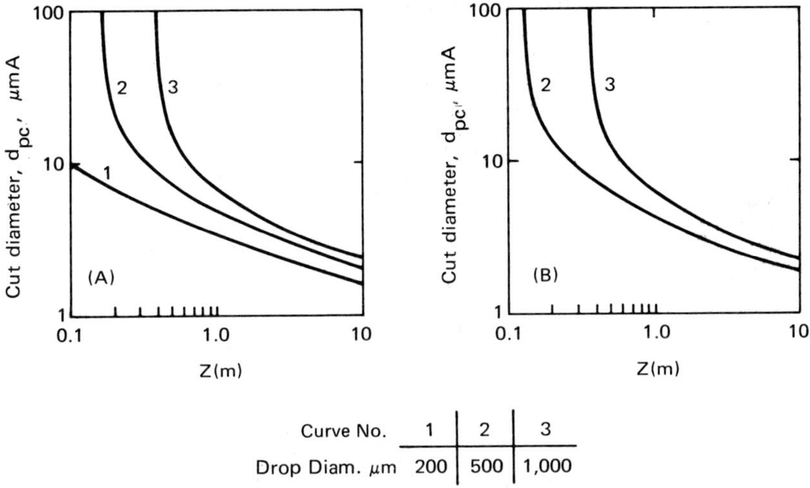


Figure 9. Performance cut diameter predictions for typical vertical countercurrent spray. (A) $Q_L/Q_G = 1$ liter/m³; $\mu_G = 0.6$ m/second. (B) $Q_L/Q_G = 1$ liter/m³; $\mu_G = 0.9$ m/second.

## 2. Cross Flow—Inertial Impaction

In the cross-flow case, the water is sprayed at the top of the spray chamber while the gas flows horizontally. For collection by inertial impaction in a spray chamber, Equation (34) predicts the penetration:

$$Pt = \exp - 3Q_L h\eta/4Q_G r_d = \exp - [0.25(A_d u_d \eta/Q_G)] \tag{34}$$

Some solutions of the equation for inertial collection in a crosscurrent spray chamber are plotted in Figure 10. The same procedure advocated for countercurrent spray chambers also applies here.

### *E. Gas Atomized (Cocurrent) Sprays*

An important class of scrubbers in which the liquid is atomized by a high velocity gas stream includes venturi, flooded disc, orifice, and similar types. Geometric design appears to have a minor effect on the performance of these cocurrent spray scrubbers. Several investigators (*16–19*) utilize the following equation to describe particle collection in a gas atomized spray:

$$-\frac{dc}{c} = \frac{u_r}{u_G}\frac{3Q_L\eta}{2(u_G - u_r)Ad_d}\,dZ \tag{35}$$

From Equation (35) the investigators cited take different paths to a final design relationship. The one by Calvert *et al.* (*7*) leads to an explicit

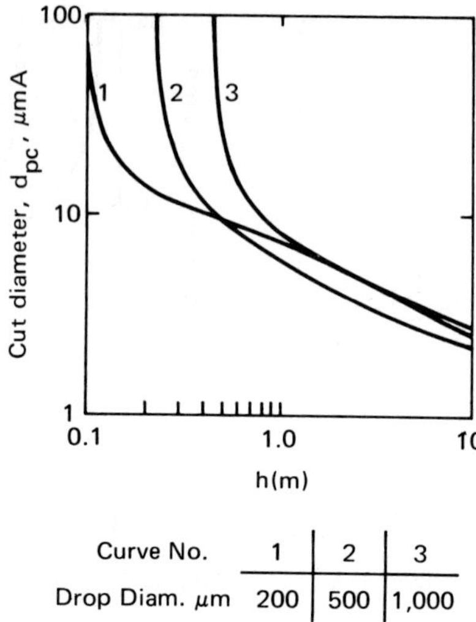


Figure 10. Performance cut diameter predictions for typical crosscurrent spray. $Q_L/Q_G = 1$ liter/m³.

form which can be solved without a digital computer and yields the following relationships:

$$Pt = \exp\left[\frac{2Q_L u_G \rho_L d_d}{55 Q_G \mu_G} F(K_{pt}, f)\right] \tag{36}$$

where $K_{pt}$ is the inertia parameter evaluated at the velocity of the gas in the throat, $u_G$:

$$K_{pt} = d_{pa}{}^2 u_G / 9\mu_G d_d \tag{37}$$

and

$$F(K_{pt}, f) = \frac{1}{K_{pt}}\left[-0.7 - K_{pt}f + 1.4 \ln\left(\frac{K_{pt}f + 0.7}{0.7}\right) + \left(\frac{0.49}{0.7 + K_{pt}f}\right)\right] \tag{38}$$

The factor $f$ is an empirical factor which absorbs the influence of various parameters not included explicitly in Equation (36). These parameters include collection by means other than impaction, particle growth due to condensation or other effects, drop sizes other than those predicted, loss of liquid to the venturi walls, maldistribution, and other effects.

For conservative design purposes in the case of hydrophobic particles it is recommended an $f$ value of 0.25, which is about average for the available data, be used. For hydrophilic aerosols, such as soluble compounds, acids, and fly ash with $SO_2$ and $SO_3$, values of $f$ will be significantly higher; perhaps 0.4 to 0.5. The situation is further complicated by the fact that $f$ increases at liquid-to-gas ratios below about 0.2 liter/m$^3$. Large scrubber tests show $f = 0.5$.

Drop diameter is taken to be the Sauter (surface) mean diameter computed from the Nukiyama and Tanasawa (*20*) correlation for pneumatic atomization:

$$d_s = \frac{58{,}600}{u_r}\left(\frac{\sigma}{\rho_L}\right)^{1/2} + 597\,\frac{\mu_L^{0.45}}{\sqrt{\rho_L \sigma}}\left(1000\,\frac{Q_G}{Q_L}\right)^{1.5} \tag{39}$$

In this empirical equation, units are as follows: $u_r$ = cm/sec; $\sigma$ = dyn/cm; $\rho_L$ = gm/cm$^3$; $\mu_L$ = poise; $Q$ = volume/second; $d_d = \mu$m.

Hesketh *et al.* (*21*) have experimentally discovered a different type of atomization in a venturi scrubber when liquid injection nozzles larger than 1 mm i.d. are used, and when critical air velocities are above that given by

$$v_{\text{critical}} = (0.795/d_{\text{nozzle}})^{1/2} \times 10^3 + 466 \tag{40}$$

where $v_{\text{critical}}$ = cm/second; $d_{\text{nozzle}}$ = cm. The authors have termed it cloud-type atomization. The average drop size produced is roughly less than one-half the size predicted by Equation (39).

Venturi performance is shown in terms of the predicted aerodynamic

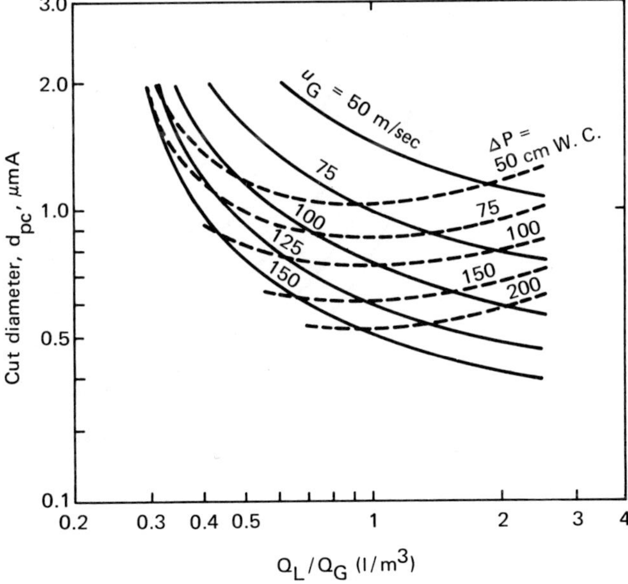


Figure 11. Performance cut diameter predictions for venturi scrubber. $f = 0.25$.

cut size against gas velocity, with liquid-to-gas ratio as parameter, and with predicted constant pressure drop lines indicated on Figure 11 based on mean drop diameter.

Once the required separation cut diameter for a given application has been computed, the approximate operating region can be found from Figure 11. This does not tell the whole story however, because penetration depends not only on the collection efficiency of a single drop but also on the extent to which the gas is swept by drops. In other words, the drop hold-up (volume fraction drops) in the throat is a significant factor in determining particle penetration, and it cannot be accounted for by a simple power relationship with particle diameter.

Figure 12, a plot of penetration as a function of aerodynamic diameter and the liquid-to-gas flow rate ratio, shows this effect. Note that penetration reaches a limiting value as particle size increases; showing that even though the collection efficiency of one drop for that size particle ap-

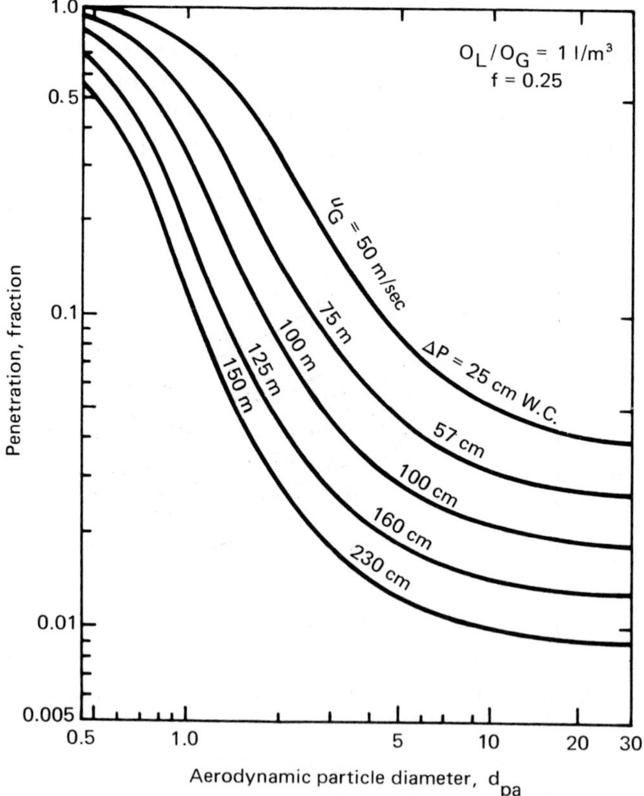


Figure 12. Predicted penetration for venturi scrubber.

proaches 100%, there are not enough drops to completely sweep the gas stream.

## IV. Scrubber Energy Requirement

The energy required for particle scrubbing is mainly a function of the gas pressure drop, except for preformed sprays and mechanically aided scrubbers. It has been shown that there is an empirical relationship between particle penetration and power input to the scrubber for a given scrubber and a specific particle-size distribution. However, this "power law" does not provide a way to predict performance versus power input for any size dust.

A relationship between $d_{pc}$ and scrubber pressure drop (*24*) is presented in Figure 13—a plot of performance cut diameter $d_{PC}$ versus gas pressure drop for sieve plates, venturi and orifice scrubbers, impingement plates, and packed columns. Predictions were made by means of performance equations such as those given previously, and pressure drop correlations for the scrubbers.

(a) Sieve plate penetration and pressure drop predictions for one plate are plotted as lines 1a and 1b for perforation diameters of 0.5 and

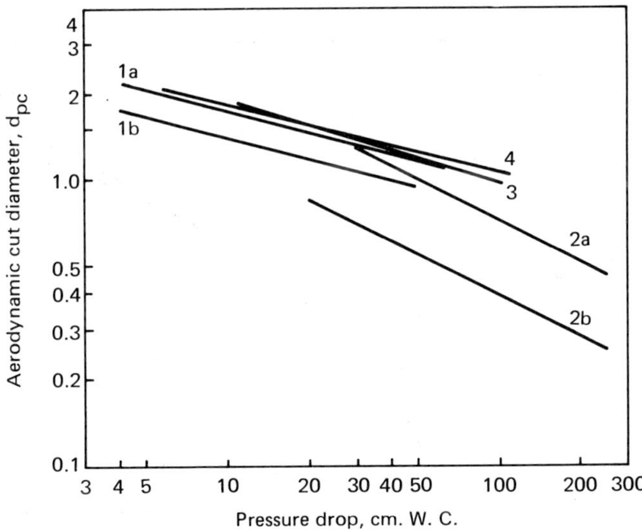


Figure 13. Representative cut diameters as a function of pressure drop for several scrubber types. 1a: Sieve, $F = 0.4$, $d_h = 0.5$ cm; 1b: sieve, $F = 0.4$, $d_h = 0.3$ cm; 2a: venturi, $f = 0.25$; 2b: venturi, $f = 0.5$; 3: impingement plate; 4: packed column, $d_c = 2.5$ cm.

0.3 cm, respectively, and $F = 0.4$. Cut diameters for other froth densities can be computed from the relationship that they are inversely proportional to $F$. Cut diameters for two and three plates in series would be 84 and 80%, respectively, of those for one plate at any given pressure drop. Note that these predictions are for wettable particles and that both froth density and pressure drop are dependent on plate design and operation.

(b) Venturi penetration and presure drop data are given for $f = 0.25$ and $f = 0.5$ in lines 2a and 2b, respectively. The predictions are for a liquid-to-gas ratio $Q_L/Q_G \approx 1$ liter/m$^3$ corresponding to about the minimum pressure drop for a given penetration. Data recently obtained for several large scrubbers fit a value of $f = 0.5$.

(c) Impingement plate data used for line 3 were predicted for one plate. Cut diameters for 2 and 3 plates in series are 88 and 83%, respectively, of those shown in line 3.

(d) Packed column performance as shown by line 4 is representative of columns from 1 to 3 m high and packing of 2.5 cm nominal diameter.

To estimate the penetration for particle diameters other than the cut size, under a given set of operating conditions, one can use the approximation of Equation (24) with $B = 2.0$. Alternatively, one could use more precise data or predictions for a given scrubber. Figure 14 is a plot of

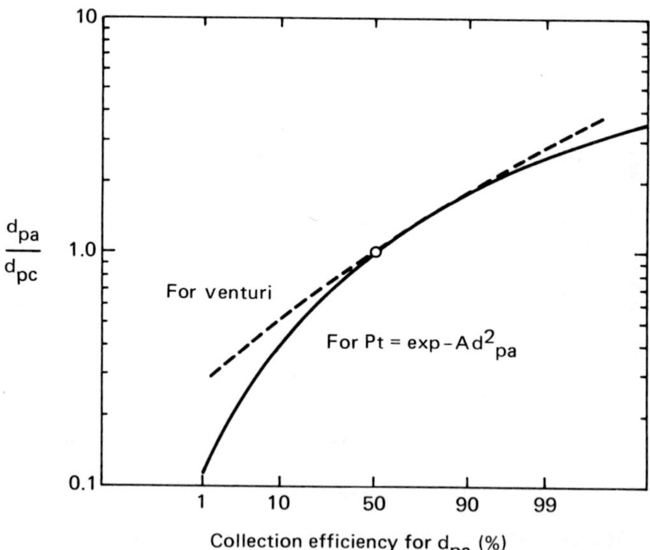


**Figure 14.** Ratio of particle diameter to cut diameter as a function of collector efficiency.

the ratio of particle aerodynamic diameter to cut diameter versus penetration for that size particle ($d_{pa}$), on log-probability paper. One line is for Equation (24) and the other is based on data for a venturi scrubber.

### Performance Limit for Inertial Impaction

The limit of what one can expect of a scrubber utilizing inertial impaction is clearly indicated by Figure 13. If a cut diameter of 1.0 μmA, or smaller is required, the necessary pressure drop is in the medium-to-high energy range. High efficiency on particles smaller than 0.5 μmA diameter would require extremely high pressure drop if inertial impaction were the only mechanism active.

High efficiency scrubbing of submicron particles at moderate pressure drop is possible, but it requires either the application of some particle separation force which is not dependent on gas velocity or the growth of particles so that they can be collected easily. Particle separation phenomena which have these characteristics are the "flux forces" due to diffusiophoresis, thermophoresis, and electrophoresis. Brownian diffusion is also useful when particles are smaller than about 0.1 μm diameter. Particle growth can be accomplished through coagulation, chemical reaction, condensation on particles, ultrasonic vibration, and electrostatic attraction.

## V. Diffusional Collection

Particle collection by Brownian diffusion can be described by relationships for mass transfer, and it is possible to outline the magnitude of efficiency which can be attained with typical scrubbers. The general relationship which describes particle deposition in any control device in which turbulent mixing eliminates any concentration gradient normal to the flow outside the boundary layer and in which the deposition velocity is constant is

$$Pt = \exp - (u_{PD} A_d / Q_G) \tag{41}$$

The particle deposition velocity for Brownian diffusion, $u_{BD}$, can be estimated from penetration theory [see Eqs. (13) and (14)] as

$$u_{BD} = 1.13 (D_p / t_g)^{1/2} \tag{42}$$

For packed columns the penetration time $t_g$ can be taken as the time required for the gas to travel one packing diameter. For plate scrubbers which involve bubbles rising through liquid, the penetration time for a

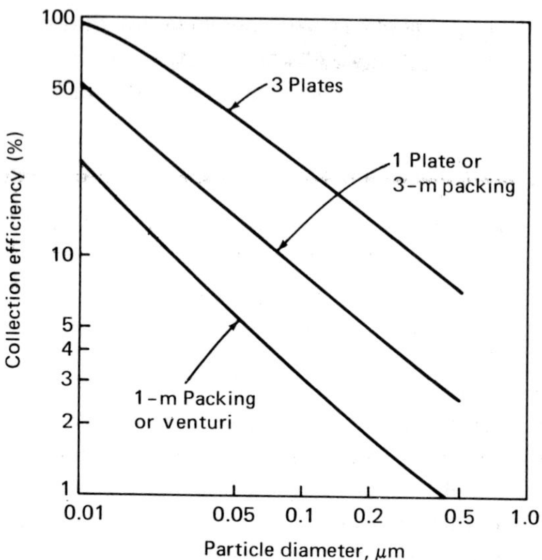


**Figure 15.** Predicted particle collection by diffusion in plates, packing, and venturi scrubbers.

circulating bubble is about that for the bubble to rise one diameter, as shown by Taheri and Calvert (*14*). For spray scrubbers the penetration time is that for the gas to travel one drop diameter.

Predictions of particle penetration due to Brownian diffusion only were made by means of Equations (41) and (42) for typical sieve plate and packed columns. The prediction for a venturi scrubber was made by means of an equation (*7*) for gas phase controlled mass transfer.

The results are plotted on Figure 15 as collection efficiency versus particle diameter. It can be seen that high efficiency collection of 0.01 μm diameter particles is readily attainable with a three plate scrubber, typical of a moderately effective device for mass transfer. Collection efficiency for particles a few tenths micron diameter is poor, as is well known.

## VI. Flux Force/Condensation

Particle separation by flux force mechanisms is not amenable to such simple treatment as Brownian diffusion because of the variation of deposition velocity with heat and mass transfer rates or electrical charging and field within the scrubber. The magnitude of flux force/condensation (F/C) effects in a variety of scrubbers (*25*) is illustrated in Figure 16 (*26–36*), a plot of particle penetration against water vapor condensa-

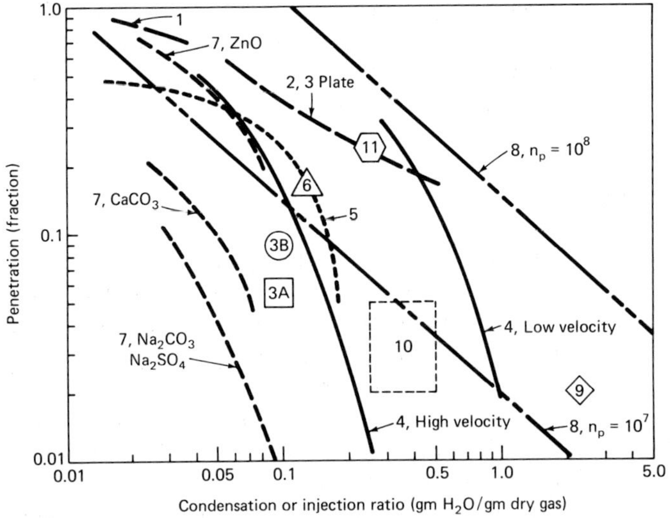


Figure 16. Condensation scrubbing performance (see tabulation below).

| Curve No. | Investigator(s) | Scrubber type | $d_{pg}$ ($\mu$m) | Particle material | $n_p$ #/cm³ |
|---|---|---|---|---|---|
| 1 | Calvert *et al.* (*26*) | Sieve plate (1 cold plate) | 0.7 | Dibutyl phthalate | $5 \times 10^5$ |
| 2 | Calvert and Jhaveri (*27*) | Sieve plate (3 cold plates) | 0.4 | Ferric oxide | $10^5 - 10^6$ |
| 3 | Fahnoe *et al.* (*28*) | (A) Cyclone, or (B) Peabody (1 plate) | <2.0 | NaCl | $10^3$ |
| 4 | Goldsmith and May (*29*) | Tubular condenser | ? | Nichrome and others | ? |
| 5 | Lancaster and Strauss (*30*) | Steam nozzle + spray + cyclone | 1.0 | ZnO | $10^5 - 10^6$ |
| 6 | Litvinov (*31*) | Venturi + 2 sieve plates | 1.7 | Apatite | $\sim 10^5$ |
| 7 | Prakash and Murray (*32*) | Steam nozzle + dry duct | | ZnO, $CaCO_3$, $Na_2CO_3$, $Na_2SO_4$ | |
| 8 | Rozen and Kostin (*33*) | Sieve plate with alternate hot and cold plates | 0.3 | Oil | $10^5 - 10^6$ |
| 9 | Schauer (*34*) | Steam nozzle + Peabody (5 plates) | 0.3 | Dioctyl phthalate | $2 \times 10^7$ |
| 10 | Stinchombe and Goldsmith (*35*) | Tubular condenser | 0.1 | Iodine | $10^3 - 10^4$ |
| 11 | Terebenin and Bykov (*36*) | Vertical netted | 0.05 | Tin fume | $5 \times 10^7$ |

tion ratio (grams water vapor per gram of dry gas). It can be seen that both the condensation ratio and the number concentration of particles are important parameters. Particle surface properties, apparatus design, and operating conditions are also significant.

## VII. Economics

Costs are the ultimate criterion of the optimum system in gas scrubbing, as in any industrial operation. One must decide whether one type of equipment is better than another when both are capable of the desired performance; whether to use less expensive equipment and more power; whether to use more expensive materials or to have higher maintenance costs; or whether to use a higher stack and less efficient collection equipment. The rationalization of these various trade-offs requires the use of a single method of evaluation: total cost.

Cost estimation methods range from quick and dirty predesign approximations to elaborate compilations of firm bids on completely designed systems. Even the latter will only give results with a probable error of between 7% over and 15% under actual capital costs. Other elements such as labor and maintenance will be more inaccurate. Nevertheless, a decision based on approximate costs is better than one based on no cost considerations. The situation is eased considerably because the comparison of cost estimates generated in the same way is likely to be more accurate than their absolute magnitudes and because sometimes the most doubtful items may contribute only a small fraction of the total cost.

The following list includes average capital investment costs in chemical plants.

(a) Delivered equipment cost; updated to the present is the starting point for calculations

(b) Installed equipment cost is taken from publications or is computed as 1.43 times delivered cost

(c) Process piping is equal to 30–60% of installed cost [item (b)]

(d) Instrumentation is equal to 3–20% of installed cost, depending on amount of automatic control

(e) Buildings and site preparation is equal to 10–30% of installed cost for outdoor and 60–100% for indoor plants

(f) Auxiliaries are equal to 0–5% of installed cost for minor additions to 25–100% for new facilities

(g) Outside lines (i.e., not with the process) = 5–15% of installed cost for intermediate cases

(h) Fixed capital investment is the total of items (b) through (g) plus costs for engineering and construction, contingencies, and a size factor

The methods presented here are those commonly used in the chemical process industry and can serve as a reasonable basis for preliminary decision making. It is important that all relevant items under the general headings of capital and operating costs be recognized and included. Capital investment covers the plant site and buildings, utilities, storage facilities, and emergency facilities, in addition to the process equipment cost. The amount of nonmanufacturing investment such as for site, shop, and warehouse facilities to be assessed against the project will depend upon circumstances such as whether it is a plant addition or a new plant. Process equipment costs include the basic collection equipment plus installation, piping, instrumentation, insulation, fans, foundations, and supporting structure and may be estimated by use of published data. The total system cost is the sum of items (b)–(g) in the above list. To this must be added cost for engineering and construction (20–60% of total system cost), contingencies (30% of total), and a size factor (0 for routine plants costing \$2,000,000, 5–15% of total for routine plants costing \$500,000 to \$2,000,000, and 15–35% for experimental plants costing less than \$500,000) to give the fixed capital investment. An overall average for fluid processes is that fixed capital investment is 4.74 times delivered equipment cost.

For extrapolation purposes one may use the following generalizations:

(a) Equipment cost varies with about the 0.6 power of capacity

(b) Total plant cost varies with about the 0.7 power of capacity

(c) Costs may be updated by means of the Marshall and Stevens (M & S) Equipment Cost Index as published in "Chemical Engineering" magazine. The M & S Index at the end of 1973 was about 350

(d) Cost variation with material can be computed as the cost for plain carbon steel equipment times a multiplying factor. These factors are: copper = 1.4, aluminum = 1.5, lead = 1.6, 304 stainless = 2.3, 316 stainless = 2.7, Monel or nickel = 3.0, Hastelloy = 3.5.

A survey of scrubber users conducted in 1970 (7) yielded some cost data on six types of scrubber systems. These data are presented in Table III in terms of the approximate mean of the cost range at various gas flow capacities, the magnitude of the highest reported cost relative to the mean, and the magnitude of the lowest reported cost relative to the mean. Thus, for example, the total installed cost of a venturi scrubber system for 50,000 actual cubic feet per minute (ACFM) capacity would range from a mean of \$3.00/ACFM to a high of \$9.00/ACFM and a low of \$1.00/ACFM. The wide range of costs is due to differences in materials, power requirements, and other specific features such as ducting and structural requirements.

**Table III Reported Costs of Complete, Installed Scrubber Systems**[a]

| *Scrubber type* | *Mean cost/ACFM*[b] *1000* | *10,000* | *50,000* | *100,000* | *High/mean* | *Low/mean* |
|---|---|---|---|---|---|---|
| Venturi | $14.00 | $5.50 | $3.00 | $2.20 | 3 | ⅓ |
| Packed bed | $14.00 | $3.00 | $0.80 | — | 3 | ⅓ |
| Spray | $50.00 | $5.00 | $1.00 | $0.70 | 2 | ½ |
| Centrifugal | $3.00 | $1.30 | $0.70 | — | 2 | ½ |
| Impingement and entrainment | $8.00 | $3.50 | $2.00 | $1.50 | 1.5 | 0.7 |
| Mobile bed | — | $3.00 | $2.00 | — | 1.5 | 0.7 |

[a] Costs are for a Marshall and Stevens Index of about 280.
[b] ACFM: Actual cubic feet per minute.

## REFERENCES


1. J. H. Perry, ed., "Chemical Engineers Handbook," 5th ed. McGraw-Hill, New York, New York, 1973.
2. A. L. Kohl and F. C. Riesenfeld, "Gas Purification." McGraw-Hill, New York, New York, 1960.
3. G. Astarita, "Mass Transfer with Chemical Reaction." Amer. Elsevier, New York, New York, 1967.
4. P. V. Danckwerts, "Gas Liquid Reactions." McGraw-Hill, New York, New York, 1970.
5. T. Hobler, "Mass Transfer and Absorbers." Pergamon, Oxford, England, 1966.
6. W. S. Norman, "Absorption, Distillation, and Cooling Towers." University Press, Aberdeen, Scotland, 1961.
7. S. Calvert, J. Goldshmid, D. Leith, and D. Mehta, "Scrubber Handbook," NTIS No. PB-213-016. U.S. Department of Commerce, NTIS, Springfield, Virginia, 1972.
8. R. Higbie, *AIChE J.* **31,** 365 (1935).
9. H. L. Shulman, C. F. Ullrich, N. Wells, and A. Z. Proulx, *AIChE J.* **1,** 247 and 259 (1955).
10. "Proceedings of Second International Lime/Limestone Wet-Scrubbing Symposium," APTD-1161. United States Environmental Protection Agency, Research Triangle Park, North Carolina, 1972.
11. S. Calvert, D. M. Mehta, and R. R. Russell, *Amer. Ind. Hyg. Ass., J.* **30,** 57 (1969).
12. A. V. Slack, *Environ. Sci. Technol.* **7,** 110 (1973).
13. J. S. Eckert, *Chem. Eng. Progr.* **57,** 54 (1961).
14. M. Taheri and S. Calvert, *J. Air Pollut. Contr. Ass.* **18,** 240 (1968).
15. S. Jackson and S. Calvert, *AIChE J.* **12,** 1075 (1966).

16. R. H. Boll, *Ind. Eng. Chem., Fundam.* **12,** 40 (1973).
17. S. W. Behie and J. M. Beeckmans, *Can. J. Chem. Eng.* **51,** 430 (1973).
18. H. E. Hesketh, *J. Air Pollut. Contr. Ass.* **24,** 938 (1974).
19. S. Calvert, *in* "Air Pollution" (A. Stern, ed.), 2nd ed., Chapter 46, p. 457. Academic Press, New York, New York, 1968.
20. S. Nukiyama and Y. Tanasawa, *Trans. Soc. Mech. Eng., Tokyo* **4,** 86 (1938).
21. H. E. Hesketh, A. J. Engel, and S. Calvert, *Atmos. Environ.* **4,** 639 (1970).
22. R. Ingebo, *NASA Tech. Note* **TN D-3762** ((1956).
23. K. T. Semrau, *J. Air Pollut. Contr. Ass.* **10,** 200 (1960).
24. S. Calvert, *J. Air Pollut. Contr. Ass.* **24,** 929 (1974).
25. S. Calvert and N. C. Jhaveri, *J. Air Pollut. Contr. Ass.* **24,** 947 (1974).
26. S. Calvert, J. Goldshmid, D. Leith, and N. Jhaveri, "Feasibility of Flux Force/Condensation Scrubbing for Fine Particulate Collection," NTIS No. PB 227 307. U.S. Department of Commerce, NTIS, Springfield, Virginia, 1973.
27. S. Calvert and N. C. Jhaveri, "Study of Flux Force/Condensation Scrubbing of Fine Particles," EPA-600/2-75-018. A.P.T., Inc. San Diego, California, 1975.
28. F. Fahnoe, A. E. Lindroos, and R. J. Abelson, *Ind. Eng. Chem.* **43,** 1336 (1951).
29. P. Goldsmith and F. G. May, *in* "Aerosol Science" (C. N. Davies, ed.), pp. 163–194. Academic Press, New York, New York, 1966.
30. B. W. Lancaster and W. Strauss, *Ind. Eng. Chem., Fundam.* **10,** 362 (1971).
31. A. T. Litvinov, *Zh. Prikl. Khim.* **40,** 353 (1967).
32. C. B. Prakash and F. E. Murray, "Particle Conditioning by Steam Condensation," Preprint of paper. Dept. Chem. Eng., University of British Columbia, Vancouver, British Columbia, 1973.
33. A. M. Rozen and V. M. Kostin, *Int. Chem. Eng.* **7,** 464 (1967).
34. P. J. Schauer, cited by A. Bralove, *in* "Air Pollution" (L. C. McCabe, ed.), Chapter 40. McGraw-Hill, New York, New York, 1952.
35. R. A. Stinchombe and P. Goldsmith, *J. Nucl. Energy* **20,** 261 (1966).
36. A. N. Terebenin and A. P. Bykov, *Zh. Prikl. Khim.* **45,** 1012 (1972).

# 7

## Mist Elimination

Werner Strauss

## I. Introduction

The word *mist* will be used in this chapter as a generic term for liquid droplets which range in size from less than 1 μm to several millimeters. The smaller droplets, i.e., those less than 1 μm are formed by nucleation and condensation of vapor, while the larger droplets are formed by the bursting of bubbles, by entrainment from surfaces, by sprays, or by splash-type liquid distributors.

Small drops tend to be spherical, maintaining their shape because of liquid surface tension. Large droplets of the order of a millimeter tend to be spheroidal, changing between oblate to prolate relative to their direction of motion.

The most common anthropogenic source of mists is the air passing through cooling towers, where droplets are virtually pure water, although they may contain traces of additives used for algae and corrosion control. However, if, owing to an accident, there is a broken pipe or tube in the tower, appreciable quantities of the material being cooled, which in an oil refinery could be a low vapor pressure hydrocarbon, can be entrained. Nonetheless, the main problem with cooling tower mists is the loss of water and the nuisance value of the "light but objectionable local drizzle" (*1*).

More critical are the condensation mists from phosphoric and sulfuric acid plants, where the droplets are much smaller and are highly corrosive. Effective mist elimination is also essential for shielding high efficiency particulate air (HEPA) filters in containment vessels for pressurized water reactors (PWR) in nuclear power generation and for the removal of droplets from the air entering surface ships in zones near nuclear explosions.

Generally speaking however, mist eliminators are used in chemical engineering operations for separating droplets such as tars from the vapors from the boilers of distillation columns for preventing entrainment and carry over from the top of these columns; and from steam drums in boilers to ensure the discharge of dry steam. The degree of separation required depends on the particular application. Thus, while there are stringent air pollution control restrictions for acid plants specifying the quantity and concentration of the emission of mist droplets, design and economic considerations predominate for mist eliminators used as part of an industrial process.

Early types of mist collectors in acid plants were packed beds—usually coke. To be effective they used very low velocities, and therefore were large, with high capital costs. In addition, they suffered frequent blockage and required cleaning and repacking with consequent high maintenance costs (*2, 3*). More recently, a wide range of mist eliminators has been developed which are fairly specific for particular applications. Thus for coarse droplets there are simple disengagement chambers and louver demisters. Smaller droplets require smaller impingement obstacles, and a multiplicity of these in series. They range from knitted wires to fine fibers, as well as combinations of these.

Other mist eliminators widely used commercially are centrifugal and electrostatic collectors. Sintered ceramic tubes, closed at one end and referred to as "candles" have been used, but are fragile and tend to block up. Venturi scrubbers have been used for fine mists in some acid plants, and fluidized beds and sonic agglomerators have been tested extensively but with few commercial applications.

## II. Size Distribution and Stability of Mists

While solid particles are relatively stable and independent of atmospheric conditions, the size and stability of mists is very dependent on their method of formation, temperature and humidity, and the presence and frequency of solid nuclei. The formation of droplets by breakup of liquids, or by homogeneous or heterogeneous nucleation has been covered in depth by Hidy and Brock (*4*).

For sulfuric acid mists, Kapcznski and Generalczyk (*5*) give Amelin's theory in which mist formation is associated with critical supersaturation of the vapor. Empirical observation confirms these theoretical predictions. Thus, Fairs (*2*) reports very different mist droplet size distributions for similar sulfuric acid plants operating under different conditions. In one case the droplets ranged from 0.8 to 13 $\mu$m, and in the other from 3 to 26 $\mu$m. He also reports that mists from directly fired acid concentrators with more nuclei present are both smaller and more numerous than mists from indirectly fired concentrators.

Brink and Contant (*6*) found that phosphoric acid mist droplets from an acid plant ranged from 0.6 to 3 $\mu$m. After passage through a venturi scrubber, the smallest droplets were only 0.3 $\mu$m in size.

The droplets from cooling towers are an order of magnitude larger. Chilton (*1*) found that for an up-flow cooling tower (2.1 m $sec^{-1}$) with a splash distributor, 6% of the doplets were less than 200 $\mu$m, 23% less than 300 $\mu$m, 50% less than 400 $\mu$m, and all less than 500 $\mu$m. Bell and Strauss (*7*) found that droplets from a packed cross-flow tower tended to vary from 40 $\mu$m to 1 mm, with maxima between 100 and 400 $\mu$m, depending on liquid loading and gas velocity (Fig. 1).

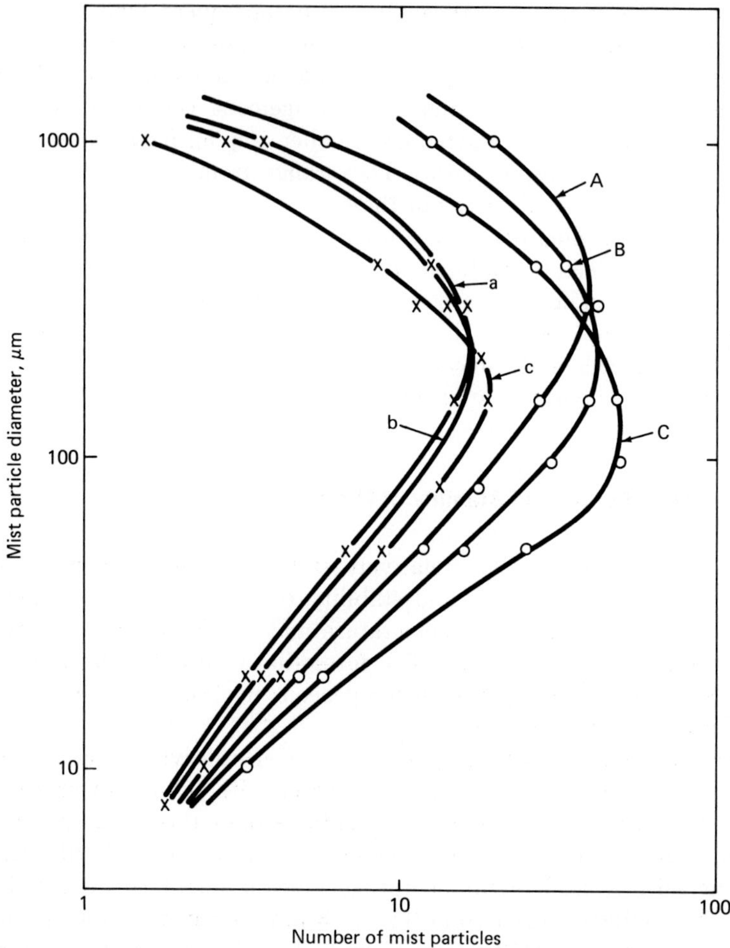


Figure 1. Droplet size distribution for mists entering and leaving a *W* mist eliminator after a cross-flow packed tower with different gas velocities and liquid loadings. A. Entering mist, 3 m $sec^{-1}$; 0.75 kg $sec^{-1}$ liquid (water) loading. a. Leaving mist, 3 m $sec^{-1}$; 0.75 kg $sec^{-1}$ liquid (water) loading. B. Entering mist, 3 m $sec^{-1}$; 1.50 kg $sec^{-1}$ liquid (water) loading. b. Leaving mist, 3 m $sec^{-1}$; 1.50 kg $sec^{-1}$ liquid (water) loading. C. Entering mist, 4.6 m $sec^{-1}$; 0.75 kg $sec^{-1}$ liquid (water) loading. c. Leaving mist, 4.6 m $sec^{-1}$; 0.75 kg $sec^{-1}$ liquid (water) loading.

## III. Theory of Mist Elimination

Mist droplet collection from a gas stream is basically the same as solid particle collection, but there are some simplifying features. First, small mist droplets are always spherical because of the surface tension of the liquid. Therefore, theories developed for spherical particles can be applied without modification. Second, when a droplet has landed on a surface, its subsequent shape and behavior depends on the hydrophobic or hydrophilic nature of the surface and the droplet. Thus, if the attractive forces between the liquid droplet molecules and the surface molecules are stronger than between the liquid molecules, the droplet will wet the surface, spreading out in a film. Water therefore wets a clean metal or glass surface, but not one coated with a hydrophobic oil or resin. It has been observed (*8*) that droplets collected by hydrophobic fibers such as glass wool treated with silicone resin, or garnetted polyester fibers, tend to resist reentrainment by agglomerating and running out of the filter.

### *A. Inertial Collectors*

The essential collection mechanism for droplets in inertial collectors is inertial impaction, and to a lesser extent interception. Droplets are generally too large for diffusion to play a major role. The collecting bodies are cylinders, spheres, or ribbons, and the theory of the mechanisms is reviewed fully elsewhere (*9*). (See also Chapter 3, this volume.) The equations of motion of the droplets in gas streamlines which divert around an obstacle have been solved numerically, and determined experimentally. The efficiency $E$ of collection by inertial impaction is found as a function of the inertial impaction parameter or separation number $\psi$ which is, numerically, half the Stokes' number:

$$\psi = D_p^2 V \rho_p / 18 \mu D_b \tag{1}$$

where

$\rho_p$ = droplet (liquid) density
$V$ = upstream velocity of the gas stream relative to the target
$D_p$ = droplet diameter
$D_b$ = target diameter or width of ribbon
$\mu$ = gas viscosity

The Cunningham correction (*9*, p. 182) is assumed to be 1, because of the relatively large size of the droplets, and is not included. The part of the inertial impaction parameter represented by $D_p^2 V \rho_p / 18\mu$ is the stopping distance, i.e., the distance that the droplet would travel horizontally if it were injected into still air with a velocity $V$, meeting Stokes' law resistance.

The experimentally determined relation between $E$ and $\psi$ is shown in Chapter 3, this volume, Figure 20. Inspection shows that with the inertial impaction parameter $\geq 6$, collection efficiencies over 80% are obtained with a cylindrical target body. On the basis of this Soole *et al.* (*10*) developed the rotary impaction filter, in which filaments set in a disk are spun across the gas containing the mist droplets, which impact on the filaments.

For a mist eliminator with a multiplicity of obstacles (targets) in series, such as a wire mesh mist eliminator or a fiber pack, the total efficiency $E_T$ for a number of $n$ of noninteracting targets in series is

$$E_T = 100[1 - (1 - E/100)^n] \tag{2}$$

A slightly modified form of this equation, proposed by Carpenter and Othmer (*11*) gives excellent agreement with the measured performance for a wire mesh eliminator:

$$E_T = 100[1 - (1 - E/100c)^n] \tag{3}$$

where $c = n/k'F$. $k'$ is the product of the wire diameter and total length divided by the cross-sectional area of the eliminator perpendicular to the gas stream, and $F$ is the total frontal area occupied by the target (i.e., wire) in each layer. In the example quoted, there were 48 layers of wire ($n$) of total length 11,234 m (from weight of wire, density of wire mesh, and wire diameter). The cross-sectional diameter of the eliminator was 0.6223 m giving an area of 0.304 m$^2$. The diameter of the wire was 0.287 mm and $F$ was found by inspection to be 0.67. Thus $c = 6.748$.

For the theoretically more complex case of the louver demister which consists of a series of parallel baffles, inclined at an angle $\theta$ to the direction of flow, Calvert *et al.* (*12, 13*) suggest the following equation for the efficiency $E$ of a series of $n$ sets of baffles:

$$E = 100\{1 - \exp - [(u_t/V)t(nw\theta/b \tan \theta)]\} \tag{4}$$

where

- $u_t$ = terminal velocity of the droplet with gravity acceleration, i.e., $D_p{}^2\rho_p g/18\mu$ for $D_p < 100\ \mu$m
- $w$ = width of baffle
- $b$ = distance between baffles, measured at right angles to the direction of flow
- $\theta$ = angle of inclination of baffle to the direction of flow, radians
- $n$ = number of rows of baffles of series

While the inertial impaction parameter gives the minimum velocity of the gas stream relative to the mist eliminator for a certain separation efficiency, the upper limit to the gas stream velocity is determined by the flooding and reentrainment characteristics of the packing. This can be calculated from the empirical Souders–Brown equation (*14*):

$$V = k\sqrt{(\rho_p - \rho)/\rho} \tag{5}$$

where $\rho$ is the gas stream density and $k$ is an empirical constant, the value of which depends on the units used. York *et al.* (*15, 16*) recommend that values of $k$ of 0.35 ft/second (i.e., 0.10 m sec$^{-1}$) be used for most vertical flow applications, although this should be reduced for high loadings and dirty liquids. If wetting agents are present in a water system, lower $k$ values should be used. York (*17*) further suggests that good performance is obtained for all velocities between 30 and 110% of the optimum, as calculated from Equation (5). In a cross-flow system, where there is no liquid hold up, an optimum value of $k$ of 0.40 ft/second (0.12 m sec$^{-1}$) is applicable (*7*).

Sorokin *et al.* (*18*) point out that Equation (5) is related to the Kutateladze number (Ku), which is

$$\text{Ku} = V\sqrt{(\rho)}/\sqrt[4]{\{g^2\sigma(\rho_p - \rho)\}} \tag{6}$$

where $g$ is the gravity acceleration and $\sigma$ the surface tension of the liquid. Comparing Equations (5) and (6) it can be seen that the "constant" $k$ in Equation (5) is

$$k = \phi\sqrt{g^2\sigma/(\rho_p - \rho)} \tag{7}$$

where $\phi$ is a coefficient independent of the physical properties of the liquid or gas. Sorokin *et al.* (*18*) show that the function $[g^2\sigma/(\rho_p - \rho)]^{1/2}$ is similar for many systems, and varies only slightly with pressure from subatmospheric (0.04 kg/cm$^2$) to several atmospheres (several kg/cm$^2$). However, large changes in the surface tension can affect the value of $k$.

Poppele (*19*) found experimentally in conventional ft-lb-second units that the value of $k$ varies with the liquid loading on knitted wire entrainment separators according to the empirical relation

$$k = (\log G - B)/m \text{ ft/second} \tag{8}$$

where $G$ is the liquid loading (lb/hour ft$^2$) and $B$ and $m$ are constants specific to an entrainment separator (Table I).

For baffles in a louver-type eliminator, flooding or partial flooding is unlikely to occur before reentrainment. The liquid flows across the baffle and is then reentrained from the edge. For both vertical and horizontal baffles, Calvert *et al.* (*12, 13*) give the velocity $V$ of the gas stream for the onset of reentrainment as

$$V = \sqrt{\sigma \cos\theta/\rho\delta} \tag{9}$$

**Table I Calculation of the Variation of the Souders–Brown Constant $k$ with Liquid Loading, Using Equation (8)[a]**

| *Eliminator type* | *York[b] type 421* | *York[b] type 931* |
|---|---|---|
| Mesh density $(\mathrm{lb/ft^3})$[c] | 12 | 5 |
| Wire surface area $(\mathrm{ft^2/ft^3})$[d] | 132 | 55 |
| Flood point | | |
| Liquid loading $(\mathrm{lb/hour\ ft^2})$[e] | 25–930 | 25–930 |
| $B$ | 3.71 | 7.22 |
| $m$ | −3.18 | −8.70 |
| Load point | | |
| Liquid loading $(\mathrm{lb/hour\ ft^2})$[e] | 5–930 | 5–930 |
| $B$ | 4.64 | 7.64 |
| $m$ | −7.15 | −11.1 |

[a] $k = (\log G - B)]/m$ ft/second. $B$ and $m$ are the constants in this equation.

[b] Designation of eliminators manufactured by O. H. York, Inc., New Jersey.

[c] To convert a metric density kg $\mathrm{m^{-3}}$ to $\mathrm{lb/ft^3}$, divide by 16.018.

[d] To convert metric surface area $\mathrm{m^2m^{-3}}$ to $\mathrm{ft^2/ft^3}$, divide by 3.28.

[e] To convert metric liquid loading kg/hour $\mathrm{m^2}$ to lb/hour $\mathrm{ft^2}$, divide by 4.882.

where $\delta$ is a calculated film thickness (*20*). In Equation (9) it is assumed that liquid collected is uniform across baffles that are wetted on both sides, interfacial stress is uniform, and film flow across the baffles occurs.

### B. Cyclone Collectors

Cyclone mist eliminators have virtually the same efficiency for liquid droplets as for solid particles. The equations given by Caplan (Chapter 3, this volume) can be used for predicting efficiency. However, to avoid reentrainment of the collected liquid on the walls, there is an upper limit to the tangential velocity (spinning speed) that can be used. Even at somewhat lower spinning speeds the liquid film may creep to the edge of the exit pipe, from which droplets are then entrained.

Stairmand (*21*) has suggested that the highest spinning speed, which is only slightly less than the inlet velocity because of surface friction, that can be used before reentrainment occurs at the walls is a function of the shear stress at the liquid film surface, represented by the product $\rho V^2$. With $\rho$ in kilograms per cubic meter and $V$ in meters per second this product should be less than 1330 for the air/water system, or twice this (2660) for the air/oil system. These values, which are 28% less than those originally given by Stairmand, are from Stearman and Williamson (*22*). For the air/water system at ambient conditions (with $\rho_{air} = 1.185$ kg $m^{-3}$) this represents a maximum velocity of 33 m $sec^{-1}$. However, Calvert *et al.* (*12, 13*) found no reentrainment with velocities of 45 m $sec^{-1}$, indicating that the Stearman and Williamson values are conservative, and Stairmand's original value is more realistic.

These velocities are well above the optimum spinning speed of about 15 m $sec^{-1}$ recommended by Stairmand (*21*). With increasing gas pressure, the density rises, and the velocity calculated from the relation, $\rho V^2 = \text{const}$, falls off. Thus, at pressures above 5 atm, the optimum velocity of 15 m $sec^{-1}$, which is the maximum recommended by Stearman and Williamson (*22*), should be about the limit before reeentrainment.

## IV. Single-Stage Mist Eliminators

### A. Disengagement Chambers

The simplest system is a chamber in which droplets are removed from the gas stream by gravity. A simple sloped channel for draining the collected liquid is required. Since the reentrainment velocity for liquid droplets is high (about 3 m $sec^{-1}$), a small chamber, with relatively high velocity is feasible for removal of spray droplets. The maximum velocity is the same as for reentrainment of liquid from the walls of cyclones (Section III,B).

For simple settling chambers of length $L$, breadth $B$, and gas volume flow $Q$ (all in uniform units of meters and cubic meters per second), the minimum droplet size that theoretically will be collected is

$$d_{min} = \sqrt{18Q\mu/\rho_p gBL} \qquad (10)$$

The effectiveness of a horizontal flow chamber can be improved by inserting a number of horizontal plates into the chamber, thereby decreasing the vertical distance a droplet has to fall before capture. By using waved plates and sloping them so that the droplets drain to one end, an inexpensive, low pressure drop collection system for relatively fine droplets can be constructed.

The other common type of disengagement chamber, usually referred to as a "knock-out" drum (Fig. 2) will remove all droplets with a falling speed greater than the upward flow of gases. Since a 400-$\mu$m-diameter droplet has a falling speed in air of 1.6 m $sec^{-1}$ (at 20°C), a 100-$\mu$m droplet

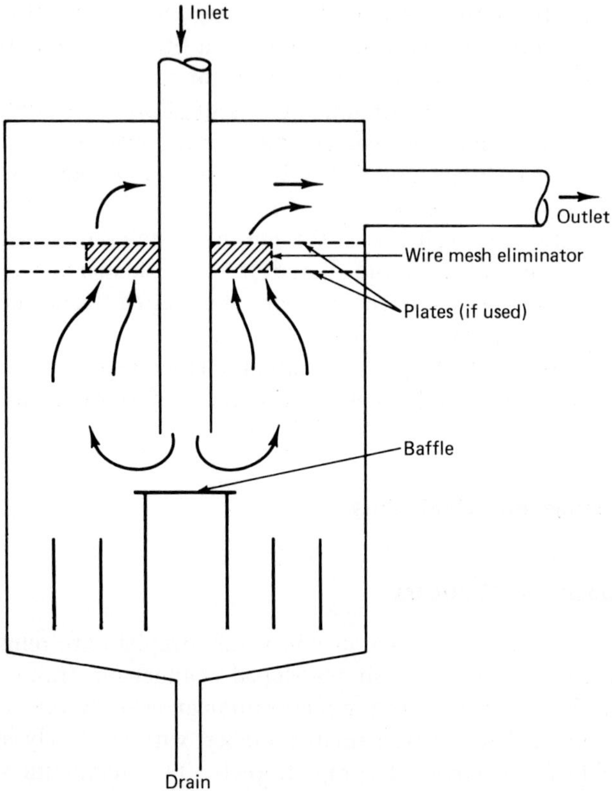


**Figure 2. Knock-out drum, showing position of wire mist eliminator if used. After Stearmann and Williamson (*22*).**

a falling speed of 0.25 m sec$^{-1}$ and a 40-$\mu$m droplet a falling speed of 0.048 m sec$^{-1}$, the knock-out drum is useful for droplets over 100 $\mu$m. Stearman and Williamson (*22*) recommend that the diameter of the vertical inlet pipe should be such that the velocity $V$ (meters per second) is not greater than that given by the product $\rho V^2 = 185$. In the case of a gas at ambient pressure and temperature $V$ is approximately 11 m sec$^{-1}$. Thus a separation chamber, approximately 3.5 m diameter with an 0.5 m diameter inlet pipe will treat about 7800 m$^3$ hr$^{-1}$ (4000 ft$^3$/minute), removing droplets greater than 100 $\mu$m.

To avoid reentrainment, a knock-out drum should contain an impingement disk, one inlet pipe diameter below the inlet pipe, and twice the diameter of the inlet pipe. This protects the liquid surface from the impinging air jet. Stearman and Williamson (*22*) also recommend a number of cross baffles (four is suggested) to further reduce any reentrainment.

These workers also show combinations of knock-out drums with other separators, such as wire mesh demisters, to remove finer droplets.

### B. Inertial Separators: Venetian Blind, V, W, and Wave Separators

These simple eliminators are widely used for relatively coarse mists and sprays from cooling towers, because of their small space requirements and low pressure drops. They operate by diverting the gas stream and ejecting droplets onto collector baffles. In their simplest form they employ parallel strips set at angles of about 30° to 45° in the direction of the gas stream (Fig. 3a). Chilton (*1*) investigating this design in a vertical upward-flow gas stream, using 75-mm-wide (12-mm-thick) wooden slats, found that 45° slats were somewhat more efficient at lower velocities (1 to 1.5 m $sec^{-1}$) when compared with 30° slats. At higher velocities (over 2 m $sec^{-1}$) there was little difference in the measured collection efficiency. However, at the 45° angle there was higher flow resistance. In order to obtain optimum performance with such a system it would therefore be desirable to vary the angle of the baffles with gas stream velocity.

Chilton also studied the effectiveness of adding a second row of parallel baffles, at 20° and 30° to the direction of flow (Fig. 3a). Removal efficiency was improved from 60% without the second row to 85% at low velocities (1 m $sec^{-1}$), but only marginal improvement (96% without the second row to 97% with the second row) was obtained at higher velocities. The pressure loss with a second row added was only slightly (10%) greater than with a single row.

The next stage of complexity is the double *V* or *W* pattern mist eliminator, which Bell and Strauss (*7*) and Calvert *et al.* (*12, 13*) have studied in a cross-flow system. Detail of the pattern used by Bell and Strauss is shown in Figure 3c. Calvert *et al.* used a broader blade (76 mm cf 63 mm) and a closer spacing (25 mm cf. 32 mm). The maximum efficiency achieved by Bell and Strauss was 76% at superficial velocities of 4.8 m $sec^{-1}$. Calvert *et al.* report almost 100% effectiveness with droplets somewhat coarser (84 $\mu$m to 1.2 mm) in the range of gas velocities from 3 to 6 m $sec^{-1}$, although at 7 m $sec^{-1}$ the efficiency fell off to 78%.

Bell and Strauss (*7*) investigated a range of mist droplet loadings at several gas stream velocities, and determined the droplet size distribution of the mists entering and leaving the eliminator (Fig. 1). Although the range of droplet size was from about 40 to 500 $\mu$m, doubling the water input to the scrubber preceding the mist eliminator increased the number of droplets entering the eliminator, but reduced their mean size. The *W* type proved more efficient in removing larger droplets than smaller ones. At lower water loadings (0.75 kg $sec^{-1}$) the mean droplet size was 400

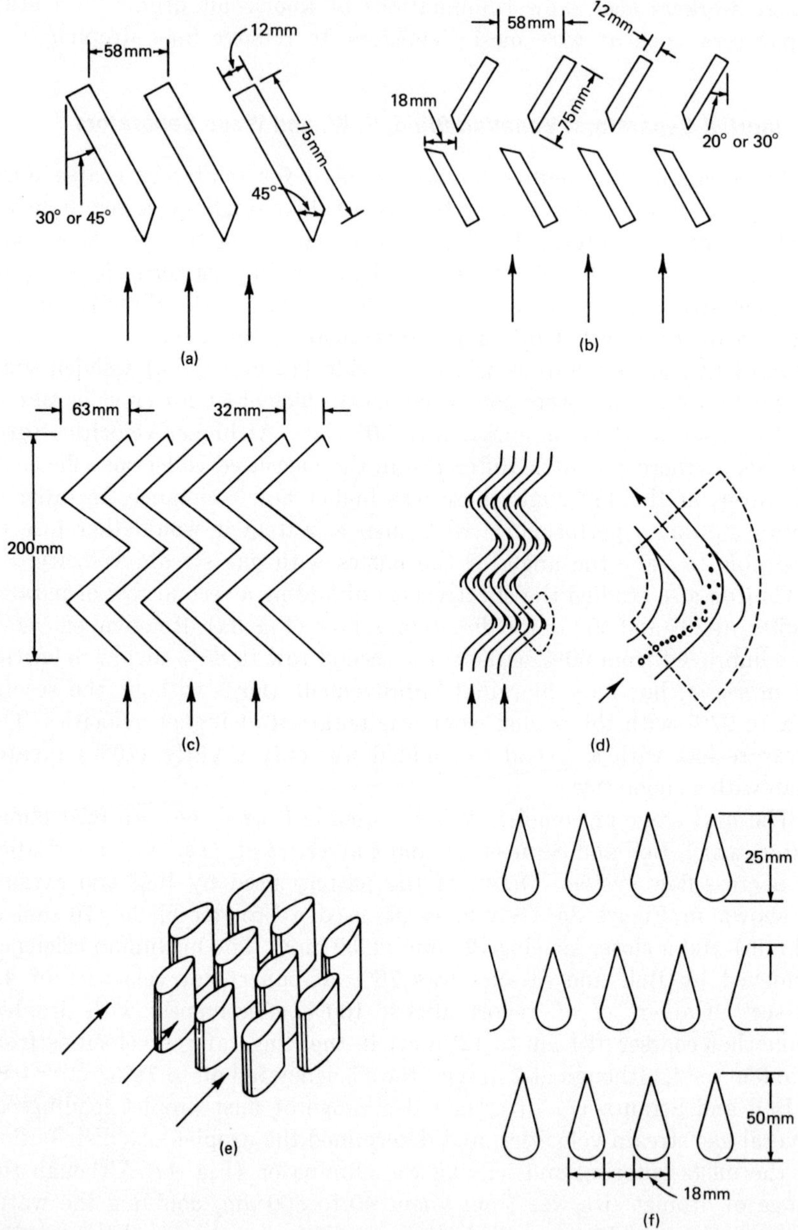


Figure 3. Simple inertial separators. (a) Single row of baffles (*1*), (b) two rows of baffles (*1*), (c) *W* pattern separator (*7*), (d) wave separator (Euroform pattern) (*24*), (e) vertical rod tear-drop shape in horizontal flow system (*31*), (f) horizontal rod with tear-drop shape in vertical flow system (*1*).

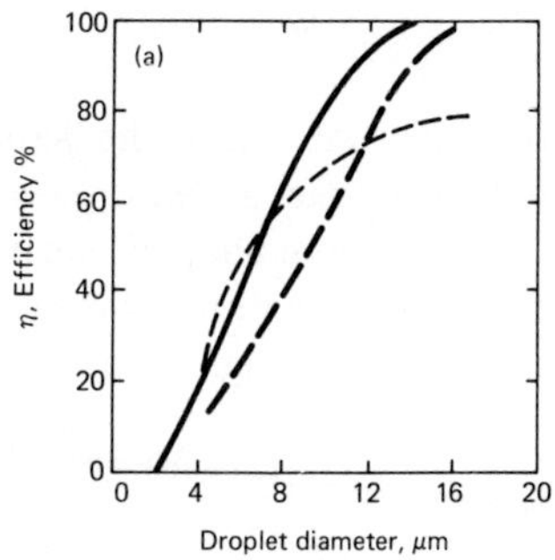


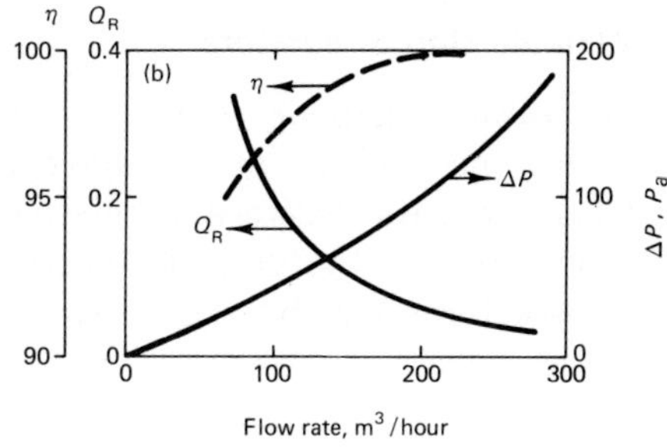


Figure 4. Performance of Euroform separator (a) Fractional efficiency curves for Euroform and simple wave separator (*23, 24*). Solid line: Euroform type TS 5; long dash line: Euroform type TS 5/2; short dash line: simple wave eliminator. (b) Performance of a separator on a scrubber on a fertilizer plant with gas throughput of 250,000 m$^3$/hour: correlation of flow rate with efficiency $\eta$ (%), pressure drop, $\Delta P$ (Pa), and residual moisture, $Q_R$ (ml/m$^3$). Efficiency was 99.9% under normal operating conditions.

μm before and 300 μm after the eliminator. At higher loadings (1.5 kg sec$^{-1}$) the mean was 200 μm before and 300 μm after the eliminator. With constant liquid loading (0.75 kg sec$^{-1}$) and a 50% increase in superficial gas velocity (3 to 4.5 m sec$^{-1}$) the total number of droplets entering the eliminator increased by 20% and the mean size fell to 100 μm, while on leaving, the number increased by 30% with a mean size of about 150 μm, showing the effect of increased entrainment at the higher velocity, and the need for better design to prevent this.

The pressure drop through these demisters at 4.8 m sec$^{-1}$ did not exceed 50 Pa* [0.2 in. water gauge (W G)] (see also Fig. 10). Calvert *et al.* (*12, 13*) obtained a pressure drop of 120 Pa (0.476 in. WG) at 4.8 m sec$^{-1}$ and 280 Pa (1.1 in. WG), at 6 m sec$^{-1}$ due to the closer spacing of their baffles.

Brink (*23*) reports that the efficiency of a wave-shaped demister made up of 11-mm 90° arcs, with 3.8-mm clearance between sheets was 75% for droplets over 15 μm. For droplets from a bubbling source, Sorokin *et al.* (*18*) measured 68% removal efficiency on a similar design. Regehr (*24*) has placed a special collecting slot on the convex side of the wave (Fig. 3d) which propels the droplets to the concave side of the eliminator and reduces droplet entrainment from this point, which is a major source of inefficiency in conventional *V* and *W* eliminators.

Figure 4a shows fractional efficiency for simple and Regehr-modified wave eliminators, while Fig. 4b gives efficiency and pressure drop data for the modified collector used on a scrubber in a fertilizer plant. The modified collector is best used at velocities from 4.5 to 6.5 m sec$^{-1}$.

* Pa = Pascal = 0.1 mm water gauge (WG) (pressure), 1 kPa = 4 in. WG.

### C. Variable Gap Separators

The variable gap or "pressure jump" separator (*25*) uses the same operating mechanisms as the wave separator. The gas stream passes through the gaps between successive turns of a wave-shaped spring thereby throwing the droplets together (Fig. 5). The distance between successive turns can be decreased by tightening the spring. Efficiencies are given in Table II.

### D. Rotary Impaction Filters

Instead of passing the gas stream at high velocity through a series of stationary collecting bodies, the collecting bodies can be moved at high

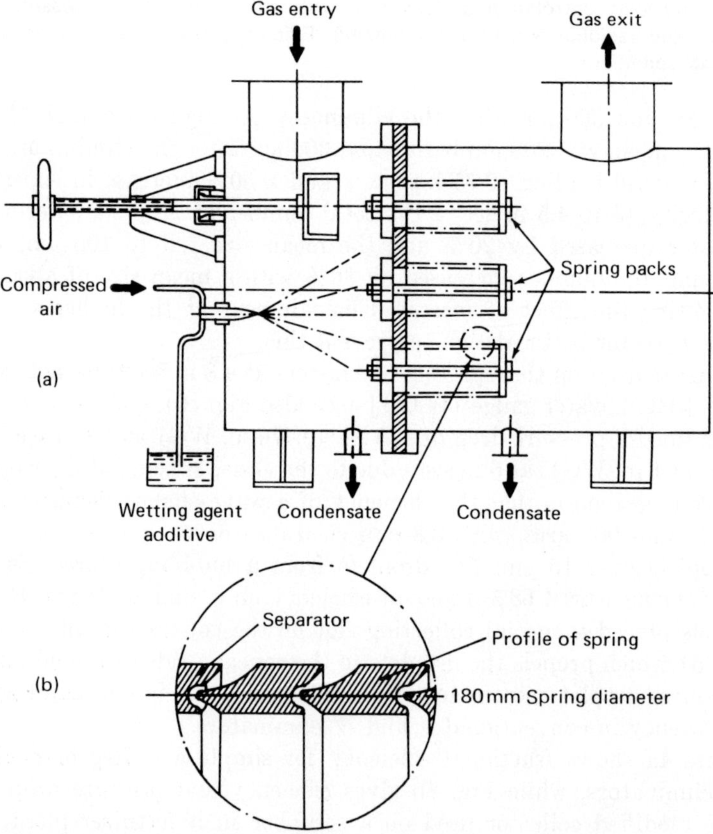


Figure 5. Variable gap separator (*25*). (a) Variable gap mist eliminator assembly; (b) detail of shape of elements.

**Table II Efficiency of the Variable Gap Mist Eliminator** (*25*)

| | *Inlet concentration* ($gm/m^3$) | *Outlet concentration* ($gm/m^3$) | *Efficiency* (%) |
|---|---|---|---|
| Phosphoric acid mist from acid, manufactured as $P_2O_5$ | 52 | 0.06 | 99.9 |
| Sulfuric acid mist from acid, manufacture as sulfur trioxide | 18 | 0.15 | 99.16 |
| Hydrogen chloride in waste gases from $CaCl_2$ manufacture | 17.7 | 0.0004 | 99.998 |
| Hydrogen fluoride in waste gases from superphosphate manufacture | 1.8 | 0.005 | 99.73 |
| Hydrogen chloride waste gases from pickling | 0.08 | 0.0002 | 99.75 |
| Waste gases from sulfuric acid, manufacture—ammonia added | | | |
| Sulfuric acid | 0.8 | 0.05 | 93.75 |
| Sulfur dioxide | 6 | 0.2 | 96.7 |

velocity through a slow moving gas. One example of this is the rotary impaction filter, in which collecting wires are spun at high speed. In initial designs, the wires were loose at their outer ends but they tended to break because of vibration. In later designs they are supported by an outer rim connected to the central disk by spokes.

It was first suggested that the velocity of the wires, and their size relative to the droplets must be such that the Stokes' number (which is twice the inertial impaction parameter) is equal or greater than 12 (i.e., $\psi \geq 6$). Subsequent work has shown that $\psi = 6$ is, in fact, an optimum value, giving collection efficiencies of 100% for 10-$\mu$m droplets. This decreases linearly to 92% for $\psi = 20$ (10-$\mu$m drops) (*25*). Tapered elements, which produce a consistent inertial impaction parameter along their length, are more effective than cylindrical wires. For 50-$\mu$m droplets the increased efficiency of using tapered elements was from 99.80 to 99.96%, although for 10-$\mu$m droplets the effect was far less (*27*).

In an experimental unit tested by Pyne *et al.* (*10*), which was capable of treating 1.18 $m^3\ sec^{-1}$ (2500 $ft^3$/minute), the outer diameter $D_2$ was 0.76 m and the filaments had a length of 0.1 m. At 1900 rpm, the linear speeds at the inner and outer ends of the filaments are 56 and 76 m $sec^{-1}$, respectively (*28*). For droplets with diameter $D_p$ (sufficiently large that the Cunningham correction is unity) and density $\rho_p$, approaching a collecting body (diameter $D_b$) with velocity $V$ and an inertial impaction parameter of 6, the following relation between $D_p$ and $D_b$ can be deduced from Equation (1):

$$D_b = (V\rho_p/108\mu)D_p^2 \tag{11}$$

For the unit investigated, for $V$ at the inner end equal to 56, $\mu$ equal to $1.8 \times 10^5$ N-second/m$^2$ (i.e., air at 18°C) and $\rho_p$ equal to 1000 kg m$^{-3}$:

$$D_b = 2.88 \times 10^7 D_p^2 \tag{12}$$

Therefore, for 4-$\mu$m droplets, the wire filaments should have a diameter of 0.460 mm [26 Birmingham wire gauge (BWG)]. The number of filaments required $N$ is based on the swept volume at $n$ rpm, which must be equal to the volume flow $Q$ if all droplets are to have a chance of being collected. The gas passes through the annulus swept by the filaments, the area of which is $(\pi/4)\ (D_2^2 - D_1^2)$, where $D_1$ and $D_2$ are the inner and outer diameters of the annulus, so that in unit time the volume swept is

$$Q = D_b N(D_2^2 - D_1^2)n(\pi/4) \tag{13}$$

In the above example, $N$ is 390. The number of filaments cannot exceed the space available for them on the inner disk. A further design parameter is therefore $ND_b < \pi\ D_1$. In this case $ND_b$ is $390 \times\ 460 \times 10^{-6}$ m, i.e., 1.8 m, compared with $\pi D_1$, which is 1.76 m. While, in general, pressure loss through a rotary impaction filter is low (less than 100 Pa), pressure drop rises sharply with closer spacing of the filaments.

A later model, able to handle four times the volume (10,000 ft$^3$/minute) was investigated intensively. The disk could be spun at 500 and 1000 rpm, the outer diameter was 1.12 m, and the inner diameter could be varied from 0.266 to 0.742 m. Wire filaments ranged from 0.914 mm (20 BWG) to 2.032 mm (14 BWG), with typical values for an inner disk of 0.688 m and 1.829 mm (15 BWG) wires. At 500 rpm the pressure loss was 25 Pa and about twice this at 1000 rpm.

The device has the theoretical advantage that by choosing the right combination of speed of rotation and filament diameter, it can be designed to collect at very low pressure loss virtually all droplets above a certain size. Some anomalies became apparent after further experimental work, notably that contrary to expectation, a decrease in filament diameter led to a decrease in effectiveness. Soole (*28*) points out that the assumption of $\psi = 6$ for 100% collection, while a useful preliminary value, is not correct, as it is based on the measured single wire collection efficiencies of Wong, Ranz, and Johnstone (*29*). These are 10% greater at $\psi = 6$ than the values of May and Clifford (*30*). That efficiencies greater than predicted from single wire measurements, and calculations based on potential flow are obtained, is due to deposition in the lee of the filaments from their turbulent wake (*28*).

## V. Multiple-Stage Mist Eliminators (Few Stages)

### A. Simple Impingement Separators

The simplest mist eliminator pattern uses rows of vertical tubes in a horizontal flow system. Calvert *et al.* (*12, 13*) tested an arrangement of 19-mm tubes, spaced equidistant to one another in six staggered rows, with stream velocities from 1 to 7 m $sec^{-1}$ and droplets greater than 84 $\mu$m. Collection efficiencies of 95 to 100% were achieved.

Vertical rods, which have a streamlined teardrop cross section, are used commercially (Fig. 3e). The manufacturers claim (*31*) that for an Equation (5) coefficient $k$ of unity, which corresponds to a flow velocity of 8.5 m $sec^{-1}$ almost 100% effectiveness is achieved for 11-$\mu$m droplets of water in air at 1 atm, with 46% removal for 5-$\mu$m droplets.

A similar pattern (Fig. 3f) with two and three rows was tested by Chilton (*1*) in a vertical flow system. With two rows 66% efficiency was attained at 1 m $sec^{-1}$ and 85% at 3 m $sec^{-1}$. With three rows over 90% efficiency was obtained.

### B. Calder–Fox Scrubber

The device known as the Calder–Fox scrubber was the first mist eliminator developed for collecting acid mists by using one or two stages of collection by impingement. It was developed by Calder, Fox, and Palmer in 1915 to replace coke boxes (*3*).

In the earliest designs, small and large holes were punched in successive lead plates, the first acting to accelerate the gas stream and the second acting as an impact plate. In later designs glass strips were used. Typically the glass strips were 3.2 mm × 6.4 mm, set 0.8 mm apart, and the impact plate strips were 15.8 mm × 5.5 mm, set 1.6 mm apart, both on 7.14-mm centers. Alternative designs are given by Fairs (*3*). In a more recent design, the initial (main) plate assembly is followed by a collector plate assembly of equal sized plates. Fairs (*2*) offers procedures for calculating the relation between particle size, reentrainment and pressure loss, as well as a number of other factors, such as gas density and velocity. He suggests that the Calder–Fox scrubber is relatively cheap and effective for mist droplets above 2.5 to 3 $\mu$m, if sufficient pressure is available. However, knitted wire and fiber bed eliminators have largely replaced the Calder–Fox scrubber.

## VI. Multiple-Stage Impingement Separators (Many Stages)

### *A. Wire Mesh Mist Eliminators*

Knitted wire mist eliminators are formed from meshes of wire, knitted into a cylindrical open weave stocking, which is then crimped to give a stable wire configuration, with a very large number of successive impingement targets. The wire can be of a wide range of materials,* depending on the application, temperature, and corrosive nature of the medium in which it operates. The wire has to be flexible, so that it can be woven, yet sufficiently rigid so that the meshes, after crimping, remain in a stable configuration during the life of the eliminator. The stockings are woven from round or laminated filaments between 0.11 and 0.33 mm in diameter (29–42 BWG), are then flattened to give a double layer, crimped (if metal) or pleated (if polymeric material), and then are formed into pads by rolling or otherwise placing the crimped layers together, to give a shape which will fill the required cross section of the tower, duct, or housing (Fig. 6).

The depth of pad used varies from 50 to 300 mm (2–12 in.), with 100- to 150-mm pads (4–6 in.) being the most common. In small cylindrical vessels the eliminator is a simple spiral wound pad, while in larger vessels it is made of a number of sections fitted together. The pad is tied to a support frame, as shown in Figure 6. If the quantities of liquid to be removed are very high and many of the droplets are in the coarser size ranges, the wire mesh eliminator can be placed in a knock-out drum (Fig. 2). However, to get adequate velocity through the mesh generally requires about three to four times the inlet pipe cross section. This requires a reduction in cross section in a knock-out drum originally designed as a gravity settling chamber.

Although knitted wire eliminators are widely used, there is surprisingly little detailed information published on their effectiveness, particularly in vertical flow systems. Carpenter and Othmer (*11*) measured the efficiency of collecting brine droplets from a vertical tube evaporator with a wire mesh eliminator. The eliminators were 622 mm ($24\frac{1}{2}$ in.) and 149 mm ($5\frac{7}{8}$ in.) in diameter and 100 mm deep, and were made of 0.287-mm-diameter copper wire. The eliminator had a free volume of 98%. At a superficial gas velocity of 0.6 m $sec^{-1}$ (2 ft/second) they were 79.3% efficient, exceeding 99% at 3.34 to 4.25 m $sec^{-1}$ (11–14 ft/second). This

* The following have been used: copper, aluminum, mild steel (carbon steel), stainless steel (different grades), Monel, Inconel, tantalum, titanium, Carpenter alloy, Hastalloy, polypropylene, polytetrafluorethylene, and polyvinylchloride.

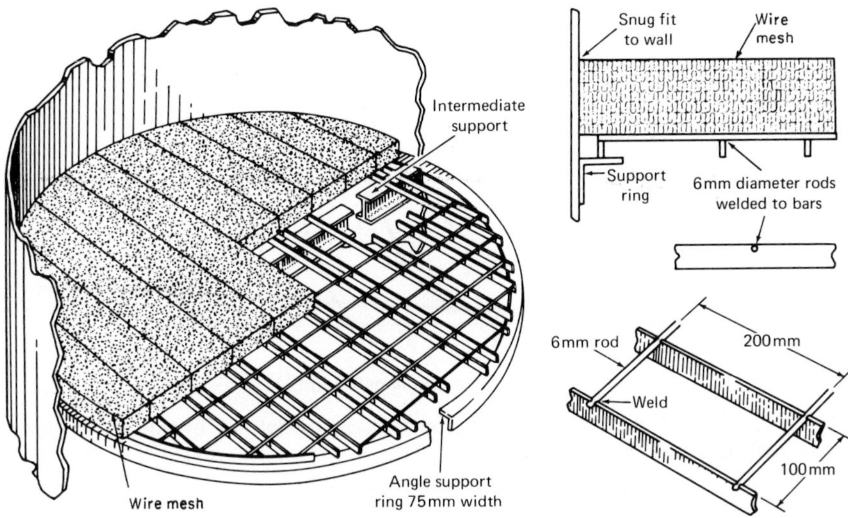


Figure 6. Wire mesh mist eliminator assembly in a large round column.

fell due to reentrainment when the gas velocity exceeded 5.47 m $sec^{-1}$ (18 ft/second).

Three detailed investigations have been published on horizontal flow systems with vertical eliminators; two (*7, 12, 13*) with commercial sized units, and one (*32*) on a laboratory system. Bürkholz (*32*) used a 150 × 150-mm eliminator with 0.27-mm wire, arranged to give 30 successive stages of collection, with velocities from 0.35 to 20 m $sec^{-1}$ (Fig. 7).

Bell and Strauss (*7*) used a 600 × 600-mm (2-ft-square) eliminator, 100 mm deep, made of 0.25-mm stainless-steel wire, with a packing density of 160 kg $m^{-3}$ (10 lb/$ft^3$). The water droplet size on entering the eliminator was as shown in Figure 1, and on leaving as shown in Figure 8. The highest efficiency obtained (92%) was at 4 m $sec^{-1}$ (13 ft/sec) velocity above which entrainment occurred (Fig. 9). Bell and Strauss (*7*) also tested a 150-mm-deep eliminator made of crushed aluminum turnings (25 μm thick, averaging 2.5 mm diameter) packed between wire mesh to a packing density of 64 kg $m^{-3}$ (4 lb/$ft^3$). In this eliminator the maximum efficiency was only 84%, due to channeling of the gases through irregularities in the packing. The efficiency/velocity correlation and droplet size distribution for this eliminator is also shown in Figures 8 and 9. Figure 8 shows clearly that the wire mesh eliminator, and to a lesser extent the packed aluminum turning eliminator, is more efficient at collecting droplets less than 50 μm than the *W* baffle eliminator.

Calvert *et al.* (*12, 13*) used a 610-mm-long eliminator made of 0.28-mm

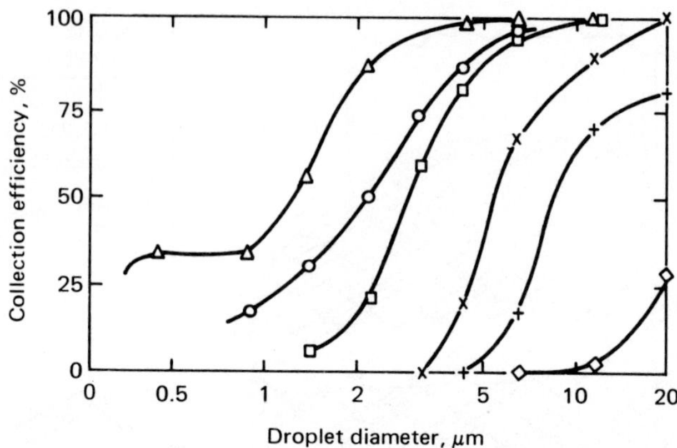


Figure 7. Effect of gas stream velocity on fractional efficiency of wire mesh eliminator (*32*). See following tabulation.

| | *Flow velocity, m sec$^{-1}$* | *Pressure drop, Pa* |
|---|---|---|
| ◇ | 0.35 | 0 |
| + | 0.70 | 0 |
| × | 1.40 | 0 |
| □ | 2.8 | 15 |
| ○ | 6.0 | 120 |
| △ | 12 | 300 |

wire and in a 100-mm-deep bed, with a packing density of 144 kg m$^{-3}$ and a void volume of 98.2%. With water droplets 84 μm and larger, almost 100% efficiency was obtained for velocities from 1 to 7 m sec$^{-1}$. At velocities above 7 m sec$^{-1}$, some entrainment was apparent, and efficiency fell to about 90%.

Pressure drop is a complex function of velocity in mist eliminators, particularly horizontal units operating in a vertical gas stream, because of loading, and subsequent blocking of the eliminator by the collected liquid, similar to the behavior of packed towers. In normal operation, an eliminator working below the loading point is unlikely to have a pressure drop of over 250 Pa (1 in. WG). Figure 10 shows the loading points, i.e., the change in the direction of the pressure drop/velocity curves, observed visually as penetration of liquid into the mesh; and the flooding or reentrainment points for two wire mesh eliminators (York type 421 and 931).

First-stage flooding has been used as a method of droplet agglomeration for very fine droplets, such as sulfuric acid mists, in two-stage eliminators (*33*). Here the lower first stage, which acts as an agglomerator, is a high

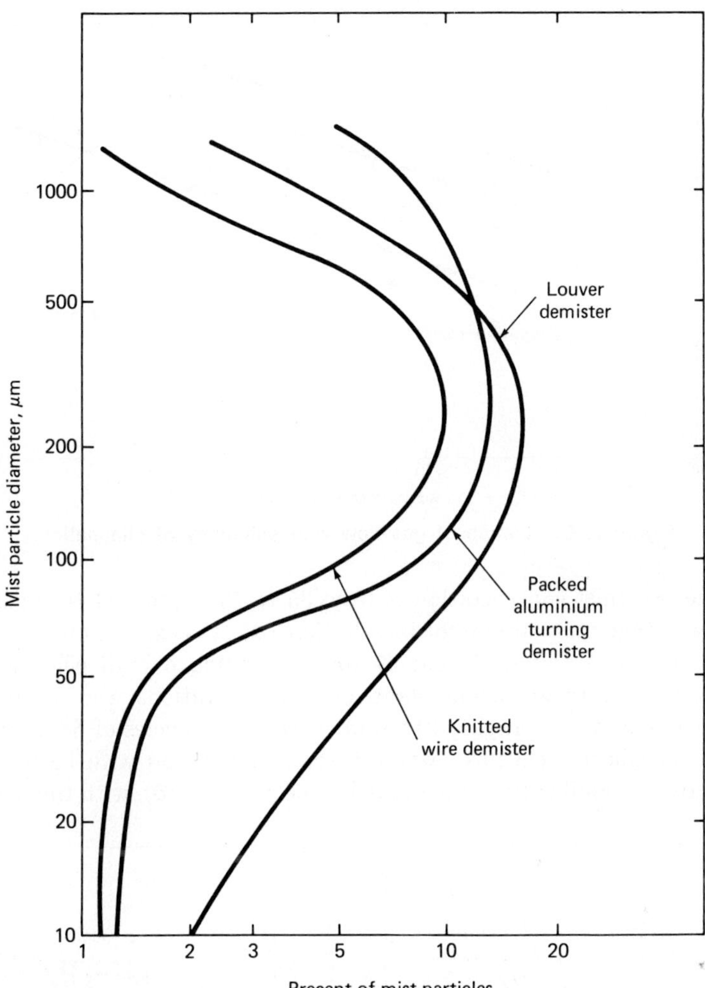


**Figure 8. Droplet size distribution for droplets leaving knitted wire, packed aluminum turnings and *W* louver mist eliminators (*7*).**

density mesh, in practice about 192 to 224 kg $m^{-3}$ (12–14 lb/$ft^3$), while the upper second stage has a much lower packing density, 80–113 kg $m^{-3}$ (5–7 lb/$ft^3$). Each pad is from 100 to 200 mm thick, and the two are separated by about $\frac{3}{4}$ equivalent tower diameters. For sulfuric acid stacks the optimum velocity is from 4.6 to 5.5 m $sec^{-1}$ (15 to 18 ft/second), but under turn-down conditions, the mist eliminator remains effective for velocities as low as from 3 to 3.3. m $sec^{-1}$ (10 to 11 ft/second). The unit has a total pressure drop of from 370 to 500 Pa (1.5–2 in. WG),

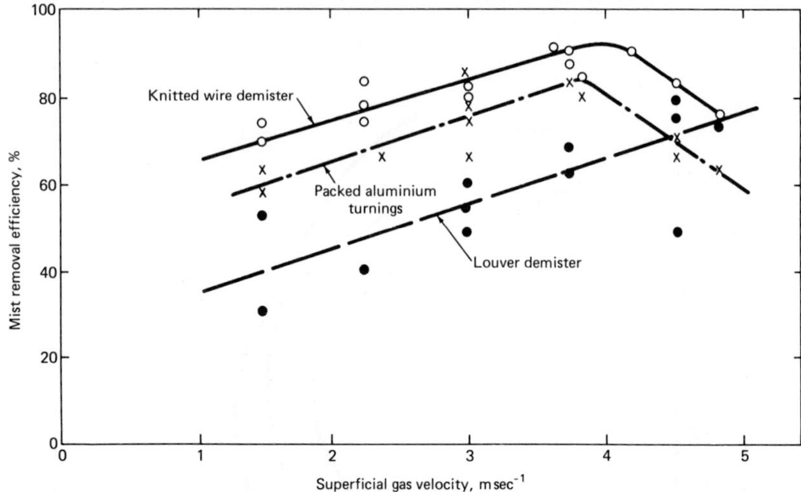


Figure 9. Correlation of gas flow with efficiency of elimination.

and the exit mist concentration is from 35 to 70 mg m$^{-3}$ (1 to 2 mg/ft$^3$). In a two-stage mist eliminator installation (*34*) the exit acid mist concentration varied between 37 and 57 mg m$^{-3}$ giving overall efficiencies of 97.4 to 99.1%. In an oleum plant, exit concentrations were even lower (5.7 to 45 mg m$^{-3}$, averaging 19 mg m$^{-3}$), with efficiencies of 88 to 96%.

For phosphoric acid mist, where there is a problem with blocking the eliminator, a similar two-stage unit has been used (*35*) with the addition

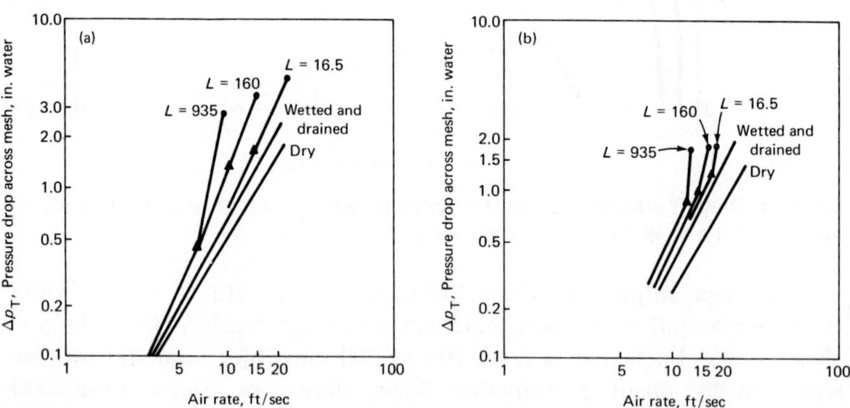


Figure 10. Correlation of pressure drop with gas flow at different liquid loadings for York type mist eliminators (*16*). *L*: liquid loading in lb/hour-ft$^2$. Thickness: 6 in.(a) Type 421 mesh style; ▲: 1 in. penetration depth of water into mesh; ●: reentrainment point. (b) Type 931 mesh style; ▲: 2 in. penetration of water into mesh; ●: reentrainment point. For details of mesh see Table I.

of an auxillary water spray on the agglomerator section. Efficiencies of 99.96 to 99.98% were obtained with exceptionally high pressure drops of 10 kPa (40 in. WG) and a velocity of 8.4 m $sec^{-1}$ (27.5 ft/second). Inlet loadings were of the order of from 112 to 165 gm $m^{-3}$ and outlet concentrations were from 0.038 to 0.025 gm $m^{-3}$. In general, the liquid loading for wire mesh eliminators should not exceed 115 gm $m^{-3}$ (50 grain/$ft^3$).

The pressure drop through a vertical wire mesh eliminator in a horizontal gas stream tends to be less than in a vertical countercurrent flow system. Thus at the maximum velocity used by Bell and Strauss (*7*) (4.8 m $sec^{-1}$, 16 ft/second) it was only 140 Pa (0.55 in. WG) (Fig. 11). The pressure drop/velocity relation for the knitted wire and packed aluminum

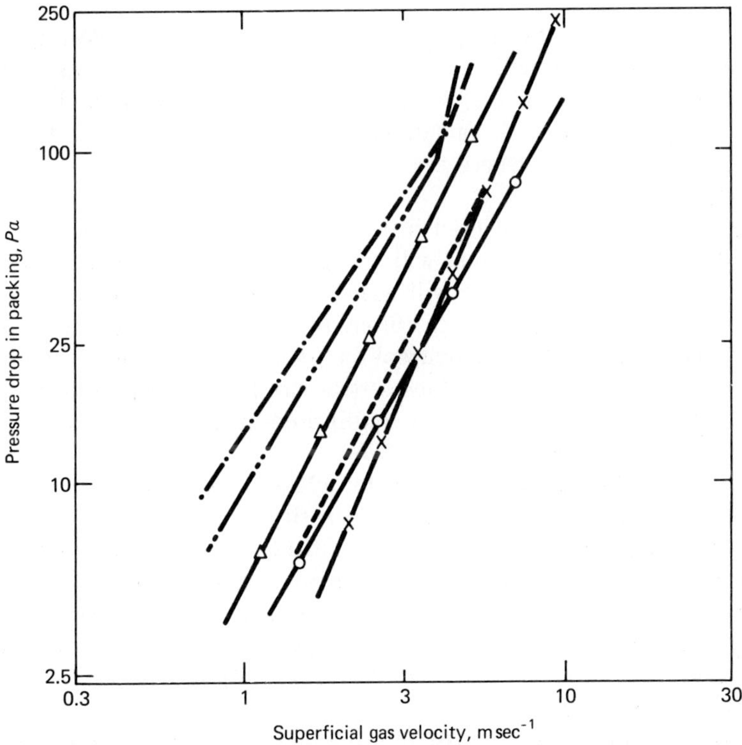


Figure 11. Correlation of pressure drop with superficial gas velocity for cross-flow mist eliminators, investigated by Bell and Strauss (*7*), Jashnani and Calvert (*13*), and Burkholz (*32*). Dot-dashed line: packed aluminum turnings (150 mm), Bell and Strauss (*7*); double dot-dashed line: packed aluminum turnings (150 mm), Bell and Strauss (*7*, double dot-dashed line); dashed line: multi-louvers, Bell and Strauss (*7*); △: zig-zag baffles, Jashnani and Calvert (*13*, Fig. 11); mesh (100 mm), Jashani and Calvert (*13*, Fig. 11); ×: mesh (60 mm, regular pattern), Burkholz (*32*); O: Jashnani and Calvert (*13*, Fig. 11).

turnings, on a log–log plot, consist of two intersecting straight lines, the intersection coinciding with the optimum velocity for maximum collection efficiency, i.e., the reentrainment points. The pressure drops measured by Calvert *et al.* (*12, 13*) are somewhat lower at high velocities, and appear to lie parallel to the line drawn through the Bell and Strauss points before the entrainment point. On the other hand, the pressure drops measured by Bürkholz (*32*), again for very high velocities, lie parallel to the extension of the line drawn through the Bell and Strauss points after the point of reentrainment. It is therefore necessary to consider droplet behavior when designing wire mesh eliminators for optimum velocity.

### B. Fiber Mist Eliminators

Inertial impaction theory predicts that the smaller the collecting body, the more effective it will be in collecting finer droplets. Thus, a filter constructed from fine fibers, 30 to 15 μm diameter, should be better for collecting fine mists, such as sulfuric acid or phosphoric acid, than one made from wire mesh where the wires are usually from 100 to 350 μm diameter.

Many fine fibers, when used in a mist eliminator, will become "wet" with the liquid, i.e., the liquid will spread over the fiber, unless the fiber is hydrophobic. Such wetting will lead to matting of the fibers, retention of the liquid, and blocking of the filter. It is therefore necessary to select hydrophobic fibers, or treat hydrophilic fibers to give them hydrophobic surface properties, and to produce a structure from which liquid droplets will readily drain after an initial equilibrium amount of liquid has been collected.

Fairs (*8*) describes a treatment for glass fibers which establishes a stable structure at the right packing density. First the fibers are compressed from an original packing density of 32–48 kg $m^{-3}$ to about 112–160 kg $m^{-3}$. The packed fibers are then heated at 500°C for 1 hour, which relieves the stresses in the compressed fibers and stabilizes the structure. To produce hydrophobic properties, the glass wool pack is treated with a silicone resin obtained from the partial hydrolysis and polymerization of methyl chlorosilane.

Fairs used an eliminator, 762 mm diameter and 50 mm thick, on the tail gases of an acid plant. The pad rested on coarse, resin-coated steel gauze. The average gas flow through the filter was 0.078 m $sec^{-1}$, with a pressure drop of 2.5 kPa (10 in. WG). Initially, satisfactory operation was obtained, with exit concentrations averaging 0.6 mg $m^{-1}$ and never exceeding 0.83 mg $m^{-3}$ sulfur trioxide. At close to 2000 hours operation, bisulfite deposits blocked the filter, and the supporting gauze failed. After repairs, a total of 4000 hours was established before retreatment of the fiber bed with the silicone resin was required. After retreatment, the exit concentration averaged 0.4 mg $m^{-3}$. A normal service life of 5000 hours between maintenance periods is usual.

Fairs also tested a filter pad made of garnetted polyester fiber (Terylene)* which gave slightly poorer performance, averaging exit concentrations of 0.8 mg $m^{-3}$ with flow rates of 0.068 m $sec^{-1}$. However, the polyester fiber does not require pretreatment with a resin to make it hydrophobic, and is resistant to acid concentrations up to 70% sulfuric acid. These filters should therefore have much longer service life, possibly over 2 years.

On the basis of the above work, commercial units have been built where acid resistant materials such as polyvinylcholride sheets or wire mesh have been used as support and the filter medium is formed around them. When mists are accompanied by fine particles, which tend to block the filter, the filter must be irrigated with water. This raises the pressure drop to 4.5 kPa (18 in. WG). Such an application, where fine titanium dioxide particles and sulfuric acid are collected simultaneously, has been studied by Morash, Krouse, and Vosseller (*36*) using a two-stage unit, in which the first stage was a flat, irrigated filter pad and the second, with a larger surface area, was wrapped around a supporting cylinder. Fibers of polyester (average diameter 23 $\mu$m) as well as of polypropylene and polyacrylonitrile were tried in the pilot plant. Polyester fibers gave the best performance. The efficiency of titanium dioxide dust removal (inlet concentration 1 gm $m^{-3}$) was virtually 100%, and of acid removal (inlet concentration 4 gm $m^{-3}$) was between 83 and 94%, giving 0.05 to 0.20 gm $m^{-3}$ acid mist in the exhaust. The pressure drop was from 1.4 to 2.5 kPa and the gas velocity from 0.3 to 1.5 m $sec^{-1}$. A water spray rate of 1 kg $m^{-3}$ was required for high mist removal efficiencies.

Simultaneous removal of acid mists and fumes was also reported in an investigation by Billings, Kurker, and Silverman (*37*), who used 25- to 50-mm pads of slag wool and other mineral fibers to collect sulfuric, phosphoric, and nitric acid mists. Very high velocities (1 to 2.5 m $sec^{-1}$) were used with comparatively loosely packed fibers (48–80 kg $m^{-3}$). The measured efficiencies ranged from 26% (for the lowest packing density) to about 80%, which is well below the efficiencies required for a commercial plant.

The most extensive series of commercial fiber mist eliminators has been developed by Brink (*23, 38, 39*) and his co-workers (*40–44*). The eliminators are similar to those described above, i.e., a pad of fiber on a support,

* Registered Trade Name, Imperial Chemical Industries Ltd., England.

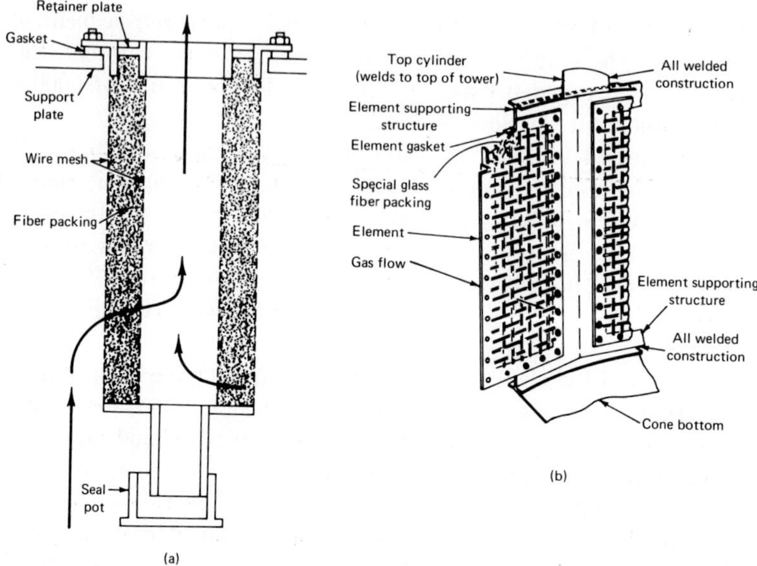


Figure 12. Brink mist eliminator. (a) High efficiency type; (b) high velocity type (*39*).

either cylindrical or flat (Fig. 12). In some cases the fibers are placed in the annulus between two open-weave baskets made of fiber reinforced plastic.

## 1. Sulfuric Acid Plants

The mist eliminators operate at almost 100% efficiency on droplets greater than 3 $\mu$m, while for smaller droplets the cylindrical eliminator in some circumstances will give efficiencies over what can be achieved by the flat type. Operating parameters are shown in Table III, and experimental efficiencies are given in Table IV (*38, 41, 43*).

**Table III Operating Parameters of Fiber Mist Eliminators on Sulfuric Acid Plants (*43*)**

| | *Cylindrical* | *Flat* |
|---|---|---|
| Superficial velocity (m sec$^{-1}$) | 0.076–0.200 | 2.01–2.50 |
| Efficiency on particles greater than 3 $\mu$m | 100% | 100% |
| Efficiency on particles less than 3 $\mu$m | 95–99+% | 90–98% |
| Pressure drop (Pa) | 1250–3750 | 1500–2000 |

**Table IV Collection efficiencies for Sulfuric Acid Mist from a Sulfur-Burning Contact-Acid Plant with Fiber Mist Eliminators** (*38*)

| *Contact plant production* | *Mist loading of gases leaving absorber and entering mist eliminator*[a] *(gm $m^{-3}$ $H_2SO_4$)* | *Mist loading of gases leaving mist eliminator (gm $m^{-3}$ $H_2SO_4$)* | *Collection efficiency with droplets less than 3 μm (%)* |
|---|---|---|---|
| Mist eliminator A[b] | | | |
| 99% $H_2SO_4$ and 65% oleum at full capacity | 1.10 | 0.053 | 95.1 |
| | 1.13 | 0.052 | 95.3 |
| | 1.42 | 0.058 | 95.9 |
| 99% $H_2SO_4$ and 25% oleum at 75% capacity | 0.23 | 0.0044 | 98.1 |
| | 0.31 | 0.0060 | 98.1 |
| | 0.24 | 0.0044 | 98.1 |
| Mist eliminator B[b] | | | |
| 99% $H_2SO_4$ and 25% oleum at full capacity | 0.51 | 0.0030 | 99.4 |
| | 0.65 | 0.0040 | 99.4 |
| | 0.69 | 0.0033 | 99.5 |
| 99% $H_2SO_4$ and 25% oleum at 60% capacity | 0.24 | 0.0016 | 99.3 |
| 99% $H_2SO_4$ only at 60% capacity | 0.075 | 0.0037 | 99.3 |

[a] The mist loading values given are for droplets 3 μm diameter or less.
[b] Mist eliminator A was designed for 100% efficiency for droplets larger than 3 μm and for 95% efficiency for droplets less than 3 μm. Mist eliminator B was designed for 100% efficiency on droplets larger than 3 μm and 99% on droplets less than 3 μm.

## 2. Chlorine Gas

Commercial fiber mist eliminators have been used extensively on chlorine plants, particularly wet and dry mercury cell gases, treating up to 3500 $m^3$/ hour (*40*).

## 3. Phosphoric Acid Mist

A cylindrical unit on a pilot plant, after 2521 hours operation produced efficiencies from 95 to 99%, with inlet loadings of 1.9 gm $m^{-3}$ and outlet loadings of from 0.15 to 0.06 gm/$m^{-3}$ A flat unit, after 7400 hours operation had efficiencies over 92% (*41*).

### 4. Combined Sulfuric and Phosphoric Acid Mists

Brink (*39*) reports that a cylindrical mist eliminator reduced loadings of from 2.8 to 7.0 gm $m^{-3}$ $P_2O_5$, 1 gm $m^{-3}$ sulfuric acid ($<3$ $\mu$m) to from 0.30–0.074 gm $m^{-3}$ $P_2O_5$, 0.010 gm $m^{-3}$ sulfuric acid.

### 5. Compressed Gases

The cylindrical eliminators have performed well at temperatures to 67°C and pressures up to $3.7 \times 10^5$ kPa (375 atm) removing oils and other impurities from a whole series of gases (*42*) (mist in parentheses): acetylene (water), air (oil, water, sulfuric acid, etc.), ammonia (oil), carbon dioxide (oil and water), chlorine (sulfuric acid, chlorinated organics), helium (oil), hydrogen (oil, water), hydrogen chloride (biphenyl, dichloroethane, benzene, phenol, benzyl chloride, etc.), methanol (oil, etc.), nitrogen (oil, nitric acid), natural gas (water, monoethanolamine).

### 6. Higher Temperatures

Although the filters reported on have operated at very high pressures, their maximum temperature of operation was 67°C. However, service at higher temperatures is required, and a specific application of this is the reduction of the load on HEPA filters in a nuclear reactor housing in a possible accident situation. Griwatz *et al.* (*45*) tested a number of commercial fiber filters made of combinations of fibers and wire for their ability to operate at 320 kPa (3.2 atm) and 133°C against monodisperse droplets of 0.3, 0.6, 10, and 100 $\mu$m. The filters tested were

(a) tetrafluoropolyethylene (Teflon, DuPont) fibers (20 $\mu$m) combined with wire (152 $\mu$m), in a bed 67 mm deep, held between 16-gauge wire grids (York type 321 SR)

**Table V Operating Characteristics and Efficiencies for Fiber Mist Eliminators (*45*)**

| *Type (see text)* | *a* | *b* | *c* | *d* | *e* |
|---|---|---|---|---|---|
| Bed depth (mm) | 67 | 610 | 125 | 125 | 100 |
| Flow velocity (m $sec^{-1}$) | 2.0 | 1.44 | 2.0 | 2.0 | 2.26 |
| Pressure drop (Pa) | 322–555 | 200–300 | 250–475 | 250–425 | 65–87 |
| Efficiency (%) | | | | | |
| at 100 $\mu$m | 26 | 100 | 100 | 100 | 100 |
| at 10 $\mu$m | 36 | 100 | 100 | 100 | 90 |
| at 0.6 $\mu$m | 31 | 7 | 20 | 22 | 1 |
| at 0.3 $\mu$m | 7 | 5 | 4 | — | 0 |

(b) bonded fiberglass, 610 mm deep (American Air Filter, AAF type T)

(c) bonded fiberglass, 125 mm deep (Mine Safety Appliances)

(d) fiberglass (9 $\mu$m) and stainless-steel wire (127 $\mu$m), mixed and knitted, 125 mm deep (Mine Safety Appliances)

(e) knitted wire mesh, 100 mm deep (Farr type 68-44 MHZ)

The operating characteristics—gas flow rate, pressure drop—and efficiencies for the different types are shown in Table V.

## VII. Cyclone Mist Eliminators

Cyclones are used for collecting very heavy liquid loadings of droplets over 10 $\mu$m, such as following venturi scrubbers. As stated in Section III,B, their design follows the principals of cyclone design for particulates (Chapter 3, this volume). There are however two problems connected with the use of cyclones in mist elimination: first, at very high spinning speeds, entrainment of liquid by shear off the wall is possible. This has been dealt with in Section III,B. Second, in a cyclone of the design shown in Figure 13a, the normally downward flowing liquid will flow countercurrent to the gas stream. There is, therefore, a tendency for the wall film to creep upwards. This can be prevented by having a circular gap opening about three-fourths up the wall. This is surrounded by a weir with drains

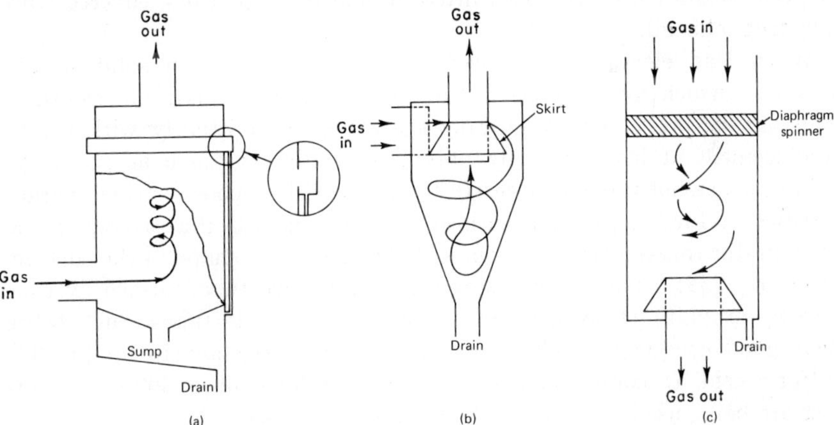


Figure 13. Cyclone mist eliminator. (a) Upward flow cyclone with annular gap for preventing creep of liquid (*46*); (b) reverse flow cyclone with skirt around exit pipe to prevent creep (*21*); (c) down-flow, in-line cyclone with skirt around exit pipe to prevent reentrainment (*21*).

to the sump. The upward creeping liquid is pushed through the gap by the centrifugal force in the cyclone, and then drains to the sump (*46*). Two other types of centrifugal mist eliminators have been suggested by Stairmand (*21*). The first is to use a conventional reverse flow cyclone, and provide a skirt around the exit pipe to prevent creep to the exit (Fig. 13b), and the second is to reverse the flow in the cyclone so that it becomes cocurrent. Again, a skirt around the exit pipe prevents reentrainment (Fig. 13c).

## VIII. Electrostatic Precipitators

Electrostatic precipitators for liquid droplets are essentially the "wetted wall" designs shown in Chapter 5, this volume, Figures 30 and 31. However, special factors must be considered for the use of these precipitators for sulfuric and phosphoric acid mists. In the case of sulfuric acid, the tubes in the older design are of lead and the discharge wires are star shaped, also of lead with a Monel core (for strength), their effective size being 10 mm (*47*). A lead grid keeps the wires taut. The bottom of the precipitator is sloped to one side for draining the liquid. A serious problem is the sealing of the high tension leads, because condensation of acid mist can cause shorting and breakdown. Current techniques for preventing this are the use of a closed chamber for housing the insulators, and then feeding a stream of dried and purified gas into this chamber. Alternatively steam coils have been fitted around the insulators, covered with a hydrophobic oil.

When lead electrodes are used, they get coated with solid sulfate ($PbSO_4$), which forms layers 1–3 mm thick, limiting the life of the tube. This can be prevented by flushing the tubes intermittently with water. Replacement of lead tubes by polymeric materials (such as polyvinyl chloride) has not proved successful because of their poor wetting characteristics. A further problem with organic polymers is that arcing in the precipitator causes carbonization, and changes in the shape of the surface. Stopperta (*48*) reports work using a graphite-lined tube, formed by condensing graphite (Korobon) on phenol formaldehyde tubes. Such tubes have good conductance characteristics and are resistant to up to 50% sulfuric acid. In commercial installations 4-m long tubes, 300 mm diameter are being used, with an applied voltage of 70 kV.

Brink (*39*) gives details of a large electrostatic precipitator for phosphoric acid mists. It treats 34,000 $m^3\ hr^{-1}$ (20000 $ft^3$/minute) at 106°C. It is an upward flow unit with carbon tubes and stainless-steel discharge wires, which are hung from lead-covered steel pipes. The housing is of acid-proof brick, and the electrical leads entering the unit pass through lead-lined oil seals containing the insulators. It is pointed out that the losses of acid were 180 kg per day, giving an overall efficiency of 99.4%. This, however, was inadequate for air pollution control, so that the tail gases from the precipitator had to be treated by a scrubber before discharge to the atmosphere through a 60-m stack.

## IX. Ceramic Candles

Thimble-shaped, porous, acid-resistant ceramic tubes have been used for collecting acid mists. However, Fairs (*2*) points out that there is high gas velocity in the pores of the filter which leads to reentrainment. The superficial velocity per unit area of candle is about 0.3 m $sec^{-1}$. The overall superficial velocity for a nest of tubes is of the order of 1.5 m $sec^{-1}$.

Massey (*33*) has reported excellent performance with these tubes, with overall efficiencies of 72 to 98.6%, consistently reducing outlet concentrations to less than 30 to 150 mg $m^{-3}$. The pressure drop is from 1.7 to 2.7 kPa (7–11 in. WG). Massey comments that maintenance is very costly because the candles are fragile and therefore subject to breakage.

## X. Venturi Scrubbers

Venturi scrubbers have been used for the collection of sulfuric and phosphoric acid mists. Their design and operation is discussed in Chapter 6, this volume.

In a commercial unit Jones (*49*) successfully reduced the sulfuric acid concentration in the tail gases from a contact plant concentrator from 1.78–26 gm $m^{-3}$ to 0.018–0.106 gm $m^{-3}$, using 0.58–0.77 kg $m^{-3}$ water at a pressure drop of 1.75–3.75 kPa (9–15 in wg). Increasing the water flow reduced the mist, but increased the pressure drop. Subsequently, Eckmann and Johnstone (*50*) reported similar results, using 0.81 kg $m^{-3}$ and velocities in the throat of 60–90 m $sec^{-1}$, with pressure drops of 3.75 kPa (15 in. WG).

Brink and Contant (*6*) used a venturi scrubber, rectangular in cross section ($150 \times 863$ mm) on a phosphoric acid mist. The rectangular section was 300 mm long, and was approached by a 25° convergent section. It was followed by a 2.2° divergent section for 1.5 m, which diverged to 15°. The inlet mist concentration was 50 gm $m^{-3}$. The efficiency of collection at droplet diameter 0.5 $\mu$m was 79%, at 0.6 $\mu$m 91%, at 0.8 $\mu$m 90%, at 1.0 $\mu$m 98%, and at 1.5 $\mu$m 99%. Brink (*39*) states that at a 10-kPa (40 in. WG) pressure drop, with wide ranges of liquid-to-gas ratios, the unit had overall efficiencies of 98%.

## XI. Sonic Agglomerators

Sulfuric acid mist has been collected using a sonic agglomeration technique. In early experiments *(51)* the gas stream was first passed down a large empty tower, where it was exposed to intense sound waves (150 dB at 2.25 kHz) for 4 seconds, and then through multicyclones. The tower was 2.43 m diameter (8 ft) and 7.6 m (25 ft) tall. The gas flow was 40,000 $m^3$ $hr^{-1}$ (24,000 $ft^3$/minute). The inlet acid mist concentration was 3.5 gm $m^{-3}$, which was reduced to 0.14–0.018 gm $m^{-3}$, which corresponds to efficiencies of 96 to 99.5%.

A device has been tested in which a wire mesh mist eliminator (120-mm-deep bed) is placed below a sound source *(52)*. The gases enter at the base of the unit and are submitted to a water spray before passing to the mist eliminator where they are exposed to a sonic field of 60–80 W at 9.8 kHz. Experiments showed that this was a more effective mist eliminator than a two-stage unit, which had twice the pressure drop *(53)*.

However, no commercial units seem to have resulted from this work. Fairs *(2)* points out that sonic agglomeration only works well at high mist concentrations, and the total energy requirement is considerably greater than for venturi scrubbers of similar capacity.

## XII. Measurement of the Effectiveness of Mist Eliminators

Because mists are liquid droplets, which change their nature in a saturated gas stream when the temperature changes, it is very important when determining the droplet size distribution, that the gases be not only sampled isokinetically, since they are generally over 5 $\mu$m, but also kept virtually constant in temperature. In practice the present author has found that keeping the sampling system 2°C above that of the mist eliminator system compensates for heat losses and reduces condensation effects.

If the aim is to measure overall efficiency (without determining droplet size distribution) it is necessary to collect droplets in bulk, before and after the mist eliminator. While it appears reasonable to adsorb mist droplets on a chemical adsorbent such as calcium chloride, silica gel, or magnesium perchlorate, in granular form, packed into an adsorbent tube, usually a U-tube, this gives poor results because the channeling which occurs gives incomplete adsorption. In practice, it is more satisfactory to use U-tubes packed with glass wool. These are weighed before and after use as sample collectors, and are stored at constant vapor pressure. To check on their efficiency, U-tubes packed with a chemical adsorbent are placed in the train after the glass wool packed units, and checked for weight changes. Similarly, Stairmand (*54*) found filters packed with fine (2–5 μm) glass wool, to a depth of 25 mm in a 50-mm filter, a satisfactory method of trapping mists.

Fairs (*2*) describes a small electrostatic precipitator for mist sampling. The discharge electrode is a platinum wire with a spiral gauze welded to it, charged to 20–50 kV. It is hung centrally in a glass tube (25 mm diameter), surrounded by a copper tube or one covered with tin or aluminum foil. In a 500-mm-long sampler, from 1 to 4.25 $m^3$ $hr^{-1}$ of gas is completely demisted.

If the mist is pure water (or other solvent) precautions have to be taken that the collected water (or other solvent) does not evaporate, or, if it is hygroscopic, attract more moisture between the time of collection and the time when the collected liquid is measured, usually by weighing. If the mist consists of acid droplets [or of a salt solution, as in the case of the Carpenter and Othmer experiments (*11*)], it is possible to wash the collected droplets out of collection tube, and determine their acid (or salt) content chemically.

Determining droplet size distributions is difficult. Coarse droplets, i.e., those above 10 μm, can be sampled by briefly exposing a glass microscope slide, covered with a hydrophobic oil, to the gas stream. Brief exposure is accomplished either manually or mechanically, depending on the speed with which it has to be carried out to get a representative sample. In this method allowances have to be made for the proportion of smaller droplets carried around the slide in the gas stream. Size distribution is then evaluated by microscopic sizing. For finer droplets a multistage cascade impactor can be used, as the one described by Brink (*55*). This separates out droplets smaller than 3 μm into five groups on the five stages of the impactor, from 0.33 to 3.14 μm. In a later modification (*56*), a preliminary glass cyclone is used to collect the droplets above 3 μm. Bürkholz (*57*) has also described a cascade impactor assembly with nine stages. This has enabled him to cover the range from 0.5 to 10 μm.

## ACKNOWLEDGMENTS

The author wishes to acknowledge the work of his students at the University of Melbourne and at the University of North Carolina, particularly Mr. C. G. Bell and Mr. J. Floyd. The chapter was compiled and written while the author was visiting Professor of Air Hygiene at the University of North Carolina, and was supported by the Triangle Universities Consortium on Air Pollution through a grant from the United States Environmental Protection Agency.

## REFERENCES

1. H. Chilton, *Trans. Inst. Chem. Eng.* **30,** 235 (1952).
2. G. L. Fairs *in* "Gas Purification Processes" (G. Nonhebel, ed.), 2nd ed., p. 520. Butterworth, London, England, 1972.
3. G. L. Fairs, *Trans. Inst. Chem. Eng.* **22,** 110 (1944).
4. G. M. Hidy and J. R. Brock, "The Dynamics of Aerocolloidal Systems," Chapters 8–10. Pergamon, Oxford, England, 1970.
5. J. Kapcznsky and K. Generalczyk, *Ochr. Powietrza* (English transl. *Air Conservation*) **6,** 27 (1972) [quoting A. G. Amelin, "Theory of Fog Condensation" (in English). Davey, New York, New York, 1966].
6. J. A. Brink and C. E. Contant, *Ind. Eng. Chem.* **50,** 1157 (1958).
7. C. G. Bell and W. Strauss, *J. Air Pollut. Contr. Ass.* **23,** 967 (1973).
8. G. L. Fairs, *Trans. Inst. Chem. Eng.* **36,** 475 (1958).
9. W. Strauss, "Industrial Gas Cleaning," 2nd ed. Pergamon, Oxford, England, 1976.
10. H. W. Pyne, R. B. Wilson, and B. W. Soole, *Brit. J. Appl. Phys.* **18,** 1177 (1967).
11. C. L. Carpenter and D. F. Othmer, *AIChE J.* **1,** 549 (1955).
12. S. Calvert, I. L. Jashnani, and S. Yung, "Entrainment Separators for Scrubbers," APT Fine Particle Scrubber Symp. United States Environmental Protection Agency, San Diego, California, 1974.
13. I. L. Jashnani and S. Calvert, Annu. Meet. Paper No. 74-230. Air Pollution Control Association, Pittsburgh, Pennsylvania, 1974.
14. M. Souders and G. G. Brown, *Ind. Eng. Chem.* **26,** 98 (1934).
15. W. D. Matthews and O. H. York, "Wire Mesh Mist Eliminators in the Gas Industry," 1963 Gas Cond. Conf. University of Oklahoma, Norman, Oklahoma, 1963.
16. O. H. York and E. W. Poppele, *Chem. Eng. Progr.* **59**(6) (1963).
17. O. H. York, *J. Teflon (du Pont, Wilmington, Delaware)* **2**(12), 1 (1961).
18. Yu. L. Sorokin, L. N. Demidova, and N. P. Kuz'min, *Khim. Neft. Mashinostr.* **8,** 20 (1968).
19. E. W. Poppele, "Correlation of Maximum Air Velocity with Liquid Entrainment Loading for Wire Mesh Mist Eliminators." M.S. Thesis, Newark College of Engineering, Newark, New Jersey, 1958.
20. I. L. Jashnani, APT Inc., Riverside, California (private communication, to be published).
21. C. J. Stairmand, *Trans. Inst. Chem. Eng.* **29,** 356 (1951).
22. F. Stearman and G. J. Williamson, *in* "Gas Purification Processes" (G. Nonhebel, ed.), 2nd ed., p. 564. Butterworth, London, England, 1972.
23. J. A. Brink, *in* "Chemical Engineers Handbook" (R. H. Perry and C. C. Chilton, eds.), 5th ed., p. 18.82. McGraw-Hill, New York, New York, 1973.
24. U. Regehr, *VDI (Ver. Deut. Ing.) Ber.* **149,** 344 (1970).
25. H. Peterson, "Achema Jahrbuch," Vol. II. Dechema, Frankfurt, Germany, 1971/1973 (special reprint).

26. B. W. Soole and H. C. W. Meyer, *Aerosol Sci.* **1,** 147 (1970).
27. B. W. Soole, *Aerosol Sci.* **3,** 321 (1972).
28. B. W. Soole, *Filtr. Separ.* **11,** 483 (1974).
29. J. B. Wong, W. E. Ranz, and H. F. Johnstone, *J. Appl. Phys.* **26,** 244 (1955).
30. K. R. May and R. Clifford, *Ann. Occup. Hyg.* **10,** 83 (1967).
31. Karbate Entrainment Separators, Type MV, Publ. K 26. British Acheson Electrodes Ltd., Wincobank, Sheffield, England; see also Cat. Sect. S 6900. Union Carbide Corp., New York, New York, 1960.
32. A. Bürkholz, *Chem.-Ing.-Tech.* **42,** 1314 (1970).
33. O. D. Massey, *Chem. Eng.* **66**(14), 143 (1959).
34. O. H. York and E. W. Poppele, *Chem. Eng. Progr.* **66**(11), 67 (1970).
35. J. W. Coykendall, E. F. Spencer, and O. H. York, *J. Air Pollut. Contr. Ass.* **18,** 315 (1968).
36. N. Morash, M. Krouse, and W. P. Vosseller, *Chem. Eng. Progr.* **63**(3), 70 (1967).
37. C. E. Billings, C. Kurker, and L. Silverman, *J. Air. Pollut. Contr. Ass.* **8,** 185 (1958).
38. J. A. Brink, *Can. J. Chem. Eng.* **41,** 134 (1963).
39. J. A. Brink, *in* "Gas Purification Processes" (G. Nonhebel, ed.), 2nd ed., Chapter 15B. Butterworth, London, England, 1972.
40. J. H. Nichols and J. A. Brink, *Electrochem. Technol.* **2,** 233 (1964).
41. J. A. Brink, W. F. Burggrabe, and J. A. Rauscher, *Chem. Eng. Progr.* **60**(11), 68 (1964).
42. J. A. Brink, W. F. Burggrabe, and L. E. Greenwell, *Chem. Eng. Progr.* **62**(4), 60 (1966).
43. J. A. Brink, W. F. Burggrabe, and L. E. Greenwell, *Chem. Eng. Progr.* **64**(11), 82 (1968).
44. J. A. Brink, E. D. Kennedy, C. N. Dougald, and T. R. Metzger, *Environ. Pollut. Manage.* **1,** 79 (1971).
45. G. H. Griwatz, J. V. Friel, and J. L. Creehouse, Report 71-45, United States Atomic Energy Commission Contact AT(45-1)-2145. Mine Safety Appliances Res. Corp., Evans City, Pennsylvania, 1971.
46. W. Strauss, Australian Patent 403,736 (1970).
47. J. S. Lagarias, *J. Air Pollut. Contr. Ass.* **10,** 271 (1960).
48. K. Stopperta, *Staub-Reinhalt. Luft* **25,** 508 (1965).
49. W. P. Jones, *Ind. Eng. Chem.* **41,** 2424 (1949).
50. F. O. Eckmann and H. F. Johnstone, *Ind. Eng. Chem.* **43,** 1358 (1951).
51. H. W. Danser and G. P. Neumann, *Ind. Eng. Chem.* **41,** 2439 (1949).
52. S. H. V. Asklöf, U.S. Patent 3,026,966 (1962).
53. R. M. G. Boucher and G. R. Koehler, "Ultrasonic Demister Project," Preliminary Report. Macrosonics Corp., Carteret, New Jersey (n.d.)
54. C. J. Stairmand, *Trans. Inst. Chem. Eng.* **29,** 15 (1951).
55. J. A. Brink, *Ind. Eng. Chem.* **50,** 645 (1958).
56. W. F. Patton and J. A. Brink, *J. Air. Pollut. Contr. Ass.* **13,** 162 (1963).
57. A. Bürkholz, *Chem.-Ing.-Tech.* **42,** 299 (1970).

# Adsorption

Amos Turk



## I. General Principles

The forces that hold atoms, molecules, or ions together in the solid state exist throughout the body of a solid and at its surface. The forces at the surface may be considered to be "residual" in that they are available for binding other molecules which come in contact with it. Any gas, vapor, or liquid will, therefore, adhere to some degree to any solid surface. This phenomenon is called adsorption, or sorption, the adsorbing solid is called the adsorbent, or sorbent and the adsorbed material is the adsorbate, or sorbate. Adsorbed matter may also condense in the submicroscopic pores of an adsorbent; this phenomenon is called capillary condensation. A molecule that moves to and is held at the surface of a solid loses the energy of its motion; adsorption is therefore always an exothermic, or energy-releasing process. Since this chapter is concerned with control of air pollutant sources, only adsorption from the gaseous state is within its scope.

Adsorption is useful in air pollution control because it is a means of concentrating gaseous pollutants, thus facilitating their disposal, their recovery, or their conversion to innocuous or valuable products.

When an adsorbate is chemically bonded to the adsorbent, it is said to be chemisorbed. Chemisorption is typified by the following characteristics: (a) the energy released is greater than that which occurs in physical adsorption; it is in the range associated with chemical reactions, usually over 10 kcal/mole, (b) the process is often irreversible, for example, part of the oxygen adsorbed on activated carbon at ambient temperatures is recoverable only as CO and $CO_2$, (c) the rate increases with a rise in temperature, (d) the process is more highly selective than physical adsorption, and (e) the capacity of the adsorbent is limited to that of the active sites on its surface, which cannot exceed a unimolecular layer.

More important in air pollution control are phenomena in which adsorbates react with each other; the adsorbent, by serving as a concentrating medium, speeds up the reaction rate. Material in the adsorbed state may be especially reactive; the adsorbent then functions as a true catalyst. In some cases a specially selected catalyst or reactant is incorporated into an adsorbent prior to the use of the adsorbent as a gas purifier. The adsorbent is then said to be impregnated with the material. Such impregnation may increase the rate, capacity, or selectivity of the adsorbent for gas purification.

For practical air pollution control applications, chemical reaction of the pollutant with the adsorbent (chemisorption) or with another adsorbate (reactive impregnation) is a mixed blessing. The irreversibility and positive temperature dependence that are associated with chemical reactions are advantageous, and in some cases so is the selectivity. But the total capacity of the adsorbent for nonselective retention of gaseous pollutants is reduced by these effects, and in most cases this reduction is an overriding disadvantage.

The rate of gas purification by adsorption depends on the rate of transfer of gas molecules to the surface ($s$), on the fraction of molecules ($k$) reaching the surface that are retained there, and on the rate of desorption ($d$):

$$\text{Rate} \propto (ks - d) \tag{1}$$

When a polluted airstream is passed through a fresh adsorbent bed, practically all the vapor molecules that reach the surface are adsorbed ($k \approx 1$), and desorption is very slow ($d \approx 0$). Furthermore, if the bed consists of closely packed granules, the distance that the molecules must travel to reach some point on the surface is small, and the transfer rate is therefore high. In practice, the half-life of airborne molecules streaming through a packed adsorbent bed is of the order of 0.01 second, which means that a 95% removal can occur in about four half-lives, or around 0.04 seconds (*1*). By contrast, equivalent gas removal efficiencies by flame or thermal incineration (Chapter 9, this volume) require contact times at least an order of magnitude greater (tenths of seconds), and most ambient temperature oxidizing agents, such as ozone or potassium permanganate, are considerably slower than flame reactions (*2*). These rate comparisons imply that an adsorber may occupy less space than any other gas cleaning devices of comparable efficiency. This conclusion is indeed true for the initial performance of a fresh bed, but it ignores the requirement for sufficient adsorbent to enable gas purification to continue for a specified time. The role of adsorbent bed depth will be treated in Section III,A,1.

The quantity of material that can be physically adsorbed by a given weight of adsorbent depends on the following factors: (a) the concentration of the material in the space around the adsorbent, (b) the total surface area of the adsorbent, (c) the total volume of pores in the adsorbent whose diameters are small enough to facilitate condensation of adsorbed gases, (d) the temperature, (e) the presence of other gases in the environment which may compete for a place on the adsorbent, (f) the characteristics of the molecules to be adsorbed, especially their weight, electrical polarity, size, and shape, and (g) the electrical polarity of the adsorbent surface. Maximum capacity for adsorption of a given substance is favored by a high concentration of the substance is the space adjoining the adsorbent, a large adsorbing surface, freedom from competing substances, low temperature, and by aggregation of the substance in large molecules which fit and are strongly attracted to the receiving shapes of the adsorbent.

The net rate of adsorption of a substance depends on the rate at which molecules reach the adsorbing surface, the fraction of those making contact that are adsorbed, and the rate of removal of molecules from the surface (desorption). Therefore, to favor rapid adsorption, the adsorbing equipment should be designed with a view to providing ample duration of contact (detention time) between the gas to be purified and an adsorbent which is sufficiently retentive to the contaminants that are to be removed.

Disposal of pollutant gases that have been concentrated by adsorption may be effected in any of the following ways: (a) The adsorbent with its adsorbate may be discarded. Since even the saturated adsorbent is relatively nonvolatile, this step seldom involves difficult problems. An exception occurs when the sorbate is radioactive. (b) The adsorbate may be desorbed and either recovered, if it is valuable, or discarded. In either case, the adsorbent is recovered. (c) The adsorbate may be chemically converted to a more easily disposable product, preferably with preservation and recovery of the adsorbent.

## II. Adsorbents

### A. Comparative Properties

Adsorbents are most significantly characterized by their chemical natures, by the extent of their surfaces, and by the volume and diameters of their pores. The most important chemical differences among adsorbents are in their electrical polarity.

Activated carbon, consisting largely of neutral atoms of a single element, presents a surface with a relatively homogeneous distribution of electrical charge. As a result, there are no significant potential gradients of molecular dimensions on the surface which would selectively orient and bind polar in preference to nonpolar adsorbate molecules.

All commercially important sorbents other than carbon are simple or complex oxides; their surfaces contain inhomogeneous distribution of charge on a molecular scale and hence are polar. These sorbents show considerably greater selectivity than does activated carbon, and overwhelmingly greater preference for polar than for nonpolar molecules. As a result, they are more useful than carbon when separations are to be

**Table I Surface Areas and Pore Sizes of Adsorbents**

| | *Activated carbon* | *Activated alumina* | *Silica gel* | *Molecular sieve* |
|---|---|---|---|---|
| Surface area ($m^2/gm$) | 1100–1600 | 210–360 | 750 | — |
| Surface area ($m^2/cm^3$) | 300–560 | 210–320 | 520 | — |
| Pore volume ($cm^3/gm$) | 0.80–1.20 | 0.29–0.37 | 0.40 | 0.27–0.38 |
| Pore volume ($cm^3/cm^3$) | 0.40–0.42 | 0.29–0.33 | 0.28 | 0.22–0.30 |
| Mean pore diameter (Å) | 15–20[a] | 18–20 | 22 | 3–9 |

[a] Refers to micropore volume (<25 Å diameter); macropores (>25 Å) not included.

made among different types of pollutants, but much less useful when overall decontamination of an airstream is to be accomplished. They are essentially ineffective for direct decontamination of a moist air or gas stream. Since the latter tasks greatly predominate in problems of control of air pollutant sources, polar sorbents are of less general applicability.

Table I shows ranges of surface areas and pore volumes for several different adsorbents. Among these, activated carbon is generally highest in surface area and pore volume, the properties which primarily determine overall adsorptive capacity.

### *B. Activated Carbon*

Activated carbon (also called active carbon, or activated charcoal) consists of particles of moderately to highly pure carbon which have a large surface area per unit weight or volume of solid. For use in a fixed bed for air or gas purification, the particles must be so sized that they impose little difference to flow for a given sorption efficiency; the range of 4–20 mesh (U.S. Sieve Series) encompasses the predominant portion of carbon for such use. To minimize mechanical attrition during transportation and use, the activated carbon should be hard. Hardness is determined in part by the nature of the raw material used for manufacture of the activated carbon, and in part by the manufacturing process. Raw materials include coconut and other nut shells, fruit pits, bituminous coal, hard woods, and petroleum residues.

As stated in the previous section, activated carbon is effective is adsorbing molecules of organic substances with less selectivity than is exhibited by other, more polar sorbents. Activated carbon is effective in adsorbing organic molecules even from a humid gas stream. The water molecules, being highly polar, exhibit strong attractions for each other, which compete with their attractions for the nonpolar carbon surface; consequently the larger, less polar organic molecules are selectively adsorbed.

The total adsorptive capacity of a sample of activated carbon may be measured by its activity or retentivity for a standard vapor. The activity is the maximum amount of a vapor which can be adsorbed by a given weight of carbon under specified conditions of temperature, concentration of the vapor, and concentration of other vapors (usually water). The retentivity is the maximum amount of adsorbed vapor which can be retained by the carbon after the vapor concentration in the ambient air or gas stream has been reduced to zero. Because an adsorbent may be required to retain its adsorbate even in pure air, the retentivity represents the practical capacity of the carbon in service.

Typical specifications for activated carbon to be used for air purification are given in Table II.

The pore-size distributions of activated carbons are important determinants of their adsorptive properties. Pores less than about 25 Å in diameter are generally designated as micropores; larger ones are called macropores. The distinction is important because most molecules of concern in air pollution range in diameter from about 4 to about 8.5 or 9 Å. If the pores are not much larger than twice the molecular diameter, opposite-wall effects play an important role in the adsorption process by facilitating capillary condensation. Maximum adsorption capacity is determined by the liquid packing that can occur in such small pores. The opposite-wall effect was shown by Stoldt and Turk (*3*), who found that bromine undergoes *trans*-addition to alkenes on brominated activated carbon, which means that both sides of the alkene molecules are exposed to the adsorbed bromine.

At very high vapor pressures, multilayer adsorption can lead to capillary condensation even in large pores ($> 25$ Å), but since many odorous pollution problems are associated with vapors at relatively low partial pressures, the small-pore volume is the more important property. Figure 1 shows pore-size distribution curves for (a) a carbon used for water purification and not suitable for use in gas phase, (b) a typical air purification carbon, and (c) a high performance "superactivated" air purification carbon.

The plot of adsorption capacity versus pressure of the adsorbate at a given temperature is called the adsorption isotherm. Figure 2 shows how adsorption capacity increases with increasing partial pressure, as

**Table II Typical Specifications for Activated Carbon Used for Air Purification**

| *Property* | *Specification* |
|---|---|
| Activity for $CCl_4$[a] | At least 50 % |
| Retentivity for $CCl_4$[b] | At least 30 % |
| Apparent density | At least 0.4 gm/ml |
| Hardness (ball abrasion)[c] | At least 80 % |
| Mesh distribution | 6–14 range (Tyler Sieve Series) |

[a] Maximum saturation of carbon, at 20°C and 760 Torr in an airstream equilibrated with $CCl_4$ at 0°C.

[b] Maximum weight of adsorbed $CCl_4$ retained by carbon exposure to pure air at 20°C and 760 Torr.

[c] Percent of 6–8 mesh carbon which remains on a 14-mesh screen after shaking with 30 steel balls of 0.25–0.37 in. (0.635–0.940 cm) per 50 gm carbon, for 30 minutes in a vibrating or tapping machine.

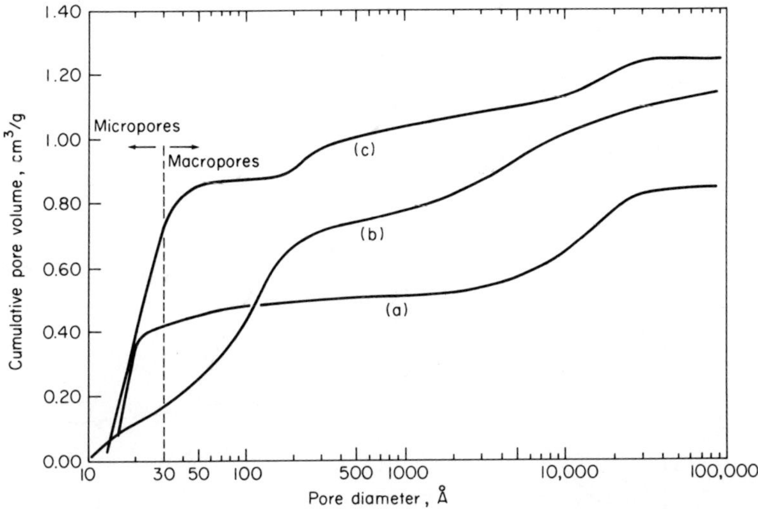


Figure 1. Pore volumes of activated carbons. (a) Air purification carbon; (b) water purification carbon; (c) "superactivated" air purification carbon.

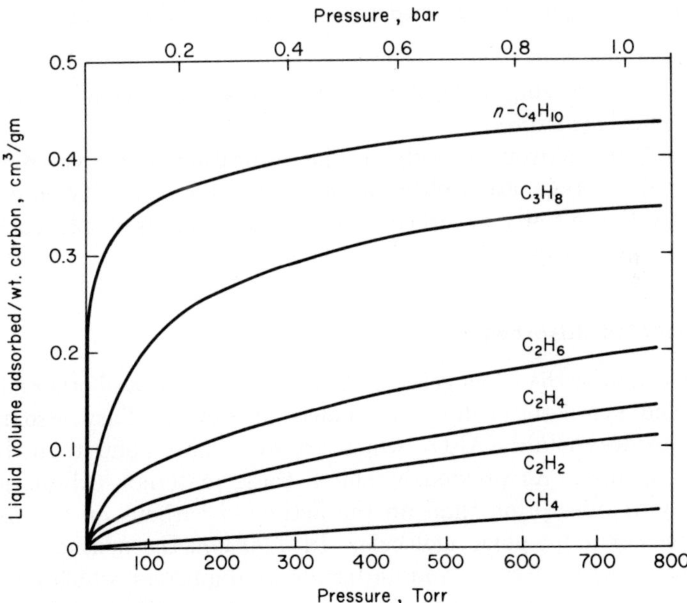


Figure 2. Adsorption isotherms of hydrocarbon vapors at 100°F on air purification activated carbon. Liquid volumes measured at boiling points of the hydrocarbons.

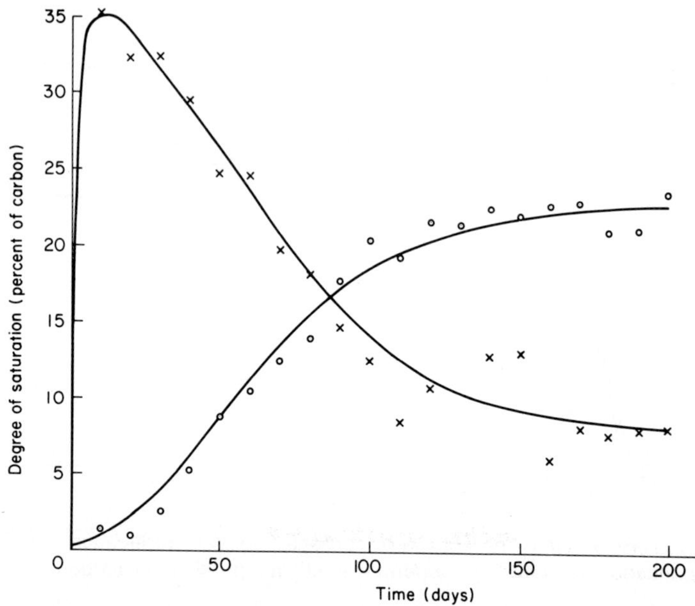


Figure 3. Saturation of coconut shell activated carbon in commercial apple storage (Entiat, Washington). X: Adsorbed water; ○: adsorbed organic vapors.

well as with increasing molecular weight in a series of compounds of related chemical structure.

Figure 3, taken from a study of the saturation of activated carbon in an apple storage atmosphere at 85% relative humidity and 35°F, shows how the initially adsorbed moisture is gradually displaced by the adsorbed organic vapors (*4*).

### C. Oxygenated Adsorbents

These comprise the silica gels, fuller's, diatomaceous, and other siliceous earths, and synthetic zeolites or "molecular sieves." They also include metallic oxides, notably $Al_2O_3$, which are even more polar than the siliceous compounds. As previously noted, these materials exhibit greater specificity of adsorption than do the activated carbons on the basis of preference for more polar molecules. In addition, the synthetic zeolites can be made with specified and uniform pore diameters, which give them outstanding properties of adsorbent specificity on the basis of size or shape of adsorbate molecules. Even this structural uniqueness, however, does not obviate the preference of the adsorbent for polar rather than nonpolar molecules; as a result, these materials will not adsorb organic

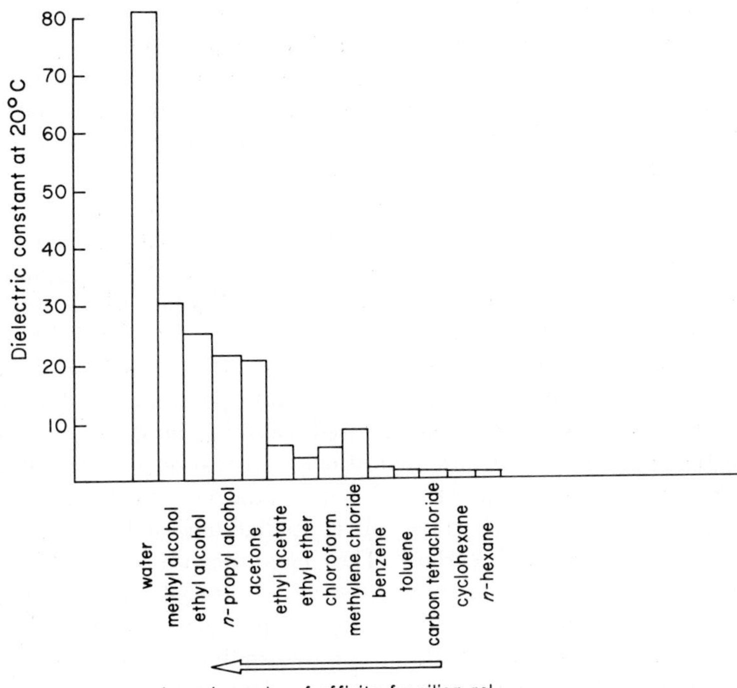


**Figure 4.** Relationship between dielectric constant of solvent and affinity for silica gel.

molecules, even of sizes which match their pores, from a moist airstream, the water molecules being adsorbed in preference.

The relationship between the polarity of an adsorbate and its affinity for polar adsorbants is shown in Figure 4, which displays various substances in the order of their affinity for silica gel (*5*). The dielectric constant, which is a good index of electric polarity, closely follows the affinity order. These relationships imply that small, polar molecules, such as methanol or formaldehyde, can be effectively removed from dry gas streams by strongly polar adsorbents like $Al_2O_3$, and that the saturated adsorbent can be effectively regenerated with water.

## *D. Adsorbent Impregnations*

The effectiveness of adsorbent impregnations may be related to any of the following modes of action:

1. The impregnant may be a reagent that chemically converts a pollutant to a harmless or adsorbable product. As an example, carbon may

be impregnated with 5–15% of its weight of bromine. The adsorbed bromine reacts with alkenes to yield the brominated addition products. Ethylene, which causes significant plant damage and which, because of its low molecular weight, is not significantly removed from an airstream by physical adsorption alone, is thus converted to 1,2-dibromoethane, which is readily adsorbed, and remains on the surface (*6*).

2. The impregnant may be a catalyst that acts continuously. Since the only reacting materials available are the carrier gas (air) and the pollutant itself, the only reactions to be catalyzed are oxidation and decomposition. For continuous oxidative catalysis, limitations are imposed by the following factors: if activated carbon is the sorbent, a highly active oxidation catalyst will tend to make the carbon pyrophoric; if the catalyst activity is low, only easily oxidizable pollutants will be converted; if a nonoxidizable carrier is used, selection must be made from among the polar sorbents whose limitations were discussed above. Within this framework, applications of continuous oxidative catalysis have been made by using chromium, copper, silver, palladium, and platinum impregnations on activated carbon. The metal depositions are usually effected by *in situ* decomposition of complex salts which were used in the impregnating solutions. At ambient temperatures the impregnated carbon itself is stable to oxidation, but its kindling temperature is significantly lowered and hence it must be treated as a potentially combustible substance. Catalytic oxidations on nonoxidizable carriers are generally conducted at elevated temperatures; these are the processes of catalytic combustion described in Chapter 9, this volume.

When the pollutant character of a substance can be abated by decomposition, the acceleration of this action by a catalyst is a valid objective in source control. This application is limited to inherently unstable substances; most important among these are the molecules which contain oxygen-to-oxygen linkages: ozonides, peroxides, hydroperoxides, and ozone itself. The untreated surface of activated carbon acts as a catalyst for rapid decomposition of these substances. Polar adsorbents impregnated with metallic oxides also catalyze their decomposition (*7*).

3. Finally, the impregnant may be a catalyst that acts intermittently. This action would be applied to a pollutant that is first collected for an interval of time by physical adsorption. When the capacity of the sorbent has been used up, the temperature may be raised to initiate a catalytic surface oxidation of the collected sorbate.

The catalysts used in an early study of this application were oxides of Cr, Mo, and W (*8*). Since then, more extensive studies have shown that $V_2O_5$, Pd, and Pt can be effective for the cycling adsorption-oxidation of a variety of hydrocarbons and oxygenated adsorbents (*9–12*).

**Table III Adsorbent Impregnations**

| *Adsorbent* | *Impregnant* | *Pollutant* | *Action* |
|---|---|---|---|
| Activated carbon | Bromine | Ethylene; other alkenes | Conversion to dibromide, which remains on carbon |
| | Lead acetate | $H_2S$ | Conversion to PbS |
| | Phosphoric acid | $NH_3$; amines | Neutralization |
| | Sodium silicate | HF | Conversion to fluorosilicates |
| | Iodine | Mercury | Conversion to $HgI_2$ |
| | Sulfur | Mercury | Conversion to HgS |
| | Sodium sulfite | Formaldehyde | Conversion to addition product |
| | Sodium carbonate or bicarbonate | Acidic vapors | Neutralization |
| | Oxides of Cu, Cr, V, etc; noble metals (Pd, Pt) | Oxidizable gases, including reduced sulfur compounds such as $H_2S$, COS, and mercaptans | Catalysis of air oxidation |
| Activated alumina | Potassium permanganate | Easily oxidizable gases, especially formaldehyde | Oxidation |
| | Sodium carbonate or bicarbonate | Acidic gases | Neutralization |

Table III lists some adsorbent impregnations and summarizes their modes of action.

## III. Equipment and Systems

### *A. Equipment*

#### 1. Design Principles

The general requirements that sorption phenomena impose on equipment design are (a) long enough duration of contact (detention time) between airstream and sorbent bed for adequate sorption efficiency, (b) sufficient sorption capacity to provide the desired service life, (c) small enough resistance to airflow to allow adequate operation of the air moving devices being used, (d) uniformity of distribution of airflow over the sorbent bed to ensure full utilization of the sorbent, (e) adequate pretreatment of the air to remove nonadsorbable particles which would impair the action of the sorbent bed, and (f) provision for renewing the sorbent after it has reached saturation. The action of a sorbent bed on a moving airstream as it affects system design is illustrated by the schematic "adsorption wave" in Figure 5.

Curve a shows the vapor concentration within the bed shortly after the start of the adsorption process. (The bed is fresh.) The inlet concentration of pollutant is $C_i$. This level drops off sharply with increasing distance through the bed, and reaches "zero" after some finite distance. (Theoretically, zero is never reached, but we assume that a nondetectable concentration is a practical "zero.") Let us assume that the objective of the process is to reduce $C_i$ to some target concentration $C_t$ which may be an odor threshold level, or perhaps some arbitrary fraction of $C_i$. The minimum bed thickness that could achieve this objective initially is called the critical bed depth, designated $L_c$ in Figure 5. As time goes on and adsorption continues, the upstream portion of the bed becomes partly saturated and the vapor therefore penetrates more deeply, as shown in curve b. Finally the upstream section of adsorbent becomes completely saturated (curve c). The section of bed between saturation and "zero" concentration is the region in which adsorption is taking place, or the mass transfer zone. At the time the system has reached the condition of curve c, the length of this transfer zone is $T_c$. Now, as adsorption continues further, the transfer zone progresses down-

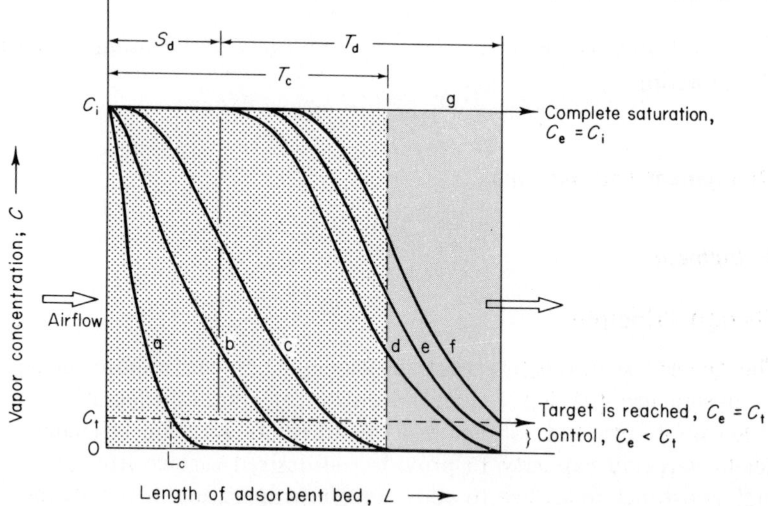


Figure 5. The adsorption wave.

stream, but its length remains the same. Thus, when the system has reached the condition of curve d, the length of the transfer zone $T_d$ is the same as it was before, but it is displaced to the right. The upstream section of the bed $S_d$ is now saturated and inactive. We see now that as the curve continues to advance, $C_e$ becomes greater than zero (curve e), and will reach the target concentration (curve f) long before the entire bed is saturated. Complete saturation ($C_e = C_i$) is represented by curve g. The entire pattern of moving curves is called the "adsorption wave."

Design requirements imposed by the objectives set forth above and by the nature of the dynamic adsorption wave are met by several types of equipment and of system arrangements.

## 2. Adsorbent Disposed on a Carrier

Considerable flexibility in sorbent bed design may be obtained when the sorbent, usually in the form of fine powder, is disposed on an inert

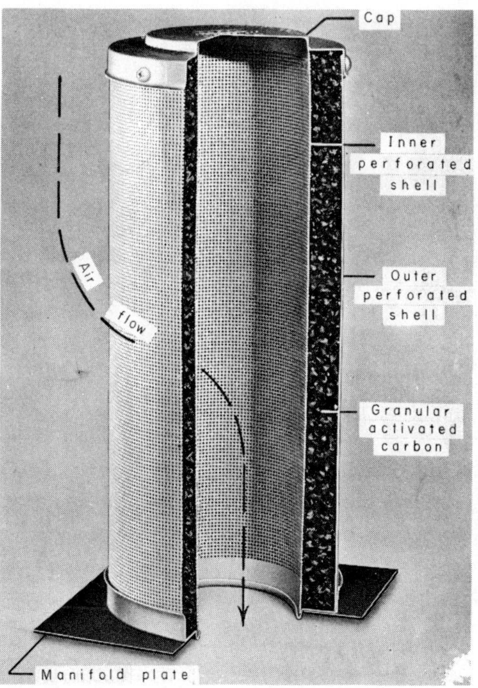


Figure 6. Cylindrical thin-bed canister adsorber. (Courtesy of Connor, Inc., Danbury, Connecticut.)

carrier. The latter may be paper, organic or inorganic textiles, or extruded plastic filaments. Papers have been developed that contain 50–75% carbon by weight; cellulose monofilaments extruded with activated carbon powder may effectively reach a level of about 80% of activated carbon. Even at best, however, the limitations in sorption capacity imposed by carriers, and their limitations of rigidity, have excluded such media from application to industrial source control of atmospheric pollution.

## 3. Stationary Thin-Bed Granular Adsorbers

The great advantage of thin-bed adsorbers is the low resistance which they impose to airflow. Since even small differences in depth of thin-bed adsorbers constitute a significant portion of the total depth and may promote channeling, bed depth uniformity is important. Flat, cylindrical, and pleated bed shapes have been used. Beds may be retained by porous barriers, screening, or perforated sheet metal. Requirements for rigidity have led to the overwhelming use of perforated sheet metal retainers. Equipment is illustrated in Figures 6, 7, and 8. Commerically available

Figure 7. Pleated cell thin-bed adsorber. (Courtesy of Barneby-Cheney Co., Columbus, Ohio.)

Figure 8. Aggregated flat cell thin-bed adsorber. The small test element located on the upstream side of the cell contains carbon that is to be analyzed after some period of service for degree of saturation, and thereby to predict the remaining capacity of the cell. (Courtesy of Connor, Inc., Danbury, Connecticut.)

cylindrical canisters are designed for about 25 $ft^3$/minute of air; the larger pleated cells handle 750–1000 $ft^3$/minute and cells comprising aggregates of flat bed components handle 2000 $ft^3$/minute.

## 4. Stationary Thick-Bed Granular Adsorbers

These are used when large adsorbing capacity is needed. The concomitant increase in efficiency is generally not, in itself, sufficient justification for use of thick adsorbent bed equipment. Even with this type of equipment, reasonable bed uniformity is desirable for effective use of the sorbent. Adsorbers for regenerative systems (see below) are usually in the range of 1–6 ft (0.3–1.8 m) in bed depth, with a downward gas flow to minimize bed lifting. Design air flow capacities are up to 40,000 cfm (67,960 $m^3$/hour). The ratio of weight of carbon to design airflow capacity is typically about 0.5 lb/cfm (0.27 kg/$m^3$/hour). Typical thick-bed adsorbers, such as are used in solvent recovery systems are shown in Figures 9, 10, and 11.

Figure 9. Thick-bed adsorbers used in a solvent recovery system. (Courtesy of Union Carbide Corp., New York, New York.)

Stationary adsorbers impose two inherent disadvantages: (a) The existence of a saturated zone ($S_d$, in Fig. 5) represents idle carbon, which occupies space and wastes energy in the form of work done by the gas stream that flows through it. The portion of bed ahead of the transfer zone is completely fresh, and this carbon, too, is therefore idle. (b) When the performance criterion has been reached ($C_e = C_t$), and the carbon bed must be replaced or reactivated, the upstream section $L - (S + T)$ comprises an unsaturated portion of the bed, and the carbon in this portion is therefore underutilized. The moving-bed systems described below are designed to circumvent these drawbacks.

## 5. Fluidized Adsorbers

When a gas is passed upward through a bed of granular adsorbent, the pressure drop imposed by the bed (see Fig. 17) is in opposition to its own weight. If the gas velocity is increased to a sufficient value, the

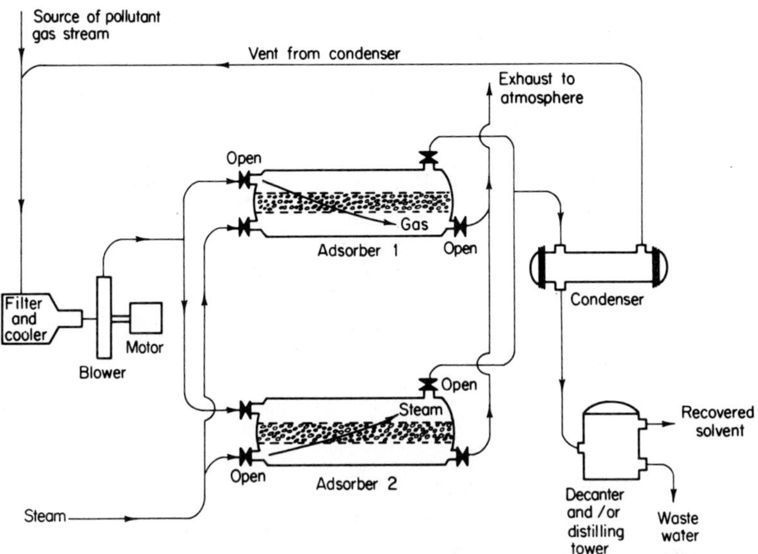


Figure 10. Two-stage regenerative system. Adsorber 1 is adsorbing; adsorber 2 is steaming. After number 1 is saturated and number 2 is clean, their functions are reversed.

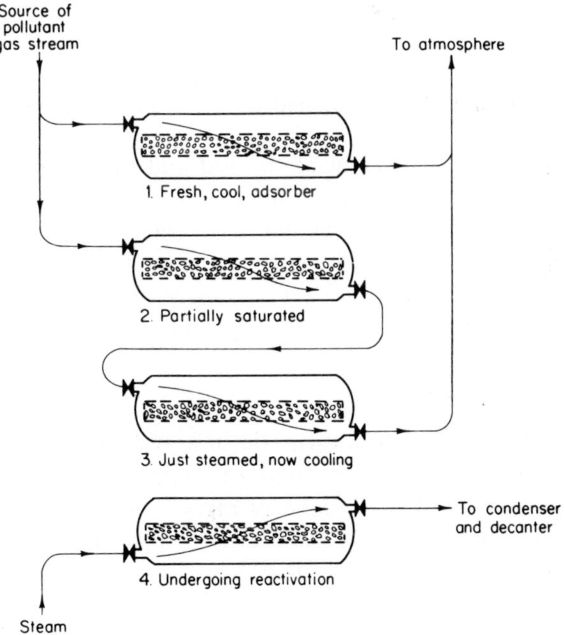


Figure 11. Four-stage regenerative system.

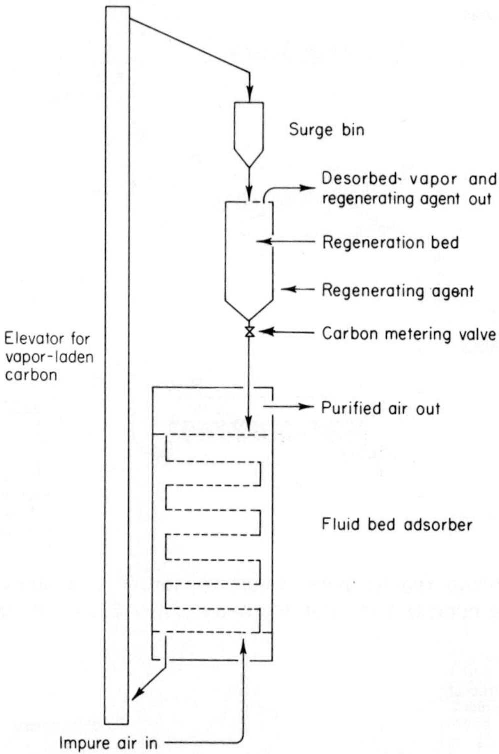


**Figure 12. Fluidized bed solvent recovery system.**

pressure drop equals the weight of the bed and the solids begin to move. This motion is the beginning of fluidization. At higher gas velocities (about 250 ft/minute), the granular adsorbent particles may be maintained in constant motion. The pressure drop required for fluidization depends on the bed depth and the densities of the gas stream and of the solid particles. A typical system is shown in Figure 12.

## 6. Rotating Bed System

To reduce the amount of idle carbon in a system and to keep pressure drop to a minimum, it would be helpful to direct the airflow only through the operating portion of the bed ($T$ in Fig. 5) while the saturated carbon is being reactivated for future service. One approach (*13*) to such a design is the rotating bed shown in Figure 13. There are four cylinders in all. The carbon bed is in the annular space between two inner concentric perforated cylinders (numbers 2 and 3) and is pertitioned into segments. The outside cylinder (number 1) is blank except for slots near

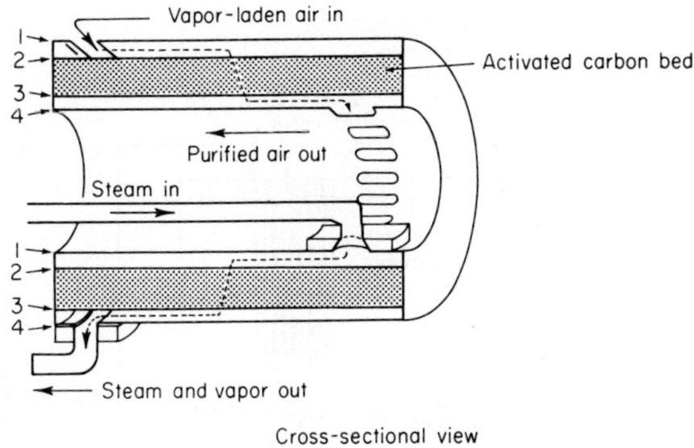


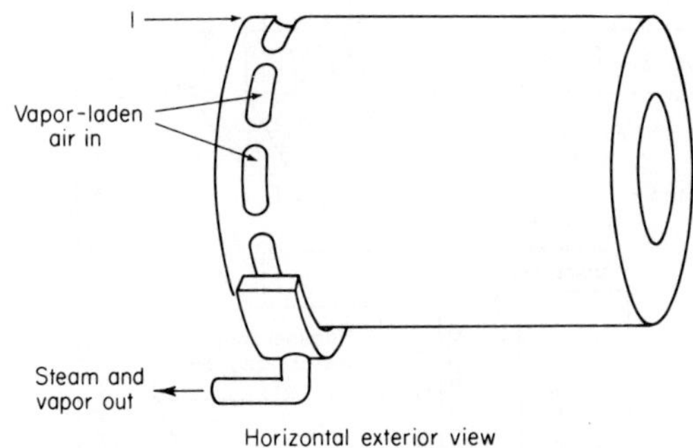


**Figure 13. Continuous rotary bed. (Based on Sutcliffe Speakmen Co., Ltd illustrations, Bronxville, New York.)**

the left end. The uncovered slots serve as inlet ports for polluted air; the covered ones as outlets for steam and desorbed vapors during reactivation. The inside cylinder (number 4) is also blank except for slots near the right end which serve as outlet ports for purified air or as steam inlets. The entire assembly rotates at a rate such that each segment of the bed is switched from adsorption to desorption as soon as it reaches the required level of saturation.

Although carbon utilization is thus improved, steam utilization becomes less efficient. The maintenance of seals and moving parts is also more expensive than that for a stationary bed.

## 7. Falling Bed

Figure 14a shows a falling bed adsorber "house" composed of 12 parallel adsorber compartments, each containing 2400 lb (1089 kg) of activated carbon, and each capable of handling 10,500 cfm (17,834 $m^3$/hour) of gas (*14*). As the carbon becomes saturated, a portion is dropped to the

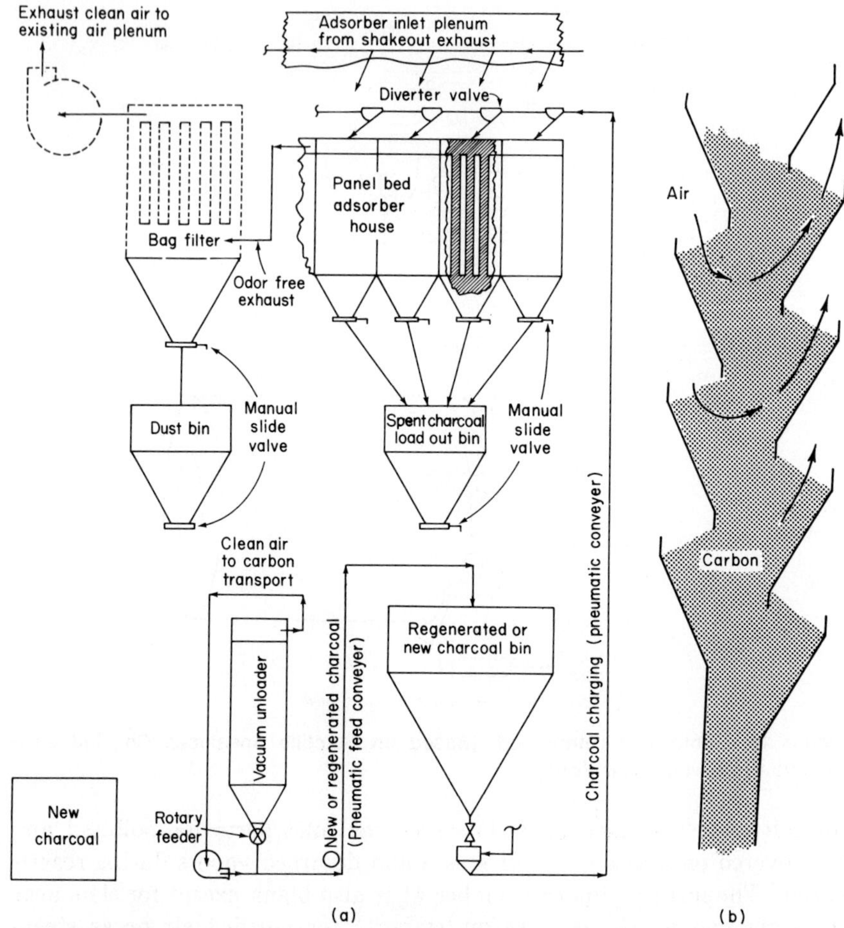


**Figure 14. Falling bed. (a) Adsorber system; (b) individual bed.**

spent carbon loadout bin and fresh carbon is added to the top. Thus, the carbon moves downward through the panel. Spent carbon is treated thermally and reused. Each panel (Fig. 14b) is 2 in. (5.08 cm) thick and is completely filled with activated carbon. The effective bed depth, however, is more than 4 in. (10.16 cm), since shells extend out from the panel sides to support the carbon and to provide a path for the gas flow. Panels are arranged in a series of "V" configurations so that the gas coming from one side must go through the carbon before exiting through a "V" on the other side of the chamber. As the carbon flows down through the bed, it is continuously turned and exposes new surfaces for maximum adsorption and particulate filtration.

## B. Systems

### 1. Principles

The contaminated air may be passed through the sorbent bed partially (the remainder being bypassed), completely (single-pass), or with some recirculation (multiple-pass).

When an adsorbent is used in recirculation to purify air in an enclosed space, the system imposes some inherent limitations on the rate of air purification and on the ultimate air purity attainable, regardless of the efficiency of the adsorber. Assuming a constant rate of generation of contaminants, and a constant rate of their removal by the adsorber, the concentration in the space undergoes an exponential falloff to an equilibrium value in which generation and removal rates are equal (*15, 16*):

$$C = C_0 e^{-EQt/V} + (G/EQ)(1 - e^{-EQt/V}) \quad (2)$$

and

$$C_\infty = G/EQ \quad (3)$$

where

$V$ = volume of chamber
$t$ = time
$C$ = concentration of vapor in chamber at any time
$C_0$ = initial concentration of vapor in chamber
$C_\infty$ = concentration of vapor in chamber at equilibrium
$E$ = efficiency of vapor reduction by the adsorber
$Q$ = volume rate of air delivered by the adsorber
$G$ = quantity rate of generation of vapor within (or injected into) chamber

In the special case where generation of contaminant has ceased ($G = 0$), the initial concentration falls off according to the equation

$$C = C_0 e^{-EQt/V} \quad (4)$$

and the final concentration ($C_\infty$) equals zero.

When the contaminated gas stream is to be released to the atmosphere, only a single pass is made through the adsorber. Before the adsorbing step, however, the gas stream may have to be treated in any of the following ways: (a) Particles so large that they may build up an obstructive coating on the adsorbent bed must be removed by filtration. (b) Sorptive capacity for the contaminant can be increased by any action that preconcentrates contaminants; this includes operating under increased pressure. (c) Moisture droplets, which may also act as an obstructant to gas adsorption, may be removed by electrical or mechanical means such as are elaborated in Chapter 7, this volume. As an alternative, the capacity of the carrier airstream for removing the droplets by vaporizing them may be increased by diluting the stream with ambient air if the relative humidity of the latter is low. Excessively high humidities may be reduced by cooling the air sufficiently with fin-tube-type coils so that condensation occurs and then by reheating the air until the relative humidity falls below about 50%. (d) If the gas stream is too hot (above about 100°F), it may have to be cooled to avoid excessive reduction of carbon capacity. Cooling by dilution with ambient air necessitates an increase in the size of the adsorbers and in the power cost for moving the air. Nonetheless, for thin-bed equipment, dilution is sometimes the most economical approach to cooling. The increased cost of larger adsorbers is partially compensated by a longer service life. (e) Excessively high gas concentrations are disadvantageous because the accumulated heat of adsorption may raise the temperature of the carbon bed to a degree that significantly impairs its adsorbing capacity. This problem can be remedied by dilution, but at the expense of the increased costs cited above. More significant, however, is the fact that high gas concentrations bring about rapid saturation of the adsorbent, and therefore impose high costs for its regeneration or replacement. Under these circumstances, it is often advantageous to use another vapor-control device before the carbon. For example, a liquid scrubber (see Chapter 6, this volume) may be used to remove soluble vapors of low molecular weight, such as light amines or oxygenates, leaving the less soluble gases of higher molecular weights, which are often strongly odorous even in small concentrations, to be removed by the activated carbon adsorbent.

A dual-stage system can be operated so as to utilize the total capacity of the adsorbent. Curve a, Figure 15, represents breakthrough from adsorber I, but complete control by adsorber II, which has been barely penetrated by the adsorption wave. The system may be operated until the effluent from adsorber II reaches $C_t$, at which time the carbon in adsorber I will be completely utilized (curve b). Adsorber I may then be removed for regeneration and replaced by adsorber II, while adsorber II is replaced by a fresh bed. This system will thus have been returned to the condition of curve b. Continual repetition of this process utilizes all the carbon to saturation while purifying the airstream to within the established criterion, $C_t$.

The service life of the adsorbent is limited by its capacity and by the contaminating load. Provision must therefore be made for determining when the adsorbent is saturated and for renewing it. The weight of an adsorbent is not a valid measure of its saturation because its moisture

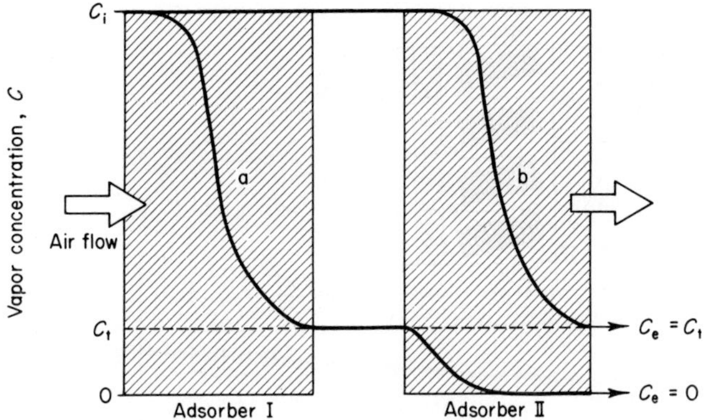


Figure 15. Dual-stage adsorbing system.

content depends on the relative humidity of the gas streaming through it, and is therefore apt to be variable. If mechanically feasible, a representative element or portion of the adsorbent bed may be removed and chemically analyzed to determine the degree of saturation of the entire bed (*1*). In many cases, a schedule for renewal of adsorbent is determined by actual deterioration of performance (breakthrough), or is based on a time schedule calculated from previous performance history or predicted from values assumed for the various saturation factors shown in Equation (5), Section IV,A,4.

To estimate the degree of saturation of a carbon bed or its residual capacity by chemical analysis requires a method that does not itself destroy or saturate the bed and that is insensitive to moisture. One such technique (*17*) subjects the bed to a pulse of $CO_2$, and the residual capacity is estimated by the degree of broadening of the eluting $CO_2$ peak.

## 2. Replacement of Adsorbent Bed

In some systems the adsorbent is discarded or removed elsewhere for reactivation after its performance has deteriorated. Either thin- (Fig. 16) or thick-bed equipment may be used, depending on factors previously described.

Such systems are used when the service life is so long that the cost of on-site regenerating equipment is unwarranted. This time span is typically 1 month or longer, depending on various other factors to be discussed in Section IV,A.

The types of pollutant sources that normally fall into this category are (a) those whose vapor concentrations are low (below about 2 ppm),

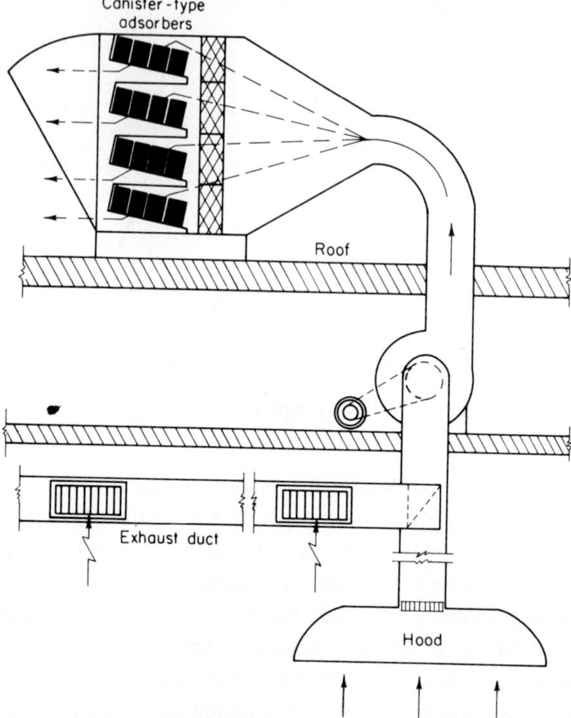


Figure 16. Thin-bed exhaust purification system.

such as gas streams that contain traces of highly odorous vapors, and (b) those that are potential single-episode releases, such as accidental leakages of radioactive or chemically toxic vapors.

## 3. On-Site Regeneration

a. Recovery of Sorbent and Sorbate. It is often economically advantageous to recover adsorbed vapors (usually solvents) from gas streams in which their concentrations are above about 700 ppm (0.07%) but not higher than about 25% of the lower explosive limit (LEL). Since the LEL of organic vapors is typically about 2%, this upper limit is about 0.5% or 5000 ppm. In operation, vapor-laden air, free from gross particulate matter and not warmer than about 100°F, is driven through one or more thick-bed stages of activated carbon, and the effluent is released to the atmosphere. When the sorbent reaches a given level of saturation, the vapor-laden pollutant source is directed to another, fresh adsorber, so that pollution abatement is continuous. Meanwhile, the saturated bed is regenerated by blowing a regenerating gas or vapor through it in the direction counter to that which the polluted gas stream had taken.

The regenerating agent of choice is low temperature steam. Steam is advantageous for several reasons: (a) At atmospheric pressure, steam temperature (100°C) is high enough to desorb most solvents of interest but not so high that it damages the carbon or the desorbed vapor. Some of the steam condenses in the bed, and its high heat of condensation aids in the desorption. (b) The effluent steam is easily separated from the recovered vapor by condensation followed by decantation or distillation. (c) The residual moisture in the carbon bed is readily removed by a stream of cool air, either pure or containing some organic vapor.

There are, however, a few disadvantages: (a) The system involves the capital and operating expenses of steam generation. (b) Vapors of high molecular weight, which may exist as an impurity in the solvent to be recovered, are not effectively desorbed by low-pressure steam, and may accumulate as a permanent resinous residue in the carbon. (c) Although the recovered vapor may not react with low pressure steam in the gas phase, it may do so on the carbon surface, which is an effective catalyst for many reactions, including oxidation. As an example, the action of hot activated carbon on an airstream containing methylethyl ketone yields an objectionable effluent that contains formaldehyde, acetic acid, and other irritating oxidized components.

Typical systems are illustrated in Figures 10 and 11. Figure 11 shows a scheme for a regenerative system designed to make full use of the sorbent's capacity. Absorber 1 is fresh and its effluent may be discharged to the atmosphere. Adsorber 2 is partly saturated, but its residual capacity may be used if its effluent is passed through a second stage (adsorber 3). The latter adsorber, having just been steamed, is simultaneously being cooled. The final adsorber (4) is undergoing reactivation. When the cycle is complete, the adsorbers change functions according to the schedule: $1 \rightarrow 2$; $2 \rightarrow 4$; $3 \rightarrow 1$; $4 \rightarrow 3$.

As noted above, caution must be exercised before any contaminated airstream is committed to a hot carbon bed, because such contact may promote decomposition or partial oxidation and thereby result in the discharge of odorous or irritating gases to the atmosphere. It is not usually necessary to have separate cooling and drying cycles because the vapor-laden air cools and dries the carbon bed so rapidly that there is always enough cool dry carbon to handle the adsorption. In such cases, the cooling wave travels through the bed considerably more rapidly than the adsorption wave. It is when the rates of travel of the two waves at high vapor concentrations are essentially equal, that precooling is necessary.

b. Recovery of Sorbent and Destruction of Sorbate. When the adsorbate is not worth recovering, either because its intrinsic value is low or because the recovery procedure is too difficult or expensive, it may nonetheless pay to regenerate the adsorbent at the site. The desorbed matter is then disposed of or destroyed. Such conditions are frequently encountered in the vapor concentration range above about 1 or 2 ppm but below about 700 ppm.

In such applications, the preferred regenerating gas is hot air, either at atmospheric or reduced pressure. (The data of Figure 18 show the advantages of evacuation.) The desorbate may then be removed from the effluent stream by incineration or scrubbing. In effect, the adsorber serves as a vapor concentrating medium. For example, benzene at a concentration of 150 ppm in air can be effectively stripped by a carbon bed and returned to a regenerating airstream at concentrations up to about 3%, or 30,000 ppm (*18*). This ratio represents a 200-fold magnification, which greatly reduces the cost of any subsequent treatment.

The oxidation of the adsorbate by air may also occur on the adsorbent surface, preferably in the presence of a catalyst, as mentioned in Section II,D. It has been shown (*9–12*) that various oxide and noble metal catalysts are effective for such applications, that hydrocarbons and oxygenates can be completely oxidized before the carbon bed itself starts to oxidize, and that repeated cycles of adsorption and catalytic oxidation can be carried out without impairing the function of the carbon. The method offers the advantage that the formation of polymer (e.g., styrene → polystyrene) on the adsorbent surface, which renders desorption much more difficult, is not a barrier to surface oxidation.

## IV. Applications to Source Control

### A. Effect of Process Variables

#### 1. Airflow

The rate of airflow is in inverse proportion to the detention time, or duration of contact, between airstream and sorbent bed. The possible effects of rapid airflow in uneven channeling through the sorbent bed, with resultant displacement or mechanical attrition, must also be considered. In the great majority of cases, however, the prime determinant of permissible flow rate is the pressure drop imposed by the sorbent bed on the moving airstream. Figure 17 presents pressure drop information as a function of flow rate for several typical granulations of activated carbon. Representative airflows in a commercial sorption equipment range from about 25 ft/minute (0.125 m/second) in thin-bed adsorbers up to about 80 ft/minute (0.4 m/second) in the thicker solvent recovery beds.

The distribution of particle sizes in a granular adsorbent bed is an important factor in determining the bed efficiency. Adsorption efficiencies exceeding 95% can be attained in a well-packed bed of carbon granules in the 6–16 mesh size range, even with thin beds (0.5–0.7 in.) (1.27–1.78 cm) at linear velocities of 30–50 ft/minute (0.15–0.25 m/second) and residence times approximating 0.03 seconds. For the thick beds used in recycling systems, linear air velocities as high as 200 ft/minute (1.0 m/second) have been used without adversely affecting bed efficiencies. The limiting factor in such cases is usually the increased power costs for driving the air through the bed.

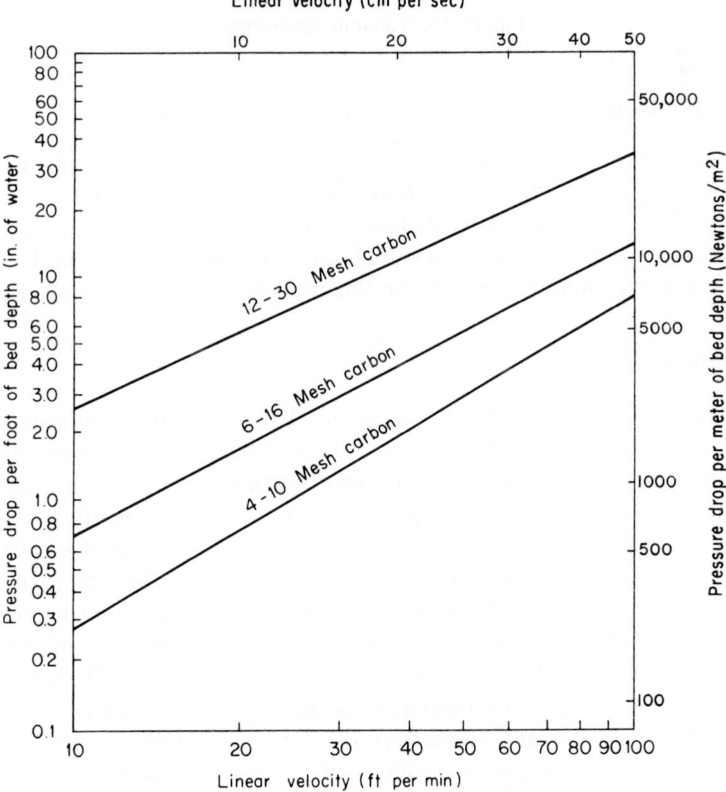


Figure 17. Pressure drop versus flow rate through granular carbon beds.

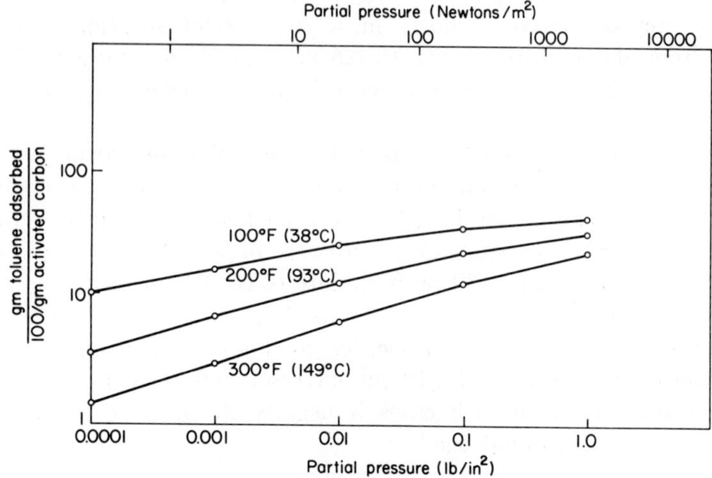


Figure 18. Toluene isotherms.

## 2. Temperature

The capacity of an adsorbent increases with decrease of temperature. As mentioned in Section III,B,1, temperatures above about 100°F (30°C) (see Fig. 18) should be avoided. Excessive temperature fluctuations such as may be caused by process cycling are also disadvantageous because they may induce massive desorption during periods of temperature rise.

## 3. Humidity

Activated carbon will adsorb moisture from a humid airstream, but will release this moisture to the atmosphere during the process of adsorbing larger, less polar molecules such as are usually common to pollutant gas streams. The presence of some moisture, in fact, is desirable because the vaporization of the water from the carbon dissipates heat of adsorption and provides a more uniform bed temperature. The presence of a moderate amount of moisture in the airstream from a polluting source, therefore, is not significantly deleterious to the performance of gas-adsorbing activated carbon. When the relative humidity of the airstream exceeds about 50%, its detraction from the carbon's capacity begins to be significant, but is not necessarily disabling. At higher moisture contents, the loss of capacity and efficiency increases. This effect has been found to be largely independent of temperature in the ranges normally encountered.

### 4. Vapor Concentration

The duration of service of the adsorbent is determined by the prevailing vapor concentration according to the relationship (*19*)

$$t = 6.43(10)^6 SW/EQMC \tag{5}$$

where

$t$ = duration of adsorbent service before saturation (hours)
$S$ = proportionate saturation of sorbent (fractional). Refer to Table IV for typical maximum values (retentivities)
$W$ = weight of adsorbent (pounds)
$E$ = sorption efficiency (fractional)
$Q$ = airflow rate through sorbent bed ($ft^3$/minute)
$M$ = average molecular weight of sorbed vapor (gm/mole)
$C$ = entering vapor concentration, ppm by volume

When weight ($w$) is in kilograms and airflow rate ($Q$) is in $m^3$/hour, the constant [$6.43(10)^6$] becomes $2.41(10)^7$.

For a typical average retentivity value, in source control by adsorption, we may take $S = 0.20$ (see Table IV). Note that this assumption implies complete saturation of the bed, and therefore applies to systems of the type shown in Figures 11 and 14. For a vapor with $M = 100$, and complete adsorption ($E = 1$), this relationship then reduces to

$$t = 1.29(10)^4 W/QC \tag{6}$$

**Table IV Rent- entivity of Vapors by Activated Carbon[a]**

| *Substance* | *Formula* | *Molecular weight* | *Normal boiling point, °C* | *Approx. retentivity[a] in % at 20°C 760 Torr* |
|---|---|---|---|---|
| Acetaldehyde | $C_2H_4O$ | 44.1 | 21 | 7 |
| Amyl acetate | $C_7H_{14}O_2$ | 130.2 | 148 | 34 |
| Butyric acid | $C_4H_8O_2$ | 88.1 | 164 | 35 |
| Carbon tetrachloride | $CCl_4$ | 153.8 | 76 | 45 |
| Ethyl acetate | $C_4H_8O_2$ | 88.1 | 77 | 19 |
| Ethyl mercaptan | $C_2H_6S$ | 62.1 | 35 | 23 |
| Eucalyptole | $C_{10}H_{18}O$ | 154.2 | 176 | 20 |
| Ozone | $O_3$ | 48.0 | −112 | Decomposes to oxygen |
| Putrescine | $C_4H_{12}N_2$ | 88.2 | 158 | 25 |
| Skatole | $C_9H_9N$ | 131.2 | 266 | 25 |
| Toluene | $C_7H_8$ | 92.1 | 111 | 29 |

[a] Percent retained in a dry airstream at 20°C, 760 Torr, based on weight of carbon.

### 5. Degree of Regeneration

The degree of saturation of an adsorbent is defined as

$$\text{Saturation capacity} = \frac{\text{weight of adsorbate}}{\text{weight of adsorbent}} \tag{7}$$

However, when a saturated adsorbent is regenerated, especially in an on-site installation, it is often uneconomical to remove all of the adsorbate. Instead, the regeneration is carried to the point where the cost per unit quantity of desorbate reaches a minimum value. This point is always short of complete desorption; the residual material is commonly called a "heel." The practical capacity of the carbon is then defined as

$$\text{Working capacity} = \frac{\text{weight of adsorbate} - \text{weight of ``heel''}}{\text{weight of adsorbent}} \tag{8}$$

One advantage of a heel is that it reduces the heat of adsorption in subsequent cycles. Working capacities of activated carbon may be as low as one-half of the total saturation capacity, depending on the cost of reactivation per unit weight of desorbate at different stages of desorption.

The effectiveness of heat and vacuum for the complete regeneration of saturated carbon depends in large measure on the molar volume of the adsorbate. The method is inadequate for adsorbates of molar volume above that of *n*-octane (162 ml/mole). For the range between acetone (75 ml/mole) and *n*-octane, desorption conditions of 200°C and $10^{-4}$ Torr for about 2 hours are generally adequate. Adsorbates of lower molar volumes can be completely desorbed at still lower temperatures.

## *B. Specific Processes*

### 1. Vapor Recovery in Process Industries

Major applications are in the recovery of solvents from air streams that are free of particulate matter and whose vapor concentrations are above about 700 ppm. These criteria are met in various solvent-using operations such as dry cleaning, degreasing, surface coating (including paper), rubber processing, and flexographic and gravure printing. Other applications occur in chemical manufacturing that involves vapor-phase reactions in an airstream. An example is the air oxidation of cumene to produce phenol; an activated carbon system is used to recover unreacted cumene.

## 2. Deodorization of Odorous Emissions

Many odorants in low concentrations, such as $10^{-7}$ (100 ppb) or less, are detectable and objectionable. The deodorization of such gas streams by solid adsorption is, in many instances, an effective and economical procedure. When odorous gases are discharged in high concentration, even the effluents from reasonably efficient cleanup methods may still be objectionably odorous. Solid adsorption methods may then be used effectively as a final deodorizing stage.

Examples of unpleasant and pervasive odorous effluents discharged to the atmosphere, which can be controlled in a final cleanup stage by solid adsorption, include those shown in Table V.

## 3. Control of Gaseous Radioactive Emissions

The adsorption of radioactive substances follows the principles set forth for nonradioactive gases and vapors. Applications have been made in which activated carbon systems serve as active or standby installations to prevent the emission of radioactive gases from nuclear reactors or other sources (*20, 21*). Radon and radioiodine are examples of adsorbable radioactive gases.

**Table V Processes That Involve Atmospheric Discharge of Adsorbable Odors**

| *Industry* | *Process* |
|---|---|
| Food processing | Dehydration, canning, cooking, frying, baking, coffee roasting; processing of fish, poultry, and meats; handling and blending of spices; fat and scrap rendering and other waste digestions; fermentation processes |
| Manufacturing and use of chemicals | Processes involving discharge to atmosphere of waste or recoverable by-product, solvent, or plasticizer; loss of small quantities of highly odorous materials, as in manufacture of pesticides, glues, cements, adhesives, fertilizers, and pharmaceutical products (especially those extracted from natural sources such as glands, urine, and blood); paint and varnish production; release of odorous vapors by displacement from storage tanks during filling and transfer operations |
| Miscellaneous processes | Gas odorizing sites, including containers, storage tanks, and odorant injection points; paper and pulp manufacturing; tannery operations; foundries; manufacturing of asphaltic products such as roofing; discharge of odorous exhausts from animal laboratories; and many others |

## 4. Evaporative Loss of Gasoline

Starting with the 1970 model year, new automobiles in the United States have been equipped with an activated carbon canister which traps the vapors generated from the fuel tank and in some cases from the carburetor during the "hot soak" period after the engine has been turned off. Subsequently, when the engine is restarted, the adsorbed vapors are desorbed from the carbon by a controlled air purge and the vapors are returned to the intake manifold for combustion in the engine (see also Chapter 14, this volume).

In the absence of controls, considerable losses of gasoline vapor occur during transfer and storage at marketing terminals and service stations. Activated carbon adsorbers can be used to concentrate and control these vapors, obviating the need for large vapor-holding tanks (*22*).

## 5. Removal of Sulfur-Containing Gases

Activated carbon serves as a contact catalyst for various reactions of sulfur compounds, including air oxidation of $H_2S$ to sulfur, of $SO_2$ to $SO_3$

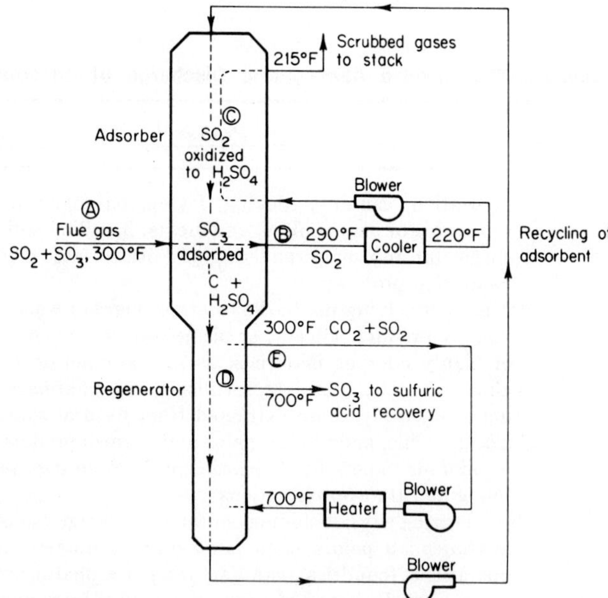


Figure 19. The Reinluft process. Simplified schematic representation (see Table VI).

**Table VI Steps in the Reinluft Process**

| *Step* | *Chemical or physical action* |
|---|---|
| A. Flue gas, containing $SO_2$ and $SO_3$, enters the adsorber at 300°F (149°C) | The $SO_3$ is adsorbed on the carbon; the $SO_2$ passes through |
| B. The flue gas is drawn off, cooled, and returned to the adsorbent bed at a higher level | Cooling of $SO_2$ from 300°F (149°C) to 220°F (104°C) |
| C. The $SO_2$ is oxidized to $SO_3$, and sulfuric acid is produced. The sulfuric acid remains on the carbon | $SO_2 + \frac{1}{2}O_2 \rightarrow SO_3$<br>$SO_3 + H_2O \rightarrow H_2SO_4$ |
| D. The downward motion of the bed carries the carbon with sulfuric acid to the regenerator section (700°F) (371°C), where the sulfuric acid dissociates. Some of the $SO_3$ is recovered; some reacts with the carbon | $H_2SO_4 \rightarrow SO_3 + H_2O$<br>$2SO_3 + C \rightarrow CO_2 + 2SO_2$ |
| E. The product gas that leaves the regenerator at 300°F (149°C) is reheated to 700°F (371°C) and returned to base of the regenerator | |

[a] The letters A–E refer to steps shown in Figure 19.

or to sulfuric acid, and reduction of $SO_2$ or sulfuric acid by $H_2S$ to produce sulfur:

$$H_2S + 1\tfrac{1}{2}\,O_2 \rightarrow SO_2 + H_2O \tag{9}$$
$$SO_2 + \tfrac{1}{2}\,O_2 \rightarrow SO_3 \tag{10}$$
$$SO_2 + \tfrac{1}{2}O_2 + H_2O \rightarrow H_2SO_4 \tag{11}$$
$$2\,H_2S + SO_2 \rightarrow 2\,H_2O + 3\,S \tag{12}$$
$$3\,H_2S + H_2SO_4 \rightarrow 4\,H_2O + 4\,S \tag{13}$$

One example of the use of activated carbon to remove sulfur-containing gas is the Reinluft process (*23*), which involves both adsorption and catalytic oxidation by means of a moving bed of activated carbon in a regenerative system. The separate steps involved are listed in Table VI and are illustrated in the flow diagram of Figure 19.

## V. Comparative Costs of Adsorbers and Other Systems

Figure 20 compares various systems for the removal of a typical solvent, methylisobutyl ketone (MIBK) from an airstream. Note that all sorption systems cost much less than incineration below vapor concentrations of about 500 ppm (which is 0.05% by volume, or about 4% of the LEL). At still higher concentrations, the heat content of the vapor lowers the cost of incineration below that of adsorption if there is no credit for

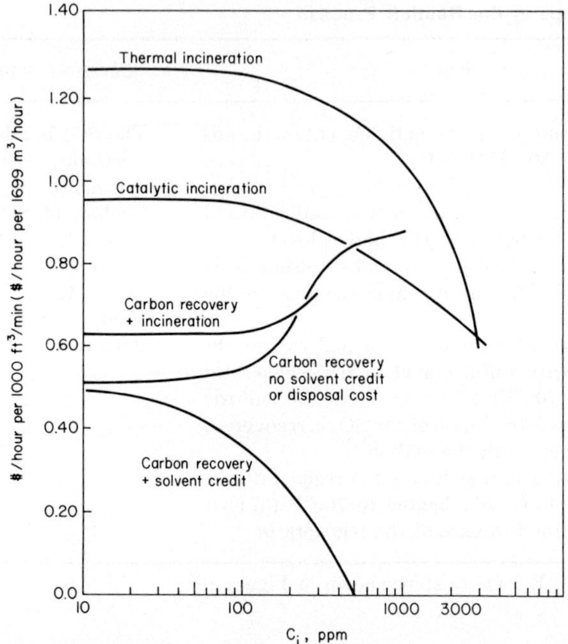


**Figure 20. Comparative costs of pollution control with carbon recovery, catalytic incineration, and thermal incineration, 3800 ft³/minute capacity system. Calculations based on methyl isobutyl ketone (MIBK), 2080 hour/year operating time, 1973 prices.**

the recovered solvent. With such credit, however, the costs of adsorption systems rapidly become self-sustaining.

## GENERAL REFERENCES

### Books and Reports


S. Brunauer, "The Adsorption of Gases and Vapors." Princeton Univ. Press, Princeton, New Jersey, 1945.

A. K. Doolittle, "The Technology of Solvents and Plasticizers," Chapter 9. Wiley, New York, New York, 1954.

R. L. Goldsmith, K. J. McNulty, G. M. Freedland, A. Turk, and J. Nwankwo, "Contaminant Removal from Enclosed Atmospheres by Regenerable Adsorbents," Nat. Aeron. and Space Admin. Report of Contract NAS2-7896. Ames Research Center, Moffett Field, California, 1974.

J. W. Hassler, "Purification with Activated Carbon." Chem. Publ. Co., New York, New York, 1974.

MSA Research Corp., "Package Sorption Device Systems Study," EPA-R2-73-202. United States Environmental Protection Agency, Washington, D.C., 1973.

T. M. Olcott, "Development of a Sorber Trace Contaminant Control System Including Pre- and Post-Sorbers for a Catalytic Oxidizer," Rep. No. NASA CR-2027. Nat. Aeron. Space Admin., Washington, D.C., 1972.

H. Sleik and A. Turk, "Air Conservation Engineering," 2nd ed. Connor Eng. Corp., Danbury, Connecticut, 1953.

H. L. Barnebey, *Heat., Piping, Air Cond.* **30,** 153 (1958).

H. L. Barnebey, *J. Air Pollut. Cont. Ass.* **15,** 422 (1965).

R. S. Joyce, J. R. Lutchko, R. K. Sinha, and J. E. Urbanic, *Ann. N.Y. Acad. Sci.* **237,** 389 (1974).

N. A. Richardson and W. C. Middleton, *Heat., Piping, Air Cond.* **30,** 147 (1958).

H. H. Todd, *Air Eng.* **4,** 26 (1962).

A. Turk and K. A. Bownes, *Chem. Eng.* **57,** 156 (1951).

## REFERENCES

1. A. Turk, H. Mark, and S. Mehlman, *Mater. Res. Stand.* **9,** 24 (1969).
2. A. Turk, S. Mehlman, and E. Levine, *Atmos. Environ.* **7,** 1139 (1973).
3. S. Stoldt and A. Turk, *J. Org. Chem.* **34,** 2370 (1969).
4. A. Turk and A. Van Doren, *Agr. Food Chem.* **1,** 145 (1953).
5. W. Trappe, *Biochem. Z.* **305,** 150 (1940).
6. A. Turk, J. Morrow, P. F. Levy, and P. Weissman, *Int. J. Air Water Pollut.* **5,** 14 (1961).
7. W. D. Ellis and P. V. Tometz, *Atmos. Environ.* **6,** 707 (1972).
8. A. Turk, *Ind. Eng. Chem.* **47,** 966 (1955).
9. J. Nwankwo and A. Turk, *Ann. N.Y. Acad. Sci.* **237,** 397 (1974).
10. J. Nwankwo and A. Turk, *Environ. Sci. Technol.* **9,** 846 (1975).
11. J. Nwankwo and A. Turk, *Carbon* **13,** 495 (1976).
12. A. B. Stiles, U.S. Patent 3,658,724.
13. "Solvent Recovery with Activated Carbon." Sutcliffe Speakman & Co., Bronxville, New York, 1963.
14. J. Schaum, *Mod. Cast.* (1973).
15. A. Turk, (*ASHRAE*) *Amer. Soc. Heat., Refrig. Air-Cond. Eng. J.* (Oct. 1963).
16. A. Turk, *in* "Basic Principles of Sensory Evaluation," ASTM STP 433. Amer. Soc. Test. Mater., Philadelphia, Pennsylvania, 1968.
17. A. Stamulis, J. K. Thompson, and H. F. Bogardus, *J. Air Pollut. Contr. Ass.* **21,** 709 (1971).
18. W. D. Faulkner, W. G. Schuliger, and J. E. Urbanic, Pap., 74th Nat. Meet. Amer. Inst. Chem. Eng., New York, New York, 1973.
19. A. Turk, H. Sleik, and P. J. Messer, *Amer. Ind. Hyg. Ass., Quart.* **13,** 23 (1952).
20. T. T. Porembski, *Air Cond., Heat., Vent.* **57,** 97 (1960).
21. R. E. Adams and W. E. Browning, Jr., "Removal of Radioiodine from Air Streams by Activated Carbon," At. Energy Comm. Rep. ORNL-2872, UC-70-Radioactive Waste, TID-4500, 15th ed. United States Atomic Energy Commission, Washington, D.C., 1960.
22. R. A. Fusco and R. L. Poltorak, Annu. Meet., Sept. 12–13. Nat. Petrol. Refiners Ass., Huston, Texas. 1973.
23. D. Bienstock, J. H. Field, S. Katell, and K. D. Plants, *J. Air Pollut. Contr. Ass.* **15,** 459 (1965).

# Combustion

Chad F. Gottschlich



## I. Introduction

Combustible pollutants are emitted to the atmosphere as gases, vapors, or particulates in many industrial processes. Solvent evaporation processes cause approximately half of the pollution, while petroleum refining and gasoline marketing produce another sixth. A number of other sources such as fat rendering and coffee roasting can produce severe local nuisances, even though they produce an insignificant fraction of the total weight of combustible airborne pollutants (*1*).

There are four well-developed rapid oxidation devices for the destruction of these pollutants: thermal afterburners, catalytic afterburners, furnaces, and flares. The choice of the method to be used in a given application depends upon such factors as the concentration of the combustible in the waste gas (or fume), the steadiness of its flow rate, and the presence of other contaminants. Because the handling of vaporized or particulate combustibles involves the hazards of fire and explosion, the design of equipment for the disposal of combustible wastes requires a knowledge of combusion and related safety equipment.

The related problem of air pollution control from the incineration of liquid and solid wastes is covered in Chapter 13, this volume. Control of air pollution from mobile combustion sources by means of afterburners is covered in Chapter 14, this volume.

## II. Principles of Combustion

Combustion is a rapid oxidation process that we normally associate with a flame. Edwards (*2*) describes this process in terms of mixing, precombustion reactions, combustion, and postflame reactions.

Mixing can occur before combustion begins (premix combustion) or while combustion is in progress (nozzle mix combustion or the diffusion flame). The chemistry of combustion is dependent upon the presence of free radicals, which are electrically neutral molecular fragments containing an unpaired valence bond. Free radicals are extremely reactive and unstable. Even though their life is short and their concentration is low, they profoundly increase the speed of the oxidation reaction. In the precombustion zone, formation of free radicals is initiated. The temperature in this zone is brought to about 500°C by heat conduction and radiation from the combustion zone.

In the combustion zone, the free radicals combine rapidly with the oxygen and the fuel molecules by a complex sequence of chemical reactions. These include reactions which continue to produce free radicals; reactions which produce such partial oxidation products as aldehydes, carbon monoxide and hydrogen; and reactions which produce such final products of combustion as carbon dioxide and water. Along with these chemical changes, the heat of combustion is released and causes the temperature to rapidly approach 1700°C and higher. Because of the high temperature of the combustion zone, the combustion reactions cannot go to completion. Equilibrium concentrations of a few percent of hydrogen, carbon monoxide, and oxygen remain. Also, concentrations of the order of a fraction of a percent of hydroxyl radical and atomic hydrogen exist at temperatures of 1700°C and higher.

The postflame zone follows. In it, the combustion products cool as heat is transferred from them. As the temperature drops, the residual partial combustion products and free radicals combine with oxygen and each other to bring the oxidation reaction to completion. As the time available for the total oxidation process is finite, a small amount of oxidation products, of the order of a few parts per million, will remain unreacted and be vented to the atmosphere. If the flame is cooled too quickly (a symptom of improper operation), soot may be formed and the concentration of aldehydes can be high enough to cause obvious effects such as odor nuisance and eye irritation.

Combustion can proceed only when the concentration of a combustible in an oxidant, e.g., air, is within a limited range (*3*). This range is bounded by concentrations called the lower and upper explosion (or flammable) limits. As an example, for a mixture of natural gas and air at room temperature, the lower explosion limit (LEL) is 5.3 vol% of natural gas. The upper flammable limit (UEL) occurs at 15.0 vol%. A flame will be self-propagating for any concentration within this range. A flame that is initiated in a natural gas–air mixture having a concentration below 5.3% or above 15.0% will be quenched by the mixture. The limits stated above are not absolute because they are affected by the temperature and the pressure of the mixture, the geometry of the containing vessel, and the presence of other contaminants.

The flammable range is controlled primarily by the amount of energy that is released by the combustion process. On comparing 31 common combustible chemicals mixed with air, it was found that the heats of combustion of the mixtures at the LEL were in a range of 330 to 610 kcal/Nm$^3$ (normal cubic meters), with an average value of 500 kcal/Nm$^3$. This fact may be used to estimate the LEL of a combustible, if its molar heat of combustion is known or can be estimated. The phrase fume energy concentration is often used to express the heat of combustion per unit volume of a mixture.

Oxygen and combustible content may be used to define four types of waste gas streams to be treated by combustion (*4*). Types I and II are below 25% of the LEL. For type I, the oxygen level is above 15% and for type II, below 15%. For type III, the combustible concentration is above the UEL. In type IV, the combustible and oxygen concentrations are within the flammable range. The last two definitions differ slightly from Hemsath's (*4*). Type IV is unsafe. Process conditions should be changed to reduce either the oxygen or the combustible content to convert the waste stream to one of the other three types. Types I and II are usually treated in afterburners and occasionally in furnaces. Type III waste streams are normally treated in flares.

## III. Afterburners

### *A. Thermal Afterburners*

A thermal afterburner is used for the destruction of pollutants in a waste stream having a combustible concentration below the LEL. It usually consists of a refractory-lined combustion chamber containing a burner at one end. It is generally operated at temperatures of 550° to 850°C and has sufficient volume to hold the waste stream for 0.3 to 1.0 seconds. The burner can be oil fired, but is more commonly gas fired. If the waste stream has enough oxygen (type I), it can be used as the oxygen source for the burner fuel. If not (type II), supplementary air must be supplied along with the fuel. Supplementary air is avoided when possible because its use increases the fuel requirement by up to 50%.

If enough fuel is added to the waste stream so that the mixture temperature will be 820°C after completion of the combustion reactions, the fume energy concentration of this mixture is only 280 kcal/$Nm^3$, which is well below the LEL. A flame initiated in this mixture will not propagate, but will be quenched. If flame quenching occurs, the unburned fuel and partial oxidation products formed will often increase, rather then reduce, the contaminants in the waste stream. This problem is solved by mixing only part, usually about one-half, of the waste stream with the fuel; such a mixture will support flame combustion. The remainder of the waste stream bypasses the burner and then is mixed with the flame combustion products far enough downstream to be sure the flame is not quenched.

As the hot combustion products mix with the remainder of the waste stream, rapid oxidation of the rest of the combustibles occurs, even though the flame is no longer present. At temperatures above 700°C, the oxidation reaction rate is usually much faster than the rate at which the two streams mix together. Thus, the rate of oxidation becomes controlled by fluid mechanical rather than chemical processes.

The burners used may be divided into two classes: the distributed burner and the discrete burner (*1*). In the distributed burner (Fig. 1), a large number of small burners are distributed over the flow cross section at the entrance of the afterburner. With discrete burners, only one or a few burners are used. The advantage of the distributed burner is that because all parts of the waste stream come within a few inches of the flame, only a short distance (10 to 20 times the burner spacing) is required to complete the mixing of the burner products and the bypassed waste stream. Relatively greater mixing lengths are required for discrete

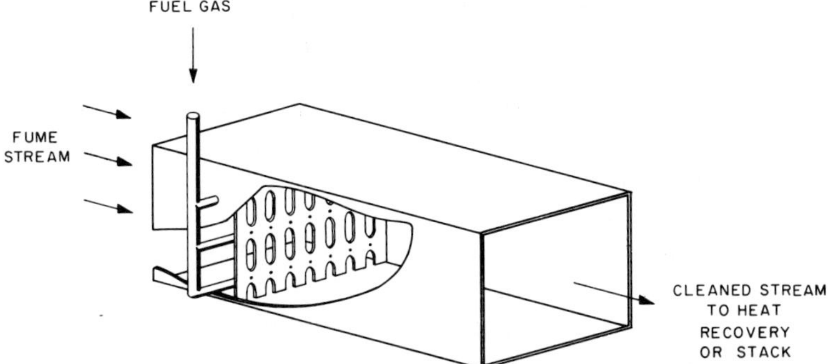


Figure 1. Thermal afterburner with a distributed burner.

burners because the cross-channel distances are greater between the bypassed waste stream and the burner products. Baffles and other turbulence inducing devices are commonly needed to promote cross-channel mixing when using discrete burners. Such burners, however, have the advantage of being more tolerant of a fouling waste stream. In addition, they can be oil fired while distributed burners are presently limited to gaseous fuels. Each general type of burner is available from several burner manufacturers.

### B. Catalytic Afterburners

The catalytic afterburner (Fig. 2) usually has two parts. In the first, the waste gas is preheated, usually by a burner. Indirect heat exchange with another hotter stream can also be used. In the second, the waste stream is brought into contact with a bed of catalyst. The most popular catalyst is a finely divided noble metal, such as platinum supported on another solid. The catalyst support can be in a variety of forms, e.g.,

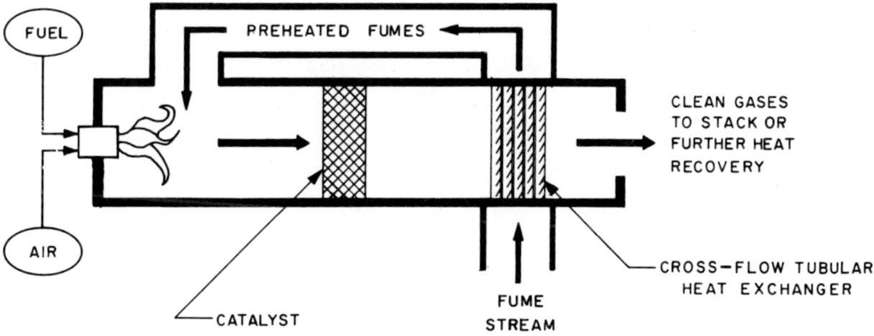


Figure 2. Catalytic afterburner using its waste heat to preheat the fume stream.

wire mesh, ceramic honeycomb, rods, beads, etc. In addition to the noble metals, certain metal oxides are effective catalysts. Some of these are $Co_3O_4$, $CoO \cdot Cr_2O_3$, $MnO_2$, $LaCoO_2$, and CuO (*2*). The volume of catalyst required is about 0.03 to 0.12 $m^3$ per $Nm^3$ of fume/second. The volume includes the space occupied by the air passages surrounding the catalyst particles. Even if the entire catalyst volume is considered as available for gas flow, the residence time is only 0.01 to 0.05 seconds. Thus, most of the total volume of the catalytic burner is taken up by the preheat zone.

Catalytic materials are expensive, but the amount of catalyst used is small so that the purchase cost of a catalytic afterburner is about the same or just a little more than the cost of a thermal afterburner. However, because of the catalytic afterburners' lower operating temperature, lighter construction is possible and installation becomes easier and cheaper than for thermal afterburners.

A catalytic afterburner operating in the 400° to 500°C range can usually oxidize a waste gas as effectively as a thermal afterburner operating in the 700° to 800°C range. The lower temperature can result in a fuel savings of the order of 40 to 50%. As an example, consider an afterburner treating 4.7 $Nm^3$/second of waste gas with an inlet temperature of 205°C, an outlet temperature of 760°C, and using natural gas as the fuel at \$0.053/$Nm^3$. The fuel cost can be as high as \$20/hour if the fume energy concentration is negligible. At 8000 operating hours per year, this is about \$160,000/year. The fuel savings possible with a catalytic afterburner are, in fact, the main incentive for their use.

Even though the operating cost is substantially less and the installed cost moderately less for catalytic afterburners, they are not as widely used as thermal afterburners. There are several reasons for this. If the gas stream has particulates in it, they will gradually coat the catalyst and make it unavailable for the oxidation reaction. The catalyst activity can usually be recovered by washing, but this adds to maintenance cost. The catalyst can be easily poisoned by compounds containing sulfur, heavy metals, phosphorous, and halogens. (Sulfur and the halogens are less serious poisons because the catalyst can recover most of its activity after they are removed from the gas stream.) Even in the absence of poisons and fouling materials, the catalyst gradually loses its activity as it ages. Deactivation by aging is faster at higher temperatures. In a poorly controlled waste stream, there may be high levels of combustibles at times that raise the catalyst to excessive temperatures. Such incidents can severely shorten the useful life of the catalyst. If the maximum bed temperature is maintained below 600°C, a useful life of 3 to 5 years can be expected. At 675° to 700°C, catalyst life drops to 1 year. However, even if the process conditions are such that the catalyst has a useful life of only 1 year, the catalytic afterburner will be more economical than the thermal afterburner. Used noble metal catalysts can normally be returned to the manufacturer to recover their metal value.

The rate at which oxidation occurs on the catalyst is dependent upon the chemical nature of the pollutant. For the normal amounts of catalyst used, 90% conversion of most common pollutants can be achieved with temperatures of 425° to 480°C. Carbon monoxide is more easily oxidized and requires a temperature of about 370°C; methane is particularly difficult to oxidize, requiring a temperature of about 650°C.

Just as with thermal incinerators, a high fume energy concentration can significantly reduce the supplemental fuel requirement. The fuel replacement by fume energy concentration cannot be one-for-one however, because a high fume energy concentration means that there will be a large temperature rise through the catalyst bed. The lower inlet temperature to the bed then requires a higher outlet temperature to maintain the same percent reduction of the pollutant. Using data from Rolke *et al.* (*1*), suppose a fume with a negligible fume energy concentration requires a catalyst temperature of 450°C to achieve a 90% reduction of its pollutant concentration. If the pollutant is increased to a fume energy concentration of 44.5 kcal/Nm$^3$ (about 10% LEL), the inlet temperature can be dropped to about 340°C for the same percent reduction of pollutant, but the outlet temperature rises to about 490°C. As a result, the efficiency of the fume in replacing preheat energy is about 75%. This comes about because the oxidation rates are slower at lower temperatures. The slow start that the reaction has at a low inlet temperature must be compensated by a higher outlet temperature.

### C. Heat Recovery

The use of a catalyst in an afterburner was motivated by the accompanying reduced fuel requirement. Fuel can also be saved by using a heat exchanger to transfer heat from the recuperator's exit stream to its inlet stream. This may not be the most attractive use for the waste heat from the afterburner, however. For example, it is generally more economical to generate steam with the heat than to preheat a gas stream because steam generators have much higher heat transfer coefficients. A need for the steam has to exist, of course.

An even more attractive possibility is to transfer the heat to the process that produced the fume. We recall that solvent evaporation processes are the principal source of combustible pollutants vented to the atmosphere. In this process, large flows of hot air or hot gases supply the process heat and carry the evaporated solvent out of the oven. Generally, the oven temperature is well below that of the afterburner temperature. Therefore, when using the afterburner gases in the process, it is necessary to temper them with ambient air or a recirculation stream of oven gas. Not all of the afterburner flue gases can be returned to the process. Part must be vented because evaporating solvent, air in-leakage, and afterburner fuel are continually adding volume to the recirculating gas stream. Further, if no air is added to the recirculating stream, its oxygen content will be depleted to the point where combustion cannot be supported in the afterburner. Also in curing ovens it is often necessary to maintain some oxygen in the circulating hot gases because the curing chemistry requires it. Any air added for these two purposes requires an equal volume of afterburner flue gas to be vented from the system. The vented hot gases can still be used to preheat the fume entering the afterburner, to preheat fresh airstreams entering the process, or to generate steam. In individual applications, other uses for the hot afterburner gases may appear.

The most commonly used heat transfer equipment is the shell-and-tube exchanger. Generally the cool stream is inside the tubes while the afterburner flue gases flow over the outside of the tubes perpendicular to their axis. This flow pattern, called cross-flow (Fig. 3), leads to simpler construction and gives a substantially higher heat transfer coefficient on the outside of the tubes than is possible when the outside flow is parallel to the tubes. By using a multipass arrangement, a cross-flow exchanger can be made to approximate the performance of a counter-flow exchanger.

The cyclic regenerator is a second kind of heat transfer device that is commonly used. In its most conventional form, it consists of two stationary packed beds through which the hot and cold streams flow alternately (*5*). The packing, which acts as a heat storage medium, usually consists of ceramic shapes such as balls, saddles, and bricks. In some

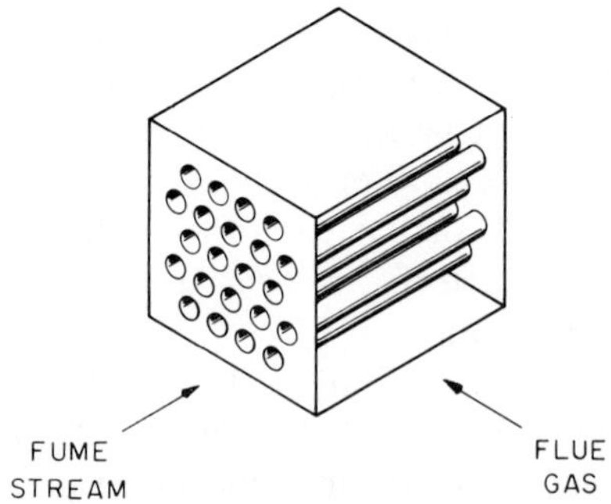


**Figure 3. Cross-flow heat exchanger.**

cases, stones graded to a narrow size range may be used. A variation of the regenerator is the rotary regenerator (*5, 6*). It is also known as the Lungstrom air heater and consists of a cylindrical heat storage matrix that rotates at a constant speed. As it rotates, each part of it alternately allows the hot and the cold stream to flow through it. The matrix is usually built as a honeycomb structure made of ceramic, glass, or metal. The cells of the honeycomb are the flow passages, and the walls store the heat.

Both the shell-and-tube heat exchanger and the regenerator are easily fouled by dirty fluids. The symptoms of fouling differ, though. In the shell-and-tube exchanger, fouling primarily reduces the heat transfer rate; in the regenerator, increasing pressure drop is the first effect observed.

To compare the performances of the shell-and-tube exchanger and the regenerator, it is useful to use the effectiveness factor. It is defined as the amount of heat transferred to the cold stream, divided by the amount of heat that could be recovered from the hot stream, if it were cooled from its inlet temperature to the inlet temperature of the cold stream. Because it is less expensive to create heat transfer surface area in a regenerator than in a shell-and-tube exchanger, regenerators usually have much higher effectiveness factors. A 95% effectiveness factor is practical in a regenerator, where 60 to 70% may be an economically practical limit for tubular exchangers.

The use of fume preheat can create a hazard because the fume's additional energy content may cause the fume to oxidize significantly before it enters the afterburner. The additional temperature rise could produce thermal stresses beyond the design capability of the exchanger or other parts of the system. It is particularly important to consider this possibility if the flow rate or the fume can vary over wide limits, since the temperature rise experienced by the fume in the heat exchanger increases as its flow rate decreases. For example, if the fume flow rate varies over a 4 to 1 range, the fume temperature rise will be approximately 50% greater at the minimum rate than at the maximum rate. Even if no precombustion occurs, this additional temperature rise may overheat the afterburner. One method of overcoming this problem is to add tempering air to the fume stream to control its exit temperature from the exchanger.

## IV. Furnaces

Many processes that generate a fume are part of a plant having utility and process furnaces. As these furnaces generally operate at temperatures even higher than those of afterburners and usually have adequate residence times, they would seem to offer an attractive potential as a disposal device for all three types of fume. The advantage of using the fume as a supplementary fuel or air supply in existing equipment is apparent. Further, the fuel requirement of the afterburner is eliminated and recovery of the fuel value of the fume is easily possible.

Such an application has only occasional practicality, however. The furnace has to be tolerant of an additional stream having a fluctuating flow rate determined by an external process rather than the needs of the furnace. This means that the fume stream should be a small contribution to the furnace's total fuel–air requirements. Additional furnace controls may be needed because of the fume. Ducting must be added to bring the fume to the furnaces. The fume generating process now becomes dependent upon the reliability of operation of the furnace, which means that if the furnace is shut down, either the process must also be shut down or polluted gases will have to be vented to the atmosphere. There may be contaminants in the fume that can damage the furnace or the material being heated by it.

## V. Flares

Process plants that handle combustible gases are subject on occasion to either off-design operating conditions or emergencies which require the release of these gases to the atmosphere. Flares are used to do this safely and without pollution. The flare is usually mounted from about 5 to 100 m above ground level and far enough from other structures to provide protection to equipment and personnel from the heat, flame, light, and noise emitted by the flare. During the past few years, ground-level flares have become popular.

Although a flare is basically a burner, it differs from conventional industrial burners because of the special performance characteristics demanded of it. In particular it must provide safe, complete combustion of the waste stream despite rapid, large changes in the flow rate of that stream. Flares are expected to handle a turndown ratio (i.e., flow rate range) of 1000:1 whereas conventional industrial burners are seldom designed for more than a 10:1 turndown ratio. Flares have been designed and built for flow rates in excess of 30,000 $Nm^3$/hour and still have the ability to maintain combustion at flow rates as low as 30 $Nm^3$/hour.

Because of these extreme turndown requirements, flares are normally nozzle mix burners. Steam jets are frequently used to promote good mixing and an adequate flow of combustion air. The steam requirements range from 0.05 to 0.30 kg/kg of waste gas. An example of a steam injection flare is shown in Figure 4. Alternatively, the air can be supplied by fans. Lower operating cost and less noise are claimed for fans. If the waste gas has enough pressure, it can be used to aspirate its air just as in a bunsen burner. A danger with the last method is that a substantial volume of explosive mixture exists upstream of the ignition point. Thus, the flare must be carefully designed in order to prevent flashback over its wide operating range With all three methods, good design can ensure smokeless combustion over the operating range of the flare.

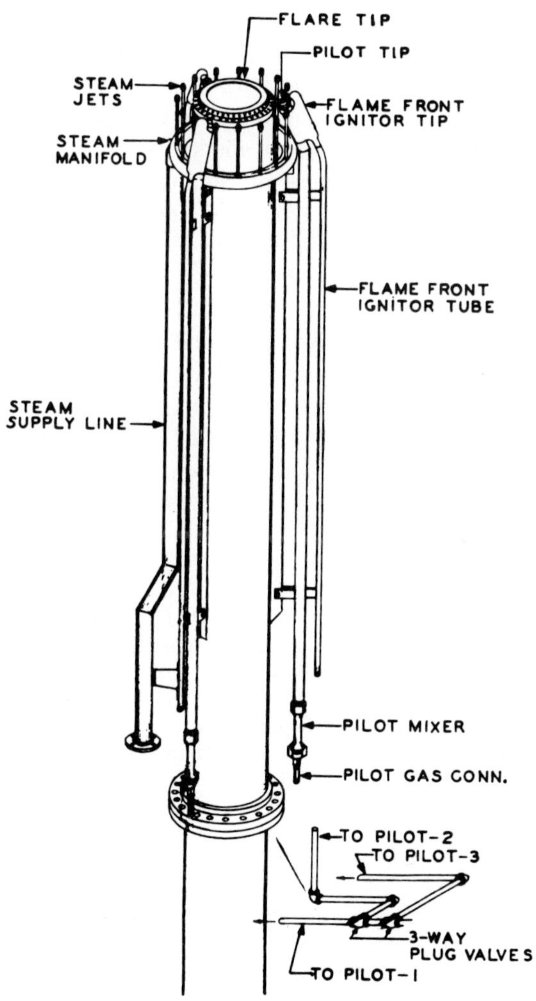


Figure 4. Steam injection flare (*6a*).

Elevated flares often have a shroud surrounding the burner. It promotes better mixing of the waste combustible stream with air and protects the flue and its pilot lights from being blown out by winds of upward of 160 km/hour.

Ground-level flares use a multijet method to introduce the waste gas stream. A stack consisting of a steel shell lined with refractory is mounted above the jet assembly. The combination of momentum exchange from the jets and a stack effect supplies the combustion air required with relatively low noise levels. The stack effectively shields the surroundings from heat. Further, the light from the flare is almost totally blocked.

A flare can tolerate inert materials in the waste gas; but, it may be necessary to use preheat to maintain stable combustion. This can be done by surrounding the flare with a ring burner in addition to the pilot burners. The fume energy concentration of the waste gas, air, and steam (if present) mixture should be kept above 450 kcal/$Nm^3$.

As flares are designed to dispose of combustible gases, any entrained oil mist should be continuously separated and drawn away at the base of the stack. Also flares should be designed to prevent back drafts into the waste gas piping.

The design of flares has evolved as an art because the complex interactions of fluid mixing, heat transfer, and combustion chemistry have been difficult to quantify. This has gradually changed with the steady development of the science of combustion. Swithenbank (*7*) has shown how combustion science can be used in the design of a flare to provide for wide turndown, complete combustion, low emissions of nitrogen oxides and low noise levels.

## REFERENCES

1. R. W. Rolke, R. D. Hawthorne, C. R. Garbett, E. R. Slater, T. T. Phillips, and G. D. Towell, "Afterburner Systems Study," Report EPA-R2-72-062. United States Environmental Protection Agency, Washington, D.C., 1972.
2. J. B. Edwards, "Combustion, the Formation and Emission of Trace Species." Ann Arbor Sci. Publ., Ann Arbor, Michigan, 1974.
3. B. P. Mullins and S. S. Penner, "Explosions, Detonations, Flammability and Ignition." Pergamon, Oxford, England, 1959.
4. K. H. Hemsath and A. C. Thekdi, 75th Nat. Meet. Amer. Inst. Chem. Eng., New York, New York, 1973.
5. R. H. Perry, C. H. Chilton, and S. D. Kirkpatrick, "Chemical Engineer's Handbook." McGraw-Hill, New York, New York, 1963.
6. W. M. Kays and A. L. Londen, "Compact Heat Exchangers." McGraw-Hill, New York, New York, 1964.

6a. R. J. Ruff, "Air Pollution" (A. C. Stern, ed.), 1st ed., Vol. II. p. 360. Academic Press, New York, New York, 1962.

7. J. Swithenbank, *AIChE J.* **18,** 553 (1972).

# Part C

# PROCESS EMISSIONS AND THEIR CONTROL

# 10

## Fuels and Their Utilization

Richard B. Engdahl and Richard E. Barrett

## I. Introduction

A characteristic of civilization since its beginning has been the combustion of wood and fossil fuel for useful heat, usually with the emission of smoke and some fine ash. More recently, combustion has been used extensively for industrial steam and thermal power generation—practically always accompanied in the early applications by emissions to the atmosphere of smoke, ash, odors, and noxious and benign gases. After the Industrial Revolution transformed many ways of life, the smoking industrial chimney came to be widely accepted as a welcome sign of the new prosperity. Then, slowly, as living in industrial smoke became almost intolerable, industrial centers, one by one, began breaking free from acceptance of the tradition that smoke and other pollution is a necessary concomitant of prosperity. More recently, governmental authority has been used as the mechanism for enactment and enforcement of increasingly stricter air pollution regulations.

Although much fuel has been used wastefully, and some of it even discharged unburned into the air, fuel users have had interest in improved fuel utilization. Thus, economic pressures have resulted in gradual improvements in the efficiency of most fuel-burning equipment and reductions in air pollution emissions have occurred as a result. However, it must be emphasized that clear flue gas emanating from a chimney does not automatically mean high efficiency or low pollutant emissions. Health effect studies have pointed an accusing finger at some invisible gases (sulfur oxides, nitrogen oxides, etc.) as contributors to the adverse impact of pollution on life. Probably the most wasteful fuel-burning installations are those which have clear effluents, but in which the fuel gases are needlessly diluted with tremendous volumes of air which carry wasted heat away from the point of heat release.

## II. Fuels

The predominant fuels being utilized today (Table I) are fossil in origin, having been formed by processes of nature and stored in the earth for many thousands of years. Wood wastes, bagasse, and municipal refuse are minor, nonfossil fuels consisting primarily of cellulose. Although growing shortages of other fuels has generated increased interest in utilization of waste fuels, many waste materials contain substances which can cause undesirable emissions to the air when used as fuels (*1*).

The atmosphere of many of the industrialized countries has greatly improved since World War II by a strong trend toward use of "clean"

**Table I Common Fossil Fuels**

| *Solid fuels (coals), as received* | *Fixed carbon* | *Volatile matter* | *Ash* | *Moisture* | *C* | *H* | *O* | *N* | *S* | *Higher heating value* *kJ/kg* | *Btu/lb* |
|---|---|---|---|---|---|---|---|---|---|---|---|
| Anthracite | 88 | 1 | 8 | 3 | 85 | 2 | 4 | 1 | 1 | 30,000 | 13,000 |
| Semianthracite | 75 | 10 | 12 | 3 | 78 | 4 | 4 | 1 | 1 | 30,000 | 13,000 |
| Bituminous | 60 | 25 | 10 | 5 | 78 | 5 | 7 | 1 | 3 | 28,000 | 12,000 |
| Subbituminous | 45 | 25 | 5 | 25 | 55 | 6 | 34 | 1 | 1 | 21,000 | 9,000 |
| Lignite | 22 | 35 | 8 | 35 | 42 | 7 | 42 | 1 | 1 | 16,000 | 7,000 |

| *Liquid fuels* | | *Higher heating value* *MJ/m³* | *Btu/gal* |
|---|---|---|---|
| No. 2 fuel oil | Specification for various grades of fuel oils are based on required characteristics for firing in different types of burners. Fuel composition varies with crude source and refining. | 39,000 | 140,000 |
| No. 4 fuel oil | | 40,400 | 145,000 |
| No. 5 fuel oil | | 41,800 | 150,000 |
| No. 6 fuel oil | | 42,100 | 151,000 |

| *Gaseous fuels* | *Composition* | *Higher heating value* *kJ/m³* | *Btu/scf*[a] |
|---|---|---|---|
| Natural gas | 86–95% methane, remainder ethane, propane, carbon dioxide, and nitrogen | 37,200 | 1,000 |
| Commercial propane | 90–95% propane, remainder propylene | 93,100 | 2,500 |
| Commercial butane | 95% butane, remainder butylene | 119,200 | 3,200 |
| Fuel gas from coal gasification | See Table II (p. 400) | 3,700 to 13,000 | 100 to 350 |

[a] scf: Standard cubic feet.

fuels—oil and gas for residential, commercial, and industrial purposes, and in some countries, by the use of electricity for comfort heating. This developing fuel-use pattern has now been arrested by a worldwide shortage of available oil and gas. For the short run, ample oil and gas are in reserve for immediate needs if economics and regulations permit their recovery, but in the long run these reserves are not inexhaustible.

Owing to growing oil and gas shortages the trend must once more swing back, at least partly, to bituminous coal, still abundantly available in many areas of the world. However, the national determination of many countries has well established that their populations will not again accept smoky, heavily sulfurous atmospheres. Accordingly, until the massive research now accelerating to produce clean fuels from coal at an acceptable cost bears fruit, conversions from oil and gas to high-sulfur coal will be confined primarily to those large users who either can afford to install and maintain the expensive pollution abatement equipment required to utilize coal in an environmentally acceptable manner, or are located where sulfur dioxide dispersion by tall stacks is practical. Smaller heat sources such as industrial, commercial, and residential heating systems will move toward the use of low-sulfur coal, often at considerable cost penalty.

## III. Pollutants Emitted by Combustion Processes

Although approximately 99% of the products generated by combustion are innocuous gases—namely, nitrogen, carbon dioxide, and oxygen—combustion processes do emit significant quantities of pollutants. The pollutants generated by combustion processes can be categorized as (a) gases and vapors, (b) particulate matter (solid particles and liquid droplets), (c) visible smoke, and (d) toxic and hazardous pollutants.

The first two categories are by definition mutually exclusive. Their sum is equal to the total mass of pollutants emitted by a source. Visible smoke is a portion of the particulate matter and should not be considered when quantifying the mass of emissions. However, smoke is important (a) for esthetic reasons, (b) as an indicator of incomplete combustion, (c) as an indicator of heat-exchange fouling, and (d) because it is easily evaluated at low-cost by relatively unskilled observers. Smoke readings may be more indicative of the potentially more harmful fine particulate emissions than is total mass of emissions. Toxic and hazardous pollutants are materials that are emitted in such small quantities that they are a barely perceptible fraction of the total emissions on a mass basis, but which may be very harmful to the environment.

### A. Gases and Vapors

Gases and vapors in the flue gases from combustion sources that are generally considered to be the principal pollutants are the sulfur oxides, $SO_2$ and $SO_3$; the nitrogen oxides, NO and $NO_2$ or $NO_x$ ($NO + NO_2$); carbon monoxide, CO; and gaseous hydrocarbons, HC.

Additionally, concern has been expressed over the release of large quantities of carbon dioxide into the atmosphere by combustion of fossil fuels because $CO_2$ accumulates and is not cleaned from the air by natural processes. There are $2100 \times 10^9$ metric tons (about $2300 \times 10^9$ tons) of carbon dioxide in the atmosphere and about $140{,}000 \times 10^9$ metric tons (about $150{,}000 \times 10^9$ tons) dissolved in the sea (*2*). The great increase in the use of carbonaceous fuels has resulted, over the years, in the release of about $180 \times 10^9$ metric tons (about $200 \times 10^9$ tons) of carbon dioxide to the air. This is approximately 10% of the total amount of carbon dioxide now in the atmosphere. All of the fuel used in the world today adds an additional $8 \times 10^9$ metric tons (about $9 \times 10^9$ tons) of carbon dioxide to the atmosphere each year (*3*). Photosynthesis is estimated to consume about $54 \times 10^9$ metric tons (about $60 \times 10^9$ tons) per year, but respiration and decaying vegetable and animal life returns carbon dioxide to the atmosphere at about the same rate. The present average atmospheric concentration of carbon dioxide is approximately 300 ppm, and it is rising at the rate of about 1 ppm per year (*4*). Although the most obvious sources of the additional $CO_2$ are combustion processes, very slight increases in ocean water temperature can produce similar effects.

#### 1. Sulfur Oxides

Sulfur oxides are emitted from combustion process due to conversion of most of the sulfur in the fuel to sulfur dioxide and sulfur trioxide during combustion. Sulfur is present in significant quantities in coal (up to about 6% sulfur) and oil (up to about 4.5% sulfur). Except possibly for coal burned on grates, at least 90% of the sulfur in the fuel is emitted as sulfur oxides in the flue gas. About 99% of the sulfur oxides are emitted as sulfur dioxide and from 0.5 to 2.0% as sulfur trioxide.

Raw natural gas may contain significant quantities of hydrogen sulfide but it is generally removed from the gas before distribution and widespread utilization. Hence, combustion of natural gas emits essentially no sulfur oxides.

#### 2. Nitrogen Oxides

Nitrogen oxides (generally referred to as $NO_x$ but really the sum of nitric oxide and nitrogen dioxide) are emitted from all combustion processes. About 90–95% of the nitrogen oxides are emitted as nitric oxide (NO) and the remainder as nitrogen dioxide ($NO_2$).

The source of the nitrogen oxide emissions is twofold: by thermal fixation, where the high temperature flame environment promotes the reaction of nitrogen and oxygen to form $NO_x$, and by conversion of the organic nitrogen in the fuel to $NO_x$. Natural gas contains essentially no fuel nitrogen so that all $NO_x$ emitted by gas-fired sources is thermal. Fuel oil and coal contain up to about 1.0 and 2.8% fuel nitrogen, respectively, and combustion of these fuels produces $NO_x$ by both mechanisms.

Figure 1 shows the influence of temperature on the formation of thermal $NO_x$ (*4a*). The precise reactions by which thermal fixation occurs are still subject to question. However, models by Zeldovich (*5*) seem to explain the mechanism.

If the hot products of combustion cool slowly the nitrogen oxides formed in the flame will have time to decompose to oxygen and nitrogen. However, in the usual case the gases are cooled rapidly by heat absorbing surfaces which extract useful heat, hence, nitric oxide decomposition is arrested and the flue gases retain high concentrations of $NO_x$.

Conversion of fuel-bound nitrogen to $NO_x$ has been well documented. Also documented is the fact that the proportion of fuel nitrogen converted to $NO_x$ is highest when the quantity of nitrogen in the fuel is low. As the

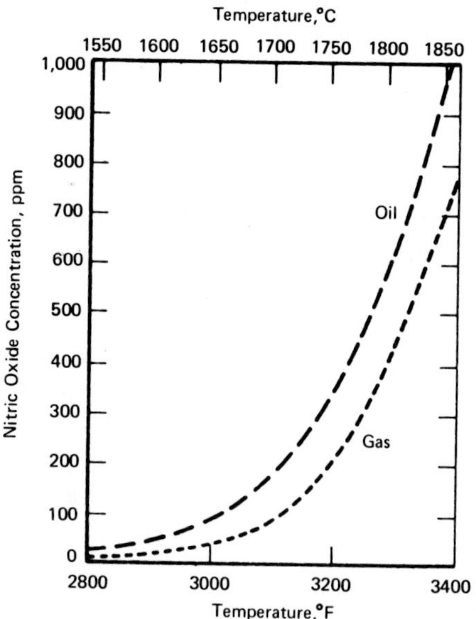


Figure 1. Calculated nitric oxide concentration from oil- and gas-firing with typical excess air levels and 0.5 second residence time (5).

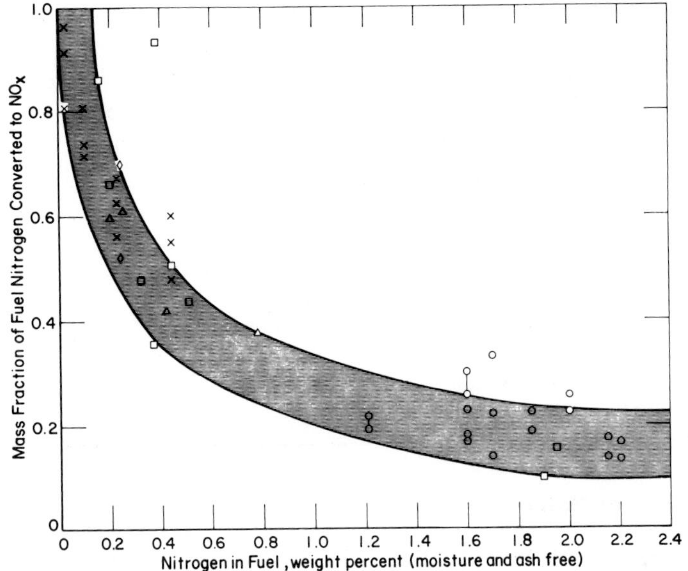


Figure 2. Conversion of fuel-bound nitrogen to $NO_x$ in combustion systems. ◇: Dykema (*5a*); ○: Crawford (*5b*); △: Turner (*5c*); □: Cato (*5d*); ×: Hazard (*5e*).

nitrogen in the fuel increases, the proportion of the fuel nitrogen converted to $NO_x$ decreases. Figure 2 shows that at fuel nitrogen levels below 0.05% as much as 90% of the fuel nitrogen is converted to $NO_x$. As fuel nitrogen levels reach 0.5 and 1.0% only about 40 and 20% of the fuel nitrogen is converted to $NO_x$, respectively.

## 3. Carbon Monoxide and Gaseous Hydrocarbons

Carbon monoxide and gaseous hydrocarbons are emitted by combustion processes due to the incomplete combustion of the fuel. Incomplete combustion may result from incorrect air-to-fuel ratio, insufficient mixing of air and fuel, insufficient residence time in the flame, or quenching by cold surfaces. Although carbon monoxide and hydrocarbons are present in flue gas from any combustion source, in properly designed and operated units the quantities of these gases emitted are so low as to be insignificant—carbon monoxide less than 100 ppm for gas-, 20 ppm for oil-, and 200 ppm for coal-fired furnaces; and hydrocarbons less than 20 ppm for gas-, oil-, and coal-fired furnaces. However, in poorly designed or operated systems, emissions of carbon monoxide and hydrocarbons may be increased by 10 or 100 times.

One problem encountered in reducing emissions from combustion sources is the fact that modifications that reduce carbon monoxide and hydrocarbon emissions generally increase $NO_x$ emissions and vice versa. The expected effect of various modifications in operation of combustion sources on emissions is shown in the following tabulation.

| *Change in operation* | *Effect on CO and HC emissions* | *Effect on $NO_x$ emissions* |
|---|---|---|
| Increase excess air | Decrease | Increase, then gradually decrease |
| Increase flame temperature | Decrease | Rapid increase |
| Increase residence time at high temperature | Decrease | Small increase |

### B. Particulate Matter

Particulate matter may be defined as the solid particles and liquid droplets present in flue gas.* Solid particles are generally either ash from the fuel and/or carbonaceous material (soot) resulting from incomplete combustion. Liquid droplets generally consist of unburned and/or partially combusted fuel droplets. Sulfur trioxide ($SO_3$) may be present as droplets of sulfuric acid if the flue-gas temperature is below about 150°C (300°F); however, air preheater and stack corrosion problems generally dictate that sufficiently high flue-gas temperatures be maintained to avoid formation of these acid droplets.

### C. Smoke

Smoke is a submicrometer particulate aerosol, from a combustion source, which obscures vision. It may, and usually does, contain comparatively little particulate matter by weight. However, it may appear to be an impenetrable mass because of the light-scattering properties of materials in the size range 0.3 to 0.5 $\mu$m *(6)*. On the other hand, the effluent from a stack emitting a much larger tonnage of large particles, say 100 $\mu$m, may appear clear because such particles do not scatter light and, moving at high speed, they may be invisible. Smith and Gruber (7)

* The United States Environmental Protection Agency has defined particulate matter as "any material, other than uncombined water, which exists in a finely divided form as a liquid or solid at standard conditions" (*Federal Register,* August 17, 1971). However, this definition was not included in a later publication of the same standard (*Federal Register,* December 23, 1971).

conclude from the extensive measurements of Hangebrauck *et al.* that "no direct relationship seems to exist between the total particulate loading and the opacity of the smoke plumes." However, for some noncombustion processes, if the emitted particle-size distribution is constant, opacity has been shown to be related to the particulate mass loading of the flue gas (*8*).

A source of large particle emission is the accumulation of fine particles of smoke and fly ash on boiler heating surfaces and their subsequent flaking-off as agglomerates. Within a few hours of operation the insulating effect of these deposits reduces heat transfer to such an extent that their removal is necessary by a process called "sootblowing." Most of the agglomerates thus removed are so coarse that they are readily collected in conventional dust collectors. Particulate emissions from oil-fired units are more than doubled during sootblowing (*9*).

### D. Toxic and Hazardous Pollutants

As more has been learned about the pollutants making up the mass of total emissions and how to control them, increasing interest has been directed to pollutants that are classified as toxic or hazardous, but are emitted in only trace quantities. Most of the pollutants of interest here have attracted attention because either (a) they are not rapidly excreted from the body and tend to accumulate, (b) they are carcinogenic, or (c) they are retained in the lungs. Among the pollutants of interest are (a) metals (mercury, cadmium, zinc, lead, etc.); (b) polycyclic organic matter (benzo-$\alpha$-pyrene, benzo-$\epsilon$-pyrene, etc.); and (c) fine particulate matter (0.1–0.6 $\mu$m).

Metals are present in liquid and solid fuels. Metals in liquid fuels are generally entrained in the flue gas upon combustion and may be collected in particulate control devices or emitted to the air.

For coal firing, it is expected that most of the trace elements will remain associated with the bulk inorganic combustion residues (ash). This is not true, however, for mercury and the volatile Periodic Table Group V and Group VI oxides, e.g. selenates and arsenates. For example, preliminary results by Oak Ridge National Laboratory (*10*) indicate poor collection of selenium by electrostatic precipitators, and also the fact that as much as 90% of the mercury escapes from a combustion system because of its high volatility. Also, preliminary information (*11*) indicates component segregation as a function of particle size in residual fly ash. This observation implies dependence of trace metal emissions on combustion conditions, as well as on the nature and effectiveness of particulate matter cleanup.

Among the "hazardous pollutants" identified by the United States National Academy of Sciences and the United States Environmental Protection Agency (USEPA) is the general class of polycyclic organic matter. This class of materials, including polynuclear aromatic hydrocarbons and nitrogen–heterocyclic materials, is poorly defined, both from the standpoint of collection and analysis, and of the effect of combustion conditions on their generation. It is known that (a) they are abundantly present in fuel oil and coal, (b) some of them, e.g., benzo-$\alpha$-pyrene, are carcinogenic, and (c) combustion conditions do affect both the gross emission of polynuclear hydrocarbons and the specific distribution of individual compounds (*12*) (see also Chapter 3, Vol. I; and Chapter 6, Vol. II).

Since production of polynuclear hydrocarbons is affected by combustion conditions, it must be considered separately for different combustion systems. Conditions leading to relatively low combustion temperatures are especially suspect, since it is believed that compounds such as benzo-$\alpha$-pyrene form more rapidly than they decompose at temperatures below 815°–870°C (1500°–1600°F). Appreciable residues of these materials may persist at much higher temperatures, but detailed kinetic information on their decomposition is not available.

For base-load utility power plants that operate under relatively steady-state conditions and with high flame temperatures and good air–fuel mixing, benzo-$\alpha$-pyrene emission levels are believed to be low. Smaller combustion systems, with inherent quenching by cold surfaces and short residence times in the flame, tend to emit larger quantities of polycyclic organic matter. Combustion systems with poor air–fuel mixing (such as small coal-fired systems) are relatively high emitters of polycyclic organic matter. Consideration of these emissions is especially important for combustion in systems where combustion modifications, like flue-gas recirculation or staged combustion, are employed to reduce peak temperatures for $NO_x$ control purposes. Other combustion systems that may emit high concentrations of polycyclic organic matter are fluidized beds (with lower combustion temperatures) and systems operating in on–off models (with poor mixing at start-up and shutdown).

## IV. Fuel Modification to Reduce Emissions

The most desirable way to reduce emissions from fuel impurities is to clean the fuel. In the case of "sour" natural gas, containing sulfurous impurities, the desulfurization of the gas is a relatively simple, low-cost process which has been widely applied (*13*). More costly, but also in wide practice, is the production by petroleum refineries of low-sulfur distillates from high-sulfur crude oil. In addition, hydrodesulfurization of residual oil is coming into use despite its significant addition to the final cost to the consumer. This added cost is generally less than the cost of subsequent combustion gas cleanup (*14*, *15*).

Coal is not so readily separated from its sulfur and ash. Some desulfurization and deashing of coal is practical by means of well-established coal-washing techniques which add very little to the cost of the coal. Other coal cleaning methods (physical separation and solvent and chemical coal cleaning and refining) are under development. However, few of these processes have progressed beyond laboratory or pilot scale operations. In addition, the conversion of coal to a liquid or gaseous fuel has as one of its primary motivations, the removal of contaminants from the coal.

Removal of most of the sulfur from coal, or the substitution of a low-sulfur coal in place of a high-sulfur one, usually has a markedly deleterious effect on the ash-collection efficiency of electrostatic precipitators because of the increased resistivity of the fly ash produced. Conditioning of the flue gases by small amounts of sulfur trioxide or other additives, or redesign for larger and higher temperature precipitators, can correct for this deterioration. Operation of a precipitator at a temperature above 425°C (about 800°F) results in a decrease of the resistivity of the fly ash to such a low level that high collection efficiency can be maintained regardless of the sulfur trioxide content of the gases. Unfortunately for existing plants, modifying precipitators from low temperature to high temperature operation can be very costly.

### A. Coal Washing

Coal washing is a commercially practiced coal preparation technique, with about 50% of the coal mined in the United States being washed. The inorganic portion of the sulfur is usually in the form of chunks or finely divided particles of $FeS_2$, iron pyrites, which if separable by crushing can be removed because of the high density of $FeS_2$ (more than twice that of coal). The organic sulfur is chemically bound into the coal substance and cannot readily be separated. Some coals benefit much more from washing than others, hence, a closely defined statement of the benefits of coal cleanup by washing cannot be made. United States Bureau of Mines tests on the washability of many different coals (*16*) indicate that washing can reduce the sulfur content of bituminous coal by 15–30% and the ash content by 20–50%.

In a coal preparation plant, the raw coal is subjected to (a) size reduction and screening, (b) separation of coal from ash and pyrite by devices utilizing differences in specific gravity or surface properties, and (c) dewatering the product coal. The amenability of the coal for sulfur removal by coal washing is established by "float and sink" analyses or washability tests. In these tests, the coal is crushed to various sizes, and each narrow size range is introduced into vessels containing liquids in the specific gravity range of 1.2 to 1.8. The upper part of the range approaches the specific gravity of impurities such as shale or rock, and the lower part of the range approaches the specific gravity of pure coal. The specific gravity of pyrite is about 4.90 and of shale or rock is 1.6 to 2.2; that of pure coal is about 1.2. The float material is termed the yield, while the sink material is called the refuse.

In practice; sulfur and ash removal are increased by using a medium (a magnetite or sand slurry) with a lower effective specific gravity, however, the total yield of product coal in the float material is reduced. Typical coal cleaning plants operate with media with an effective specific gravity of 1.6 and yieds vary between 80 and 90% of the total heat input.

In the modern coal-cleaning plant shown in Figure 3 (*17*), incoming raw coal is first crushed to a top size of 127 mm (5 in.). An initial size separation of the 6.4 mm × 0 mm ($\frac{1}{4}$ in. × 0) fraction is achieved over double-decked screens. The 127 × 6.4 mm (5 × $\frac{1}{4}$ in.) fraction is sent to heavy-media vessels where coal is separated from refuse with at least 99% recovery of the coal possible on a heat content basis. The yield from the primary heavy residue separation is screened to separate the 127 × 32 mm (5 × $1\frac{1}{4}$ in.) and the 32 × 6.4 mm ($1\frac{1}{4}$ × $\frac{1}{4}$ in.) size ranges. The former is crushed to a top size of 32 mm ($1\frac{1}{4}$ in.) while the latter is centrifuged. These two streams are combined and sent to the load-out facilities. The yield from the secondary separation vessel is crushed and treated with the 6.4 mm × 0 ($\frac{1}{4}$ in. × 0) portion of the raw coal.

The 6.4 mm × 0 ($\frac{1}{4}$ in. × 0) size range resulting from the initial separation is separated into two fractions, 6.4 × 0.6 mm ($\frac{1}{4}$ in. × 28 mesh) and 0.6 mm × 0 (28 mesh × 0), using sieve bends and vibrating screens. Separation of refuse from the 6.4 × 0.6 mm ($\frac{1}{4}$ in. × 28 mesh) fraction is achieved in heavy-media cyclones. Froth flotation is used for the 0.6 mm × 0 (28 mesh × 0) fraction. The large-size fraction is dewatered by centrifugation, while the small-size fraction is dewatered by a vacuum disk filter. Both fractions are combined and dried in a coal-fired fluidized-bed dryer equipped with a cyclone and a flooded-disk venturi scrubber for particulate control.

This plant is designed for closed-loop operation. Make-up water is needed mainly for losses in the fluidized bed dryer. A slurry of magnetite in water is the heavy medium employed. Elaborate schemes are used for recovery and conservation of the magnetite. The circuit for this purpose consists of (a) a sump wherein the dilute medium collects, (b) classifying cyclones for separation of the concentrated slurry, (c) a magnetite thick-

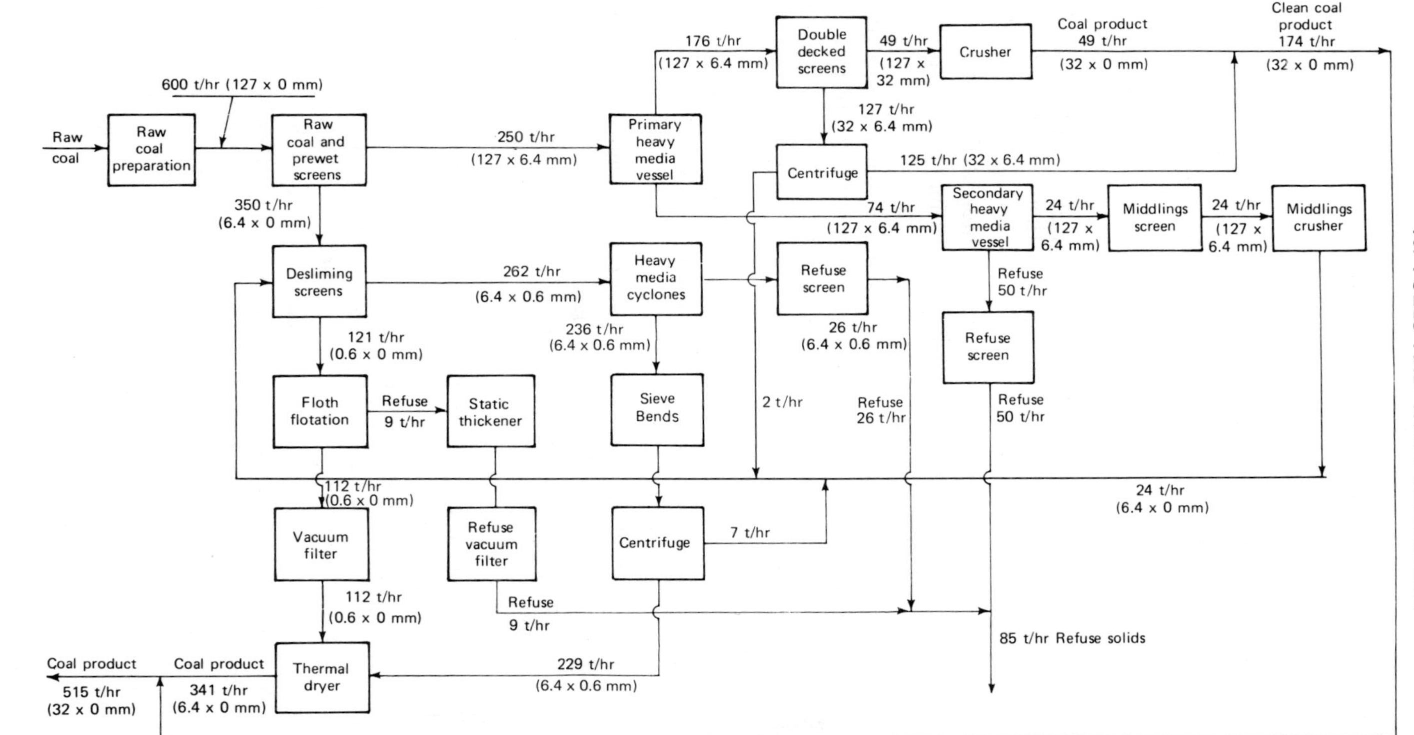


Figure 3. Schematic flow diagram of a 600 ton/hour metallurgical coal cleaning plant. (Only solids containing streams are shown.)

ener for the same purpose as (b) above, and (d) a magnetic separator for separation of magnetite from refuse.

### B. Other Physical Separation Processes

High-pyritic-sulfur raw coal can be fractionated by utilizing the difference in the electrophoretic mobilities of pyrite and coal in water. This has been done on a laboratory scale to remove a siliceous as well as a pyritic refuse (*18*).

A dry-removal process can clean fine-sized coal by removing the lightweight, pyrite-depleted portions centrifugally and then electrostatically concentrating the pyrite from the remaining pyrite-enriched portion. On a pilot scale, the dry-separation process, when integrated with stage grinding, is nearly as effective as float-sink separation at 1.6 specific gravity (*19*).

Several types of electrostatic separators, including triboelectric and nonuniform field devices, have been shown to remove pyrite from pulverized coal (*20*). In general, concentrating the pyrite is a relatively slow process and many cycles are necessary to concentrate it into a fraction that can be economically discarded.

Thermal-magnetic processes produce selective heating of pyritic impurities in coal by dielectric means to convert the pyrite to magnetic pyrrhotite while leaving the coal constituents undisturbed (*21*). Conventional magnetic separation can be used to selectively remove the inorganic sulfur after the dielectric treatment. Significant reductions of inorganic sulfur in coal have been shown possible by steam heating in the range of 190° to 260°C (375° to 500°F) followed by conventional magnetic separation methods (*22, 23*).

An immiscible liquid separation technique can be used for fine-coal separation. This involves the selective migration of one or more compounds from a water suspension to an oil phase or to an oil–water interface. Patents for oil flotation date back to 1860. Interest in this method has been renewed in recent years, e.g., the so-called "Convertol Process" (*24*). Processes in which one of the constituents is concentrated by using an immiscible collecting oil phase are typified by the Shell pelletization process (*25*) for removing fine carbon black particles from dilute aqueous suspensions. Meadus and co-workers (*26*), using this principle, were able to reduce the ash and sulfur content of a fine coal. The coal formed spherical agglomerates while the hydrophilic ash and sulfur-containing components remained in aqueous suspension. In the above cases, the naturally hydrophobic material was agglomerated with the oil phase as a bridging agent.

If the particles of one species are selectively aggregated into flocs while maintaining the other species fully dispersed, the flocs can then be separated by conventional means, such as ordinary gravity settling. Studies have shown that reversible flocculation can be obtained in ultrafine particulate systems (630 Å) by means of thermodynamic steric effects (*27*). Flocculation-dispersion stability is controlled by the free energy of interpenetration of specifically adsorbed organic groups which, in many systems, is a function of the temperature of the system. Therefore, particular attention is given to long-chained surfactants that exhibit a sharp reversal in their free energy of interpenetration with temperature. The separation mechanism is based on close control of the temperature of the system and is reversible within a narrow range of temperatures. This feature of the method allows spontaneous reversibility of suspension stability with temperature, and is a major advantage over current flocculation technology.

Among the "trace" metals that may be found in raw coal are cadmium, chromium, copper, lead, manganese, mercury, nickel, selenium, and zinc (*28*). These metals tend to separate out with the pyrite and ash. Thus, coal preparation by physical separation processes may significantly reduce the amounts of these metals in the fuel.

### *C. Chemical Refining*

Several chemical refining processes for converting raw coal into clean solid fuel have been developed to various stages of feasibility over the past 50 years or so. One method is treatment with aqueous solvents to preferentially leach out pollutants and leave a reatively clean solid carbonaceous residues. Another is solvent refining which consists of heating coal in an organic solvent at a temperature and pressure sufficient to dissolve most of the organic material in the coal, leaving the majority of the pollutants as the residue. A third method involves the treatment of raw coal with molten salt baths to reduce the pollutant content of raw coal and produce a relatively clean solid fuel. Finally, a variety of miscellaneous processes for chemically cleaning coal have attracted interest.

#### 1. Aqueous Leaching Processes

Pyrite ($FeS_2$) is found in coal as nodules or as tiny discrete particles and is, therefore, subject to attack by a wide variety of chemical solvents or reagents, generally at an elevated temperature. Leaching of coal, as opposed to physical cleaning, involves the separation of the impurities—sulfur, nitrogen, and ash, which includes the toxic or hazardous metals—

by treating or leaching the coal with a solvent or reagent to convert the impurities to a soluble form, generally water soluble. Some of the impurities are removed as a gas, e.g., hydrogen sulfide. In addition, some leaching processes are capable of removing some or all of the organically bound sulfur (which is not separable by physical techniques).

Although leaching of coal is not yet practiced commercially in the United States, many methods for leaching sulfur and the mineral matter from coal have been reported to be successful on a laboratory or pilot-plant scale.

In the Bureau of Mines process, coals are treated at 225°C (437°F) for 2 hours using a 10% NaOH aqueous solution as the leachant. In some cases, the hydrothermally leached coal is treated further by an acid leach at ambient temperature. Pyritic sulfur is removed by the NaOH treatment (*29*). The leached coal has a somewhat higher ash content than the original coal. However, if the NaOH treatment is followed by an acid leach, most of the ash is removed. For example, an Illinois No. 6 coal originally contained 9.7% ash and 1.1% pyritic sulfur; the leached coal contained 0.7% ash and 0.1% pyritic sulfur. The yield of coal was 91.5%. Similar results were obtained with Indiana No. 5 coal. The depyriting–leaching procedure usually produces a coal product of higher heating value than the raw coal. The free-swelling index changes only slightly and little or no organic sulfur is removed.

It has been observed that the sulfur content of coal extracts is reduced by pressure leaching of extracts with aqueous NaOH and ZnO at temperatures greater than 150°C (302°F) (*30*). The alkali-soluble material from the extraction of a Pittsburgh seam coal with NaOH contained 1.16% sulfur (*31*).

The Southern Research Institute process has established the feasibility of oxidizing and removing the pyritic-sulfur by the treatment of coal with steam and air at pressures of 10 to 13 atm (150 to 200 psi) and temperatures up to 120°C (250°F) (*32*). High-temperature steam–air treatment of Romanian and Indian coals has resulted in sulfur reductions of 30% and approximately 57%, respectively. The addition of ammonia with the steam has resulted in the removal of 88% of the sulfur from Indian coals (*33*).

Leaching of pulverized coal with aqueous solutions of $FeCl_3$ or $Fe_2(SO_4)_3$ at temperatures up to 130°C (266°F) has resulted in the extraction of the majority of the pyritic sulfur. The pyritic sulfur is converted to iron sulfate and elemental sulfur (*34*). The majority of the iron sulfate is soluble and is removed in the leach liquor, whereas the elemental sulfur is insoluble and must be removed with the leached coal product. The elemental sulfur is separated from the coal by vacuum distillation or extraction with a solvent such as toluene or kerosene. The leached coal product is washed to remove any remaining soluble iron sulfate. By-products from the process are elemental sulfur, iron oxide, and iron sulfate.

The Battelle Hydrothermal Coal process has been shown in bench-scale tests to remove up to 99% of the pyritic sulfur and 70% of the organic sulfur (*34a*). This process uses an aqueous slurry of coal and leachant at moderate pressures and temperatures. The most promising leachants are NaOH and $Ca(OH)_2$.

Since pyrite and mineral sulfides are not part of the coal matrix, they can, in theory, be considered sulfide ores and, thus, hydrometallurgical technology developed for treating sulfide ores should be applicable for treating the mineral sulfides in coal.

Inorganic sulfide ores can be leached (a) in the absence of oxidizing agents, and (b) in the presence of oxidizing agents (*35*). In the absence of oxidizing agents, dilute acids dissolve some metal sulfides with the liberation of hydrogen sulfide. For example, iron pyrite reacts with mineral acids to produce the corresponding iron salt and hydrogen sulfide. Ferrous sulfide (FeS) is completely soluble in mineral acids. In principle, the resulting hydrogen sulfide can be recovered and converted to elemental sulfur (*36*). Other sulfides which are soluble in mineral acids include CoS, ZnS, and NiS.

In the presence of oxidizing agents, pyrite as well as a number of other sulfides react to liberate elemental sulfur or soluble sulfate. For example, $FeS_2$ reacts quantitatively with $Fe_2(SO_4)$ to produce $FeSO_4$ and elemental sulfur (*37*) whereas NaCl converts pyrite to the corresponding sulfate. Elemental sulfur is not formed in the latter case. Other oxidizing media include nitric acid, nitrates, and oxygen in the presence of water or other solvents such as aqueous ammonia solutions.

Leaching of coal may remove ash as well as sulfur. Leaching with ferric salts such as $Fe_2(SO_4)_3$ has resulted in the removal of approximately 10% of the ash from coal (*34*). The ash content of an Illinois No. 6 coal, which had been leached in caustic was reduced from 9.8 to 0.7 wt% by treating the leached coal with HCl at ambient temperature (*29*).

Deashing may also be achieved with other reagents. Mineral (ash) matter has been leached from coal using aqueous solutions of $HNO_3$, $Cl_2$, and HF (*35*). In another study, the ash content of a low-sulfur coking coal was reduced from 0.76 to 0.28% by heating the coal in an aqueous caustic soda paste at 120 atm and 250°C (482°F) for 20 minutes (*38*).

Nitrogen, like organic sulfur, is believed to be linked to coal substance. Therefore, the removal of nitrogen from coal must be achieved by chemical rather than physical cleaning. Coal nitrogen is subject to extraction by weak organic and strong inorganic acids (*39*). In one case, up to 90% of the nitrogen was extracted by heating coal in 85% $H_3PO_4$ at 100° to 150°C (212° to 302°F) (*40*).

## 2. Solvent-Refining Processes

Solvent refining of coal differs from the leaching approach in that the carbonaceous value of the coal is extracted and converted to a solid fuel. The potential air pollutants either remain in the residue from the extraction, or are subsequently removed from the carbonaceous content of the coal after the extraction operation. Solvent refining was initiated in the United States with the limited objective of producing a low-cost alternative to desulfurized residual oil and natural gas for use in boilers (*40a*).

The fuel product is prepared by dissolving coal in a coal-derived solvent in the presence of hydrogen at 68 atm or greater and 440°C (825°F). Only limited hydrogenation is required since the process is not primarily designed to produce lighter oil products. The hydrogenation step is followed by ash separation and conversion of sulfur to a removable form. The coal-solvent solution is then filtered and evaporated to yield a fuel product with ash and sulfur both substantially removed. The fuel product can be utilized as either a liquid or solid. The solvent-refined coal process can use any rank coal below anthracite, from lignite to low-volatile bituminous. Using the process, an ashless, low-sulfur fuel having a heating value of about 37,200 kJ/kg (16,000 Btu/lb) can be produced. The development is in the large pilot-plant stage.

## 3. Molten-Salt Processes

Treatment in a molten-salt bath has the potential for the removal of sulfur, nitrogen, and mineral matter from coal. These potential pollutants are converted to water-soluble salts which are removed from the coal by a series of washing operations.

For the extraction of pyritic sulfur from high-sulfur coals, a molten bath of NaOH/KOH can be employed (*34*). One method employs molten caustic (Na/KOH in a ratio of 1:1) to dissolve pyritic sulfur from conventionally cleaned Pittsburgh seam coal (*33*). Complete removal of pyritic sulfur is achieved by treating 1 part of less than 0.4 mm (—40 mesh) coal with four parts of caustic at between 150° and 250°C (302° and 482°F). Below 150°C (302°F) no removal of sulfur is achieved. Between 250° and 400°C (482° and 752°F) extraction is rapid, requiring approximately 5 minutes. Organic sulfur is not attacked in this system.

## 4. Miscellaneous Processes

Other leaching treatments said to decompose pyrites, that might be useful in treating coals are (a) sodium bicarbonate and heavy metal carbonate in a closed system at 185°C (365°F); (b) steam at 300° to

400°C (572° to 752°F); (c) sulfur monochloride vapors at 140°C (284°F); (d) moist $CO_2$ at 250°C (482°F); and (e) carbon tetrachloride at 250°C (482°F) (*41*). Leaching Illinois No. 5 and Lower Kittaning coals with *p*-cresol removed 10–50% of the organic sulfur (*34*).

The use of somewhat higher temperatures in combination with various atmospheres has been found to be effective in reducing the sulfur content of coal; however, the final product is "char" rather than coal (*42*). For example, when 0.4 to 0.85 mm (20- to 40-mesh) size coal was treated for 4 hours at 1000°C (1832°F) in atmospheres of $N_2$, $CO_2$, $CH_4$, or $C_2H_4$, approximately 50–60% of the sulfur was removed. Similarly, treatment with water–gas removed 76%, anhydrous ammonia removed 82%, hydrogen removed 87%, and steam removed 84% of the original sulfur present.

Biological processes may also be used in aqueous coal-cleaning systems. Bacteria are divided into two classes: (a) autotrophic—organisms that live on inorganic matter; and (b) heterotrophic—organisms that live on organic matter. Autotrophic bacteria use atmospheric carbon dioxide as their sole source of the carbon necessary for cellular growth. These organisms have adapted themselves to live and grow in a strongly acidic environment (pH 1.5–3) and in the presence of heavy metal ions which are extremely toxic to most other forms of life. Some bacteria also convert sulfides to sulfates and sulfates to sulfides; others oxidize hydrogen sulfide to sulfur, and nitrogen to nitrates.

The reduction of sulfur in coal by the employment of bacterial action to oxidize and remove pyrite represents a specialized form of chemical benefication (*43*). By immersing coal in acidified water (pH 3.5) containing *Ferrobacillus ferrooxidans* bacteria and maintaining the water at room temperature for 4 days, considerable pyrite reduction can be achieved. Approximately 50% was removed from a Kentucky No. 11 coal while 47.1 to 76.2% reduction was achieved (in all but a few tests) with Pittsburgh coal which was treated in a more acidic, bacteria-laden water of pH 2.6.

### D. Liquefaction

Liquid fuels may be prepared from coal by three principal generic schemes, synthesis, pyrolysis, and hydrogenation. All of these remove sulfur, nitrogen, and ash to a greater or lesser extent. Removal in this context refers to the component being absent (or in reduced concentration) in liquefied products, though probably present in a more concentrated form in secondary products.

Synthesis refers to production of hydrocarbons (Fischer–Tropsch synthesis) or methanol by catalytic reaction of carbon monoxide and hydrogen (synthesis gas). Synthesis gas may be formed by the gasification of coal with steam. The synthesis products must be essentially free of sulfur and nitrogen, as well as other contaminants; for the most part these will have been removed by a gas-purification step following gasification. The major drawback to production of fuels by synthesis is high cost.

Liquid fuels can be made in low yield (25% of the moisture and ash-free coal) by pyrolysis. An example is the FMC COED (Char-Oil Energy-Development) process (*44*). If the liquid (tar) is filtered to remove entrained solids, and then hydrocracked, the product will contain about 0.1% sulfur, 0.2–0.4% nitrogen, and 0.01% ash. A portion of the sulfur and nitrogen are removed as elemental sulfur and ammonia from the treatment of the carbonizer and hydrocracker gases. The remaining portion of these contaminants and most of the ash will appear in the char by-product. Filtration is a key step in the process—very low filtration rates are common.

The most likely way by which liquefaction will be applied to produce clean fuel from coal is hydrogenation. Such processes differ with respect to the severity of treatment and, hence, the quantity of hydrogen which is reacted. The more severe the treatment, the higher the proportion of coal which is converted to distillate products. Distillate products are free of ash and contain 0.1–0.5% sulfur and about 0.2–0.5% nitrogen. If desired these may be reduced to 0.1% or less by conventional petroleum refinery practice. Residual products contain 0.5–1.5% sulfur and 0.1–1% ash, even after having been purified by filtration. Removal of ash and, hence, mineral sulfur (FeS not degraded at liquefaction conditions) from the residual products is a problem in liquefaction processes. The unconverted coal, removed from the liquified products by filtration, will contain high levels of ash and sulfur. The disposition of the unconverted coal from liquefaction processes is uncertain, but in most cases probably will be by gasification.

### E. Coal Gasification

Coal gasification was widely practiced in industry before the discovery of substantial natural gas resources and additional gasification processes are under development. These produce either a low-heating-value fuel gas with a heat content of about 5600 kJ/m$^3$ (about 150 Btu/scf), an intermediate-heating-value fuel gas with a heat content of 11,000 to 24,000 kJ/m$^3$ (300 to 650 Btu/scf), or a high-heating-value fuel gas, equivalent to natural gas, with a heat content of about 37,000 kJ/m$^3$ (about 1,000 Btu/scf), often referred to as synthetic or substitute natural gas (SNG). The processing alternatives consist of (a) adding oxygen (either pure or as air) to the gasifier, (b) adding hydrogen to the gasifier, or (c) using an externally fired circulating heat carrier. Processes that produce low- or intermediate-heating-value fuel gas will, in most cases, be located in close proximity to a power plant or industry that will use the gas as fuel. For this purpose, coal will be transported to the gasifier site. Processes that produce high-heating-value gas (SNG) will most likely be located at the coal mine and the product gas will be transported by pipeline to its users.

In the production of low- and intermediate-heating-value fuel gas, the process consists of devolatization of the coal:

$$\text{Coal} \rightarrow \text{C(char)} + CH_4 + \text{a mixture of liquids and gases} \qquad (1)$$

and reacting steam with coal at elevated temperatures:

$$C + H_2O \rightleftharpoons CO + H_2 \qquad (2)$$

Reaction (2) is endothermic, the heat for it being supplied by burning a portion of the coal in air or oxygen:

$$C + \tfrac{1}{2}O_2 \rightleftharpoons CO \qquad (3)$$

$$C + O_2 \rightleftharpoons CO_2 \qquad (4)$$

In addition, the water–gas shift reaction occurs:

$$CO + H_2O \rightleftharpoons CO_2 + H_2 \qquad (5)$$

Thus, the product fuel gases are a mixture of $H_2$, $CO$, $CO_2$, $CH_4$, inerts (such as $N_2$), and minor amounts of other hydrocarbons and impurities. If an air–steam mixture is used directly to gasify the coal, the product has a heat content of approximately 5600 kJ/m$^3$ (about 150 Btu/scf) and is called a low-heating-value gas. With proper cleanup, this gas can be used as an energy source either for power generation or industrial application.

If oxygen is used in place of air for Reactions (3) and (4), the gas produced will have a heating value of approximately 11,000 kJ/m$^3$ (about 300 Btu/scf). Depending upon the amount of methane contained in the gas, the heating value can be as high as 17,000 to 24,000 kJ/m$^3$ (about 450 to 650 Btu/scf). After cleanup, this intermediate heating value gas is suitable for the production of chemicals (such as ammonia), synthetic liquid fuels (such as methanol), gaseous fuels (such as methane or SNG), or for direct use as an energy source. Table II gives typical compositions of gas from gasifiers for producing low- and intermediate-heating-value fuel gases.

The production of high-heating-value gas (mostly methane) requires that the intermediate heating value gas be methanated to increase its heating value and reduce the CO content to an acceptable limit. The basic reaction of the methanation step is

$$C + 2\,H_2 \rightleftharpoons CH_4 \qquad (6)$$

**Table II Typical Compositions of Fuel Gas Produced by Commercial Gasifiers**

| Gasifier | Gasifying medium | Type of fuel | $CO_2$ | $CO$ | $H_2$ | $CH_4$ | $N_2$ | Higher heating value | |
|---|---|---|---|---|---|---|---|---|---|
| | | | | | | | | $kJ/m^3$ | $Btu/scf$[a] |
| Lurgi | Oxygen–steam | Brown coal | 32 | 13 | 36 | 15 | 1 | 12,000 | 322 |
| | Air–steam | Brown coal | 14 | 16 | 25 | 5 | 40 | 6,820 | 183 |
| Koppers–Totzek | Oxygen–steam | Bituminous coal | 7 | 56 | 35 | 0 | 2 | 10,810 | 290 |
| Winkler | Oxygen–steam | Lignite | 20 | 35 | 40 | 3 | 2 | 10,170 | 273 |
| | Air–steam | Lignite | 10 | 22 | 12 | 1 | 55 | 4,360 | 117 |
| Wellman–Galusha | Oxygen–steam | Bituminous coal | 12 | 52 | 33 | 1 | 2 | 10,510 | 282 |
| | Air–steam | Coke | 3 | 29 | 15 | 3 | 50 | 6,410 | 172 |

[a] scf: Standard cubic feet.

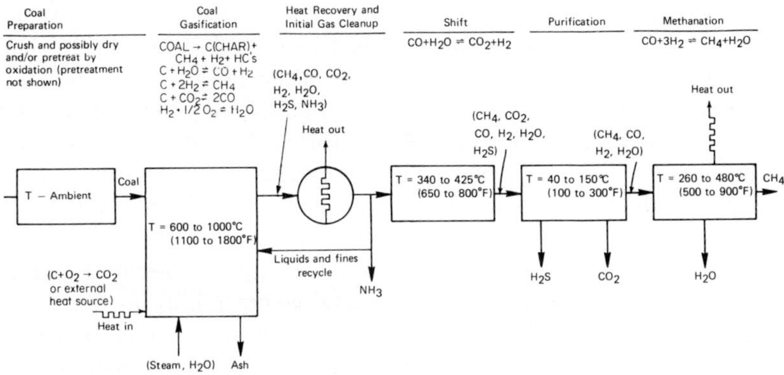


**Figure 4. Generalized flow sheet for coal gasification processes. *T*: Temperature.**

which is exothermic. The methanated gas is dried and, if necessary, compressed to pipeline pressure. Methanation processes for large-scale application are still under development.

Whether high- or low-heating-value gas is produced depends upon the conditions of temperature, pressure, and residence time, under which the gasifier is operated and whether the methanation step [Eq. (6)] is incorporated in the process. Gasification of coal can be conducted at atmospheric or elevated pressure in a fixed-, fluidized-, entrained-, or molten-bed reactor.

A generalized flow sheet for gasification* to produce either low- or high-heating-value gas is shown schematically in Figure 4. After the coal is pretreated to prevent caking, it passes to the gasifier where oxygen is added [for Reactions (3) and (4)] or an external heat source is used to bring about the gasification reactions.

The mixture leaving the gasifier requires further treatment or cleanup to convert it into usable low-heating-value gas or SNG. Steps in the gas cleanup are essentially the same regardless of the gasification method, although design details, heat recovery, sulfur removal, etc., differ. The gas mixture passes through cyclone separators to remove fine particles and dust. It is then scrubbed to remove condensible materials and sent through a shift converter (if needed) to obtain the proper ratio of hydrogen to carbon monoxide for methane formation. The stream is then scrubbed with activated potassium carbonate, methanol, isopropanol, or other solvent (*45*), to remove carbon dioxide and

* This flow sheet is not representative of the two-stage gasification process, e.g. HIGH GAS process.

hydrogen sulfide and, perhaps, passed over activated carbon to remove organic compounds. Iron oxide or equivalent absorbents are being developed to remove remaining traces of sulfur (*46*). At this point the heat content of the gas is about 3700 to 19,000 kJ/m³ (100–500 Btu/scf), and the option is available to use this material directly as low- or intermediate-heating-value gas, or to further process it into a high-heating-value gas with a heat content of about 37,000 kJ/m³ (about 1000 Btu/scf).

Coal gasification processes can be divided into two groups, commercially available and advanced processes. Of commercially available processes, the Wellman–Galusha gasifier is the only one in operation in the United States. The Koppers–Totzek, Winkler, and Lurgi processes (*47*) are in active use outside the United States, and Lurgi and Koppers–Totzek gasifiers will soon be used for both high- and low-energy gas applications in the United States. Advanced processes which have operated are all less than 90 metric tons/day (100 tons/day).

## Pollution Control in Gasification

Coal gasification is, of itself, a pollution control method and represents an alternative to flue gas cleaning, solvent refining, chemical cleaning, coal washing, fuel switching, etc., to meet emission regulations.

Figure 5 shows a simplified block diagram of the major coal gas treatment steps which may be required to clean the fuel gas and to protect

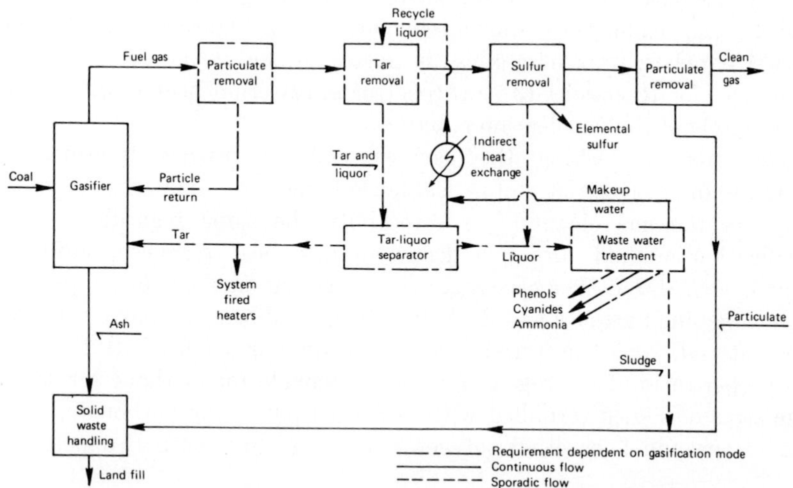


Figure 5. Simplified diagram of a fuel gas cleanup system.

the environment. The existence and/or severity of the pollution problem areas are determined by gasifier design and operating parameters, the coal selected for gasification, coal-gas treating steps selected, and the end use of the gas.

The bulk of the ash in the coal is usually removed in front-end processing equipment. For example, in the Lurgi process it is removed from the bottom of the gasifier, while in the $CO_2$ Acceptor process it is removed along with unburned char and spent acceptor (reacted dolomite) from the acceptor-stripper unit downstream from the gasifier. However further cleanup of the fuel gas to reduce particulate loading is necessary if clean fuel gas is required.

In the usual form of gasifier, sulfur is volatilized under reducing conditions to form hydrogen sulfide along with small quantities of COS and $CS_2$, mercaptans, and thiophenes. Table III (*48*) gives typical concentrations of nonhydrocarbon impurities in fuel gas produced by coal gasification. These sulfur compounds are then scrubbed from the fuel gas, stripped from the absorption liquor as a concentrated hydrogen sulfide stream, and converted to sulfur using Claus plant technology (see below). Fuel gas scrubbing and stripping to remove and concentrate these sulfur compounds (mainly, hydrogen sulfide) is accomplished by any one of many processes: (a) amine scrubbing systems; (b) hot potassium carbonate process; (c) Sulfinol process (Shell Oil Company); (d) Seaboard and vacuum carbonate processes (Koppers Company); (e) phosphate process (Shell Development Company); (f) Wet Iron Box process; (g) Thylox process; (h) hot-ferric oxide process; and (i) dolomite acceptor process (*45*).

**Table III Typical Concentrations of Nonhydrocarbon Impurities in Fuel Gas Produced by Coal Gasification** (*48*)

| *Constituent* | *Concentration, vol %* |
|---|---|
| $H_2S$ | 0.3–3.0 |
| $CS_2$ | 0.016 |
| COS | 0.009 |
| Thiophene | 0.010 |
| Mercaptans | 0.003 |
| $NH_3$ | 1.1 |
| HCN | 0.10–0.25 |
| Pyridine bases | 0.004 |
| NO | 0.0001 |

The concentrated hydrogen sulfide stream is usually then sent to a Claus plant. In the Claus process, hydrogen sulfide is burned under precisely controlled conditions so that one-third of it is converted to sulfur dioxide. The cooled combustion products are catalytically reacted to form elemental sulfur:

$$2\ H_2S + SO_2 \underset{}{\overset{\text{Cat}}{\rightleftharpoons}} 3\ S + 2\ H_2O \tag{7}$$

after which the sulfur is condensed and separated. However, recovery is not complete, and as much as 5% of the entering sulfur is contained in the exit gas. Tail gas treating technology, e.g., the Beavon, Cleanair, and IFP processes (*45*), can achieve recovery of almost all of the remaining sulfur. Consequently sulfur removal of 99.5–99.9% can be achieved on the hydrogen sulfide in the gases where the investment and operating cost can be justified.

The nitrogen in the coal can be converted to ammonia, cyanides, and phenols. As these compounds serve as sources of fuel-bound nitrogen for production of nitrogen oxides ($NO_x$) in combustion systems, they are removed from the fuel gas in the tar removal and sulfur removal steps. They can be recovered from the wastewater and may be useful by-products.

In most processes, aqueous waste waters from gasifiers will be treated using conventional technology, i.e., biological oxidation, gravity separation, and a wide variety of physical–chemical unit operations (ion exchange, reverse osmosis, adsorption, filtration, chlorination, etc.).

## V. Combustion Technology for Clean Fuels from Coal-Conversion Processes

The advent of clean fuels from coal processing will bring some new combustion problems not encountered in present practice, and new firing techniques may be needed to utilize these fuels. Characterizing new fuels and investigating expected problems on a small scale as the fuels become available in laboratory or pilot quantities could provide feedback to influence the choice of processing variables. This could improve the overall success of these fuel developments by linking processing and utilization early in their development.

### A. Firing Low-Heating-Value Gas

Of the gasification processes for producing sulfur-free gas from coal, on-site generation of low-heating-value gas having a heating value in the range of about 5200–6000 kJ/m$^3$ (140 to 160 Btu/scf) appears most economical and suitable for firing base-load steam power plants. Some promising processes using oxygen enrichment are expected to produce an intermediate-heating-value gas of approximately 11,000 kJ/m$^3$ (about 300 Btu/scf) (*49, 50*) which may be used to fire industrial processes.

Gasification plants will operate over a range of production rates as needed to match fuel demand, and will include the process steps of gasification, particulate removal, and removal of sulfur as hydrogen sulfide. Most of the older sulfur-removal processes require cooling of the gas to about 150°C (about 300°F) or less, with consequent loss of sensible heat. Processes now under development are intended to remove sulfur at gasification temperatures. For close-coupled gasifier-user systems, this would permit boiler firing with gas at above 540°C (about 1000°F) to utilize the sensible heat in the fuel gas as manufactured.

Present gasifier art is based largely on fixed bed Wellman–Galusha (*51*) or Lurgi (*52*) producers, which are relatively small and must be used in multiples to serve high-demand uses. For central-station power plant applications there is much interest in large pulverized-coal-fired gasifiers of the entrainment type that would operate more like present boiler equipment.

Special considerations involved in firing low-heating-value gas compared to natural gas include: (a) gas volume, (b) burner area, (c) furnace volume, (d) heat exchanger pressure drop, and (e) control of $NO_x$ emissions. The following comments are specifically directed toward power plant operations but are generally true for a wide range of combustion applications.

## 1. Gas Volume

Low-heating-value gas has about six times the volume of natural gas per unit of heat content at standard conditions. If fired at a temperature of about 540°C (about 1000°F), the volume of low-heating-value gas would be 17 times as great as the volume of cold natural gas per unit of heat content. Thus, gas piping for low-heating-value gas must be much larger than that for natural gas, and gas-flow areas in gas burners must be similarly larger.

## 2. Burner Area

The flame speed of low-heating-value gas is somewhat lower than that of natural gas, and it is usual to design for lower burner velocities. Thus, substitution of low-heating-value gas in an existing boiler might require boiler derating, use of more burners, or use of larger burners.

### 3. Furnace Volume

The furnace volume presently used for firing low-heating-value gas is comparable with that needed for pulverized coal. Thus, direct substitution of low-heating-value gas for coal appears feasible. However, many boilers designed for burning gas or oil have small furnaces, and would not be suitable for burning low-heating-value gas without derating. Based upon experience with other types of combustion systems used in industry, it appears possible to substitute additional burner pressure drop for furnace volume, although this would require a major change in burner design.

### 4. Heat Exchanger Pressure Drop

The mass flow of combustion products from cooled low-heating-value gas is about 20% greater than that for other fuels for the same boiler output (assuming the fuel gas is cooled for cleaning). Thus, when cool low-heating-value gas is considered as a fuel for substitution in existing boilers, the pressure drop through boiler tube banks may be a limiting factor. For operation at the full boiler rating, boiler pressure drop would increase about 40%. Alternatively, to maintain the same pressure drop, derating of about 20% would be necessary.

To avoid derating the entire plant because of boiler pressure drop, it would be feasible to increase the boiler gas-flow area by removing a few rows of boiler tubes. This would restore capacity, but at lower boiler efficiency. With further modification to add boiler surface, efficiency could be restored.

When low-heating-value gas is fired hot, boiler output could be maintained with no increase in flue-gas volume or in boiler pressure drop.

### 5. Control of $NO_x$ Emissions

The *thermal* $NO_x$ emission when firing low-heating-value gas should be very low. In effect, the high content of molecular nitrogen is equivalent to very effective flue-gas recirculation. The slow mixing and combustion of low-heating-value gas in the furnace reduces maximum flame temperature. However, few experimental data on emissions of $NO_x$ from combustion of low-heating-value gas are available at this time.

Ammonia and other nitrogen compounds are produced in gasification of coal-containing organic nitrogen. If these are not removed from the low-heating-value gas, they would be expected to be partially converted to $NO_x$ during combustion of the gas. Accordingly, it appears necessary to explore means of burning low-heating-value gas containing nitrogen compounds without emitting excessive $NO_x$. This may be preferable to the alternative of removing nitrogen compounds from the gas. Although low temperature, wet processes for sulfur removal would also remove ammonia, it is not clear that high temperature sulfur-removal techniques would remove it.

### 6. Overall Assessment of Low-Heating-Value Gas Firing

The use of low-heating-value gas in existing boilers may require extensive modification to provide suitable gas ducting, burners, and superheat control. In addition, derating of up to 20% may be necessary to accomodate boiler pressure drop and furnace volume requirements. However, sulfur dioxide should be low, but conversion of ammonia and organic nitrogen compounds to $NO_x$ during combustion is a potential problem if these compounds are not removed from the gas.

## *B. Firing Liquid Fuels from Coal or Shale*

Liquid fuels derived from coal are of interest as a substitute for petroleum-based fuels, and as low-sulfur fuels to replace coal. Liquid fuels from shale are also under development, and problems in firing them are expected to be similar to problems in firing liquids from coal for corresponding service.

Several coal liquefaction processes are under development for making synthetic crude oil for refinery feed and for liquid fuels of all kinds (*49*, *53*). Two types of low-sulfur liquid fuels appear of greatest interest for electric power generation: (a) a heavy, low-ash fuel that can be substituted for No. 6 fuel oil or coal in boilers, and (b) a lighter, cleaner fuel suitable for firing gas turbines and combined-cycle plants.

In these processes, coal is dissolved, hydrogenated, modified by catalytic treatment, and treated to remove sulfur and, if necessary, fuel nitrogen. One of these processes, catalytic coal liquefaction (*54*), can provide a range of products suitable for use as refinery feedstock, boiler fuels, or gas-turbine fuels. These products have lower hydrogen content than petroleum fuels of similar viscosity, sulfur content below 0.2%, and high nitrogen content.

In the various coal liquefaction processes, ash is removed by filtration from the liquid, and sulfur is reduced to 0.2% or less. However, most of the organic nitrogen in the original coal appears in the liquefied product, and nitrogen is more difficult and expensive to remove than is sulfur.

Thus, it appears necessary to define the need for nitrogen removal, considering the alternative of burning the fuel under conditions to minimize fuel nitrogen conversion to $NO_x$.

### 1. Boiler Fuels

Liquid boiler fuels are expected to be somewhat similar to No. 6 fuel oil. Boiler slagging and corrosion should be minimized because nearly all of the ash is removed in making the fuel; this should also make the fuel suitable for boilers without electrostatic precipitators, but this will depend on how thoroughly the ash is removed in coal refining.

### 2. Gas-Turbine Fuels

Gas-turbine fuels must be much cleaner than boiler fuels in terms of ash constituents that can cause erosion of turbine blades. Production of both a light gas-turbine fuel suitable for aircraft-type gas turbines and a heavier gas-turbine fuel suitable for industrial-type gas turbines appears desirable.

### 3. Methanol

Methanol can be made from coal by combining carbon monoxide and hydrogen from coal gasification in a catalytic process (*54*). It should cost about the same as methane produced from coal, can be stored in tanks, and can be burned readily in any power cycle. It can also be used in fuel cells. The heating value of methanol is only about 18,600 $MJ/m^3$ (about 66,775 Btu/gal), compared to 39,000 to 44,600 $MJ/m^3$ (140,000 to 160,000 Btu/gal) for petroleum fuels. Thus, use of methanol would require about twice as much storage-tank and transportation capacity as petroleum-based fuels. Methanol fuels need not be pure methanol, but can include a mixture of higher alcohols to minimize cost and improve heating value.

No particular combustion or emission problems are anticipated in burning methanol in either boilers or gas turbines. However, some rearrangement of burners and boiler surfaces may be required for superheat control, as for other fuel substitutions.

## *C. Firing Chemically Refined Coal*

Solvent-refined coal can be treated as solid fuel and, with proper techniques, fired in a crushed or pulverized form, or it can be treated as a liquid fuel and fired in atomized form after preheating to about 400°C (about 750°F) under pressure, which is feasible with care. For pulverization, its high grindability index (Hardgrove 164) indicates that it is easily pulverized, but the pulverizer and conveying lines must operate at low temperatures to prevent the fuel from melting and agglomerating.

The fuel appears to be similar to a high-volatile coal in ignition and ignition stability. However, to obtain substantially complete carbon burnout, the burners and furnaces should be similar to those designed for low-volatile coals, because the fuel is less reactive than the usual high-volatile bituminous coal.

## VI. Combustion Modification to Reduce Emissions

### A. Nitrogen Oxides

Nitrogen oxides form during combustion by two processes: (a) the *thermal* fixation of nitrogen from combustion air at high flame temperature; and (b) the *conversion* of organic nitrogen compounds in the fuel to $NO_x$, which can take place at much lower temperature. Combustion of gas and No. 2 fuel oil, which contain little or no organic nitrogen, primarily produces $NO_x$ by the thermal mechanism. Combustion of residual fuel oils and coal, which contain large concentrations of organic nitrogen (0.4–1.0% for residual oil, and 1–2% for coal) produce $NO_x$ by both mechanisms. For the high-nitrogen fuels, as the fuel nitrogen content increases the *conversion* mechanism becomes increasingly predominant as the source of $NO_x$ emissions.

Nitrogen found in fuels in the form of organic nitrogen compounds is a significant source of $NO_x$ produced during combustion. Between 20 and 90% of the nitrogen in liquid fuel can be converted to $NO_x$ in combustion (*5e, 55, 56a, 56b, 57*). When the nitrogen content is less than 0.1%, the percent *conversion* can approach 90%, but this percentage declines as fuel nitrogen content increases, reaching about 40–50% with fuel nitrogen content of about 0.5%. As the fuel nitrogen content increases above 1%, the efficiency of *conversion* of fuel nitrogen to $NO_x$ decreases further.

About 20–25% of the organic nitrogen in coal is converted to $NO_x$ under normal combustion conditions (*58, 59*). With coal nitrogen content ranging from 0.7 to 2.8%, this level of conversion could account for all of the $NO_x$ output permitted by United States Federal Standards. It appears necessary to develop coal combustion techniques specifically oriented to minimizing conversion of fuel nitrogen to $NO_x$ in order to minimize total $NO_x$ emissions.

The formation of thermal $NO_x$ can be reduced considerably by combustion modifications which reduce flame temperatures (see Fig. 1); however, reduction of $NO_x$ formed by conversion is more difficult to achieve and is generally accomplished by maintaining reducing conditions during the early part of the combustion process. Hence, decreasing flame temperature is an effective means of decreasing $NO_x$ emissions for firing natural gas and distillate oils, but it not nearly as effective when residual oil or coal are fired.

Techniques that have been developed or proposed for decreasing $NO_x$ emissions include those listed in the tabulation below.

| *Techniques for reducing flame temperatures to decrease production of thermal $NO_x$* | *Techniques for reducing conversion of fuel nitrogen to $NO_x$* |
|---|---|
| Flue-gas recirculation | Biased or two-stage combustion |
| Water addition | Low excess air combustion |
| Biased or two-stage combustion in combination with heat removal | |

Techniques for reducing flame temperature and, thus, *thermal* $NO_x$ involve diluting the combustible fuel/air mixture with an inert gas. The inert gas absorbs some of the energy released during the combustion process and, thus, reduces peak flame temperatures. The two most common inert diluents are recirculated flue gas and water vapor. For flue-gas recirculation, if the cooled flue gas is returned to the burner and mixed with the combustion air prior to combustion, it can be as effective as excess air for achieving low smoke levels *(60, 61)*. Commercial application of flue-gas recirculation has been limited, except for its use in controlling $NO_x$ emissions from automobile engines. Since flue-gas recirculation does not increase the mass of flue products, its use does not decrease equipment efficiency.

The use of water injection to reduce $NO_x$ formation has been applied to gas turbines with success *(61a)*.

Several attempts to commercially use water vapor as an inert diluent have involved firing water-in-oil emulsions containing up to 30% water *(62, 62a)*. Use of water vapor as a diluent increases flue losses and, thus, potentially decreases the efficiency of the combustion system. Claims have been made that the firing of water-in-oil emulsion improves combustion sufficiently to permit operation of the combustion system at lower excess air levels and, thus, avoids the loss in efficiency.

Two-stage combustion consists of burning the fuel with only a portion of the stoichiometric air (about 80%) in the first stage, and then adding the remaining air and completing combustion in a second stage. Two-stage combustion reduces *thermal* $NO_x$ by lowering peak temperatures, and reduces *conversion* $NO_x$ by maintaining reducing conditions during the early burning. A variation of this is biased firing of multiburner furnaces where some burners are run fuel-rich and others admit only air to the furnace. This achieves a form of two-stage combustion without requiring separate combustion zones. Two-stage burning may thus be accomplished in various ways including designing the burner to control air/fuel mixing, biased firing, and the use of two distinct combustion zones.

For some combustion systems, *conversion* $NO_x$ can be reduced by firing the system at a lower excess air level. Firing at low excess air decreases the concentration of oxygen and, thus, decreases the tendency for $NO_x$ to form. However, firing at low-excess air without smoke requires better mixing of air and fuel and, thus, may increase flame temperatures and *thermal* $NO_x$. Therefore, decreasing excess air to decrease $NO_x$ emissions has only limited application, usually to fuels with high nitrogen content.

### 1. Gas and Oil Firing

Considerable research has been carried out in reducing $NO_x$ emissions from gas- and oil-fired utility boilers by biased firing and flue-gas recirculation. This was begun in California, and resulted in sufficient reduction of $NO_x$ levels for boilers fired with natural gas and oil to meet emission standards of 160 ppm for gas and 250 ppm for oil firing.

### 2. Pulverized Coal Firing

The response of pulverized coal flames to combustion modifications differs significantly from the response of gas and oil flames. For example, flue-gas recirculation reduces $NO_x$ from oil flames by 50% or so, but has almost no effect on $NO_x$ from coal flames (*63, 64*). This implies that recirculation reduces *thermal* $NO_x$ but has little effect on *conversion* $NO_x$, leading to the conclusion that most of the $NO_x$ formed in burning coal comes from fuel nitrogen. $NO_x$ levels increase continuously for coal flames in utility boilers as excess air is increased from 0 to 60% (*65*). The response of $NO_x$ from coal flames to increasing excess air infers that conversion of fuel nitrogen to $NO_x$ is limited by the availability of oxygen in the flame and that the conversion percentage continues to increase with increasing excess air, even though *thermal* $NO_x$ may be declining.

In existing coal-fired units, lowest $NO_x$ emission levels are obtained with tangential firing, higher $NO_x$ levels are obtained with front-wall firing or horizontally opposed firing, and the highest $NO_x$ levels are obtained in hot slag-tap and cyclone furnaces. Furnace shape, furnace volume, heat release per unit boiler surface in the burner zone, and the air preheat temperature all affect $NO_x$ emission levels significantly. For new units, furnaces should be designed which provide more effective cooling of the flame in the immediate vicinity of the burner.

Although $NO_x$ emissions from pulverized-coal-fired boilers can be reduced significantly by biased firing, staged combustion, and water injection; each technique may involve undesirable results of unknown magnitude. Biased firing and staged combustion depend upon maintaining large zones of reducing atmosphere. These may alter furnace slagging characteristics and fly ash characteristics. In addition, if reducing zones occur along walls, experience indicates that severe wall-tube corrosion (tube wastage) may result *(66)*. Thus, the application of these two techniques must be approached with care, and consideration must be given to protection of wall tubes. It may be possible to adjust burners and flow patterns to provide a barrier of air or inert combustion products along the walls, thus avoiding slagging and corrosion. Alternatively, corrosion resistance may be increased by use of more corrosion-resistant tube alloys in local areas subject to attack, or by surface treatments or coatings.

### 3. Cyclone Furnace Firing

At the time of writing this chapter, there are about 100 cyclone-fired boilers in service, fired by about 500 cyclone units. These are particularly appropriate where high-ash highly abrasive coal of low grindability and low-ash fusion temperature must be burned. Most units are in Illinois, southern Ohio, and West Virginia. The $NO_x$ emission level for cyclone furnaces using bituminous coals is high, in the range of 800 to 1000 ppm for normal operation.

Although $NO_x$ emission regulations would seem to restrict further application of cyclone furnaces for firing bituminous coal, cyclone furnaces have some advantages for burning lignite. Lignite is hard to pulverize and has ash-fusion characteristics suitable for cyclone furnaces.

It appears possible to greatly reduce $NO_x$ emission from cyclone furnaces by two-stage combustion. The cyclone would be fired with about 90–95% of theoretical air, and the combustion products cooled considerably in the secondary furnace before adding secondary air. Short-term experiments of this kind indicate that operation is satisfactory in that good slag tapping can be maintained and carbon loss is low. $NO_x$ level is reduced to 500 ppm or less (*67*). However, there is concern about excessive tube corrosion with such operation because of the formation of iron sulfide which can attack the tubes which cool the cyclone.

The United States Bureau of Mines has operated a small laboratory-size cyclone furnace at outlet temperatures of up to about 2425°C (about 4400°F) for magnetohydrodynamic (MHD) studies, by using air preheat temperature and oxygen enrichment (*68*). At this temperature, the $NO_x$ level for firing with 110% theoretical air is extremely high, 4400 ppm. However, by firing the laboratory cyclone furnace with only 95% of theoretical air, cooling the combustion products to about 1100°C (2000°F) and completing the combustion at 106% theoretical air, $NO_x$ emission of only 150 ppm was obtained. With cyclone firing at 92–94% theoretical air, and with combustion products cooled to about 1350°C (about 2450°F), the $NO_x$ level was 550–575 ppm.

### 4. Stoker-Fired Furnaces

There is little data on $NO_x$ emission from stoker-fired furnaces. Normally, stoker-fired furnaces are limited to sizes below the energy input level of 73,000 MJ/second (about 250 million Btu/hour) and, therefore, are not required to meet United States Environmental Protection Agency $NO_x$ emission standards for power plants.

The basic problems in reducing $NO_x$ from stoker-fired units is that complete combustion takes place within the stoker fuel bed with relatively high local temperatures. Thus, there appears to be no opportunity to utilize concepts such as biased firing or staged combustion. Flue-gas recirculation and air humidification may be feasible and should reduce $NO_x$ emission significantly.

## *B. Sulfur Oxides*

Although many emission regulations have been aimed at limiting sulfur dioxide emissions, there is relatively little that can be done to the combustion process to affect sulfur oxide emissions. Generally, it is a matter of most of the sulfur in the fuel entering the furnace being emitted as $SO_2$. From 5 to 10% of the sulfur in the coal may be retained in the ash, but the remaining 90 to 95%, or more, of the sulfur is emitted as $SO_2$. Eventually this $SO_2$ is oxidized in the atmosphere to $SO_3$ which will react to form sulfates or be washed out of the air by rain. The fluidized-bed combustor is the only unit offering a good chance for limiting sulfur oxide emissions by design or operation of a combustion process.

## Fluidized-Bed Combustion for $SO_2$ Control

Fluidized-bed combustion is a relatively new concept for combined combustion and sulfur control, insofar as simultaneous combustion and desulfurization takes place in a limestone or dolomite fluidized bed operating at atmospheric or elevated pressures. A schematic diagram of a simple coal-burning fluidized-bed combustion boiler concept is shown in Figure 6. It is formed by an enclosure consisting of waterwalls—abutting boiler tubes in which water and steam flow. The pressure in this enclosure may be in the 1- to 25-atm range. A distributor plate with holes or bubble cap devices distributes the air flow uniformly over the base on the enclosure. The air then passes at superficial velocities in the range of 0.6 to 4.6 m/second (2 to 15 ft/second) through a bed of particles at temperatures from about 760° to 1040°C (about 1400° to 1900°F). These parti-

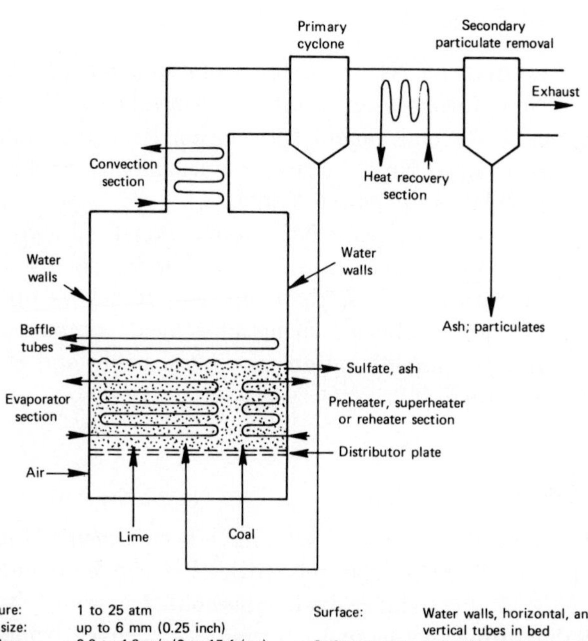


Figure 6. Fluidized-bed combustion boiler. Pressure: 1–25 atm; coal size: up to 6 mm (0.25 in.); airflow: 0.6 to 4.6 m/second (2 to 15 ft/second); temperature; 760 to 1040°C (1400 to 1900°F); surface: water walls, horizontal, and vertical tubes in bed; sulfur removal: $CaO + SO_2 + \frac{1}{2}O_2 \rightarrow CaSO_4$.

cles, whose top size is 1.6 to 6.4 mm ($\frac{1}{16}$ to $\frac{1}{4}$ in.), consist of ash or other inert material, lime sorbent, and small quantities (less than 1%) of unburned coal or carbon. Coal and sorbent are fed to the bed by feeders extending through the air distributor or the waterwalls. From 60 to 70% of the heat released in burning the fuel with air is transferred to the water/steam in the tubes surrounding and submerged in the bed.

The air and combustion gases passing through the bed cause considerable agitation: particles are thrown from the bed into the empty volume above—the freeboard. Larger, heavier particles fall back into the bed; smaller, lighter particles are carried out of the enclosure by the gases. Much of the ash in the coal is eventually carried from the bed in the combustion gases. Larger ash particles may accumulate in the bed and require periodic removal.

Splashing of the particles into the freeboard can be minimized by locating a screen of horizontal water/steam tubes as a baffle over the surface of the agitated bed. The baffle tubes increase the heat surface exposed to the bed and reduce the height in the freeboard required to minimize the loss of large particles from the enclosure.

Convection heat transfer surface can be located in the path of the combustion products to generate more steam from the sensible heat of these gases. A reasonably high heat transfer coefficient from the combustion gases to the tube surface requires high gas velocities. These velocities can be achieved by narrowing or restricting the gas passage in the convection section or by packing tubes into a broad gas passage with small clearances between tubes.

Particles carried from the fluidized bed enclosure by the combustion gases can be captured by inertial separation apparatus such as a cyclone. These particles comprise ash, fragments of the limestone/dolomite sorbent, and typically 20 to 40% by weight unburned coal char or carbon. To obtain high combustion efficiencies, the particles captured in this primary cyclone can be recycled to the boiler bed or to a separate fluidized bed combustor where the burning of the char is completed. Additional heat can be extracted from the combustion gases in the heat recovery section, which may be either an economizer, a boiler feedwater heater, or an air preheater.

Final cleanup of the combustion gases is carried out in a secondary particulate removal system, using cyclones or an electrostatic precipitator.

Fluidized-bed combustion has a high volumetric heat release rate of 5 MJ/second $m^3$ (500,000 Btu/$ft^3$hour), as compared to 0.2 MJ/second $m^3$ (20,000 Btu/$ft^3$hour) in a pulverized-coal-fired boiler. Also the rapid movement of the solid particles passing over tubes immersed in the bed results in a high rate of heat transfer of about 280 J/second m$^2$K (about 50 Btu/ft$^2$hour °F) compared to about 50 to 85 J/second m$^2$K (about 10–15 Btu/ft$^2$hour °F) in a conventional boiler, although the temperature difference will be about one-half. Thus, smaller boilers with less tube surface should be possible for fluidized-bed combustion systems.

The fluidized-bed temperatures of about 760°–1040°C (1400°–1900°F) are selected to achieve maximum $SO_2$ capture by limestone (over 90% removal). At these low temperatures, $NO_x$ emission is reduced (250–600 ppm) and clinkering problems are minimized. Experimental evidence indicates that the reaction of $NO_x$ with CO to form $N_2$ is promoted by CaO. Therefore, when combustion is carried out in two stages, one in reducing conditions (oxygen deficient) and one in oxidizing condition (oxygen sufficient), $NO_x$ emission has been reduced to 70 ppm. Pressurized operation also favors $NO_x$ reduction (50–200 ppm at 5 atm) (*69*). By virtue of the low combustion temperature employed, emission of alkalies has been found to be lower in fluidized-bed combustion than in conventional plant operation.

### *C. Particulate Matter*

#### 1. Coal Firing

a. Pulverized Coal. The ash particles produced by pulverized coal firing are the spherical fused remnants of the high-temperature combustion of fine (5–100 μm) coal particles. During combustion, which occurs in much less than 1 second, the particle of an eastern United States bituminous coal softens due to heating, puffs up as volatile gases bubble out of the sticky mass, and burns rapidly as a lacy, fragile, puffed-up mass called a cenosphere (*70, 71*).

The coal ash is distributed nonuniformly throughout the cenosphere. As the high temperature combustion proceeds rapidly, the ash becomes liquid and the lacy ash remnant agglomerates into a solid sphere or breaks up into a series of true spheres of varying smaller sizes. If the temperature of the particle decreases before agglomeration or liquid breakup can occur, a hollow spherical particle results. Practically all of the carbon in the coal is completely oxidized to carbon dioxide in this intense combustion reaction.

Although most of the ash particles become liquid during combustion, they cool and are solidified before they reach the walls of dry-bottom furnaces. In slagging or wet-bottom furnaces some of these particles are liquid or sticky when they reach the walls where they combine to form a viscous slag layer which slowly drains down to the slag-tap opening in the furnace bottom. Because of their inflexibility in coping with the varying ash properties of many different coals, slag-tap furnaces are disappearing.

A small amount of the volatile ash constituents such as mercury, sodium, and potassium compounds are driven off as vapors by the high temperature of particle combustion. Most of these vapors, except mercury, condense rapidly into ultrafine droplets and immediately solidify as submicron spheres that are extremely difficult to capture. Although these condensed solids are a small proportion of the total ash, they may constitute an important fraction of the ash which is so fine that it is difficult to collect.

The size of the fly ash discharged from a conventional pulverized coal-fired furnace is affected primarily by the size of coal fired. The finer the grind, the finer the fly ash. While fine grinding helps combustion, it makes the resulting ash harder to capture. Also, high temperature flames promote combustion but increase ash vaporization and subsequent condensation into finer particles. Figure 7 shows typical fly ash distributions for pulverized-coal-fired boilers (*72*).

Western United States coals tend to produce finer ash because the coal particles do not puff up into a tarry mass but burst into finer particles during the heating period.

b. Cyclone Furnaces. In a cyclone-fired boiler furnace, the coal particles are burned in a horizontal, water-cooled chamber before flue gases

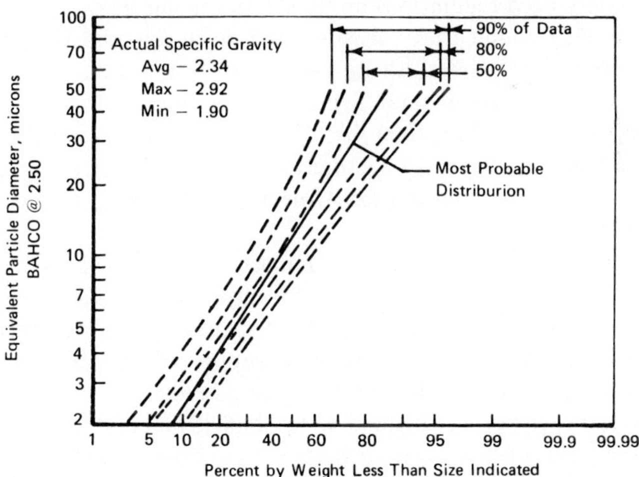


Figure 7. Particle-size distribution for pulverized-coal-fired boilers (*72*). Actual specific gravity: average, 2.34; maximum, 2.92; minimum, 1.90. BACHO method is described in References (*72a*) and (*74b*).

are discharged into the larger furnace cavity. The coal and combustion air are introduced into the cyclone chamber with a high swirl that throws the coal particles against the water-cooled, ash laden walls of the cylinder. The coal particles are retained in the ash where the combustible portion is burned. About 85% of the coal ash is retained in a cyclone furnace as slag. Some ash will not melt and form slag in a cyclone furnace. This depends on the chemical composition of the ash. Cyclone furnaces are therefore not as flexible as conventional dry-bottom furnaces in coping with the ash from different coals.

Another drawback of the cyclone furnace is that its high temperature vaporizes more of the ash than does a pulverized coal-fired furnace. The resulting condensed submicron ash particles are very difficult to capture. In addition, the high temperature in the cyclone produces higher nitric oxide emissions. Accordingly, the cyclone boiler, originally a very attractive combustion modification, has turned out to have problems which can be troublesome.

c. Spreader Stokers. The spreader stoker employs either a mechanical spreader or jets of steam or air to throw the solid fuel into the furnace where it burns partially in suspense, and then falls onto a grate that is traveling or stationary to complete its burning. Overfire air jets have been found essential to smoke-free operation (*73*). Overfire air jets reduce dust emission significantly but not enough to meet most ordinances (*74*). Without a dust collector, dust emission at all burning rates is well above most commonly used regulations on fly ash in the flue gas.

Kuhlman (*75*) found that spreader stokers in firetube boilers emitted little more flue dust, about 1.8 kg per metric ton (about 4 lb/1000 lb) of steam generated, than hand firing or traveling grates, about 1.3 kg per metric ton (about 3 lb/1000 lb). The former figure is roughly equivalent to dust emissions of 15 kg/metric ton (30 lb/ton) of coal fired, 2.5 kg/MJ (1.2 lb per million Btu), or 1.4 kg/1000 kg (1.4 pounds per 1000 lb) of flue gas.

Smoke emission is characteristic of most spreader stokers if they are operated at less than approximately 25–50% of full load, because low furnace temperature causes incomplete suspension burning. Accordingly, a spreader stoker-fired system should be supported by suitable auxiliary heat-generating means so that at low plant loads the stoker can either be operated to carry all of the load or be shut down completely. Modulating controls have been shown to help reduce dust emission (*76*).

The unburned gases, carbon monoxide and hydrocarbons, emitted by spreaders are in extremely low concentration when operation is smokeless, similar to values characteristic of pulverized-coal firing.

## 2. Residual Oil Firing

A successful combustion modification to reduce sulfur trioxide formation and thus minimize boiler corrosion from high-sulfur residual oil is by firing with very low excess air (*77*). However, a disadvantage of this technique is that the emission of finely divided unburned carbon particulate matter is increased (*78*).

## 3. Particle Collectability

The efficiency of collection in electrostatic precipitators is enhanced somewhat by *decreasing* the resistivity of the ash particles. The electrical conductivity (and its reciprocal resistivity) is influenced by small amounts of acid gas present such as sulfur trioxide. Accordingly, numerous techniques have been tried to "condition" the particles by adding sulfur trioxide, sulfuric acid, and ammonia to the gases entering precipitators (*79*). The results have been mixed. Usually some improvement has been noted but, in general, a precipitator that is seriously deficient owing to design, construction, or maintenance faults will not be aided appreciably by flue-gas conditioning.

**ACKNOWLEDGMENTS**

The authors would like to express their thanks to others who contributed to the material presented in this section including Herbert R. Hazard (combustion modification), Dr. Harvey S. Rosenberg and Dr. Eugene J. Mezey (coal cleanup), Herman Nack (fluidized-bed combustion), all of the Battelle-Columbus staff, and Dr. J. M. Genco (gasification) formerly of the Battelle-Columbus staff.

**REFERENCES**

1. W. R. Niessen and A. F. Sarofim, "Incinerator Air Pollution: Facts and Speculation," Proc. Nat. Incinerator Conf., Cincinnati, Ohio, pp. 167–181. Amer. Soc. Mech. Eng., New York, New York, 1970.
2. Conservation Foundation, "Implications of Rising Carbon Dioxide Content of the Atmosphere," Conference Report. Conserv. Found., New York, New York, 1964.
3. H. W. Seuss, *Bull. At. Sci.* **17,** 374 (1961).
4. J. R. Garratt and G. I. Pearman, "Atmospheric Carbon Dioxide," Proc. Int. Clean Air Conf., Melbourne, Australia, pp. 17–22, 1972.

4a. "Control Techniques for Nitrogen Oxides from Stationary Sources," Publ. AP-

67. U.S. Department of Health, Education and Welfare, Washington, D.C., 1970.

5. Y. B. Zeldovich, *Acta Physicochimica U.R.S.S.* **21**, 577–628.

5a. O. M. Dukema, "Analysis of Test Data for $NO_x$ Control in Gas- and Oil-Fired Utility Boilers," EPA-650/2-75-012. U.S. Environmental Protection Agency, Research Triangle Park, North Carolina, 1975.

5b. A. R. Crawford, E. H. Manny, and W. Bartok, "Field Testing: Application of Combustion Modifications to Control $NO_x$ Emissions from Utility Boilers," EPA-650/2-74-066. U.S. Environmental Protection Agency, Research Triangle Park, North Carolina, 1974.

5c. D. W. Turner and C. W. Siegmund, "Staged Combustion and Flue Gas Recycle: Potential for Minimizing $NO_x$ from Fuel Oil Combustion," Paper presented at 1st American Flame Days, Chicago, Ill. The American Flame Research Committee, 1972 (unpublished).

5d. G. A. Cato, H. J. Buening, C. C. DeVivo, B. G. Morton, and J. M. Robinson, "Field Testing: Application of Combustion Modifications to Control Pollutant Emissions from Industrial Boilers—Phase I," EPA-650/2-74-078a. U.S. Environmental Protection Agency, Research Triangle Park, North Carolina, 1974.

5e. H. R. Hazard, "Conversion of Fuel Nitrogen to $NO_x$ in a Compact Combustor." ASME Paper No. 73-WA/GT-2. Annu. Meet. Amer. Soc. Mech. Eng., New York, New York (1973).

6. H. L. Green and W. R. Lane, "Particulate Clouds, Dusts, Smokes, and Mists," p. 96. E. & F. N. Spon, London, England, 1957.

7. W. S. Smith and C. W. Gruber, "Atmospheric Emissions from Coal Combustion," Publ. 999-AP-24, p. 45. U.S. Public Health Service, Department of Health, Education and Welfare, Washington, D.C., 1966.

8. H. Breuer, J. Gebhart, K. Robock, and U. Yeichert, *Staub-Reinhalt. Luft* **33**(4), 187–190 (1973).

9. R. W. Gerstle, S. T. Cuffe, A. A. Orning, and C. H. Schwartz, *J. Air Pollut. Contr. Ass.* **15** (2), 59–64 (1965).

10. N. E. Bolton, W. S. Lyon, R. I. Van Hook, A. W. Andren, W. Fulkerson, J. A. Carter, and J. F. Emery, "Trace Element Measurements at the Coal-Fired Allen Steam Plant," Progress Report, June, 1971–January, 1973, Report No. ORNL-NSF-E.P.-43. Oak Ridge Nat. Lab., Oak Ridge, Tennessee, 1973.

11. J. W. Kaakinen, R. M. Jorden, and R. E. West, "Trace Element Study in a Pulverized Coal-Fired Plant," Paper No. 74-8, 67th Annu. Meet. Denver, Colorado. Air Pollut. Contr. Ass., Pittsburgh, Pennsylvania, 1974.

12. B. T. Commins, *Atmos. Environ.* **3**, 565–572 (1969).

13. J. S. Conners, C. L. Perkins, and F. E. Vandaveer, "Gas Engineering Handbook," Chap. 9, pp. 4-82 and 4-92. Industrial Press, New York, New York, 1965.

14. F. L. Robson, A. J. Giramonti, G. P. Lewis, and J. Gruber, "Technological and Economic Feasibility of Advanced Power Cycles and Methods of Producing Nonpolluting Fuels for Utility Power Stations," PB 198392. Nat. Tech. Inform. Serv., U.S. Dept. of Commerce, Springfield, Virginia, 1970.

15. E. A. Hall, P. Choi, and E. Kropp, "Assessment of the Potential of Clean Fuels and Energy Technology," EPA-600/2-74-001, p. 85. U.S. Environmental Protection Agency, Research Triangle Park, North Carolina, 1974.

16. A. W. Duerbrouck and E. R. Palowitch, "Survey of Sulfur Reduction in Appalachian Region Coals by Stage Crushing," Inform. Circ. 8301. U.S. Bureau of Mines, Washington, D.C., 1966.

17. E. Hurst, J. O. Lively Manufacturing and Construction Company, Glen White, West Virginia, 1974 (personal communication).
18. K. J. Miller and A. F. Baker, "Electrophoretic-Specific Gravity Separation of Pyrite from Coal, Laboratory Studies," Report RI-7440. U.S. Bureau of Mines, Washington, D.C., 1970.
19. W. T. Abel, M. Zulkoski, G. A. Brady, and J. W. Eckerd, "Removal of Pyrite from Coal by Dry Separation Methods," Report RI-7732. U.S. Bureau of Mines, Washington, D.C., 1973.
20. R. B. Reif, "Electrostatic Separation of Pyrite from Coal," Second Summary Report to Bituminous Coal Research, Inc., Monroeville, Pennsylvania. Battelle's Columbus Laboratories, Columbus, Ohio, 1962.
21. Anonymous, *Chem. Week* **103** (7), 64 (1968).
22. A. G. Yurovsky and I. D. Remesnikov, *Koks Khim.* **13,** 8 (1958).
23. W. M. Kester, Jr., *Mining Eng.* **17** (55), 72 (1965).
24. S. C. Sun and W. L. McMorris, *Mining Eng.* (*New York*) **11** (11), 1151 (1959).
25. R. J. Quideriveg and N. V. Campagne, *Chem. Eng.* (*London*) **220,** 223 (1968).
26. F. W. Meadus, G. Paillard, A. F. Sirianni, and I. E. Puddington, *Can. Mining Met. Bull.* **61,** 736 (1968).
27. D. H. Napper and A. Netschey, *J. Colloid Interface Sci.* **37** (3), 528 (1971).
28. A. W. Deurbrouck, *Mining Congr. J.* **60,** (2), 65 (1974).
29. L. Reggel, R. Raymond, I. Wender, and B. D. Blaustein, "Preparation of Ash-Free, Pyritic-Free Coal by Mild Chemical Treatment," Prepr. Pap., 164th Nat. Meet, pp. 44–48. Amer. Chem. Soc., Washington, D.C., 1972.
30. H. Dreyfus, U.S. Patent 2,221,866 (1940).
31. H. H. Lowry, ed., "Chemistry of Coal Utilization" Vol. 1, p. 1134. Wiley, New York, New York, 1945.
32. Anonymous, *Environ. Sci. Technol.* **4** (9), 718 (1970).
33. J. W. Leonard and C. F. Cockrell, *Mining Congr. J.* **56** (12), 65–70 (1970).
34. J. W. Hamersma, E. P. Koutsoukos, M. K. Kraft, R. A. Meyers, and G. J. Ogle, "Chemical Desulfurization of Coal: Report of Bench-Scale Developments, Vol. 1," EPA-R2-73-173a. U.S. Environmental Protection Agency, Research Triangle Park, North Carolina, 1973.
34a. E. P. Stambaugh, J. F. Miller, S. S. Tam, S. P. Chauhan, H. F. Feldman, H. E. Carlton, J. F. Foster, H. Nack, and J. H. Oxley, "Environmentally Acceptable Solid Fuels by the Battelle Hydrothermal Coal Process," Paper No. 116. Amer. Power Conf., Chicago, Illinois, 1976. (To be published in conference proceedings.)
35. P. X. Masciantino, *Fuel* **44,** 269–275 (1965).
36. K. Ono, M. Kameda, A. Yazawa, and K. Koike, *Bull. Res. Inst. Miner. Dressing Metallurgy, Sendai* **18** (2), 147–158 (1962).
37. I. N. Kuzminykh and Ye L. Yakhontoya, *Zh. Prikl. Khim.* (*Leningrad*) **23,** 1121–1126 (1950).
38. A. Crawford, "Leaching of Coal by Combined Jig Washings, Froth Flotation and Extraction with Caustic Soda," BIOS Final Report No. 522. Central Air Documents Office, Dayton, Ohio, 1946 (now Defense Documentation Center, Alexandria, Virginia).
39. R. A. Meyers, German Patent No. 2,108,786 (to TRW, Inc., Redondo Beach, California) (1971).
40. R. A. Myers, J. S. Land, and C. A. Flegal, "Chemical Removal of Nitrogen and Organic Sulfur from Coal," TRW, Inc., Redondo Beach, California,

APTD-0845. U.S. Environmental Protection Agency, Research Triangle Park, North Carolina, 1971.

40a. D. L. Kloepper, T. F. Rogers, C. H. Wright and W. C. Bull, "Solvent Processing of Coal to Produce a De-Ashed Product," Res. Develop. Rep. No. 9. U.S. Department of the Interior, Office of Coal Research, Washington, D.C., 1965.

41. P. H. Given and W. F. Wyss, *Brit. Coal Util. Res. Ass., Mon. Bull.* **25,** 165–179 (1961).

42. P. D. Snow, *Ind. Eng. Chem.* **24** (8), 903–909 (1932).

43. J. W. Leonard, C. T. Holland, and E. U. Syed, "Unusual Methods of Sulfur Removal from Coal: A Survey," Report No. 30, p. 14. Coal Research Bureau, West Virginia University, Morgantown, West Virginia, 1967.

44. A. H. Strom and R. T. Eddinger, *Chem. Eng. Progr.* **67** (3), 75–80 (1971).

45. Anonymous, *Hydrocarbon Process.* **52,** 87–116 (1973).

46. W. T. Abel, F. G. Schulz, and P. F. Langon, "Removal of Hydrogen Sulfide from Hot Producer Gas by Solid Absorbents," R.I. 7947. U.S. Bureau of Mines, Washington, D.C., 1974.

47. L. K. Mudge, G. F. Schiefelbein, C. T. Li, and R. H. Moore, "The Gasification of Coal," Report to Battelle Energy Program. Battelle's Columbus Laboratories, Columbus, Ohio, 1974.

48. A. V. Slack, "Sulfur Dioxide Removal from Waste Gases," Pollution Control Review, No. 4. Noyes Data Corp., Park Ridge, New Jersey, 1971.

49. Office of Coal Research, "Clean Energy from Coal—A National Priority," Annual Report for Calendar Year 1972. U.S. Department of Interior, Washington, D.C., 1972.

50. "Evaluation of Coal Gasification Technology: Part II-Low Btu Fuel Gas." National Academy of Science, National Academy of Engineering, Washington, D.C., 1973.

51. T. E. Ban and G. M. Hamilton, "Low-Btu Gas from Coal," Paper presented at the Energy Resource Conference. University of Kentucky, Lexington, Kentucky, 1973.

52. J. Agosta, H. F. Illian, R. M. Lundberg, and O. G. Tranby, *Chem. Eng. Progr.* **69** (3), 65–66 (1973).

53. H. C. Hottel and J. B. Howard, "New Energy Technology—Some Facts and Assessments." MIT Press, Cambridge, Massachusetts, 1971.

54. F. H. Kant, "Feasibility Study of Alternative Automotive Fuels," ESSO Research and Engineering Status Report Prepared for Alternative Automotive Power Systems Coordination Meeting. U.S. Environmental Protection Agency, Ann Arbor, Michigan, 1974 (unpublished).

55. D. E. Turner, R. L. Andrews, and C. W. Siegmund, *Combustion* **44** (2), 21–30 (1972).

56a. R. E. Barrett, S. E. Miller, and D. W. Locklin, "Field Investigation of Emissions from Combustion Equipment for Space Heating," EPA-R2-73-084a. U.S. Environmental Protection Agency, Research Triangle Park, North Carolina, 1973.

56b. G. B. Martin, D. W. Pershing, and E. E. Berkau, "Kinetics of the Conversion of Various Fuel Nitrogen Compounds to Nitrogen Oxides in Oil-Fired Furnaces," AIChE Paper No. 37f, presented at the 70th Nat. Meet., Atlantic City, New Jersey. Amer. Inst. Chem. Eng., New York, New York, 1971.

57. C. V. Sternling and J. O. L. Wendt, "Kinetic Mechanisms Covering the Fate

of Chemically Bound Sulfur and Nitrogen in Combustion," Final Report by the Shell Development Co. U.S. Environmental Protection Agency, Research Triangle Park, North Carolina, 1972.

58. W. Bartok, A. R. Crawford, and G. J. Piegari, "Systematic Field Study of $NO_x$ Emission and Combustion Control Methods for Power Plant Boilers," Air Pollution and its Control, Ser. 68 (126), pp. 66–74. Amer. Inst. Chem. Eng., New York, New York, 1972.
59. A. H. Rawdon and S. A. Johnson, *Proc. Amer. Power Conf.* **35,** 828–837 (1973).
60. P. W. Cooper, R. Kamo, C. J. Marek, and C. W. Solbrig, "Recirculation and Fuel–Air Mixing as Related to Oil-Burner Design," Publ. 1723. Amer. Petrol. Inst., Washington, D.C., 1964.
61. R. L. Andrews, C. W. Siegmund, and D. G. Levine, "Effects of Flue Gas Recirculation on Emissions from Heating Oil Combustion," Paper No. 68-21, 61st Annu. Meet., St. Paul, Minnesota. Air Pollut. Contr. Ass., Pittsburgh Pennsylvania, 1968.

61a. H. Shaw, "The Effect of Water on Nitric Oxide Production in Gas Turbine Combustors," Paper 75-GT-70, Gas Turbine Conf., Amer. Soc. Mech. Eng., New York, New York, 1975.

62. R. E. Hall, "The Effect of Water/Distillate Oil Emulsions on Pollutants and Efficiency of Residential and Commercial Heating Systems," Paper 75-09.4, 68th Annu. Meet., Boston, Massachusetts. Air Pollut. Contr. Ass., Pittsburgh, Pennsylvania, 1975.

62a. R. E. Hall, "The Effect of Water/Residual Oil Emulsions on Air Pollutant Emissions and Efficiency of Commercial Boilers," Paper 75-WA/APC-1, Annu. Meet. Amer. Soc. Mech. Eng., New York, New York, 1975.

63. W. J. Armento and W. L. Sage, "The Effect of Design and Operation Variables on $NO_x$ Formation in Coal-Fired Furnaces," Proceedings of the Coal Combustion Seminar, pp. 193–205. U.S. Environmental Protection Agency, Research Triangle Park, North Carolina, 1973.
64. D. W. Pershing, J. W. Brown, and E. E. Berkau, "Relationship of Burner Design to the Control of $NO_x$ Emissions Through Combustion Modification," Proceedings of the Coal Combustion Seminar, pp. 87–139. U.S. Environmental Protection Agency, Research Triangle Park, North Carolina, 1973.
65. A. R. Crawford, E. H. Manny, and W. Bartok, "$NO_x$ Emission Control for Coal-Fired Utility Boilers," Proceedings of the Coal Combustion Seminar, pp. 214–285. U.S. Environmental Protection Agency, Research Triangle Park, North Carolina, 1973.
66. W. T. Reid, "External Corrosion and Deposits in Boilers and Gas Turbines," Fuel and Energy Science Series, pp. 133–134. Amer. Elsevier, New York, New York, 1971.
67. Personal communications with representatives of Babcock & Wilcox Company, Barberton, Ohio.
68. D. Bienstock, "Air Pollution Aspects of MHD Power Generation," Paper VI 1.1, presented at the 13th Symposium on Engineering Aspects of MHD. Stanford University, Stanford, California, 1973.
69. S. J. Wright, "The Reduction of Emissions of Sulfur Oxides by Additions of Limestone or Dolomite During the Combustion of Coal in Fluidized Beds," Proc. Third International Conference on Fluidized Bed Combustion held at Hueston Woods, Ohio, pp. 135–154. U.S. Environmental Protection Agency, Research Triangle Park, North Carolina, 1972.

70. M. A. Field, D. W. Gill, B. B. Morgan, and P. G. W. Hawksley, "Combustion of Pulverized Coal," p. 211. Brit. Coal Util. Res. Ass., Leatherhead, England, 1967.
71. A. R. Ramsdan, *J. Inst. Fuel* **41,** 451–454 (1968).
72. Anonymous, "Criteria for the Application of Dust Collectors to Coal-Fired Boilers." Industrial Gas Cleaning Institute—American Boiler Manufacturers Association, Joint Technical Committee, Industrial Gas Cleaning Institute, Stamford, Connecticut, 1967.
72a. "Determining the Properties of Fine Particulate Matter," Power Test Code, PTC 28-1965, pp. 23–25 Amer. Soc. Mech. Eng., New York, New York, 1965.
72b. A. Wolf, *Staub-Reinhalt, Luft* **27** (4), 30–33 (1967) (in English).
73. C. Morrow, W. C. Holton, and H. L. Wagner, *Trans ASME* **75,** 1363–1372 (1953).
74. W. C. Holton and R. B. Engdahl, *Trans. ASME* **74,** 207–214 (1952).
75. A. Kuhlman, *Staub* **25** (6), 121–131 (1964).
76. E. J. Boer and C. W. Porterfield, "Dust Emissions from Small Spreader-Stoker-Fired Boiler," ASME Paper 53-S-26. Amer. Soc. Mech. Eng., New York, New York, 1953 (unpublished).
77. F. Glaubitz, *Combustion* **7,** 31–35 (1963).
78. G. C. Jefferis and J. D. Sensenbaugh, *Mech. Eng.* **82,** 111 (1960) (abstr.).
79. J. Dalman and D. Tidy, *Atmos. Environ.* **6,** 721–734 (1972).

# 11

## Space Heating and Steam Generation

R. E. Barrett, R. B. Engdahl, and D. W. Locklin

## I. Introduction

This chapter is concerned with three combustion sources of air pollution, namely, residential space heating, commercial space heating, and industrial steam generation for processing and space heating.

Residential space heating may conveniently be defined as the application of combustion equipment for the purpose of heating single-family residences, individual apartments, condominium units, multifamily units of up to about four families, and mobile homes. The size of heating unit required for these applications ranges up to about 90,000 J/second (about 300,000 Btu/hour) and includes warm-air systems, hot-water systems, and steam systems. The load on combustion equipment in this category is very seasonal.

Commercial space heating includes such applications as larger multifamily apartments, commercial business operations, schools, governmental operations, and some small industrial operations. The size of combustion unit required for most of these applications ranges from 90,000 J/second to 3,000,000 J/second (about 300,000 Btu/hour to 10,000,000 Btu/hour). Warm-air, hot-water, and steam heating systems are included. Obviously, there are commercial operations that are much larger (e.g., large universities served by a single steam plant), but the vast majority of commercial operations will fit within the above-defined size range. It is a convenient range to use to define combustion equipment, because equipment within this range generally has similar design and operational features. The prime load in this category is space heating and, therefore, is seasonal in nature.

Industrial steam generation includes boilers used to generate steam for industrial processing and manufacturing and for space heating and cooling at industrial facilities. Large industrial boilers may also be used to generate on-site electrical power. A predominant feature of these boilers is the fact that the load is related to the manufacturing process and is relatively constant through the year; space heating generally provides a small portion of the load. The size of boilers included in this category generally ranges from 3 MJ/second to 75 MJ/second (about 10 million Btu/hour to 250 million Btu/hour).

Combustion equipment and applications exist that do not fit within the above definitions, but most of the equipment and firing practices for such applications will be similar to that for equipment included in the above categories. For instance, residential and commercial water heaters utilize equipment similar to that used for space heating but with lower firing capacities.

## II. Fuel-Use Patterns

Before the 1930's, the predominant fuel for residential and commercial space heating and industrial steam generation was coal, mainly because of its availability. However, in the United States the development of large pipelines, accelerated by the needs of World War II, made large quantities of natural gas and fuel oil available to many segments of the country. In recognition of the convenience of firing gaseous and liquid fuels, most residential and commercial space heating and a significant portion of industrial steam generation gradually shifted to firing natural gas and fuel oil. Electric heating also emerged during the late 1950's and 1960's as an option to fuel firing for residential and commercial space heating. The relatively high energy cost has limited the acceptance of electric heating for residential use in areas with high heating loads. Electric heating is a somewhat more viable option for those commercial applications where first cost is relatively more important than operating cost.

Figure 1 and Table I show the recent history of the relative use of various energy sources for residential heating, commercial heating by hot-water or steam systems, and industrial steam generation in the United States.

The shift in fuel from coal to natural gas and fuel oil for residential and commercial space heating has had a dramatic effect on air quality in urban areas. Photographic evidence and the memories of many bring to mind the smoke emissions of the coal-fired units and the smoke and haze conditions that prevailed during the winter months in many cities

**Table I Fuel Selection For Firing Boilers (United States)** (3)

| *Boiler size* | *1930* | *1950* | *1970* |
|---|---|---|---|
| Commercial heating | 70% coal | 20% coal | 5% coal |
| boilers | 20% oil | 40% oil | 30% oil |
| 0.09 to 3.0 MJ/sec | 10% gas | 30% gas | 40% gas |
| (0.3 to 10 × $10^6$ Btu/hr) | | 10% oil/gas | 25% oil/gas |
| Small industrial boilers | 80% coal | 15% coal | 5% coal |
| 3.0 to 30 MJ/sec | 15% oil | 40% oil | 30% oil |
| (10 to 100 × $10^6$ Btu/hr) | 5% gas | 25% gas | 3% gas |
| | | 20% oil/gas | 30% oil/gas |
| Large industrial boilers | 90% coal | 45% coal | 15% coal |
| 30 to 150 MJ/sec | 5% oil | 20% oil | 20% oil |
| (100 to 500 × $10^6$ Btu/hr) | 5% gas | 20% gas | 25% gas |
| | | 10% oil/gas | 25% oil/gas |
| | | 5% coal/oil/gas | 5% coal/oil/gas |

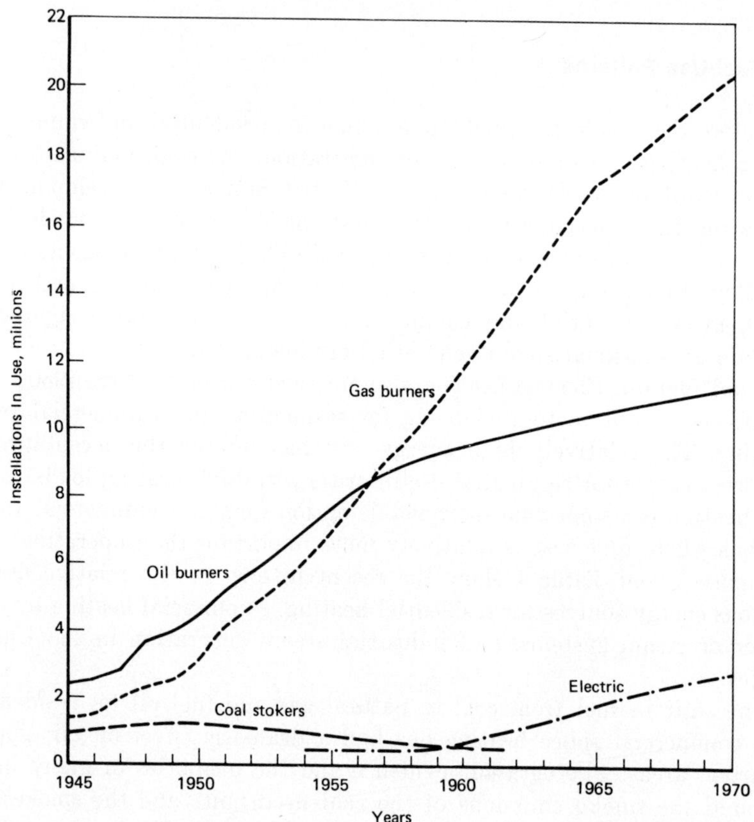


Figure 1. Automatic central-heating installations in the United States (*1*, *2*).

through the 1940's. Although all of the credit for the reduction in urban ambient particulate levels cannot be credited to fuel switching for space heating, such fuel switching has been a major factor in the large reduction in particulate emissions discharged at low elevations. Less obvious is the reduction in ambient sulfur dioxide in cities due to the conversion from coal-fired space heating to natural gas and distillate oil.

## III. Types of Combustion Equipment

### *A. Residential Heating*

Gas- and oil-fired residential heating systems distribute heat to the residence either directly as warm air or indirectly by supplying steam or hot water to heat transfer surfaces. In the United States gas-fired resi-

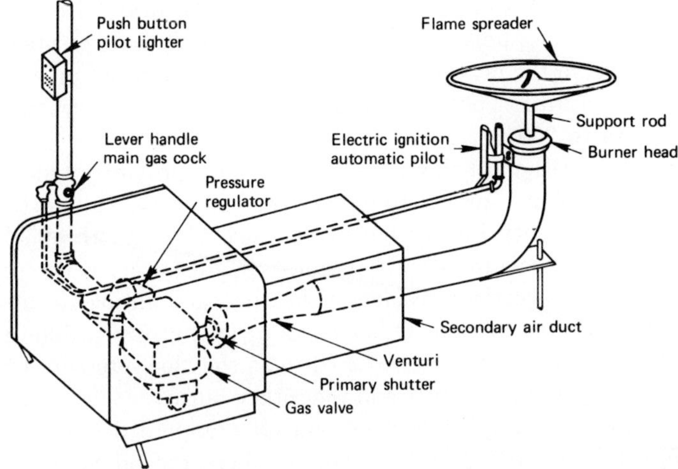


Figure 2. Typical single-port gas conversion burner (*4*).

dential heating systems are predominantly of the warm-air type because natural gas first became widely available in the Midwest and Far West where warm air heating systems predominated. Conversely, in the United States, oil-fired residential heating systems are nearly evenly divided between warm air systems and steam or water systems, with the largest concentration of steam or water systems occurring along the East Coast. Both gas- and oil-fired warm-air furnaces may be of the forced-air type (using a fan to move air over the heat exchanger) or of the gravity type (generally for older and two-story houses).

Gas-fired furnaces and boilers generally are equipped with atmospheric pressure or natural-draft burners (Figs. 2 and 3). The fuel gas passing through a venturi induces a flow of primary air which mixes with the fuel gas. This mixture is then fed through a burner into a combustion

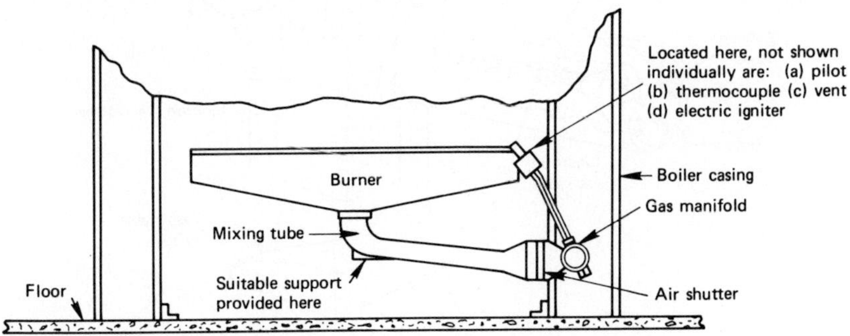


Figure 3. Atmospheric gas burner (*5*).

space where the fuel is burned. The secondary air is drawn from the room in which the heating unit is located by the natural draft created by the hot exhaust gas in the heating unit. The air–fuel ratio is not precisely controlled in these units. They therefore tend to operate at very high excess air levels—up to 200%.

Oil-fired furnaces and boilers are nearly always provided with a burner that includes a blower or fan to deliver the combustion air to the flame with sufficient velocity and turbulence to mix with and throughly burn the oil. The high-pressure-atomizing gun-type oil burner is the predominant type of oil burner for residential heating, accounting for over 95% of the United States market. Other types of oil burners include the air-atomizing gun burner (usually referred to as the "low-pressure burner"), the vaporizing pot-type burner, and the rotary burner.

The gun burner (Fig. 4) includes an oil pump to deliver the oil at about 700 kN/m$^3$ (100 psi) to a nozzle where the oil is sprayed into the combustion chamber as fine droplets. The oil droplets mix with the combustion air supplied by the fan and burn. Appropriate ignition and control systems are required. Because the air is supplied by a fan, it is possible to adjust oil-heating units to operate at lower excess air levels than natural-draft gas burners. However, to avoid smoke and heat exchanger fouling, most are still fired at moderately high excess air levels—about 60 to 120%.

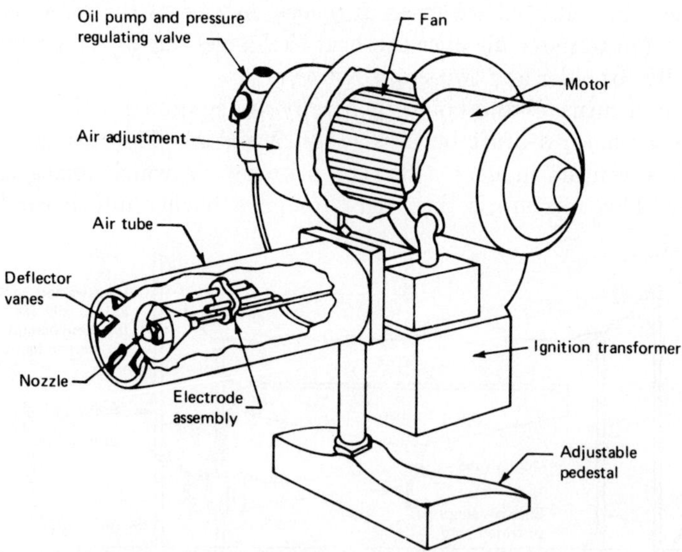


Figure 4. High-pressure gun-type oil burner (*4*).

### B. Commercial Boilers

Commercial heating boilers are primarily of three types; the firebox–firetube boiler, the package Scotch boiler, and the cast-iron boiler. The firebox–firetube boiler unit (Fig. 5) consists of an internal furnace and firetubes, i.e., tubes surrounded by water through which combustion products pass. The first pass of firetubes extends from the furnace to the rear header and the second pass extends between the front and rear headers. Firebox–firetube boilers are suitable for firing gas, oil, or, if grates are added, coal. The Scotch boiler (Fig. 6) consists of a shell-type boiler

Figure 5. Firebox type of boiler. (Courtesy of Kewanee Boiler Corp.)

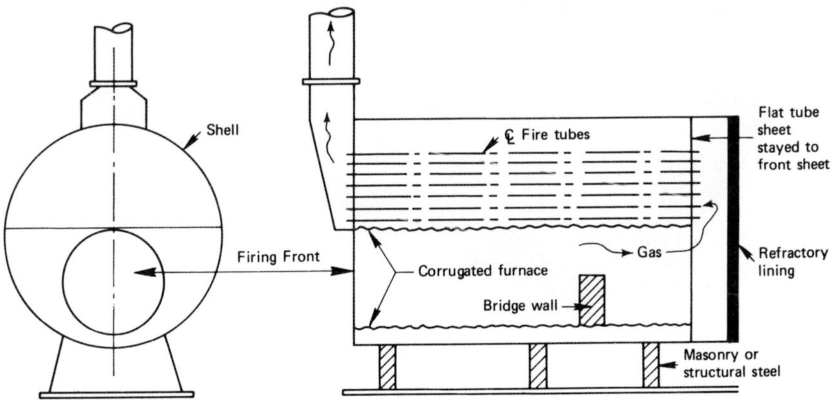


Figure 6. Scotch boiler.

with a large furnace tube and one or more passes of smaller diameter firetubes contained within the shell. The fuel is burned within the furnace tube. Scotch boilers are generally limited to firing of gas or oil. Cast-iron boilers (Fig. 7) are constructed of a number of identical (except for end sections) cast-iron sections. Since these boilers are assembled on site, they are particularly useful when installing a boiler in a room that does not have sufficient access to permit installing a package boiler. Cast-iron boiler sections are designed so that, when assembled, a tunnel-shaped combustion space is provided. Cast-iron boilers can be fired with gas and oil, and, if sufficient space is provided for grates or a stoker, coal.

Figure 7. Gas-fired commercial cast-iron boiler for hot-water or steam systems. (Courtesy of Weil-McLain Hydronic Div.).

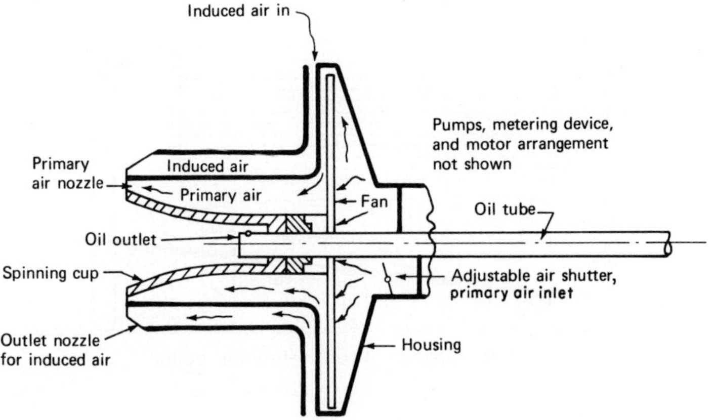


Figure 8. Rotary oil burner (5).

Standard practice has been to provide commercial-sized burners with forced-draft fans. The use of forced-draft fans permits the entire boiler package to be preassembled prior to delivery and reduces installation costs. For gas-fired boilers, the use of power burners permits easier conversion to oil (if necessary) and permits operating gas-fired boilers at lower excess air due to better mixing between air and fuel. Commercial gas-fired boilers generally operate at 10–30% excess air.

Small oil-fired commercial heating boilers (2–10 gal/hour) generally fire No. 2 fuel oil [fuel oil grades are defined by ASTM Standard D-396 (*6*)] and have burners much like the burners of residential heating units. For larger oil-fired boilers, the air-atomizing oil burner (using a two-fluid atomizer) and the rotary oil burner (Fig. 8) are more common than the pressure-atomizing oil burner. Commercial oil-fired boilers generally operate at 20 to 40% excess air.

### *C. Industrial Boilers*

Small industrial boilers 3 MJ/second to 60 MJ/second (about 10 to $200 \times 10^6$ Btu/hour) are generally of the firebox–firetube, packaged Scotch, horizontal-return-tubular (HRT), or watertube type. The HRT boiler (Fig. 9) consists of a shell containing firetubes mounted over a refractory-wall furnace. Small watertube boilers (Fig. 10) consist of a large number of water- and steam-filled tubes arranged as walls surrounding a center cavity that serves as the combustion chamber or fur-

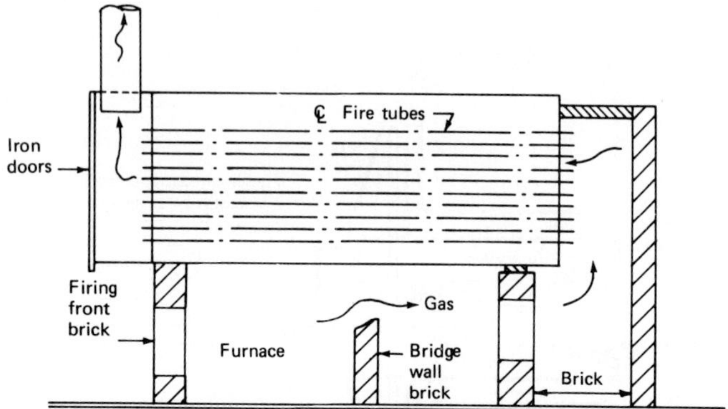


Figure 9. Horizontal-return-tubular boiler.

nace. The fuel is fired into this combustion space. Industrial boilers with firing rates above 60 MJ/second (about $200 \times 10^6$ Btu/hour) are mostly field-erected watertube boilers.

Nearly all types of industrial boilers, except the Scotch units, can be adapted to fire oil, gas, and, if a grate or a stoker is provided, coal. Pulverized coal can be fired in watertube boilers with inputs above about 60 MJ/second (about $200 \times 10^6$ Btu/hour). Oil and gas are fired with power burners, large boilers having multiple burners. Oil-fired industrial boilers are generally fired with heavier fuel oils (Nos. 5 and 6) which require heating for pumping and atomizing. Steam atomization is used for most oil burners on industrial boilers.

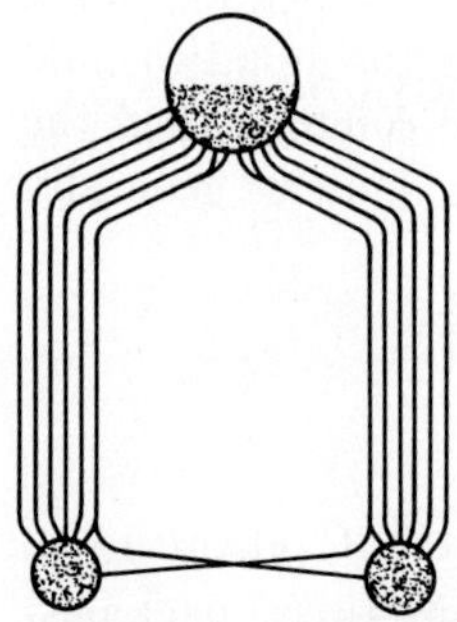

A-type has two small lower drums or headers. Upper drum is larger to permit separation of water and steam. Most steam production occurs in center furnace-wall tubes entering drum.

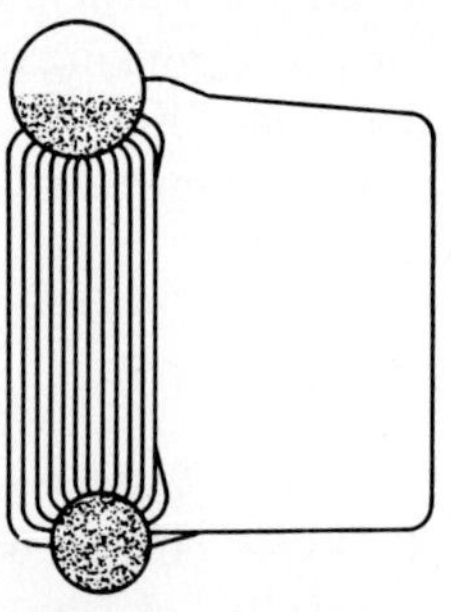

D-type allows much flexibility. Here the more active steaming risers enter drum near water line. Burners may be located in end walls or between tubes in buckle of the D, right angles to drum.

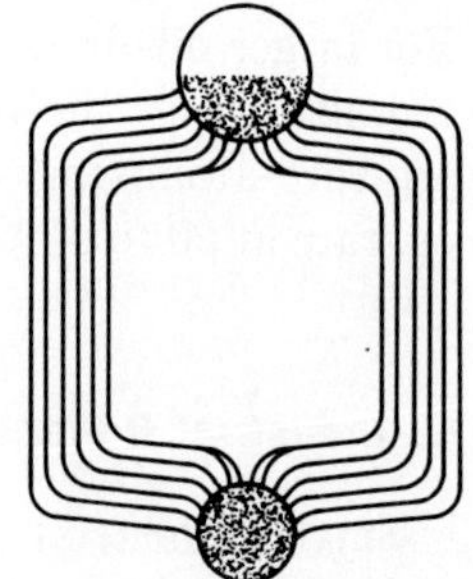

O-type is also a compact steamer. Transportation limits height of furnace so, for equal capacity, longer boiler is often required. Floors of D and O types are generally tile-covered.

Figure 10. Types of bent-tube packaged watertube boilers.

### D. Characterization of Stoker Types

Stokers of various types are designed mechanically to feed coal uniformly onto a grate within a furnace, to supply combustion air to the fuel bed, and to remove ash from the zone of combustion. The development of the mechanical stoker has advanced so the modern stoker is still considered an important element in the combustion of coal as well as other types of solid fuel. The modern stoker–boiler system incorporates controls to coordinate air and fuel supply to changing loads, and dust collecting equipment to minimize emissions. Fly carbon return systems have been used to increase efficiency but they are losing favor because they increase particulate emissions.

Stokers are classified according to the method of feeding fuel to the furnace, namely, chain-grate or traveling grate, water-cooled vibrating grate, spreader, and underfeed.

The type of stoker used depends upon the capacity required and the type of fuel burned. In general, the spreader stoker is the most widely used in the capacity range from 34,000 to 180,000 kg (about 75,000 to 400,000 lb) of steam per hour because it responds quickly to load changes and can burn a wide range of coal. The underfeed stokers are principally utilized with boilers generating less than 14,000 kg (about 30,000 lb) of steam per hour. The larger size underfeed units, as well as the chain and traveling grate stokers, are being displaced by the spreader and the vibrating-grate stokers in the intermediate range.

#### 1. Chain or Traveling Grate

The designations chain grate and traveling grate (Fig. 11) are often used interchangeably as these types of stokers are fundamentally the same except for grate construction. The one essential difference is that links of the chain-grate stokers are assembled so that they move with a scissorlike action at the return bend of stoker, while in most traveling grates there is no relative movement between adjacent grate sections. Accordingly, the chain grate is more suitable to handling coals with clinkering-ash characteristics than is the traveling grate.

The operation of each type is similar. Coal, fed from a hopper onto the moving grate, enters the furnace after passing under an adjustable gate to regulate the thickness of the fuel bed. The layer of coal on the grate entering the furnace is heated by radiation from the furnace gases and is ignited as its volatile matter is driven off by this rapid radiative heating. The fuel continues to burn as it moves across the furnace and

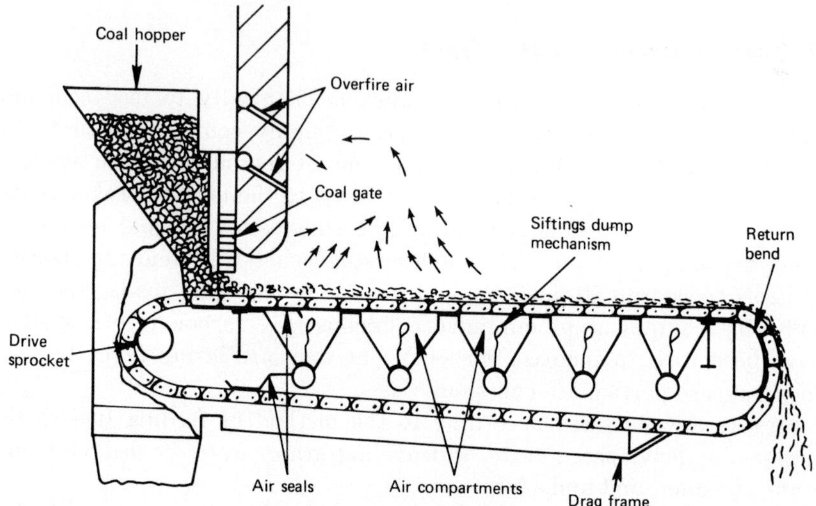


Figure 11. Chain-grate stoker.

the fuel bed becomes progressively thinner. At the far end of the grate, where the combustion of coal is completed, the ash is discharged into the pit as the grates pass downward over a return bend. Often the stokers use furnace arches, front and/or rear, to improve combustion by reflecting heat to the fuel bed. The front arch also serves as a bluff body to mix rich streams of volatile gases with air to reduce unburned hydrocarbons. As shown in Figure 11, the stoker may be zoned or sectionalized and is equipped with individual zone dampers to control air pressure and quantity to the various sections as the fuel travels along with the grate.

The chain and traveling-grate stoker can burn a variety of fuels including peat, lignite, subbituminous, free-burning bituminous, anthracite, and coke, providing the fuel is sized properly. Strongly caking bituminous coals may have a tendency to mat and prevent proper air distribution to the fuel bed. Also, a bed of strongly caking coal may not be responsive to rapidly changing loads.

Fuel-bed thickness varies with the type and size of coal burned. For bituminous coal, a 125 to 175 mm (about 5- to 7-in.) bed is common, and for small-sized anthracite, the fuel bed is reduced to 90 to 125 mm (about $3\frac{1}{2}$ to 5 in.).

Chain- and traveling-grate stokers are offered for a maximum continuous burning rate of between 1.1 and 1.6 MJ/second-m$^2$ (about 350,000 to 500,000 Btu/hour-ft$^2$ or approximately 40 lb coal/hour-ft$^2$) depending on the ash and moisture content and type of fuel.

## 2. Vibrating-Grate Stoker

The vibrating-grate stoker (Fig. 12) is similar to the chain-grate stoker in that both are overfeed, mass burning, continuous-ash discharge units. However, in the vibrating stoker, the sloping fuel bed is supported on equally spaced vertical plates that oscillate back and forth in a rectilinear motion causing the fuel to move from the hopper through an adjustable coal gate onto the active grates. Air is supplied to the stoker through lateral passages beneath the stoker formed by the flexing plates. Ash is automatically discharged into a shallow or basement ash pit. The grates are water-cooled and are connected to the boiler circulating system. The rate of coal feed and fuel-bed movement are controlled by frequency and duration of vibrating cycles and are regulated by automatic combustion controls that proportion the air supply to optimize heat release rates.

The vibrating-grate stoker has found increasing acceptance since its introduction because of simplicity, low maintenance, wide turndown ratio (10 to 1), and adaptability to multiple-fuel firing.

The water-cooled vibrating-grate stoker is suitable for burning a wide range of bituminous and lignite coals. Because of the gentle agitation and compaction caused by the vibratory actions, coal having a high free-swelling index can be burned, and a uniform fuel bed can be maintained without blowholes and thin spots. The uniformity of air distribution and resultant fuel-bed condition produces good response to load swings, and smokeless operation over the entire load range. Fly ash emission is prob-

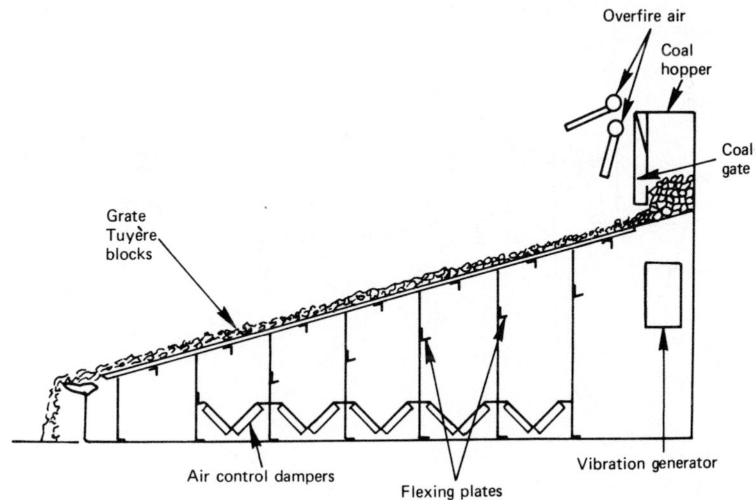


Figure 12. Vibrating-grate stoker.

ably greater than from the traveling-grate stoker because of the agitation of the fuel bed.

Burning rates of these stokers vary with the type of fuel burned, but, in general, the maximum heat release rates should not exceed 1.25 MJ/second-$m^2$ (about 400,000 Btu/hour-$ft^2$ or approximately 32 lb coal/hour-$ft^2$) to minimize fly ash carryover.

## 3. Spreader Stokers

Spreader stokers utilize a combination of suspension burning and grate burning. As illustrated in Figure 13, coal is continually projected into the furnace above an ignited fuel bed. The coal fines are partially burned in suspension and the larger particles are burned in a thin fast-burning fuel bed on the grate. This method of firing provides extreme flexibility for load fluctuations as ignition is almost instantaneous as firing rate is increased.

Because the spreader stoker burns approximately 50% of the fuel in suspension, much higher particulate loadings are generated than is usual from other types of stokers. Dust collectors are therefore required to col-

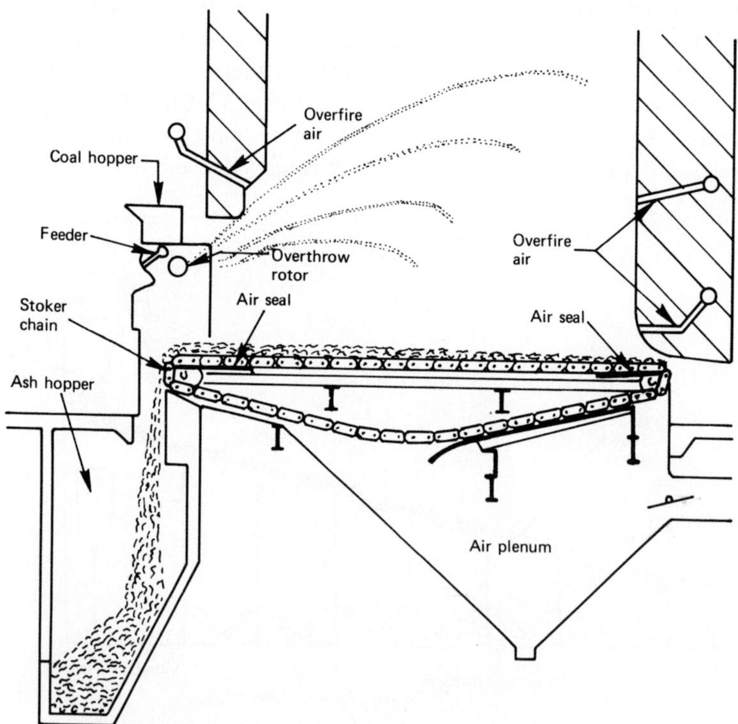


Figure 13. Traveling-grate spreader stoker.

lect particulate in the flue gas before discharge to the stack. To minimize carbon loss, fly carbon reinjection systems are utilized to reinject the carbon-containing fly ash into the furnace for complete burnout, although this increases particulate emission.

Grates for spreader stokers may be of several different types. All grates are designed with high flow resistance to avoid blowholes through the thin fuel bed. The earliest grate designs were simple stationary grates from which ash was removed manually by raking through doors. Later designs provided for intermittent dumping of the grate (either manually or power-cylinder operation) to remove ash. These units are frequently used for the small and medium-sized boilers. Both types of grates are sectionalized with separate undergrate air chambers for each grate section and a grate section for each spreader unit. Consequently, both air and fuel supply to one section can be temporarily discontinued for cleaning and maintenance without affecting the operation of other sections of the stoker.

For high efficiency operation, a continuous-ash discharge grate, such as the traveling grate, is necessary. The introduction of the spreader stoker with a traveling grate increased burning rates by about 70% over the stationary- and dumping-grate types.

All spreader stokers, and in particular ones with traveling grates, are able to burn fuels with a wide range of burning characteristics (including coals with caking tendencies because the rapid surface heating of the coal in suspension destroys its caking tendency). High moisture, free-burning bituminous and lignite coals are commonly burned, while coke breeze can be burned in a mixture with high volatile coal. Anthracite, because of its low volatile content, is not a suitable fuel for spreader-stoker firing.

Ideally, the fuel bed of coal-fired spreader stoker is from 50 to 100 mm (about 2 to 4 in.) thick. The maximum heat release rates are from 1.4 MJ/second-$m^2$ (about 450,000 Btu/hour-$ft^2$ or approximately 35 lb coal/hour-$ft^2$) on stationary or dumping grates to 2.4 MJ/second-$m^2$ (about 750,000 Btu/hour-$ft^2$ or 60 lb coal/hour-$ft^2$) on traveling-grate spreader types. Heat release rates of up to 3.2 MJ/second-$m^2$ (1,000,000 Btu/hour-$ft^2$) are practical with certain waste fuels in which a greater portion of fuel can be burned in suspension than is possible with coal.

## 4. Underfeed Stokers

Underfeed stokers, as their name implies, introduce the raw coal into a retort beneath the burning fuel bed. The underfeed stokers are classified into horizontal and gravity feed types. In the horizontal type, coal travels in a retort within the furnace parallel to the floor, while in the gravity-feed type, the retort is inclined 25°.

In the horizontal underfed retort stoker, coal is fed to the retort by a screw (for small stokers) or a ram (for larger units, see Fig. 14). Once the retort is filled, the coal is forced upward and spills over the retort to form and feed the fuel bed. Air is supplied through tuyeres at each side of the retort and through air ports in the side grates. As the coal rises within the retort, heat is conducted downward from the actively burning bed above, and the volatile gases are distilled off and burned. The rising fuel bed then ignites from contact with the burning bed. The incoming raw coal displaces the fuel bed gradually over the tuyeres and side grates. The coal is completely burned by the time it reaches the dump grates at both sides from which the ash and clinker are dropped into pits. Overfire air is used to provide additional combustion air to the flame zone directly above the bed to reduce smoke emissions. Operation of the gravity-feed-type underfeed stoker is similar in principle. These units consist of sloping multiple retorts and have rear ash discharge.

Either type of underfeed stoker can burn a wide range of coal, although the horizontal type is better suited for free-burning bituminous. In general, these types of units can burn caking coal providing there is not an excessive amount of fines. The ash-softening temperature is an important factor in selecting coals because the possibility of clinkering increases with lower ash-softening temperature coals.

Single or double retort horizontal types are generally used to service boilers with capacities up to 11,000 kg steam/hour (about 30,000 lb/hour). These units are designed for heat-release rates of 1.9 MJ/second-$m^2$ (about 50 lb coal/hour-$ft^2$).

A multiple-retort stoker consists of a series of inclined single-retorts placed adjacent to each other. Coal is fed into each retort where it is moved slowly to the rear while simultaneously being forced upward over the retorts.

## 5. Stoker Selection

Table II summarizes the characteristics, capacity range, and the maximum burning rates for each type of stoker discussed.

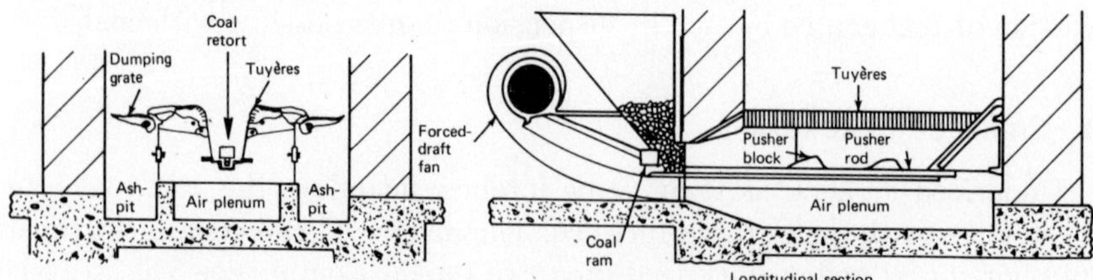


Figure 14. Horizontal underfeed stoker with single retort.

**Table II Characteristics of Various Types of Stokers**

| Stoker type | Capacity range (steam-generating rate) | | Maximum coal burning rate | | Characteristics |
|---|---|---|---|---|---|
| | *kg/hr* | *lb/hr* | *MJ/sec-m*$^2$ | *Btu/hr-ft*$^2$ | |
| Chain grate and traveling grate | 9,000 to 45,000 | 20,000 to 100,000 | 1.6 | 500,000 | Similar characteristics as vibrating grate except these stokers experience difficulty in burning strongly caking coals |
| Vibrating grate | 14,000 to 150,000 | 30,000 to 150,000 | 1.3 | 400,000 | Low maintenance, low fly ash carryover, capable of burning wide variety of coals, smokeless operation over entire range |
| Spreader | 35,000 to 180,000 | 75,000 to 400,000 | | | Capable of burning a wide range of coals, best ability to follow fluctuating loads, high fly ash carryover, low load smoke |
| Stationary and dumping grate | | | 1.4 | 450,000 | |
| Traveling grate | | | 2.4 | 750,000 | |
| Vibrating grate | | | 1.3 | 400,000 | |
| Underfeed | | | | | Capable of burning caking coals and a wide range of coals (including anthracite), high-maintenance, low fly ash carryover, suitable for continuous load operation |
| Single or double retort | 11,000 to 14,000 | 25,000 to 30,000 | 1.3 | 425,000 | |
| Multiple retort | 14,000 to 225,000 | 30,000 to 500,000 | 1.9 | 600,000 | |

## IV. Emissions from Residential Heating Units

### A. Oil-Fired Units

Manufacturers of oil-fired heating units have been concerned about smoke emissions for many years in order to control deposits on heat transfer surfaces. The Bacharach smoke scale was developed as a measure of the soot-forming potential of flue gas from such units. Manufacturers originally designed oil burners to achieve a Bacharach No. 5 smoke, until the 1960's when Bacharach No. 2 smoke became the industry standard (*7*, *8*). (Depending on the stack size and viewing background, a smoke number of 4 to 8 on the Bacharach scale is barely visible at the stack.)

Figure 15 shows the effect of excess air on emissions for a typical oil-fired residential heating unit fired with a pressure-atomizing gun-type burner (*9*). If the excess air level is set either below or above a critical range (which varies from unit to unit), smoke, carbon monoxide, and

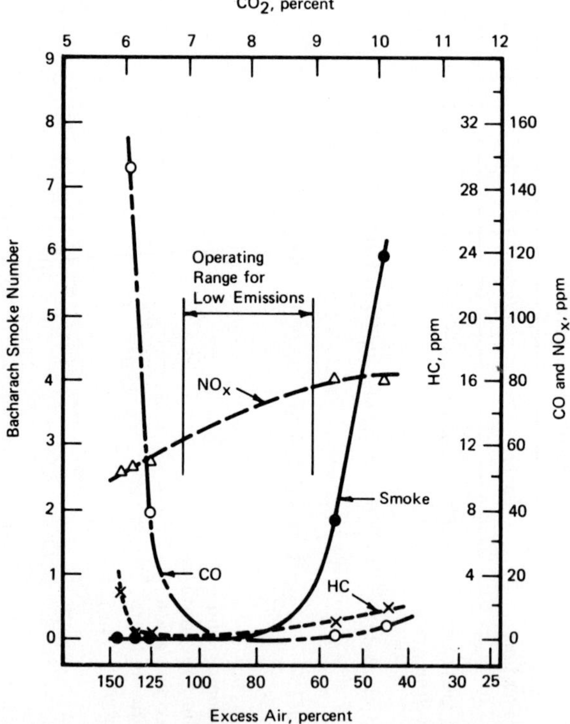


Figure 15. Typical smoke and gaseous emission characteristics for a residential unit in the tuned condition (*9*). △: $NO_x$; ○: CO; ×: HC; ●: smoke.

**Table III Comparison of Mean Emissions for Cyclic Runs on Residential Oil-Fired Units (9)**

| Units | Condition | Mean Bacharach smoke number | Mean emission factors[a] kg/1000 liters (lb/1000 gal) | | | |
|---|---|---|---|---|---|---|
| | | | *CO* | *HC* | $NO_x$ | *Filterable particulate* |
| All units | As-found | *b* | >2.65 (>22.1) | 0.68 (5.7) | 2.3 (19.4) | 0.35 (2.9) |
| | Tuned | *b* | >1.96 (>16.4) | 0.36 (3.0) | 2.3 (19.5) | 0.28 (2.3) |
| All units, except those | As-found | 3.2 | 0.93 (7.8) | 0.086 (0.72) | 2.3 (19.6) | 0.29 (2.4) |
| in need of replacement | Tuned | 1.3 | 0.52 (4.3) | 0.068 (0.57) | 2.3 (19.5) | 0.26 (2.2) |

[a] For cyclic runs of 10 minutes on and 20 minutes off.
[b] Not meaningful as units in need of replacement had raw oil on smoke spot.

hydrocarbon emissions climb rapidly due to incomplete combustion. The typical operating range of oil-fired units is 6 to 9% carbon dioxide, or 60 to 110% excess air. From the heat exchanger sooting viewpoint, the desired air setting is at a carbon dioxide value 0.5 to 1.0% less than that at which the smoke curve starts increasing rapidly.

The results of one field study of emissions from 33 residential oil-fired furnaces and boilers determined that $NO_x$ emissions were relatively insensitive to all burners or system design variables and to burner operating condition (tuned, untuned, in need of replacement, etc.) (*9*). Similarly, particulate emissions were only slightly lower for tuned units than for untuned units (Table III). Particulate emissions from the newer flame-retention head burners were lower than for older burner designs.

Smoke, carbon monoxide, and hydrocarbon emissions were found to be quite sensitive to burner operating conditions. The few oil-fired units in poor operating condition (and judged in need of overhaul or replacement) were found to contribute a disproportionately large share

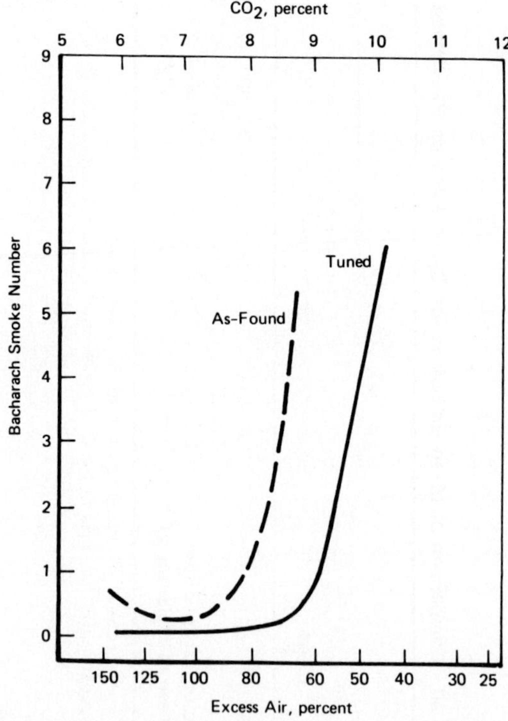


Figure 16. Typical smoke vs carbon dioxide curves for a residential unit in the as-found and tuned conditions (9).

of carbon monoxide and hydrocarbon emissions from the group of units. For instance, of 33 units tested in this program, 3 poorly performing units were responsible for in excess of 65% of the carbon monoxide emissions and for 87% of the hydrocarbon emissions generated by the entire group of units. Once the poorly performing units were eliminated, tuning of the remaining units substantially reduced smoke (Fig. 16) but achieved marginal benefits in reducing already low emissions of carbon monoxide, hydrocarbons, and particulates. Thus, the major factor in any program to control emissions from such units should include the screening of units to identify the poorly performing ones and to effect their renovation or replacement.

Another study to compare emissions from various burner designs (in the laboratory) concluded that the modern flame-retention-type oil burners are cleaner burning and produce lower emissions of smoke, carbon monoxide, hydrocarbons, and particulates than older burner designs (*10, 11*).

The air pressure available from blowers on residential oil burners is insufficient to generate the furnace turbulence required to completely burn the fuel at low excess air. Reducing excess air for oil-fired residential units from the 60 to 110% range to the 10 to 15% range while maintaining the same flue-gas temperature would increase unit efficiency by 5 to 10 percentage points (e.g., from 70 or 75% to 80%).

### B. Gas-Fired Units

Natural gas-fired residential heating units, when properly adjusted, are low emitters of nearly all pollutants (except possibly $NO_x$). These units are equipped with atmospheric burners and operate at high excess air ratios. For a properly adjusted burner, the presence of a generous supply of secondary air and a gaseous fuel which mixes rapidly with combustion air results in essentially clean combustion, very low emission of carbon monoxide and hydrocarbons, and a nearly total absence of smoke or particulate emissions (Table IV) (*12*). However, improper adjustment of the gas burner primary air can cause incomplete combustion and high levels of carbon monoxide, hydrocarbons, and smoke emissions.

Because of low flue gas temperatures the overall thermal efficiencies of gas-fired units with atmospheric burners are typically about 75% in spite of high excess air (6% carbon dioxide or 80% excess air). Increased efficiency can be achieved by significantly reducing excess air but this requires the use of fanpowered rather than atmospheric burners. Emissions from gas-fired fanpowered burners should not be higher than those from atmospheric units.

**Table IV Average Emissions of 34 Gas-Fired Forced-Air Furnaces with Atmospheric Burners** (*12*)

| | | | *Mean flue-gas concentration, ppm, air free* | | | | *$NO_x$ Emission factor (as $NO_2$)* | |
|---|---|---|---|---|---|---|---|---|
| *Average* | *Flame*[a] | *Mean $CO_2$, %* | *CO* | *NO* | *$NO_2$* | *Aliphatic aldehydes* | *kg/1000 MJ* | *lb/$10^6$ Btu* |
| Overall | Blue | 5.8 | 15 | 87.8 | 4.6 | 0.18 | 0.042 | 0.097 |
| | Yellow | 5.9 | 359 | 73.4 | 9.7 | 0.60 | 0.037 | 0.087 |
| Multiport | Blue | 5.9 | 18.2 | 93.8 | 4.8 | 0.20 | 0.045 | 0.104 |
| burner | Yellow | 6.0 | 338 | 79.9 | 8.8 | 0.62 | 0.040 | 0.093 |
| Single port | Blue | 5.5 | 8.7 | 73.4 | 4.2 | 0.14 | 0.035 | 0.082 |
| burner | Yellow | 5.8 | 410 | 57.9 | 11.6 | 0.60 | 0.045 | 0.104 |

[a] Blue flame—proper adjustment; yellow flame—improper adjustment.

## V. Emissions from Commercial and Industrial Boilers

### *A. Oil- and Gas-Fired Boilers*

Emissions from commercial and small industrial boilers were measured in a field program that included 13 boilers—the smallest being a 40 boiler horsepower (0.47-MJ/second or $16 \times 10^5$-Btu/hour input) boiler and the largest a 600 boiler horsepower (70-MJ/second or $24 \times 10^6$-Btu/hour input) (*9*). Results of that study are summarized below.

Figures 17 and 18 illustrate typical curves of smoke and $NO_x$ emissions versus excess air for a boiler fired at 80% load with three types of fuel oil and natural gas. The smoke curves are generally flat over a range of excess air levels above 30% but increase sharply as the combustion air is reduced below that percentage. For oil firing, as excess air is reduced, smoke levels tend to increase before carbon monoxide and hydrocarbons increase. With gas firing, carbon monoxide and hydrocarbons increase before smoke density increases.

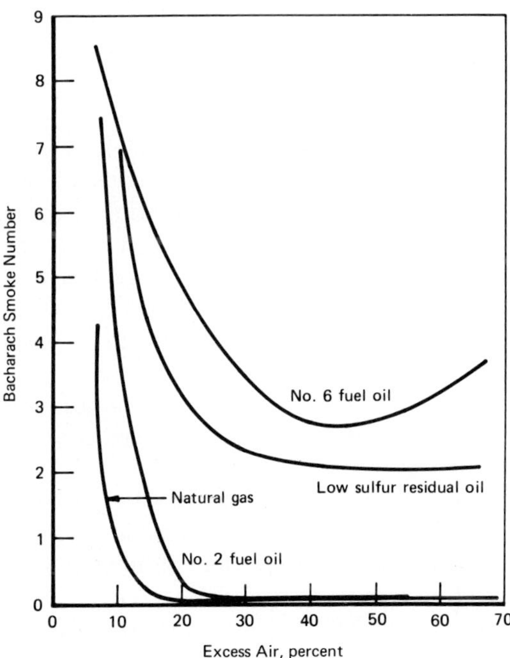


Figure 17. Typical characteristics of smoke vs air for a commercial boiler firing different fuels at 80% load (9).

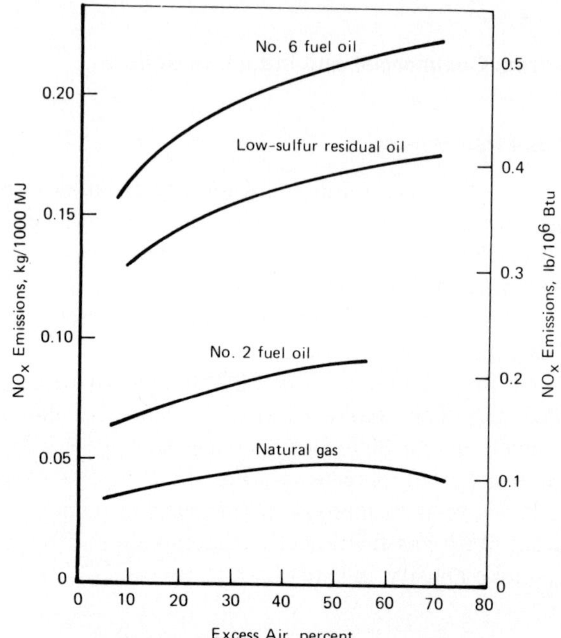


**Figure 18. Typical nitrogen oxide emissions vs excess air for a commercial boiler firing different fuels at 80% load (9).**

To adjust the air setting to acceptable smoke (or minimum smoke for that boiler/fuel combination) requires combustion air settings between 30 and 40% excess air for most boilers operating with heavier fuels. It is possible to fire natural gas at 10% excess air without exceeding Bacharach No. 1 smoke; however, 20% excess air is required to avoid rapidly increasing carbon monoxide emissions. Smoke increased slightly at high excess air in only a few cases with oil firing and in only one case with gas firing.

$NO_x$ emissions are relatively insensitive to the air setting. Emissions of carbon monoxide and hydrocarbons are consistently low for commercial boilers in normal operating ranges.

Load generally has little influence on emission levels, including $NO_x$. In some cases smoke and particulate emissions increased with load when firing No. 6 oil, suggesting that limitations in mixing or combustion volume can be reached at high load.

Smoke, particulates, and $NO_x$ emissions are strongly influenced by fuel grade. Particulate emissions and smoke increase with increasing fuel-grade designations (heavier oils) when several fuel oils are fired in the same boilers at a given excess-air setting; these emissions are consistently low when firing natural gas at normally excess-air ratios. Figure 17 shows the influence of fuel grade on smoke for a typical boiler.

The effect of fuel grade on filterable particulate emissions is summarized in Figure 19. Particulate emissions are plotted against API (American Petroleum Institute) gravity (*13*) as an indicator of fuel grade and burning characteristics—the heaviest fuels having lowest API gravity. The approximate ranges of API gravity for different fuel grades are shown on the figure.

Ash content tends to be higher for fuels of low API gravities but not sufficient to account for higher particulate levels with heavier fuels. The band of ash content for the fuels in this investigation is shown in Figure 19.

A low-sulfur residual oil was similar in performance to a No. 4 or No. 5 oil; it yielded filterable particulate levels about equal to those from No. 4 oil and only one-third of those from No. 6 oil.

Figure 20 shows the strong effect of fuel nitrogen on $NO_x$ emissions from boilers in which different grades of fuel oil were fired. The zero

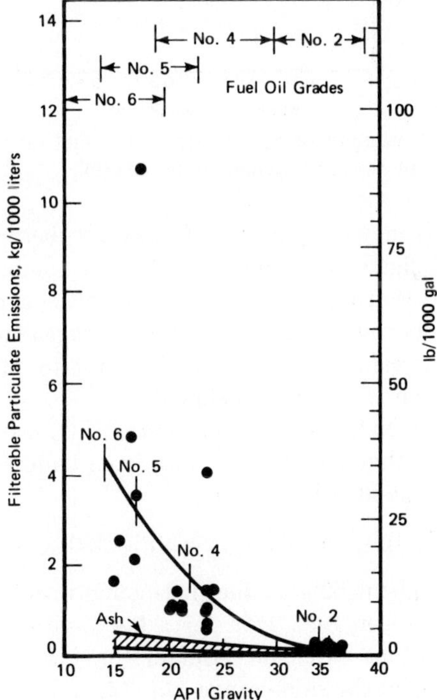


Figure 19. Relation of filterable particulate matter and API gravity for commercial boilers firing different fuels at 80% load (*9*).

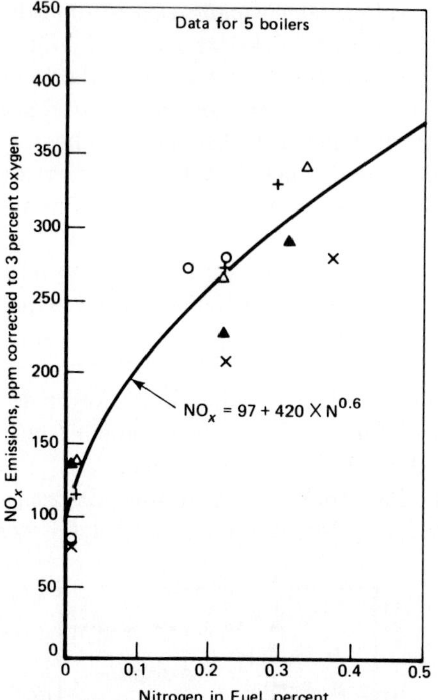


Figure 20. Relation of nitrogen oxide emissions and fuel nitrogen for a commercial boiler firing different fuels at baseline conditions (*9*).

nitrogen intercept is indicative of $NO_x$ formed by thermal fixation, which is dependent upon flame temperature and other combustion parameters influenced by boiler/burner design. The slope of the curve of $NO_x$ versus fuel nitrogen reflects the conversion of fuel nitrogen to $NO_x$. The range of fuel nitrogen in residual fuel oils is from 0.1 to 1.0%; for distillate fuel oils, fuel nitrogen is generally below 0.2%.

Both the *thermal* $NO_x$ and the *conversion* $NO_x$ vary from boiler to boiler. The equation that best fits the data from boilers in which fuels of different nitrogen content were fired is

$$NO_x \text{ (ppm at 3\% } O_2) = 97 + 420 \times N^{0.6} \tag{1}$$

This indicates that about 62% of the fuel-bound nitrogen is converted to $NO_x$ for a 0.2% nitrogen fuel, 47% for a 0.4% nitrogen fuel, and 38% for a 0.7% nitrogen fuel. Other investigators have reported conversion of fuel nitrogen to $NO_x$ in the same range (*14, 15*).

It should be pointed out that other fuel properties, such as gravity and viscosity, also vary from fuel to fuel. Thus, the $NO_x$ levels shown in

Figure 20 may indirectly include the effects of factors other than fuel nitrogen.

$NO_x$ emissions from firing of natural gas in commercial boilers are in the same range as $NO_x$ emissions from firing No. 2 fuel oil. Both of these fuels have low fuel nitrogen and produce mostly *thermal* $NO_x$.

Table V presents suggested emission factors for oil-fired boilers by fuel grade (*9*). Emission factors intended for use in compiling emission inventories or in evaluating control strategies should discriminate between fuel-oil grades where these are known.

Table VI summarizes emissions as measured from seven commercial boilers firing natural gas and gives suggested emission factors (*9*).

### *B. Coal-Fired Boilers*

Utilization of solid fuels has been, in the past, frequently associated with various degrees of smoke emission. The advent of completely automatic equipment to replace hand-firing methods greatly reduced and, in most cases, eliminated visible smoke from solid fuel burning. Increasingly widespread enforcement of air pollution control ordinances has caused users of solid fuels either to install practically smoke-free equipment or to convert to firing the more easily utilized, but often more costly, gaseous or liquid fuels. In modern applications of solid fuels, provisions for fly ash collection must be made to keep the final solids emission within the limits of local ordinances.

Sulfur dioxide emission continues to be a problem when burning most coals. However, considerable effort is being expended to develop methods for producing clean solid fuels (see Chapter 10, this volume).

Emissions of polynuclear hydrocarbons by combustion equipment has become of increasing concern because of the carcinogenic characteristics of some compounds in this class. Coal-fired combustion units, especially small coal-fired units, are generally acknowledged to be one of the prime sources of these emissions, Table VII (*16, 17*).

Tables VIII and IX (*18, 19*) give emissions of combustible gases and $NO_x$ for various types of coal-fired units, respectively.

#### 1. Traveling-Grate and Vibrating-Grate Stokers

The traveling grate generally works best with noncaking fuels because air distribution is most uniform through a uniform noncaked bed. For many years, mixing of the burning gases above the traveling grate was achieved by the use of elaborate suspended arches which forced the upward flow of flame into a narrow throat past incandescent refractory.

**Table V Suggested Emission Factors for Oil-Fired Commercial Boilers (9)[a]**

| Fuel oil grade | Emission factors kg/1000 liters (lb/1000 gal) | | | | |
|---|---|---|---|---|---|
| | *CO* | *HC* | $NO_x$[b] | $SO_2$[c] | *Filterable particulate* |
| No. 2 | 0.06 (0.5) | 0.020 (0.17) | $2.40 + 9.35 \cdot N^{0.6} (20 + 78 \cdot N^{0.6})$ | $17.0 \cdot S (142 \cdot S)$ | 0.14 (1.2) |
| No. 4 | 0.11 (0.9) | 0.029 (0.24) | $2.40 + 10.19 \cdot N^{0.6} (20 + 85 \cdot N^{0.6})$ | $18.4 \cdot S (154 \cdot S)$ | 1.68 (14.0) |
| No. 5 | 0.13 (1.1) | 0.034 (0.28) | $2.40 + 10.43 \cdot N^{0.6} (20 + 87 \cdot N^{0.6})$ | $19.1 \cdot S (159 \cdot S)$ | 3.24 (27.0) |
| No. 6 | 0.14 (1.2) | 0.035 (0.30) | $2.40 + 10.67 \cdot N^{0.6} (20 + 89 \cdot N^{0.6})$ | $19.4 \cdot S (162 \cdot S)$ | 4.31 (36.0) |
| LSR[d] | 0.11 (0.9) | 0.028 (0.23) | $2.40 + 10.07 \cdot N^{0.6} (20 + 84 \cdot N^{0.6})$ | $18.3 \cdot S (153 \cdot S)$ | 1.44 (12.0) |

[a] These values are based on mean emission data for the identified fuel grades having typical API (American Petroleum Institute) gravity as follows: 34 deg API for No. 2; 22 for No. 4; 17 for No. 5; 14 for No. 6; and 23 for LSR. Where actual API gravity is known, interpolated values should be used. Assuming st eady-state base-line operating conditions: 80% load and air adjustment for 12% $CO_2$ flue-gas concentration.

[b] N: multiplication factor equal to percent nitrogen in fuel oil. If concentration is unknown, the following values are suggested: 0.01% for No. 2; 0.2% for No. 4; 0.3% for No. 5; 0.4% for No. 6; and 0.2% for LSR.

[c] S: multiplication factor equal to sulfur content in fuel.

[d] LSR: low sulfur residual oil (1.0% S).

**Table VI Suggested Emission Factors for Gas-Fired Commercial Boilers (9)**

| | Emission factors, $kg/10^6m^3$ $(lb/10^6\ ft^3)$[a] | | | | |
|---|---|---|---|---|---|
| | *CO* | *HC* | $NO_x$ | $SO_2$ | *Filterable particulate* |
| Mean emission from 7 boilers: operated at 80% load and 10% $CO_2$ in flue gas | 268 (16.7) | 59 (3.7) | 1680 (106) | Nil | 91 (5.7) |
| Suggested emission factors | 321 (20.0) | 64 (4.0) | 1600 (100) | 9.6 (0.6) | 96 (6.0) |

[a] To convert to emission factors in lb/$10^6$ Btu, multiply lb/$10^6$ $ft^3$ values by 0.00098.

**Table VII Polynuclear Hydrocarbons Emitted by Bituminous Coal with Various Firing Methods** *(16, 17)*[a]

| | *Type of unit* | | | | | | | | |
|---|---|---|---|---|---|---|---|---|---|
| | | | *Spreader stoker*[b] | | | *Underfeed stoker*[a] | | | |
| | *Pulverized firing* | *Cyclone* | *A* | *B* | *Chain-grate stoker* | *C* | *D* | *E* | *Handfired* |
| Benzo(*a*)pyrene | 0.03–0.13 | 0.49 | 0.04 | 0.057 | 0.082 | 22 | 0.26 | 8.4 | 880 |
| Pyrene | 0.20–0.40 | 2.25 | 0.23 | 1.30 | 0.860 | 35 | 3.70 | 17 | 1320 |
| Benzo(*e*)pyrene | 0–0.58 | 0.87 | 0.13 | 0.770 | 0.290 | 17 | 0.510 | 11.9 | 220 |
| Perylene | 0–0.15 | 0.07 | | | | 3.5 | | | 132 |
| Benzo(*ghi*)perylene | 0.02–1.42 | 0.44 | | | | 9.9 | | 1.28 | 660 |
| Anthanthrene | 0–0.02 | | | | | 0.64 | | | 198 |
| Coronene | 0–0.15 | 0.01 | 0.02 | 0.057 | | 0.73 | | 2.64 | 66 |
| Anthracene | | | | | | 1.9 | | | 880 |
| Phenanthrene | | | | | | 22 | 2.2 | 64 | 2200 |
| Fluoranthene | | 0.17 | 0.11 | 0.790 | 1.50 | 83.9 | 7.1 | 103 | 2200 |
| Benz(*a*)anthracene | | | | | | 8.6 | | 1.23 | |

[a] lb/$10^6$ Btu input (multiply by 0.430 to obtain kg/1000 MJ).
[b] Letter designations refer to different units.

**Table VIII Combustible Gases Emitted by Bituminous Coal with Various Firing Methods (*17*)[a]**

| | *Pulverized firing* | *Cyclone* | *Spreader stoker* | *Chain-grate stoker* | *Underfeed stoker* | *Hand-fired* |
|---|---|---|---|---|---|---|
| Single atom hydrocarbons | 1–10 | — | — | 5 | 36–120 | 730 |
| Carbon monoxide | 5–44 | — | 29 | 510 | 160–1100 | 3500 |
| Formaldehyde | 0.1–9.25 | 0.17 | 0.06 | 0.14 | 0.21–0.38 | — |

[a] lb/10 $10^6$ Btu input—multiply by 0.430 to obtain kg/MJ.

However, the recent trend has been to use only rear arches or simple open furnaces and to attain mixing by means of high-velocity jets of secondary air directed from a number of nozzles located in one or more of the furnace walls. These are effective in eliminating smoke which may result from poor mixing (*20*). Zoned control of the air admitted to various portions of the fuel bed is an effective means for reducing smoke and fly ash emission. Fly ash may be slightly higher on vibrating-grate stokers than on traveling-grate stokers because of increased agitation of the fuel bed.

Measurements on a single power-generating traveling-grate stoker, operating at near rated load, showed negligible hydrocarbon emissions, even with smoke between 20 and 40% opacity, but particulate emissions were 0.99 kg per metric ton of gas (0.99 lb per 1000 lb) at 50% excess air (*18*).

**Table IX Emission of Nitrogen Oxides (*18, 19*)**

| *Type of unit* | *kg/1000 MJ (input)* | *lb/$10^6$ Btu (input)* |
|---|---|---|
| Pulverized coal | | |
| Vertical firing | 0.16 | 0.38 |
| Corner firing | 0.41 | 0.95 |
| Front wall firing | 0.29 | 0.68 |
| Horizontal opposed firing | 0.28 | 0.65 |
| Cyclone | 1.1 | 2.5 |
| Stoker | | |
| Spreader stoker | 0.28 | 0.65 |
| Commercial underfeed | 0.13 | 0.30 |
| Residential underfeed | 0.15 | 0.36 |
| Hand-fired | 0.05 | 0.11 |

## 2. Spreader Stokers

The spreader stoker employs overfeed burning, an inherently smoky method, plus suspension burning, an inherently smoke-free fly ash-producing method. Overfire jets have been found essential to smoke-free operation.

Figure 21 shows the effect of burning rate on dust emission from spreader and underfeed stokers (*21–23*) when the dust was measured ahead of the dust collector. The two upper curves indicate that, without a dust collector, dust emission from a spreader stoker is well above most commonly used limitations on fly ash in the flue gas at all burning rates. Overfire jets reduce dust emission significantly but not enough to meet most ordinances. On the other hand, the cluster of points in the lower right of the figure are dust-emission data after the collector for a similar unit having a dust collector of high efficiency.

As a part of extensive studies in Germany on a number of different fuels, boilers, and firing systems, Kuhlman (*24*) found that spreader stokers in firetube boilers emitted little more flue dust, about 1.8 kg per metric ton of steam generated, than hand firing or traveling grates, about 1.3 kg per metric ton. The former figure is roughly equivalent to 30 lb of dust per short ton of coal fired, 1.2 lb per million Btu, or 1.4 lb per 1000-lb flue gas.

Smoke is characteristic of most spreader stokers if they are operated at less than approximately 25% of full load, because low furnace temperature causes incomplete suspension burning. Accordingly, a spreader stoker-fired system should be supported by suitable auxiliary heat-generating means so that, at low plant loads, the stoker can either be operated to carry all of the load or be shut down completely. Modulating controls have been shown to help reduce dust emission (*25*).

The quantity of unburned gases emitted by spreaders are extremely low when operation is smokeless, similar to values characteristic of pulverized-coal firing.

## 3. Underfeed Stokers

Overfeed fuel beds have been shown to be inherently smoky in operation because burning gases rise through fresh fuel, thus resulting in rapid devolatilization of the new fuel in a zone having a deficiency of oxygen. On the other hand, underfeed beds are inherently smoke free. The air and fresh fuel flow concurrently, usually upward; hence, the zone of ignition, which is near the point of maximum evolution of combustible gases, is supplied with ample air and well mixed, which promotes complete combustion.

Because high-velocity jets of burning gas escape through fissures in the underfed fuel bed, some fly ash will be produced (Fig. 21). Measurements on small- and medium-sized underfeed stokers show particle emissions ranging from 0.24 to 0.68 kg per metric ton flue gas (0.24 to 0.68 lb per 1000 lb) at 50% excess air (*18*). At burning rates of 1.6 MJ/second-m$^2$ (approximately 500,000 Btu/ft$^2$-hour or about 40 lb coal/ft$^2$-hour) the amount of fly ash may be great enough to require dust collectors to clean the flue gases. Data on dust emission are available for two large multiple-retort stokers. Measurements on one unit, made in 1932 (*26*), showed emissions ranging from 0.7 to 3.4 gm/m$^3$ (about 0.3 to 1.5 gr/ft$^3$) for burning rates of between 1.0 and 1.8 MJ/second-m$^2$ (about 25 and 45 lb coal/ft$^2$-hour). Later measurements on a similar unit using low-volatile coal (*27*) showed emissions of 0.46 to over 3.4 gm/m$^3$ (0.2 to over 1.5 gr/ft$^3$) for approximately the same range of burning rates. The amounts of unburned gases which escape from underfed stokers have been shown, in limited tests, to be low when operation is smokeless; moderately high when smoky (*18*). Flue-gas analyses on a single retort-underfeed stoker applied to a small industrial boiler showed no carbon monoxide, hydrogen, methane, ethane, or illuminants within the limits of detection of the combustion gas analyzer used, i.e., 0.1%, even when smoke was produced deliberately by firing a coarse, uniformly sized coal (*28*).

From the smoke emission viewpoint, the usual underfeed stoker is bet-

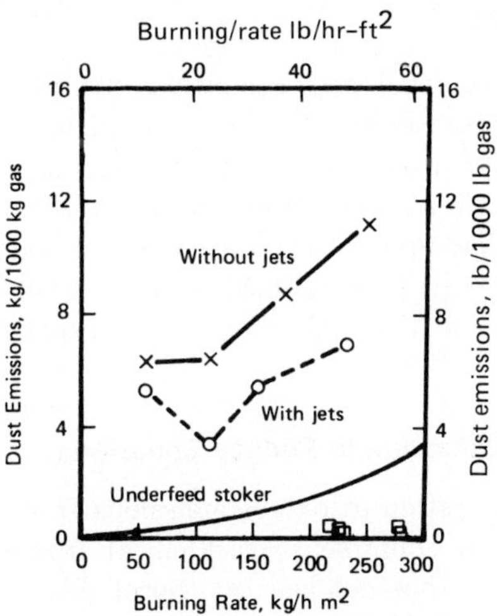


Figure 21. Effect of burning rate, overfire jets, and collector efficiency on dust emission from spreader stokers and one large underfeed stoker burning eastern bituminous coal (*21–23*) ×, ○: No reinjection, no collector; □: with reinjection from high efficiency collector.

ter suited for caking coals than for noncaking or free-burning coals because the volatiles are released more slowly from caked masses than from a porous bed of coals. However the caked masses formed when coking coals are fired can cause uneven air distribution and result in the ash containing appreciable unburned carbon. Consequently, there is the possibility of smoke from misapplication of coal to such equipment, but no data are available on the amount of pollution which can result from such a condition.

### 4. Hand Firing

The only instances in which hand firing of high-volatile fuels has been acceptably and smokelessly accomplished has been in small domestic equipment utilizing the downdraft principle (*29*). In the usual open fireplace, stove, or furnace the hot products of combustion rise upward through fresh fuel that is fired on top of the incandescent bed, thereby producing rapid devolatilization and smoke emission and carry-off fly ash. Downdraft combustion requires a downward flow of air and gases, so that complete combustion is readily attained with a wide variety of fuels. The method produces practically no fly ash. Unfortunately, successful application of this principle has been limited to small units, owing to problems of fuel flow and of heat transfer through the fuel mass.

### 5. Overfire Air Jets

Jets of air over the fuel bed, sometimes introduced into the furnace by jets of high pressure steam, have been used for over a hundred years (*30*) to abate smoke from solid fuel-fired furnaces. The action of the jets is to cause intense mixing of the stratified streams of air and burning gas which usually flow upward from a solid fuel bed. Methods for application of overfire jets have been thoroughly described (*20, 31*), and data have been obtained on their effects on some types of firing systems.

## VI. Combustion Modification to Reduce Emissions

Combustion modification to reduce emissions from small- and intermediate-size combustion equipment (residential furnaces and commercial and industrial boilers) has not been as thoroughly studied as has its application to utility boilers. However, combustion modification offers the opportunity for reducing emissions from the smaller oil- and gas-fired equipment. About the only effort that has been devoted to reducing emissions from small coal-fired combustion equipment has been the work in industrial-size fluidized bed combustors. (Chapter 10, this volume contains additional information on combustion modification to reduce emissions.)

In addition to the work discussed below, a considerable amount of research has been directed toward developing low-emission burners for various types of continuous combustion–Rankine cycle or gas turbine engines for automobiles. The principles and techniques for reducing emissions that have been developed for these applications are also applicable to oil- and gas-fired burners for nonautomotive applications.

### A. Improved Burners

The effect of combustion–air swirl has been studied by the International Flame Research Foundation (*32, 33*) and others (*34*). Results of these studies have shown that varying the swirl intensity can increase or decrease $NO_x$ emissions by at least 30%. However, no simple rules for preselecting proper swirl to minimize $NO_x$ emissions have been developed.

Howekamp and Hooper have investigated the emissions from combustion-improving burner heads for domestic oil burners (*10*). They found one burner head that could achieve low smoke emissions (Bacharach No. 1) at excess air levels as low as about 15% excess air. However, at this low excess air level, the $NO_x$ emissions from this burner were among the highest of the six burners tested. Other new burner concepts are reported to offer low smoke or $NO_x$ emissions, but are reported in less detail (*11*).

### B. Recirculation

Cooper *et al.* (*35*) have shown that flue-gas recirculation can be an effective technique for firing oil at low excess air while maintaining low emissions of smoke. Andrews *et al.* (*36*) demonstrated that for two recirculation configurations (one involving a blower to return exhaust gas to the blast tube of a gun burner and the other a new burner with internal recirculation), low $NO_x$ emissions could be achieved by flue-gas recirculation while maintaining low levels of carbon monoxide, hydrocarbon, and smoke emissions. $NO_x$ emissions were reduced from about 1.44 kg/m$^3$ (12 lb/1000 gal) with no recirculation to as low as 0.48 kg/m$^3$ (3 lb/1000 gal) with recirculation. Also, with recirculation, a Bacharach No. 1 smoke could be achieved with 20% excess air, versus the about 60% excess air required to achieve this smoke level without recirculation.

Muzio, Wilson, and McComis have studied the application of flue-gas recirculation to reduce emissions from package boilers fired with distillate and residual oil and natural gas (*37*). They reported that $NO_x$ reductions of up to 25% for firing No. 6 fuel oil, 40% for No. 2 fuel oil, and 60% for natural gas could be accomplished with recirculation without increasing soot emissions excessively. They found that the recirculated flue gas must be intimately mixed with the combustion air in the vicinity of the burner to be effective in controlling $NO_x$ emissions.

### C. Staged Combustion

True staged firing with two or more separate combustion zones generally is not practical in most small combustion systems owing to space requirements and other considerations. However, it has been found that some of the benefits of staged combustion (lower $NO_x$ emissions) can be achieved by limiting combustion air to the burners to less than the stoichiometric air required for combustion. After the fuel is partially combusted the remaining combustion air is added and combustion is completed. This technique is variously referred to as biased firing, delayed air admission, and staged combustion.

Muzio *et al.* (*37*) examined the use of staged combustion to reduce $NO_x$ emissions from a package boiler firing No. 6 fuel oil. They found that, for their burner, only a 25% reduction in $NO_x$ could be accomplished before smoke emissions become excessive. Combining flue-gas recirculation and staged combustion resulted in accomplishing a 45% reduction in $NO_x$ emissions for firing No. 6 fuel oil, while maintaining a smoke no greater than Bacharach No. 2. Thus, the combination of staged combustion and flue-gas recirculation was more effective than either separately.

Cato *et al.* (*38*), have examined off-stoichiometric firing of multiburner industrial boilers. They found that $NO_x$ emissions would be reduced by up to 29% for oil-fired industrial boilers and by up to 40% for gas-fired industrial boilers. Similar results have been reported for utility boilers (*39, 40*).

### D. Fuel Additives

The use of fuel additives to reduce emissions is a fertile field for extravagant claims, few of which withstand rigorous technical evaluation. Claims of $NO_x$ and sulfur dioxide reductions are especially subject to question. However, additives have been shown to be effective in reducing smoke emissions from gas turbines (*41*) and, thus, there may be some basis for their use to reduce emissions of smoke, soot, carbon particulate, carbon monoxide, and hydrocarbons associated with incomplete combustion.

Martin conducted tests on about 300 additives in a residential-heating gun-type oil burner (*42*). More extensive tests were conducted with a small number of additives in three domestic oil burners and in a package boiler (*43*). These tests showed that $NO_x$ or sulfur dioxide emissions could not be reduced by additives. Although some additives decreased smoke or soot emission, it was generally shown that the improvement in performance achieved by the use of additives was less than what could be achieved by use of improved burner design.

Giammar *et al.* evaluated the performance of additives in a commercial Scotch boiler firing residual fuel oil and found that certain alkaline-earth and transition metals were effective in reducing emissions of smoke, carbon particulate, and polycyclic organic matter (*45*).

### E. Oil/Water Emulsions

The firing of water-in-oil emulsions to reduce emissions from oil combustion has received attention due to the marketing of several emulsification burners or devices. The concept is twofold: first, the water within the individual oil droplets flashes to steam and explodes the droplets when the oil is sprayed into the hot furnace, thus producing a secondary atomization and finer droplets and cleaner combustion, and, second, the water acts as a diluent and cools the flame, thus reducing thermal $NO_x$ formation. Tests of several emulsification devices by the United States Environmental Protection Agency determined that they reduced emissions of smoke, particulate, and $NO_x$ under some conditions (*44*).

## REFERENCES

1. D. W. Locklin, "Recommendations for an Industry-Wide Oil-Burner Research Program," Publ. 1700, p. 19. Amer. Petrol. Inst., Washington, D.C., 1960.
2. "Oilheating Gains in 1970," *Fueloil Oil Heat* **30** (1), 38–41 (1971).
3. D. W. Locklin, H. H. Krause. A. A. Putnam, E. L. Kropp, W. T. Reid, and M. A. Duffy, "Design Trends and Operating Problems in Combustion Modification," EPA-650/2-75-032, p. A-20. U.S. Environmental Protection Agency, Research Triangle Park, North Carolina, 1974.
4. W. R. Johnson, in "ASHRAE Handbook and Product Directory—1975," Chapter 23. Amer. Soc. Heat, Refrig. Air-Cond. Eng., New York, New York, (1975).
5. R. H. Emerick, "Heating Handbook." McGraw-Hill, New York, New York, 1964.
6. "Specifications for Fuel Oils," ASTM Standard D-396 (latest edition). Amer. Soc. Test. Mater., Philadelphia, Pennsylvania, 1974.
7. "Safety and Performance Standard for the Installation of Residential Boiler and Furnace Units for Distillate Fuel," Standard B-58. Oil-Heat Institute of America, Inc., New York, New York, 1959.
8. "Standard for the Installation of Distillate Fuel Burners for Residential Application," Standard B-59. Oil-Heat Institute of America, Inc., New York, New York, 1959.

9. R. E. Barrett, S. E. Miller, and D. W. Locklin, "Field Investigation of Emissions from Combustion Equipment for Space Heating," EPA-R2-73-084a. U.S. Environmental Protection Agency, Research Triangle Park, North Carolina, 1973.
10. D. P. Howekamp and M. H. Hooper, "Effects of Combustion-Improving Devices on Air Pollutant Emissions from Residential Oil-Fired Furnaces," Proceedings of the New and Improved Oil Burner Workshop. Nat. Oil Fuel Inst., Inc., New York, New York, 1969.
11. D. P. Howekamp, "Flame Retention-Effects on Air Pollution," presented at the 9th Annu. Meet. Nat. Oil Fuel Inst., Inc., New York, New York, 1970.
12. D. W. De Werth, "An Investigation of Emissions from Domestic Natural Gas-Fired Appliances," Proceedings, How Significant are Residential Combustion Emissions, Publ. SP-8, pp. 42–58. Air Pollut. Contr. Ass., Pittsburgh, Pennsylvania, 1974.
13. "Standard Method of Test for API Gravity of Petroleum Products," ASTM Standard D 287-67. Amer. Soc. Test. Mater., Philadelphia, Pennsylvania, 1974.
14. W. Bartok, A. R. Crawford, A. R. Cunningham, H. J. Hall, E. H. Manny, and A. Skopp, "Systems Study of Nitrogen Oxide Control Methods for Stationary Sources," Vol. II, Report No. PB-192789. U.S. Dept. of Commerce, Nat. Tech. Inform. Serv., Springfield, Virginia (1969).
15. G. A. Cato, H. J. Buening, C. C. De Vivo, B. G. Morton, and J. M. Robinson, "Field Testing: Application of Combustion Modification to Control Pollutant Emissions from Industrial Boilers—Phase I," EPA-650/2-74-078-a. U.S. Environmental Protection Agency, Research Triangle Park, North Carolina, 1974.
16. R. D. Hangebrauck, D. J. von Lehmden, and J. E. Meeker, "Sources of Polynuclear Hydrocarbons in the Atmosphere," Publ. No. 999-AP-33. Nat. Air Pollut. Contr. Admin., U.S. Dept. of Health, Education, and Welfare, Washington, D.C., 1967.
17. W. S. Smith and C. W. Gruber, "Atmospheric Emissions from Coal Combustion—An Inventory Guide," Publ. 999-AP-24. Public Health Service, U.S. Dept. of Health, Education, and Welfare, Cincinnati, Ohio, 1966.
18. R. P. Hangebrauck, D. J. von Lehmden, and J. E. Meeker, *J. Air Pollut. Contr. Ass.* **14,** 267–278 (1964).
19. S. T. Cuffe and R. W. Gerstle, "Summary of Emissions from Coal-Fired Power Plants." Amer. Ind. Hyg. Ass., Akron, Ohio, 1965.
20. "Applications of Overfire Jets to Prevent Smoke from Stationary Plants." Bituminous Coal Research, Pittsburgh, Pennsylvania, 1944 (revised 1957).
21. W. C. Holton and R. B. Engdahl, *Trans. ASME* **74,** 207–215 (1952).
22. C. Morrow, W. C. Holton, and H. L. Wagner, *Trans. ASME* **75,** 1363 (1953).
23. P. H. Hardie and W. S. Cooper, *Trans. ASME* **56,** 833–849 (1953).
24. A. Kuhlman, *Staub* **24,** 121–131 (1964).
25. E. J. Boer and C. W. Porterfield, "Dust Emissions from Small Spreader-Stoker-Fired Boilers," ASME, Paper 53-S-26 (unpublished) (available from Engineering Societies Library, United Engineering Center, New York, New York).
26. A. C. Stern, *Combustion* **5,** 35–47 (1933).
27. C. E. Miller, *Proc. Midwest Power Conf.* **9,** 97 (1949).
28. R. B. Engdahl and J. H. Stang, *Nat. Eng.* (May, 1947).
29. B. A. Landry and R. A. Sherman, *Trans. ASME* **72,** 9–17 (1950).
30. E. D. Benton and R. B. Engdahl, *Trans. ASME* **69,** 35 (1947).
31. W. Gumz, *Combustion* **22,** 39–48 (1951).
32. M. P. Heap, T. M. Lowes, R. Walmsley, and H. Bartelds, "Burner Design Prin-

ciples for Minimum $NO_x$ Emissions," Doc. No. K20/a/67. Int. Flame Res. Found., IJmuiden, Netherlands, 1973.

33. T. M. Lowes, M. P. Heap, and R. B. Smith, "Reduction of Pollution by Burner Design," Doc. No. K20/a/74. Int. Flame Res. Found., IJmuiden, Netherlands, 1974.
34. D. R. Shoffstall and D. H. Larson, "Aerodynamic Control of Nitrogen Oxides and Other Pollutants from Fossil Fuel Combustion," EPA-650/2-73-033a. U.S. Environmental Protection Agency, Research Triangle Park, North Carolina, 1973.
35. P. W. Cooper, R. Kamo, C. J. Marek, and C. W. Solbrig, "Recirculation and Fuel–Air Mixing as Related to Oil-Burner Design," Publ. 1723. Amer. Petrol. Inst. Washington, D.C., 1964.
36. R. L. Andrews, C. W. Siegmund, and D. G. Levine, "Effect of Flue Gas Recirculation on Emissions from Heating Oil Combustion," Paper 68-21, 61st Annu. Meet., St. Paul, Minnesota. Air Pollut. Contr. Ass., Pittsburgh, Pennsylvania, 1968.
37. L. J. Muzio, R. P. Wilson, Jr., and C. McComis, "Package Boiler Flame Modifications for Reducing Nitric Oxide Emissions—Phase II of III," EPA-R2-73-292-b. U.S. Environmental Protection Agency, Research Triangle Park, North Carolina, 1974.
38. G. A. Cato, H. J. Buening, C. C. DeVivo, B. G. Morton, and J. M. Robinson, "Field Testing: Application of Combustion Modifications to Control Pollutant Emissions from Industrial Boilers—Phase I," EPA-650/2-74-078-a. U.S. Environmental Protection Agency, Research Triangle Park, North Carolina, 1974.
39. W. Bartok, A. R. Crawford, and G. J. Piegari, "Systematic Field Study of $NO_x$ Emission Control Methods for Utility Boilers," APTD-1163. U.S. Environmental Protection Agency, Research Triangle Park, North Carolina, 1971.
40. A. R. Crawford, E. H. Manny, and W. Bartok, "Field Testing: Application of Combustion Modifications to Control $NO_x$ Emissions from Utility Boilers," EPA-650/2-74-066. U.S. Environmental Protection Agency, Research Triangle Park, North Carolina, 1974.
41. N. J. Friswell, *in* "Emissions from Continuous Combustion Systems" (W. Cornelius and W. G. Agnew, eds.), p. 170–172. Plenum, New York, New York, 1972.
42. G. B. Martin, D. W. Pershing, and E. E. Berkau, "Effects of Fuel Additives on Air Pollutant Emissions from Distillate-Oil-Fired Furnaces," Publ. AP-87. Office of Air Programs, U.S. Environmental Protection Agency, Research Triangle Park, North Carolina, 1971.
43. D. W. Pershing, G. B. Martin, E. E. Berkau, and R. E. Hall, "Effectiveness of Selected Fuel Additives in Controlling Pollution Emissions from Residual-Oil-Fired Boilers," EPA-650/2-73-031. U.S. Environmental Protection Agency, Research Triangle Park, North Carolina, 1973.
44. R. E. Hall, "The Effect of Water/Distillate Oil Emulsions on Pollutants and Efficiency of Residential and Commercial Heating Systems," Paper 75-09.4, 68th Annu. Meet., Boston, Massachusetts. Air Pollut. Contr. Ass., Pittsburgh, Pennsylvania, 1975.
45. R. D. Grammar, H. H. Krause, and D. W. Locklin, "The Effect of Additives in Reducing Particulate Emissions from Residual Oil Combustion," Paper 75-WA/CD-7, Annu. Meet. Amer. Soc. Mech. Eng., New York, New York, 1975.

# 12

## Power Generation

F. E. Gartrell

## I. Electric Power Demand

Ever-increasing energy use has been indispensable to developing modern society. The substitution of manufactured energy for human and animal energy has led to widespread availability of goods and services. The quality of life of the peoples of the world depends increasingly upon the availability of large amounts of energy in useful form and at reasonable cost. In developed countries this has resulted in an energy-dependent society which demands increasing quantities of manufactured energy to meet the needs of population growth and changing per capita demands for goods and services. In developing countries, progress toward national goals of adequate food supplies, education and health services, improved housing, and economic self-sufficiency is geared closely to expansion of energy supplies. Thus, both developed and developing countries have expanding energy requirements.

Because of its utility, convenience, and environmental advantages at point of use, electric energy is an essential and major component of the worldwide expanding energy requirement. Many estimates have been made of world electrical generating capacity (Fig. 1) (*1*). As indicated by Table I and Figure 2 (*2*), a phenomenal growth in electrical generat-

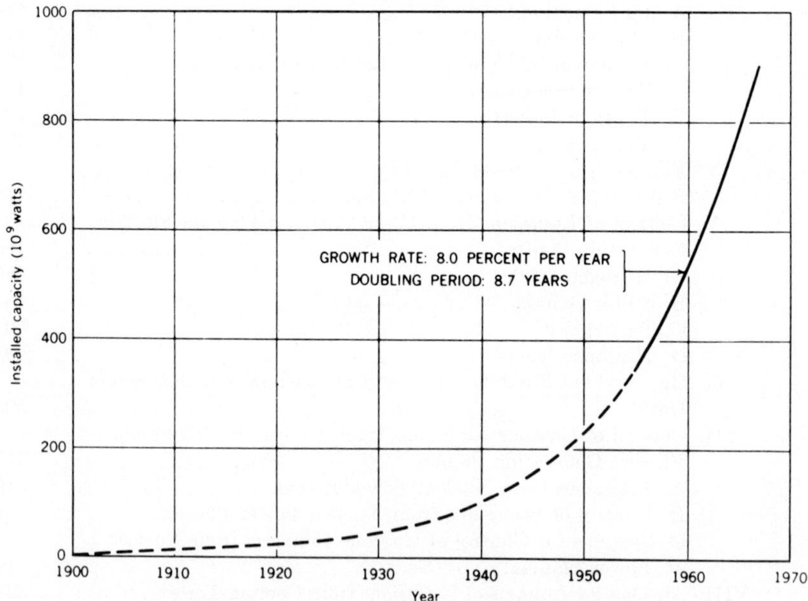


Figure 1. World electrical generating capacity (*1*).

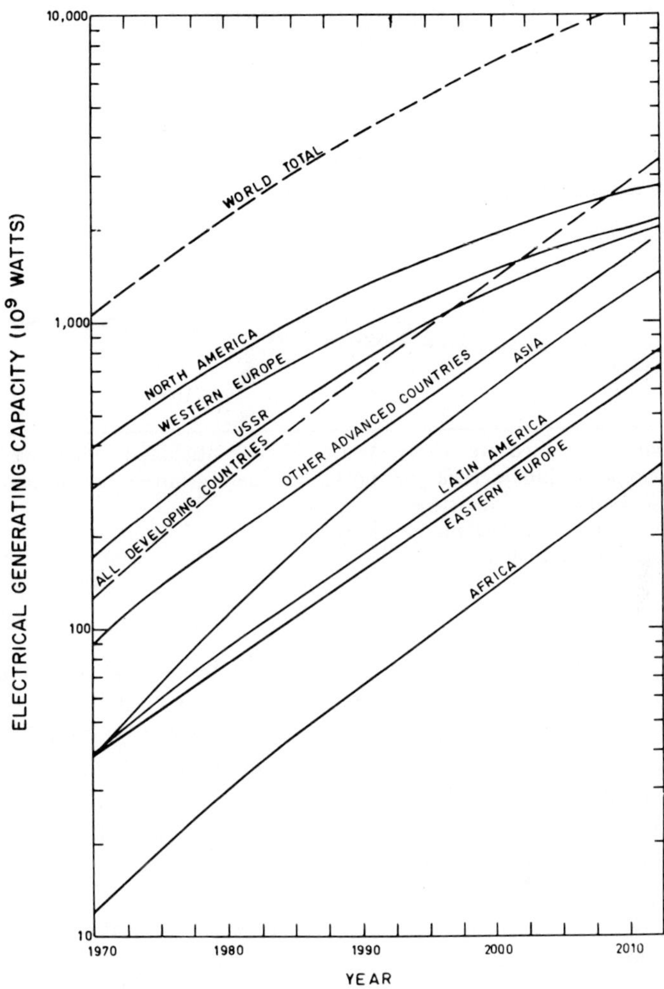


Figure 2. Projected growth of electrical generating capacity (2).

ing capacity has occurred in the past several decades and is expected to continue for the immediately foreseeable future.

## II. Meeting the Demand

On the basis of present technology of energy conversion systems, electric energy demands for the foreseeable future will be met primarily by power plants based on fossil fuels and nuclear fission (*3*). Gas- and oil-

**Table I Expected Electrical Capacity ($10^9$ watts) for Period 1970–1985 (2)**

| | *Developing regions*[a] *(total)* | *Advanced regions* | | | | *World*[c] *total* |
|---|---|---|---|---|---|---|
| | | *North America* | *Western Europe* | *USSR* | *Other*[b] *advanced* | |
| 1970 Population (millions of people) | 1845 | 230 | 345 | 240 | 150 | 2810 |
| 1970: Capacity | 0.5 | 10 | 10.5 | 1.7 | 1.3 | 24 |
| 1975: Capacity | 4.4 | 65 | 38 | 9 | 7 | 123 |
| 1980: Capacity | 22.5 | 155 | 100 | 43 | 24 | 345 |
| 1985: Capacity | 52.5 | 320 | 210 | 160 | 65 | 810 |

[a] The regions designated are Africa, Asia (excluding Mainland China and Turkey), Eastern Europe (including Turkey), and Latin America, but do not necessarily conform to standard political divisions.
[b] Australia, Japan, New Zealand, South Africa.
[c] Excluding Mainland China.

fired turbines and diesel-powered generating units will play important roles in power systems, but will make only minor contributions to total generation required. Direct conversion of solar energy is the only significant long-range alternative to nuclear power. However, there appears to be no economically feasible concept yet available for substantially tapping that continuous supply of energy. Prior to the advent of nuclear power, the largest source of energy for power production was fossil fuels. Hydroelectric power amounted to only 2.4% of that contributed in 1965 by coal, lignite, and crude oil (*2*). Of the world's total potential water power capacity only 8.5% is developed at present, and this principally in industrialized areas of North America (23% developed), Western Europe (57% developed), and Far East (48% developed). Thus, water power is not expected to play a significant role in meeting future power demands, except possibly in Africa and South America where a large potential water-power capacity is relatively undeveloped. Tidal energy and geothermal energy are expected to play only a minor role in meeting the world's electric energy demands.

While the pattern of development may vary in different parts of the world, the projected electric power production by primary source in the United States for the period 1950–2000 shown in Figure 3 illustrates the rapidly expanding role of nuclear power expected in developed countries (*4*).

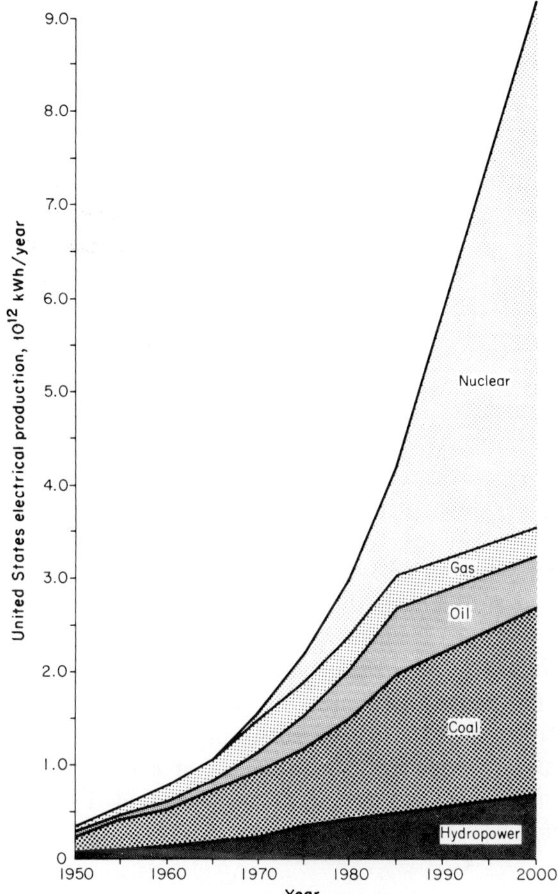


Figure 3. Net United States production of electricity by primary source, 1950–2000 (*4*).

Because of the stringent environmental controls that have been imposed from the beginning of nuclear power plant development, control of air pollution from fossil fuel-fired generating plants is the major air pollution problem associated with electric power production, with gas- and oil-fired plants generally presenting much less of a problem than coal-fired plants.

In the following sections control of air pollution from fossil fuel-fired power plants, including gas and oil turbine and diesel-powered units, and from nuclear power plants will be reviewed. Also, impacts of condenser cooling systems on air quality, common to both types of plants will be discussed.

## III. Fossil Fuel-Fired Steam Electric Power Plants

### A. Fuel Burning Systems

Basic to the planning, design, operation, and maintenance of air pollution control systems for fossil fuel-fired power generating plants is a knowledge of the engineering features of the plant itself. Air pollution control systems are required to prevent adverse effects on air quality from the discharge of waste products of combustion to the atmosphere. While some power generating plants, such as those with gas turbine or diesel-powered generating units, convert chemical energy of the fuel directly to mechanical energy, most utilize fuel burning systems to generate steam for power generation. The descriptions of fuel burning systems and boiler types which follow are from "Combustion Engineering" (*5*) except the information on cyclone furnaces, which is from "Steam, Its Generation and Use" (*6*).

The primary function of fuel burning systems in the process of steam generation is to provide controlled efficient conversion of the chemical energy of the fuel into heat energy which is then transferred to the heat absorbing surfaces of the steam generator.

Fuel burning systems function to do this by introducing the fuel and air for combustion, mixing these reactants, igniting the mixture, and distributing the flame envelope and the products of combustion. An ideal fuel burning system fulfilling these functions would have the following characteristics:

(a) No excess oxygen or unburned fuel in the end products of combustion

(b) A low rate of auxiliary ignition energy input to initiate and sustain continuity of the combustion reaction

(c) A satisfactory and economical reaction rate between the fuel and oxygen

(d) A compatible and effective method of handling and disposing of the solid impurities introduced with the fuel

(e) Uniform distribution of the product weight and temperature in relation to the parallel circuits of heat absorbing surface

(f) A wide and stable firing range

(g) Fast response to demands for a change in firing rate

(h) A high degree of equipment availability with low maintenance requirements

A fuel burning system prepares fuel and air for combustion, properly directs their introduction into the furnace, provides conditions for stable ignition and combustion and suitably removes the products of combustion from the furnace.

The fuel burning system should function so that the fuel and air input is ignited continuously and immediately upon its entry into the furnace. The total fuel burning system required to do this consists of *subsystems* for air handling, fuel handling, ignition, and combustion product removal plus the main burners and the boiler furnace.

## 1. Air Handling

This subsystem supplies properly prepared air to the main burners on a continuous and uninterrupted basis. It should be capable of providing any required air–fuel ratio over the entire operating range of the fuel burning system.

## 2. Fuel Handling

This subsystem supplies properly prepared uncontaminated fuel to the main burners on a continuous and uninterrupted basis. It should be capable of providing any required air–fuel ratio over the entire operating range of the fuel burning system.

## 3. Ignition

This subsystem provides initial ignition of any fuel input and air–fuel ratio over the entire operating range of the fuel burning system. Its primary function is to ignite the furnace input continously and immediately.

## 4. Main Burners

These are designed and constructed so as to be capable of properly regulating and directing the flow of air and fuel into the furnace to provide stable combustion. The main burners function to ensure self-ignition of the furnace input on a continuous and immediate basis.

## 5. Boiler Furnace

In terms of fuel burning systems and subsystems, the boiler furnace is designed and constructed so as to promote a flow pattern of incoming fuel and air such that stable combustion can be achieved over the entire operating range.

### 6. Combustion Product Removal

This subsystem must be capable of removing furnace gases over the entire operating range of the fuel burning system while maintaining furnace pressures within design limitations. A primary function is to remove inert combustion products so that the furnace fuel and air input can be continuously and immediately ignited. The capacity of this subsystem must allow for removal of combustion products of the fuel being used or else it can impose a limitation on the rate of furnace input.

## *B. The Combustion Process*

The rate and degree of completion of a chemical reaction such as the combustion process are importantly influenced by temperature, concentration, preparation, and distribution of the reactants, catalysts, and mechanical turbulence. All of these factors have one effect in common: to increase contacts between molecules of the reactants. Higher temperatures, for instance, increase the velocity of molecular movement and increase the possible rate of reaction. At a given pressure, three basic factors limit the maximum temperature that can be realized for increased opportunity for contact. These are the heat absorbed by the combustion chamber enclosure, by the reactants in bringing them to ignition temperature, and by the nitrogen in the air used as a source of oxygen.

The concentration and distribution of the reactants in a given volume is directly related to the opportunity for contact between interacting molecules. In an atmosphere containing 21% oxygen, the amount present in air, this rate is much less than it would be with 90% oxygen. This is another harmful effect of the presence of nitrogen in the air used as the oxygen supply for oxidizing the fuel. Distribution and concentration of reactants assume even greater importance as the reaction proceeds toward completion. Because of the dilution of reactants by the inert products of combustion, the relative distribution, and hence opportunity for contact, approaches zero.

Preparation of the reactants and mechanical turbulence greatly influences the reaction rate. It is primarily these factors that are available to the designer of a fuel burning system to provide the desirable reaction rate. The beneficial effect of mechanical turbulence on the combustion reaction becomes apparent when it is realized that agitation can improve relative distribution and impart energy permitting greater opportunity for molecular contact. Agitation assumes greater significance when it can be achieved in the later stages of the combustion process at a time when the relative concentration of the reactants is approaching zero.

## C. Fundamental Concepts of Ignition Energy

Before the combustion reaction can occur, the mixture of reactants must, of course, be ignited. The mechanism of establishing ignition of any fuel–air mixture is to elevate its temperature to the point where its rate of oxidation and subsequent release of heat equals or exceeds the rate at which heat is extracted from the products of combustion.

The functional requirement of a fuel burning system is to supply an uninterrupted flammable furnace input and ignite it continuously as fast as it is introduced and immediately upon its appearance in the furnace. Thus, no explosive mixture can accumulate in the furnace, since the furnace input is effectively consumed and rendered inert. *Ignition* takes place when the flammable furnace input is heated above the ignition temperature.

Supplying correct ignition energy for the furnace input is a substantial task. There are many factors which establish the range of combinations of ignition energy quantity, quality, and location that can provide a satisfactory furnace input ignition rate at any instant. Unfortunately, these factors are of rapidly changing value, and what constitutes sufficient ignition energy at one instant may be insufficient the next. From the rate-igniting standpoint, ignition energy is never in excessive supply. The more stable a fire is, the more likely that ignition energy is being supplied in substantial excess of the minimum necessary to maintain input ignition.

Six major factors determine the total ignition energy required.

### 1. Fuel Quality

Low-grade fuels can require substantially more heat (auxiliary and inherent ignition energy) to raise the total fuel mass up to the required ignition temperature than do high-grade fuels.

### 2. Fuel Preparation

If the fuels are not suitably prepared (too coarse or cold, for example), they can require much more ignition heat on a rate basis.

### 3. Air Preparation

If the air is not suitably prepared (too cold or diluted with inert materials, for example), it can require much more ignition energy to bring the total oxygen-bearing mass ($O_2$ + inerts) up to the ignition temperature.

### 4. Burner Product Distribution

If the main burner distributes fuel and air so that a noncombustible product is provided where the auxiliary or inherent ignition energy is still adequate to ignite a combustible furnace input, the best ignition energy will be ineffective.

### 5. Total Fuel–Air Ratio

If the main burner fuel–air ratio is far off from premix flammable ratios, even though the diffusion firing technique for furnace combustion is used, it may require much more ignition energy because the excess ingredient (fuel or air) acts as an inert, consuming heat without being able to react chemically and release heat in return.

### 6. Main Burner Mass Flow Rate

If the main burner mass flow rate is substantial, it will require a comparable ignition energy input rate to provide satisfactory ignition, since the same percentage of the flammable main burner fuel and air mixture must be raised to the ignition temperature regardless of flow rates.

## *D. Burner Design*

In a practical burner and fuel burning system as applied to a boiler, all of the fundamental factors influencing rate and completeness of combustion must be considered in conjunction with obtaining a high degree of heat transfer efficiency.

There are two basic methods of producing a total flow pattern in a combustion chamber to provide successful reactant molecular contacts through mechanical turbulence. One method of providing multiple flame envelopes is to divide and distribute fuel and air into a multiplicity of similar streams, and treat each pair of fuel streams and airstreams independently of the others. The second method, resulting in a single flame envelope, is to provide interaction between all streams of air and fuel introduced into the combustion chamber.

The first of these methods requires accurate subdivision of the total fuel and air supplied to a common combustion chamber and provides little opportunity for sustained mechanical mixing or turbulence throughout the entire combustion chamber volume. The necessity of obtaining and sustaining good distribution of fuel and air is a technical as well as an operating problem. However, it permits more time for contact between all fuel and air molecules, and mechanical turbulence is sustained throughout the entire volume of the combustion chamber. This avoids stringent distribution accuracy requirements.

## Vertically Fired System

This type of fuel burning system was developed initially for pulverized coal before the advent of water-cooled combustion chamber-wall surfaces. Pulverized coal suspended in the primary air is injected downward in the furnace, and secondary air is supplied through furnace wall openings below the nozzles to produce a U-shaped vertical pulverized coal flame. Combustion air ducts are provided around the coal nozzles to provide tertiary air. Because a large percentage of the total combustion air is withheld from the fuel stream until it projects well down into the furnace, this arrangement has the advantage that the fuel stream is heated separately from the main body of combustion air, providing good ignition stability. The delayed introduction of the main body of combustion air provides needed turbulence at a point in the fuel stream where partial dilution has taken place. This type of firing is particularly well suited for solid fuels which are difficult to ignite, such as those with less than 15% volatile matter. Physical factors limited the maximum capacity of each fuel nozzle to approximately 50,000,000 Btu/hour, and the furnace width becomes too great for high capacity central station boilers.

## Horizontally Fired System

As the name implies, this system utilizes burners which produce horizontal flames in the furnace. This type of burner was developed to provide a method of introducing fuel and air to obtain higher capacities per fuel nozzle than was practical with the vertically fired system. With this type burner, the main fuel supply is introduced in the burner, and the air is introduced into a throat area with a rotation which produces a secondary flow in the central zone. The fuel concentrated in the vortex has a relatively long residence time to reach ignition temperature with a relatively low mass of air to be simultaneously heated. While the mechanical turbulence in the immediate area of the throat is intensive, it does not persist for any appreciable distance out into the vastly greater volume of the combustion chamber. Rapid ignition and combustion must be obtained to utilize this turbulence before reduction in reactant concentration occurs. With this type of firing system it is impossible for the fuel introduced in one burner to mix suitably with the air from another burner. The proportioning of fuel and air to each nozzle and air assembly must be uniform if opportunity for contact and uniform product mass and temperature in relation to the combustion chamber are to be established. Because of the limited time and volume through which mechanical turbulence is established and maintained, the peripheral distribution of air and fuel at each burner and the throat area is of extreme importance. Peripheral distribution of air is accomplished by the yoke adjustment of the ring dampers, by the area reduction from the inlet vane area to the throat area, and by centering the fuel introduction nozzle in the throat opening.

Peripheral fuel distribution of liquid and gaseous fuels is usually accomplished through the geometry of the burner tip. In many burners this is done by arranging the tip orifices uniformly around the circumference of a circle on the tip or nozzle. Mechanical oil atomizers with a single tip orifice utilize an internal design that imparts a high angular velocity spin to the fluid which provides the quality of atomization and distribution required. In the case of pulverized coal, which is transported in mechanical suspension by a portion of the combustion air, adjustable vanes in the fuel nozzle promote good peripheral fuel distribution.

## Tangentially Fired System

This burner system was developed to incorporate the advantages of the single flame envelope technique previously described for an extremely wide range of fuels. With this system, the fuel is admitted at the corners of the combustion chamber through alternate compartments. Distribution dampers proportion the air to the individual fuel and air compartments. The fuel streams and airstreams from each corner are aimed tangent to the circumference of a circle in the center of the furnace. Control over the furnace flow pattern allows adjustment of the rate at which air is supplied to the fuel as well as the energy then available to mix these reactants. Fuel streams are essentially injected into the furnace between airstreams. Control of the input distribution and velocity provides control over the furnace flow pattern and thus the combustion process. The fuel is normally fired on a level basis so that proper interaction of the separate streams is obtained. However, selective furnace utilization and steam temperature control can be achieved by tilting tangential burners.

## Cyclone Furnaces

When coal is used to fuel the previously described burning systems, grinding to produce a fine power, "pulverized coal" (70% passing a 200-mesh screen), is required. Pulverized-coal firing, introduced in the 1920's, is the best way to burn many types of coal, particularly the higher grades and ranks. However, the cyclone furnace introduced some 20 years later is particularly suitable for the lower grades and ranks of high-ash, low-fusion temperature coal. The cyclone furnace is a water-cooled horizontal cylinder in which fuel is fired, heat is released at extremely high rates, and combustion is completed. Its water-cooled surfaces are studded, and covered with refractory over most of their area. Coal crushed in a simple crusher, so that approximately 95% will pass a 4-mesh screen, is introduced into the burner end of the cyclone. About 20% of the combustion air, termed primary air, also enters the burner tangentially and imparts a whirling motion to the incoming coal. Secondary air with a velocity of approximately 300 ft per second (fps) is admitted in the same direction tangentially at the roof of the main barrel of the cyclone and imparts a further whirling or centrifugal action to the coal particles. A small amount of air (up to about 5%) is admitted at the center of the burner. This is known as "tertiary" air. The combustible is burned from the fuel at high heat release rates and gas temperatures exceeding 3000°F are developed. These temperatures are sufficiently high to melt the ash into a liquid slag, which forms a layer on the walls of the cyclone. The incoming coal particles (except for a few fines that are burned in suspension) are thrown to the walls by centrifugal force, held in the slag, and scrubbed by the high-velocity tangential secondary air. Thus the air required to burn the coal is quickly supplied, and the products of combustion are rapidly removed.

The gaseous products of combustion are discharged through the water-cooled reentrant throat of the cyclone into the gas-cooling boiler furnace. Molten slag continually drains away from the burner end and discharges through a slag tap opening in the boiler furnace, from which it is tapped into a slag tank, solidified, and disintegrated for disposal.

By this method of combustion the fuel is burned quickly and completely in the small cyclone chamber, and the boiler furnace is used only for cooling the flue gases. Most of the ash is retained as a liquid slag and tapped into the slag tank under the boiler furnace. Thus, the quantity of fly ash is low and its particle size so fine that erosion of boiler heating surfaces is not experienced even at high gas velocities.

## Additional Burner Functions

In practice a burner design may be utilized to accomplish other ends aside from efficient combustion of fuel. For instance, the burner system may be used to control slag fluidity or steam temperature in an operating steam generator. Tilting tangenital burners permit selective utilization of the furnace heat absorbing surfaces. Another example of regulation of combustion chamber effectiveness is the use of gas recirculation at various points and in variable quantities in proper relation to the overall mass flow pattern.

### E. Types of Boilers

In modern steam generators, various components are arranged to absorb heat efficiently from the products of combustion. These components are usually described as boiler, superheater, reheater, economizer, and air preheater. The term "boiler" may be used to refer to the overall steam-generating unit, with the term "boiler surface" being used to denote the actual parts of the circulatory system which are in contact with hot gases on one side and water or a mixture of water and steam on the other.

Boilers may be classified as shell, fire-tube, and water-tube types. Modern boilers are the water-tube type. Water-tube boilers are of the drum type or "once through" pressure type. In the latter type, there is no recirculation of water within the unit, and for this reason, a drum is not required to separate water from the steam. The water-tube construction facilitates obtaining greater boiler capacity and the use of higher pressure. High pressure and high temperature single-boiler, single-turbine combinations have been generally adopted for steam electric generating plants. There are a few installations where two approximately half-size boilers per turbine are provided in an effort to avoid the very large single boilers which otherwise would be required. Since heat rate, investment, and labor costs decrease as size increases, economic incentives have led to increasingly large steam electric generating units. For example, the largest fossil-fired unit added to United States systems in 1960 was 500 megawatts and in 1973, 1300 megawatts.

### F. Types of Fuels

Steam electric generating plants are designed to burn the fuel or fuels expected to be available at the particular site at reasonable cost for the life of the plant. For existing plants, changes in fuel or types of fuel for air pollution control or other reasons may require major plant modification or in some cases might be completely impractical.

The usual fuels burned in central generating stations are bituminous coal, lignite, natural gas, and oil. Characteristics of these fuels are described in Chapter 10, this volume. Coal is the most complicated and troublesome of major fuels. Its use involves storage and handling facilities, preparation before firing using crushers or pulverizers, ash collectors, soot blowers, and ash disposal equipment. Since there is a wide variation in properties of coal and ash, the steam generating unit must be designed for optimum performance burning the particular coals available. Natural gas has the fewest design restrictions since it is clean and easy to burn. If only natural gas is burned, fuel storage facilities and ash handling equipment are unnecessary; soot blowers can be omitted; and dust collectors are not needed. Fuel oil has many of the desirable features of natural gas including ease of handling and elimination of ash handling equipment. It does require storage, heating, and pumping facilities. Oils with high sulfur and vanadium content can cause troublesome deposits on surfaces throughout a steam generating unit. These deposits may be minimized by arranging the heating surfaces for optimum cleaning by soot blowers. Air heater protection devices (a steam coil or hot water coil) are required to prevent condensation of gases and acid attack on the gas side of the air heater surfaces.

## IV. Emissions from Fossil Fuel-Fired Steam Electric Generating Plants

The products of combustion most commonly released by fossil fuel-fired steam electric generating plants are solid incombustible ash particles of various types and sizes, and gases which include carbon dioxide, carbon monoxide, water vapor, sulfur dioxide, and nitrogen oxides. Of these materials, except for attention currently being directed to nitrogen oxides, the particulates and sulfur compounds are the compounds requiring major attention in control of air pollution from thermal power stations. Trace quantities of uranium and thorium and their products of radioactive decay are released in fly ash from large fossil fueled steam electric stations, but are of negligible public health significance (*7*). Other trace substances such as mercury, arsenic, copper, iron, lead, etc., and polycyclic organic matter (products of incomplete combustion of organic fuels) also may be present. The air pollution significance, if any, of the quantities of these substances found in stack gases of modern power plants is yet to be determined.

The rate at which pollutants are emitted from a power plant burning fossil fuel depends upon the type, quality, and quantity of fuel burned, the design of the boiler and furnace, and the combustion system used. Average emission rates from conventional equipment by type of fossil fuel are listed in Table II (*8*). Nitrogen oxides and particulate emission are affected greatly by boiler and furnace design and the combustion system used, whereas sulfur oxide emissions are not significantly influenced thereby.

**Table II Emission of Pollutants at Electric Power Plants Average Rate by Type of Fossil Fuel[a] (8)**

| | *Coal* (*lb/ton*) | *Oil* (*lb/1000 gal*) | *Gas* (*lb/$10^6$ $ft^3$*) |
|---|---|---|---|
| Nitrogen dioxide | 20 | 104 | 390 |
| Sulfur dioxide | $38S^b$ | $157S^b$ | 0.4 |
| Sulfur trioxide | $0.6S^b$ | $2.5S^b$ | Negligible |
| Carbon monoxide | 0.5 | 0.04 | Negligible |
| Hydrocarbons as methane | 0.2 | 3.2 | Negligible |
| Aldehydes as formaldehyde | 0.005 | 0.6 | 1 |
| Particulate matter | $17A(1\text{-}E^c)$ | $10(1\text{-}E)^c$ | 15 |

[a] Btu Equivalent: 25,000,000 per ton; 150,000,000 per 1000 gal; 1,044,000,000 per $10^6$ ft³. [Btu (British thermal unit) is the quantity of heat required to raise the temperature of 1 lb of water 1°F at or near its point of maximum density. Btu × 0.2520 = kilogram-calories.]

[b] $S$ equals percent sulfur in the fuel. For example, coal with 2% sulfur will emit 76 lb of $SO_2$ and 1.2 lb of $SO_3$ per ton of coal burned, assuming no removal of $SO_x$ from the flue gases. In coal-fired boilers much of the $SO_3$ is removed with the ash.

[c] Emissions of fly ash are a function of the ash content of the fuel, type of furnace, and efficiency of the control equipment. For a dry bottom, pulverized coal unit, fly ash emissions in pounds per ton of coal burned would be $17A(1\text{-}E)$, where $A$ is the ash content of the coal expressed in percent and $E$ is the efficiency of the precipitator expressed as a decimal. For coal having an ash content of 10% and a precipitator operating at an efficiency of 97%, the rate of emissions would be $17 \times 10\ (1 - 0.97) = 5.1$ lb per ton of coal.

## V. Control of Emissions from Fossil Fuel-Fired Steam Electric Generating Plants

### A. Nitrogen Oxides

Nitrogen oxides ($NO_x$) are formed by oxidation of atmospheric nitrogen at high temperatures in power plant furnaces and by partial combustion of the nitrogenous compounds contained in the fuel. Using the average emission rates presented in Table II, the relative $NO_x$ emissions in pounds per million Btu are 0.37 for natural gas, 0.69 for oil, and 0.80 for coal. Allowable emissions under the new source performance standards for stationary sources promulgated by the United States Environmental Protection Agency (EPA) (December, 1971) are 0.20 lb per million Btu or 175 ppm dry at 3% excess $O_2$ when gaseous fuel is burned; 0.30 lb per million Btu or 230 ppm dry at 3% excess $O_2$ for liquid fuel; and 0.70 lb per million Btu or approximately 500 ppm dry at 3% excess $O_2$ for solid fossil fuel (except lignite). (See Chapter 12, Volume V.)

For a specific type fuel the formation of nitrogen oxides in the combustion zone of a furnace depends not only on flame temperature, but also on excess air supplied to support combustion; the position of burner flame relative to furnace walls, and the configuration of the furnace. Investigations for control of emissions of nitrogen oxides have been directed principally to modifications in combustion processes. Tangentially fired units consistently have significantly lower emissions of $NO_x$ than wall-fired and cyclone units for all types of fuel.

Blakeslee and Burbach have reported on tests on 45 tangentially fired units conducted as part of a continuing program to develop analytical methods for prediction and reduction of $NO_x$ emissions (*9*). The units tested represent a nearly equal number of gas-, oil-, and coal-fired units, covering the size range, from approximately 50 to 565 megawatts. Of the 14 gas-fired units tested, only five met the standard. There is a discernible trend toward higher emissions in larger units, most likely due to higher operating temperatures. Reduction to levels below the standard was achieved on all of the other nine units by shutting off fuel to the upper level of the fuel compartments while continuing to supply air through the upper fuel and air compartments. This type of operation is described as "overfire air simulation," and is generally referred to as two-stage combustion. This was accomplished without sacrifice of boiler efficiency. Flue gas recirculation was also demonstrated as an effective means of controlling $NO_x$ emissions with gas firing.

Of the 16 oil-fired units tested, nine met the standard without modification. Of these, seven are furnaces designed for coal firing but converted to oil burning. It appeared that with tangential firing of oil, $NO_x$ is produced primarily by conversion of fuel nitrogen. Thus, high nitrogen fuels will present special $NO_x$ control problems. Combustion modifications such as overfire air, gas recirculation through the windbox, and low excess air operation were effective in reducing $NO_x$ emissions. No reduction in boiler efficiency was reported. There was no discernible trend showing increased $NO_x$ emissions with larger units as was the case with natural gas firing.

Of the 16 coal-fired units reported, ten met the standard under normal operation. Overfire simulation by taking the pulverizer supplying the upper elevation of fuel nozzles out of service reduced $NO_x$ emissions on all units tested under these conditions. The amount of reduction is primarily dependent on the reduction in air supplied to the fuel ignition zone. On units not designed to operate with one pulverizer out of service, it was necessary to reduce load when operating in this mode. Fuel types represented in the tests were lignite, and western, midwestern, and eastern United States bituminous coals. The slagging tendencies of the individual coals coupled with furnace heat rates appear to be more closely related to $NO_x$ emissions than general coal types. The complex relationships existing among the operating variables in coal-fired furnaces preclude establishing any trend of $NO_x$ emission with increasing fuel nitrogen from the data presently available.

With the expectation that $NO_x$ emission control would be required, many tangentially fired coal, oil, and gas units have been designed with overfire air systems. Gas recirculation through the windbox has also been included on a number of gas- and oil-fired designs. Tangentially fired units are currently being designed to meet the EPA $NO_x$ standards. On oil and gas fired units, larger furnaces and both gas recirculation through the windbox and overfire air systems are used to assure meeting the standard in day-to-day operation. The overfire air system is normally designed for 15% of the total air requirement and is supplied only with air.

For coal-fired units overfire systems alone are used. Overfire air is provided by adding two air compartments at the top of the windbox in each corner of the furnace. The compartments are normally sized for 15% of the windbox airflow and have manual dampers for airflow control and have manual nozzle tilt control. The position of the overfire air dampers and nozzle tilt are optimized after initial operation to give the lowest $NO_x$ emissions consistent with satisfactory furnace performance. The designs are necessarily based on tests of relatively short duration and overall effects of long-term operation have not been evaluated. Burning coals at low excess air or with overfire air, which reduces air in the primary combustion zone, may result in excessive slag accumulation on furnace walls. There is insufficient experience with overfire air on large furnaces burning a wide variety of coals to predict ash deposits on walls, carbon loss, etc. Operational procedures using overfire air systems in coal firing have to be worked out on each unit after startup.

From tests conducted on 16 different boilers at Tennessee Valley Authority (TVA) power plants, four under modified combustion conditions and 12 under normal full-load conditions, wall-fired boilers are consistently higher in $NO_x$ emissions than tangentially fired units (*10*). Highest emissions were from cyclone-fired units. The data from the tangentially fired units using coal were consistent with the findings of Blakeslee and Burbach (*9*) referred to previously and indicate that the tangentially fired units tested meet the emission standard without combustion modification. $NO_x$ emission levels under normal full-load operation were relatively constant for tangentially fired units with capacities ranging from 120 to 900 megawatts.

All wall-fired units exceeded the standard, and in some cases, emissions were twice the standard. Significant reductions were demonstrated to be possible for all units tested (except cyclone-fired units) by using a combination of low excess air with staged firing, i.e., operating the top row of burners on air only and shutting down the pulverizer mill supplying coal to the top row. In some cases this resulted in a drop in boiler load because the pulverizers remaining in service could not supply fuel at the required rate for full-load operation. Staged firing resulted in reductions of 24–50% in $NO_x$ emissions. These were short-term tests. Long-term tests to determine whether burning coal under modified combustion conditions has adverse effects on the boiler are needed before the practicality of this approach to $NO_x$ emission control can be demonstrated. Only limited modification in combustion conditions could be achieved on the cyclone unit (rated capacity 704 megawatts) tested. Excess air could not be varied over a wide range and "staged firing" could only be simulated by operating the top cyclone burners under highly fuel-lean conditions.

Load ranged from 545 to 668 megawatts. Reducing load was the only significant control variable. A 20% load reduction reduced $NO_x$ emission by 25%, 1170 to 880 ppm, still greatly in excess of the standard of 500 ppm. The other variables had no significant effect on $NO_x$ emission. At the time of writing this chapter, there are no promising leads as to a practical method for controlling $NO_x$ emissions from coal-fired cyclone boilers to levels required to meet the USEPA standards.

### B. Sulfur Oxides

The sulfur oxides in the stack emissions from fossil fuel-fired power plants are directly proportional to the sulfur content of the fuel. For gas-fired plants the quantity is usually insignificant. Fuel oils used in power plants vary in sulfur content from less than ½% to more than 4% by weight, and coals vary from about ½% to more than 5%. Methods for producing low sulfur fuel oils from high sulfur crude oil and for reducing the sulfur content of coal, and for producing clean fuel (coal gasification and solvent refining) for power plant use are discussed in Chapter 10, this volume. Methods to be discussed here for control of sulfur oxides will be limited to those applicable at the power plant.

Sulfur is present in coals as organic sulfur, sulfate, and pyrite and in oils as sulfides, mercaptans, polysulfides, and thiophenes. A small fraction of the sulfur as pyrite in coals is removed with the rejects from the coal processing equipment. When coal and fuel oils are burned in power plants, 90–95% of the sulfur appears as $SO_2$ and 1–3% as $SO_3$ in the stack gas. The remainder leaves the furnace with the bottom ash or with the collected ash from the stack gas-cleaning equipment. Because of the large capacity of most modern fossil fueled power plants and the high sulfur content of fuel used in many locations, $SO_2$ emissions from many plants exceed 500 tons/day, with some plants on occasion exceeding 2000 tons/day.

## 1. High Stacks

Large power plants built in recent years designed to burn high sulfur fossil fuels have used high stacks, and in some cases supplementary controls have been applied when dispersion conditions are unfavorable to control ground level $SO_2$ to acceptable levels for protection of public health and welfare. The efficacy of high stacks for this purpose has been widely demonstrated. In addition to providing a dependable method for control of ground-level concentrations of $SO_2$, high stacks also provide dispersion of other products of combustion and serve as a reliable backup to mitigate effects of malfunction of air pollution control equipment and variability in fuel quality.

In Great Britain during the late 1950's, as elsewhere, to meet growing electric energy demands, progressively larger fossil fueled power plants with progressively higher stacks were planned (*11*). In 1959, the first 1000-megawatt power plant was commissioned at High Marnham and this was equipped with two stacks 450 ft high. By 1960, the first of a series of nine plants, each containing four 500-megawatt units, was being planned and attention was directed to single stacks serving multiunits and to multiflue stacks. Plant capacities and stack arrangements for some of the large power plants planned in Britain during the period 1960–1964 are presented in Table III (*11*). During the period October 1963–March 1965 a comprehensive continuous ambient $SO_2$ monitoring program was conducted at the High Marnham station. For the 12-month period October 1963–September 1964 the plant burned 2.36 million tons of coal of about $1\frac{1}{2}$% sulfur content and emitted over 65,000 tons of $SO_2$. There are no other nearby single sources of $SO_2$ of any magnitude. The 450-ft stacks were demonstrably effective in dispersion of $SO_2$ from the power plant. The overall effect of the plant on ambient $SO_2$ concentrations was equivalent to adding only 0.1 to 0.2 pphm to the average hourly mean background concentration. The mean background concentrations of 3–5 pphm recorded were consistent with other long-term average measurements in rural areas in Britain. The High Marnham studies suggested that most $SO_2$ recorded originated in cities and industrial areas 20–30

**Table III Chimney Stack Arrangements at Recent Fossil Fueled Stations of the Central Electricity Generating Board, Great Britain** (*11*)

| | Installed capacity | | Chimney stacks | | |
|---|---|---|---|---|---|
| *Station and year planned (coal-fired unless shown)* | *Unit size (megawatts)* | *Total capacity (megawatts)* | *Number* | *Height (ft)* | *Number of internal flues per stack* |
| 1960 West Burton "A" | 500 | 2000 | 2 | 600 | 1 |
| Tilbury "B" | 350 | 1400 | 2 | 550 | 1 |
| 1961 Ferrybridge "C" | 500 | 2000 | 2 | 650 | 1 |
| Eggborough | 500 | 2000 | 1 | 650 | 4 |
| Fawley (Oil) | 500 | 2000 | 1 | 650 | 4 |
| 1962 Kingsnorth (Coal/Oil) | 500 | 2000 | 1 | 650 | 4 |
| Ratcliffe | 500 | 2000 | 1 | 650 | 4 |
| Ironbridge "B" | 500 | 1000 | 1 | 650 | 1 |
| Aberthaw "B" | 500 | 1500 | 1 | 650 | 3 |
| Fiddler's Ferry | 500 | 2000 | 1 | 650 | 4 |
| 1963 Cottam | 500 | 2000 | 1 | 650 | 4 |
| Didcot | 500 | 2000 | 1 | 650 | 4 |
| Rugelye "B" | 500 | 1000 | 1 | 600 | 1 |
| Pembroke (oil) | 500 | 2000 | 1 | 700 | 4 |
| 1964 Drax | 660 | 3960 | 1 | 850 | 3 |

miles away. The background alternated between long periods of low $SO_2$ pollution and short periods of high pollution which affected all survey gauges simultaneously.

Subsequently similar studies were conducted at Tilbury-Northfleet (1080 megawatts) 1963–1968, High Marnham (1000 megawatts) 1965–1966, West Burton (2000 megawatts) 1966–1969, and Eggborough (2000 megawatts) 1967–1973 (*12*). (This statement on Central Electricity Generating Board experience was prepared at the request of TVA and included in testimony filed by TVA.) All surveys confirm the minimal effect of power plants which typically add no more than 4–5% to the average ambient $SO_2$ levels even in rural areas of low pollution.

The American Electric Power Service Corporation has conducted extensive monitoring of ambient $SO_2$ levels in the vicinity of its large coal-fired power plants some of which burn high sulfur fuel (*13*). At its Muskingum plant, which burns 5% sulfur coal, plans for the addition of a 580 megawatt unit included provision of a single new stack for the existing four units equal in height to the one to be designed for the new unit. Units 1 and 2 (205 megawatts each) were served by individual stacks 273 ft high and units 3 and 4 (215 megawatts each) by a single stack 435 ft high. A height of 825 ft was selected for the two new stacks. Conversion to a single tall stack, from three moderate height stacks, resulted in halving the peak ambient $SO_2$ concentrations for 4 unit plant operation. At the Mitchell plant which was constructed near the Kammer plant, two 800 megawatt units are served by a single 1200-ft stack in contrast to the three 225 megawatt units at the Kammer plant served by two 600-ft stacks. Sulfur in the fuel for both plants averages 3.5 to 4.0%. Based on 19 months of sampling, the addition of the Mitchell plant has not caused an increase in ambient $SO_2$ concentrations, but it is expected that further sampling under a full range of meteorological conditions will possibly show a slight increase.

TVA operates 12 large coal-fired power plants, with one to ten units each, totalling 63 units. As shown in Table IV the units range in size from 60 megawatts to 1300 megawatts, and total plant capacity ranges from 240 to 2600 megawatts. The plants burn over 34 million tons of coal per year, with an average sulfur content of coal on a plant-by-plant basis that ranges from approximately 1 to over 4%.

Beginning with its first large steam electric generating plant, the Johnsonville plant, the first unit of which went into commercial operation in 1951, TVA conducted extensive air pollution studies at each plant. The studies have included operation of particulate and $SO_2$ monitoring networks, full-scale atmospheric dispersion studies, and surveillance of effects of plant emissions on vegetation. Experience was used in planning air pollution control at succeeding plants and for additions to existing plants. As shown in Table IV, concurrent with the increasing unit size and total capacity at TVA's recent plants there has been an increase in stack height to control ambient $SO_2$ concentrations to levels that will protect vegetation. With the few exceptions of some of the older plants which have relatively short stacks, the use of high stacks has been effective in preventing significant effects on vegetation (*14*). As a part of the plan to meet ambient air quality standards for $SO_2$ applicable to the TVA power plants, new, high stacks are to be installed at the Kingston, Shawnee, and Windows Creek plants.

Power plant stack gas dispersion models are discussed in Chapter 9, Vol. I. The progressive trend toward larger units and higher stacks at TVA power plants has been accompanied by a change in the plume dispersion models associated with maximum ground-level concentration of stack gases (*15*). At TVA power plants, maximum concentrations for principal plume dispersion models do not differ significantly for the early small unit plants. However, the coning model was considered the critical model because its frequency was appreciably greater than other models.

**Table IV Major TVA Steam Plants**

| Name | First unit in operation | Unit numbers | Rated capacity: Per unit (megawatts) | Rated capacity: Total plant (megawatts) | Height of stacks (ft) | Fuel burned (FY 1973)[a]: Tons (1000) | Fuel burned (FY 1973)[a]: Average sulfur content (%) |
|---|---|---|---|---|---|---|---|
| Cumberland | 1973 | 1–2 | 1300 | 2600 | 1000 | 647 | 3.6 |
| Paradise | 1963 | 1–2 | 704 | 2558 | 600 | 6668 | 4.2 |
| | | 3 | 1150 | | 800 | | |
| Widows Creek | 1952 | 1–2 | 140.6 | 1978 | 270[b] | 3669 | 3.1 |
| | | 3 | 150 | | 270 | | |
| | | 4 | 140.6 | | 270 | | |
| | | 5–6 | 140.6 | | 270 | | |
| | | 7 | 575 | | 500 | | |
| | | 8 | 550 | | 500 | | |
| Kingston | 1954 | 1–4 | 175 | 1700 | 250[c] | 4282 | 2.2 |
| | | 5–9 | 200 | | 300 | | |
| Shawnee | 1953 | 1 | 150 | 1675 | 250[d] | 4704 | 2.7 |
| | | 2–7 | 175 | | 250 | | |
| | | 8 | 150 | | 250 | | |
| | | 9 | 175 | | 250 | | |
| | | 10 | 150 | | 250 | | |
| Johnsonville | 1951 | 1–4 | 125 | 1485 | 270 | 3192 | 3.7 |
| | | 5–6 | 147 | | 270 | | |
| | | 7–10 | 173 | | 400[e] | | |
| Colbert | 1955 | 1–3 | 200 | 1373 | 300 | 2933 | 4.0 |
| | | 4 | 223 | | 300 | | |
| | | 5 | 550 | | 500 | | |
| Gallatin | 1956 | 1–2 | 300 | 1255 | 500[e] | 3050 | 2.7 |
| | | 3–4 | 327.6 | | 500[e] | | |
| Allen | 1959 | 1–3 | 330 | 990 | 400 | 1554 | 3.2 |
| Bull Run | 1966 | 1 | 900 | 900 | 800 | 2375 | 1.0 |
| John Sevier | 1955 | 1 | 223 | 823 | 350[e] | 1822 | 1.8 |
| | | 2–4 | 200 | | 350[e] | | |
| Watts Bar | 1942 | 1–4 | 60 | 240 | 150 | 518 | 3.8 |

[a] FY: Fiscal year

[b] New high stack or stacks to be installed to serve units 1–6. Height not yet determined.

[c] Two 1000-ft stacks to be installed to serve all 9 units.

[d] Two new high stacks to be installed to serve all 10 units. Height not yet determined.

[e] One stack serves 2 units.

As the size of units and height of stack increased, the maximum surface concentrations from the coning model decreased and the concentrations (relative to the coning model) from the inversion breakup model increased. With the newer, larger units with high stacks the trapping model also referred to as the limited mixing layer dispersion model, became the critical model. For the trapping model, the actual stack height is a relatively minor determinant of surface concentration when there is little or no penetration into the capping layer by the plume. When significant penetration occurs, a corresponding decrease in surface concentration results. Thus, the advantage of high stacks during trapping conditions lies in the enhanced probability that the plume will enter, partially or completely, the layer of stable air aloft. However, because of practical considerations there may be situations, such as the case of very large plants burning high sulfur coal, where it is not feasible to achieve the desired level of control under trapping conditions by high stacks alone.

In the mid 1960's plans were made to add a 1150-megawatt unit to the two existing 704-megawatt units at the Paradise plant. With three unit plant operation and the high sulfur coal to be used, average $SO_2$ emissions at normal full load of 60–70 tons per hour were projected. Although not demonstrated at the time, TVA recognized the possible limitations of stacks to provide effective dispersion of such quantities of $SO_2$ under all meteorological conditions. While the third unit was under construction intensive dispersion studies were undertaken in 1966 to identify the specific meteorological conditions which caused the higher ground-level concentrations of $SO_2$ and to develop precedures that could be applied after the third unit went into operation so maximum ground-level concentrations of $SO_2$ would not exceed levels for two-unit operation. The trapping or limited mixing layer dispersion model proved to be the critical model. Maximum ground-level concentrations from limited mixing layer dispersion at the Paradise plant were found to persist from 2 to 5 hours beginning between 9 a.m. and 11 a.m. local time. These concentrations occur relatively close to the plant, approximately 3 miles. Nine meteorological criteria and parameter values were established to determine the critical atmospheric conditions when plant operations could cause ground concentrations of $SO_2$ to exceed desired control levels. Meteorological input information required for these determinations are vertical temperature profile, vertical wind speed profile, sky condition—percent cloud cover and forecast maximum surface temperature for the day. On-site meteorological data, wind speed, and vertical temperature profiles are taken daily at 7:00–7:30 a.m. local time. These data, along with sky condition and forecast maximum surface temperature, are processed by computer to determine whether all parameters of the nine criteria fall within the critical range of values. If all nine criteria requirements are met, allowable $SO_2$ emission, or load generation, is determined by computer using the equation for limited mixing layer dispersion so as to not exceed the ambient $SO_2$ control level. The emission limitation is scheduled beginning about 9 a.m. and ending no later than 2 p.m. The program was initiated in 1969 when the third unit was put into operation and $SO_2$ emission at normal full-load operation of the total plant was increased approximately 80%. Ambient $SO_2$ standards have been met consistently, whereas standards were not met with two-unit operation before the program was initiated. On the basis of the success of the Paradise program, similar programs are scheduled for eight other TVA power plants where monitoring data, dispersion calculations, or both, indicate that ambient $SO_2$ standards are not being met. Estimates of the impacts of the emission limitation programs on power production at each of the plants are presented in Table V (*16*). Development, including installation of air monitors, of an emission limitation program costs approximately $500,000 per plant. Annual cost for operation of the program is approximately $200,000 per plant, exclusive of power replacement or increased fuel costs. Possible modifications in the fuel systems at the TVA plants to permit quick switching to low sulfur coal are being studied. This would

**Table V Estimate of Impact on Power Production of $SO_2$ Emission Limitation Program at TVA Steam Plants (*16*)**

| *Plant* | *Estimated average frequency of emission reductions (days per year)* | *Estimated average duration of emission reductions (hours per day)* | *Estimated average power lost by emission reductions (megawatt-hours per year)* |
|---|---|---|---|
| Allen | 1 | 6.0 | 2,000 |
| Colbert | 24 | 5.5 | 25,000 |
| Cumberland | 10 | 3.0 | 15,000 |
| Gallatin | 7 | 4.5 | 9,000 |
| Johnsonville | 7 | 6.0 | 12,000 |
| Kingston | | | |
| Existing stacks | 55 | 7.0 | 100,000 |
| Tall stacks | 0 | — | — |
| Paradise | 13 | 3.5 | 22,000 |
| Shawnee | | | |
| Existing stacks | 42 | 6.0 | 87,000 |
| Tall stacks | 3 | 3.0 | 4,500 |
| Widows Creek | | | |
| Existing stacks | 25 | 6.0 | 75,000 |
| Tall stacks | 12 | 3.0 | 18,000 |

permit reductions in $SO_2$ emissions as required without loss of generation and would provide for optimum use of scarce low sulfur coal for air quality protection. Oil-fired plants and plants with dual firing systems, for example, oil and coal, have much greater flexibility for fuel switching without major changes than do coal-fired plants. For coal-fired plants, in addition to the physical modifications required in the fuel handling system, it is necessary to assure also that the characteristics of the low sulfur fuel are compatible with efficient boiler operation and that the electrostatic precipitator is adequately designed to maintain necessary fly ash collection efficiency when the low sulfur fuel is being fired.

## 2. Low Sulfur Fuels

Historically, regulatory efforts for power plant $SO_2$ control have been directed to control of emissions at the source. These efforts generally have been in the form of limitations on the sulfur content of power plant fuels. As long as sufficient quantities of low sulfur fuels compatible with the combustion systems of the plants involved are obtainable, compliance by fuel switching is usually a practical and feasible control measure—although serious disruptions of fuel markets may occur. The proliferation of strict $SO_2$ emission limitations for power plants, notably in the United States, coupled with growing electric energy requirements, has created a great demand for replacement low sulfur fuels and for development of stack gas desulfurization processes that would meet emission control requirements when high sulfur fuels are used. Stack gas desulfurization processes will be discussed in a following section.

While combustion and fuel handling systems would obviously require modification, fuel switching from coal to natural gas or oil usually does not present major conversion problems. This is not the case when changing from high sulfur coal or oil to low sulfur coal. As is discussed in Section V,C, electrostatic precipitators are designed for a specified range of sulfur content of coal to be burned. Lowering of sulfur content below design values may reduce efficiency to the extent that major modifications will be required to meet particulate emission standards. Experience of two utilities (TVA and Commonwealth Edison Company, Chicago, Illinois) in burning low sulfur subbituminous coals in furnaces designed for high rank bituminous coals with high sulfur content have been reviewed (*16, 17*).

As early as 1965, Commonwealth Edison Company concluded that on the basis of available technology, nuclear energy offered the best solution to air pollution problems. By 1974, 40% of its generation will be by nuclear units. From 1969 to 1974, it is projected that on the basis of fuel consumption, the percent generation by coal is expected to drop from 85 to 47%. Since $SO_2$ control regulations will not permit continued use of Illinois high sulfur coals normally used, sources of low sulfur coals were sought to meet continuing needs.

Efforts of Commonwealth Edison Company to obtain low sulfur coal led to extensive exploration of sources in Montana and Wyoming. Commitments to purchase low sulfur Western United States coals totalled over 5,000,000 tons per year by mid-1972. In addition to problems of transport and related problems of reliability of delivery associated with haul distances of 1200 or more miles in contrast to a maximum of 300 miles for Illinois sources, the characteristics of the coals themselves presented significant burning problems and loss of generating capability on some units. The Montana and Wyoming coals are classified as subbituminous coals. In addition to their sulfur content, they are very high in moisture content, 10–20% lower than Illinois coals in heating value, and have different ash fusion characteristics. Furthermore, the sulfur content varies from 0.3 to 1.7%. The upper range barely meets the 1970 regulatory sulfur limit of 1.8% and is considerably in excess of the 1972 limit of 1.0%. Thus, even use of "low sulfur" western coals would require careful selection to meet regulatory requirements. Problems with burning these low sulfur coals as reported by Lundberg (*17*) include serious ignition problems, longer coal handling times, and reductions in effective bunker capacity due to lower Btu content and density, reductions in pulverizer capacity due to inability to handle the drying load, increase in carbon loss to fly ash, reductions in capacity up to 20%, reductions in electrostatic precipitator performance, and carbon or coke carryover and explosions within the slag tank on slagging furnaces equipped with cyclone burners.

TVA conducted test burnings of Western United States low sulfur subbituminous coals at its Johnsonville power plant during the period September–November 1972. The coals used were typical of coal from the States of Wyoming and Montana, having a low heat content (8000–8700 Btu/lb) and high moisture content (26–30%). While no difficulty was encountered in firing the coals in tangentially fired pulverized fuel boilers at the Johnsonville plant, the low Btu content and high moisture content resulted in reducing maximum generating capability from 15 to 30%. The furnaces used for the tests are equipped only with mechanical ash collectors, so no data on effect of firing low sulfur coal on electrostatic precipitators could be obtained.

The American Electric Power Service Corporation (AEP), one of the largest in the United States, has about 14,480 megawatts of generating capacity of which 13,500 megawatts is coal fired. Compliance with $SO_2$

emission standards applicable at each of its plants is by burning conforming coal (*13*). Tests by AEP have demonstrated that many western coals can be burned satisfactorily in cyclone furnaces. To supplement the limited supplies of low sulfur coal available for its plants, AEP acquired western coal lands and contracted to purchase large tonnages of western coal. The company also ordered towboats, barges, and rail hopper cars to get the coal to its Ohio River plants as needed (*18*).

## 3. Stack Gas Desulfurization

Stack gas desulfurization has a long history. As early as 1934–1936, a full-scale gas washing installation for $SO_2$ removal was made at the Battersea and Bankside plants in London, England, and similar equipment was installed in 1963 to serve the 150-megawatt oil-fired capacity added at Bankside. The process used alkaline water from the River Thames with chalk added. Only the Bankside installation remains in full operation.

In the early 1960's, increasing worldwide emphasis on air pollution control led to an upsurge of research to develop practical processes for desulfurization of power plant stacks with potential for universal application where needed. Initially, major emphasis was upon processes for recovery of the sulfur in usable form. By 1967 ten such processes were being tested or planned in large units tied into actual power plants or other plants (process steam or chemical) that emit sulfur oxides (*19*). Four of these were in the Federal Republic of Germany (mixed metal oxides Grillo process, sulfacid Lurgi process, lignite ash Still process, and carbon absorption Reinluft process); one in England (alkalized alumina United States Bureau of Mines process); four in Japan (manganese dioxide Mitsubishi process, ammonia scrubbing Showa Denko process, catalytic oxidation Kiyoura Tokyo Institute of Technology process, and carbon adsorption Hitachi process); and one in the United States (catalytic oxidation Monsanto process). Of the three process types involved, absorption was the most popular method accounting for five of the test units; of the remainder, three involved adsorption and two catalytic oxidation. The processes are described in detail in the reference cited. Concurrently, work on "throwaway processes" includes extensive investigations on the dry limestone injection process in the Federal Republic of Germany, Japan, Poland, and the United States, and on the limestone injection–wet scrubbing process in the United States.

Extensive full-scale tests of the dry limestone injection process at the TVA Shawnee power plant were completed in 1972. It was concluded that, because of the inherently low $SO_2$ removal efficiencies and the potential for major power unit reliability problems, this process probably will not play an important role in controlling $SO_2$ emissions from power plants (*20*).

While interest in recovery processes continues, because of the need in the short term for desulfurization processes to meet stringent $SO_2$ emission standards on tight time schedules, particularly in the United States, emphasis and major developmental efforts have shifted to alkaline scrubbing throwaway processes. The technical, operational, and economic aspects of the throwaway processes are generally considered to be less complex than the recovery processes and thus more amenable for short-term development and application.

Developmental work continues worldwide, but only in the United States is there commitment for nationwide $SO_2$ abatement with primary attention focused on fossil fuel-fired power plants. A report issued April 1973 by a United States interagency expert panel, Sulfur Oxide Control Technology Assessment Panel (SOCTAP), summarized the status of control technology for sulfur oxides (*21*). The Panel concluded that stack gas cleaning is technologically feasible but that "reliability of currently available systems has been the subject of some question." Although it was projected that engineering barriers to applications of stack gas-cleaning systems could be removed by the end of 1974, widespread applications to the nation's coal-fired power plants would not be expected prior to 1980. The four processes described in the report considered sufficiently developed to potentially desulfurize flue gas on a full-scale commercial basis within the subsequent 5 years and a fifth system, called the double-alkali process—considered potentially important, are discussed below.

a. WET LIME/LIMESTONE SCRUBBING. Limestone scrubbing basically consists of combining the sulfur dioxide in flue gas with a limestone slurry to produce solid calcium sulfite and calcium sulfate (Fig. 4) (*22*). Since there is no profitable market for these materials, they present a solid waste disposal problem. The great majority of full-size power plant desulfurization systems in both the planning and operational phases involve scrubbing with limestone or lime slurries.

The primary reasons for this are that these processes are more fully developed than other first-generation systems, have relatively low capital and operating costs as compared with more complex removal systems, and have high potential removal efficiencies. Process problems include chemical scaling, erosion–corrosion, solid waste disposal, and plume heating requirements. Limestone scrubbers can remove particulates very efficiently from the flue gas along with the sulfur oxides.

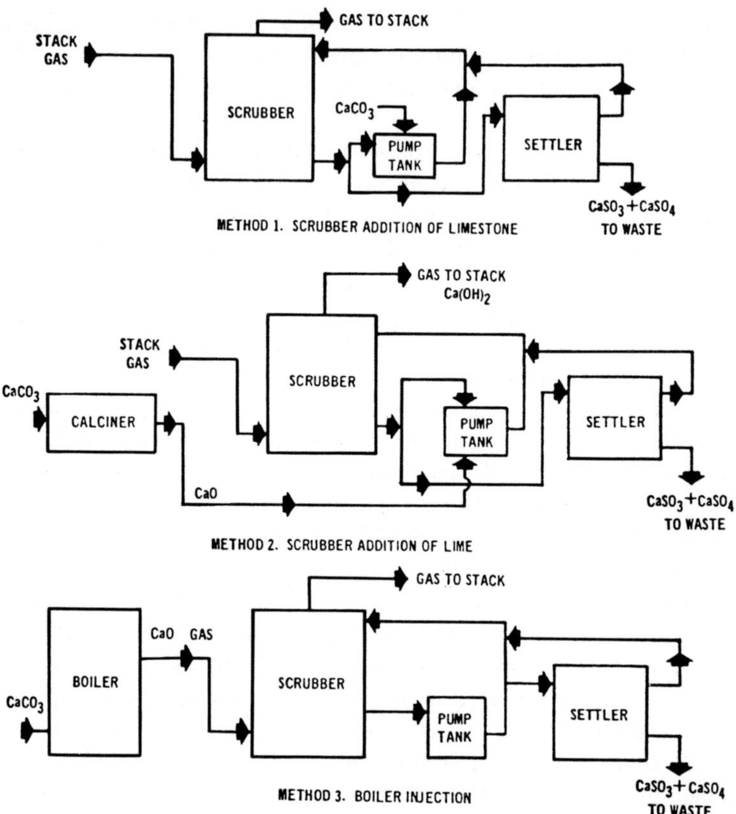


**Figure 4.** Major process variations in use of lime or limestone for removal of $SO_2$ from stack gases (*22*).

b. MAGNESIUM OXIDE (MGO) SCRUBBING. In many respects, magnesium oxide scrubbing is similar to limestone scrubbing (Fig. 5) (*22*). The principal difference is that the spent magnesium sulfite and sulfate salts are regenerated, producing a concentrated stream of 10–15% $SO_2$ and regenerated magnesium oxide for reuse in the scrubber. The concentrated sulfur dioxide gas is then conventionally converted to either concentrated sulfuric acid or elemental sulfur. Therefore, this system does not produce a solid waste problem, provided there is a firm market for these products.

Since the reactant is recycled, it must be protected from contamination by particulates. It is, therefore, necessary that the process be applied on an oil-fired boiler or that particulate matter be sufficiently removed from the flue gas prior to passing it into this system.

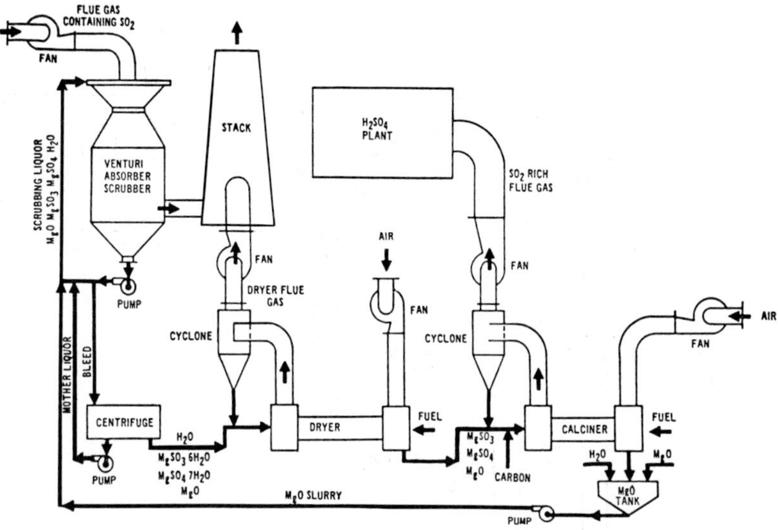


Figure 5. Magnesium oxide (MgO) slurry process for desulfurization of flue gas free of particulate matter (*22*).

c. Catalytic Oxidation. This process basically consists of passing the flue gas through a fixed catalyst bed where $SO_2$, in the presence of $O_2$, is converted to $SO_3$ (Fig. 6) (*22*). The $SO_3$ is then absorbed in recirculated sulfuric acid in an absorption tower. Since sulfuric acid is a salable product, there is no solid waste product, assuming that a market can be found. As with the magnesium oxide system, particulates must be removed from the flue gas prior to entering the system.

Catalysis requires a high temperature, about 850°F, and the fly ash must be removed before the gas enters the catalyst. Thus the catalyst and the gas cleaning equipment must be inserted ahead of the economizer or, as would be the case for installation in existing plants, the gas must be reheated.

d. Wellman–Lord Process. This process basically involves the absorption of sulfur dioxide into a solution of sodium sulfite, bisulfite, and sulfate, resulting in the production of a concentrated stream of $SO_2$ which can be further processed to liquid $SO_2$, sulfur, or sulfuric acid (Fig. 7) (*22*). Thus, no solid waste problems are produced by application of this system, if a market can be found. The specific advantage of this process is the simplicity of its unit operations. The main disadvantage is its sensitivity to buildup of contaminants, necessitating bleed.

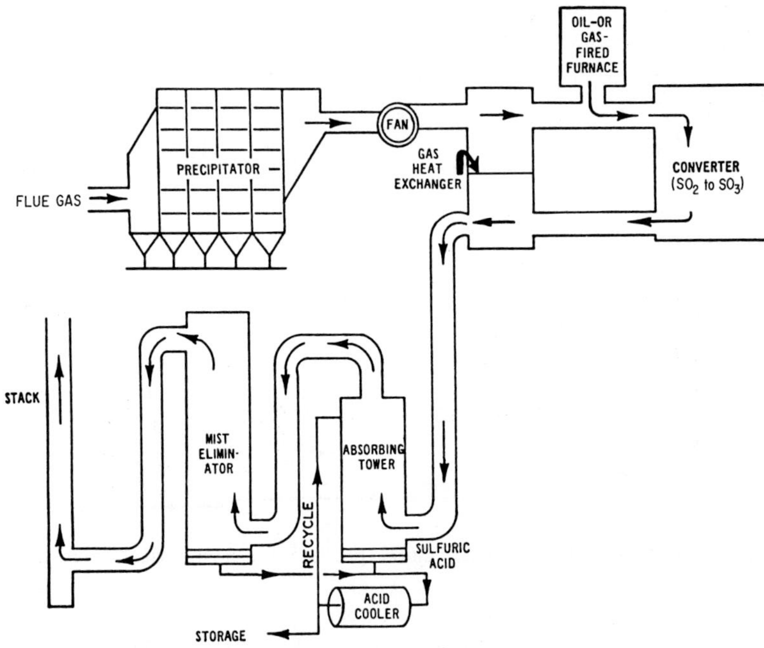


Figure 6. Catalytic oxidation process—with reheat—for desulfurization of flue gas (22).

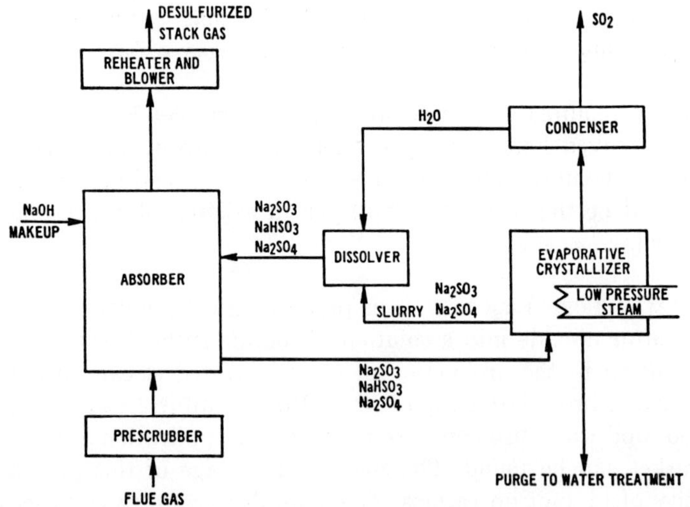


Figure 7. Wellman–Lord process for desulfurization of stack gas (22).

e. Double-alkali Scrubber Process. The many double-alkali process variations involve the scrubbing of flue gases with a clear liquor containing dissolved sodium or ammonium salts, followed by treatment of the depleted liquor with lime or limestone in a reaction producing throwaway sludge for disposal and alkali liquor for scrubbing (Fig. 8) (*22*). In the United States, primary developmental attention has been placed on sodium-based double-alkali systems using lime regeneration. In Japan, emphasis has been on systems with limestone used as the input alkali and gypsum produced as a salable product. A comparison of these systems as taken from the SOCTAP report (*21*) is given in Table VI.

A report by the USEPA in October 1973 summarized the status of flue gas technology (*22*). In the United States almost 18,000 megawatts of flue-gas desulfurization is planned with about 2000 megawatts in operation. A summary of planned and operating flue-gas desulfurization units in United States power plants as of September 1973 is presented in Table VII (*22*). In Japan more than 60 commercial and prototype flue-gas desulfurization plants are now in operation. Most of these are of relatively small capacity and are designed to treat waste gas from industrial boilers, and chemical and smelting plants. However, several large systems have been installed. Table VIII (*22*) presents a summary of selected $SO_2$ control systems on fossil fuel-fired boilers in Japan as of September 1973. None of the systems reviewed has demonstrated the high degree of reliability required by the utility industry. The two systems reported as most nearly meeting this requirement were the Chemico-Mitsui calcium hydroxide scrubbing systems in Japan and the Combustion Engineering calcium hydroxide scrubbing system at the Paddy's Run station of the Louisville (Kentucky) Gas and Electric Company. Experience at other

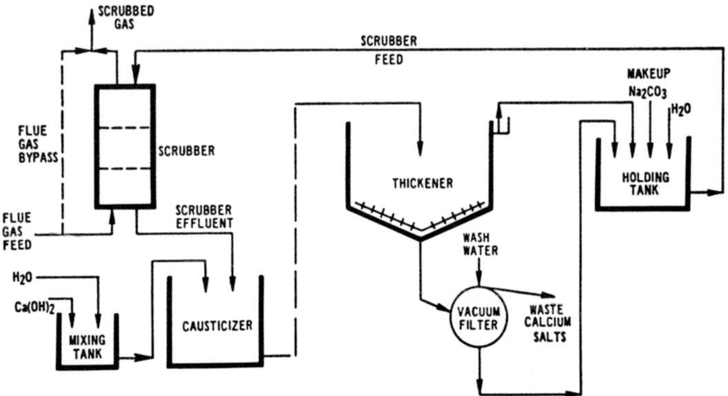


Figure 8. Double–alkali process (one variation)—sodium scrubbing with lime regeneration—for desulfurization of stack gases (*22*).

**Table VI Comparisons of $SO_2$ Control Process Systems (21)**

| Processing method | Reactant input requirements | Throwaway or recovery | Approximate investment costs[a] for coal-fired boilers ($ per kilowatt) | Approximate annual costs[b] (mills per kilowatt-hour): No credit for S recovery | Approximate annual costs[b] (mills per kilowatt-hour): With credit for S recovery | $SO_2$ Removal efficiency (%) |
|---|---|---|---|---|---|---|
| Wet lime/limestone/$Ca(OH)_2$ slurry scrubbing | Lime (100–120 % stoich)[c] Limestone (120–150 % stoich)[c] | Throwaway $CaSO_3/CaSO_4$ | 27–46 | 1.1–2.2 | N.A.[d] | 80–90 |
| Magnesium oxide scrubbing | MgO alkali; carbon and fuel for regeneration and drying | Recovery of concentrated $H_2SO_4$ or elemental sulfur | 33–58 | 1.5–3.0 | 1.2–2.7 | 90 |
| Monsanto catalytic oxidation (add-on) | Catalyst $V_2O_5$ (periodic replacement) and fuel for heat | Recovery of dilute $H_2SO_4$ | 41–54 | 1.5–2.6 | 1.3–2.4 | 85–90 |
| Wellman–Lord process (soluble sodium scrubbing with regeneration) | Sodium makeup and heat for regeneration requirements | Recovery concentrated $H_2SO_4$ or sulfur | 38–65 | 1.4–3.0 | 1.1–2.7 | 90 |
| Double-alkali process | Sodium makeup plus lime/limestone (100–130 % stoich)[c] | Throwaway $CaSO_3/CaSO_4$ | 25–45 | 1.1–2.1 | N.A. | 90 |

[a] Generally, where a cost range is indicated, the lower end refers to a new unit (1000 MW), while the high end refers to a 200 MW retrofit unit. Costs include particulate matter removal.
[b] Assumptions: costs calculated at 80 % load factor; fixed charges per year 18 % of capital costs.
[c] Percent of stoichiometric amount required.
[d] N.A. = Not applicable.

**Table VII Summary of Planned and Operating Flue-Gas Desulfurization Units on United States Power Plants as of September 1973** *(22)*

| | *Planned* | | *Operating* | | *Planned and operating* | |
|---|---|---|---|---|---|---|
| | *Number of units* | *Total* (MW) | *Number* | *Megawatts* | *Number of units* | *Total* (MW) |
| Limestone (LS) | 6 | 2855 | 3 | 1076 | 9 | 3931 |
| Lime (L) | 8 | 3220 | 4 | 725 | 12 | 3945 |
| L/LS—Not selected | 10 | 5929 | | | 10 | 5929 |
| Magnesium oxide | 1 | 120 | 2 | 250 | 3 | 370 |
| Other $SO_2$ control systems | 5 | 640 | 1 | 110 | 6 | 750 |
| Process not selected | 5 | 2960 | | | 5 | 2960 |
| | 35 | 15724 | 10 | 2161 | 45 | 17885 |

facilities indicates that reliability can be a problem if systems are not carefully designed and operated. The boiler injection mode is particularly prone to serious operating problems. System bypasses and/or extra modules are needed to permit scrubber maintenance without power unit shutdown.

**Table VIII Summary of Selected $SO_2$ Control Systems on Japanese Fossil Fuel-Fired Boilers as of September 1973** *(22)*

| | *Planned* | | *Operating* | | *Planned and operating* | |
|---|---|---|---|---|---|---|
| *Process* | *Number of units* | *Total* (MW) | *Number of units* | *Total* (MW) | *Number of units* | *Toal* (MW) |
| Wellman–Lord | 2 | 250 | 2 | 295 | 4 | 545 |
| Dilute sulfuric acid–gypsum (Chiyoda) | 2 | 484 | 1 | 52 | 3 | 536 |
| Double-alkali–gypsum (limestone) | 2 | 300 | 1 | 156 | 3 | 456 |
| Mitsubishi-JECCO lime/limestone-gypsum | 4 | 530 | | | 4 | 530 |
| Wet limestone | 1 | 100 | | | 1 | 100 |
| Carbon adsorption | | | 1 | 139 | 1 | 139 |
| DAP-Manganese | | | 1 | 110 | 1 | 110 |
| Lime scrubbing | | | 1 | 156 | 1 | 156 |
| | 11 | 1664 | 7 | 908 | 18 | 2572 |

Despite the fact that a number of large-scale plants are scheduled for operation in the United States, many engineering and process problems remain to be solved if successful operation of flue-gas desulfurization systems is to be achieved, and widespread application is to become feasible.

Closed-loop (no liquid discharge) operation is required if gas-cleaning systems are to avoid contributing to water quality problems. This means that all liquid streams should be recycled. Recycle is a particularly serious complication with the calcium-based scrubbing systems because of the tendency of calcium salts to supersaturate (higher dissolved solids content than the equilibrium concentration). The allowable makeup water in a closed-loop system is low, about 0.75 gallons/minute (gpm) per megawatt, and fresh water for cleaning is limited. To the extent that scrubber effluent can be purged from the system, the fresh water addition can be increased, and reliability is improved. The capability of scrubbing systems to operate for long periods under closed-loop conditions is still to be demonstrated.

Accumulation of solids in the mist eliminator is one of the major mechanical problems. In lime and limestone scrubbing processes, a high solids concentration is maintained in the circulating slurry in order to avoid scaling. To the extent that slurry is entrained in the gas leaving the scrubber, solids are carried over to the mist eliminator and impaction leads to buildup of insoluble solids. The buildup is removed by washing, but the limited amount of water available in a closed-loop system may not be adequate. Admixture of recycle water can lead to scaling if the solution reaches saturation. One method that shows promise is use of a wash tray ahead of the mist eliminator to remove some of the entrained solids. More work on this and alternative methods is needed.

The high solids concentration of the slurry also causes erosion of spray nozzles, pumps, and scrubber internals. Additional materials evaluation is needed for intelligent design of components.

For new power plants, a gas cleaning system can be integrated into the initial design. In retrofit installations, the equipment must be designed to occupy the space available, and in many situations this can result in a serious economic penalty; in others, it might be virtually impossible. A gas bypass system is required for large-scale test facilities in order to avoid outages for the generating unit when the gas cleaning system is out of service. Tight fitting dampers to divert the flow of gas from one duct system to another present major design problems. Use of spare scrubbing trains to improve reliability also is subject to availability of dependable large dampers.

Extensive application of lime-limestone throwaway processes would create large quantities of sludge for disposal. The quantities are 2.5 to

3.0 times the average coal ash disposal tonnage and for a 1000-megawatt plant, an area of approximately 1.5 square miles would be filled to a 10-ft depth over a 20-year period.

A further complication results because the sludge does not settle into a compact solids layer. Instead, it is thixotropic and even after prolonged periods of storage, assumes the consistency of wet cement when subjected to mechanical working. The implication is that the material would be unsuitable for use as permanent landfill and would have to be retained indefinitely in storage ponds. Several organizations are conducting research on methods for stabilizing the sludge, but neither the techniques nor economics have been established. Until suitable methods are developed, disposal of waste solids will be a major deterrent to widespread use of lime–limestone scrubbing systems, particularly in urban areas.

### C. Particulates

Gas cleaning is the most common technique used for control of particulate emissions from stationary combustion sources. Among gas-cleaning methods are settling chambers, large- and multiple small-diameter cyclones, wet scrubbers, electrostatic precipitators, and fabric filters. Electrostatic precipitators installed on the flue-gas exit side of the air preheater are the most commonly used devices for removal of particulates from coal-fired power plants (*23*). For new installations designed to burn low sulfur coal, the electrostatic precipitators may be installed on the hot side of the air preheaters. Electrostatic precipitators may be used in combination with cyclone collectors, which usually precede the electrostatic precipitators. Cyclone collectors following high-efficiency electrostatic precipitators help to sustain the highest possible efficiency by cleaning, up puffs due to rapping, soot blowing, and minor slight malfunctions of the precipitators.

Interest in use of wet scrubbers on coal-fired power plants for particulate removal has been generated primarily by potential use singly or in combination to remove both particulates and $SO_2$. While scrubbers may offer some advantage where low sulfur–high ash coal is used and high efficiency of ash removal is required, there is no indication that scrubbers will replace electrostatic precipitators as the gas cleaning method of choice where only particulate matter removal is involved.

The cost of precipitators increases rapidly at higher collection efficiencies. For example, on a 500- to 800-megawatt plant, a precipitator of 95% efficiency may cost between $800 and $1200 per megawatt; one of 95% efficiency may cost in excess of $2500 per megawatt (*24*). The total installed cost may range up to 10 times this amount. Costs for "hot side" precipitators would be substantially higher because of the greater volume of gas involved. The total cost of a precipitator is directly related to its size, which in turn depends upon the volume of gas to be cleaned and the efficiency of particulate removal required. The efficiency of removal required will determine the plate area per 1000 actual cubic feet per minute (acfm) required for a specified sulfur content of coal to be fired. The effect of sulfur content on plate area needed to obtain various precipitator efficiencies is shown in Figure 40 of Chapter 5, this volume. For coal-fired power plants gas flow per kilowatt of generation will average approximately 3.0 acfm, with temperatures of 260° to 280°F, and 15 to 20% excess air.

In the design of a precipitator to meet a specified emission standard, allowances must be made for normal variations in ash and sulfur content that occur in various coal seams, as well as for possible changes in fuels. The precipitators should be designed conservatively for the lowest operating gas temperatures expected, while burning coal with the least favorable combination of sulfur–ash content. In the design, emphasis should be upon reliability and availability to assure that design efficiencies will be maintained throughout the life of the equipment without significant outages being required for major repairs and maintenance.

The greatest reliability problem associated with precipitators is the integrity of the charging wires (*25*). A tendency to wire failure is related to competing demands for good mechanical and electrical properties. For example, the finer the wire, the better the electrical field; however, wires are required to withstand vibrations caused by electrical wind and arcing. When vibration is inhibited by fixing the top and bottom, mechanical fatigue eventually produces broken electrode wires. Designs utilizing rigid discharge electrodes to overcome or minimize these problems should be preferred for power plant installations (*26*). Further, mechanical rapping systems are preferable to vibration rapping systems. Multiple series and parallel electrical sections must be provided for consistently acceptable performance. Gas velocities must be uniform and quite limited—preferably not more than 5 ft/second.

In spite of the vast experience of the electric utility industry with electrostatic precipitators and the theoretically high level of performance that should be attainable, results from commercial operation of precipitators have not been universally satisfactory. This may be due in some cases to lack of maintenance or changes in fuel or operating conditions outside the design range of the precipitators. Potential problems related to fuel changes have already been cited. Another problem is related to operating gas temperatures. Power plant precipitators are designed for operation with gas temperatures above the acid dew point and with uniform gas distribution. While great care is taken to assure uniform gas distribution, little attention is given to the problem of maintaining temperatures above the acid dew point throughout the precipitators. With regenerative air preheaters, and to a much lesser extent with tubular-type air preheaters, wide variations in temperature distribution (as much as 30° to 40°F) occur in the precipitator inlet duct. With the downward trend in exit gas temperatures for improvement in boiler efficiency, this presents a problem in maintaining proper precipitator performance due to altered electrical characteristics of fly ash in the cold spots and acid corrosion of the precipitator itself. Such problems of course are not encountered in "hot side" precipitators, an advantage in addition to their superiority for use with low sulfur fuels.

Studies of the effects of exit gas temperature on precipitator performance conducted on two 550-megawatt units at TVA's Widows Creek plant and on one 575-megawatt unit at the Colbert plant have been reported (*27*). Each unit has four electrostatic collectors designed to remove 90% of the particulates when burning coal containing 2.5% sulfur with a gas flow of 400,000 cfm (at 280°F and an inlet dust loading of 4.1 grains/ft$^3$). Performance of the collectors on Widows Creek unit 8 and Colbert unit 5 was far below the guarantee and efforts by the manufacturers to resolve the poor efficiency were unsuccessful. Under normal operating conditions, no sparking occurred in the collectors even though set at maximum ampere rating on the transformer-rectifier set, and tests showed efficiencies of 20–60%. Collectors on Widows Creek unit 7 met the performance guarantee during early operation, but efficiency dropped after several months of service. Elevation of exit-gas temperature had a pronounced effect on collector performance. Results are shown in Figure 9. As gas temperature is raised, efficiency increases and reaches design value at 320°F. As shown in Figure 10, at elevated gas temperatures, collector efficiency is 90% or better at gas flows up to 410,000 cfm, slightly above the design flow of 400,000 cfm.

Elevation of gas temperature changed the electrical behavior of the collector. Below the acid dew point, the automatic control system maintained primary voltage at 300–310 V with no evidence of sparking. Above the acid dew point, the control system maintained voltage at 320–340 V with a spark rate of 25–50/minute. With the higher voltages, increased efficiencies were obtained as shown in Figure 11.

Further studies established a relationship between collector performance and condensed sulfuric acid in the flue gas, as shown in Figure 12. Thus, with operation below the acid dew point (which means the presence of condensed $H_2SO_4$) low efficiencies could be expected, possibly because of alteration in the electrical properties of the flue gas or the fly

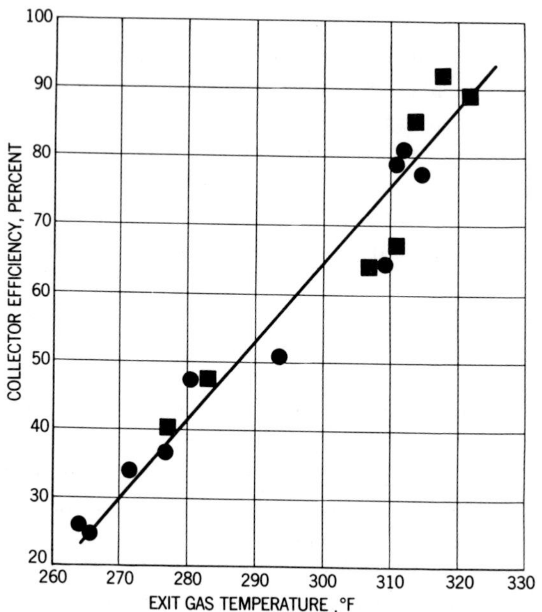


Figure 9. Relationship of collector efficiency with exit-gas temperature (*27*). ■: Widows Creek 7; ●: Widows Creek 8.

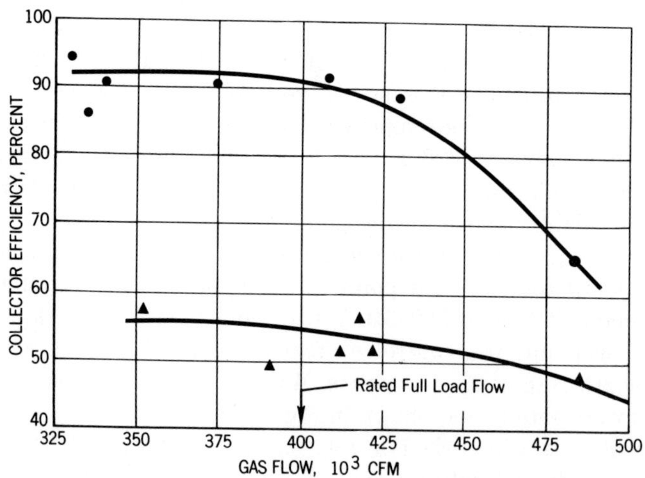


Figure 10. Comparison of collector efficiency at normal and elevated temperatures with varying gas flows (*27*). ●: Gas temperature 310°F above acid dew point); ▲: gas temperature 270°F (below acid dew point).

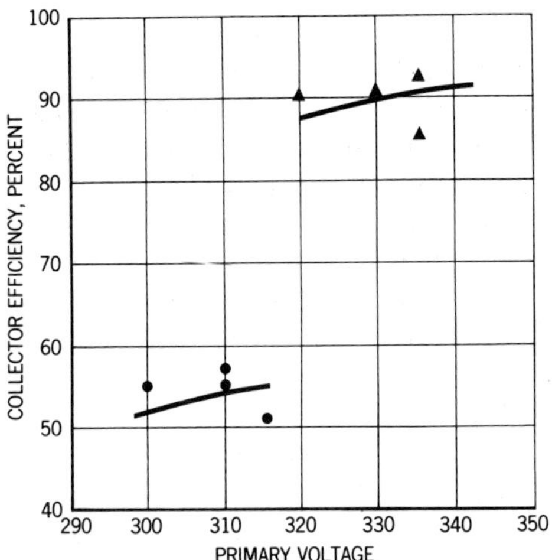


Figure 11. Relationship of collector efficiency with power input (*27*). ▲: Gas temperature 310°F (above acid dew point); ●: gas temperature 270°F (below acid dew point).

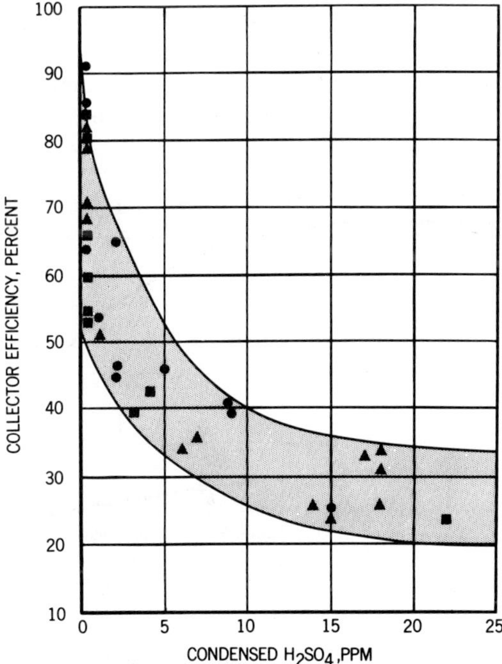


Figure 12. Collector performance as a function of the amount of condensed sulfuric acid in flue gas (*27*). ●: Widows Creek 7; ▲: Widows Creek 8; ■: Colbert 5.

ash. The presence of less than 1 ppm condensed acid reduced efficiency from 90 to 50%. Calculated acid dew points ranged from 260° to 285°F for the three boilers having the deficient precipitators. In order to raise all sections of the gas stream above the acid dew point, it would be necessary to carry an exit-gas temperature of about 320°F (some 60°F above normal operating temperature). This increase in temperature would result in a loss of 1.2% in boiler efficiency, and thus would be economically unattractive as a way to improve collector efficiency. As an alternate to this approach, the use of ammonia as a gas conditioning agent to neutralize the condensed sulfuric acid was investigated, with the results shown in Figures 13 and 14. A level of 15 ppm $NH_3$ in the flue gas, injected downstream of the air heater was selected as the design basis for the full-scale installation to bring the precipitator up to design efficiency.

Flue-gas conditioning with sulfur trioxide (or $H_2SO_4$) has been widely used for improving electrostatic precipitator performance. At the TVA Bull Run plant, completed in 1966, the electrostatic precipitators were designed for a collection efficiency of 99%. The original coal supply for this plant had a sulfur content of 2.0 to 2.5%. When a shift was made to low sulfur coal, averaging 1.0%, the efficiency of the precipitators dropped to 80 to 85%. Pilot-plant studies showed that gas conditioning with $SO_3$ increased the precipitator efficiency to between 98 and 99% with an optimum injection rate of 20 to 25 ppm. A schematic of the full-scale $SO_3$ gas conditioning system at the Bull Run is presented in Figure 15.

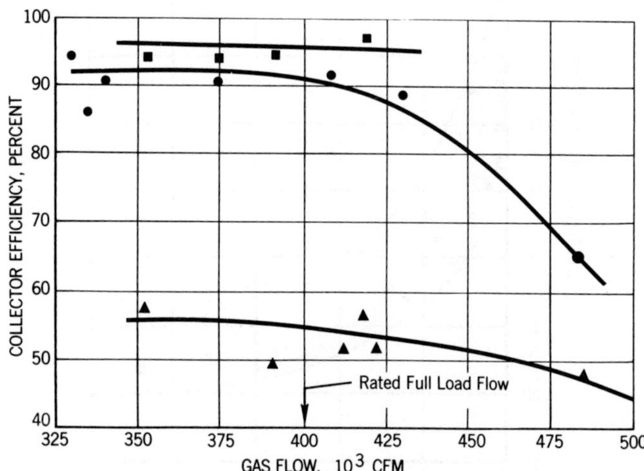


Figure 13. Comparison of collector efficiency with normal gas temperature, with elevated gas temperature, and with ammonia injection at normal gas temperature (27). ■: Gas temperature 270°F (below acid dew point with $NH_3$ feed); ●: gas temperature 310°F (above acid dew point); ▲: gas temperature 270°F (below acid dew point).

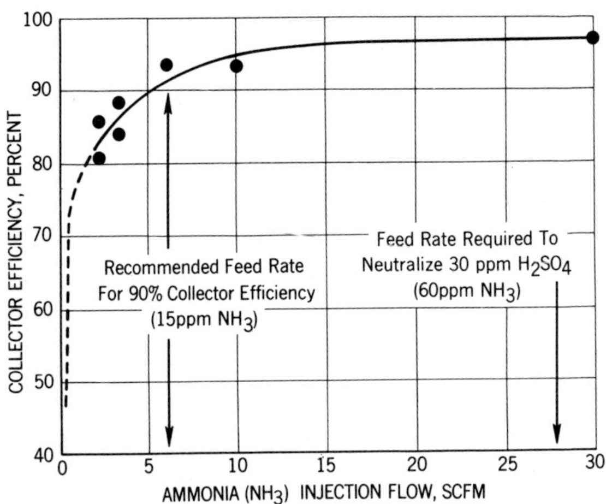


Figure 14. Collector efficiency as a function of ammonia feed rate (27).

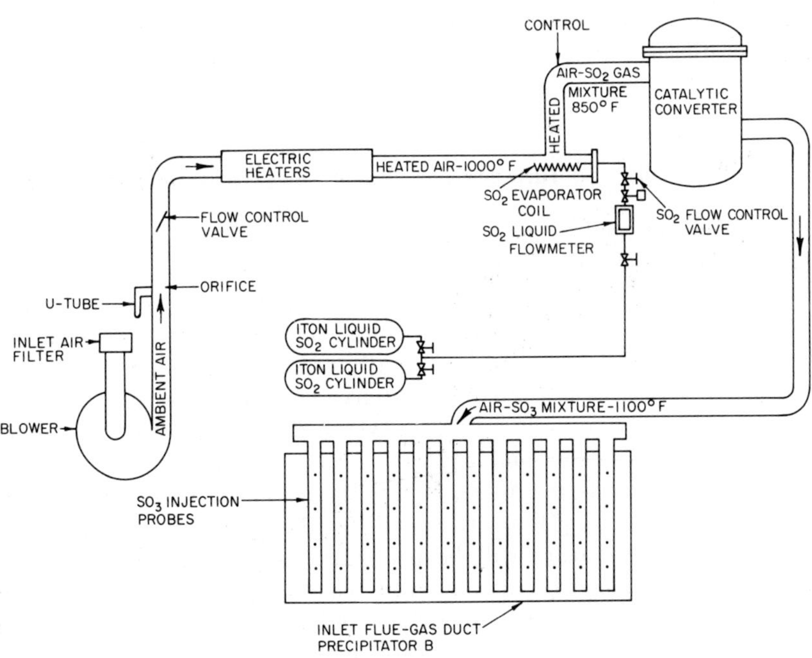


Figure 15. Schematic—$SO_3$ flue-gas conditioning system—TVA Bull Run steam plant. Note: 5 thermocouples (TC) on converter system to 6 point recorder on converter.

Flue-gas conditioning provides a means for offsetting to some degree the adverse effects of lower sulfur content of fuel and lower flue-gas exit temperatures on precipitator efficiency but is not a substitute for adequate design.

An essential part of the particulate control system is the necessary equipment to remove the collected fly ash and transport it for disposal. The ash removal system must be adequately sized to preclude buildup of fly ash in the collectors with resulting reentrainment which could significantly reduce overall removal efficiency.

### D. Nuisance Dusts

In some situations where power plants are located in proximity of residential areas, dusts from coal-handling operations and dewatered ash ponds may give rise to nuisance complaints. Use of covered conveyors, wetting sprays at transfer points, sprinkling of haulways, treatment of storage piles with polymeric compounds, and enclosure of live piles may be required singly or in combination to control fugitive dusts from coal-handling operations. Filled ash ponds, after dewatering, may require a surface layer of top soil with vegetative cover to control dust generation from ash disposal areas where such areas are not converted to industrial or other use.

## VI. Gas- and Oil-Fired Turbines and Diesel-Powered Generating Units

The principal applications of gas turbines and diesels on large power systems include peaking, standby service, emergency power for safe rundown or black start at steam electric plants, end-of-line regulation, and reserve (*28*). The combustion gas turbine consists typically of an axial-flow air compressor, one or more specially constructed combustion chambers where liquid or gaseous fuel is burned in an excess of air, and a power turbine of two or more stages where the hot gases are expanded to drive the compressor and generator. Nearly two-thirds of the total turbine horsepower is required to drive the compressor. Temperature of the gas stream entering the turbine will be about 1800°F at peak rating and exhaust temperature will be 950°F or slightly higher. The sulfur dioxide emission depends on the sulfur content of the fuel being used, but is normally low. Nitrogen oxide emissions are slightly higher than those of the comparable-sized fueled steam electric unit.

The simple open-cycle gas turbine generator unit is the gas turbine design most widely used by the electric utility industry. Early applications beginning in the 1950's were in sizes of 5 to 10 megawatts. Since about 1960, combustion turbine installations have increased in size to a range of 18 to 60 megawatts on a single turbine drive, and to 160 megawatts, using multiple turbines on a single generator shaft. Developmental work is in progress on single units of up to 100 megawatts with gas turbine inlet temperatures of 2000°F and above. Combined gas–steam turbine systems increase power plant thermal efficiency and are expected to find increased use as intermediate cycling units filling longer term use. In a combined-cycle plant, fuel is first burned in combustion turbines that spin generators to produce electricity. The hot exhaust from the gas turbine then goes to a special heat recovery boiler to make steam, which turns a conventional steam turbine that drives an additional generator. A novel application of the combined system was made by one company. In lieu of rebuilding existing boilers and installing electrostatic precipitators for old units at one of its plants, the company decided to install combustion turbine generators for peaking power and waste heat boilers to supply steam to the old steam turbine generators. This permitted retirement of several coal-fired boilers (*29*).

Since combustion turbines use large volumes of air and compressor speeds are very high, sound suppression treatment is required to keep noise levels within acceptable limits. Commonly, silencers are employed on both intake and exhaust. In some cases the units are completely housed by a sound enclosure.

Diesel engine generator units for central station power service generally fall into two broad categories—those of the slow-speed heavy-duty base-load design, or those of the intermediate-speed, lighter weight, prepackaged building-block type. The latter are generally about 2 megawatts in size. The former are regularly manufactured in sizes up to approximately 6 megawatts and are reportedly offered in sizes of 20–25 megawatts by some manufacturers (*28*).

A report of the National Academy of Engineering in 1972 summarized approaches to control of emissions of $NO_x$ from stationary gas turbines and diesel engines (*30*). Some of the simplest approaches to control of emissions from diesel engines are retard the timing to the range in which both the smoke and $NO_x$ emissions decrease; and use a longer period of injection for the same amount of fuel so that the rate of heat release in the cylinder is decreased. Rate of injection and retarded timing tend to be more of a problem on high-swirl chambers and less of a problem on quiescent chambers. Another method is staged injection—pilot injection followed by main injection. These techniques have been shown to produce a fairly high reduction in $NO_x$ from existing engines, although there is a penalty of some loss in thermal efficiency. Water injected into the intake manifold or emulsified in the fuel for diesel engines has also proven successful, as has exhaust-gas recirculation and lowering of the compression ratio.

On turbines the techniques of exhaust recirculation, water injection, and combustor redesign are also applicable. A combination of fuel-injection techniques and lean operation appears to offer the most promise for reduction $NO_x$ from gas turbines. This involves changing the mixing patterns, adding excess air at various points along the length of the combustor, and recirculating a portion of the exhaust gases. Various fuel additives may possibly reduce emissions, but no demonstrated technology or promising research data exist.

## VII. Control of Gaseous Emissions from Nuclear-Fueled Steam Electric Generating Plants

### A. Emissions from Nuclear Power Plants

The quantity of gaseous effluents from a nuclear fueled steam electric generating plant is insignificant compared to the amount from a fossil fueled plant of equal capacity. For example, the quantity of gaseous waste from a 2-unit pressurized water reactor nuclear plant with a total capacity of 2664 megawatts has been estimated to be 94,400 $ft^3$/year (*31*). In contrast, gaseous emissions from the Tennessee Valley Authority (TVA) Cumberland 2-unit coal-fired plant with total capacity of 2600 megawatts amount to 4,698,000 $ft^3$/minute (*26*). Because nuclear plant effluents include radioactive gases and particulates with serious ecological and public health implications, they are subject to stringent control and are closely regulated and monitored.

During the operation of nuclear power reactors, radioactive materials are produced by fission of the nuclear fuel and by neutron interaction with metals and materials in reactor systems and impurities in reactor coolant water. Most of the radioactive materials produced are retained in the fuel element; only a small fraction of these materials enter the plant waste systems and are channeled into the various effluent streams.

Pursuant to implementation of its licensing requirement that nuclear power reactors keep radioactive materials in effluents as low as practicable, the United States Atomic Energy Commission, in July 1973, issued a statement concerning its proposed rulemaking action providing "Numerical guides for design objectives and limiting conditions for operation to meet the criterion 'as low as practicable' for radioactive material in light-water cooled nuclear power reactor effluents" (*32*).

Schematics of a boiling water reactor (BWR) and a pressurized water reactor (PWR) power plant showing both gaseous and liquid radwaste effluents are presented in Figures 16 and 17 (*32*). A comparison of annual radioactive gaseous effluents from an operating BWR and PWR power

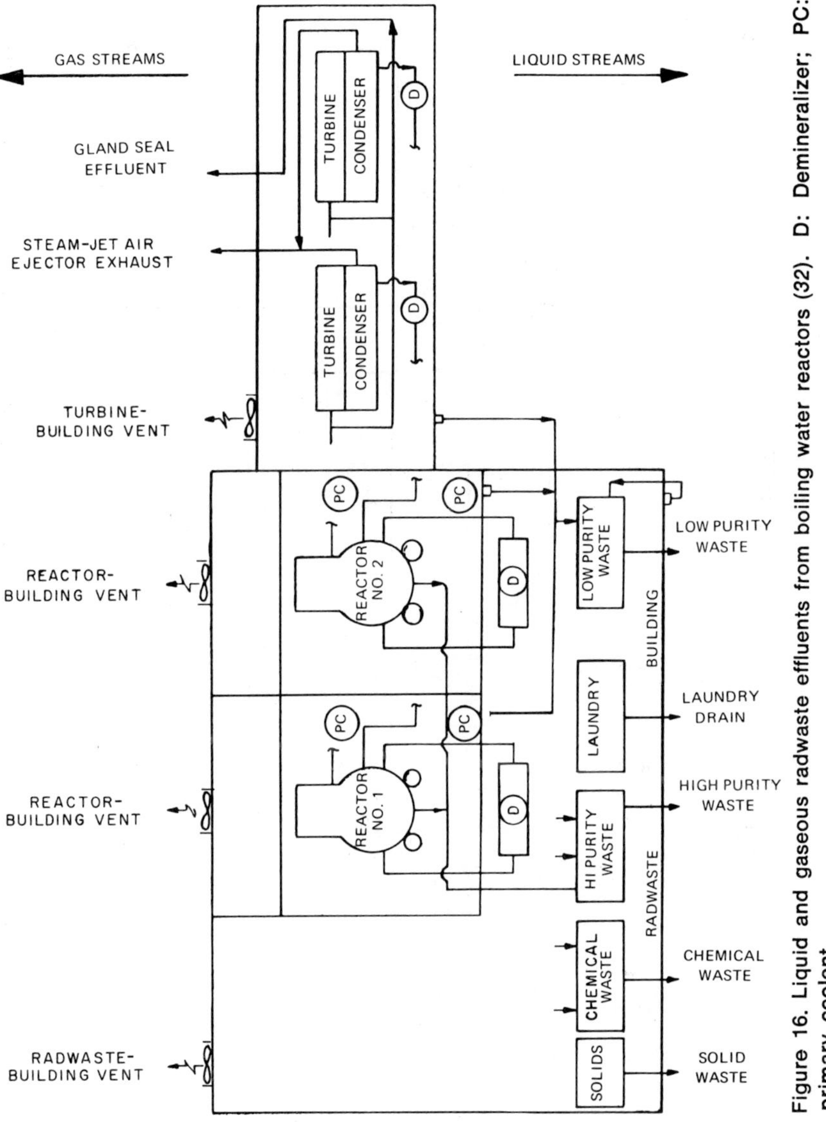


Figure 16. Liquid and gaseous radwaste effluents from boiling water reactors (32). D: Demineralizer; PC: primary coolent.

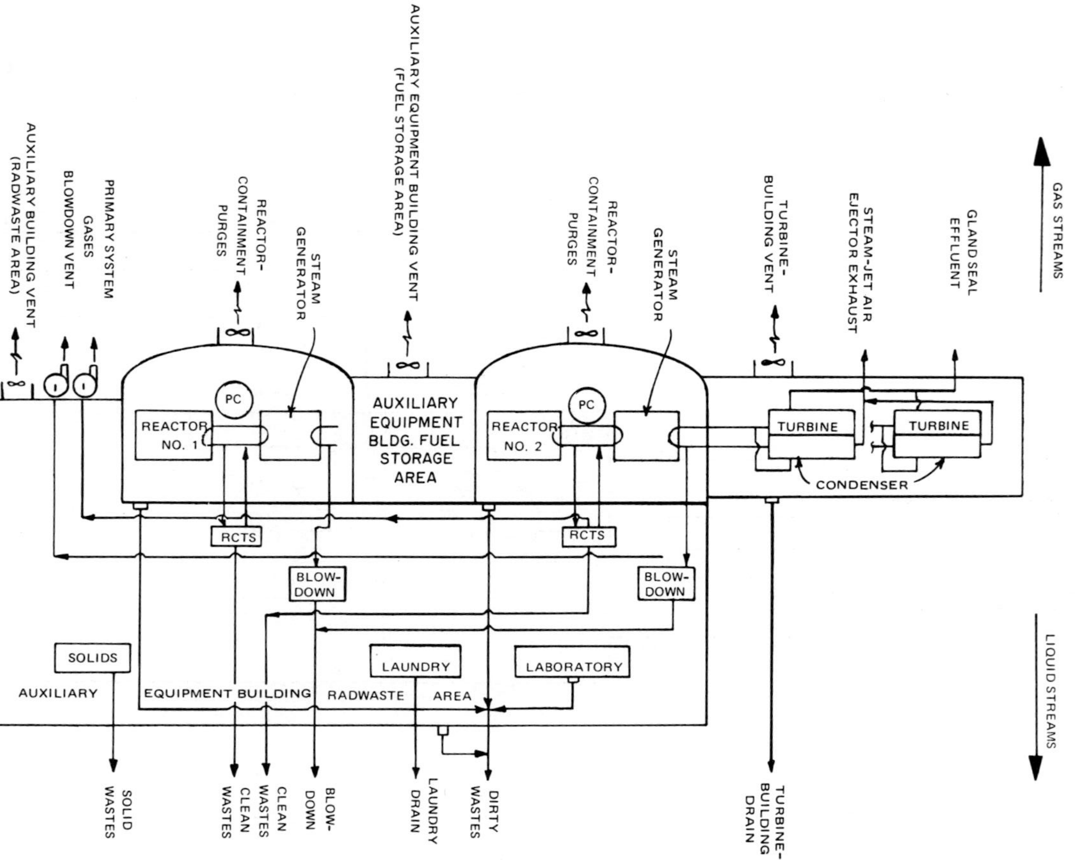


Figure 17. Liquid and gaseous radwaste effluents from pressurized water reactors (32). RCTS: Reactor coolant treatment systems; PC: primary coolant.

**Table IX Summary of Radioactive Airborne Effluent From a Boiling Water Reactor and Pressurized Water Reactor Power Plant 1972 (*33*)**

| | *Boiling water reactor*[a] | *Pressurized water reactor*[b] |
|---|---|---|
| Licensed power level (megawatts thermal) | 1670 | 1518 |
| Initial criticality | 12/10/70 | 11/2/70 |
| Power generation (megawatt-hours) | | |
| Gross thermal | $1.09 \times 10^{7}$ | $9.96 \times 10^{6}$ |
| Net electrical | $3.56 \times 10^{6}$ | $4.67 \times 10^{6}$ |
| Airborne effluents—curies released | | |
| Noble gases—total | $7.51 \times 10^{5}$ | $2.81 \times 10^{3}$ |
| Krypton-85 | $5.79 \times 10^{3}$ | $9.7 \times 10^{1}$ |
| Xenon-133 | $1.14 \times 10^{5}$ | $1.95 \times 10^{3}$ |
| Krypton-88 | $1.40 \times 10^{5}$ | $8.61 \times 10^{1}$ |
| Krypton-87 | $1.03 \times 10^{5}$ | $3.51 \times 10^{1}$ |
| Krypton-85m | $5.4 \times 10^{4}$ | $1.14 \times 10^{2}$ |
| Xenon-138 | $6.34 \times 10^{4}$ | — |
| Xenon-135 | $1.85 \times 10^{5}$ | $3.84 \times 10^{2}$ |
| Argon-41 | $<1.65 \times 10^{2}$ | $2.13 \times 10^{1}$ |
| Halogens—total[c] | $5.76 \times 10^{-1}$ | $1.15 \times 10^{-2}$ |
| Iodine-131 | $5.76 \times 10^{-1}$ | $9.7 \times 10^{-3}$ |
| Iodine-133 | 1.16 | |
| Iodine-135 | 2.26 | |
| Particulates—total | $1.25 \times 10^{-2}$ | $1.82 \times 10^{-2}$ |
| Cesium-137 | $<3.7 \times 10^{-4}$ | — |
| Barium–lanthanum-140 | $<1.3 \times 10^{-3}$ | — |
| Strontium-90 | $4.4 \times 10^{-5}$ | — |
| Cesium-134 | $<2 \times 10^{-4}$ | — |
| Strontium-89 | $5.5 \times 10^{-5}$ | — |
| Cobalt-58 | — | $6.5 \times 10^{-3}$ |
| Cobalt-60 | — | $6.9 \times 10^{-4}$ |
| Rubidium-88 | — | $1.1 \times 10^{-2}$ |

[a] Monticello Plant, Northern States Power Company.
[b] Point Beach-1,2, Wisconsin Michigan Power Company and Wisconsin Electric Power Company.
[c] Totals for halogens and particulates: only nuclides with half-life >8 days.

plant, based on 1972 data, is presented in Table IX (*33*). The much greater quantity of noble gases emitted from the BWR plant than from the PWR plant is due principally to the difference in basic reactor design, as described in Chapter 5, Vol. I.

### B. Control of Emissions from Nuclear Power Plants

Radwaste treatment systems are designed to control, measure, and reduce the amount of radioactive materials in effluents from a nuclear power station. Effluents must be monitored (a) to verify that the radioactivity released is within acceptable limits and (b) to provide an indication that remedial action is required when the radioactivity in the effluent stream increases above a predetermined level.

The waste treatment processes used in nuclear power stations depend upon (a) the amounts and types of radioactive materials present in each effluent stream, (b) the total volume to be treated in each stream, and (c) the degree of reduction required to bring the releases into accordance with the applicable guidelines.

Methods used to reduce the radioactivity in effluents include (a) holding up of the waste to permit radionuclides to decay to an acceptable level, (b) reducing the source of radioactivity entering the effluent stream, and (c) selectively removing radioactive materials prior to discharge. Holdup pipes providing a 2-minute delay are used for BWR gland seal condenser vent releases. This holdup system provides sufficient time for only the short-lived radionuclides such as N-16 (half-life $\approx$7 seconds) to decay prior to release. The main condenser offgas holdup line for early BWR's was usually designed for a 30-minute holdup to allow the decay of radionuclides such as Xe-137 (half-life $\approx$3.8 minutes). A 30-minute holdup provides little reduction of radioactive gaseous effluents with half-lives exceeding a few minutes, resulting in large noble gas releases estimated to be more than $1 \times 10^6$ Ci/year for an early type BWR with a capacity of 2400 megawatts electric (MWe) (*32*).

In contrast to this, an early (PWR) with a 7-day pressurized holdup tank to delay primary system gases results in less than 5% as many curies per year of noble gas being released as for an early version BWR. The practical difference in delay times between the BWR and PWR cases is the result of the difference in volumetric flow rates. PWR primary system gas stripping flow is approximately 0.05% of BWR offgas flow.

For newer PWR power plants, pressurized storage tanks are used to provide up to a 60-day delay for primary system gases. For BWR offgas systems, additional treatment processes are necessary to increase holdup times from a few minutes to hours or days. A catalytic recombiner can be used to decrease the volume of offgases from the steam–jet air ejector by 90% to reduce the size of components used for subsequent treatment. Subsequent treatment processes may range from charcoal adsorbers to large charcoal delay systems and cryogenic distillation systems to hold up the noble gases and to provide almost complete removal of iodine.

Charcoal delay trains and cryogenic distillation units provide delay times up to 90 days. After 90 days, Kr-85 (half-life $\approx$10.8 years) is the only radionuclide released in significant quantities.

Catalytic recombiners reduce process volumes and increase operating safety by forming water from hydrogen and oxygen. In a BWR, hydrogen and oxygen formed from the radiolytic decomposition of water in the reactor are processed through the recombiner, leaving the turbine condenser air in-leakage as the carrier gas stream. In a PWR the excess hydrogen used as a primary coolant additive is mixed with oxygen and recombined into water. Charcoal delay systems are large charcoal beds which remove iodine and selectively delay the flow of noble gases. Their performance is dependent upon the flow rate, temperature, and moisture content of the stream treated. Their performance is also influenced by the type, mass, and physical characteristics of the charcoal used, the impurities in the carrier gas (e.g., CO, $CO_2$, NO, $NO_2$), the type of carrier gas (e.g., air, hydrogen), the system pressure, and the carrier gas velocity. The length of time a nuclide is delayed increases with (a) the amount of charcoal used, (b) decreasing temperature and humidity of the charcoal bed, and (c) decreasing carrier gas flow rate. Charcoal delay systems provide nearly complete removal of iodine, a delay of 1 to 2 days for Kr, and a delay of 13 to 35 days for Xe.

Deep-bed charcoal adsorbers are capable of reducing the iodine concentrations by a factor of 10. Charcoal tray filters are not suitable for this purpose because of their low efficiency at the low iodine concentrations encountered during normal reactor operations.

Cryogenic distillation units selectively remove radionuclides. In a cryogenic system the waste gas is first cooled to low temperature and dried to remove water, and finally stream temperature is reduced to approximately —320°F with liquid nitrogen. The Kr, Xe, and I are separated from the carrier gas (normally nitrogen, oxygen, or hydrogen) by distillation. These liquefied radionuclides are stored for radioactive decay prior to release.

Owing to the vapor pressure of iodine, krypton, and xenon, 0.01% of the Xe and I and 0.025% of the Kr are vented with the overhead gases. Adequate holdup time is provided for the activation gases that normally flow overhead with the distilled nitrogen. Cryogenic distillation offers the advantage of reduced storage volume requirements.

In the early BWR facilities the radioactive material released in the turbine-building ventilation air was insignificant compared to the total radioactive gases released from the facility. In the BWR facilities employing more advanced treatment systems, the radioactive materials present in the turbine-building ventilation air become significant. The radioactivity released in turbine-building ventilation air can be reduced by either treatment prior to discharge or by measures taken to reduce the radioactivity entering the building atmosphere.

A principal source of airborne radioactivity in the turbine building results from radioactive gases escaping with main steam. The steam leakage, and thus the radioactive gas leakage, can be reduced by using more efficient seals on valve stems in steam service.

### C. *Systems for Control of Gaseous Effluents from Nuclear Power Plants*

#### 1. Browns Ferry Nuclear Plant

The TVA Browns Ferry Nuclear Plant consists of three BWR units with a total electric generating capacity of 3456 megawatts. The first unit was placed in commercial operation in early 1974, with the other two units scheduled for commercial operation by 1975. The original plant design (Fig. 18) included 30-minute holdup and elevated-stack (600 ft) release of gases from the air ejector offgas system (*34*). The air ejector offgas system continuously removes from the reactor coolant system gases formed inside and leaked from the fuel elements, gases formed in the coolant, and in-leakage air in the condenser. For the gland seal offgases a large diameter pipe provides a 1.75-minute holdup time to allow decay of short-lived radioactive N-16 and O-19 from the gland seal and mechanical vacuum pumps.

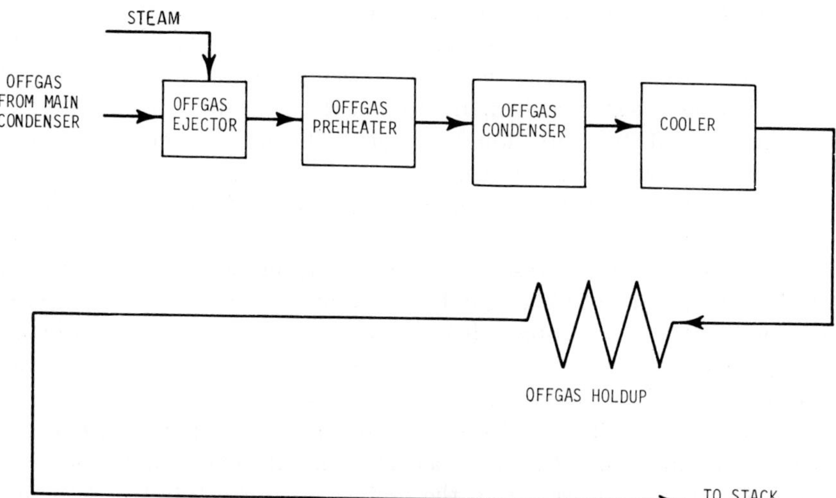


Figure 18. Original gaseous radwaste treatment system—Browns Ferry nuclear plant (*34*).

**Table X Benefits and Costs for Alternate Gaseous Waste Treatment Systems, Browns Ferry Nuclear Plant** (*34*)

| | *Annual external dose to any individual at the site boundary (mrem) three-unit plant* | *Reduction in annual external dose (mrem)*[a] | *Cost* | *Dollars per mrem reduction* | *Incremental cost of mrem reduction ($ per mrem)* |
|---|---|---|---|---|---|
| 30-Minute holdup | 290 | — | Base case | — | — |
| Recombiners only | 50.0 | 240 | $ 6,000,000 | $25,000 | $ 25,000 |
| Recombiners and 6 charcoal beds | 2.8 | 287.2 | 9,000,000 | 31,300 | 63,600[b] |
| Recombiners and 12 charcoal beds | 0.6 | 289.4 | 10,500,000 | 36,300 | 682,000[c] |
| Cryogenic system | less than 0.1 | 290 | 9,700,000 | 33,500 | 250,000[a] |

[a] Three-unit plant—0.5% failed fuel.
[b] Incremental cost over recombiners only.
[c] Incremental cost over recombiners and six charcoal beds.

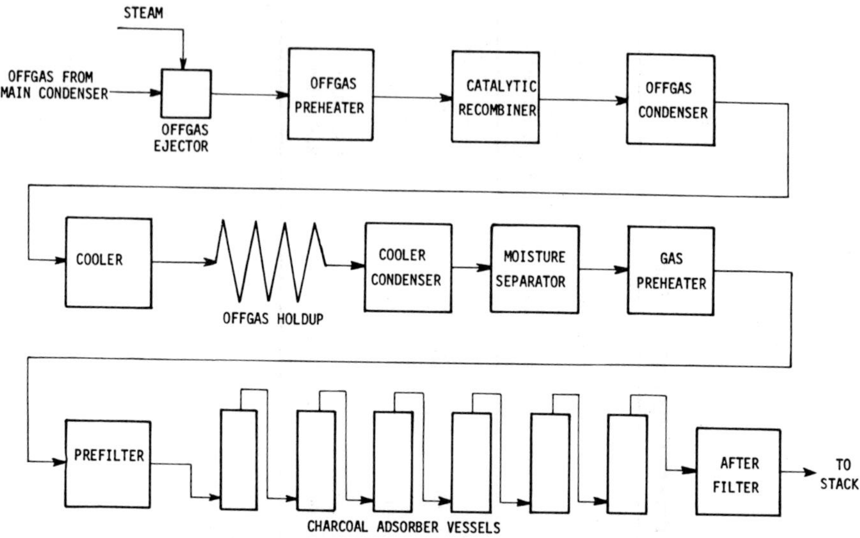


Figure 19. Extended gaseous radwaste treatment system—Browns Ferry nuclear plant (*34*).

As design and construction of the plant progressed, ways to provide additional treatment of gases were investigated to keep radioactivity in gaseous effluents as low as practicable. The results of these investigations are summarized in Table X. A recombiner and 6-charcoal bed system was selected as representing the best balance of cost, reduction in environmental impact, and feasibility (Fig. 19). In addition to gaseous waste handled by this system and discharged through the stack, ventilation air from the reactor and turbine buildings which may contain low levels of radioactivity is discharged through roof vents after filtration through high efficiency particulate absolute (HEPA) filters and charcoal filters. Discharges through the stack and all roof vents are monitored continuously. If predetermined levels of radioactivity for normal operations are exceeded, alarms are actuated and corrective measures indicated.

## 2. Bellefonte Nuclear Plant

The Bellefonte Nuclear Plant is a 2-unit PWR nuclear plant with a planned total electric generating capacity of 2664 megawatts. It is scheduled for commercial operation in 1979–1980.

The gaseous waste treatment and principal exhaust system for this plant are shown in Figure 20 (*31*). Most of the radioactive gas released from the reactor coolant system is handled in the gas decay system. Gases from the following sources are handled:

Reactor coolant bleed holdup tank
Reactor coolant bleed evaporators
Makeup tanks
Vents from pumps, filters, demineralizers, coolers in reactor coolant, makeup and purification, and chemical addition and boron recovery system
Relief valves
Reactor coolant drain tanks
Tritiated waste holdup tank
Tritiated auxiliary building sump tank
Waste evaporator feed tank
Waste evaporator
Waste evaporator distillate test tanks
Spent resin tank
Distillate storage tanks
Boric acid storage tanks
Reactor coolant bleed evaporator feed tanks
Reactor coolant bleed evaporator test tanks

Vents from each of these sources are connected to a compressor suction header. Two gas compressors are connected to this header. The compressors discharge to one of two 3000-ft$^3$ gas decay tanks. A recycle line between the compressor discharge and suction has a pressure-controlled valve which responds to pressure in the suction header.

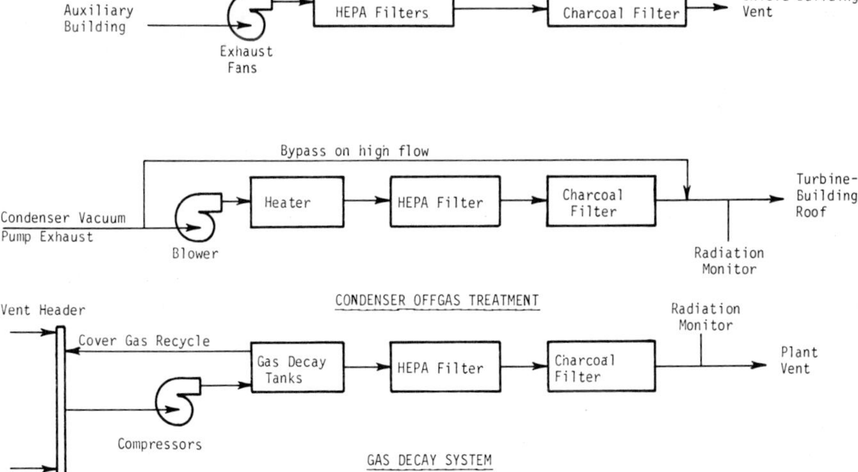


**Figure 20. Gaseous treatment and exhaust systems—Bellefonte nuclear plant (*31*).**

One compressor is normally in operation, with the other on standby. When the pressure in the suction header reaches 2.0 lb/in.$^2$ gauge, the compressor starts and runs until the header pressure drops to 0.5 lb/in.$^2$ gauge. When the header pressure drops below 0.5 lb/in.$^2$ gauge, the pressure-controlled valve in the recycle line opens, recycling gas from the storage tank to the header. Typically, when a tank such as a reactor coolant bleed tank is being filled with liquid, gas above the liquid is transferred by the compressors to the gas decay tanks. When the liquid level in the tank is lowered, gas is returned from the decay tank to the holdup tank by way of the recycle line. This arrangement allows a maximum decay time while minimizing gas consumption.

When the decay tank in service approaches a pressure of 100 lb/in.$^2$ gauge, the contents of the other tank are released. The tank vent line includes a radiation-controlled valve, a flow-controlled valve, a prefilter, high efficiency particulate filter, a charcoal filter (for iodine removal), and a radiation monitor. The monitor closes the radiation-controlled valve when a concentration of 5 $\mu$Ci/ml is exceeded. Downstream of the monitor, the line branches into two lines, each of which terminates at the top of a reactor shield building. Releases are made at a time when meteorological conditions are favorable and from the one of the two vents that will provide the best dispersion for the gases.

Other gaseous releases occur for which decay storage is not provided. Tanks and equipment in the radioactive liquid waste disposal system, other than those listed above, are vented to the auxiliary-building ventilation exhaust system. Radioactive gases and airborne particulates resulting from liquid leaks and spills in the auxiliary building are also picked up by the exhaust system. The exhaust is passed through HEPA and charcoal filters for particulate and iodine removal prior to discharge from the shield-building vent.

Because of primary to secondary leakage during operation, gases from the primary system are transported with the steam to the turbine and condenser. The gases, along with air that leaks into the condenser, are removed from the condenser by vacuum pump exhausters. The vacuum pump exhaust is passed through a blower, a heater, a HEPA filter, and two charcoal filters in series before being discharged to the turbine-building roof. This path is in operation except when vacuum pump flows in excess of 50 ft$^3$/minute occur, as during unit startup. A high-flow bypass is provided which automatically opens when there is high pressure drop across the HEPA filter.

Radioactive gases and particulates accumulate in the containment atmosphere during normal operation as a result of small leaks in the reactor coolant system. The containment atmosphere is continuously monitored for radioactivity during operation. The containment atmosphere is purged prior to entry of personnel or in order to equalize the containment pressure with the outside atmosphere. A containment auxiliary charcoal system is operated for about 8 hours prior to purging. This system consists of three fan-filter units, each located on a separate level of the building and arranged to enhance mixing of the total containment volume. HEPA and charcoal filters are provided in each unit for reduction of airborne particulates and iodine. The three units can circulate the containment atmosphere about two times during an 8-hour period. During purging, outside air is supplied to the containment. Purged air is exhausted through two 50% capacity fans and filter networks in parallel (high-efficiency particulate air filters and charcoal filters) to the plant vent where it is monitored during release to the atmosphere. The containment purge system has a capacity of approximately 1.5 complete changes of air per hour. Venting capacity is controlled by variable dampers.

In addition to purge of the full containment, the instrument room is purged separately for entries made about once per 2 weeks. The air from this room is exhausted through the containment purge exhaust system. Leakage of steam and feedwater during periods of operation with primary to secondary leakage introduces some radioactivity into the turbine-building atmosphere. Airborne radioactivity is exhausted, without treatment, at the turbine-building roof. All gaseous release systems are rigorously monitored. Whenever levels of radioactivity exceed predetermined limits, alarms are actuated, and corrective measures instituted.

In planning this gaseous radwaste treatment system, alternatives considered included the addition of recombiner and noble gas removal systems. Because of costs involved and the minimal incremental benefits to be provided, it was concluded that use of these systems was not justified. However, the present system can be modified to allow for future addition of such systems if, as a result of experience at this or similar plants or technological developments, subsequent evaluations should demonstrate their need and practicability.

## VIII. Air Quality Impacts of Emissions from Cooling Towers

The quantity of waste heat from large thermal power plants, mostly in the form of heat rejection to the condenser from the turbine exhaust is enormous. For a large modern fossil fuel-fired steam electric generating plant, heat rejection to the condenser is approximately 50% of the heat input to the boiler. This is equivalent approximately to about $1\frac{1}{3}$ kw hours of energy in the form of heat rejected to the condenser for each kilowatt hour of net electric generation. For nuclear fueled units, approximately

66% of heat input to the cycle is rejected to the condenser which is equivalent to about 2 kW hours of heat for each kilowatt hour of net electric generation (*35*). The availability of naturally occurring waters suitable for receiving the quantity of heat rejected from the condenser of large thermal power stations is limited. Where sufficient quantities and suitable discharge conditions do exist, their use as cooling water may be restricted because of water temperature standards or effluent standards which cannot be met by a once-through condenser system. These conditions lead to increasing use of cooling towers for condenser cooling for large thermal electric generating units. Cooling towers discharge heat removed from the condenser cooling water to the atmosphere. Cooling ponds and canals, with or without spray systems also may be used for this purpose, but these have relatively minor impacts on air quality as compared with impacts of cooling towers.

Cooling towers may be generally classified as either evaporative (wet) or nonevaporative (dry) or combination evaporative–nonevaporative (wet–dry). Schematics of major types of cooling towers used for major industrial installations such as power plants are presented in Figure 21 (*35*).

In the wet tower, water comes in intimate contact with the cooling air and transfers by evaporation about 75% of the total heat removed from the circulating water. Air–water contact occurs in the fill packing. Most of the remainder of the heat removed is by convection. Radiation and advection are negligible in a cooling tower.

The dry type of cooling tower uses either air-cooled condensers or air-

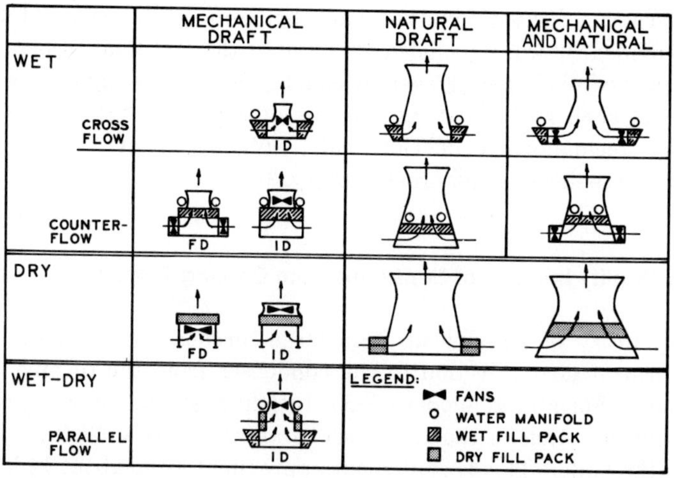

| | MECHANICAL DRAFT | NATURAL DRAFT | MECHANICAL AND NATURAL |
|---|---|---|---|
| WET CROSS FLOW | ID | | |
| COUNTER-FLOW | FD ID | | |
| DRY | FD ID | | |
| WET–DRY PARALLEL FLOW | ID | LEGEND: FANS; O WATER MANIFOLD; WET FILL PACK; DRY FILL PACK | |

Figure 21. Major types of cooling towers (35). FD: Forced draft; ID: induced draft.

cooled heat exchanges, and there is no contact between air and water. The wet–dry tower has both wet fill and dry heat exchange surfaces. The terms "cross flow" and "counterflow" describe wet towers with respect to the relative flow directions of the air and falling water in the fill.

Towers are further classified by the method of creating air movement through them. In natural draft towers air movement is induced by the large chimneylike tower and the density difference between the air inside and outside the tower. This results in an elevated (300 to 500 ft) point source of discharge for the cooling tower effluent to the atmosphere. In a mechanical draft unit, air is moved by fans, either the induced (ID) type that pulls air through the tower, or the forced (FD) type that pushes the air through. Typical large mechanical draft tower units consist of a series of connected cells about 50 ft high, 70 ft wide, and 40 ft deep, and are generally 300 to 350 ft in length (*36*). Mechanical draft cooling tower design results in low-level line source discharge of the tower discharge to the atmosphere. As shown in Figures 22 and 23 both natural draft and mechanical draft wet towers are equipped with drift (i.e., mist) eliminators which control the discharge of water droplets to a fraction of a percent of the circulating water flow. (See Chapter 7, this volume for further information on mist eliminators.)

The environmental impact of a cooling tower installation depends, in addition to local meteorology and the total quantity of heat released, upon

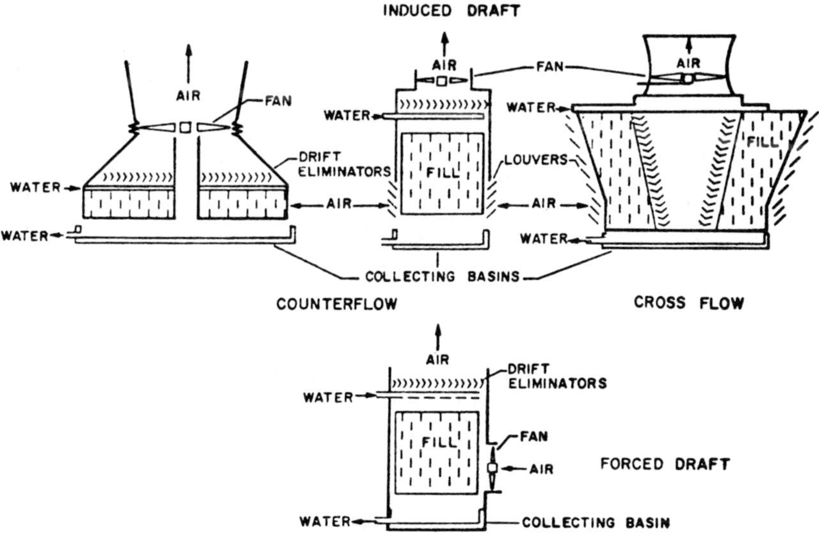


Figure 22. Mechanical-draft wet-type cooling towers showing locations of drift or mist eliminators (*39*).

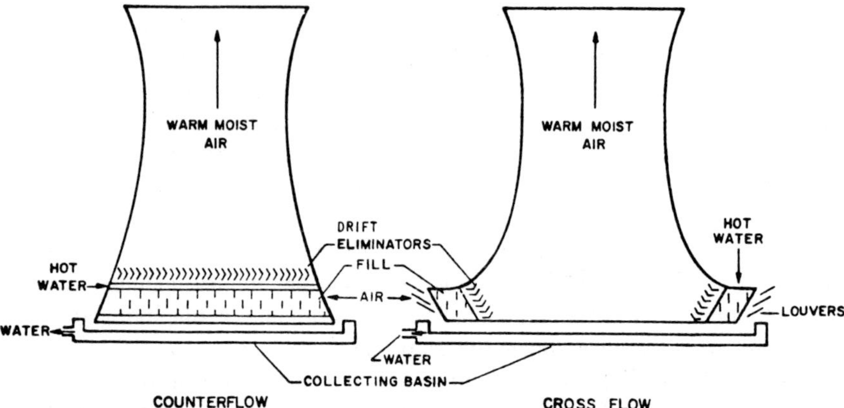


Figure 23. Hyperbolic natural-draft wet-type cooling towers showing locations of drift or mist eliminators (*39*).

the type and design of the cooling system, and, for evaporative cooling systems, the quality of the makeup water, allowable dissolved solids buildup in the recirculating water, and chemicals used to control scaling, corrosion, biological fouling, and other problems associated with the recirculation water.

Information on the impacts of thermal discharges on air quality has been summarized by Peterson (*37*). The principal impacts may be classified as those related to possible meteorological problems that may result from water vapor and heat releases and those related to salt deposition that may result from drift losses. The air quality significance of drift losses from cooling tower and associated evaporated salts that are formed in the atmosphere is yet to be determined. With the rapid increase in the number of large cooling tower installations being made, potential impacts of drift losses on air quality, in addition to those associated with salt deposition, can be expected to receive increased attention.

Approximately 1% evaporation is required to achieve a 10°F temperature reduction. This evaporation results in a buildup of concentrations of salts and other chemicals in the recirculation water. To control these concentrations to levels that will permit the tower to operate effectively, substantial volumes of the recirculation water must be discharged from the system. The discharge of salts and other chemicals to the atmosphere (aerial blowdown) depends upon the amount of drift losses and the operating concentrations in the circulating water. Normally aerial blowdown is small compared to blowdown discharge to a receiving water body. At one plant, however, because of unique operational restrictions contained in the water rights contract, design included a cooling water system for zero liquid blowdown under normal operating conditions (*38*). This was accomplished by extensive makeup water pretreatment and reliance upon "aerial blowdown" or drift from the towers to maintain concentrations of salts and other chemicals in the circulation water within control levels.

Factors affecting dispersion and deposition of drift from natural-draft and mechanical-draft towers, as summarized by Roffman (1973), are presented in Table XI (*39*). Roffman's report also illustrates mathematical models for calculating dispersion and deposition of drift from cooling towers and describes the effects of salt on biota and water bodies. Because of the problem of salt drift and deposition, if sea water or brackish water, is used, the number of saltwater cooling tower installations is quite limited. However, saltwater and brackish water cooling tower installations have been operated successfully (*39*). Improved designs of drift eliminators have resulted in manufacturers guarantees that drift will not exceed 0.002% of the circulating water in the towers (*35*). With these improved designs, more extensive use of saltwater and brackish water cooling towers can be expected.

The most common meteorological problem associated with wet cooling towers is an increase in the frequency of occurrence and the amount of

**Table XI Factors Affecting Dispersion and Deposition of Drift from Natural-Draft and Mechanical-Draft Cooling Towers (*39*)**

| *Factors associated with the design and operation of the cooling tower* | *Factors related to atmospheric conditions* | *Other factors* |
|---|---|---|
| Volume of water circulating in the tower per unit time | Atmospheric conditions including humidity, wind speed and direction, temperature, Pasquill's stability classes, which affect plume rise, dispersion, and deposition | Adjustments for non-point source geometry |
| Salt concentration in the water | Tower wake effect which is especially important with mechanical draft towers | Collection efficiency of ground for droplets |
| Drift rate | Evaporation and growth of drift droplets as a function of atmospheric conditions and the ambient conditions | |
| Mass size distribution of drift droplets | Plume depletion effects | |
| Moist plume rise influenced by tower diameter, height, and mass flux | | |

fog in the vicinity of towers (*37*). When the atmosphere is relatively humid, calm, and cooler than the tower emissions, fog or icing results. In contrast, wet tower emissions into a dry, warm and windy atmosphere generally will not cause local fog. With mechanical towers, because the effluent is emitted at relatively low height, the moisture plumes occasionally diffuse to ground level where practical problems can result, for example, where towers are located near heavily traveled highways. When the ambient temperature is below freezing, ice may form on objects upon which water droplets impinge. Such problems can be prevented or controlled by properly designed and operated mechanical wet–dry towers which combine wet and dry cooling sections in a single tower to provide flexibility in managing the discharge of a visible plume (*35*). Because of their height, emissions from natural draft towers reach ground level far less frequently than those from mechanical units, and hence fogging is correspondingly reduced. Only a few programs have been organized to systematically determine frequencies of occurrence of ground fogging for various meteorological circumstances. Consequently most quantitative estimates of ground fogging have come from mathematical model calculations of the dispersion of moist plumes. The descriptive probability of fogging or visibility obstructing potential from various types of cooling systems has been summarized by Peterson as shown in Table XII (*37*).

Emission of large amounts of heat and water vapor at times may cause

**Table XII Descriptive Probability of Visibility Obstructing Potential for Various Types of Cooling Systems** (*37*)

| *Type of cooling system* | *Descriptive probability of obstructing visibility* |
|---|---|
| Tall, natural-draft towers standing alone, fully equipped with drift eliminators | Extremely low, virtually zero |
| Tall, natural-draft towers alone but without drift eliminators | Low, but likely to occur with high humidity and stagnation |
| Tall, natural-draft towers, close to fossil fuel smokestacks emitting acid-producing stack gases | Low to substantial, depends on prevalence of wind, direction, and spacing of stack and tower(s) |
| Mechanical-draft towers emitting at low level | Substantial, but highly variable, depending on wind and orientation or grouping of units |
| Slack water, ponds, and spray ponds | Low to substantial, depending on the stagnation of the atmosphere and confinement of humidified air |

increased cloudiness in the vicinity of power plants. One study of this phenomenon was reported on by Niemeyer (1970) (*40*) and Stockham (1971) (*41*). In this study of water vapor from a group of four large natural-draft cooling towers at a coal-fired power plant, consideration was given to formation of ground fog, icing, reduction of visibility, abnormal precipitation, and possible interaction between water vapor plumes and local clouds, particularly convective clouds. Comingling or mixing of vapor plumes from cooling towers with the stack plumes containing $SO_2$ was observed.

Results from the study indicate that humidity increases can be detected for many miles downwind from the tower and that on frequent occasions the tower plumes can be seen to evaporate and then recondense to some extent at higher altitudes further downwind under higher relative humidities. In the presence of stable atmospheric conditions, the water vapor plumes persist after leveling off and appear as stratus clouds or merge with and reinforce existing cloud cover. Cumulus clouds initiated by the heat and water vapor from the plant have also been observed; generally these "man-made" clouds appear shortly before natural cumulus activity begins. Cloud initiation during periods of otherwise clear skies is infrequent. In summary, results from the study indicate that water vapor from cooling towers does have an effect on the environment but that this effect is limited and reinforces natural phenomena, i.e., formation of cumulus clouds that would occur in any event. If a cooling tower is located in a deep valley or in other topographically restricted regions, these effects may be greater than the study indicates.

As reported by Stockham the humid plumes from the cooling towers under study become mixed with the stack plumes at downwind distances ranging from 200 to 1000 m (*41*). Because the cooling tower water supplies were neutralized, the presence of acid droplets sampled in the converged plumes is considered *a priori* evidence that sulfur dioxide from the stack gases had reacted with moisture in the cooling tower plume. While a relationship apparently exists between humidity of the air and the ratio of acid to neutral droplets, no relationship was found between the sulfur dioxide content and the presence of acid droplets. Almost without exception, droplets in the plume, both acid and neutral, contained a large number of small solid particles, usually less than a few micrometers in size. The particles were presumed to be fly ash.

## REFERENCES

1. M. K. Hubbert, *in* "Environmental Aspects of Nuclear Power Stations," Paper IAEA-SM-146-1. IAEA, Vienna, Austria, 1971.

2. B. I. Spinarad, *in* "Environmental Aspects of Nuclear Power Stations," Paper IAEA-SM-146-2. IAEA, Vienna, Austria, 1971.
3. C. Starr, *Sci. Amer.* **225** (3), 37–49 (1971).
4. United States Council on Environmental Quality, "Energy and the Environment—Electric Power," Stock No. 411-00019. U.S. Govt. Printing Office, Washington, D.C., 1973.
5. G. R. Frying, ed., "Combustion Engineering," pp. 17–1 to 17–18. Combust. Eng., Inc., New York, New York, 1966.
6. "Steam, Its Generation and Use" pp. 10-1 to 10-3. Babcock & Wilcox, New York, New York, 1972.
7. J. E. Martin, E. D. Harward, D. T. Oakley, J. M. Smith, and P. H. Bedrosisian, *in* "Environmental Aspects of Nuclear Power Stations," Paper IAEA-SM-146-2. IAEA, Vienna, Austria, 1971.
8. United States Federal Power Commission, "The 1970 National Power Survey" (December 1971), Part I, p. I-11-3. U.S. Govt. Printing Office, Washington, D.C., 1971.
9. C. E. Blakeslee and H. E. Burbach, *J. Air Pollut. Contr. Ass.* **23**, No. 1, 37–42 (1973).
10. G. A. Hollinden, S. S. Ray, and N. D. Moore, Proc. Air Pollut. Contr. Div. Nat. Symp., Philadelphia, Pennsylvania, pp. 8-1 to 8-30 Amer. Soc. Mech. Eng., New York, New York, 1973.
11. G. N. Stone and A. J. Clarke, *Proc. Amer. Power Conf.* **39**, 540–556 (1967).
12. Tennessee Valley Authority Status Report for United States Environmental Protection Agency, "Control of Sulfur Oxides," Vol. 1, Part 3. Tennessee Valley Authority, Knoxville, Tennessee, 1973. (Presented at United States Environmental Protection Agency Public Hearing on Sulfur Oxide Compliance of Utility Power Plants, Washington, D.C., October 18–November 2, 1973.)
13. Testimony of American Electric Power Systems at United States Environmental Protection Agency Public Hearing on Sulfur Oxide Compliance of Utility Power Plants, Washington, D.C., October 18–November 2, 1973.
14. F. E. Gartrell, G. F. Stone, and T. A. Wojtalik, *in* "Environmental Aspects of Nuclear Power Stations," Paper IAEA-SM-146-78. IAEA, Vienna, Austria, 1971.
15. S. B. Carpenter, T. L. Montgomery, J. M. Leavitt, W. C. Colbaugh, and F. W. Thomas, *J. Air Pollut. Contr. Ass.* **21**, No. 8 (1971).
16. Status Report, "Control of Sulfur Oxides," Vol. I, Part 2. Tennessee Valley Authority, Knoxville, Tennessee, 1973. (Filed with the United States Environmental Protection Agency at Public Hearings on Sulfur Oxide Compliance of Utility Power Plants, Washington, D.C., October 18–November 2, 1973.)
17. R. M. Lundbergh, Elec. World Conf. Sulfur in Utility Fuels—Growing Dilemma, 1972 (unpublished).
18. Coal Bug on a Lonely Road, *Fortune,* January 1974, p. 19.
19. A. V. Slack, *Chem. Eng.* (*New York*) **74**(25), 188–196 (1967).
20. Office of Research and Development, "Full Scale Desulfurization of Stack Gas by Dry Limestone Injection," EPA-650/2-73-019a, pp. 28–29. United States Environmental Protection Agency, Washington, D.C., 1973.
21. "Sulfur Oxide Technology Assessment Panel (SOCTAP) on Projected Utilization of Stack Gas Cleaning Systems by Steam-Electric Plants," Report No. (EPA-APTD 1569). United States Environmental Protection Agency, Washington, D.C., 1973.
22. "Presentation on Status of Flue Gas Desulfurization Technology," given at

public hearings on sulfur dioxide compliance of utility plants. U.S. Environmental Protection Agency, Washington, D.C., 1973.
23. "Control Techniques for Particulate Air Pollutants," AP-51. Nat. Air Pollut. Contr. Admin., U.S. Department of Health, Education, and Welfare, Washington, D.C., 1969.
24. United States Federal Power Commission, "The 1970 National Power Survey," Part I, p. 1-11-15. U.S. Govt. Printing Office, Washington, D.C., 1971.
25. Committee on Air Pollution and Control, Division of Engineering, "Abatement of Particulate Emissions from Stationary Sources," Report No. COPAC-5. Nat. Res. Counc., Nat. Acad. Eng., Washington, D.C., 1972.
26. J. A. Hudson and J. Greco, *Proc. Amer. Power Conf., 1974,* pp. 454–463 (1974).
27. J. T. Reese and J. Greco, *J. Air Pollut. Contr. Ass.* 18(18) pp. 523–528 (1968).
28. U.S. Power Commission, "The 1970 Power Survey," Part I, pp. 1-8-1 to 1-8-7. U.S. Power Commission, Washington, D.C., 1971.
29. F. R. Jackson, J. T. Pavel, and J. N. Kovacik, *Combustion* 43, No. 3, 17–26 (1971).
30. Committee on Air Pollution and Control, Division of Engineering, "Abatement of Nitrogen Oxides Emissions from Stationary Sources," Report No. COPAC-4. Nat. Res. Counc., Nat. Acad. Eng., Washington, D.C., 1972.
31. "Draft Environmental Statement. Bellefonte Nuclear Plant." Tennessee Valley Authority, Knoxville, Tennessee, 1973.
32. "Final Environmental Statement concerning Proposed Rulemaking Action: Numerical Guides for Design Objectives and Limiting Conditions for Operation to Meet the Criterion As Low As Practicable for Radioactive Material in Light-Water-Cooled Nuclear Power Reactor Effluents." U.S. Atomic Energy Commission, Washington, D.C., 1973.
33. Directorate of Regulatory Operations, "Report on Releases of Radioactivity in Effluents and Solid Waste from Nuclear Power Plants for 1972." U.S. Atomic Energy Commission, Washington, D.C., 1973.
34. "Final Environmental Statement. Browns Ferry Nuclear Plant, Units 1, 2, and 3." Tennessee Valley Authority, Chattanooga, Tennessee, 1972.
35. T. D. Kolflat, *Power Eng.* 78, 32–39 (1974).
36. Woodson, R. D., *in* "Thermal Considerations in the Production of Electric Power" (M. Eisenbud and G. Gleason, eds.). Gordon & Breach, New York, New York, 1969.
37. J. T. Peterson, *In* "Effects and Methods of Control of Thermal Discharges" (report to Congress by the U.S. Environmental Protection Agency), pp. 571–613, Serial No. 93-14. U.S. Govt. Printing Office, Washington, D.C., 1973.
38. P. B. Christiansen and D. R. Colman, "Industrial Process Design for Water Pollution Control," Vol. II. Amer. Inst. Chem. Eng., New York, New York, 1970.
39. A. Roffman, "The State of the Art of Salt Water Cooling Towers for Steam Electric Generating Plants," Report No. WASH-1244 (prepared for the U.S. Atomic Energy Commission by Westinghouse Electric Corporation, Pittsburgh, Pennsylvania, February 1973). National Technical Information Service, U.S. Dept. of Commerce, Springfield, Virginia, 1973.
40. L. E. Niemeyer, R. A. McCormick, and J. H. Ludwig, *in* "Environmental Aspects of Nuclear Power Stations," pp. 711–722. IAEA,.Vienna, Austria, 1971.
41. J. Stockham, "Cooling Tower Study," Report APTF-0702 by IIT Research Institute, Chicago, Illinois, 1971, for Environmental Protection Agency, Durham, North Carolina. U.S. National Technical Information Service, U.S. Dept. of Commerce, Springfield, Virginia, 1971.

# 13

## Incineration

Richard C. Corey

## I. Introduction

Incineration is a waste-disposal process in which combustible solid, liquid, and gaseous wastes are burned *completely*, under controlled conditions. Ideally, incineration will reduce solid combustible wastes to a comparatively small volume of inert, nonputrescible, odorless residue, which can be disposed of in the same manner as any other clean fill material. Concomitantly, the gases discharged from the ideally operating incinerator stack will not contain products of incomplete combustion, such as smoke, tars, carbon monoxide, hydrogen, hydrocarbons, and charred particles.

Incineration is an exceedingly complex combustion process. It includes pyrolytic decomposition; surface and gas combustion; conductive, convective, and radiative heat transfer through heterogeneous media; and gas flow through randomly packed beds of material whose size, shape, and orientation are changing continually. The physical and chemical complexity of the burning processes, in conjunction with the heterogeneous nature of solid waste, results in incinerators frequently trading one environmental problem for others that may be more serious. For example, municipal incinerators, which disposed of about 18 million of the 200 million tons of household, institutional, and commercial solid wastes generated in the United States in 1972, often discharge to the atmosphere objectionable amounts of smoke, fly ash, and odors, and produce a residue that contains relatively large amounts of unburned, putrescible material. Such residues are difficult to dispose of in an environmentally acceptable manner; that is, without breeding rodents and insects, liberating offensive odors, and defacing the landscape.

## II. Background

The first furnaces designed specifically to burn solid wastes were operated in England about 100 years age (*1*). Since then, incinerators have slowly evolved from little better than an enclosed bonfire, to large sophisticated plants that are virtually nuisance free and aesthetically acceptable (*2, 3*). Niessen and others (*4*) have prepared a comprehensive inventory of American incinerators, giving their location, year built, capacity, air pollution equipment, and type of furnace and grate.

Among recent incinerator improvements has been a marked trend toward utilizing the heat generated to produce revenue. From 1930 through about 1965 heat utilization fell into disrepute. Since then there has been a growing trend toward utilizing the heat generated in incinerators, because, with growing national energy problems in the United States, Europe, and Japan, it makes sense to use combustible wastes as an important national energy resource.

In the United States, landfills handle most of the wastes, usually at a much lower average cost than incineration. However, acceptable landfill sites are steadily being exhausted. Composting, biodegradation of the organic matter in wastes, has not been successful on a municipal scale in the United States because of marketing problems with the humuslike product. The outlook, therefore, is that incineration will probably become the only suitable means of disposing of residential and commercial wastes in large municipalities.

Incinerator technology has been accelerated by recent theoretical studies of combustion in fuel beds, which allow better understanding of the heat, mass, and momentum transfer processes within and above a burning bed. It will take time, however, to reduce these theoretical findings to manageable equations and graphs that incinerator designers can use with confidence. Until this happens, each new large incinerator continues to be built as essentially a full-scale pilot plant, to the extent that it incorporates previously untried improvements. This is neither a new nor an undesirable technological phenomenon; modern steam generators, which are highly efficient and reliable, have been developed in this manner because there has been no satisfactory substitute for getting the data necessary to formulate definitive empirical relationships for future designs.

## III. The Urban Waste Problem

Municipal solid waste has been increasing at a rate of about 7 million tons annually, while the area available for economical disposal by landfill has been dwindling rapidly. The economic attractiveness of incineration to urban planners increases as land values and transportation costs continue to rise, because incineration facilities can be located near population centers, and require less land area than landfills.

**Table I Refuse—Composition Data**[a,b]

| Location | Notes | Food wastes | Yard wastes | Miscella-neous | Glass, ceramics | Metal | Paper products | Plastics | Leather, rubber | Textiles | Wood | Oil, paint, chemicals, etc. | Total |
|---|---|---|---|---|---|---|---|---|---|---|---|---|---|
| 4-City, New Jersey Region | Average for Paterson, Clifton, Passaic, Wayne | 8.3 | 13.3 | 8.96 | 6.44 | 9.44 | 43.87 | 2.66[f] | | 4.52 | 2.96 | | 100.49 |
| Composite | As collected, includes 9.05% adjusted moisture | 8.40 | 6.88 | 10.01[d] | | 6.85 | 52.70 | 1.52 | 0.76 | 0.76 | 2.29 | 0.76 | 99.98 |
| Hempstead, Long Island, New York | Predominantly residential, as received | 10.9 | 17.6 | | 9.6 | 8.5 | 42.6 | 4.6[f] | | 3.1 | 3.2 | | |
| Hempstead, Long Island, New York | Including residential and commercial, excluding bulky and industrial | 12.0 | 20[c] | | 8[e] | | 46.0 | 4[f] | | 3.0 | 7.0 | | 100.0 |
| Johnson City, Tennessee | Residential, 10/67 | 26.1 | 1.6 | 1.0 | 11.0 | 10.9 | 45.0 | 1.7 | 1.0 | 1.4 | 0.4 | | 100.01 |
| Johnson City, Tennessee | Municipal, 7/68 | 34.6 | 2.3 | 0.2 | 9.0 | 10.4 | 34.9 | 3.4 | 2.4 | 2.0 | 0.8 | | 100.0 |
| Weber County, Utah | Residential and commercial, 4/68 | 8.5 | 4.2 | 5.9 | 4.6 | 8.4 | 61.8 | 2.5[f] | | 2.0 | 2.2 | | 100.1 |
| Cincinnati, Ohio | Residential, 10/66 | 28.0 | 6.4 | — | 7.5 | 8.7 | 42.0 | 1.6 | 1.0 | 1.4 | 2.7 | | 99.3 |
| Memphis, Tennessee | Residential, 7/68 | 19.7 | 12.1 | 12.5 | 9.8 | 6.6 | 29.8 | 3.0[f] | | 4.8 | 1.7 | | 100.0 |
| Alexandria, Virginia | Residential and commercial, 5/68 | 7.5 | 9.5 | 3.4 | 7.5 | 8.2 | 55.3 | 3.1[f] | | 3.7 | 1.7 | | 99.9 |
| San Diego, California | Residential and commercial, 1967 | 0.8 | 21.1 | — | 8.3 | 7.7 | 46.1 | 0.3 | 4.7 | 3.5 | 7.5 | | 100.0 |
| Genesee County, Michigan | As collected, includes commercial, industrial domestic, and demolition wastes | 7.11 | 1.99 | 23.62 | 3.34 | 4.64 | 20.39 | 1.49[f] | | 3.01 | 22.41 | 12.00 | 100.0 |
| Flint, Michigan | Annual average | 32.6 | 13.5 | 0.3 | 17.9 | 14.5 | 17.5 | 2.3[g] | | 0.5 | 0.9 | | 100.0 |
| Genesee County, Michigan | Domestic | 26.0 | 10.8 | 0.2 | 14.3 | 11.8 | 34.0 | 1.8[g] | | 0.4 | 0.7 | | 100.0 |
| Santa Clara, California | As collected, domestic, average | 2.3 | 23.8 | | 12.7 | 7.6 | 47.5 | 1.0 | 1(av.) | 1.2 | 1(av.) | | 98.1 |
| Philadelphia, Pennsylvania | Includes significant quantities of industrial wastes, as collected | 5.0 | | 16.4 | 9.1 | 8.4 | 54.4 | 0.2 | 1.5 | 2.6 | 2.4 | | 100.0 |
| Jefferson County, Kentucky | Ar collected. residential average 66/67 | 19.8 | | 1.3 | 10.5 | 9.3 | 59.1 | | | | | | 100.0 |
| New Jersey | As collected | 10.0 | | | 4.0 | 8.0 | 51.0 | | 4.0[h] | | 4.0 | | 81.0 |
| Ohio | As collected | 28.0 | | | 8.0 | 9.0 | 42.0 | | 3.0[h] | | 3.0 | | 93.0 |
| Arizona | As collected | 22.0 | | | 8.0 | 10.0 | 43.0 | | 1.0[h] | | 2.0 | | 86.0 |
| California | As collected | 15.0 | | | 2.0 | 7.0 | 54.0 | | 2.0[h] | | 2.0 | | 82.0 |
| Tennessee | As collected | 26.0 | | | 11.0 | 11.0 | 46.0 | | 5.0[h] | | 0.3 | | 99.3 |
| Generalized analysis | From study made by Purdue University | 12.0 | 12.0 | 14.5 | 6.0 | 8.0 | 42.0 | 0.7 | 1.0 | 0.6 | 2.4 | 0.8 | 100.0 |
| Hamilton, Ontario | June 28–July 26 | 31.0 | 13.0 | — | 7.0 | 5.0 | 33.0 | 1.3 | 1.0 | 2.0 | 6.0 | | 99.0 |

[a] Niessen *et al.* (5).
[b] Weight percent, on an as-fired basis.
[c] Categories of yard wastes and miscellaneous grouped together for experimental purposes.
[d] Categories of miscellaneous and glass, ceramics grouped together for experimental purposes.
[e] Categories of glass, ceramics and metal grouped together for experimental purposes.
[f] Categories of plastics, leather, and rubber grouped together for experimental purposes.
[g] Categories of plastics, leather, and rubber, are grouped together for experimental purposes.
[h] Categories of plastics, leather, rubber, and textiles grouped together for experimental purposes.

**Table II Estimated Annual Average American Refuse Composition**[a,b]

| Component | Mean weight percent | Description |
|---|---|---|
| Glass | 9.9 | Bottles mainly |
| Metal | 10.2 | Cans, wire, foil |
| Paper | 51.6 | Various types, some with fillers |
| Plastics | 1.4 | PVC, polyethylene, styrene, etc. |
| Leather and rubber | 1.9 | Shoes, tires, toys, etc. |
| Textiles | 2.7 | Cellulosic, protein, and woven synthetics |
| Wood | 3.0 | Packaging, furniture |
| Food wastes | 19.3 | Garbage |
| Total | 100.0 | |

[a] Based on Niessen *et al.* (*5*).
[b] On an as-fired basis, without yard and miscellaneous wastes.

**Table III Estimated Proximate and Ultimate Analysis of American Urban Refuse (in weight percent)**

| | *Kaiser et al.* (*6*) | *Niessen et al.* (*5*) |
|---|---|---|
| Proximate analysis | | |
| Moisture | 28.0 | — |
| Volatile matter | 43.4 | — |
| Fixed carbon | 6.6 | — |
| Glass, ash, metal | 22.0 | — |
| Ultimate analysis | | |
| Moisture | 28.0 | 28.3 |
| Carbon | 25.0 | 25.6 |
| Hydrogen, net[a] | 3.3 | 3.4 |
| Oxygen | 21.1 | 21.2 |
| Nitrogen | 0.5 | 0.6 |
| Sulfur | 0.1 | 0.1 |
| Glass, ceramics, stones | 9.3 | — |
| Metals | 7.2 | 20.8 |
| Ash, other inerts | 5.5 | — |
| High heating value (HHV), Btu/lb | 4500 | 4450 |
| Stoichiometric air required, lb air/lb refuse[b] | 3.2 | 3.1 |

[a] Hydrogen, net = total hydrogen − (oxygen/8).
[b] Stoichiometric air as lb air/lb refuse can be approximated by multiplying the heating value by 0.0007.

### A. Urban Refuse Composition

The characteristics of refuse, particularly its chemical composition, volatile and ash content, and its heating value, have a direct effect on the design, number, and capacity of incinerators, and the kinds and quantities of air pollutants generated.

Table I shows the variation in refuse components for 23 American sources (*5*). The variations in yard wastes and miscellaneous categories are generally largest because they depend on local conditions. These data demonstrate clearly that each community must analyze its particular refuse for local planning, and cannot rely on national averages.

Table II is an average of the 23 sources given in Table I, with yard wastes and miscellaneous materials omitted. These data are useful for national planning, enabling one to predict changes in the average refuse composition based on anticipated changes in the amounts and types of materials discarded nationally.

For incinerator design and operation, the chemical composition, heating value, and air requirements for wastes (*5, 6*) are needed (Table III). These data are useful for comparative purposes only. For the design or performance calculation of a specific incinerator these data must be obtained for the specific wastes being burned.

The Incinerator Institute of America classifies wastes to be incinerated as shown in the following tabulation.

| *Type* | *Description* | *Principal components* | *Approximate composition, percent by weight* |
|---|---|---|---|
| 1 | Rubbish | Paper, cartons, rags, woodscraps, floor sweepings; domestic, commercial, industrial sources | Rubbish 100 (garbage up to 20) |
| 2 | Refuse | Rubbish and garbage; residential sources | Rubbish 50<br>Garbage 50 |
| 3 | Garbage | Animal and vegetable wastes, restaurants, hotels, markets; institutional, commercial, and club sources | Garbage 100 (rubbish up to 35) |
| 4 | Animal solids and organic wastes | Carcasses, organs, solid organic wastes, hospital, laboratory, abbatoirs, animal pounds, and similar sources | Animal and human tissue 100 |
| 5 | Gaseous, liquid or semiliquid wastes | Industrial process wastes | Variable |
| 6 | Semisolid and solid wastes | Combustibles requiring hearth, retort, or grate-burning equipment | Variable |

**Table IV Average Per Capita American Rate of Refuse Generation**

| *Source* | *lb/capita/day* | *Reference* |
|---|---|---|
| Wegman, 1964 | 4.4 | (*7*) |
| Combustion Engineering, Inc., 1969 | 5.1 | (*8*) |
| Fife, 1970 | 5.1 | (*9*) |
| Niessen *et al.*, 1970 | 5.6 | (*5*) |
| Warner *et al.*, 1970 | 5.3 | (*10*) |
| Drobny *et al.*, 1971 | 4.5 | (*11*) |

### *B. Urban Refuse Generation*

For making waste management plans, knowledge of the per capita rate of refuse generation is essential. Such a number depends heavily on local conditions (Table IV). These numbers are for generated refuse that potentially would be disposed of by municipal and private solid waste disposal systems, and do not include industrial and agricultural wastes (*5–11*).

### *C. Trends in Refuse Composition and Generation*

A comprehensive report of the expected overall urban refuse situation in the United States is available (*5*). The report predicts the urban refuse composition and characteristics through the year 2000 for three general climate regions by considering the growth rate of each of the categories in the refuse analysis (Tables V and VI).

It is important to note that unanticipated major changes in packaging, legislation, or consumption could affect any of the categories. The projections indicate that paper and plastics will be the fastest growing components of refuse, with growth rates of 35 and 400%, respectively. These changes would have a marked effect on future design of incinerators and their air and water pollution control systems. Other significant trends include lower bulk density, higher heating value, and lower moisture content of urban wastes.

The report predicts that three factors will combine to increase the solid waste burden in urban areas by the year 2000: (a) the per capita rate of waste generation will increase 72%; (b) population will increase 54%; and (c) the percentage of solid waste generated that is actually collected for final disposal will increase from the current value of about 70 to 95%.

The great increase in the total urban refuse generated will, of course, necessitate increased facilities for handling the refuse. A Combustion

**Table V Projected United States Refuse Properties and Statistics[a]**

| Refuse properties and statistics | 1968[b] | | | 1970[b] | | | 1975[b] | | |
|---|---|---|---|---|---|---|---|---|---|
| | *A* | *B* | *C* | *A* | *B* | *C* | *A* | *B* | *C* |
| High heating value (HHV), Btu/lb | 4582 | 4505 | 4628 | 4628 | 4550 | 4493 | 4719 | 4640 | 4582 |
| Percent moisture | 25.9 | 27.8 | 29.3 | 25.2 | 27.1 | 28.6 | 28.4 | 25.3 | 26.9 |
| Percent ash content | 21.8 | 20.3 | 19.1 | 22.1 | 20.7 | 19.5 | 22.9 | 21.5 | 20.3 |
| Percent ash (excluding glass, metals) | 5.5 | 5.3 | 5.1 | 5.5 | 5.2 | 5.1 | 5.3 | 5.1 | 4.9 |
| Per capita growth multiplier | 1.0 | 1.0 | 1.0 | 1.05 | 1.05 | 1.05 | 1.19 | 1.18 | 1.18 |
| National population growth multiplier | 1.0 | 1.0 | 1.0 | 1.02 | 1.02 | 1.02 | 1.07 | 1.07 | 1.07 |
| Total waste-load multiplier | 1.0 | 1.0 | 1.0 | 1.07 | 1.07 | 1.07 | 1.27 | 1.26 | 1.26 |
| Per capita heat-rate multiplier, Btu/person day | 1.0 | 1.0 | 1.0 | 1.06 | 1.06 | 1.06 | 1.23 | 1.22 | 1.22 |
| Total heat-rate multiplier, Btu day | 1.0 | 1.0 | 1.0 | 1.08 | 1.08 | 1.08 | 1.32 | 1.31 | 1.31 |

| Refuse properties and statistics | 1980[b] | | | 1990[b] | | | 2000[b] | | |
|---|---|---|---|---|---|---|---|---|---|
| | *A* | *B* | *C* | *A* | *B* | *C* | *A* | *B* | *C* |
| High heating value (HHV), Btu/lb | 4811 | 4730 | 4627 | 5040 | 4956 | 4849 | 5407 | 5271 | 5161 |
| Percent moisture | 22.1 | 24.0 | 25.7 | 20.5 | 22.4 | 24.1 | 19.3 | 21.3 | 22.9 |
| Percent ash content | 23.5 | 22.0 | 20.3 | 22.4 | 21.1 | 20.0 | 19.7 | 18.6 | 17.7 |
| Percent ash (excluding glass, metals) | 5.2 | 5.0 | 4.9 | 5.2 | 5.0 | 4.9 | 5.3 | 5.2 | 5.0 |
| Per capita growth multiplier | 1.32 | 1.32 | 1.31 | 1.52 | 1.51 | 1.50 | 1.76 | 1.74 | 1.72 |
| National population growth multiplier | 1.13 | 1.13 | 1.13 | 1.33 | 1.33 | 1.33 | 1.54 | 1.54 | 1.54 |
| Total waste-load multiplier | 1.49 | 1.49 | 1.48 | 2.02 | 2.01 | 2.00 | 2.71 | 2.68 | 2.65 |
| Per capita heat-rate multiplier, Btu/person day | 1.39 | 1.39 | 1.36 | 1.67 | 1.66 | 1.64 | 2.08 | 2.04 | 2.00 |
| Total heat-rate multiplier, Btu day | 1.57 | 1.57 | 1.54 | 2.22 | 2.21 | 2.18 | 3.20 | 3.14 | 3.08 |

[a] Niessen *et al.* (*5*).
[b] A, seasonal (e.g., Massachusetts); B, semiseasonal (e.g., Alabama); C, nonseasonal (e.g., Florida).

Engineering, Inc., study (*8*) shows that at the time of their report (1969), approximately 50% of United States cities over 25,000 in population using sanitary landfill had less than 6 years of life left in their existing landfills. About 20% of the communities over 25,000 used incineration

**Table VI Projected United States Refuse Composition**[a,b]

| | 1968[c] | | | 1970[c] | | | 1975[c] | | | 1980[c] | | | 1990[c] | | | 2000[c] | | |
|---|---|---|---|---|---|---|---|---|---|---|---|---|---|---|---|---|---|---|
| *Refuse category* | *A* | *B* | *C* | *A* | *B* | *C* | *A* | *B* | *C* | *A* | *B* | *C* | *A* | *B* | *C* | *A* | *B* | *C* |
| Glass | 8.8 | 8.1 | 7.6 | 9.1 | 8.4 | 7.9 | 9.9 | 9.2 | 8.7 | 10.3 | 9.6 | 9.0 | 9.5 | 8.9 | 8.4 | 8.1 | 7.6 | 7.2 |
| Metal | 8.7 | 8.1 | 7.5 | 8.8 | 8.2 | 7.6 | 9.0 | 8.4 | 7.8 | 9.4 | 8.7 | 8.1 | 9.0 | 8.4 | 7.9 | 7.4 | 6.9 | 6.5 |
| Paper | 38.2 | 35.1 | 32.6 | 39.1 | 35.8 | 33.5 | 40.8 | 37.6 | 35.2 | 41.5 | 38.4 | 36.1 | 45.0 | 41.7 | 39.3 | 49.7 | 46.0 | 43.5 |
| Plastics | 1.1 | 1.1 | 1.0 | 1.3 | 1.3 | 1.1 | 1.9 | 1.8 | 1.7 | 2.8 | 2.7 | 2.5 | 3.5 | 3.5 | 3.1 | 4.2 | 4.2 | 3.8 |
| Leather, rubber | 1.5 | 1.4 | 1.3 | 1.5 | 1.4 | 1.3 | 1.5 | 1.4 | 1.3 | 1.5 | 1.4 | 1.3 | 1.5 | 1.4 | 1.3 | 1.6 | 1.5 | 1.4 |
| Textiles | 2.0 | 1.9 | 1.8 | 2.0 | 1.9 | 1.8 | 2.1 | 2.0 | 1.9 | 2.1 | 2.0 | 1.9 | 2.5 | 2.3 | 2.2 | 2.8 | 2.6 | 2.5 |
| Wood | 2.7 | 2.4 | 2.3 | 2.5 | 2.3 | 2.2 | 2.2 | 2.0 | 1.9 | 2.0 | 1.8 | 1.7 | 1.6 | 1.5 | 1.4 | 1.3 | 1.2 | 1.2 |
| Food wastes | 21.1 | 19.5 | 18.2 | 20.2 | 18.7 | 17.4 | 17.9 | 16.6 | 15.5 | 16.2 | 15.0 | 14.1 | 14.0 | 13.1 | 12.3 | 12.1 | 11.4 | 10.7 |
| Miscellaneous | 1.8 | 1.7 | 1.6 | 1.7 | 1.6 | 1.5 | 1.5 | 1.4 | 1.3 | 1.4 | 1.3 | 1.2 | 1.2 | 1.1 | 1.1 | 1.0 | 1.0 | 0.9 |
| Yard wastes | 14.1 | 20.7 | 26.1 | 13.8 | 20.4 | 25.7 | 13.2 | 19.6 | 24.7 | 12.9 | 19.2 | 24.1 | 12.2 | 18.1 | 23.0 | 11.8 | 17.6 | 22.3 |
| | 100.0 | 100.0 | 100.0 | 100.0 | 100.0 | 100.0 | 100.0 | 100.0 | 100.0 | 100.0 | 100.0 | 100.0 | 100.0 | 100.0 | 100.0 | 100.0 | 100.0 | 100.0 |

[a] Niessen *et al.* (*5*).
[b] Weight percent.
[c] A, Seasonal state (e.g., Massachusetts); B, semiseasonal state (e.g., Alabama); C, nonseasonal state (e.g., Florida).

in 1966, with a total installed incinerator capacity of 75,000 tons per day (tpd).

Incinerator designers will have to cope with several problems caused by the trends in refuse generation and composition. For example, increased heating value and lower moisture content of refuse will reduce the capacity of refractory-lined incinerators, otherwise the resultant higher flame temperatures would damage the furnaces. Increased volatile matter in the refuse will increase significantly the amount of heat released in the furnace volume above the grate compared to the heat released on the grate. The greatest concern among the anticipated trends is the effect of increased content of plastics in refuse. Therefore, unless designers anticipate these trends in refuse composition, pollutant emissions will increase, and maintenance costs and availability will be impaired.

## IV. Contemporary Incinerator Technology

Contemporary incinerators and incineration practices have been described extensively by Corey (*12*), Friedrich (*13*), and Baum and Parker (*13a*). Incinerators for solid wastes are categorized broadly into two classes: *on-site* units that are used mostly in residential complexes such as apartment houses, in commercial activities such as supermarkets, and in institutions such as hospitals; and *central* units, commonly known as municipal incinerators, to which solid wastes from cities and suburbs are transported for incineration.

### A. On-Site Incinerators

This class of incinerators fulfills the need for solid waste disposal *at the source.* There are two basic types of on-site incinerators, *single chamber* and *multiple chamber.* Since the single chamber incinerator is unacceptable because of excessive air pollution, only multiple-chamber incinerators, the *retort* type, and the *in-line* type will be discussed. Both are batch-fed systems (*14*). Other incinerator configurations, including incinerators with vertically arranged chambers, L-shaped units, and units with separated chambers breeched together, have appeared as variations of these two basic types.

Figure 1 shows a cutaway of a typical batch-fed retort type of multiple-chamber incinerator. The unit derives its name from the return flow of effluent gases through the U-shaped gas path, and the side-by-side arrangement of the component chambers. Figure 2 shows a typical batch-fed in-line multiple-chamber design, so called because the various cham-

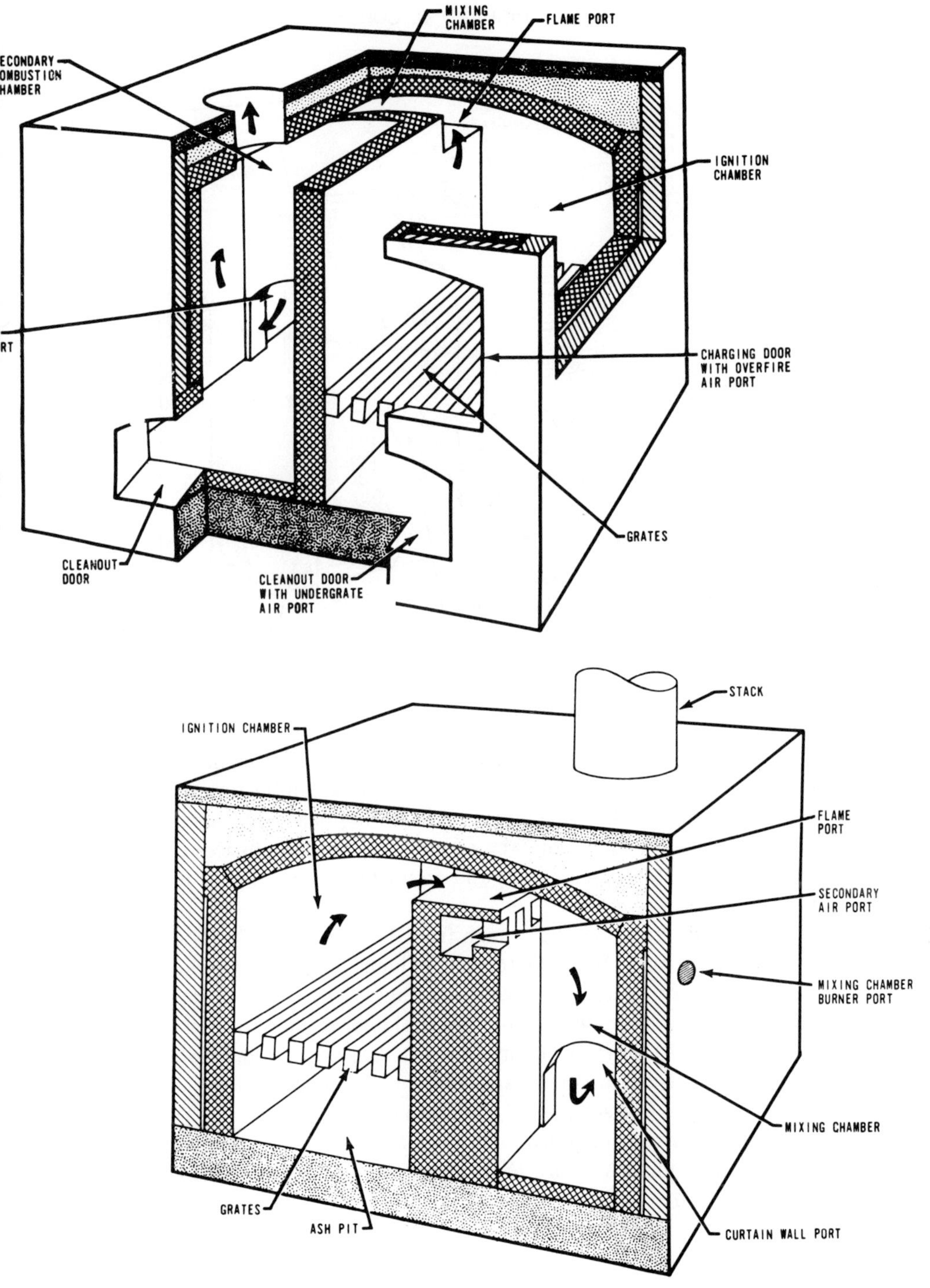


Figure 1. Cutaway of a batch-fed retort multiple-chamber incinerator (*14*).

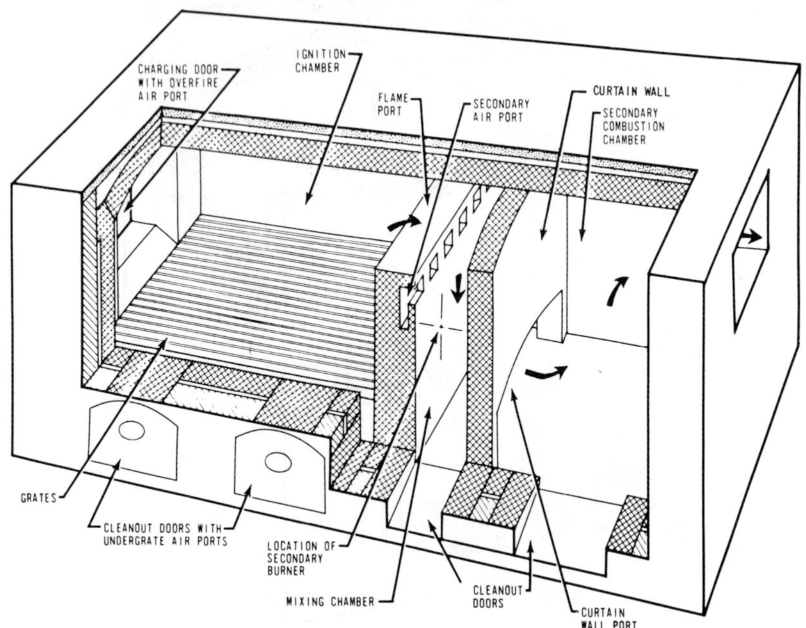


Figure 2. Cutaway of a batch-fed in-line multiple-chamber incinerator (*14*).

bers follow each other in a line. In both types the combustion process proceeds in two stages: primary, or solid-phase, combustion in the ignition chamber, followed by secondary, or gaseous-phase, combustion in subsequent chambers. The secondary combustion zone comprises two parts—a downdraft, or mixing chamber, and an up-pass expansion or final combustion chamber.

Acceptable grate loadings range from 15 to 25 lb/($ft^2$) (hour) for units with burning rates from 25 to 750 lb/hour. For burning rates in excess of 750 lb/hour acceptable grate loadings range from 25 to 35 lb/($ft^2$) (hour). Often, such incinerators are designed to operate with grate loadings of 50 to 70 lb/($ft^2$) (hour), but this has been accomplished by using underfire air as the predominant source of combustion air, a practice that leads to excessive discharge of contaminants.

Auxiliary gas- or oil-burners usually are needed in the mixing chamber if the refuse contains more than about 25% moisture. Refuse with more than about 50% moisture often requires a burner in the ignition chamber as well as in the mixing chamber.

Auxiliary burners usually are sized to provide 5000 to 10,000 Btu/hour for each hourly pound of moisture charged to the incinerator. Venturi-type gas burners, equipped with a burner block and cage, are recommended for incinerators rated up to 350 lb/hour, while nozzle-mix gas burners are recommended for larger incinerators. Burner adjustments should be made both to give a luminous flame, enhancing radiant heat transfer from the flame, and to provide flame coverage over the entire cross-sectional area of the mixing chamber.

The Los Angeles County Air Pollution Control District (LACAPCD) has developed empirical design data for on-site incinerators (*14*) based on extensive performance tests (Figs. 3 and 4). The LACAPCD makes the following recommendations concerning gas scrubbers to remove particulate matter from the stack gases:

(a) The water rate to the scrubber should be about 1 gallon per minute (gpm) for each 100 lb/hour of rated incinerator capacity.

(b) The exhaust fan should be designed to handle 700 cubic feet per minute (cfm) at 350°F for each 100 lb/hour of rated incinerator capacity.

(c) The fan should be designed to provide $\frac{1}{2}$ in. water static pressure for a 50-lb/hour incinerator, increasing uniformly to $1\frac{1}{2}$ in. for a 2000-lb/hour unit. The static pressures should be developed with the fan operating at 350°F.

The dimensional standards proposed by the LACAPCD for water scrubbers for retort and in-line multiple-chamber incinerators are given in Figures 5 and 6. Approximate costs of multiple chamber incinerators in 1966 are given in Table VII. These costs do not include foundations, electrical wiring, and gas and water piping.

Although there is a growing trend toward central incineration of community wastes, on-site incineration will continue to hold an important place in waste disposal, because, in some communities, municipal collection does not adequately serve commercial and industrial establishments and large multiple dwellings. On-site incineration can be accomplished without adverse environmental effects if the incinerators are designed and operated carefully. Since even a well-designed multiple-chamber incinerator will emit offensive discharges if it is not operated properly, the operating guide developed by the LACAPCD (Fig. 7) has proved to be very valuable in promoting clean operation.

## Multiple-Dwelling Incinerators

This type of on-site incinerator, usually called a flue-fed incinerator, fulfilled the need, many years ago, for an incinerator in apartment houses and other multistoried structures. The basic idea was to enable tenants to dump their refuse through a hopper door on each floor, each door con-

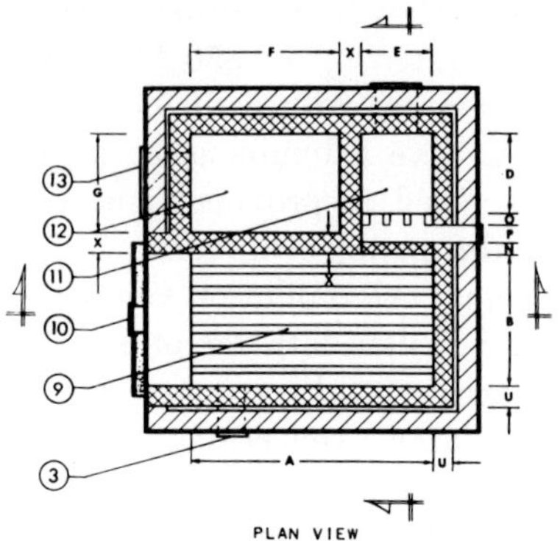


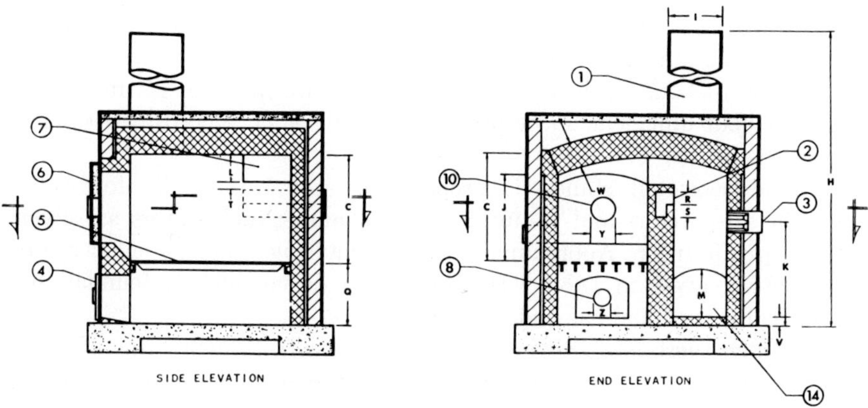


| SIZE OF INCINERATOR, lb/hr | A | B | C | D | E | F | G | H* | I | J | K | L | M | N | O | P | Q | R | S | T | U | V | W | X | Y | Z |
|---|---|---|---|---|---|---|---|---|---|---|---|---|---|---|---|---|---|---|---|---|---|---|---|---|---|---|
| | LENGTH, inches | | | | | | | | | | | | | | | | | | | | | | | | | |
| 50 | 31½ | 13½ | 22½ | 9 | 6¾ | 20¼ | 13½ | 18 | 8 | 18½ | 20 | 3¾ | 10 | 4½ | 2¼ | 2¼ | 9 | 2½ | 2½ | 2½ | 4½ | 2½ | 4½ | 4½ | 6 | 4 |
| 100 | 40½ | 18 | 28½ | 13½ | 9 | 27 | 18 | 19 | 12 | 23 | 28 | 5 | 15 | 2½ | 2½ | 4 | 14½ | 5 | 0 | 2½ | 4½ | 2½ | 4½ | 4½ | 8 | 5 |
| 150 | 45 | 22½ | 33½ | 15½ | 11½ | 29 | 22½ | 20 | 14 | 27 | 35½ | 5 | 16½ | 4½ | 2½ | 4½ | 18 | 5 | 2½ | 2½ | 4½ | 2½ | 4½ | 4½ | 9 | 6 |
| 250 | 54 | 27 | 37½ | 18 | 13½ | 36 | 27 | 22 | 18 | 30 | 40 | 7½ | 18 | 4½ | 4½ | 4½ | 20 | 5 | 2½ | 2½ | 4½ | 2½ | 4½ | 4½ | 12 | 6 |
| 500 | 76½ | 36 | 47½ | 27 | 18 | 49½ | 36 | 28 | 24 | 36½ | 48½ | 12½ | 23 | 9 | 4½ | 4½ | 26 | 5 | 5 | 2½ | 9 | 4½ | 9 | 9 | 16 | 8 |
| 750 | 85½ | 49½ | 54 | 36 | 22½ | 54 | 45 | 32 | 30 | 40 | 51½ | 15 | 28 | 9 | 4½ | 4½ | 25 | 5 | 10 | 2½ | 9 | 4½ | 9 | 9 | 18 | 8 |
| 1000 | 94½ | 54 | 59½ | 36 | 27 | 58½ | 45 | 35 | 34 | 45 | 54½ | 17½ | 30 | 9 | 4½ | 4½ | 27½ | 7½ | 12½ | 2½ | 9 | 4½ | 9 | 9 | 22 | 10 |

*Dimension "H" given in feet.

Figure 3. Design standards for batch-fed retort multiple-chamber incinerators (*14*). 1: Stack; 2: secondary air port; 3: gas burners; 4: ash pit cleanout door; 5: grates; 6: charging door; 7: flame port; 8: underfire air port; 9: ignition chamber; 10: overfire air port; 11: mixing chamber; 12: combustion chamber; 13: cleanout door; 14: curtain wall port.

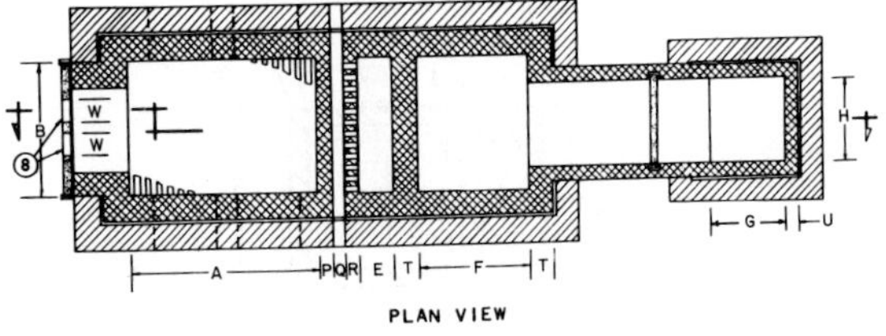


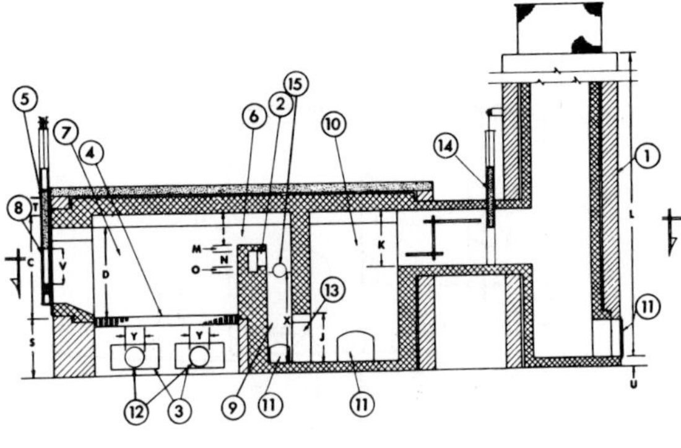


| SIZE OF INCINERATOR, lb/hr | LENGTH, inches | | | | | | | | | | | | | | | | | | | | | | | | |
|---|---|---|---|---|---|---|---|---|---|---|---|---|---|---|---|---|---|---|---|---|---|---|---|---|---|
| | A | B | C | D | E | F | G | H | I | J | K | L* | M | N | O | P | Q | R | S | T | U | V | W | X | Y |
| 750 | 85½ | 49½ | 51½ | 45 | 15¾ | 54 | 27 | 27 | 9½ | 24 | 18 | 32 | 4½ | 5 | 7½ | 9 | 2½ | 2½ | 30 | 9 | 4½ | 5 | 11 | 51 | 7 |
| 1000 | 94½ | 54 | 54 | 47½ | 18 | 63 | 31½ | 31½ | 11 | 29 | 22½ | 35 | 4½ | 5 | 10 | 9 | 2½ | 2½ | 30 | 9 | 4½ | 7 | 12 | 52 | 8 |
| 1500 | 99 | 76½ | 65 | 55 | 18 | 72 | 36 | 36 | 12½ | 32 | 27 | 38 | 4½ | 5 | 7½ | 9 | 4½ | 4½ | 30 | 9 | 4½ | 8 | 14 | 61½ | 9 |
| 2000 | 108 | 90 | 69½ | 57½ | 22½ | 79½ | 40½ | 40½ | 15 | 36 | 31½ | 40 | 4½ | 5 | 10 | 9 | 4½ | 4½ | 30 | 9 | 4½ | 9 | 15 | 63½ | 10 |

*Dimension "L" given in feet.

Figure 4. Design standards for batch-fed in-line multiple-chamber incinerators (*14*). 1: Stack; 2: secondary air ports; 3: ash pit cleanout doors; 4: grates; 5: charging door; 6: flame port; 7: ignition chamber; 8: overfire air port; 9: mixing chamber; 10: combustion chamber; 11: cleanout doors; 12: underfire air ports; 13: curtain wall port; 14: damper; 15: gas burners.

necting to a vertical duct leading to an incinerator in the building basement. At certain intervals the refuse that accumulated in the incinerator was burned, and the residue was collected and hauled away.

Because of poor design, inept operation, and tenant reluctance to follow simple rules, this type of incinerator usually has been a community nuisance. There are heavy discharges of smoke, and large particles of char, which present a serious fire hazard because the particles often are incandescent; offensive odors from the stacks; and putrescible residues that attract rodents and breed vermin.

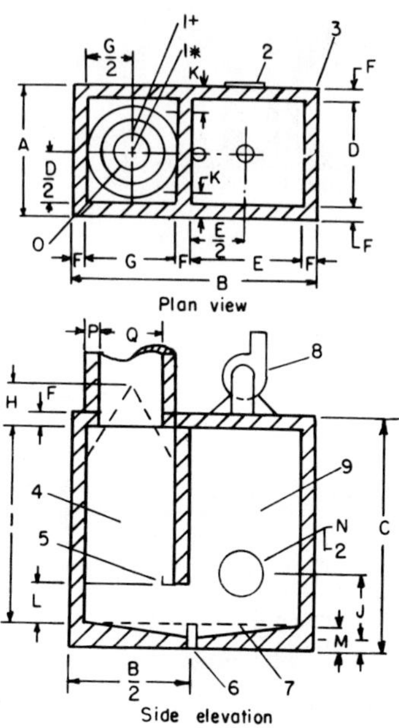


| Size of incinerator, pounds per hour | A | B | C | D | E | F | G | H | I | J | K | L | M | N | O | P | Q | No. of nozzles | Water flow GPM/nozzle | CFM | Static pressure drop (in.) | Fan (horsepower) |
|---|---|---|---|---|---|---|---|---|---|---|---|---|---|---|---|---|---|---|---|---|---|---|
| | Dimensions in inches | | | | | | | | | | | | | | | | | | | | | |
| 50 | 21 | 42 | 42 | 15 | 23 | 3 | 10 | 5 | 30 | 20 | 2 | 6 | 6 | 20 | – | $2\frac{1}{2}$ | 8 | 1 | .67 | 350 | 1/2 | 1/4 |
| 100 | 31 | 51 | 51 | 18 | 23 | 3 | 17 | 9 | 37 | 20 | 2 | 8 | 6 | 24 | – | $2\frac{1}{2}$ | 12 | 1 | 1.3 | 700 | 5/8 | 1/3 |
| 150 | 31 | 53 | 60 | 21 | 25 | 3 | 19 | 10 | 45 | 20 | 2 | 9 | 6 | 24 | – | $2\frac{1}{2}$ | 14 | 1 | 1.9 | 1050 | 3/4 | 3/4 |
| 250 | 31 | 58 | 68 | 27 | 25 | 3 | 24 | 13 | 48 | 20 | 2 | 12 | 6 | 24 | – | $2\frac{1}{2}$ | 18 | 1 | 2.0 | 1750 | 3/4 | 1 |
| 500 | 32 | 66 | 87 | 33 | 26 | 3 | 33 | 5 | 62 | 20 | 4 | 18 | 7 | 24 | 17 | $4\frac{1}{2}$ | 24 | 4 | 1.3 | 3500 | 7/8 | 2 |
| 750 | 48 | 79 | 93 | 40 | 27 | 4 | 40 | 8 | 65 | 20 | 4 | 21 | 7 | 24 | 20 | $4\frac{1}{2}$ | 30 | 4 | 1.9 | 5250 | 1 | 3 |
| 1000 | 54 | 85 | 100 | 46 | 28 | 4 | 46 | 8 | 68 | 20 | 4 | 25 | 7 | 24 | 23 | $4\frac{1}{2}$ | 34 | 4 | 2.9 | 7000 | 1 | 5 |

**Figure 5.** Design standards for batch-fed retort multiple-chamber incinerator scrubbers (*14*). 1: Spray nozzles; 2: access door; 3: castable refractory; 4: downpass; 5: channel; 6: drain; 7: water level; 8: induced draft fan; 9: up-pass.

Many efforts have been made to upgrade existing multiple-dwelling incinerators, and some guidelines are now available. Because this means of on-site disposal is a poor option in new multiple dwellings, it will not be discussed further here. However, the reader will find information on

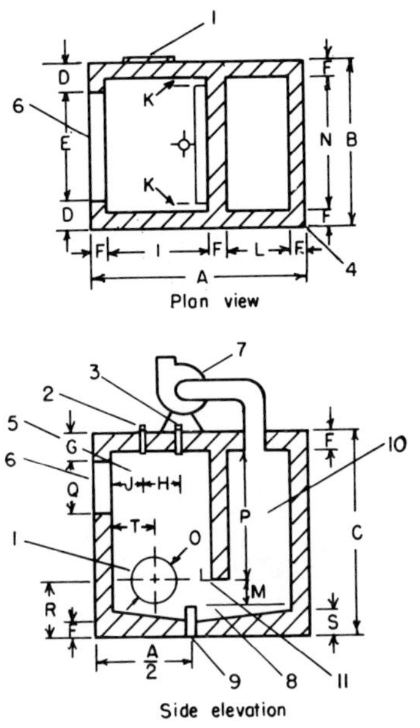


| SIZE OF INCINERATOR, POUNDS PER HOUR | DIMENSIONS IN INCHES | | | | | | | | | | | | | | | | | | | |
|---|---|---|---|---|---|---|---|---|---|---|---|---|---|---|---|---|---|---|---|---|
| | A | B | C | D | E | F | G | H | I | J | K | L | M | N | O | P | Q | R | S | T |
| 750 | 61 | 58 | 92 | 8 | 42 | 4 | $12\frac{1}{2}$ | 12 | 32 | 8 | 2 | 17 | 17 | 50 | 24 | 63 | $17\frac{1}{2}$ | 30 | 8 | 16 |
| 1000 | 75 | 62 | 102 | 7 | 48 | 4 | 18 | 15 | 39 | 10 | 2 | 24 | 20 | 54 | 24 | 70 | 20 | 30 | 8 | 20 |
| 1500 | 78 | 80 | 113 | 9 | 63 | 4 | 14 | 18 | 42 | 12 | 2 | 24 | 24 | 72 | 24 | 81 | 22 | 30 | 9 | 21 |
| 2000 | 90 | 98 | 124 | 13 | 72 | 4 | $18\frac{1}{2}$ | 21 | 50 | 12 | 2 | 28 | 25 | 90 | 24 | 105 | 27 | 30 | 10 | 25 |

| SIZE OF INCINERATOR, POUNDS PER HOUR | NO. OF NOZZLES | | WATER FLOW, GPM/NOZZLE | | GAS FLOW | STATIC PRESSURE DROP(IN.) | FAN HORSE-POWER |
|---|---|---|---|---|---|---|---|
| | PRIMARY | SECONDARY | PRIMARY | SECONDARY | CFM | | |
| 750 | 3 | 3 | 1.0 | 1.9 | 5250 | 1 | 3 |
| 1000 | 3 | 4 | 1.0 | 1.9 | 7000 | 1 | 5 |
| 1500 | 4 | 4 | 1.0 | 1.9 | 10,500 | $1\frac{1}{4}$ | $7\frac{1}{2}$ |
| 2000 | 5 | 5 | 1.0 | 2.9 | 14,000 | $1\frac{1}{2}$ | 10 |

Figure 6. Design standards for batch-fed in-line multiple-chamber incinerator scrubbers (*14*). 1: Access door; 2: primary sprays; 3: secondary sprays; 4: castable refractory; 5: downpass; 6: effluent inlet; 7: induced draft fan; 8: water level; 9: drain; 10: up-pass; 11: channel.

the various types of these incinerators, the work that has been done to upgrade existing units, and upgrading costs, in recent publications (*15, 16*).

1. Clean out grate and ash pit.
2. Close charging door (6) and ash pit door (7).
3. Open overfire air port (1), secondary air port (3) and undergrate air port (2).
4. Ignite mixing chamber burner (4) through lighter port (8) and close port (8).
5. Open charging door, charge material to fill chamber 1/2 to 3/4 full.
6. Ignite material on grate at top rear of pile and close charging door (6)
7. Turn on ignition chamber burner (5) <u>only</u> if very moist or wet material is charged.
8. Before adding more material to the burning pile in the incinerator
   a. Wait until burning pile fills less than 1/2 the chamber.
   b Push burning pile to rear of grates. (Do this gently, without causing bits of burning rubbish to fly off the pile).
   c Charge new material on front portion of grates. <u>Do not</u> put new material on top of the burning pile.
9. Operation during burndown. Close all air ports (1), (2), (3). Ignite ignition chamber burner (5), leave it on until only ash is left on grate. Leave mixing chamber burner (4) on until all smoking of material on grate is stopped, then shut it off.

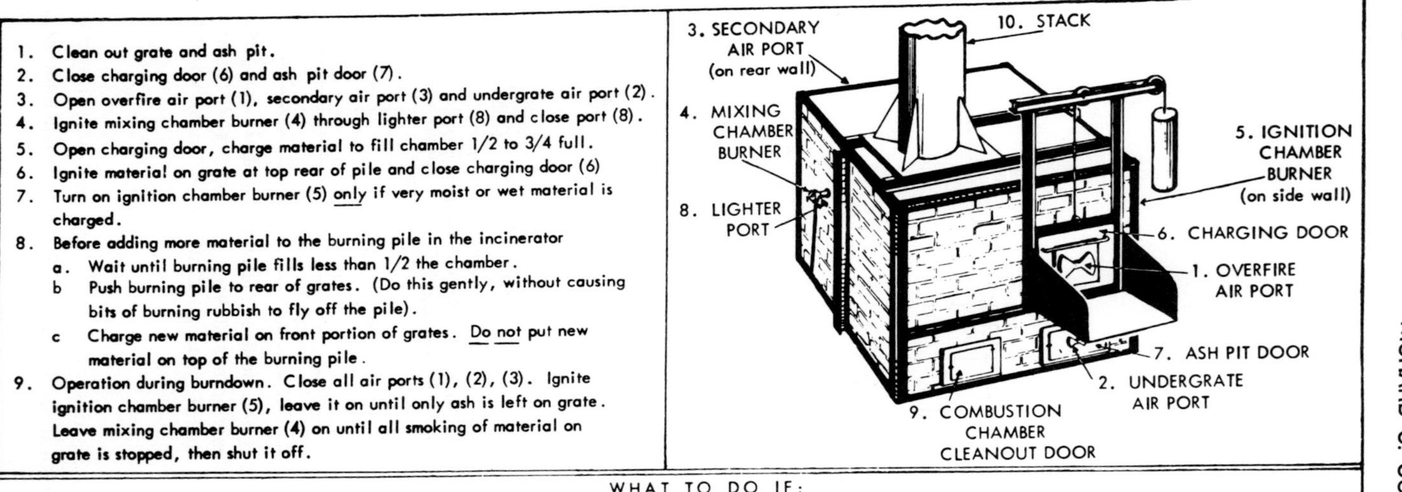


WHAT TO DO IF:

<u>Mixing chamber burner (4) flame goes up instead of down when it is first lit.</u>
1. Shut off burner.
2. Put piece of burning paper into incinerator through combustion chamber cleanout door (9), close door, light mixing chamber burner again.

<u>Smoke comes out around the charging door or ash pit door or both.</u> (This is normally the result of overcharging.) (Follow steps until it stops.)
1. Shut off ignition chamber burner (5).
2. Make sure flameport (opening at top rear inside ignition chamber), is not blocked by charged material.
3. Make sure combustion chamber cleanout door (9) is closed (and if it has a spinner or damper, be sure it is closed).
4. Do not overcharge incinerator again.

<u>White smoke comes out of the stack.</u> (Follow steps until it stops.)
1. Check mixing chamber burner (4), be sure it is burning.
2. Close secondary air port (3).
3. Close overfire air port (1).
4. Close undergrate air port (2).
5. Open gas valve of mixing chamber burner (4) fully.

<u>Black smoke comes out of the stack.</u> (Follow steps until it stops.)
1. Check mixing chamber burner (4), be sure it is burning.
2. Open secondary air port (3).
3. Open overfire air port (1).
4. Shut off ignition chamber burner (5).
5. Open charge door (6) about 1/4.

Figure 7. Operating Instructions for Multiple-Chamber Incinerators (*14*).

**Table VII Approximate Costs of Multiple-Chamber Incinerators and Scrubbers—1966[a]**

| *Capacity, lb/hour* | *Incinerator cost* | *Gas-scrubber cost* |
|---|---|---|
| 50 | $ 1,200 | $1,100 |
| 100 | 1,700 | 1,500 |
| 150 | 2,000 | 1,800 |
| 250 | 2,700 | 2,200 |
| 500 | 6,000 | 3,100 |
| 750 | 9,500 | 3,800 |
| 1,000 | 12,500 | 4,400 |
| 1,500 | 20,000 | 5,600 |
| 2,000 | 25,000 | 6,600 |

[a] George and Williamson (*14*).

### *B. Central Incinerators*

Early central municipal incinerators were batch-fed, hand-stoked burning chambers. The increasing costs of labor forced the introduction of mechanical stokers. Although incinerators with mechanical stokers had a higher initial cost, they cost less to operate and gave a lower total annual cost. However, feeding was still mainly batch. The early mechanical designs were primarily of the circular type with rabble arms for bed agitation, or of the cell-type with rocking grate. These were definitely improvements over manual stoking, but the wide variation in furnace temperature and poor combustion efficiency due to intermittent batch feeding made their operation unsatisfactory. Smoke and odors were emitted because of incomplete combustion after a fresh charge of feed had been added on top of the burning bed, thereby chilling the bed and impairing the distribution of combustion air through and above the bed.

In an effort to solve these problems, the next development was continuous feeding. These early continuous charging systems were essentially the same as those in use today, in which a vertical charging chute is kept filled with refuse, and the stoking action of the grate causes fuel to be charged through the bottom of the chute into the furnace. Some designs have used ram feeds, in which a mechanically driven piston-type device pushes a uniform quantity of feed horizontally onto the grate. The grate system moves the refuse through the incinerator to the discharge.

The earliest continuous-feed design was an inclined rotary kiln, but in the 1950's the traveling grate became the most common design for continuous-feed furnaces. The traveling grate was initially developed in the

1930's for use in coal-fired boiler plants, and its first use in a refuse incinerator was about 25 years later. While this grate does not provide continuous agitation, it is an efficient method for charge transportation through the furnace, and for efficient burning. It was not until the 1960's that grates were developed that simultaneously could provide continuous bed agitation and constant flow. Since that time several reciprocating grate designs have been developed. These include the continuous rocking grate and the reciprocating grate, of which there are numerous variations on the basic principles involved.

The engineering aspects of contemporary municipal incineration have been described by Meissner (*7*), Stephenson (*3*), and Niessen and others (*4*). Consequently, this section will deal mainly with matters that relate to air pollution from incinerators, including only enough technical background information on municipal incinerators to enable the reader to comprehend the technology.

Municipal incineration has improved considerably in recent years, owing to a certain extent from use of the *system* concept, in which each step, from collection of refuse to final disposal of the residue is considered as an essential factor in overall disposal efficiency.

## 1. Collection and Delivery of Refuse

Most refuse is delivered to the incinerator site in motor vehicles. In some instances, large compaction trailers are used to move refuse from a centrally located transfer station to a distant incinerator site. Such transfer stations receive the original refuse from a variety of pickup vehicles, with a marked trend toward trucks that compress the bulky refuse at the pickup sites from about 5–15 $lb/ft^3$ to about 15–30 $lb/ft^3$.

## 2. Handling and Storage of Refuse

Once the refuse arrives at the incinerator site it must be handled for storage, and for feeding to the next stage in the system. Usually, refuse is stored in a pit below ground level. A traveling crane and bucket is used to pile the refuse for storage and to move it from the unloading area, so that the pit can accommodate additional refuse. The crane and bucket can also be used to feed the incinerator.

## 3. Refuse Size Control and Salvage

This subsystem has grown in importance in recent years, with recognition that: (a) shearing or shredding oversize refuse achieves better burning characteristics, and (b) salvaging salable noncombustible components such as stoves and refrigerators from refuse adds operating revenue. Currently, attention is being devoted to reclaiming revenue-producing components such as glass and metals from incinerator residues (*18, 19*).

## 4. Incinerator Feed Systems

Refuse must be fed to the furnace at a controlled rate in either a batch or a continuous manner. Batch feeding directly to the furnace is usually done with a clamshell bucket or grapple on a traveling crane. In a few plants, a front-end loader operating on a paved floor charges the furnace.

Sometimes a ram-type feeding device is used. It provides an air seal at the feed to the furnace, an improvement over the bucket and front-end loader, which usually lets in uncontrolled amounts of cold air while the charging gate is open. The inrush of cold air can damage the refractory walls of the furnace, and can cause intermittent smoke evolution by cooling and quenching of the combustion process.

Newer incinerator designs usually specify continuous feeding by a hopper and gravity chute; a mechanical feeder, such as a pusher, ram, or rotary feeder; or a conveyor transporting refuse from the receiving area. The most common system is the hopper and gravity chute (Fig. 8).

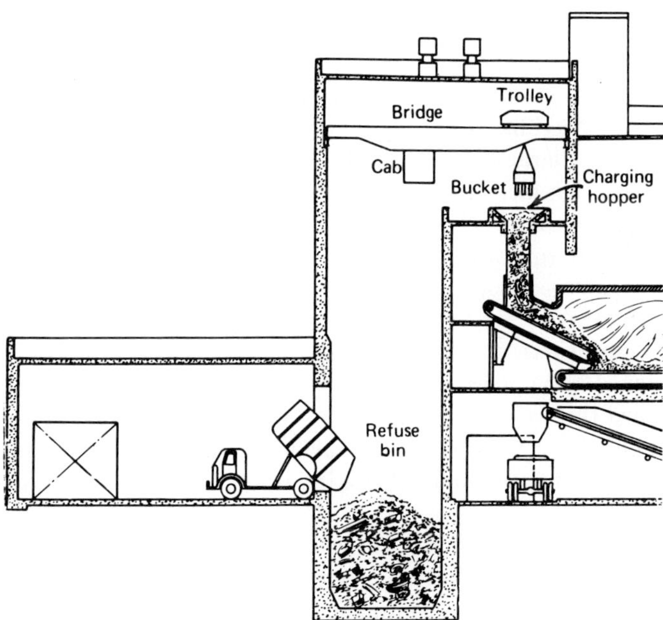


Figure 8. Hopper and Gravity Continuous Feed System.

## 5. Grates and Hearths

All contemporary incinerator furnaces use either a refractory hearth to support, dry, and burn the refuse, or a variety of grate types that move the refuse continuously through the furnace. Some grates simultaneously move and agitate the fuel bed. There are many types of hearths and grates, each having its own special features. Incinerators operating without grates include the stationary hearth, the rotary hearth, and the rotary kiln. There are two major grate types, stationary and mechanical.

a. STATIONARY HEARTHS. The stationary hearth is a refractory floor to the furnace, which may have openings for the admission of air through the fuel bed (underfire air). If underfire ports are not used, air is admitted from the sides or the top of the furnace, and burning occurs mainly at the top of the refuse (a simple form of underfeed burning). Often, combustion is assisted with auxiliary fuel burners.

b. ROTARY HEARTHS. The rotary hearth comprises a rotating refractory table that turns on a mechanically driven vertical shaft. A stationary rabble-arm system above the rotating table both moves the residue toward the table periphery, and stirs the burning bed to afford access of oxygen to the entire bed.

c. ROTARY KILNS. The rotary kiln consists basically of a refractory-lined cylinder that is inclined and rotated axially. Refuse is fed to the upper end and residue is withdrawn from the lower end. The movement of the burning charge is controlled by the speed of rotation. Rotary kilns, used *directly* as incinerators and without adequate flue-gas cleaning equipment, have failed because of excessive emissions. However, they have been used successfully in some instances to complete combustion of refuse discharged from continuous grate systems (Fig. 9).

d. STATIONARY GRATES. Stationary grates are basically frames of metal bars designed to support the fuel bed and to permit combustion air to pass upward through the grate and into the fuel bed. While some stoking or agitation of the fuel bed can be achieved by shaking the grate, such grates normally require manual stoking to obtain burnout of the residue.

e. MECHANICAL GRATES. Mechanically operated grates in batch-type furnaces evolved from stationary grate furnaces. Although batch-type incineration has yielded to continuous feeding for large municipal units, many new, small-capacity incinerators still use batch feeding (Fig. 10).

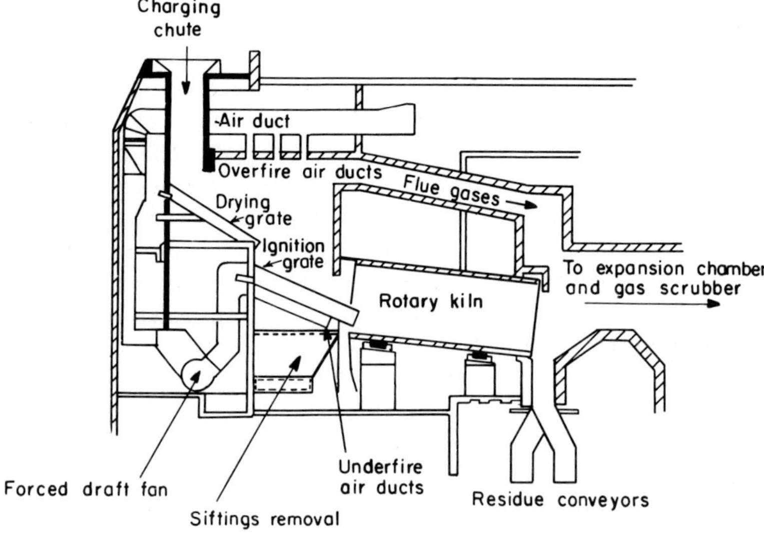


Figure 9. Combination Grate and Rotary Kiln (*15*).

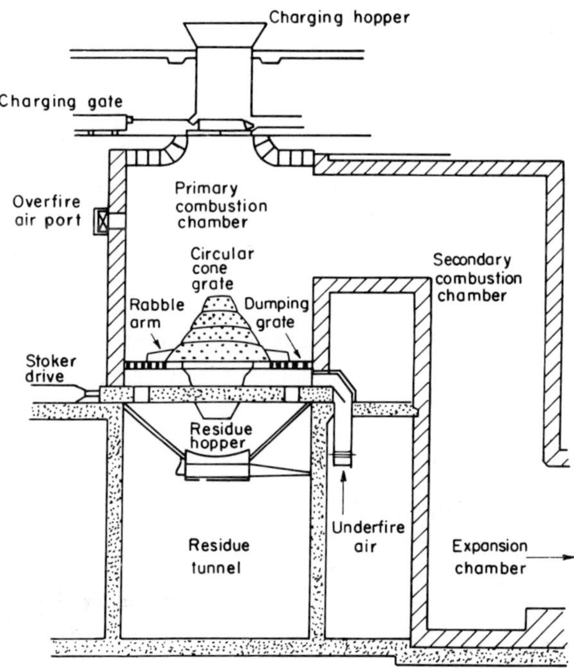


Figure 10. Batch-Feed Cylindrical Incinerator.

The grates form annuli inside the vertical cylindrical walls of the furnace. A solid grate covers the central area of the annulus. A hollow rotating hub with extended rabble arms rotates slowly above the solid grate to provide mechanical stoking. The residue works to the sides where it is dumped through the periphery of the circular grate. The hub is covered with a cone, with one or more consecutively smaller cones stacked on top of the first one. Underfire air is forced through the arms to the space above the stationary grate. Additional cone air cools the metal. Overfire air usually is introduced to the upper part of the circular furnace.

Mechanical, continuous grates function to feed the refuse from the feed chute to the furnace, move the refuse bed through the furnace, move the ash residue toward the discharge end of the grate, and to stoke and mix the burning bed of fuel. Underfire air passes in a controlled manner upward through openings in the grate to provide oxygen for combustion, and to cool the grate, protecting it from heat damage.

Some of the current grate types for large municipal incinerators include the following:

i. *Traveling Grate.* Because the traveling grate does not stoke or mix the fuel bed, it is often cascaded in two to four steps. Figure 11 illustrates two such grates in series. The refuse is carried progressively over a number of dampered air compartments, each adjusted to admit air only at the points where it is needed. First, the refuse dries; then it begins to burn. Finally, the residue is discharged to the ash pit.

ii. *Reciprocating Grate.* The *Von Roll* is one example of the reciprocating type of grate (Fig. 12). The rows move alternately to convey the refuse from the feed chute through the combustion area to the ash hopper. Another example is the *Martin* reverse-acting reciprocating grate (Fig. 13). In this system, the grate sections are stacked like inverted overlapping roof shingles. Alternate sections reciprocate uphill against the downward flow of refuse, thereby producing agitation of the refuse. Both these types of grates have a stepwise configuration, rather than an inclined

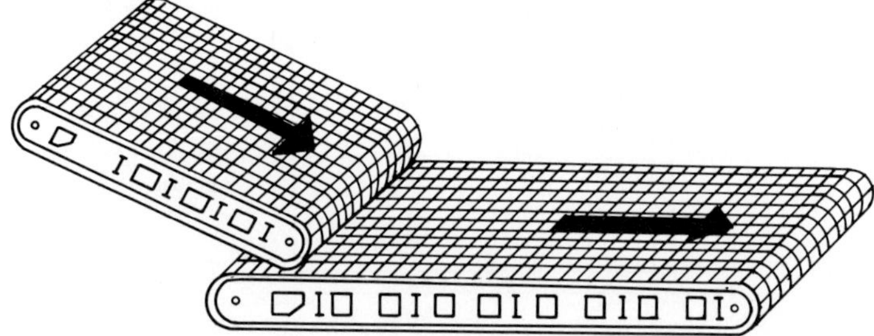

Figure 11. Cascaded Traveling Grates.

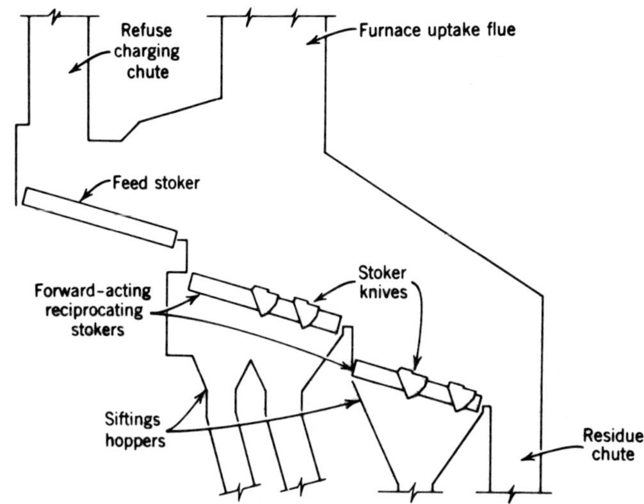


Figure 12. Von Roll Reciprocating Grate (*15*).

plane, so that additional mixing occurs as the refuse tumbles from one level of the next. Also, like traveling grates, reciprocating grates can be arranged in multiple-level series, providing additional bed agitation.

iii. *Roller or Drum Grate.* Developed in the West German city of Dusseldorf, the roller grate (Fig. 14) was introduced to counteract the

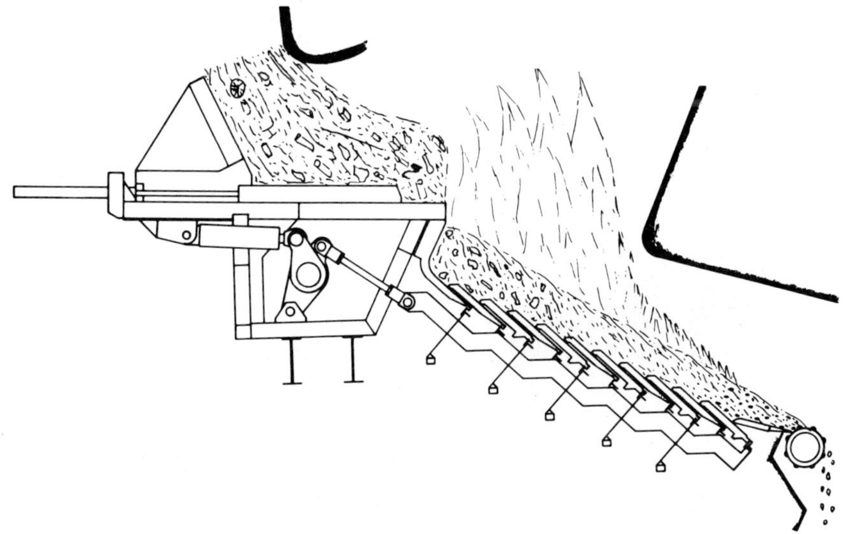

Figure 13. Martin Reverse-Acting Reciprocating Grate.

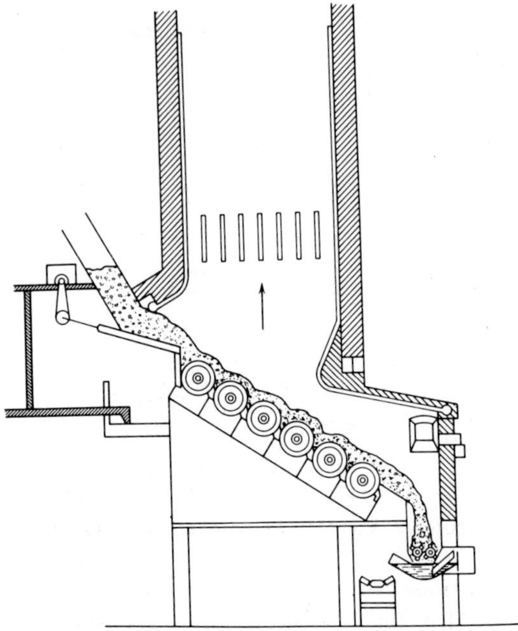

Figure 14. Roller Grate (*15*).

relatively high cost of multiple traveling grates. Each rotating drum represents a minimum length of a traveling grate section, thereby providing a maximum number of tumbling zones in the fuel bed. The slow rotation of the drums develops mixing action in the refuse between successive drums, thereby reducing the density of otherwise dense refuse components.

iv. *Rocking Grate.* A rocking grate (Fig. 15) slopes downward from the feed to the discharge end, with two, three, four, or more grate sections

Figure 15. Rocking Grate. (Courtesy of Flynn & Emrich Co.)

in series. The rocking grate includes a multiplicity of grate segments that are about quarter-cylindrical, and have openings for undergrate air. Alternate rows of grate sections are rotated 90 deg about the axis toward the discharge of the grates, with the grate face rising up into the burning mass, thus breaking it up and thrusting it forward to the discharge. These grate sections each rotate back to rest position, and alternate sections rotate as before, causing a similar stoking action and pushing the refuse bed forward.

## 6. Incinerator Furnaces

Details of furnace construction have been described by Meissner (*17*); only the design factors affecting emissions will be mentioned here.

a. Grate Loading. Loading is expressed as the weight of refuse fed per unit area of grate surface per unit time, or as the heat released per unit area of grate per unit time. The range of values for mechanical grate loadings is from 50 to 70 $lb/(ft^2)$ (hour). An average value for heat release is 300,000 $Btu/(ft^2)$ (hour).

b. Air Leakage. Control of air leakage is an important factor. Poor design and maintenance can permit inward air leakage through refractory walls and roofs, which, if excessive, will reduce furnace temperature, impair combustion, and overload pollution-control systems.

c. Furnace Geometry. The physical shape of a furnace is a factor in controlling incinerator performance. For example, hot combustion gases will tend to rise to the top of the furnace enclosure. If the furnace outlet is at the top of the enclosure the cooler gases from the discharge section of a grate will leave the furnace below the layer of hotter combustion gases, often without mixing. Thus, a large fraction of the combustion air is not used and the excess air in the flue gases may be excessive. This problem is mitigated with designs that exhaust the flue gases from the furnace either at the top near the refuse feed, or at the lower sections of the furnace near the ash discharge. Overfire air jets, refractory baffles, or bridge walls enhance the mixing of air with the products of combustion.

d. Furnace Heat Release. Volumetric heat release, $Btu/(ft^3)$ (hour), characterizes the combustion intensity through its effect on flame temperature. Most refractory-lined furnaces are designed for 15,000 to 25,000 $Btu/(ft^3)$ (hour); furnaces with welded waterwalls are designed for from 12,000 to 16,000 $Btu/(ft^3)$ (hour).

e. Combustion Air. The theoretical amount of air required for combustion is calculated from analysis of the specific refuse to be burned (*20*), or by assuming that the feed is pure cellulose or a mixture of cellulose and organics with the same heating value as the refuse. This allows calculation of the theoretical (stoichiometric) quantity of air to be supplied. Then empirical curves can be consulted to determine the percentage of excess air [i.e., (actual air supplied-theoretical air/theoretical air) $\times$ 100] required to keep the furnace temperature within the desirable range for the specific moisture content and heating value of the fuel. These values will be different for refractory and water-walled furnaces, owing to the different values of heat loss through the walls of the furnace. They range from a minimum of 70% excess air to 200% or more excess air.

It is necessary to decide how the total air should be distributed. Underfire air is needed to supply oxygen for combustion of solids in the bed, and also for grate cooling. Overfire air is required to supply oxygen for the gaseous combustion zone above the bed, to create turbulent mixing of air with combustible gases, and to provide a means for furnace temperature control by removing excess heat that may be generated. Underfire air quantities have ranged from 0 to 100% of the total air, but research and experience have shown that an excessive percentage of underfire air causes operating and pollution problems. Generally, municipal incinerators operate with 40–60% underfire air. This range has been arrived at by operating experience, although two recent installations use up to 75% underfire air.

After the quantity of overfire has been decided upon, the method of its distribution is then set. Within the past several years, a discussion has developed concerning how the overfire air should be introduced. Air nozzles can either be located in the roof of the furnace or in the sidewalls. Opinions as to how much should be introduced through the sidewalls vary from 0 to 100% of the overfire air. Since the primary function of the overfire air is to create turbulence, the distribution ratio should be that which will create the greatest turbulence in the combustion chamber without excessive entrainment of particulate matter from the fuel bed into the flue gas. Different furnace configurations, sizes, and innovations will require different overfire air distribution patterns for most efficient operation.

## V. Causes of Incinerator Emissions

### A. Combustion Principles Related to Emissions

To understand better the causes of incinerator emissions, it is essential to understand some of the principles of combustion having particular reference to refuse. (More details on combustion principles are given in Chapters 10, 11, and 12, this volume.) It is worthwhile at this point to quote an important observation—an observation that has needed attention by the incinerator industry for many years—and only recently appears to have gained limited acceptance.

> Refuse is a fuel, although a difficult one to handle. Incinerators are combustion systems, which obey the same fundamental laws as a modern power boiler or a jet engine. This view of municipal solid waste disposal by incineration is as powerful as it is obvious. As long as refuse and incineration carry a mystique, as altogether different from other fuels and combustion processes, the technology of incineration will remain primitive, and the problems of the past will remain both puzzling and unsolved (*4*).

It should not be inferred from this observation that whatever is essential to burn coal or wood efficiently is necessarily adequate for refuse, but it does mean that the combustion mechanisms are much the same, differing in degree rather than kind. For example, unlike coal, lignite, or wood, raw refuse, in even a given location, often will change substantially in composition and moisture content with time, necessitating changes in operating conditions, such as the distribution of underfire air along the length of the fuel bed, and the ratio of underfire to overfire air. Some interesting comparisons of refuse with other solid fuels are given in Table VIII. Notable resemblances are seen in theoretical air requirements in terms of pounds/$10^6$ Btu and standard cubic feet (scf)/$10^3$ Btu. A highly simplified version of what happens to solid waste in a furnace is given below.

For a substance to burn, both surface and internal moisture must be driven from the material. The vaporization of moisture present in waste material will keep the temperature of the material at or below 212°F until it is dry. Once the moisture is removed, the temperature of the substance rises to the ignition point, although the outer surface of a solid may be dried and ignited before the inner material is dried. This drying process continues throughout the entire length of the furnace, but proceeds at its greatest rate immediately following the charging of the solid waste.

**Table VIII Some Approximate Characteristics of Various Solid Fuels**

| *Fuel* | *Density, lb/ft³* | *Void fraction, percent* | *Theoretical air requirements* | | |
|---|---|---|---|---|---|
| | | | *lb/lb fuel* | *lb/10⁶ Btu* | *scf/10³ Btu* |
| Refuse | 17 | 72 | 3.2 | 725 | 9.8 |
| Wood chips | 17.5 | 50 | 3.3 | 700 | 9.5 |
| Bituminous coal | 50 | 45 | 10.0 | 750 | 10.3 |

To facilitate drying, some furnace designs direct preheated air, or radiate heat from reflecting arches to the incoming charged material, or do both. The first part of the grate system is frequently referred to as the drying zone. Ignition takes place as the solid waste is dried and continues through the furnace. The portion of the grates where ignition first occurs is called the ignition zone.

The combustion process for solid fuels as well as refuse is thought of as occurring in two overlapping stages, primary combustion and secondary combustion. Primary combustion generally refers to the physical-chemical changes occurring in proximity to the fuel bed, and consists of drying, devolatilization, ignition, and burning of the solid waste. Secondary combustion refers to the oxidation of gases and particulate matter released by primary combustion. Secondary combustion aims at achieving combustion of unburned furnace gases (carbon monoxide, hydrogen, and hydrocarbons), elimination of odors, and combustion of carbonaceous particles suspended in the gases. To promote secondary combustion a high temperature must be maintained, sufficient air must be supplied, and turbulence must be imparted to the gas stream.

The function of turbulence is to ensure mixing of each volume of gas with sufficient air to complete the burning of the unburned vapor-phase combustible matter and the suspended particulates. The turbulence must be intense and must persist long enough for mixing to be completed while the temperature is still high enough to ensure complete burning (Fig. 16).

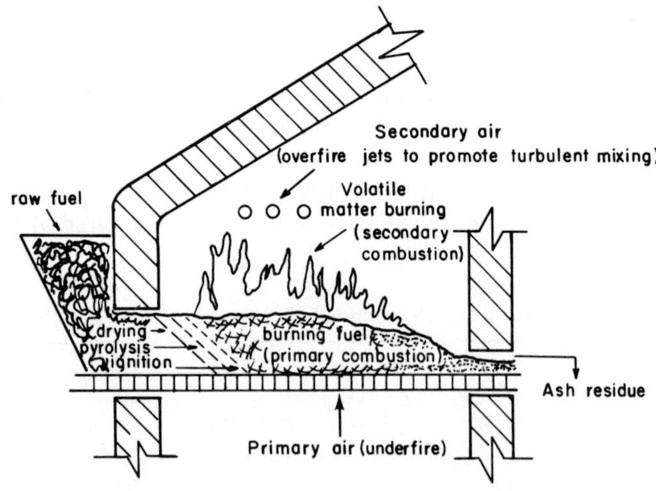


Figure 16. Diagrammatic Representation of a Burning Fuel Bed.

Often, one hears about the three T's needed for complete combustion, meaning *time, temperature,* and *turbulence.* What is difficult to ascertain is how much time, what temperature, and how much turbulence are needed? The simplified version of fuel bed combustion described above does not state these parameters quantitatively because each system has its unique values, determined from experimental empiricisms. Current research efforts hopefully will enable the future designer to quantify such basic operating parameters as *gas residence time, flame and postflame temperature,* and *turbulent mixing of air and combustible gases.* Until this goal is reached, incinerator furnace design will remain an empirical art, continuing to depend on past experience for guidance in design innovation.

### Overfire Mixing

Incinerator *overfire volume* mixing is provided by baffles, and overfire air jets, located at strategic positions in the primary combustion chamber, and sometimes beyond the primary chamber. It is particularly important to have sufficient jet energy in the region of the furnace where the pyrolysis reactions are most intense. On cold days, chilled air has been found to retard combustion.

Design equations have been developed to describe the mixing processes in the overfire volume. These equations have been based largely on the behavior of free, round, isothermal jets. Although some of these equations afford a starting point for jet design, they do not adequately represent the complex gas flow in an incinerator. Very little work has been done on the effectiveness of overfire jets for complete and rapid burnout of combustible particulates and gases in incinerators. Some broad guidelines have come from a study of overfire air jets to control smoke from a municipal incinerator (*21*). Moreover, there is a good base of technology developed for locomotives and coal-fired boiler furnaces, which provides an empirical starting point with minimum technological risks, for the use of overfire mixing techniques (*22–24*).

## *B. Stack Emissions*

The air pollutants from incinerators comprise three classes: particulate matter (fly ash and smoke), combustible gases (carbon monoxide, hydrocarbons, and polynuclear hydrocarbons), and noncombustible gases (nitrogen oxides, sulfur oxides, and hydrogen chloride).

## 1. Particulate Matter

The principal factors responsible for particulate emissions are (a) mechanical entrainment of particles from the burning fuel bed by furnace gases; (b) thermal decomposition (pyrolysis) of hydrocarbons in the furnace gases, and condensation reactions; and (c) volatilization of metallic salts or oxides.

Particulate emissions include particles of mineral matter, and of mineral matter containing unburned combustible matter that vary in size, both usually referred to as fly ash; and smoke having shades that vary from black through brown to white. If combustion in the furnace is complete, the fly ash particles will be virtually colorless. If combustion is incomplete, the fly ash particles will be gray to black, depending on the amount of carbon they contain. Often, large, thin, black flakes will be emitted. These generally are incompletely burned paper, called *char.*

Black smoke is formed when volatile matter from the burning bed is condensed into a liquid aerosol and then heated to such high temperatures in a deficiency of oxygen that the droplets are thermally decomposed, mainly to very fine carbon particles. Aggregates of smoke, called soot, usually are relatively difficult to burn, requiring more rigouous conditions of mixing, residence time, and temperature for combustion than other products of incomplete combustion.

Light-colored smoke (shades of brown to virtually white) or *distillation* smoke, is essentially volatile matter that escapes the fuel bed and has not ignited. Distillation smoke usually occurs when a cold charge of refuse is dumped onto a burning fuel bed. It consists largely of fine droplets of complex tarry materials.

Particulate emissions are reported in a wide range of units. To compare data from different sources, it has been convenient to develop conversion factors between the different sets of units. The prevailing units and the conversion factors are shown in Table IX. Unfortunately, conversion factors depend on the fuel composition and certain operating conditions in a particular incinerator, information that is not always reported. Consequently, the factors in Table IX have been derived for a representative set of conditions, with the understanding that, compared to other uncertainties in emission data, the errors will be small for data from different incinerators. Nevertheless, these factors should only be used to obtain an order-of-magnitude comparison of data from different sources, because the use of supplementary fuel, and of refuse compositions that differ significantly from the one used to derive the factors, can cause significant errors.

There appears to be some justification that any emission unit based

**Table IX Conversion Factors for Incinerator Particulate Emissions[a,b]**

| | *lb/ton of refuse (as received)* | *lb/1000 lb flue gas at 50% excess air* | *lb/1000 lb flue gas at 12% $CO_2$* | *Grains/scf flue gas at 50% excess air* | *Grains/scf flue gas at 12% $CO_2$* | *Grams/N $m^3$ flue gas at 7% $CO_2$* |
|---|---|---|---|---|---|---|
| lb/ton of refuse (as received)[c] | 1 | 0.089 | 0.10 | 0.047 | 0.053 | 0.067 |
| lb/1000 lb flue gas at 50% excess air | 11.27 | 1 | 1.12 | 0.52 | 0.585 | 0.74 |
| lb/1000 lb flue gas at 12% $CO_2$ | 10.0 | 0.89 | 1 | 0.46 | 0.52 | 0.66 |
| Grains/scf flue gas at 50% excess air[d] | 21.31 | 1.93 | 2.16 | 1 | 1.12 | 1.42 |
| Grains/scf flue gas at 12% $CO_2$[d] | 18.85 | 1.71 | 1.92 | 0.89 | 1 | 1.26 |
| Grams/N $m^3$ flue gas at 7% $CO_2$[e] | 15.0 | 1.36 | 1.53 | 0.70 | 0.79 | 1 |

[a] Based on Niessen *et al.* (*4*). Multiply items in left-hand column by factor to get quantity desired in top row.
[b] Each numerator is the weight of particulate matter.
[c] Based on refuse with a higher heating value of 4450 Btu/lb.
[d] Standard conditions are 60°F and 1 atm. There are 7000 grains/lb.
[e] Standard conditions are 32°F and 1 atm. There are 35.3 $ft^3/m^3$.

on a fixed percent of $CO_2$ in the flue gas does not provide a consistent basis, since the corrected emission will vary with the composition of the waste.

### 2. Combustible Gases

It is a characteristic of all combustible solids (refuse, coal, coke, wood, etc.) that when heated in the absence of oxygen some combustible gases and tar vapors are evolved. If a bed of such fuel is heated on a grate, with air passing upward through the bed, under certain conditions the oxygen will be consumed before it reaches the top of the bed. The unburned fuel on top, therefore, will be heated by the hot gases passing from below and volatile matter will be released from the top of the fuel bed. The volatile matter comprises unburned components, such as carbon monoxide, hydrogen, a variety of hydrocarbons, and some polynuclear hydrocarbons. The remainder of the gases is carbon dioxide, water vapor, and nitrogen. If the refuse contains halogenated polymers and sulfur-bearing components, the fuel-bed gases will also contain compounds such as hydrogen chloride, hydrogen sulfide, and sulfur dioxide.

If the combustible components are not burned completely with overfire air the gases leaving the stack will contain gaseous pollutants, including smoke, depending on the history of the hydrocarbons as they move through the incinerator. The presence of unburned or partially burned gases in stack gases is unnecessary. Such emissions can be controlled by adequate design and proper operation rather than by control devices. The key parameters are, as mentioned earlier, adequate furnace temperatures (not less than about 1800°F), sufficient air for combustion, sufficient refuse and gas residence time in the furnace, and sufficient turbulence in the combustion space (by overfire jets, baffles, or other means) to mix combustible gases with air. Gas residence time and turbulent mixing frequently are enhanced by using a secondary combustion chamber, which may be fired with auxiliary fuel (see Section IV,A on multiple-chamber incinerators).

a. Carbon Monoxide. Carbon monoxide (CO) is the most significant gaseous pollutant emitted by municipal incinerators. As explained earlier, CO in furnaces is the result of incomplete burnout of the gaseous pyrolysis products, and of the char in the fuel bed. The latter source results from carbon dioxide ($CO_2$) in the fuel bed reacting with hot carbon according to the reaction

$$C + CO_2 \rightarrow 2\,CO \qquad (1)$$

The water–gas reaction, Equation (2), may also be a source of CO, as well as of hydrogen ($H_2$):

$$C + H_2O \rightarrow CO + H_2 \quad (2)$$

the water coming mostly from the drying of the refuse.

It is important to note that CO, as well as other combustible gases, may be found in stack gases *even when free oxygen is also present.* (Theoretically, if free oxygen is present there should be no combustible gases evident.) Such a condition means that even if excess air is used, turbulent mixing in the flame and postflame region has been insufficient for complete combustion to take place. It can also mean that overfire air is being added to a region where the temperature is too low for combustion of CO to occur.

b. HYDROCARBONS. The combustion of hydrocarbons is thought to consist first of rapid oxidation to CO, followed by slow oxidation of the CO to $CO_2$. Consequently, hydrocarbon emissions always appear to be associated with CO in stack gases. Because of the faster rate of the hydrocarbon-to-CO reaction, the ratio of CO to hydrocarbon is always greater than one. Data from a 3-ton/day incinerator have been reported by Rose, Stenburg, and others (*25–27*). The authors reported 51 observations, and the derived data are shown in the tabulation below:

| | *Carbon monoxide (lb/ton)* | *Hydrocarbons (lb/ton)* | *CO/HC* |
|---|---|---|---|
| Average | 22.3 | 1.6 | 17.0 |
| Standard deviation | 21.9 | 1.2 | 12.6 |

c. POLYNUCLEAR HYDROCARBONS. The emission of polynuclear hydrocarbons* (PNH) is a very small fraction of total emissions of municipal incinerators (Table X). These limited results indicate that the smaller units emit more PNH, and that the sprays for the 50-ton/day unit reduced the PNH emission about 95% (*28*).

## 3. Nitrogen Oxides

Nitrogen oxides ($NO_x$) have two sources: nitrogen in the fuel and the reaction between atmospheric nitrogen and oxygen at high temperatures.

* The following polynuclear hydrocarbons have been identified in incinerator emission: benzo(*a*)pyrene, perylene, benzo(*ghi*)perylene, anthracene, coronene, phenanthrene fluoranthrene, benz(*a*)anthracene.

**Table X Emission Rates of Polynuclear Hydrocarbons (PNH) from Incinerators**[a]

| *Type of incinerator* | *PNH, gm/ton of refuse* | *CO, lb/ton of refuse* |
|---|---|---|
| 250 tpd[b] Municipal (before settling chamber) | 0.035 | 0.7 |
| 50 tpd[b] Municipal (before scrubber) | 0.284 | 4 |
| 50 tpd[b] Municipal (after scrubber) | 0.015 | <2 |
| 4.3 tpd[b] Commercial single chamber | 1.92 | 4 |
| 3 tpd[b] Commercial multiple chamber | 1.94 | 25 |

[a] Hangebrauck *et al.* (*28*).
[b] tpd: Tons per day.

Generally, the nitrogen content of refuse is low. The formation of nitric oxide

$$N_2 + O_2 \rightarrow 2NO \quad (3)$$

has been studied extensively, particularly in connection with high temperature systems, such as internal combustion engines, gas turbines, and large fossil fuel-fired boilers.

It is believed that NO is formed mainly at the flame front, where the temperature is high and oxygen is available, since these factors drive Equation (3) to the right. Some data on nitric oxide emissions from municipal incinerators are given in Table XI. A statistical summary of a

**Table XI Nitric Oxide Emissions from Different Sized Municipal Incinerators**

| *Plant capacity, tpd*[a] | *Emission* | |
|---|---|---|
| | *lb/ton of refuse* | *lb/10⁶ Btu* |
| 50 | 2.3[b] | 0.29 |
| 100 | 1.9 | 0.22 |
| 250 | 1.6[b] | 0.18 |
| 250 | 2.7[c] | 0.30 |

[a] tpd: Tons per day.
[b] Average of three values.
[c] Average of four values.

large number of observations (*4*) gives the results shown in the following tabulation.

| | Nitric oxide emissions | | |
|---|---|---|---|
| | *Number of observations* | *Arithmetic average ($lb/10^2$ Btu)* | *Standard deviation ($lb/10^6$ Btu)* |
| All data | 99 | 0.2126 | 0.1044 |
| Small units | 77 | 0.1879 | 0.0866 |
| Large units | 22 | 0.3122 | 0.1325 |

The NO formed in the furnace subsequently oxidizes to nitrogen dioxide ($NO_2$) in the atmosphere.

### 4. Sulfur Oxides and Hydrogen Chloride

a. Sulfur Oxides. The potential contribution of sulfur pollutants from municipal incinerators is small. In fact, studies are being made of refuse as a partial replacement for high-sulfur fuels in power plants (Section VIII,E).

b. Hydrogen Chloride. There is increasing concern about hydrogen chloride (HCl) emissions from incinerators, owing to the growing amount of halogenated polymers, notably polyvinyl chloride (PVC) and, to a much smaller extent, polyvinylidene in refuse. Pure PVC yields about 1180 lb of HCl per ton of PVC.

Two factors may limit air pollution by HCl: absorption of HCl by alkaline ash materials in the incinerator furnace, and absorption of HCl in wet scrubbers. A study of a large municipal incinerator (*29*), which measured 20 gaseous chemical species in the stack emissions, concluded ". . . water scrubbing appears to be effective in removing chlorides from flue gas. Scrubber efficiencies of 80 to 97 percent were measured." No details of the scrubber system are given.

It has also been noted that the fly ash removed by electrostatic precipitators absorbs some of the HCl and sulfur oxides. The average reduction between the inlet and the outlet of the electrostatic precipitator on the Southwest Brooklyn Incinerator was 94% for HCl and 28.5% for $SO_2$.

Fire research has provided information on the partial decomposition products of plastics (*30*). PVC may yield phosgene ($COCl_2$); other plastics may produce hydrogen cyanide (HCN) and ammonia ($NH_3$). The toxicity of some of the gases that may be emitted from incompletely burned plastics have been reported (*31*).

### 5. Other Emissions

a. Volatile Metals. Selenium, lead, zinc, and cadmium compounds volatilize quite easily, usually appearing in incinerator emissions as fumes of their respective oxides. Insufficient data are available to assess the role of these metals as air pollutants from incinerators.

b. Hydrogen Fluoride. Emissions of hydrogen fluoride (HF) arise from the combustion of fluorinated hydrocarbons, such as polytetrafluoroethylene. HF is highly soluble in water, and excellent collection efficiency can be anticipated with wet scrubber systems.

## VI. Air Pollution Control Systems for Incinerators

Although air pollution control systems incorporating water scrubbing will remove some water soluble gases, air pollution control equipment now in operation on incinerators has been designed primarily to remove particulate matter. The remainder of this section deals solely with the role of such equipment in removing particulate matter.

### *A. Settling Chambers**

A settling chamber (also called an expansion or subsidence chamber) is the simplest and least expensive particulate removal device used in incinerators. It consists of a large refractory-lined chamber wherein the velocity of the flue gases is slowed, permitting gravity settling of coarse particulates. Many older installations also employ refractory baffles extending downward from the roof and upward from the floor, causing the flue gases to change direction abruptly and additionally induce inertial separation of particulates. Some incinerator settling chambers have been supplemented by sprays to wet the walls and the bottom to inhibit reentrainment of settled ash into the gas stream. The collection efficiencies of settling chambers depend largely on the size characteristics of the particulates. Generally, efficiencies are low, in the range of 10 to 30% (*27*, *28*). Since about 1960, the settling chamber as the only means of particulate control has become obsolete.

* See also Chapter 3, this volume.

### *B. Scrubbers**

There are two basic kinds of wet systems for removing particulates from incinerator combustion products: the wetted baffle–spray system and wet scrubbers.

#### 1. Wetted Baffle–Spray System

These systems usually consist of vertical impingement baffle screens that are wetted by flushing sprays or overflow weirs. There may be one or more screens in the collection system. Particulate removal efficiency has been found to range from 10 to 50% on several different municipal incinerators (*32, 33*). Pressure drop is in the range of 0.3 to 0.6 in. water, and water consumption is in the range of 0.5 to 2.0 gal/minute for each ton per day (tpd) of refuse capacity. The installed cost of such baffles is about $0.02 to $0.04 per actual cubic foot per minute (cfm) of gas at the collector inlet temperature. As with settling chambers, wetted baffle systems alone cannot be expected to meet particulate emission standards for incinerators.

#### 2. Wet Scrubbers†

About 20% of the incinerator plants built since 1960 have been equipped with wet scrubbers. Most scrubbers in municipal incinerators operate at pressure drops of 5 to 7 in. water with efficiencies above 90%. High energy scrubbers, such as the venturi scrubber, may operate at over 99% efficiency (*34*). Water requirements range from about 5 to 15 gal/1000 actual $ft^3$/minute of gas.

The main disadvantage of scrubbers is the corrosion problems that arise from absorption of chemicals from the gas stream, causing the effluent scrubber water to become highly acidic. This requires the use of corrosion-resistant materials. Experience has shown that treatment and recirculation of scrubber water is difficult to accomplish. Some scrubber water is lost by evaporation into the effluent gas stream. The hot, saturated gas stream from the stack contacts air at ambient temperature. This results in an unsightly vapor plume. Condensation of the vapor can cause settling of droplets and particles on local structures and vehicles, with damaging results.

* See also Chapters 6 and 7, this volume.

† See also Chapters 6 and 7, this volume.

### *C. Cyclones**

The cyclone configurations that have been used for incinerators are, in order of decreasing particulate removal efficiency:

(a) The multiple-cyclone system with numerous small-diameter (less than 12 in.) cyclone units installed in a tube sheet.

(b) The multiple-cyclone system of larger diameter (over 18 in.) units installed in clusters, with flue gas manifolded to the inlets of the individual cyclones, and the outlets manifolded to a common duct.

(c) Single or double cyclone units of very large diameter (over 4 ft) with a single or split flue duct at the inlet and outlet.

Under ideal operating conditions, the smaller diameter cyclone system can attain 80% collection efficiency on incinerator ash (*35*). However, plugging of the tubes by adherent fly ash, or by particles wetted by an upstream gas-cooling system, will impair their collection efficiency disastrously. The larger diameter cyclone systems are usually free of plugging, and can achieve efficiencies up to about 70%. These systems have been the most popular for incinerators using cyclones. Very large diameter cyclone units are rarely used because of their relatively low collection efficiency.

### *D. Fabric Filters†*

At the time of writing this chapter, there was no municipal incinerator installation using bag filters as the means of air pollution control. Incinerator system designers have avoided fabric filter systems for incinerators because of their assumption that initial costs and bag replacement costs would be prohibitive; greater control of combustion would be necessary to prevent formation of soot and tarry condensates, and closer gas temperature control would be necessary to prevent thermal destruction of the fabric, or condensation on the fabric, with the possibility of fabric plugging and subsequent damage to the collector.

### *E. Electrostatic Precipitators‡*

Electrostatic precipitators have been used for a number of years with excellent results in European incinerator plants that recover heat. Efficiencies in the range of 96–99.6% have been achieved at pressure drops below 0.5 in. water. Electrical power requirements are in the range of

* See also Chapter 3, this volume.

† See also Chapter 4, this volume.

‡ See also Chapter 5, this volume.

200 to 400 W per 1000 actual cubic feet per minute (acfm) of gas treated. Inlet temperatures are usually in the range of 350° to 700°F (*36*). A number of municipal incinerators in the United States are either equipped with electrostatic precipitators, or will be so equipped in the future.

### F. Economics and Trends of Air Pollution Control

Table XII summarizes the cost and space requirements of these air pollution control systems. These data show that electrostatic precipitators have comparatively low maintenance and operating costs, which may offset their high initial investment costs (*37*).

As mentioned earlier, much of the maintenance and repair costs on wet scrubber units is due to metal corrosion from scrubber water. Many incinerator scrubbers simply dump effluent scrubber water to the nearest means of disposal, a practice that water pollution control regulations are expected to make obsolete. In general, effective water treatment and recycle has not been successfully accomplished for wet air pollution control systems, owing to the varying and corrosive characteristics of effluent water. If effective and economic water treatment methods can be developed, the wet scrubber could become a more attractive alternative to the electrostatic precipitator in incinerator air pollution control systems.

## VII. Emission Tests of Incinerators

Sampling incinerator gas requires special consideration of several characteristics unique to incinerator operation (*38–40*): (a) The need to obtain representative samples in large duct cross sections. (b) The presence of relatively large, low density particulates in the gas. (c) The high moisture content of the gas. (d) The need to sample gas in the range of

**Table XII Comparative Air Pollution Control Data for Municipal Incinerators**[a]

| | *Relative capital cost factor* | *Relative space, percent* | *Relative operating cost factor* |
|---|---|---|---|
| Multicyclones | 1.0–1.5 | 20–30 | 0.5–1.0 |
| Scrubbers with water treatment system | 3 | 30 | 2.5 |
| Electrostatic precipitators | 6 | 100 | 0.75 |

[a] Based on Fernandes (*37*).

500–2000°F, depending upon the sampling location in the incinerator system.

### A. Tests of Small (Less than 72 tpd) Incinerators

Operating conditions and stack emission data for several typical multiple-chamber incinerators, ranging in capacity from 50 to 6000 lb/hour, have been reported by George and Williamson (*14*) (Tables XIII and XIV).

Comparisons of sampling trains employing filters followed by wet impingers with sampling trains employing only filteration of dry particulate matter [especially the train prescribed by the American Society of Mechanical Engineers (ASME) Performance Test Codes 21 and 27] have been made. As might be expected, the *total* particulate catch (dry filter plus impingers) usually was higher than the dry filter catch alone (*41–43*).

The emissions of nongaseous contaminants measured by a sampling train consisting of a filter followed by a series of wet impingers are reported in Table XIV in four different ways. The first gives the concentrations of particles collected on the filter plus the materials (e.g., tars) condensed in the impingers in grains per standard cubic foot (scf) of gas at stack conditions. The second group of figures gives only the grains of particulate matter collected on the filter per standard cubic foot of dry stack gas. The third converts the total value above to the basis of 12% carbon dioxide in the flue gas. The carbon dioxide values are for the refuse only, and do not include carbon dioxide from the auxiliary fuel. The fourth is the filter collection only, calculated on the basis of 12% carbon dioxide in the flue gas.

The results for filter only collection indicate that all but three of the tests, 4a, 8, and 12, exceed the standard of the United States Environmental Protection Agency (USEPA) of 0.08 grain/scf (filter collection only) corrected to 12% $CO_2$.

### B. Tests of Medium-Sized (250 tpd) Incinerators

Walker and Schmitz (*32*) tested the incinerators shown in Figures 17 and 18. Unit no. 1 (Fig. 17) is a continuous-feed, multiple traveling grate incinerator having a capacity of 250 tons daily. It is equipped for overfire and wall-cooling air, and has a dry subsidence chamber. Furnace outlet measurements were made just before the furnace gases entered the subsidence chamber. No measurements were made at the subsidence chamber outlet or the stack. Unit No. 2 (Fig. 18) is a continuous-feed, reciprocating grate furnace, also having a capacity of 250 tpd. It is equipped with

**Table XIII Operating Conditions for Tests on Small Multiple-Chamber Incinerators**[a]

| | Test number | | | | | | | |
|---|---|---|---|---|---|---|---|---|
| | *1a* | *1b* | *4a* | *5* | *6* | *8* | *12* | *13* |
| Operating conditions | | | | all normal | | | | |
| Capacity, lb/hour | 50 | 50 | 350 | 750 | 1000 | 1000 | 2500 | 6000 |
| Charging rate, lb/batch | 2–4 | 2–4 | 30 | 20 | 75 | mech.[b] | 400 | 650 |
| Refuse composition, percent/weight | | | | | | | | |
| Paper | 100 | 69 | 0 | 71 | 83 | 0 | 100 | 65 |
| Garbage | 0 | 31 | 0 | 17 | 17 | 0 | 0 | 0 |
| Wood | 0 | 0 | 100 | 12 | 0 | 100 | 0 | 35 |
| Auxiliary fuel, standard cubic foot/hour-gas | | | | | | | | |
| Primary chamber | 0 | 165 | 0 | 0 | 0 | 0 | 0 | 0 |
| Mixing chamber | 0 | 165 | 0 | 1125 | 2850 | 0 | *c* | 0 |
| Combustion air, percent of total | | | | | | | | |
| Overfire air | 85 | 45 | 55 | 79 | 50 | 20 | 60 | 70 |
| Underfire air | 15 | 10 | 3 | 7 | 20 | 4 | 3 | 10 |
| Secondary air (mixing chamber) | 0 | 45 | 42 | 14 | 30 | 76 | 37 | 20 |
| Orsat gas analysis, percent/volume | | | | | | | | |
| Carbon dioxide | 4.8 | 6.4 | 8.4 | 6.0 | 7.4 | 5.8 | 2.2 | 6.3 |
| Oxygen | 13.8 | 6.3 | 11.2 | 12.6 | 9.9 | 14.7 | 18.3 | 9.4 |
| Carbon monoxide | 0.0 | 0.0 | 0.0 | 0.0 | 0.0 | 0.0 | 0.0 | 0.0 |
| Nitrogen | 81.4 | 87.3 | 80.4 | 81.4 | 82.7 | 79.5 | 79.5 | 84.3 |

[a] Based on George and Williamson (*14*).
[b] Mech, mechanized feed.
[c] Oil at 2.5 gallons per hour.

**Table XIV Emissions from Small Multiple-Chamber Incinerators**[a]

| | Test number | | | | | | | |
|---|---|---|---|---|---|---|---|---|
| | *1a* | *1b* | *4a* | *5* | *6* | *8* | *12* | *13* |
| Maximum opacity of stack gases, percent | 10 | 0 | 0 | 45 | 10 | 0 | 15 | 0 |
| Smoking time, minutes | 1 | 0 | 0 | 1 | 2.5 | 0 | 9 | 0 |
| Particulates, total, grains/standard cubic foot | | | | | | | | |
| At stack conditions | 0.099 | 0.058 | 0.024 | 0.075 | 0.083 | 0.052 | 0.019 | 0.092 |
| Particulates, filter collection only, grains/standard cubic foot | 0.068 | 0.039 | 0.023 | 0.055 | 0.045 | 0.025 | 0.009 | 0.066 |
| Total corrected to 12% carbon dioxide | 0.270 | 0.300 | 0.038 | 0.205 | 0.248 | 0.116 | 0.113 | 0.200 |
| Filter collection only, corrected to 12% carbon dioxide | 0.185 | 0.182 | 0.033 | 0.130 | 0.119 | 0.053 | 0.057 | 0.126 |
| Total minus filter collection, grains/standard cubic foot | | | | | | | | |
| Corrected to 12% carbon dioxide[b] | 0.085 | 0.118 | 0.005 | 0.075 | 0.129 | 0.063 | 0.056 | 0.074 |
| Gaseous emissions, grains/standard cubic foot | | | | | | | | |
| Carbon monoxide | — | — | 0.0 | — | — | 0.0 | — | — |
| Nitrogen oxides | — | — | 0.032 | <0.0001 | — | 0.028 | — | — |
| Aldehydes | — | — | 0.002 | <0.0001 | — | 0.002 | — | — |
| Organic acids | — | — | 0.021 | — | — | 0.009 | — | — |
| Sulfur dioxide or trioxide | — | — | — | — | — | — | — | — |

[a] Based on George and Williamson (*14*).

[b] Represents condensible liquid materials in sample collection system.

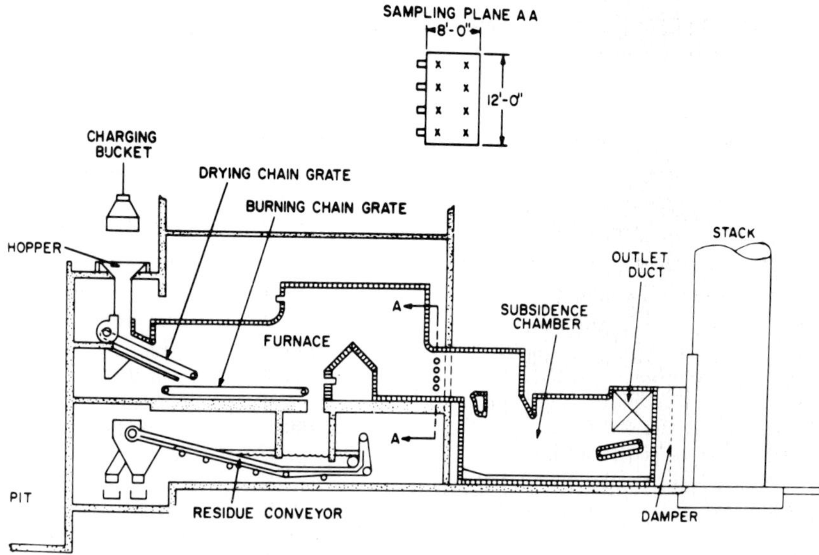


Figure 17. Traveling-Grate Incinerator with Dry Subsidence Chamber (*32*).

a wetted-baffle particulate collector, consisting of a spray manifold upstream of a series of stainless-steel baffles. The collector has a wet bottom and an auxiliary settling tank and decanting overflow system. Measurements were made at both the furnace outlet and the induced draft fan outlet breeching leading to the stack.

The furnace samples were taken as recommended by Rehm (*33*). The stack samples were taken as specified by the American Society of Mechanical Engineers Power Test Code 27, "Determination of Dust Concentrations in Gas Stream."

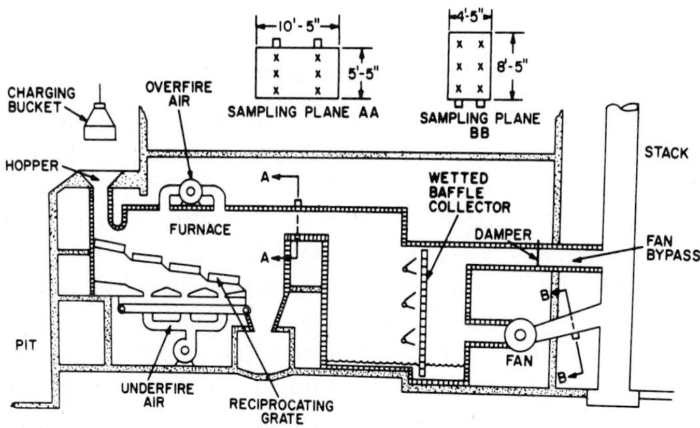


Figure 18. Reciprocating-Grate Incinerator with Wetted-Baffle Collector (*32*).

**Table XV Summary of Test Results on Two 250-tpd Municipal Incinerators**[a]

| | *Unit No.* | | |
|---|---|---|---|
| | *1 (9 tests)* | *2 (4 tests)* | |
| *Measurements, average values* | *Furnace outlet* | *Furnace outlet* | *Stack* |
| Refuse charged, tons/hour | 12.9 | 10.3 | |
| Gas volume[b] | | | |
| acfm (thousands) | 143 | 142 | 90 |
| scfm (thousands) | 36.5 | 43.2 | 47.5 |
| Carbon dioxide, percent/volume | 6.0 | 5.0 | |
| Excess air, percent | 189 | 185 | |
| Underfire air, scfm/ft² of grate | 45 | 113 | |
| Particulate emissions[c] | | | |
| Grains/scf | 1.10 | 1.65 | 0.85 |
| lb/hr | 156 | 241 | 140 |
| lb/ton refuse charged | 12.4 | 25.1 | 11.8 |
| Collection efficiency, percent (estimated) | 21 | 53 | |

[a] Based on Walker and Schmitz (*32*).
[b] Cubic foot per minute based on actual conditions of the gas at sampling point.
[c] Loading corrected to 70°F, 29.92 in. Hg, and 12% $CO_2$.

Comparing underfire air rates with particulate loadings at the furnace outlets for both units appears to confirm findings by others that furnace emissions increase with underfire air rates (Table XV), a not unexpected phenomenon. Neither unit meets the 1973 USEPA standard of 0.08 grain/scf at 12% $CO_2$.

### C. Tests of Large (Greater than 300 tpd) Incinerators

Achinger and Daniels (*44*) tested the six incinerators described in Table XVI. Units 3, 4, and 6 are shown schematically in Figure 19. A summary of the test results is given in Table XVII. To achieve comparability with test data that represent only the collection on the sampler filters, the grains/scf in Table XVII should be multiplied by 0.7, keeping in mind that this factor is a rough approximation based on limited data, which show that the sampler filter collection comprises about 70% of the total catch in a filter impinger train sampling system.

Even with such adjustment to the data in Table XVII, it is evident that the particulate loadings exceed the 1973 USEPA standard of 0.08 grain/scf (sampler filter collection only) at 12% $CO_2$. Obviously, such

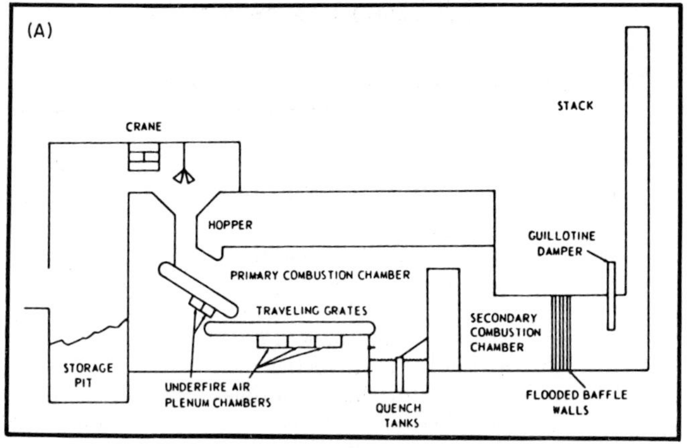


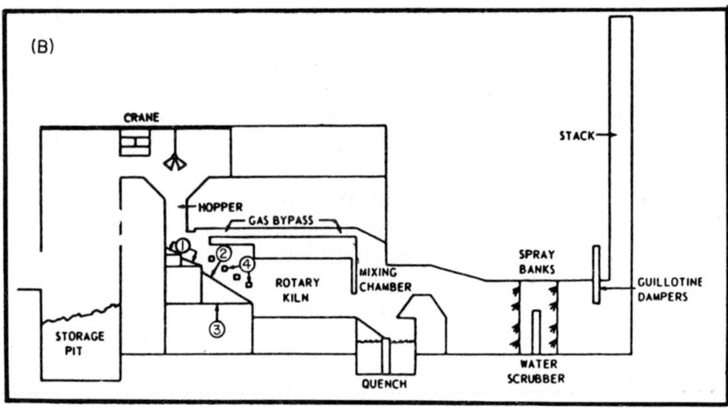


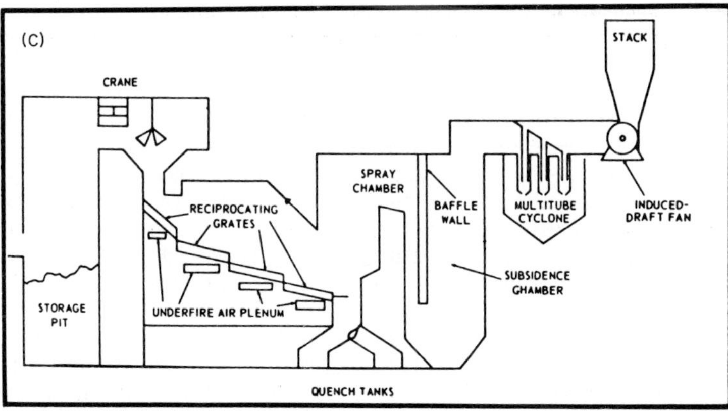


Figure 19. Incinerators Tested (*44*). (A) Traveling grate incinerator with flooded baffle walls; (B) grate-kiln incinerator with water scrubber. 1: drying grates; 2: ignition grate; 3: underfire air plenum; 4: overfire air ducts; (C) reciprocating grate incinerator with multitube dry cyclone following a wet-baffle wall.

**Table XVI Summary of Design Information on Six Large Incinerators**

| *Unit car and year built* | *Capacity, tpd*[a] | *Furnace and grates* | *Air pollution control equipment* |
|---|---|---|---|
| 1 (1966) | 300 | Two refractory-lined multiple-chamber furnaces with inclined, modified reciprocating grate sections followed by stationary grate sections | Wetted-column water scrubber |
| 2 (1966) | 300 | Two refractory-lined multiple-chamber furnaces with three sections of inclined rocking grates | Flooded baffle-wall water scrubber |
| 3 (1965) | 500 | Two refractory-lined multiple-chamber furnaces with air inclined followed by a horizontal traveling grate | Flooded baffle-wall water scrubber |
| 4 (1963) | 500 | Two furnaces with three reciprocating grate sections followed by a rotary kiln | Water sprays and a baffle wall |
| 5 (1963) | 600 | Two furnaces with three reciprocating grate sections followed by a rotary kiln | Water sprays and a baffle wall |
| 6 (1967) | 400 | Two furnaces with four sections of inclined reciprocating grates | Multitube dry cyclones following a wet-baffle wall in a subsidence chamber |

[a] tpd: Tons per day.

**Table XVII Summary of Test Results on Six Large Incinerators**

| | *Unit No.* | | | | | |
|---|---|---|---|---|---|---|
| | *1* | *2* | *3* | *4* | *5* | *6* |
| Refuse charged, tpd[a] actual | 281 | 308 | — | 660 | 645 | 482 |
| Gas volume, cfm (thousands) | 70 | 131 | 120 | 186 | 165 | 130 |
| Burning rate, lb/(hour)($ft^2$) grate | 42 | 53 | — | 59 | 50 | 62 |
| Excess air, percent | 270 | — | 260 | 220 | 320 | 500 |
| Carbon dioxide, percent/volume | 4.6 | 3.5 | 5.0 | 5.0 | 3.9 | 3.2 |
| Particulate emissions[b] | | | | | | |
| Grains/scf[c] | 0.55 | 1.12 | 0.46 | 0.73 | 0.72 | 1.35 |
| lb/hour | 122 | 186 | 173 | 238 | — | 386 |
| lb/ton refuse charged | 10.4 | 14.5 | 8.8 | 8.6 | 12.5 | 20.4 |

[a] tpd: Tons per day.
[b] Total collection in sampler filter plus that in subsequent wet impingers.
[c] Conditions: 70°F, 29.92 in. Hg, and 12% $CO_2$.

incinerators would need more efficient air pollution control equipment to meet the 1973 regulations.

The Chicago Northwest Incinerator (Fig. 20), which began operating in 1970 has four 400 tpd incineration units and a total capacity of 1600 tpd, which made it, at the time of writing this chapter, the largest incinerator installation in the Western Hemisphere (*45, 46*).

Its salient design features are

(a) A hydraulically operated Martin reverse reciprocating stoker on a 26-deg incline. It has a load limit of 10 million Btu per foot of grate width.

(b) An integral welded waterwall boiler of multipass design. The boiler is designed for 250 psig (saturated), and a steam generation rate of 3 lb/lb of refuse having a heating value of 5000 Btu/lb. The high heat-absorption rate in the water walls permits operation with comparatively low excess air.

(c) A low rear furnace arch forces unburned gases back into the high temperature combustion zone for maximum consumption of unburned combustible gases.

(d) An electrostatic precipitator designed for low gas velocities (3 feet/second), which facilitates collection of low-density paper char.

The total catch in the gases from the precipitator outlet indicated emissions were significantly less than the EPA standard of 0.08 grain/scf corrected to 12% $CO_2$ (Table XVIII).

The following conclusions are suggested by Table XVIII. There is a considerable reduction in the impinger catch while the flue gases pass through the electrostatic precipitator. It is believed that absorption of some of the condensable materials occurs on the fly ash that is removed by the electrostatic precipitator. For these particular tests, the collection efficiencies of the electrostatic precipitator were consistent, and a difference of less than 1% in efficiency was involved when comparing efficiencies on the basis of dry catch alone, and dry catch plus impinger catch.

The Harrisburg Incinerator, Harrisburg, Pennsylvania, began operation in 1973. Two furnace-boiler units provide a capacity of about 700 tpd. Each unit has a Martin reverse reciprocating stoker, and an electrostatic precipitator. Measurements were made of the sulfur oxide, hydrochloric acid, nitrogen oxide, and particulate matter concentrations at the inlet and outlet ducts of the electrostatic precipitators (*46a*). A modification of USEPA Method 8 was used for $SO_2$, $SO_3$, and HCl. USEPA Method 7, based on phenol disulfonic acid reagent, was used for nitrogen

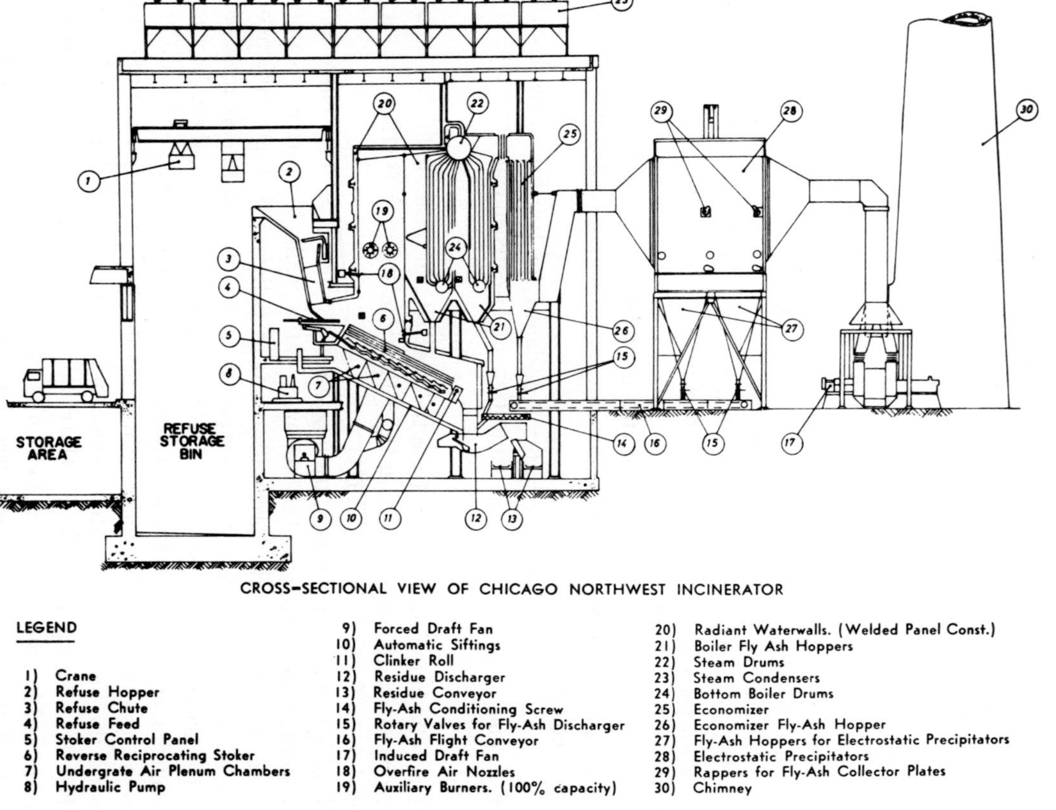


Figure 20. Chicago Northwest Incinerator. (Courtesy of Ovitron Corp. IBW-Martin Incinerator Group.)

**Table XVIII Summary of Test Results on the Chicago Northwest Incinerator**

| *Test* | *1* | *2* |
|---|---|---|
| Refuse charging rate, tons/hour | 16.7 | 16.7 |
| $CO_2$, percent/vol | 10.1 | 9.5 |
| Excess air, percent | 87.0 | 98.0 |
| Stack flow rate, dry scfm (thousand)[a] | 51.9 | 51.5 |
| Dry catch,[b] grain/scf | 0.040 | 0.027 |
| Impinger catch,[c] grain/scf | 0.009 | 0.013 |
| Total catch, grain/scf | 0.049 | 0.040 |
| Impinger/total catch, percent | 18 | 32 |
| Impinger catch, grains/scf | | |
| Inlet | 0.079 | 0.036 |
| Outlet | 0.009 | 0.013 |
| Reduction in impinger catch, percent | 88.6 | 64.0 |
| Dry catch, grain/scf | | |
| Inlet | 1.140 | 1.140 |
| Outlet | 0.040 | 0.027 |
| Electrostatic precipitator[d] efficiency, percent | 96.6 | 97.7 |
| Total catch, grain/scf | | |
| Inlet | 1.219 | 1.176 |
| Outlet | 0.049 | 0.040 |
| Electrostatic precipitator[e] efficiency, percent | 96.3 | 96.7 |

[a] Standard conditions: 70°F, 29.92 in. Hg, and 12% $CO_2$.
[b] Dry catch is the dry particulate matter from the probe, cyclone, and filter.
[c] Impinger catch is the material, except uncombined water, that is condensable at 70°F and 14.7 psia.
[d] Efficiency based on dry catch only.
[e] Efficiency based on total catch.

oxides. Both methods are described in the *Federal Register*, December 23, 1971. The average results are shown in the following tabulation:

| | *Mass flow (lb/10⁶ scf)* | | | | *Particulates (50% excess air) (lb/1000 lb gas)* |
|---|---|---|---|---|---|
| | *HCl* | *$SO_2$* | *$H_2SO_4$* | *$NO_x$* | |
| Inlet | 29 | <0.3 | 0.1 | 51 | 1.90 |
| Outlet | 24 | <0.3 | 0.2 | 72 | 0.085 |

It had been suggested that the precipitator might have an effect on the concentration of gaseous constituents in the flue gases, but these results show no significant effect.

## VIII. Novel Methods of Incineration

Considerable research and development is underway in the United States and abroad on new design concepts for incinerators, aimed at eliminating problems inherent in conventional designs. These novel concepts are still mostly in the pilot-plant stage, and have not yet been demonstrated in commercial plants. Design details are given in a study by the University of Alabama for the U.S. Bureau of Mines (*47*). The different design concepts are broadly categorized as *slagging, fluidized beds, suspension burning,* and *pyrolysis.*

### A. Slagging

The principle of slagging is to operate at a temperature sufficiently high that all incombustible materials are melted and drawn from the incinerator as a fluid slag. When quenched, the solid slag has a lower bulk specific volume (cubic foot per pound) than the residue from conventional incinerators. Some of the advantages claimed are

(a) Greater volume reduction of refuse, consequently lower hauling and land fill costs

(b) Residue has potential value as a raw material for other uses, and can provide credits for operating costs

(c) Less excess air requirements, consequently less flue gas to handle

(d) Less entrained particulate matter, consequently lower dust loading in the flue gas

(e) Designs that do not have a grate system would have less mechanical problems

Some of the advantages claimed for slagging systems may, however, lead to other problems. For example, nitrogen oxide emissions might be higher than from conventional incinerators, because of the higher combustion temperatures needed to melt the incombustibles to a slag.

Slagging incinerator systems currently under development include *Dravo/FLK* (Dravo Corp.); *Ferro-Tech* (Ferro-Tech Industries); *Melt-Zit* (American Thermogen, Inc.); *Sira* (Sira Corp.); *Torrax* (Torrax Systems, Inc.); *URDC* (Urban R & D Corp.). Baum and Parker discuss slagging systems in more detail (*48a*).

### B. Fluidized Beds

Fluidized beds have long been used as reactors in the chemical process industries. Fluidized bed incineration applications currently are limited to relatively homogeneous liquids, slurries, or semisolid mixtures, such as dewatered sewage sludges and oily sludges.

Some of the advantages claimed for fluidized bed systems are

(a) Simple internal construction with no moving parts

(b) Intense mixing in the bed, consequently low temperature gradients through the bed (no hot spots), low excess air, and a comparatively low bed temperature (1600°–1800°F)

(c) High volumetric heat release rate in the bed, possibly 100,000–200,000 Btu/(hour)

(d) Low nitrogen oxide emissions because of low bed temperature

The fluidized bed has some disadvantages, the most significant one being high particulate loadings in the effluent gases, which might require more efficient particulate removal equipment than conventional incinerators.

The only fluidized bed system developed for application to nonhomogeneous waste is the *CPU-400* (Combustion Power Co., Inc.). This design uses the hot gases from the incinerator to drive a gas turbine and generate electric power, and a waste heat boiler, through which the turbine-exhaust gases flow to raise steam. Bergin, Furlong, and Riley have described the *CPU-400* system in detail (*48b*).

### C. Suspension Burning

This system is based on first shredding the refuse to small particles, which are then blown into the combustion chamber with air, and burned in suspension. In the *Vorcinerator* (General Electric Co.) the shredded refuse is blown tangentially into a cylindrical combustion chamber where suspension burning occurs. In the *Solid Waste Reduction Unit* (Hamilton, Ontario), the shredded refuse is blown into the combustion chamber, but not tangentially. Half of the refuse burns in suspension, and the other half burns on the grate system. The *Cycloburner* (Energy, Ltd.) is a horizontal cylindrical combustion chamber into which pulverized solid wastes are fed pneumatically, and combustion air is added through a number of circumferential tuyeres (*48*). *Sira,* the previously noted slagging-type incinerator burns in suspension refuse that has been shredded to a topsize of $\frac{1}{2}$ in. By their nature, suspension-burning systems can be expected to have high particulate loadings in the effluent gases.

### D. Pyrolysis

Pyrolysis involves heating refuse to 1000°–2000°F *in the absence of oxygen* (destructive distillation), so that all volatile components are distilled off, leaving a combustible char. The volatile products comprise combustible gases and condensable liquids (tars, alcohols, ketones, and acetic acid) (Fig. 21). Baum and Parker (*48c*) have reviewed refuse pyrolysis.

Pyrolysis is not incineration, but it is a very attractive alternate because of its potential for (a) substantial reduction of air pollution; (b) production of useful products; and (c) self-sustaining operation in terms of energy.

The U.S. Bureau of Mines has published pilot-plant studies on refuse pyrolysis (*49, 50*). Monsanto Enviro-Chem Systems, Inc., has developed a hybrid combustion-pyrolysis process called *Landgard*. It is a total disposal enclosed system, including receiving, handling, and shredding

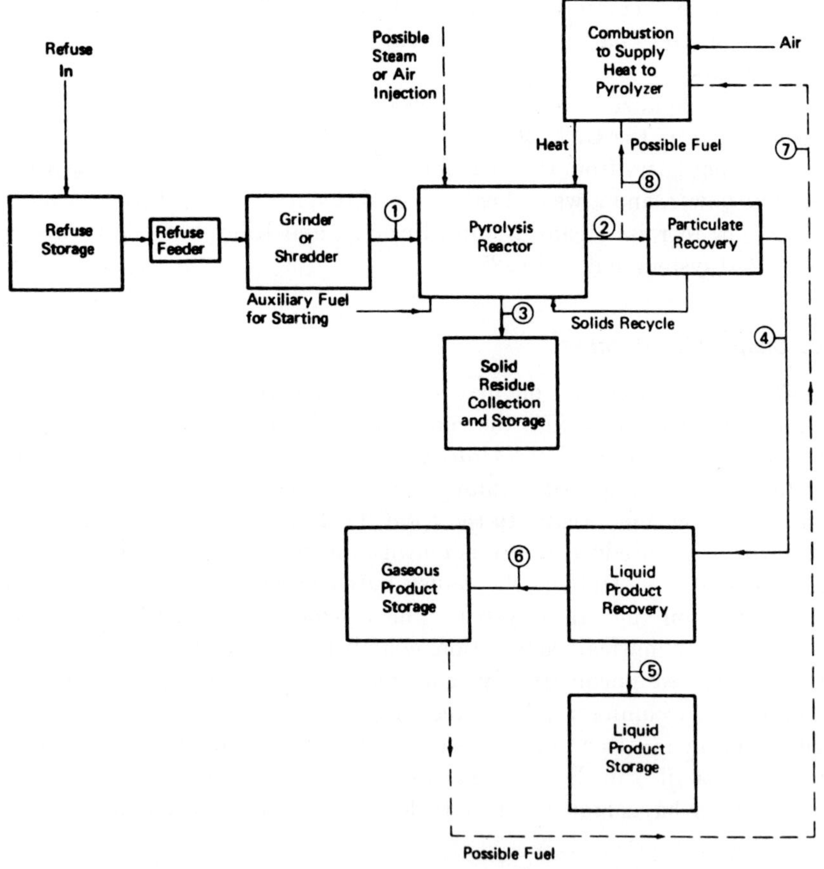


Figure 21. Schematic of a refuse pyrolysis system (*5*). 1: Solid refuse; 2: volatile products and entrained particulates; 3: solid product (char); 4: volatile product; 5: liquid product; 6: gas product; 7: gas for heating; 8: volatiles for heating.

**Table XIX Air Distribution in Various Types of Incinerators**

| | Percent of theoretical air | | |
|---|---|---|---|
| | *Conventional* | *Starved air* | *Controlled air* |
| Underfire | 40 | 48 | 48 |
| Overfire | 280 | 0 | 186 |
| Secondary | 80 | 232 | 46 |
| | 400 | 280 | 280 |

wastes; employing a rotary kiln for pyrolysis, and gas purifying and residue processing. Garrett Research and Development Co. also has developed a complete refuse pyrolysis system which maximizes the production of salable liquids. All of the gases and some of the char are burned for process heat.

Cost estimates on two pyrolysis processes show that municipal refuse can be pyrolyzed for $5 per ton of refuse taking credit for salable products. These costs are substantially less than those estimated for large municipal incinerators. However, there are differences of opinion on the credit value of some of the liquid, gaseous, and solid products from pyrolysis, and until large municipal pyrolysis units come on stream to provide a substantive basis for the market value of the products, the costs cited above should be viewed with reserve.

There are two hybrid types of refuse pyrolysis, commonly referred to as *starved-air incineration* and *controlled-air incineration*. The main aim of each is to reduce particulate emissions, by controlling the amount and the location of the combustion air (*50a*). The main difference between these types and conventional incinerators is shown in the approximate distribution of total air (*50b*) (Table XIX).

Since the underfire air to starved-air and controlled-air incinerators is less than theoretical, the refuse in the primary chamber is pyrolyzed. The unburned combustibles leaving the bed are burned at the top of the primary chamber, or in the secondary chamber by adding more air at these points.

### *E. Use of Refuse in Existing Central Station Electric Utility Plants*

The concept of utilizing municipal refuse as an *auxiliary* fuel in an *existing* electric utility plant has been evaluated in a series of papers discussing (a) the refuse energy aspects (*51*); (b) the steam generator aspects (*52*); (c) the air pollution aspects of such utilization (*53*); and all aspects (*53a*).

The following conclusions were reached regarding retrofitting existing pulverized-coal-fired utility boilers to accept municipal refuse:

(a) The concept is viable, and refuse disposal costs for retrofitted utility boiler plants would generally be lower than those for new construction of plants of corresponding size and configuration

(b) The emission of $SO_2$ from coal-fired utility plants would be reduced, since refuse is a low-sulfur fuel

An experiment was undertaken to determine, in an existing boiler furnace, the engineering and economic feasibility of using municipal refuse as an auxiliary fuel (*54, 55*) in a 125-megawatt pulverized-coal-fired boiler furnace of the Meramac Station of the Union Electric Company of St. Louis, Missouri. Other than installing refuse-burning ports in the furance, no other boiler modifications were made. Prior to firing, the refuse went through a processing system that removed magnetic materials and reduced it to particle sizes less than $1\frac{1}{2}$ in. At full load, the quantity of refuse fired was equivalent in heating value to 10% of the coal fired, or about 300 tons/day of refuse.

## IX. Disposal of Bulky Solid Wastes

Disposal of bulky solid wastes in an economically and environmentally acceptable manner has long been a troublesome problem. Bulky solid wastes are easily recognized but difficult to characterize with precision. They are often defined as pieces of waste too large to fit into the usual mechanized compaction truck, or into a conventional incinerator. This definition can be broadened to include items that burn too slowly for the residence times provided by conventional incinerators or are too large for easy compacting into a sanitary landfill.

The TS-2.4 (Refuse) Committee of the Air Pollution Control Association has issued *Informative Report No. 1,* which deals with the bulky waste problem. The report discusses comparative advantages and disadvantages of (a) *disposal on land,* including dumping and sanitary landfilling, (b) *ocean disposal,* and (c) *incineration* by special furnace designs for bulky wastes.

### X. Industrial Waste Disposal

Among the advantages of incineration for the disposal of industrial wastes are the following:

(a) It may be the most economical process available, especially if the fuel value of the waste can be utilized to generate steam for heat and power

(b) If the industrial waste contains components that can be recycled in the industry or sold for revenue, their separation from the incoming waste can be accomplished before incineration, or from the incinerator residue (*56*)

(c) Toxic materials can be destroyed in an environmentally acceptable manner

On-site single- or multiple-chamber incinerators of the type described earlier in this chapter are often used for solid wastes when the amount is no more than several hundred pounds per hour. With large amounts of wet or high-ash, materials, firing to mechanical grate-type incinerators of the types described for municipal wastes can be used. Special incinerators with integral pollution control systems, have been designed for industrial solid, sludge, liquid, and gaseous industrial wastes (*57–59*).

Today an increasing number of incinerators are being designed and built to dispose of aqueous wastes containing only traces of combustible material. These wastes may vary from liquids that are nearly all water to ones with materials that may be toxic, acidic, basic, or biologically active. Methods for incinerating such wastes have been described (*60–62*).

Incinerators have been developed especially for waste liquids and sludges from paint and resin plants (*63*). The Department of Health, Lausanne, Switzerland, developed and placed in successful operation an incinerator for waste oils and other combustible liquid wastes (*64*). Basically, the incinerator is a vertical, cylindrical chamber into which the wastes are fed tangentially.

Four types of incinerators are mainly used for incinerating sewage sludge (*65, 66*). One type, the *multiple hearth,* is a cylindrical chamber with several grates mounted axially one above the other. Each grate has an air-cooled rabble arm rotated by a common shaft. Sludge enters the top. Hot gases enter the bottom of the incinerator. The sludge progressively dries, ignites, and burns as it moves downward from one grate to another. Another type is a *fluidized* bed of hot sand into which the sludge, predried to between 40 and 80% water, is injected. The other types are the cyclone reactor and the rotary kiln. An evaluation of these types is shown in the following tabulation:

| *Type* | *Evaluation* |
|---|---|
| Multiple hearth | Maintenance costs high; frequent rabble arm replacement necessary |
| Fluidized bed | Several technical advantages. Disadvantages are high fly ash emissions, and relatively high investment cost |
| Cyclonic reactor | Small and relatively inexpensive |
| Rotary kiln | Investment and maintenance relatively low, fly ash system appears to be adequate |

Safe disposal of chlorinated and fluorinated wastes by controlled incineration has been discussed (*67*). Both low energy and high energy wet scrubbers have been used to remove the halogens and the halogen acids from the products of combustion.

Eastman Kodak Co. burns separately and in combination waste paper, chemicals, solvents, sludges, and bulky objects (*68*). A rotating-hearth incinerator, equipped with a spray washer, disposes of plant trash, waste solvents and oils, and various liquid and solid chemicals. A rotary-kiln incinerator disposes of sludge from process vacuum filters. A water scrubber and an afterburner are used to clean and deodorize the flue gases.

The conical ("teepee") burner and the silo burner are commonly used in lumber processing areas for incinerating wood chips and sawdust. Both types of burners emit excessive smoke and fly ash unless they are carefully designed and operated (*69–72*). The silo type is more amenable to better design than the conical type, which has been subject to considerable criticism and gradual replacement in many areas.

The metals in scrap cars and trucks can be salvaged by incinerators designed to burn all the combustible components (upholstery, insulation, rubber, etc.) relatively smokelessly and virtually free of fly ash emissions, without impairing the quality of the scrap metals by overheating. Kaiser has designed an incinerator that burns two car bodies simultaneously and has a capacity of 28 cars per 8 hour day (*73*). When a larger capacity is needed, twin-cart or conveyor furnaces are used. The minimum initial cost of the small incinerator is about $27,000 (in 1962 dollars), and the larger about $1200 to $1500 per unit of body capacity in 8 hours. The Bureau of Mines has designed a junk auto incinerator that will handle 32–60 cars daily, depending on the degree to which the autos are stripped prior to incineration (*74*). The construction cost is roughly $22,000 (in 1972 dollars). A summary report on air pollution from several kinds of junk auto incinerators concludes that unfavorable economics experienced by some operators of such incinerators have increased the use of fragmentizers and shredders for disposing of junk autos (*75*).

Monroe (*76*) has shown that for certain types of wastes, principally those with less than 1 or 2% ash, an open pit supplied with high velocity air jets angled downward across the pit is an effective and nuisance-free burner for solid, liquid, and gaseous wastes. If its limitations are recognized, open-pit incineration has a place, especially for liquid wastes.

A concept has been proposed for burning all kinds of industrial wastes in a central facility in an environmentally acceptable manner (*77*). This would eliminate the multiplicity of incinerators designed for specific industrial wastes. The author suggests systems that might be technically and economically feasible. Each system recovers the chemical heat in the wastes to generate steam for heat and power, and cleans the flue gases with a mechanical dust collector followed by an electrostatic precipitator.

A significant step has been taken towards the central system approach in Harrisburg, Pennsylvania, where an incinerator has been designed to burn 600 tons/hour of municipal refuse, and receive 40 tons daily of sewage sludge that has been processed in an adjacent facility. Steam generated by the incinerator is utilized at the waste disposal site, and marketed for power generation and district heating and cooling.

It appears that many difficult industrial waste disposal problems have been solved by incineration. However, the paucity of emission data in the literature suggests the possibility that in some instances a waste disposal problem was solved but an air or water pollution problem was created. Trade-offs among air, water, and ground pollution are no longer acceptable.

## XI. Pathological Waste Disposal

Pathological wastes from hospitals and other institutions should be incinerated at the source. If they are handled at a municipal incinerator, extreme care is required to prevent workers from becoming infected. Guidelines for disposing of operating room wastes, experimental animals, and cadavers are available (*78*).

Pathological wastes containing low-level radioactivity present a special problem, which has not yet been resolved to the extent that specifications for safely incinerating such wastes are available. Corey has reviewed work in this field (*79*) and concluded, "At the present time the state of the technology is not sufficiently advanced for a packaged system to be installed. Each situation must be considered as different from any other one; this means that each step from collection of the waste, through incineration, to disposal of the final residue must be examined in detail and made compatible with all the other steps."

## REFERENCES


1. R. Hering and S. A. Greeley, *in* "Collection and Disposal of Municipal Refuse," p. 311. McGraw-Hill, New York, New York, 1921.
2. F. R. Bowerman, *in* "Principles and Practices of Incineration" (R. C. Corey, ed.), Chapter 1, p. 1. Wiley (Interscience), New York, New York, 1969.
3. J. W. Stephenson, Pap. No. 69-WA/Inc-1. Amer. Soc. Mech. Eng., New York, New York, 1969.
4. W. R. Niessen, S. H. Chansky, A. N. Dimitriou, A. N. Field, C. R. LaMantie, and R. E. Zinn, "Systems Study of Air Pollution from Municipal Incineration," Vol. II, Report PB 192379. National Technical Information Service, U.S. Dept. of Commerce, Springfield, Virginia, 1970.
5. W. R. Niessen, S. H. Chansky, A. N. Dimitriou, A. N. Field, C. R. Mantie, and R. E. Zinn, "Systems Study of Air Pollution from Municipal Incineration," Vol. I, Report PB 192378. National Technical Information Service, U.S. Dept. of Commerce, Springfield, Virginia, 1970.
6. E. R. Kaiser, C. D. Zeit, and J. B. McCaffery, *in* "Proceedings of the National Incinerator Conference," p. 142. Amer. Soc. Mech. Eng., New York, New York, 1968.
7. L. S. Wegman, *in* "Proceedings of the National Incinerator Conference," p. 1. Amer. Soc. Mech. Eng., New York, New York, 1964.
8. Combustion Engineering, Inc., "Technical Economic Study of Solid Waste Disposal Needs and Practices," Report SW-7C, Pub. Health Serv. Publ. No. 1886. U.S. Govt. Printing Office, Washington, D.C., 1969.
9. J. A. Fife, *in* "Proceedings of the National Incinerator Conference," p. 249. Amer. Soc. Mech. Eng., New York, New York, 1970.
10. A. J. Warner, C. H. Parker, and B. Baum, Report for Manufacturing Chemists Assoc., Project 14402. Washington, D.C., 1970.
11. N. L. Drobny, H. E. Hull, and R. F. Testin, Battelle Memorial Institute, "Recovery and Utilization of Municipal Solid Waste," Pub. Health Serv. Publ. No. 1908. U.S. Govt. Printing Office, Washington, D.C., 1971.
12. R. C. Corey, ed., "Principles and Practices of Incineration." Wiley (Interscience), New York, New York, 1969.
13. F. D. Friedrich, *Can., Mines Br., Tech. Bull.* **TB-134** (1971).

13a. B. Baum and C. H. Parker, "Solid Waste Disposal," Vol. I. Ann Arbor Sci. Publ., Ann Arbor, Michigan, 1973.

14. R. E. George and J. E. Williamson, *in* "Principles and Practices of Incineration" (R. C. Corey, ed.), Chapter 5, p. 106. Wiley (Interscience), New York, New York, 1969.
15. H. G. Meissner, *in* "Principles and Practices of Incineration" (R. C. Corey, ed.), Chapter 4B, p. 84. Wiley (Interscience), New York, New York, 1969.
16. J. J. Sableski and W. A. Cote, *J. Air Pollut. Control. Ass.* **22**, 239 (1972).
17. H. G. Meissner, *in* "Principles and Practices of Incineration" (R. C. Corey, ed.), Chapter 6, p. 163. Wiley (Interscience), New York, New York, 1969.

18. P. M. Sullivan and M. H. Stanczyk, *U.S., Bur. Mines, Tech. Progr. Rep.* **33,** 1–19 (1971).
19. K. C. Dean, C. J. Chindren, and L. Peterson, *U.S. Bur. Mines, Tech. Progr. Rep.* **34,** 1–10 (1971).
20. A. A. Orning, *in* "Principles and Practices of Incineration" (R. C. Corey, ed.), Chapter 2, p. 9. Wiley (Interscience), New York, New York, 1969.
21. E. R. Kaiser and J. B. McCaffery, Pap. No. 69-225. Air Pollut. Contr. Ass., Pittsburgh, Pennsylvania, 1969.
22. R. B. Engdahl, *Combustion* p. 47, March 1944.
23. W. Gumz, *Combustion* p. 39, April 1951.
24. "Application of Overfire Jets to Prevent Smoke in Stationary Plants." Bituminous Coal Research, Inc., Pittsburgh, Pennsylvania, 1957.
25. A. H. Rose, R. L. Stenburg, M. Corn, R. R. Horsely, R. Allen, and P. W. Kolp, *J. Air Pollut. Contr. Ass.* **8,** 297 (1959).
26. R. L. Stenburg, R. P. Hangebrauck, D. J. von Lehmden, and A. H. Rose, *J. Air Pollut. Contr. Ass.* **11,** 376 (1961).
27. R. L. Stenburg, R. L. Horseley, R. A. Herrick, and A. H. Rose, *J. Air Pollut. Contr. Ass.* **10,** 114 (1960).
28. R. P. Hangebrauck, D. J. von Lehmden, and J. E. Meeker, *J. Air Pollut. Contr. Ass.* **14,** 267 (1964).
29. A. A. Carotti and E. R. Kaiser, Pap. No. 71-67. Air Pollut. Contr. Ass., Pittsburgh, Pennsylvania, 1971.
30. F. D. Friedrich, *Can., Mines Br., Tech. Bull.* **TB-135.**
31. M. E. Fulmer and R. F. Testin, "A Report on the Role of Plastics in Solid Waste." Battelle Memorial Institute, Columbus Laboratories, Columbus, Ohio, 1970.
32. A. B. Walker and F. W. Schmitz, *in* "Proceedings of the National Incinerator Conference," p. 64. Amer. Soc. Mech. Eng., New York, New York, 1966.
33. W. Jens and R. R. Rehm, *in* "Proceedings of the National Incinerator Conference," p. 74. Amer. Soc. Mech. Eng., New York, New York, 1966.
34. J. S. Busch, *Pollut. Eng.* p. 32. (1973).
35. J. H. Fernandes, *in* "Proceedings of the National Incinerator Conference," p. 101. Amer. Soc. Mech. Eng., New York, New York, 1968.
36. R. L. Bump, *in* "Proceedings of the National Incinerator Conference," p. 161. Amer. Soc. Mech. Eng., New York, New York, 1966.
37. J. H. Fernandes, N. Engl. Plant Eng. Maintenance Conf., Newton, Massachusetts, May 20–21, 1970, pp. 1–7.
38. F. R. Rehm, *J. Air Pollut. Contr. Ass.* **15,** 127 (1965).
39. R. B. Engdahl, *in* "Symposium on Source Sampling of Atmospheric Contaminants." Chemical Institute of Canada, Toronto, Ontario, Canada, 1971.
40. W. A. Crandall, *Mech. Eng.* **94,** 14 (1972).
41. E. F. Gilardi and H. F. Schiff, *in* "Proceedings of the National Incinerator Conference," p. 102. Amer. Soc. Mech. Eng., New York, New York, 1972.
42. F. A. Govan and L. Terracciano, Pap. No. 72-72. Air Pollut. Contr. Ass.. Pittsburgh, Pennsylvania, 1972.
43. S. J. Selle and G. H. Gronhovd, Pap. No. 72-WA/APC-4. Amer. Soc. Mech. Eng., New York, New York, 1972.
44. W. C. Achinger and L. E. Daniels, *in* "Proceedings of the National Incinerator Conference," p. 32. Amer. Soc. Mech. Eng., New York, New York, 1970.
45. G. Stabenow; *in* "Proceedings of the National Incinerator Conference," p. 178. Amer. Soc. Mech. Eng., New York, New York, 1972.

46. G. Stabenow, Pap. No. 72-WA/APC-1. Amer. Soc. Mech. Eng., New York, New York, 1971.
46a. P. R. Webb and R. B. Engdahl, "Incinerator Gas Sampling at Harrisburg, Pennsylvania," Special report. Battelle-Columbus, Columbus, Ohio, 1973.
47. D. Massey, W. E. Kelly, and E. K. Landis, "Urban Refuse Incinerator Design and Operation: State of the Art," BER Rep. No. 113-119. College of Engineering, University of Alabama, Tuscaloosa, Alabama, 1970.
48. R. G. Mills and L. G. Desmon, *in* "Proceedings of the National Incinerator Conference," p. 195. Amer. Soc. Mech. Eng., New York, New York, 1972.
48a. B. Baum and C. H. Parker, "Solid Waste Disposal," Vol. 1. Ann Arbor Sci. Pub., Ann Arbor, Michigan, 1974.
48b. T. J. Bergin, D. A. Furlong, and B. T. Riley, "A Progress Report on the CPU-400 Project," Pub. Health Service Publ. Office of Solid Waste Management, U.S. Environmental Protection Agency, Washington, D.C., 1970.
48c. B. Baum and C. H. Parker, "Solid Waste Disposal," Vol. 2. Ann Arbor Sci. Publ., Ann Arbor, Michigan, 1974.
49. R. C. Corey, *Proc. Mineral Waste Util. Symp., 2nd, 1970,* p. 299. Illinois Institute of Technology, Research Institute, Chicago, Illinois, 1970.
50. W. S. Sanner, C. Ortuglio, J. G. Walter, and D. E. Wolfson, *U.S., Bur. Mines, Rep. Invest.* **RI-7428** (1970).
50a. F. L. Cross, *Pollut. Eng.,* December 1973, pp. 30–32.
50b. F. M. Lewis, *3rd Annu. Ind. Air Pollut. Contr. Semin., 1973* p. 12.
51. R. M. Roberts and E. M. Wilson, Pap. No. 71-WA/Inc-3. Amer. Soc. Mech. Eng., New York, New York, 1971.
52. R. E. Sommerlad, R. W. Bryers, and J. D. Shenker, Pap. No. 71-WA/Inc-2. Amer. Soc. Mech. Eng., New York, New York, 1971.
53. A. P. Konopka, Pap. No. 71-WA/Inc-1. Amer. Soc. Mech. Eng., New York, New York, 1971.
53a. Electric Power Research Institute (EPRI), "Fuels from Municipal Refuse for Utilities: Technology Assessment," EPRI Rep. No. 261-1. EPRI, Palo Alto, California, 1975.
54. F. E. Wisely, G. W. Sutterfield, and D. L. Klumb, *in* "Proceedings of the National Incinerator Conference," p. 97. Amer. Soc. Mech. Eng., New York, New York 1972.
55. J. G. Singer and J. F. Mullen, Pap. No. 73-Pwr-18. Amer. Soc. Mech. Eng., New York, New York, 1973.
56. C. B. Kenahan, R. S. Kaplan, J. T. Dunham, and D. G. Linehan, *U.S. Bur Mines, Inform. Circ.* **8595** (1973).
57. R. B. Engdahl, *in* "Principles and Practices of Incineration" (R. C. Corey, ed.), Chapter 7, p. 210. Wiley (Interscience), New York, New York, 1969.
58. J. Frankel, Pap. No. 68-WA/Inc-1. Amer. Soc. Mech. Eng., New York, New York, 1968.
59. C. A. Hescheles, Pap. No. 68-PEM-10. Amer. Soc. Mech. Eng., New York, New York, 1968.
60. J. A. Challis, *in* "Proceedings of the National Incinerator Conference," p. 208. Amer. Soc. Mech. Eng., New York, New York, 1966.
61. E. S. Monroe, *in* "Proceedings of the National Incinerator Conference," p. 204 Amer. Soc. Mech. Eng., New York, New York, 1968.
62. G. B. Westall and E. G. Gjerde, Pap. No. 70-53. Pollution Air Control Assoc. Pittsburgh, Pennsylvania, 1970.

63. D. P. Bridge and J. D. Hummell, *in* "Proceedings of the National Incinerator Conference," p. 55. Amer. Soc. Mech. Eng., New York, New York, 1972.
64. H. Blanc and M. Maulaz, *in* "Proceedings of the National Incinerator Conference," p. 61. Amer. Soc. Mech., New York, New York, 1972.
65. H. Eberhardt, *in* "Proceedings of the National Incinerator Conference," p. 124. Amer. Soc. Mech. Eng., New York, New York, 1966.
66. J. F. Ferrel, *Pollut. Eng.* p. 36, March, 1973.
67. R. D. Ross and C. E. Hulswitt, Pap. No. 69-114. Air Pollution Control Assoc., Pittsburgh, Pennsylvania, 1969.
68. R. L. Merles, in "Proceedings of the National Incinerator Conference," p. 202. Amer. Soc. Mech. Eng., New York, New York, 1966.
69. R. W. Boubel, M. Northcraft, A. Van Vliet, and M. Popovich, *Oreg. Eng. Exp. Sta. Bull.* **39** (1958).
70. H. C. Johnson, "Methods and Costs of Wood Waste Disposal in the Bay Area," Inform. Bull., pp. 6–61. Bay Area Air Pollut. Contr. District, San Francisco, California, 1961.
71. H. Droege, H. C. Johnson, L. Clayton, and T. McEwen, "Performance Characteristics and Emission Concentrations from Various Type Incinerators," Inform. Bull., p. 1–63. Bay Area Air Pollut. Contr. District, San Francisco, California, 1963.
72. T. E. Kreichelt, Environmental Health Series, *U.S., Pub. Health Serv., Publ.* **999-AP-28** (1966).
73. E. R. Kaiser and J. Tolciss, *J. Air Pollut. Contr. Ass.* **12,** 64 (1962).
74. C. J. Chindgren, K. C. Dean, and J. W. Sterner, *U.S. Bur. Mines, Tech. Progr. Rep.* No. 21 (1970).
75. F. M. Alpiser, Pap. No. 68-25. Air Pollution Control Assoc., Pittburgh, Pennsylvania, 1968.
76. E. S. Monroe, *in* "Proceedings of the National Incinerator Conference," p. 226. Amer. Soc. Mech. Eng., New York, New York, 1966.
77. C. A. Hescheles, *J. Eng. Power,* p. 39 (1970).
78. "Municipal Waste Disposal," pp. 219–221. Public Administration Service, Chicago, Illinois, 1966.
79. R. C. Corey, *in* "Principles and Practices of Incineration" (R. C. Corey, ed.), Chapter 9, p. 239. Wiley (Interscience), New York, New York, 1969.

# 14

## The Control of Motor Vehicle Emissions

Donel R. Olson

## I. Introduction

Pollution control cannot be the only criterion for motor vehicle engines. Fuel economy, energy source versatility, safety, noise, cost, driveability, serviceability, size, weight, and a variety of other parameters must also be considered. The reciprocating spark-ignition carbureted gasoline engine, presently used in most motor vehicles, has favorable characteristics in all categories except emissions. Engines that are superior to spark-ignition engines in only one or a few characteristics cannot compete effectively in the United States automobile market within the existing economic system, but may be acceptable and, in many cases, desirable alternatives in other countries. As emphasis shifts from one criterion to another, as from emissions to fuel economy, new engine concepts may find acceptance, even in the United States market.

The approaches for reducing vehicle emissions include engine modification, the use of alternative fuels, exhaust treatment, periodic vehicle inspection and maintenance, traffic control, integration of mass transit systems, and various forms of rationing. Alternative engines to the spark-ignition gasoline engine include diesel, gas turbine, electric, steam (or other vapor), Stirling, flywheel, and hybrid.

## II. Mobile Sources of Emissions

Mobile pollution sources include railroad locomotives, ships, airplanes, and motor vehicles, i.e., passenger cars, buses, trucks, motorcycles, and recreational and other special purpose vehicles.

### A. Motor Vehicles

Motor vehicles are by far the major mobile pollution source. The sources of pollutants from motor vehicles are crankcase and exhaust emissions and evaporative emissions from the fuel tank and carburetor.

## 1. Crankcase Emissions

Crankcase emissions occur when some of the air–fuel mixture within the cylinder is forced past the piston rings as so-called "blowby." If uncontrolled, blowby vented to the air from the crankcase can represent about 25% of the total hydrocarbon emissions of an engine. The hydrocarbon concentration of blowby does not vary widely among vehicles and among operating modes because blowby is mainly the carbureted air–fuel mixture. Its volume, however, varies over a wide range because pressures in the cylinders change according to operating mode. Blowby rates are highest during the compression and power strokes; lowest during deceleration and idle. There are negligible amounts of pollutants other than hydrocarbons in blowby.

Since diesel engines compress only air (and small amounts of residual exhaust gas), blowby contains very low levels of pollutants, e.g., hydrocarbons.

## 2. Evaporative Emissions

In gasoline-powered vehicles, gasoline vapors can escape from the fuel tank and carburetor. The amount escaping depends upon fuel composition, engine operating temperature, and ambient temperature. Evaporative losses from the fuel tank are strongly influenced by the ambient temperature and exposure of the tank to the sun, the atmosphere, and pavements. These losses can be very high when the ambient air temperature approaches the boiling point of the gasoline.

The evaporative losses from the carburetor take place mainly during the "hot soak" period when the carburetor is heated by the hot engine after the engine has been turned off and no longer has its radiator fan in operation. Ambient temperature has less effect on evaporative losses from the carburetor. If uncontrolled, evaporative losses are about 15% of the total hydrocarbon loss from a vehicle. There are no pollutants other than hydrocarbons in the evaporative loss.

## 3. Exhaust Emissions

The exhaust is the source of most of the hydrocarbon emissions—60% from an uncontrolled engine, almost 100% from an engine with crankcase vent and evaporative emission controls—and practically all the oxides of nitrogen and carbon monoxide emissions. Exhaust products are the result of combustion in the engine under high temperature and pressure. Depending upon fuel composition and fuel additives employed, sulfur dioxide, lead, lead scavenger compounds, oxygenated compounds, particulate matter, and other compounds may also be present in the exhaust.

The type and quantity of pollutants emitted through the exhaust are strongly dependent on the engine operating modes.

### B. Railroad Locomotives and Ships

All of the pollutants resulting from combustion of coal or oil in stationary furnaces, turbines, etc., also exist when these prime movers are used in mobile sources, such as ships and locomotives. However, space constraints and operating flexibility requirements in locomotives and ships make it difficult to apply the same combustion controls as can be used in a stationary plant.

### C. Airplanes

Airplane operations represent a significant source of pollution at and near airports. As a contributor to overall air pollution, however, airplanes are a relatively small source. Many of the same control approaches used for other combustion sources are applicable, but more than any other mobile pollution source, consideration for the effect of pollution controls on safety is of overriding importance. Ground-operations management also plays a key role in airplane emissions control.

## III. Emission Control Technology

### A. Engine Design Considerations

Automotive emissions control is a compromise resulting from the superposition of emissions regulations upon an engine that was developed over a period of 60 years with practically no consideration of the composition of the exhaust. The current engine has had three primary design objectives: low cost, high power-to-weight ratio, and high power-to-size ratio. Secondary design objectives have been good durability, low maintenance, high reliability, and low fuel consumption. The diesel engine has had similar design objectives during its evolution.

Fuel economy is so strongly tied to vehicle weight that focusing upon the engine when seeking improvements in fuel mileage is dangerous (*1–3*). The concern with fuel economy, however, has fortunately resulted in improved engine efficiency.

The additional design objective of reducing pollutant emissions has been constrained in an effort to disturb existing engine designs as little as possible.

### B. Engine Types

Combustion engines are categorized by the nature and the time duration of the combustion process (Table I). Internal-combustion engines utilize combustion products as a fluid to be expanded to produce work. External-combustion engines employ a second separate working fluid, which by relieving the combustion products from the constraints of also acting as the working fluid, generally makes control of the combustion process easier and allows the attainment of very low emission levels. However, external-combustion engines suffer complexity of the second fluid and the need for efficient heat transfer surfaces in order to transfer energy from the combustion products to the working fluid. In intermittent-combustion engines, charges of fuel and air are ignited and burned repetitively, at a rate of from about 5 to 50 times per second. In continuous-combustion engines, the fuel and air are continuously supplied to a combustion chamber with no need for repeated ignitions. An advantage of the intermittent-combustion engine is that working fluid temperatures substantially in excess of engine surface temperatures are possible. In both internal- and external-type continuous-combustion engines, the temperature of the working fluid is usually set by material limitations. Many possibilities exist for combining these engine types to produce compound engines. The turbocharged diesel is an example which combines the Brayton and Diesel cycles.

**Table I Engine Classification by Combustion Characteristics**[a]

| | *Internal combustion* | *External combustion* |
|---|---|---|
| Intermittent combustion | Otto cycle (spark ignition)<br>Conventional<br>Stratified charge<br>Rotary<br>Diesel cycle (compression ignition) | |
| Continuous combustion | Brayton cycle (gas turbine)<br>Open cycle | Rankine cycle (steam engine)<br>Stirling cycle<br>Brayton cycle (closed-cycle gas turbine) |

[a] Noncombustion "engines" include electric and flywheel types.

### C. Engine Efficiency

The fraction of the energy in the high temperature combustion products which can be converted into useful energy depends upon both the particular engine cycle and the nature of the real losses, such as friction and heat transfer. As a general rule, high temperatures of the working fluid at the beginning of the expansion process and low temperatures at the end of the expansion are desired. It follows that, for good efficiency, high compression or high pressure ratios are favored and, somewhat less obviously a low-specific-heat working fluid is desirable in most cases. For internal-combustion engines, fuel-lean mixtures give the lowest specific heat ratios. It is always desirable to maximize the ratio of useful work produced to losses sustained through engine friction, engine accessories, cooling water, and the pumping of the incoming air. Actual engine efficiencies vary from 0 (at idle) to about 35% at full power for the best automotive engines (Fig. 1). The relative efficiency of engines under

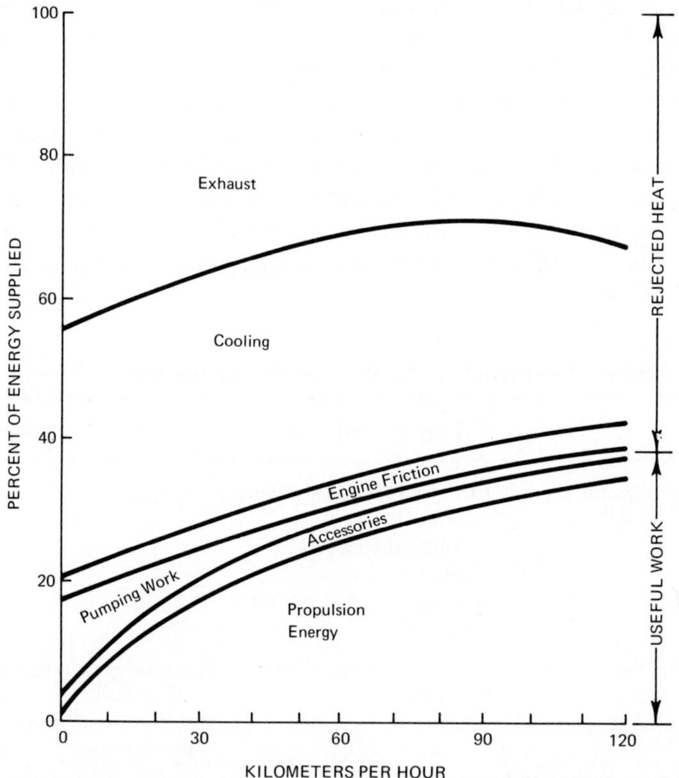


**Figure 1. Typical engine energy balance (approximate).**

actual-use conditions involves substantial part-load operation, cold starts, and transient operation, and are probably adequately inferred from vehicle fuel-economy data.

### *D. Combustion Processes*

The combustion process is extremely efficient. More than 95% of the available chemical energy is released in the combustion process. One of the most important parameters affecting combustion is the air/fuel mixture ratio (Fig. 2). If less than the required amount of air is provided, increased concentrations of carbon monoxide and hydrocarbons result. If the air supply becomes excessive, the engine will lose power and begin to misfire and the combustion will be incomplete. Without emission constraints most engines operate within the (air/fuel) ratio range of 12:1 to 16:1.

During a cold start, choking is required to make the air/fuel ratio extremely rich and, thus, compensate for the low vapor pressure of cold

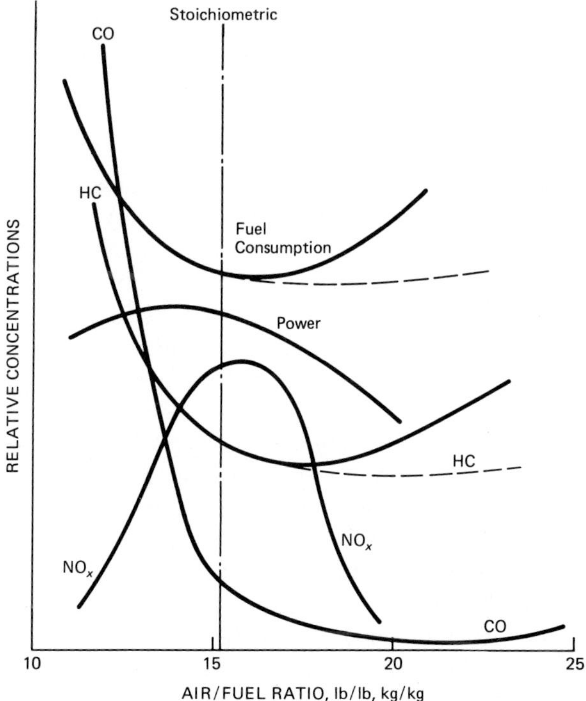


Figure 2. The relationship of typical engine emission and performance to air/fuel ratio. Solid line: conventional engine; dashed line: lean-burn engine.

gasoline. Much of the hydrocarbons and carbon monoxide is emitted during the first 2 minutes of the 23-minute United States certification test cycle. By heating the incoming air and providing hot spots for fuel evaporation, choking times have been reduced to less than 1 minute, yielding a reduction in hydrocarbon and carbon monoxide emissions.

Nitric oxide, which is formed only at high combustion temperatures, is a maximum for air/fuel ratios slightly leaner than stoichiometric. Carbon monoxide emissions decrease on the lean side of stoichiometry, but, with conventional engines, show an upturn as the air/fuel ratio is further increased because the mixture becomes too lean to support complete combustion. Conventional spark-ignition engines can be modified to operate at higher air/fuel ratios (leaner) by improving the homogeneity of the air/fuel mixture.

Diesel engines and stratified-charge engines are designed to operate with large excesses of air. Stratified-charge engines are similar in many respects to conventional gasoline burning engines, but they have the distinguishing characteristic of nonuniform distribution of the air/fuel mixture in the combustion chamber. Prior to combustion, the mixture is divided or "stratified" into fuel-rich and fuel-lean regions. Ignition is obtained in the relatively easily ignited rich mixture, and the flame front which is generated ignites the lean mixture to complete the combustion process.

By very lean overall operation (Fig. 2), these systems achieve relatively low levels of nitrogen oxide emissions without exhaust gas recirculation or catalysts and provide good fuel economy. Carbon monoxide emissions are also low. As high air/fuel ratios are achieved, hydrocarbon emissions tend to rise due to misfires in spark-ignited engines. Conventional and stratified-charge engines, therefore, still require the use of oxidizing catalysts, thermal reactors, or unique ignition systems which assure combustion ignition of very lean mixtures, to meet hydrocarbon standards.

Flame propagation, limits of flammability, and flame or wall quenching play important roles in the combustion process. Mixtures of fuel and air that lie within certain limits (flammability limits) will support combustion. The speed at which a flame will travel through a combustible mixture depends on the fuel/air ratio, the mixture temperature and pressure, and the extent of dilution by inert gases. Flame propagation in engines is a complex phenomenon, which is of extreme importance in the control of engine emissions. The reader is referred to References (*4–7*) for a more comprehensive view.

Combustion rates in engines are influenced by a variety of design and operating variables. For example, a basic characteristic of gasoline engine combustion is that increasing engine speed increases mixture swirl and turbulence, which in turn increases the combustion rates. Doubling the speed will approximately double the combustion rates, and it is this proportionality relationship which permits high speed engine operation. Generally, maximum combustion rates will occur at air/fuel ratios 10–20% richer than stoichiometric. The other major operating variable which affects the flame propagation rate is dilution of the charge by residual exhaust gases. High dilution decreases flame speed and also reduces peak combustion temperatures.

Quenching of the flame front at combustion chamber walls or in small crevices such as piston ring slots is the predominant source of exhaust hydrocarbons. A thin layer of fuel/air mixture will not burn at the relatively cool combustion chamber walls. It has been demonstrated experimentally that the "quench distance" or thickness of the unburned film is almost inversely proportional to combustion chamber pressure and to the square root of the absolute temperature. In typical modern engines, this film varies between 0.05 and 0.4 mm. The quench distance is influenced by several variables other than mixture temperature, pressure, and fuel/air ratio. Wall surface material including combustion deposits has an effect. Combustion chamber deposits act as a sponge to soak up raw fuel during the intake and compression strokes. This fuel may vaporize and enter the exhaust late in the expansion stroke, but this phenomenon is distinct from wall quenching.

Small crevices such as the piston ring areas, spark plug interior, or the crevice caused by an imperfectly fitted head gasket will also quench combustion flames. Mixtures in crevices narrower than about 1 mm probably do not burn at all.

### E. Fuel Factors

Internal-combustion engine fuels are mostly derived from crude petroleum and are a mixture of hydrocarbon components of different molecular structures. Fuels derived from other sources have been used and will be used in the future, but they presently represent a very small percentage of the total energy for operating internal-combustion engines. In this category are methanol (*8, 9*), derived from natural gas as feedstock, and pulverized solid fuels. Operation of internal-combustion engines on hydrogen has been demonstrated as a feasible method of reducing emissions, and increased emphasis on its use is being studied (*10, 11*).

The hydrocarbon composition of the components of internal-combustion engine fuel plays an important role in the achievability of emission control. Nonhydrocarbon fuel additives are helpful in achieving better combustion either directly as combustion improvers or indirectly by permitting higher compression ratios, cleaner induction systems, or longer valve life. Some compounds found in gasolines, such as sulfur, are undesirable. In the presence of exhaust gas catalytic converters and excess oxygen availability gasoline sulfur converts to sulfates and sulfuric acid mist.

The primary factors determined by hydrocarbon composition are ignition quality and volatility. Low ignition quality and relatively low volatility are necessary for diesel or compression-ignition engines. High ignition quality and more volatile fuels are necessary for spark-ignition engines.

## 1. Volatility

Emissions from spark-ignition engines are dependent on gasoline volatility characteristics (Fig. 3). This is especially true of emissions which occur during cold starts when the carburetor choke circuit is still in operation. To achieve required low emission levels, spark-ignition engines must get off the choke circuit as quickly as possible while not significantly impairing the "drive-away" capability of the vehicle. High front end gasoline volatility enhances this requirement, permitting shorter choke

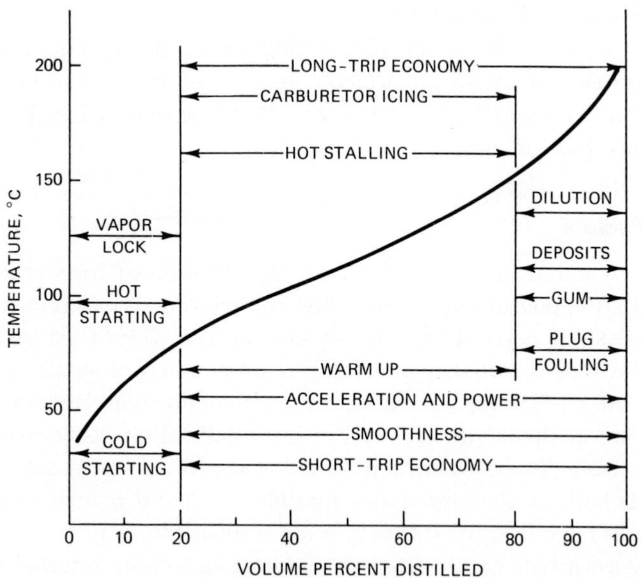


**Figure 3. Effect of gasoline volatility on vehicle performance.**

cycles and faster engine warm-up, but light front end (high volatility) gasolines aggravate the problem of evaporative emissions. With carbon canister collection and recycle systems, an over-rich mixture can occur for an undue length of time on start-up with a resulting average increase in carbon monoxide emissions.

Since compression-ignition engine fuels are relatively nonvolatile, there are no significant emissions. However, engines such as the M.A.N. using the M combustion principle depend on controlled rates of fuel vaporization following ignition initiation in the temperature-controlled precombustion chamber.

The M system was developed by Meurer (*12*) with the aim of reducing both the preignition reactions of the autoignition process and the decomposition of the fuel molecules that are associated with the heating of heterogeneous mixtures. The system is described in considerable detail by Obert (*4*) and Lichty and is summarized and compared to direct injection (D.I.) and indirect injection (I.D.I.) by Patterson and Heinen, who state

> the fuel is injected on a temperature-controlled surface of a spherical chamber in the piston. The rate of evaporation is controlled by wall temperature and the air swirl. After the fuel mixes with air, it is ignited by several ignition sources formed by injecting a small percentage of the fuel into the chamber, away from the walls (*7*).

The M system results in higher swirl and radial velocity components than in the other types of open-chamber engines. The M system seems to be the most sensitive in terms of oxides of nitrogen production to engine speed at any brake mean effective pressure (BMEP).

## 2. Alternative Fuels

Hydrogen, natural gas, and methanol can be used in slightly modified existing engines. These fuels can also be used in gas-turbine and in external-combustion engines. Their use in diesel engines may require the incorporation of a spark assist to achieve suitable combustion. This would be similar to a stratified-charge gasoline engine running on a cleaner-burning fuel.

Liquid natural gas (LNG) and liquefied petroleum gas (LPG) have been effectively used for years in specialized applications and result in low hydrocarbon and carbon monoxide emissions as a result of lean air/fuel ratios. Additional control measures, such as catalytic afterburners or exhaust gas recycle (EGR), are necessary, however, to attain satisfactorily low levels of oxides of nitrogen. Ammonia has promise as an alternate fuel, but emissions of nitrogen-containing compounds are quite high.

Hydrogen is an excellent fuel from the standpoint of exhaust emissions. Engines operated with pure hydrogen have essentially no hydrocarbon or carbon monoxide emissions and nitrogen oxide emissions can be substantially reduced through lean-mixture operation. Even if adequate supplies of hydrogen could be made available, major problems would be encountered in vehicle-hydrogen storage and engine combustion.

On a volumetric basis, almost 20 times as much compressed hydrogen as gasoline would be required for equal vehicle range. Four times as much liquid hydrogen as gasoline would be required. Solid hydride storage systems have been proposed; however, the weights involved are prohibitive. The combustion characteristics of hydrogen differ substantially from conventional fuels and would necessitate major engine and fuel-system design changes.

### 3. Water Injection and Water–Fuel Emulsions

Introduction of water into both gasoline and diesel engines, either by means of special injection systems or in the form of water–fuel emulsions, has received varying levels of attention for many years. Water can reduce audible knock in gasoline engines operating with extremely low octane fuels. However, with engines of modern design and gasolines of present octane quality, the antiknock benefits do not appear great enough to justify the costs and probable inconveniences of large-scale water use.

Water introduction reduces oxides of nitrogen emissions in the same manner as EGR; but, by comparison, water induction involves significantly more expense and inconvenience and appears to offer no functional advantage over EGR. In some cases, water–fuel emulsions have appeared to reduce carbon monoxide emissions. This generally results from the carburetor leaning effect produced when water displaces part of the metered fuel volume. In late-model vehicles, with precise carburetor calibration, this effect would be expected to impair driveability and probably increase the emission of unburned hydrocarbons.

Some emulsions will undergo microexplosions when droplet combustion proceeds. It has been argued that this will reduce all emissions, including particulates. Although possibly of some significance in spark-ignition engines, this phenomenon could be of great significance in reducing oxides of nitrogen and particulate emissions in the diesel engine. Research in this area is just beginning.

## IV. Control of Blowby and Evaporative Emissions

### A. Blowby Control

Blowby from internal-combustion engines is controlled in essentially the same way by all vehicle manufacturers. Crankcase blowby is routed to the engine induction system either by use of a sealed connecting system or a closed positive ventilating system. In the sealed system, no ventilation air is admitted to the crankcase and all the blowby passes through a tubing connection directly into the engine induction air cleaner. Blowby is predominantly fuel–air mixture, so this control approach enriches the engine mixture because a new charge of fuel is drawn in with the blowby as it passes through the carburetor a second time. The most popular control approach is the closed positive ventilating system shown in Figure

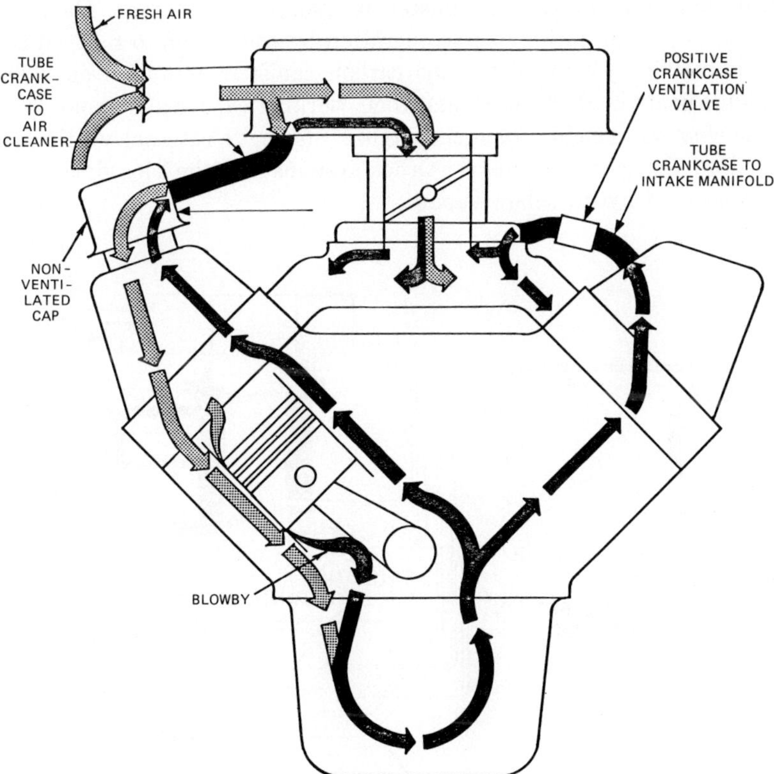


Figure 4. Closed positive ventilating system.

4. This control approach provides for ventilation airflow to the crankcase and, by properly sizing the positive crankcase ventilation (PCV) valve, mixture enrichment can be minimized. In some systems, a dual-action PCV valve will permit controlled flow to the engine induction air cleaner and the intake manifold simultaneously. A restricted oil filler cap is utilized to provide fresh air ventilation inflow to the crankcase.

### *B. Evaporative Control*

The control of evaporative losses from the fuel tank and carburetor vents requires more complicated plumbing than blowby control. A typical control system utilizing an activated carbon canister for absorption and temporary storage of hydrocarbons is shown schematically in Figure 5. Conventional systems consist of expansion tanks, liquid-vapor separators, control valves, and a carbon canister for hydrocarbon storage. The carbon canister is purged during engine operation by connection to reduced pressure induction system inlets. The carbon canisters contain from 300 to 800 gm of activated charcoal, and their working capacity is about 5–10% of the charcoal weight. Some carburetors also have external venting connected to the carbon canister. Other systems use the closed crankcase for storage of hydrocarbon vapors.

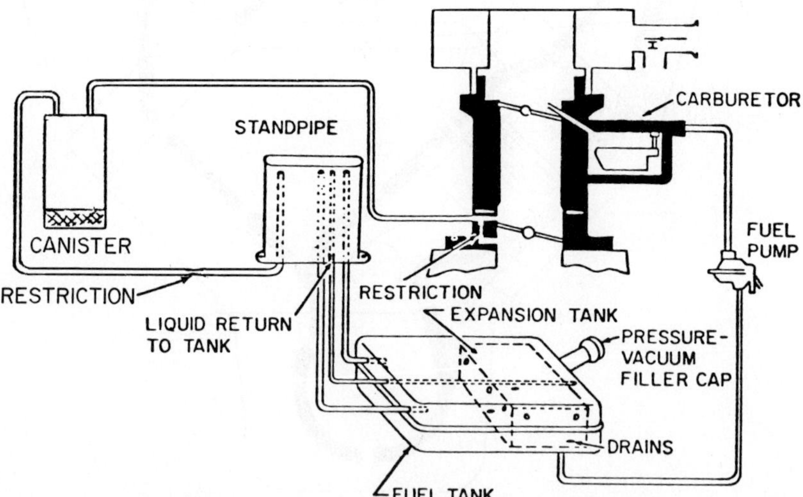


Figure 5. Typical carbon canister fuel evaporative control system with internal carburetor venting.

## V. Control Approaches for Exhaust Emissions—Conventional Spark-Ignition Engine Motor Vehicles

Exhaust emissions are controlled in conventional spark-ignition engines by air/fuel mixture control, combustion modification, exhaust treatment, or combinations of these approaches (Fig. 6). Generally, emission control is an original equipment design and engineering application problem, but retrofit or postassembly installation of vehicle emission controls provides a feasible and cost-effective approach for specific situations such as low volume foreign vehicle imports to the United States.

Exhaust treatment devices include thermal and catalytic reactors and particulate traps. Catalytic reactors fit into three categories: (a) oxidation catalyst systems which combine free oxygen with unburned hydrocarbons and carbon monoxide, (b) reduction catalyst systems which reduce nitrogen oxides to nitrogen and oxygen in a fuel-rich or reducing atmosphere, and (c) three-way-catalyst (TWC) systems which simultaneously reduce oxides of nitrogen while oxidizing hydrocarbons and carbon monoxide. Systems combining reduction and oxidation catalysts represent a fourth category.

In all cases catalytic efficiency is determined to a large extent by air/fuel mixture control. The addition of excess secondary air for oxidation catalysts causes high sulfate concentrations in the exhaust and in the case of three way catalysts requires very precise control of air/fuel mixture.

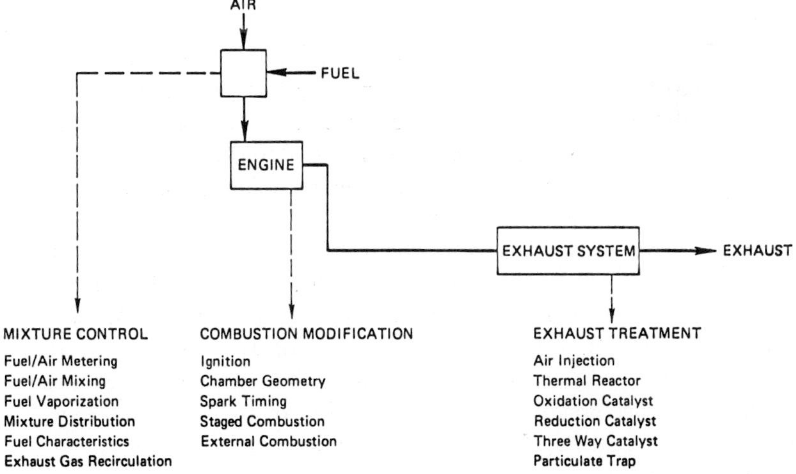


Figure 6. Approaches to the control of engine emissions.

### A. Mixture Control

Air/fuel mixture control determines vehicle emissions, fuel economy, driveability, and starting characteristics. Conventional carburetors, which are the predominant means used in the United States to maintain air/fuel ratio control, are not adequate to satisfy very low emission standards, so advanced carburetors or other mixture-control systems are required.

#### 1. Carburetors

In simple carburetors, fuel flows from a jet located in the throat (venturi) of the carburetor. These devices tend to give rich mixtures at high flows and lean mixtures at low flows. In multithroat carburetors, better mixture control is obtained by using several throats and fuel orifices to provide better flow conditions over the full operating range of the engine. These carburetors can control air/fuel ratios to within about 5%. However, the mixtures actually delivered to individual cylinders will vary greatly because of manifold design, operational transients, liquid impingement on manifold walls, and ambient changes in altitude, temperature, and humidity. As a result, cylinder-to-cylinder mixture ratios may vary as much as $\pm 20\%$ of nominal values. Such variation is grossly inadequate for the maintenance of low emissions, fuel economy, and good driveability. These problems are accentuated during cold starts and under idle or low-load operation because airflows are low, and large fuel droplets result.

One form of carburetor improvement is provided by variable-venturi and constant-depression carburetors. These more complex carburetors provide higher throat velocities, ensuring better fuel-flow control and atomization over the range of engine operating conditions. The effective application of exhaust catalyst systems as described below require in some cases very precise air/fuel ratio control. Conventional carburetors cannot adequately provide such control.

Sonic-flow carburetors, such as the Dresser (*13*), offer positive airflow control through a choked-flow diffuser, precise air/fuel ratio control, low pressure drop, fine droplet atomization, and good mixing. The ability to control homogeneity allows lean operation of the engine and reduced emissions, even without other emission-control devices. Such mixture-ratio and mixture-distribution control makes operation of a "lean-burn" engine and realization of the gains of fuel-lean operation possible. A "lean-burn" engine is one that generally operates on the lean side of stoichiometric. Table II provides typical emission and fuel economy data

**Table II Typical Emissions and Fuel-Economy Data from U.S. Environmental Protection Agency Tests of the Dresserator Induction System (*14*)**

| *Federal test procedure*[a] | *Timing* | *Mass emissions,*[b] *gm/mile (gm/km)* HC | CO | $CO_2$ | $NO_x$ | *Fuel economy, miles/gal (fuel consumption, liters/100 km)* Urban | Highway |
|---|---|---|---|---|---|---|---|
| | | 1973 Capri (259 CID)[c] | | | | | |
| 1975 average | +8 | 0.59 (0.36) | 5.50 (3.42) | 486 (302) | 1.25 (0.78) | 18.4 (12.9) | 24.4 (9.7) |
| 1972 average | +8 | 0.82 (0.51) | 8.46 (5.26) | 498 (309) | 1.51 (0.94) | 17.3 (13.6) | |
| Hot 1975 average | +8 | 0.41 (0.25) | 3.34 (2.08) | 469 (293) | 1.07 (0.67) | 18.7 (12.6) | |
| Hot 1972 average | +11 | 0.62 (0.39) | 3.43 (2.13) | 426 (265) | 1.27 (0.79) | 20.6 (11.4) | |
| | | 1973 Monte Carlo (350 CID) | | | | | |
| 1975 average | | 1.13 (0.70) | 5.11 (3.18) | 690 (429) | 1.55 (0.97) | 12.9 (18.3) | |
| 1972 average | | 1.22 (0.76) | 5.80 (3.60) | 706 (439) | 1.76 (1.09) | 12.4 (19.1) | |

[a] 1972 Values calculated from bags 1 and 2 of 1975 procedure tests; hot 1972 values calculated from bags 2 and 3 of 1975 procedure tests.
[b] HC: Hydrocarbons: $CO_2$ carbon monoxide; $CO_2$ carbon dioxide; $NO_x$: oxides of nitrogen.
[c] CID: cubic inch displacement.

obtained by the United States Environmental Protection Agency (USEPA) for two different vehicles equipped with the Dresserator system (*14*). Another type of carburetor, made by Autotronics, provides improved mixture control by promoting droplet breakup through impingement of the fuel on an ultrasonic vibrating plate.

## 2. Fuel Injection

The first electronic fuel-injection systems, developed about 1960 for gasoline-using engines, sensed airflow electronically and injected the appropriate amount of fuel. These systems offered better cylinder-to-cylinder mixture distribution, especially for engines with poorly designed intake manifolds or difficult fuel-induction problems. Such systems also improve maximum power but have been limited in their application by high cost. Electronic fuel injection alone does not have any advantage over a well-carbureted engine.

Subsequent electronic and mechanical fuel-injection systems employ air mass flow sensors. A temperature sensor enables fuel enrichment during start-up and leaning of the mixture once the engine is warm. Mechanical fuel-injection systems are simpler in operation and lower in cost than electronic fuel-injection systems. Mechanical systems are comparable to electronic fuel injection in emissions reduction and fuel economy improvement.

## 3. Feedback Systems

Conventional open-loop fuel-control systems cannot be correct under all conditions of vehicle operation. A closed-loop system which monitors an engine exhaust output parameter can maintain the air/fuel ratio within correct limits by constantly supplying a corrective signal to the primary air/fuel metering device. Such systems are in a sense self-maintaining in that they do not go out of adjustment. An example of this system is shown schematically in Figure 7 (*15*).

The combination of an oxygen sensor with electronic fuel injection can provide such a closed-loop, feedback-control system (Fig. 8) (*15*). Near stoichiometric, this sensor produces an output which varies strongly with exhaust-gas composition (Fig. 9) and provides the form and degree of mixture control necessary for the operation of three-way-catalyst emissions control systems described below. Such precise control of air/fuel mixture generally reduces fuel consumption compared to conventional carbureted systems.

The use of feedback mixture-control systems for conventional spark-ignition engines has been demonstrated. These systems provide substantial improvement in mixture control and are essential to the operation of some emissions control systems. Alone, they provide improvements in emissions, fuel economy, and driveability. They have potential applica-

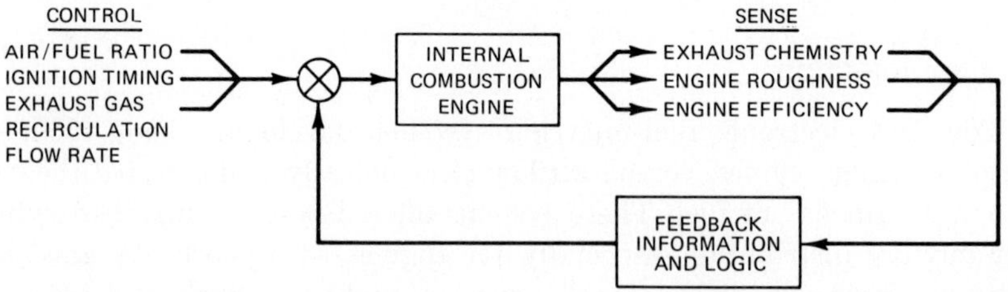


Figure 7. Concepts for closed-loop control (*15*).

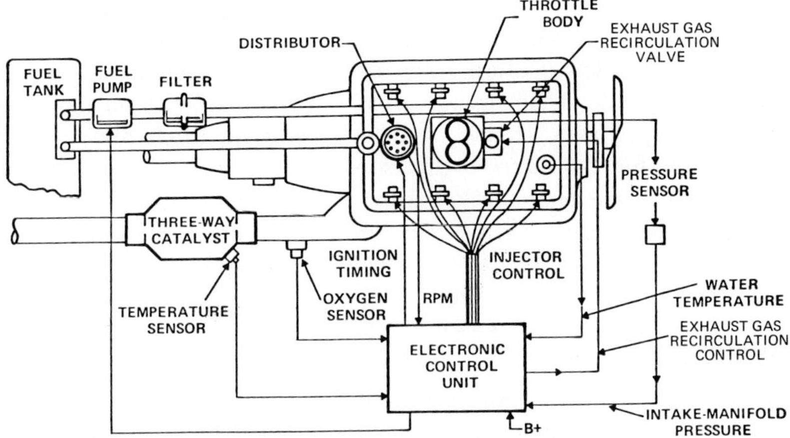


Figure 8. Closed-loop electronic fuel-injection system (*15*).

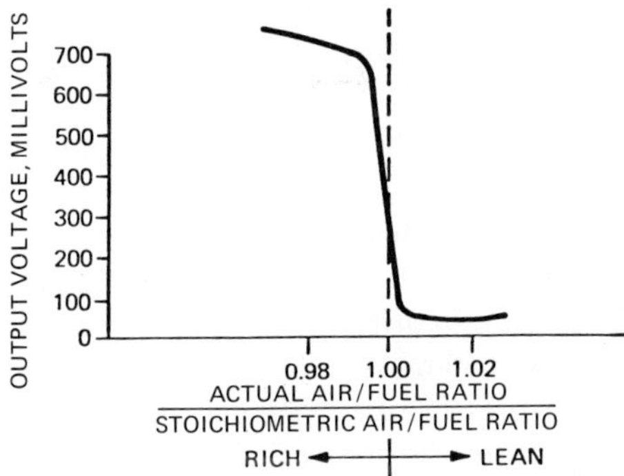


Figure 9. Typical oxygen sensor characteristics for feedback circuit (*15*).

tion to nonconventional as well as the conventional engines for which they were developed.

## 4. Exhaust Gas Recirculation (EGR)

The recirculation of exhaust gas back into the intake system is used for the control of oxides of nitrogen through reduction of the peak combustion temperature. The exhaust gas acts as an inert diluent. Small amounts of EGR have been used to improve knock resistance and improve fuel economy. However in the amounts required for oxides of nitrogen reduction, fuel consumption is usually increased and driveability is degraded. To attain oxides of nitrogen emission levels below about 0.6

gm/km, loss of fuel economy is so great as to make the use of EGR alone unattractive, even in small cars for which oxides of nitrogen emissions levels are already low because of low exhaust flows. There is some tradeoff between oxides of nitrogen and hydrocarbon control on oxidation catalyst-equipped cars. Higher engine hydrocarbon levels allow partial compensation for the adverse effects of controlling oxides of nitrogen through exhaust gas recirculation. Most EGR systems are of the proportional type which adjust gas recirculation rates to engine operating conditions so as to maximize oxides of nitrogen reduction and minimize losses of fuel economy and driveability.

### 5. Fuel Modification

Early evaporation of the fuel droplets, which are the usual form of liquid fuel after its passage through a carburetor or fuel-injection system, enhances cold-start and good-mixture distribution. A quick-heat manifold or evaporative heater, such as the Ethyl Corporation Hot-Box Manifold and the Shell Vapipe systems accomplish this result. Substantial emissions reductions have been obtained by their use even without the use of catalytic reactors.

More drastic alteration of the fuel can be accomplished by the use of an onboard catalytic fuel reformer that by a partial oxidation process, converts the gasoline to a gaseous mixture consisting primarily of hydrogen and carbon monoxide (*11*). An engine can be run entirely on such a reformed fuel or on a mixture of reformed fuel and liquid gasoline. The gains from such a system are closely related to those noted for a lean-burn engine. The presence of hydrogen is believed to assist in the combustion of lean mixtures by increasing flame speed and flammability limits.

A more direct method of fuel modification is to employ a fuel other than gasoline. As discussed earlier, if alternative fuels come into significant use, fuel availability is likely to be the controlling factor, rather than air pollution. Although some alternative fuels possess definite emission-reduction advantages, this alone will not ensure their adoption. The more important question, therefore, appears to be the effect that new motor vehicle fuels would have upon the total energy picture, rather than what fuels should be promoted for the purpose of reducing engine emissions.

## *B. Combustion Modification*

The flexibility to produce major changes in the combustion process with conventional engine combustion chamber design is limited. As a consequence, approaches which have sought to significantly modify the combustion process generally have led to substantially different engine configurations which are discussed later in this chapter. Some modification to existing combustion chambers which have proved useful in reducing emissions are mentioned below.

### 1. Spark Timing

Spark retard has been employed to reduce both hydrocarbons and oxides of nitrogen (Fig. 10) (*1, 15a*). Unfortunately, fuel economy also is reduced. The use of oxidation catalysts to control hydrocarbons and of exhaust gas recirculation to control oxides of nitrogen allows the spark to be advanced with concomitant recovery of the fuel economy losses associated with the spark retard used for emissions control on early-model-year vehicles.

### 2. Compression Ratio

Lowering of the compression ratio reduces both hydrocarbons and oxides of nitrogen, but with a loss of fuel economy of 4 to 5% in going from a 9:1 to an 8:1 compression ratio (Fig. 11). In the conventional engine, compression ratio is knock limited and, at the time of writing,

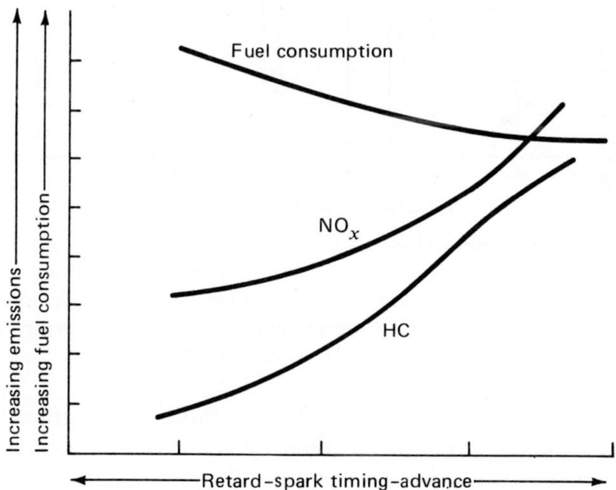


**Figure 10. Effect of ignition timing on fuel economy, oxides of nitrogen, and hydrocarbon emission (*1, 15a*).**

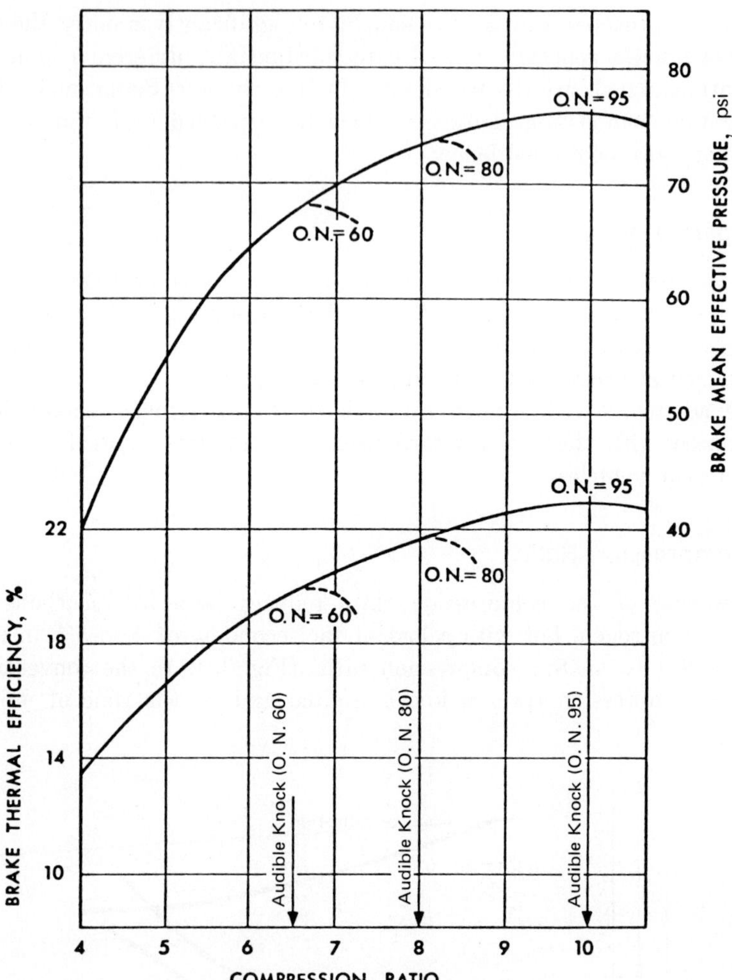


Figure 11. Effect of compression ratio on brake thermal efficiency, brake mean effective pressure, fuel economy, and oxides of nitrogen (7). Fuel economy is proportional to brake thermal efficiency. $NO_x$ is affected by compression ratio because higher chamber pressures tend to cause higher combustion temperature, which in turn, causes higher $NO_x$ emission. O.N.: Octane number.

fixed by the requirement that the vehicle be able to use unleaded 91 research octane number fuel, rather than by limitations imposed by emission requirements. With the constraints of fuel octane availability and emissions, the recovery of fuel economy by increasing compression ratio is an elusive and difficult task.

Test data for the Porsche stratified charge engine show a peculiar characteristic of periodic fluctuations in gas pressure which develops at a frequency determined by the relative volumes of the interconnected cavities—the main and auxilliary chambers. This effect contrasts with the vibratory fluctuations associated with knocking combustion. However, these fluctuations cannot be suppressed either by octane number increase or by retarding the ignition, or by variations in the physical and chemical properties of the fuels tested. Fortunately, the amplitudes of the pressure fluctuations are so small that there is no risk of mechanical damage (*18a*). The dramatic improvements in emissions are presented in Figure 12. This raises the question of whether higher compression ratios without knock limitation might not be possible with stratified charge engines as compared to conventional ones, thus providing improved fuel economy (*18a*).

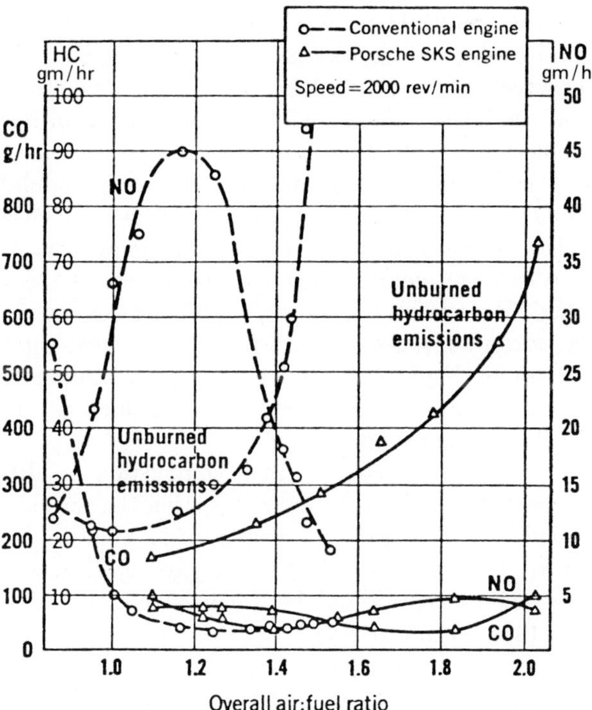


Figure 12. Comparison of combustion products for the Porsche auxiliary chamber stratified charge engine compared to conventional engines (*18a*).

### 3. Combustion-Chamber Geometry

Some relatively minor modifications to combustion-chamber geometry, such as the location of the spark plug, positioning of the piston rings, elimination of crevices, and control of turbulence, are in the category of "cleaning up" the combustion-chamber design and have yielded emission reductions. However, substantial additional gains require basically different engine configurations.

### 4. Ignition Systems

To meet even current hydrocarbon emissions levels, an engine cannot tolerate a misfire or a failure to ignite the mixture in a cylinder, even occasionally. Higher energy ignition systems give greater assurance of ignition. Breakerless systems offer increased durability and, therefore, increased reliability over the lifetime of the car. Some work has been done on the use of multiple spark plugs, multiple sparks from a single spark plug, and continuous ignition systems (*16–18*). Since reliable ignition can be ensured on conventional engines without going to such lengths, their potential application is more likely to be on lean-burn systems, which have more difficult ignition problems. A technological achievement by the Amfin Corporation uses high-frequency current modulated ignition (30–100 kHz) operating under controlled conditions of electrical resonance coupled to the secondary side of the ignition system. This ignition supplementer effectively doubles (from 1.5 to 3.0 msec) the spark duration of the plug. This system also provides a higher density discharge and effectively causes an ionization of the mixture to enhance combustion and improve flame travel. Figures 13 and 14 show the advantage of increased spark duration for leaner air/fuel ratios, $NO_x$ reduction, and increased tolerance for exhaust gas recirculation.

Ultimately the mixture control and combustion modification approaches discussed above have as their objective the formation and efficient combustion of a homogeneous lean fuel/air mixture. In a comprehensive study by the Arthur D. Little Company (*19*) it was concluded that the lean-burn concept would be the most cost-effective system to develop for the 1977–80 time period.

## *C. Exhaust Treatment*

Oxidation catalysts have been applied successfully to meet hydrocarbon and carbon monoxide emissions standards. Reduction catalysts appear

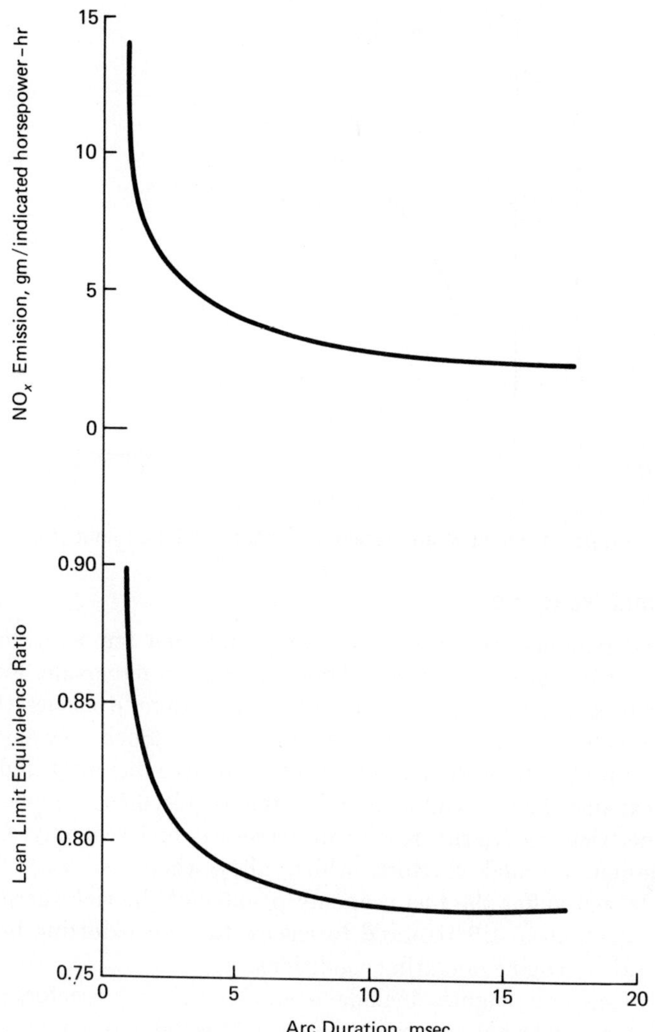


**Figure 13. Effect of arc duration on lean limit and corresponding oxides of nitrogen emission (*16*).**

to be the most likely means to allow conventional engines to meet stringent oxides of nitrogen emissions standards. Combinations of these devices, thermal reactors and particulate traps, have been proposed to overcome problems of starting or of obtaining the degree of exhaust control necessary.

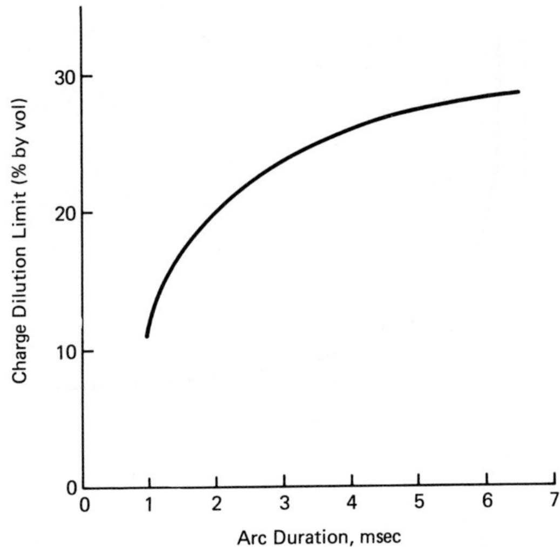


**Figure 14. Effect of arc duration on charge dilution limit (*16*).**

### 1. Thermal Reactors

Thermal reactors require excess oxygen and high temperatures to ensure efficient oxidation of hydrocarbons and carbon monoxide. Early versions required a fuel-rich exhaust and air injection to ensure that high thermal-reactor temperatures could be maintained. Such a system is particularly suited to the rotary engine because of its inherently high hydrocarbon exhaust levels. Unfortunately, the requirement to operate the engine fuel rich necessarily results in decreased fuel economy. Owing to better design, thermal reactors, which allow their use with fuel-lean engines, do not suffer the fuel economy penalty of the rich thermal reactor. Air injection is still required to ensure that an oxidizing mixture is available at all engine operating conditions.

Many lean-burn engines include a simple thermal reactor, often no more than a somewhat enlarged thermally insulated exhaust manifold. Because of the lower exhaust temperatures of lean-burn engines, thermal reactor performance is limited, but usually adequate to give approximately a 50% reduction in hydrocarbons. Since the introduction of the oxidation catalyst, thermal reactors are now found primarily on rotary, lean-burn, and stratified-charge engines.

### 2. Catalytic Reactors

There are three general categories of catalytic reactors for emissions control. The catalytic reactor system in use at the time of writing uses

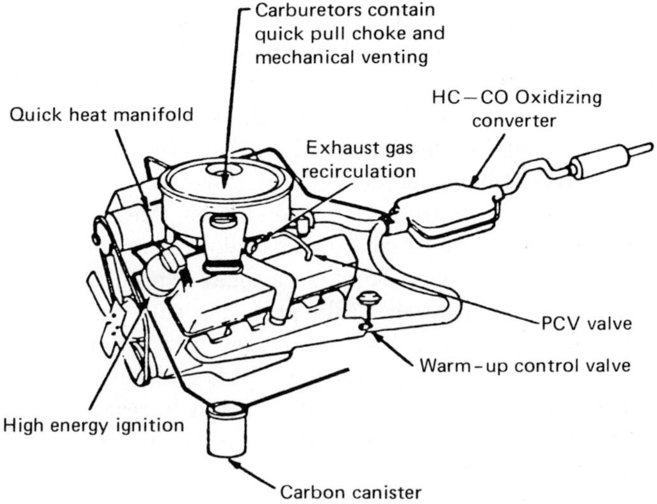


**Figure 15. Typical 1975 emission control system (*15a*).**

a noble-metal oxidation catalyst (Fig. 15). A reduction catalyst reactor for the removal of the oxides of nitrogen may not be required in the United States if the moratorium on the oxides of nitrogen standard survives. However, if the standard for oxides of nitrogen becomes as low as 0.6–0.8 gm/km, reduction catalysts will probably be used. When an oxidation and a reduction catalyst are used in series, it is known as a dual-catalyst system. When a single catalyst is used to remove hydrocarbons, carbon monoxide, and oxides of nitrogen simultaneously it is known as a three-way catalyst.

a. Oxidation Catalysts. Noble-metal catalysts, which contain platinum, palladium, and/or rhodium supported on pellets or monolithic substrates, have been demonstrated to meet a 0.41 gm/mile (0.25 gm/km) hydrocarbon standard for 50,000 miles (80,000 km) and 3.4 gm/mile (2.1 gm/km) carbon monoxide standard. Base-metal oxidation catalysts have not yet demonstrated adequate durability.

Noble-metal oxidation catalysts increase the amount of fuel sulfur emitted as sulfur trioxide, which is rapidly converted to sulfuric acid in the presence of water (*20*). The amount of fuel sulfur conversion to sulfates over the catalyst varies from less than 10 to 15–60% depending on the catalyst temperature and amount of free oxygen present. The use of an air pump to provide secondary air in the exhaust manifold causes higher $SO_3/SO_2$ ratios than occurs without it (*21, 22*). The catalyst absorbs sulfur trioxide at lower temperature but later releases it during higher temperature operation. Oxidation catalysts operate at lower temperatures than thermal reactors and, therefore, effectively decouple emissions control from engine performance, allowing the tuning of the engine for improved fuel economy. This can result in significant fuel economy gains compared to earlier model vehicles where this was not the case.

Noble metal oxidation catalysts have demonstrated durability for at least 80,000 km under controlled test conditions. They are deactivated by the presence of compounds containing lead, bromine, and phosphorus. Recovery of catalyst activity from occasional exposure to these compounds has been demonstrated. Early problems with cracking, burnout, and attrition appear to have been minimized through improvements in ignition systems and catalyst packaging. Total ignition failure at speeds above about 80 km/hour (kph) is likely to result in reactor temperatures sufficiently high to deactivate the catalysts permanently or to damage its substrate and/or container.

Oxidation catalysts essentially eliminate the emissions of polynuclear aromatics, olefins, and partially oxidized compounds, such as phenols, aldehydes, and mercaptans. These hydrocarbons are eliminated to a much greater extent than total hydrocarbons. By requiring use of nonleaded gasoline to prevent catalyst poisoning, there will be a decrease in particulate emissions attributable to lead.

Loss of noble metals from the catalyst in the exhaust gases is not a significant problem under normal operating conditions. The total amount of noble metal per vehicle is less than 0.3 gm and during 80,000 km of operation less than 10% of this material is lost. Nonetheless, ambient air monitoring for its content of noble metals, lead, sulfur, and phosphorus is proceeding so that the impact of catalyst-equipped cars can be assessed through their increase or decrease.

b. Reduction Catalysts. Reduction catalysts which are usually simply copper or steel shavings require a fuel-rich mixture and, therefore, suffer a fuel economy penalty, although this may be minimized by operating only slightly to the fuel-rich side of stoichiometric. Therefore, reduction catalysts cannot be used for oxides of nitrogen control on engines that produce fuel-lean exhaust products, e.g., lean-burn, stratified-charge, and diesel engines.

Air/fuel ratio must be controlled carefully for satisfactory operation of these catalysts. Reducing an atmosphere too much will lead to the formation of ammonia, and inadequately reducing an atmosphere will prevent the reduction of oxides of nitrogen. Excessively lean transients must be avoided when the catalyst is at operating temperature, as they lead to excessive temperatures and potential catalyst failure. In some systems an upstream oxidation catalyst or thermal reactor is employed to protect the reduction catalyst from excess oxygen.

c. Dual Catalysts. Dual catalysts also require precise air/fuel ratio control. In one configuration, a reduction catalyst followed by an oxidation catalyst is used. In another, an additional small, noble-metal catalyst is provided upstream of the reduction catalyst. Still another system uses a rich thermal reactor, followed by a reduction catalyst, which in turn is followed by a second thermal reactor. During start-up, air is injected upstream of the reduction catalyst, which, during start-up, acts as an oxidation catalyst.

d. Three-Way Catalysts. Simultaneous control of all three pollutants in a single catalytic converter is possible, but air/fuel ratio control precise to about 0.5% and an oxygen-sensor feedback-control are required. Although early oxygen sensors had very poor durability, several manufacturers now have sensors which can be guaranteed for 20,000 km or more. Bosch, for example, has reported 32,000-km durability of a three-way catalyst with an oxygen-sensor feedback-control system that meets the stringent 0.41/3.4/0.4 gm/mile (0.25/2.1/0.24 gm/km) emissions standard for $HC/CO/NO_x$ in a four-cylinder car. Engelhard Industries has reported meeting the same standards for 80,000 km with three-way-catalyst installations on a Volvo (*23*). Most three-way catalysts contain noble metals: platinum, palladium, rhodium, and, possibly, ruthenium.

Prototype systems using a three-way catalyst, an oxygen-sensor feedback-control and each of three different mixture-control systems: electronic fuel injection, mechanical fuel injection, and electronic carburetion have shown good fuel economy and driveability. The most significant problem impeding this technology is the assurance of catalyst durability.

## 3. Particulate Matter Traps

Several particulate matter trap systems have been demonstrated as means for removing lead particulate matter from conventional engine exhaust. Although the elimination of lead from gasoline obviates this potential need, such devices might be required for engines with particulate matter emissions problems, such as the diesel, should a particulate matter emissions standard be enacted. Present traps tend to be limited by their ineffectiveness for particles under 1 $\mu$m.

### *D. Combination of Control Approaches*

Control approaches are not used singly but in combination (Fig. 15). Some degrade fuel economy and some improve it (Fig. 16). The net result can be either better or worse fuel economy.

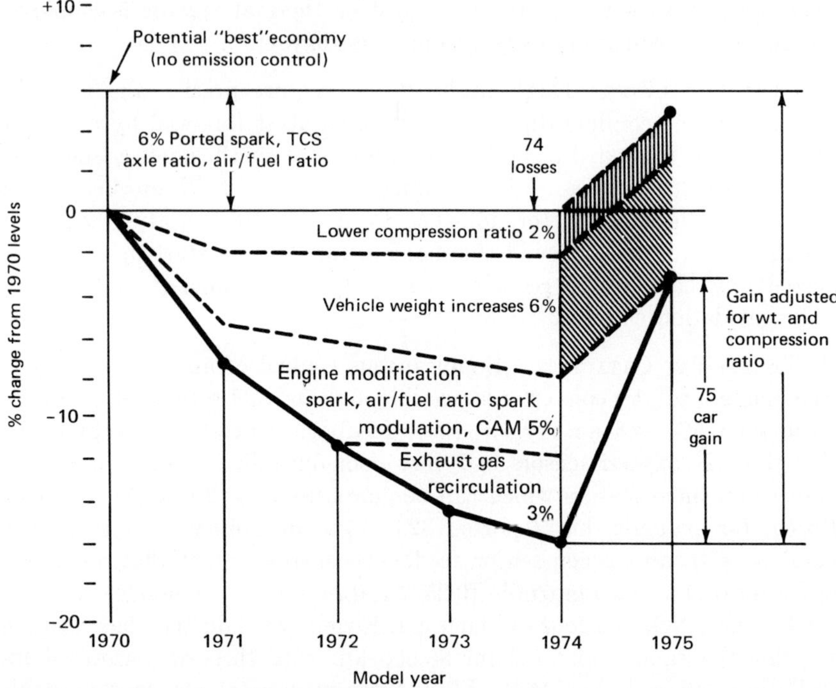


**Figure 16. Fuel economy trends [changes (gains/losses) in General Motors city-suburban schedule economy] (*15a*). Fuel economy of General Motors cars has deteriorated in recent years as a result of weight increases and emission control. Use of the catalytic converter for exhaust emission control reversed that trend for 1975. Analysis of fuel economy changes: (a) 1975 gain results from engine optimization (EGR spark and A/F ratio with catalytic converter); (b) 1975 "adjusted to 1970 weight and compression ratio approaches 1970 best economy.**

### *E. Retrofit or Postinstallation of Emission Controls*

## 1. In-Use Vehicles

The retrofit of emission controls on in-use vehicles has been successfully accomplished (*24*). Most notably, the state of California, in the United States, has required mandatory installation of retrofit emission controls on 1955 through 1970 model vehicles upon change of vehicle ownership. The devices for 1955 through 1965 models were intended to control exhaust hydrocarbons and oxides of nitrogen. Generally, they consisted of devices to disconnect original equipment vacuum advance units or to modify the ignition timing characteristics of the vehicle. Also, modulated air bleed controls were used alone or in combination with ignition-timing modifications. Controls for 1966–70 model vehicles were designed to reduce exhaust oxides of nitrogen by at least 42% compared to uncontrolled vehicles. Exhaust hydrocarbons and, to a lesser extent, carbon monoxide were also controlled by the 1966–70 model retrofit devices

The retrofit control of oxides of nitrogen was accomplished by add-on devices which provided exhaust gas recirculation to the intake manifold, vacuum ignition advance disconnect during certain operating modes or a combination of modulated air bleed to the intake manifold plus ignition-timing retard and vacuum advance disconnect. Electronic spark-timing control and water injection to the intake manifold were two additional methods of California approved control devices.

The United States Environmental Protection Agency specified various retrofit strategies to reduce vehicle pollution in localities with special problems. The strategies had varying degrees of effectiveness, but generally, political inertia and local reticence to enforce use of control devices on in-use vehicles prevailed and in-use retrofit, although technically successful, was never implemented on a wide scale.

## 2. New Motor Vehicles

Installation of emission controls in new motor vehicles after they leave the assembly line has been successfully accomplished for special situations. Since vehicle emission standards and test procedures vary in different countries, it may not be feasible to incorporate emission controls into the original vehicle design which provide significantly more control than needed in the country where most sales are made. However, it may still be desirable to export a portion of the vehicle production to other countries which have more stringent vehicle emission standards. In these cases, retrofit or postassembly installation of control devices is feasible. A specific example of this approach was accomplished with the Russian-built Lada vehicle for export to the United States and other countries. In 1976, Russia had no specific vehicle emission control standards and consequently the majority of the Lada production (called Zhiguli in Russia) did not have specific emission controls. The portion of the production scheduled for United States sales is equipped with emission controls applied in a post-assembly operation. Initially, the controls consisted of a catalytic afterburner with modulated intake air control and exhaust pulsating air control to meet the 1976 and 1977 United States standards. Also, evaporative control devices were retrofitted on the Lada vehicle to satisfy United States standards.

**Table III Effect of the Amfin Igniter on Exhaust Emissions of Typical 4-Cylinder Vehicle**[a]

| *Hot start CVS-I data*[b] | *gm/mile* *HC* | *CO* | $NO_x$ | *Carbon balance, miles/gal* |
|---|---|---|---|---|
| Baseline vehicle—standard timing (5° BTDC) | 3.20 | 43.78 | 1.45 | 23.7 |
| With: Modulating air control only at standard timing (5° BTDC) | 2.50 | 10.07 | 2.42 | 24.7 |
| With: Modulating air control and Amfin igniter at standard timing (5° BTDC) | 1.74 | 10.44 | 1.72 | 25.3 |
| With: Modulating air control and Amfin igniter at advanced timing (8° BTDC) | 2.44 | 10.46 | 2.59 | 26.3 |
| With: Modulating air control and Amfin igniter at 5° retarded timing (ATDC) | 1.50 | 8.51 | 1.55 | 23.8 |

[a] Four-cylinder, 1452 cm³ overhead cam engine, 2500 lb inertia (I), 1400 km, constant volume sampler (CVS) federal test procedure.
[b] BTDC: Before "top dead center"; ATDC: after top dead center.

Continued applied research resulted in possible elimination of the catalytic converter for four cylinder imports. The lean limit (for satisfactory driveability) was stretched about 1 to 1½ air–fuel ratio numbers through the application of new product technology using high frequency modulated current on the secondary side of the conventional ignition system. A summary of the results for this approach is shown in Table III. This application by the Amfin Corporation, previously described in Section V,B,4, also can be applied as original equipment for new vehicles. Ignition assurance for very lean mixtures is markedly enhanced with very modest current input requirements and with a retrofit package including modulated air control that costs less than $100.00 at the manufacturing level. Only about one-half hour is required for postassembly installation. Similar approaches have subsequently been successfully adapted for other limited quantity imports to the Unites States.

## VI. Stratified-Charge Engines

Like the conventional engine, stratified-charge engines are Otto cycle, spark ignition, and, mostly, gasoline fueled. Their distinguishing characteristic is that the air/fuel mixture is not uniformly distributed in the combustion chamber, but is distributed into separate fuel-rich and fuel-lean regions. Some versions of these engines are quite similar to conventional engines and, therefore, have the attractive feature of being able to draw upon existing conventional engine technology and manufacturing facilities. In several different configurations, they have demonstrated the ability to combine low emissions with improved fuel economy and good driveability.

In the following discussion, these engines have been classified into three categories: torch ignition (with prechamber), divided chamber, and open chamber (Fig. 17). Some variations are also discussed.

### A. Torch-Ignition Engines with Prechambers

The torch-ignition engine closely resembles a conventional spark-ignition engine with the spark plug replaced by a small prechamber with its own spark plug. The function of the "torch" shooting from the prechamber is to ignite a lean, homogeneous mixture in the main chamber. The mixture in the main chamber is generally prepared with a conventional carburetor. The prechamber mixture, which is usually fuel rich, may be formed by a separate carburetor or by fuel injection. Air in the prechamber may be delivered with the fuel (when a separate carburetor is used) or from the main chamber (when fuel is injected). The volume of the prechamber is usually 5 to 15% of the total clearance volume (cylinder volume when piston is at top dead center) of the main chamber. These engines are also known as "small-volume prechamber engines," "three-valve prechamber engines," and "jet-ignition engines." Historically, engines associated with the names Ricardo, Schlamann, Broderson, Heintz, and Nilov were the predecessors of the compound vortex-controlled combustion (CVCC) engine developed by Honda (*25*).

The Honda engine, in a 908-kg vehicle, met the United States 1977 emissions standards with a small lean-thermal reactor but without catalytic reactors, air injection, or exhaust gas recirculation, with little deterioration at 80,000 km. Because of the success of the Honda engine design in a variety of engine sizes, including a large V-8 engine, virtually every automobile manufacturer, worldwide, has developed a CVCC engine. The attainment of 0.4 gm/mile (0.25 gm/km) oxides of nitrogen requires the use of exhaust gas recirculation and results in a substantial loss of fuel economy. The control of hydrocarbon emissions also results in a loss of fuel economy. Fuel economy of the Honda engine appears to be comparable to a catalyst-equipped conventional engine at the 0.41 gm/mile (0.25 gm/km) hydrocarbon; 3.4 gm/mile (2.1 gm/km) carbon monoxide; 2.0 gm/mile (1.2 gm/km) oxides of nitrogen emissions level.

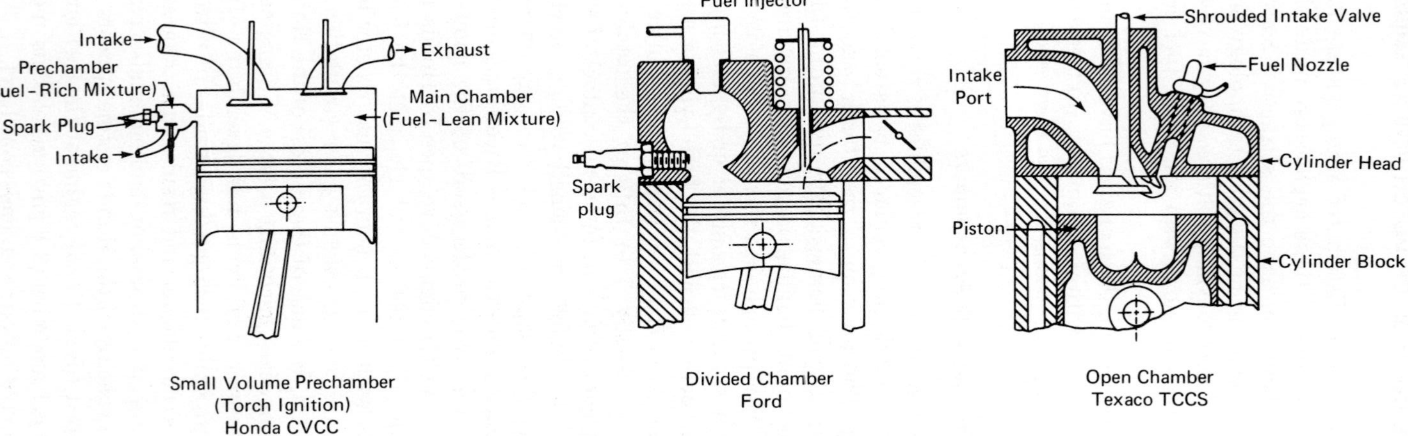


**Figure 17. Stratified-charge engine configurations.**

### B. Divided-Chamber Engines

As the prechamber volume approaches the remaining clearance volume in size, the engine falls into the divided-chamber category. Other engine names for this group are "divided-chamber staged-combustion engines," "large-volume prechamber engines," and "blind-fuel-injected prechamber engines." Direct-fuel-injected versions of these engines appear to offer oxides of nitrogen control advantages over both the torch-ignition and open-chamber engines. (Direct injection is here used to define fuel injection directly into the cylinder, as contrasted to indirect injection via the intake manifold.) The control of hydrocarbon emissions is strongly dependent on fuel-injection spray characteristics.

### C. Open-Chamber Engines

Open-chamber stratified-charge engines employ a single combustion chamber with direct injection of the fuel. In many respects, they are similar to diesel engines, but they are spark-ignited. The Ford (PROCO) and Texaco (TCCS) versions of this engine type are the most advanced. The most notable feature of these engines is their superior fuel economies, which approach those of diesel engines. Adequate control of hydrocarbon and carbon monoxide emissions requires the use of an oxidation catalyst. The Texaco version possesses a multifuel capability and provides comparable emissions levels and fuel economy whether operated on gasoline, diesel fuel, or kerosene (*26, 27*). Exhaust gas recycling is used for oxides of nitrogen control. In meeting the 0.4 gm/mile (0.25 gm/km) oxides of nitrogen emissions standard, a fuel economy loss is incurred which nearly cancels the inherent fuel economy advantage of the basic engine. These engines are still in the development stage.

## VII. Diesel Engines

For city driving, diesel-powered taxicabs and light trucks have about half the fuel consumption of comparable gasoline engine vehicles. The diesel engine is also widely used for heavy-duty trucks. As with most of the other stratified-charge engines, the diesel is an inherently low emitter of hydrocarbons and carbon monoxide. The Arthur D. Little study (*19*) referenced earlier concluded that light-duty diesels would be most cost effective for the 1980–1985 time period.

Retrofit of diesel emission controls has been demonstrated as a feasible approach on three GMC Detroit Diesel 6V-71 two-stroke cycle engines on city buses, and on three Cummins NHC-250 four-stroke cycle diesel engines as used in White Freightliner tractors in hauling produce over varied terrain and elevation from San Antonio, Texas to Colorado. As in the case of gasoline engines, retrofit can have application both for in-use vehicles and new vehicles in post-assembly.

The major air pollution problems for diesel are (a) oxides of nitrogen, (b) smoke, and (c) odors.

### A. Oxides of Nitrogen Emissions

The major emissions problem to be solved for the diesel is oxides of nitrogen emission. The use of EGR is effective only to about 0.6 gm/km oxides of nitrogen, at which point hydrocarbons and carbon monoxide emissions are increased. Retarding injection timing also reduces oxides of nitrogen, but causes an increase in carbon monoxide and often in hydrocarbons. Other methods of reducing oxides of nitrogen emissions are in the research stage.

There are fundamental reasons which prevent oxides of nitrogen control methods that are applied to conventional engines from being effective on diesel engines. Reduction catalysts require a fuel-rich exhaust for their operation and, therefore, cannot be applied to the diesel engine. Larger quantities of exhaust gas recirculation are required for the diesel engine than for the conventional engine. The different behavior and response of the diesel engine to exhaust-gas recirculation result from its fuel-lean mixture and the nature of the combustion process, which is heterogeneous (diffusion flame) rather than homogeneous (premixed flame).

Emission, fuel economy, smoke, and odor data for four typical production light-duty diesel engines are shown in Table IV (*28*). These data were obtained on the vehicles as received, and there was no special optimization for low emissions.

### B. Particulate Matter (Smoke) and Odor Emissions

Smoke and odor problems are serious for diesel engines because they are so noticeable to everyone, and the odors are so obnoxious. Unfortunately, their sources are different and so are the treatments that show promise.

#### 1. Particulate Matter (Smoke)

Smoke particulates are usually described by three categories (*7*):

(a) Liquid particulates which appear as white clouds of vapor emitted under cold starting, idling, and low loads. These consist mainly

**Table IV Emission and Fuel-Economy Data for Four Typical Production Light-Duty Diesel Engines** (*28*)

| | | *Automobile manufacturer and model* | | | |
|---|---|---|---|---|---|
| *Pollutant*[a] | *Laboratory*[b] | *Nissan Datsun* | *Mercedes 220 D* | *Peugeot 504 D* | *Opel Rekord* |
| | Emission, gm/km | | | | |
| HC | EPA | 0.14 | 0.18 | 2.07 | 0.24 |
| | SWRI | 0.22 | 0.15 | 1.22 | 0.24 |
| CO | EPA | 0.84 | 0.69 | 2.52 | 0.75 |
| | SWRI | 0.84 | 0.66 | 1.47 | 0.61 |
| $NO_x$ | EPA | 0.85 | 1.02 | 0.68 | 0.80 |
| | SWRI | 0.95 | 0.75 | 0.62 | 0.82 |
| | Fuel economy, km/liter | | | | |
| Method | | | | | |
| By carbon balance | EPA | 11.9 | 11.9 | 9.3 | 10.1 |
| | SWRI | 10.6 | 12.1 | 11.3 | 11.1 |
| Gravimetric | SWRI | 10.2 | 11.1 | 10.0 | 10.2 |
| | Odor Intensity "D"[c] | | | | |
| Six steady states | | 3.4W[d] 3.2S | 3.0 | 5.2 | 3.9 |
| Idle | | 2.9W 2.7S | 3.1 | 4.8 | 3.3 |
| Three transient conditions | | 4.7W | 3.7 | 5.7 | 4.1 |
| All ten conditions | | 3.8W | 3.2 | 5.3 | 3.9 |
| | Average exhaust smoke opacity, %[e] | | | | |
| Factors | | | | | |
| "a" acceleration | | 4.8S[d] | 3.3 | 3.7 | 5.4 |
| "b" lug-down | | 5.7S | 2.7 | 4.0 | 7.4 |
| "c" peak | | 6.1S | 5.1 | 5.4 | 8.2 |

[a] HC: Hydrocarbons: CO carbon monoxide; $NO_x$ oxides of nitrogen.
[b] EPA: Environmental Protection Agency; SWRI: Southwest Research Institute.
[c] "D" odor is the composite of: B (burnt), O (oily), A (aromatic), P (pungent).
[d] W:Winter; S:summer adjustment of air intake system for Nissan diesel car.
[e] Smoke visibility level is at approximately 3%.

of fuel and a small portion of lubricating oil, emitted without combustion; they may be accompanied by partial oxidation products. These white clouds disappear as the load is increased.

(b) Soot or black smoke emitted as a product of the incomplete combustion process, particularly at maximum loads.

(c) Other particulates include lubricating oil and fuel additives.

Soot formation or black smoke is the type of major concern. It is composed of tiny particles of irregularly shaped carbon. It is believed that these particles can be formed both in the presence of oxygen as in the lean flame region (LFR), and in the absence of oxygen by pyrolysis. They may also be formed from fuel deposited on the cylinder walls.

Black smoke formation is affected by many parameters including the following:

(a) Fuel type: For example, smoke, odor, and power are increased with DF-2 fuel (EM-139-F) as compared to DF-1 fuel (kerosene type) (*30*).

(b) Injection nozzle modification: Replacement of spherical valve injectors by needle valve and low sac injectors reduce smoke. The latter is a needle valve with the volume of the "sac" (the volume between nozzle exit and valve seat) reduced.

(c) Peak power reduction: This is usually the corrective action when other techniques fail. Unfortunately, the loss of power is often too great to be acceptable.

(d) Inlet air humidification: This has a favorable effect on smoke reduction and simultaneously permits increased power output. It also produces a moderate reduction of oxides of nitrogen emissions. It is unlikely that an added water storage system would be acceptable for passenger vehicles. However, it may be a practical solution for fleet vehicles, locomotives, and ship propulsion units, if more acceptable alternatives are not found.

Catalytic reactors have so far shown little effect on smoke reduction. More research on the problems of diesel smoke reduction is clearly needed.

Smoke opacity does not correlate with hydrocarbon, carbon monoxide, oxides of nitrogens, or odor emissions. The trends are different for each of four different light-duty vehicles tested (Table IV) (*28*).

Tests of a two-cylinder, two-cycle GMC Electromotive Division locomotive-type engine show that an effective technique to avoid the smoke point is to power derate the engine. However, in many cases the amount of power reduction required to avoid the smoke point is too great to be acceptable from an operational requirement standpoint.

## 2. Odor

Odor is not a predictable function of power, although peak power retard is a technique that is used in some cases to effect reduction of both smoke and odors (*28, 29*). Future multimodal testing will be required to quantify and correlate odor with other diesel engine test parameters. Part of the problem of correcting diesel engine odor is the difficulty of direct measurement of its quantity and magnitude. Human odor-test panels are presently the most dependable means of measurement. There appears to be some correlation between the findings of human odor-test panels and the instrumental measurements of oxygenates (Table IV).

Retrofit kits tested on three heavy-duty Cummins diesels (*29*) did not cause much change in odor results, but this type NHC-250 Cummins engine—"has always been found to be moderate in "D" odor intensity with relatively low "B" and other odor quality ratings" (*29*). (See Table IV for odor rating scale.) However, the report notes dramatic reductions in both a (acceleration) and b (lug-down) smoke factors. "All smoke values are substantially below federal limits for 1974 of 20%a, 15%b, and 50%c (peak) factors. (c is the smoke factor.) Furthermore, fuel rate, air rate, and rear wheel-power output remained consistent throughout the demonstration" (*29*). Fuel economy was unaltered, although improved power and performance at altitude was noted. These tests verify that odor and smoke emissions require separate treatment.

Retrofit kits, tested on the three GMC two-cycle Detroit Diesel 6V-71 engines in city bus service, showed an effective reduction of odors, but these engines were worse initially than the Cummins diesels.

The most effective odor reduction reported was achieved by use of the "low sac" needle (LSN) type fuel injectors. When this volume of sac is reduced, it reduces the amount of fuel which may be sucked into the cylinder during the exhaust stroke.

Smoke was also significantly reduced by substituting the LSN fuel injectors for stock crown valve or S injectors. A catalytic muffler (copper oxide coated on $Al_2O_3$ spheres) was ineffective on both odors and smoke.

## VIII. Rotary Engines

Rotary engines, predominantly of the Wankel type, are simply different mechanical configurations of the engine types already discussed (Fig. 18). Reciprocating pistons are replaced by one or more rotating drive elements. The resulting engine may be spark or compression ignited, although the spark-ignition version is by far the most common and the only version which has been placed in production. Most reciprocating stratified-charge concepts also have been applied to rotary engines. In comparison with reciprocating engines, the advantages of the rotary engines lie in their smaller size and weight, smoother operation, and potentially

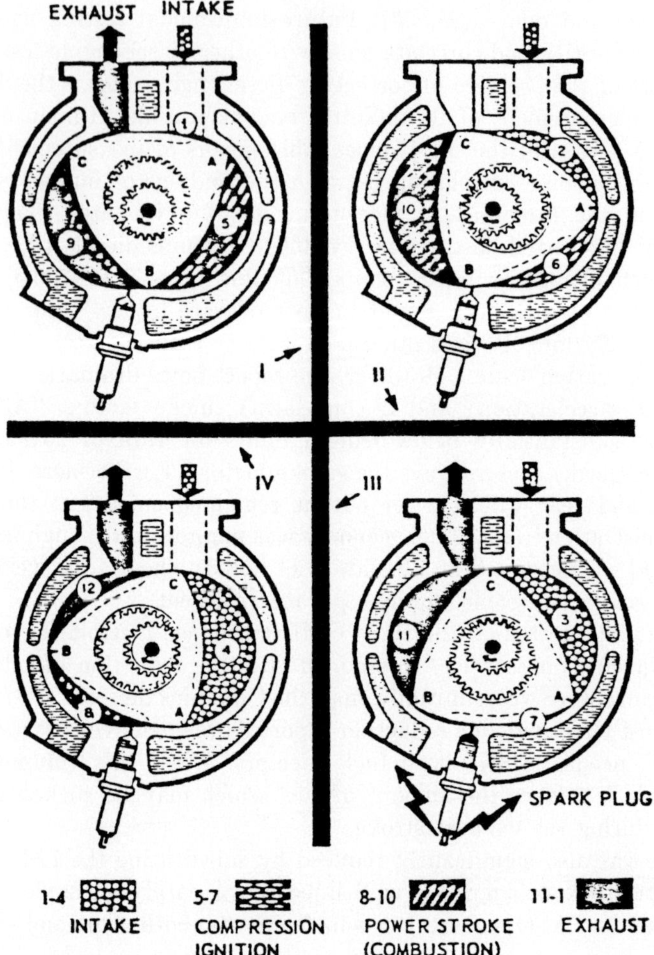


**Figure 18.** Actions in a Wankel engine during one complete rotation of the rotor.

simpler manufacturing process. The major disadvantages have been associated with seal wear and lower fuel economy. During 1975 Mazda dramatically improved the fuel economy of their engines.

## IX. Alternative Engines

Alternative engine development has focused on the gas-turbine and Rankine or steam engines (*30*). These engines have not yet been developed to the point where they are acceptable replacements for Otto cycle or diesel engines on a mass scale. They have demonstrated they can meet very low exhaust emission standards, however, such as those standards established for 1978 in the United States. Both systems incorporate continuous combustion which is inherently lower emitting than the intermittent internal-combustion engines discussed in the previous sections. The Stirling engine, which derives its power from heating and cooling a fluid within a closed volume and transmitting the resultant pressure variations to a piston, also has shown promise as an alternative power source (*30*).

Other alternative power sources with limited-use potential include electric, flywheel storage systems and hybrid systems.

### *A. Gas-Turbine Engines*

In its simplest form, the gas-turbine engine consists of a compressor, a combustion chamber, and a turbine (Fig. 19). Air is taken in by the compressor at atmospheric pressure, compressed to a higher pressure, and then delivered to the combustion chamber wherein fuel is sprayed and burned. The combustion process takes place at essentially constant pressure equal to the compressor discharge pressure. The resulting high temperature, high pressure gas is expanded in the turbine and then exhausted to the atmosphere. Part of the shaft work developed by the turbine is used to drive the compressor. The remainder is the output work used to drive the vehicle and its accessories.

All engines for on-highway vehicles have a free power turbine which uses two mechanically independent turbine stages (Fig. 20); one stage

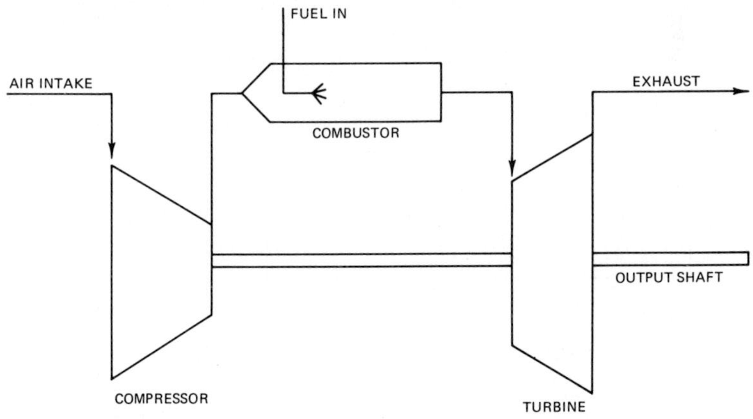


**Figure 19. Schematic diagram of simple gas-turbine engine.**

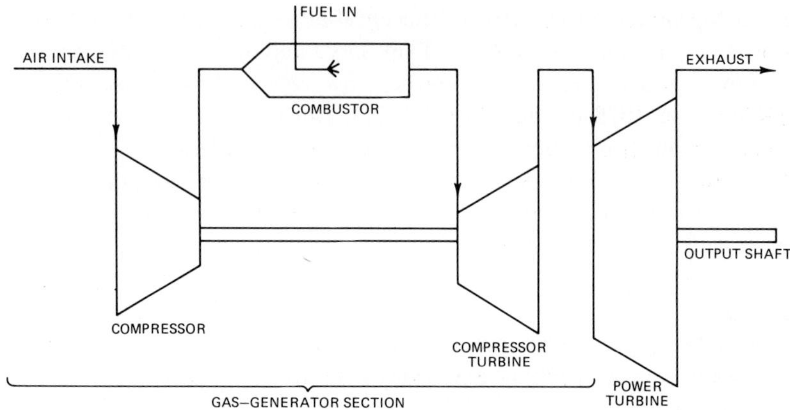


Figure 20. Schematic diagram of free-turbine engine.

drives the compressor and the other stage drives the output shaft. This approach permits use of a wide range of output-shaft speeds which are not possible with the single-shaft turbine engine.

To improve efficiency, most automobile turbine engines use a regenerative cycle which incorporates a heat exchanger that delivers heat from the turbine exhaust gas to the compressed air upstream of the combustor (Fig. 21). The regenerator decreases the exhaust temperature, but its use results in a substantial increase in engine weight and volume.

The gas turbine has demonstrated an ability to meet very low hydrocarbon and carbon monoxide emission standards, and designs incorporating variable-geometry combustors have exhibited very low oxides of nitrogen emissions. Table V shows typical emissions from two representative gas turbines. In limited cases, catalytic converters have been used

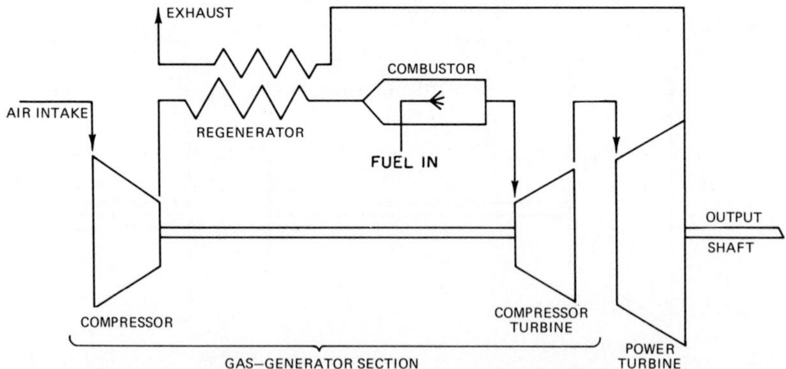


Figure 21. Schematic diagram of regenerative free-turbine engine.

**Table V Typical Emissions Performance of Alternative Engines Using the United States 1972 CVS Cold-Start Procedure[a]**

| *Vehicle* | *Emissions system* | *Emissions, gm/mile (gm/km)* HC | CO | $NO_x$ |
|---|---|---|---|---|
| General Motors GT-225 5000 lb | General Motors gas turbine, dual shaft, diesel fuel | 0.2 (0.3) | 2.7 (4.4) | 0.36 (0.58) |
| | Zwick combustor Gas turbine | 0.26 (0.42) | 2.7 (4.4) | 0.12[b] (0.19) |
| Volkswagon station wagon 2750 lb | Carter steam engine | 0.40 (0.64) | 1.0 (1.6) | 0.33 (0.53) |
| Intermediate size vehicle | Ford/Philips Stirling engine | 0.20 (0.32) | 1.2 (1.9) | 0.14[b] (0.22) |

[a] CVS = Constant volume sampler.
[b] Emissions measurements based on vehicle simulation.

in conjunction with gas turbines in successful demonstrations for reducing emissions to lower levels than shown in Table V.

Basically, the gas turbine is very simple since the rotor is often the only moving part. However, the rotor may consist of two or more very complex, precise, and expensive castings and the stator may also require precision parts. One of the primary advantages of gas turbines is the large power-to-weight ratio. Typical gas turbines for automotive application are lighter than standard automobile spark-ignition engines of equivalent power. Also, they generally require less space at equivalent power levels.

Cost of gas turbines is still higher than conventional automotive engines, but it has been estimated that by about 1980, costs in mass production could be equivalent to the conventional gasoline engine (*32a, 32b*).

Gas-turbine fuel consumption for light-duty vehicles has been shown to be similar to equivalent gasoline engines (*32*). Significantly improved fuel economy is expected as ceramic turbines are developed. Ceramic turbines will allow substantial increases in turbine inlet temperature. Fuel consumption decreases about 10% for each 100°F increase in turbine inlet temperature.

### *B. Rankine Cycle Engines*

The Rankine engine is basically a closed system operating with a working fluid that is liquefied, pumped, vaporized, expanded, and recondensed.

The expander can be a turbine or a positive-displacement device of either rotary or piston type. The pump, likewise, can be rotary, rotary positive-displacement, or reciprocating. Schematic flow systems for the various Rankine systems are shown in Figures 22–24.

The best-known working fluid is water, as used in steam engines. But for more generally useful systems, organic fluids or easy to use inorganic fluids have been applied. Depending on the fluid, the vaporizer could be either a boiler or a heater containing the fluid at supercritical pressure. Again, depending on the fluid, internal heat recovery using a recuperator can be used to increase the cycle efficiency at the expense of additional equipment.

The basic limitation of Rankine cycle efficiency is the maximum and minimum temperatures to which the working fluid can be exposed. Maximum temperature is limited by either corrosion of the high temperature

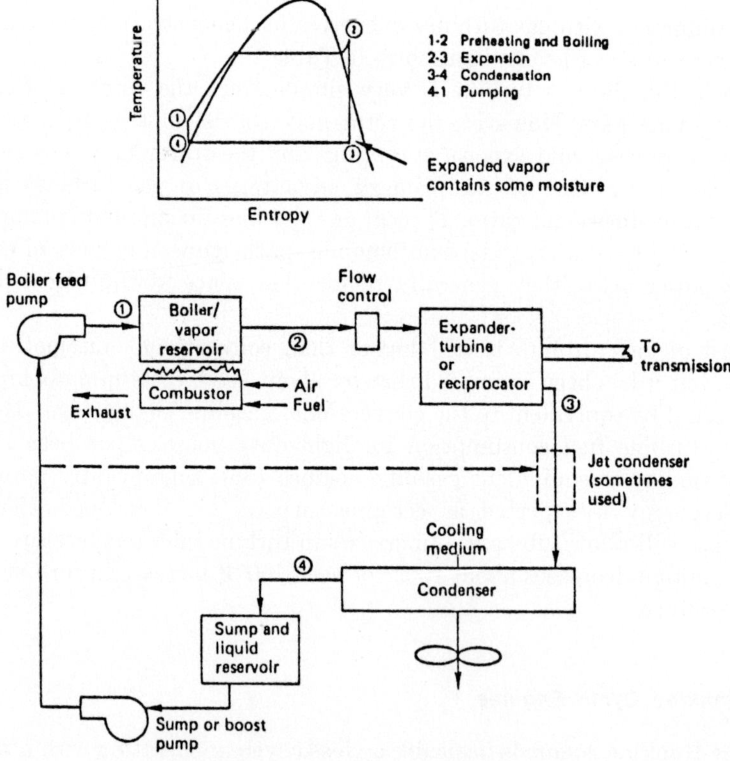


Figure 22. "Wet" vapor Rankine cycle engine.

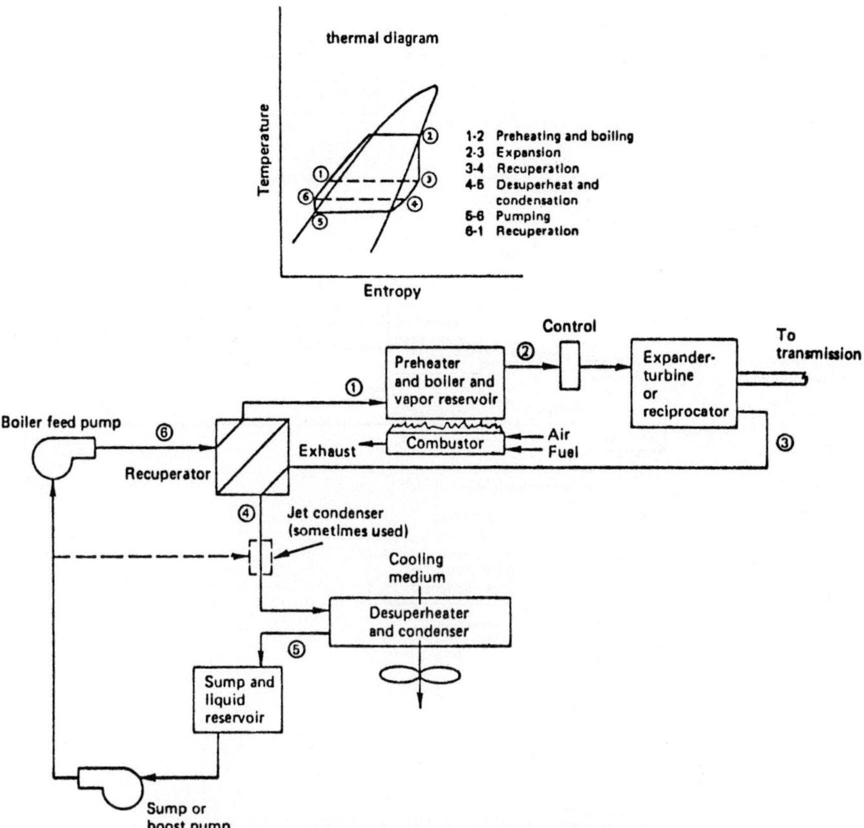


Figure 23. "Dry" vapor Rankine cycle engine.

parts or degradation of the working fluid. Minimum temperature is limited by the minimum practical vapor pressure in the condenser or by the condenser-wall temperature desired for reasonable-size radiators.

Demonstration projects have shown that it is possible to reduce exhaust emissions as shown in Table VI (*33*). Fuel consumption is higher than comparable gasoline engines, and high cost remains as a deterrent to mass use of Rankine cycle engines (*34*).

## X. Control Approaches for Railroad Locomotives and Ships

The prime movers for railroad locomotives are still large diesel engines although some experimental gas turbines have been used. Similarly, ships

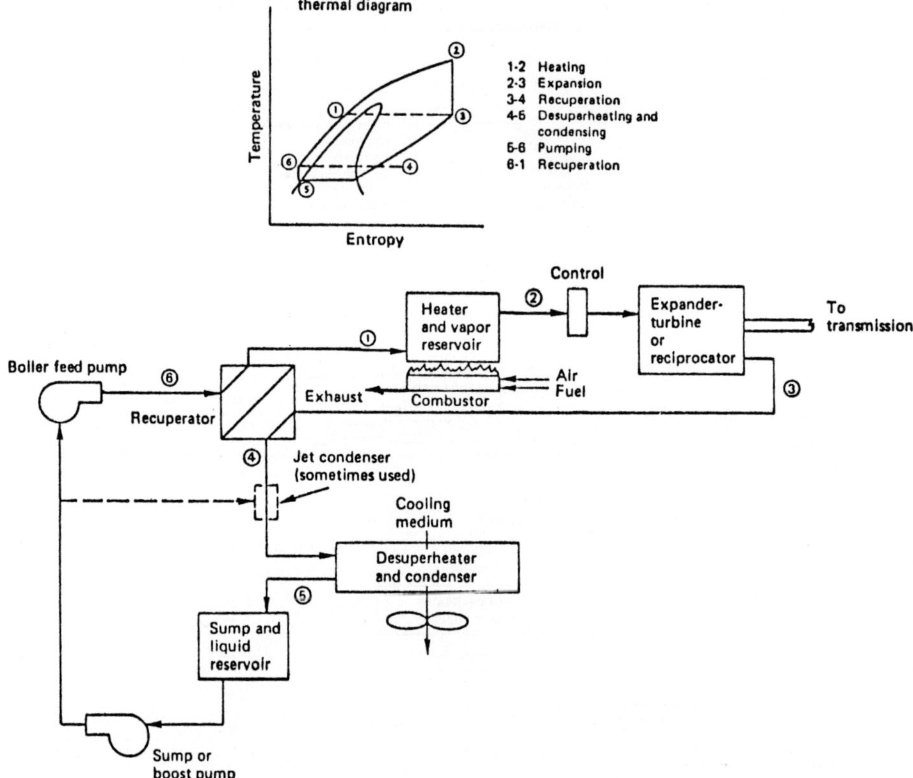


**Figure 24. Supercritical vapor Rankine cycle engine.**

use large diesel engines with an occasional gas turbine. A recent study conducted at Southwest Research Institute (*35*) characterized the emissions from three different locomotive engines under normal operating conditions. Hydrocarbons and oxides of nitrogen increase with increased load (Figs. 25 and 26). Carbon monoxide mass emissions increase with increasing load up to about midrack position and then drop off as load continues to increase for the two larger engines (Fig. 27). A smaller Roots-blown switch engine exhibited much lower carbon monoxide emissions than two large Electro-Motive Division engines. Also, carbon monoxide increased consistently for this engine as load was increased. Smoke opacity varies with rack position and for the locomotives described, maximum smoke occurred during part load operation.

The major emission control problem for locomotive-type diesel engines

**Table VI Steady-State Exhaust Emissions from Light Steam Cars (*33*)**

| *Vehicle manufacturer* | *Speed, miles/hour* | *Emissions, gm/mile*[a] | | |
|---|---|---|---|---|
| | | *HC* | *CO* | $NO_x$ |
| Aerojet Liquid Rocket Company | Idle | 0.36 | 0.13 | 0.06 |
| | 15 | 0.003 | 0.13 | 0.34 |
| | 30 | 0[b] | 0.14 | 0.22 |
| | 45 | 0.031 | 0.44 | 0.26 |
| | 60 | 0.19 | 0.69 | 0.24 |
| Steam Power Systems | Idle | 0.13 | 0.014 | 0.04 |
| | 20 | 0.32 | 0.11 | 0.2 |
| | 30 | 0.08 | 0.06 | 0.14 |
| | 40 | 0.15 | 0.25 | 0.16 |
| | 50 | 0.13 | 0.7 | 0.23 |

[a] Emission tests performed by California Air Resources Board; both cars in the 3000 lb inertia weight class; fuel used was a low volatility unleaded gasoline (6.4 lb/gal); power systems were warmed up, and stable operating conditions reached prior to taking data.
[b] A "zero" value for hydrocarbons indicates that the emission was approximately the same as the background levels in the atmosphere.

is reduction of exhaust oxides of nitrogen. To a large extent, the problem is the same for light-duty diesels, and similar control methods are effective. These include use of improved-design fuel injectors, variation in injection timing, humidification of inlet air, reduction of scavenging efficiency, and exhaust-gas recirculation. As in light-duty diesels and conventional gasoline engines, variation in operating parameters will often reduce oxides of nitrogen, but increase hydrocarbons and carbon monoxide. Fuel injectors have been modified to reduce the sac volume below the spherical or needle valve that controls fuel flow through the injector tip (*36*). The effect has been to lower hydrocarbon emissions and reduce exhaust smoke opacity, but oxides of nitrogen tend to increase. Retarded injection timing effectively reduces oxides of nitrogen and hydrocarbons, but carbon monoxide may increase. Four degrees of basic timing retard results in about 20–30% reduction in oxides of nitrogen. Water injection, reduction in scavenging efficiency, and EGR are also effective in reducing oxides of nitrogen. Combinations of approaches were shown in the referenced tests to result in oxides of nitrogen reductions of 50% or more.

Similar control approaches would be expected to have similar success in reducing emissions from diesel engines used in ship operations.

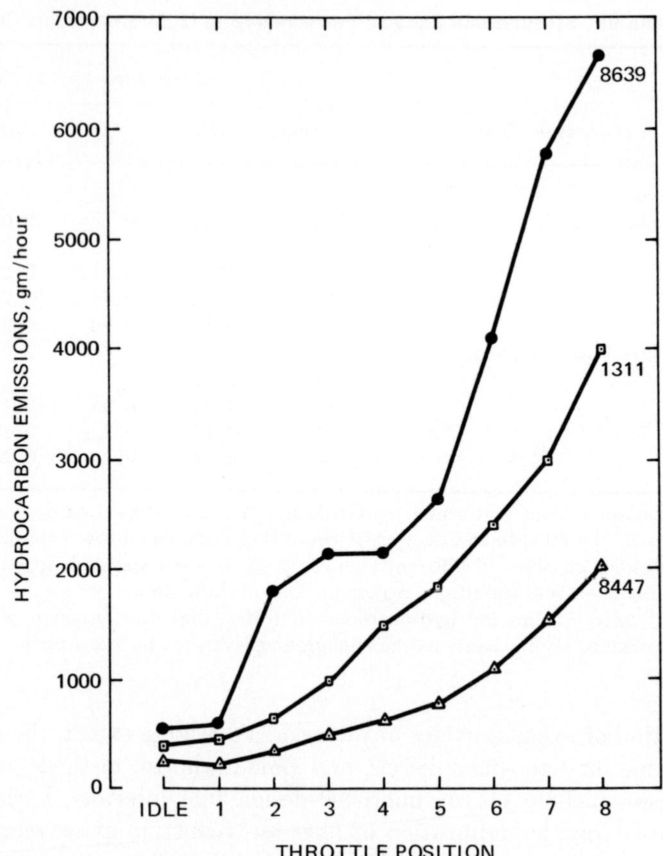


**Figure 25. Hydrocarbon emissions (gm/hour) from three locomotive diesel engines as a function of throttle position (*35*).**

## XI. Control Approaches for Aircraft Emissions

Aircraft engine emission control techniques utilize the following general approaches:

(a) Modification of ground operational procedures

(b) Improvement in maintenance and quality control procedures to minimize emissions from existing fleets

(c) Development of new combustion technology for application to new aircraft

(d) Retrofit of existing turbine engine fleets with new combustion technology

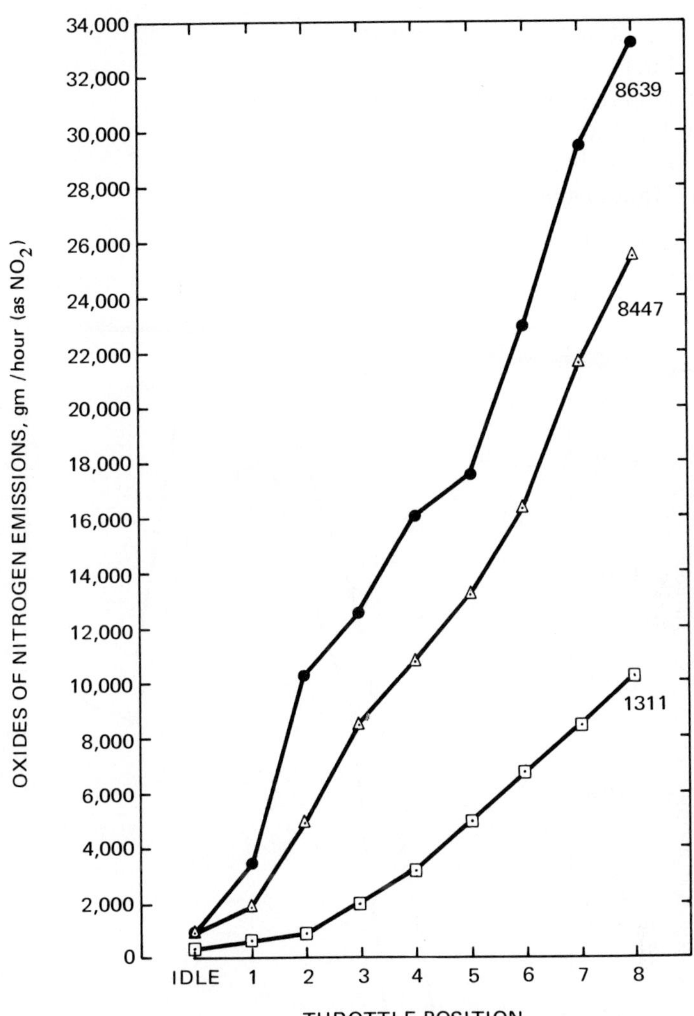


Figure 26. Oxides of nitrogen emissions (gm/hour, $NO_2$) from three locomotive diesel engines as a function of throttle position (*35*).

The various modes of aircraft operations require different control approaches. The major modes can be characterized as follows:

(a) Start-up and idle
(b) Taxi
(c) Idle at runway
(d) Takeoff

(e) Climb out to 3000 ft elevation
(f) Fuel dumping
(g) Approach from 3000 ft elevation
(h) Landing
(i) Idle and shutdown
(j) Maintenance

In assessing the feasibility of a control method, four factors must be explored: (a) effect of the method on the safety and capacity of the

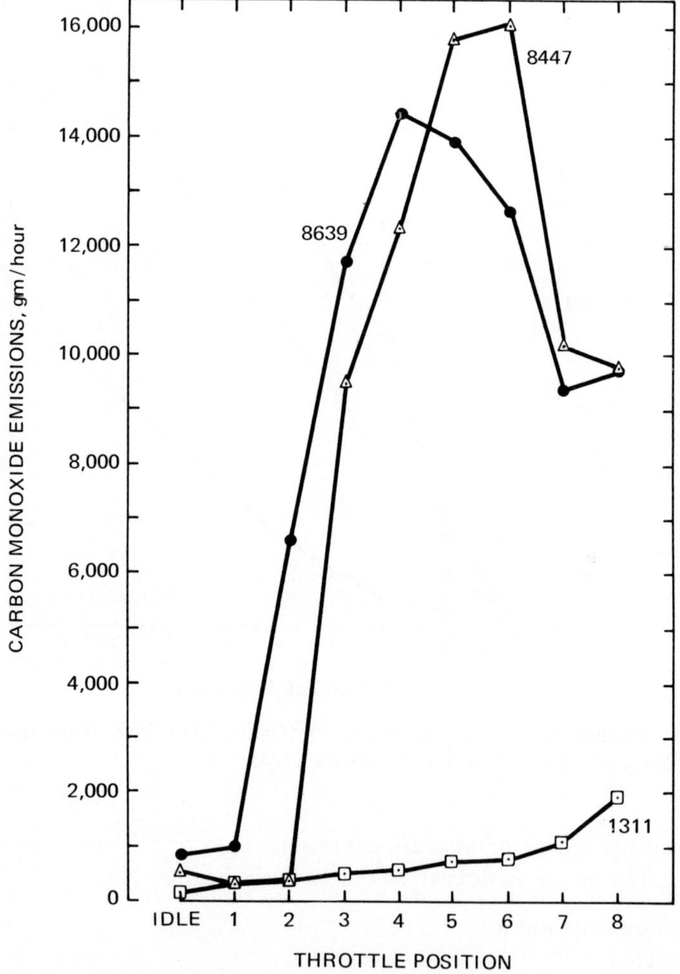


Figure 27. Carbon monoxide emission (gm/hour) from three locomotive diesel engines as a function of throttle position (*35*).

airplane; (b) effectiveness of the method in reducing emissions; (c) cost of utilizing the method; and (d) time required for implementing the method.

Emission control by fuel modification does not provide significant reduction in emissions, except for reduction in sulfur or lead content that results in proportionate reduction of sulfur oxides and lead emissions.

### A. Aircraft Classification System

The aircraft classification system groups aircraft into 12 separate types that include the currently used commercial air carriers, general aviation, and military aircraft (Tables VII and VIII). Classes 8–11 are exclusively military aircraft and are excluded from further consideration in this chapter.

### B. Emission Control by Engine Modification

Various engine modifications appear to be feasible in that they can be applied to aircraft without degrading engine reliability or seriously reducing aircraft performance (*37–39*). Costs of implementing these control methods also appear to be within reasonable limits.

#### 1. Turbine Engines

Six engine modification control methods considered feasible for turbine engines (Table IX) are, at least in principle, applicable to existing engines by retrofitting of new or modified parts, and to engines currently in production (t1–t6). Two methods (t7 and t8) are considered to be applicable only to future engines of new design, since the modifications required are too extensive to be applied to engines for which development has been completed.

The first control method (t1) consists of simple modification of the combustor and fuel nozzles to reduce all emission rates to the best levels currently attainable within each engine class. The degree of control attainable depends upon the performance of specific engines compared with those engines in the same class demonstrating the lowest emission rates. In general, this control method requires emission quality control (emission reduction to levels demonstrated by other engines of that model). Additionally, for certain high-emission engine models, it means emission reduction to the level of other engines of the same class. Reduction in emissions achievable through simple modification at the combustor and

**Table VII Aircraft Classification System**

| *Category* | *Class* | *Type* | *Examples* | *Engine model and type* | *Thrust or power*[a] | *Engines per aircraft* |
|---|---|---|---|---|---|---|
| Air carrier | 1 | Supersonic transport | Concorde<br>Tupolev TU-144 | R-R/Snecma Olympus 593 turbojet | 39,000 lb | 4 |
| | 2 | Jumbo jet transport | Boeing 747<br>Douglas DC-10 | P & WA JT9D turbofan | 43,000 lb | 4 |
| | 3 | Long range jet transport | Boeing 707<br>Douglas DC-8 | P & WA JT3D | 18,000 lb | 4 |
| | 4 | Medium range jet transport | Boeing 727<br>Douglas DC-9 | P & WA JT8D | 13,900 lb | 2.6[b] |
| | 5 | Turboprop transport | Lockheed Electra<br>Fairchild Hiller FH-227 | Allison 501-D13 turboprop | 3,750 hp | 2.5[b] |
| General aviation | 6 | Business jet | Lockheed Jetstar<br>North American Sabreliner | P & WA JT12 turbojet | 2,900 lb | 2.1[b] |
| | 7 | Piston engine | Cessna 210 Centurion<br>Piper 32-300 Cherokee Six | Continental 10-520-A, opposed piston | 292 hp | 1 |
| V/STOL | 12 | Helicopters and V/STOL | Sikorsky S-61<br>Vertol 107 | General Electric CT58 turboshaft | 1,390 hp | 2 |

[a] Equivalent shaft power.
[b] Number is derived by taking the number of engines per aircraft, times the number of aircraft of each type flying so as to get a weighted average.

**Table VIII Aircraft Engine Classification**

| Engine class | Engine type | Power range, lb thrust or eshp[a] |
|---|---|---|
| T1 | Turbine | Less than 6,000 |
| T2 | Turbine | 6,000 to 29,000 |
| T3 | Turbine | Greater than 29,000 |
| P1 | Piston | All piston engines |

[a] eshp: Estimated shaft horsepower.

**Table IX Engine Modifications for Emission Control for Existing and Future Turbine Engines**

| Control method | Modification |
|---|---|
| Existing engines | |
| t1—Minor combustion chamber redesign | Minor modification of combustion chamber and fuel nozzle to achieve best state-of-art emission performance |
| t2—Major combustion chamber redesign | Major modification of combustion chamber and fuel nozzle incorporating advanced fuel injection concepts (carburetion or prevaporization) |
| t3—Fuel drainage control | Modify fuel supply system or fuel drainage system to eliminate release of drained fuel to environment. |
| t4—Divided fuel supply system | Provide independent fuel supplies to subsets of fuel nozzles to allow shutdown of one or more subsets during low-power operation |
| t5—Water injection | Install water injection system for short duration use during maximum power (takeoff and climb-out) operation |
| t6—Modify compressor air bleed rate | Increase air bleed rate from compressor at low-power operation to increase combustor fuel–air ratio |
| Future engines | |
| t7—Variable-geometry combustion chamber | Use of variable airflow distribution to provide independent control of combustion zone fuel–air ratio |
| t8—Staged injection combustor | Use of advanced combustor design concept involving a series of combustion zones with independently controlled fuel injection in each zone |

fuel nozzles vary with the pollutant considered, the engine class, and the engine operating mode (Table X).

The estimation of emission control effectiveness for turbine engines by methods t2–t8 (Table XI) is based upon reductions attainable from "lowest current emission rates" attainable through control method t1—

**Table X Effectiveness of t1—Minor Combustion Chamber Redesign[a]—on Reduction of Emissions from Turbine Engines**

| Engine class | Pollutant[b] | *Emission rates in lb/1000 lb of fuel (kg/1000 kg of fuel)* | | |
|---|---|---|---|---|
| | | *Idle/taxi* | *Approach* | *Takeoff* |
| T1 | CO | 25 | 5 | 2 |
| T1 | THC | 10 | 1 | 0.2 |
| T1 | $NO_x$ | 3 | 7 | 11 |
| T1 | DP | 0.2 | 0.5 | 0.5 |
| T2 | CO | 45 | 6 | 1 |
| T2 | THC | 10 | 1 | 0.1 |
| T2 | $NO_x$ | 2 | 6 | 12 |
| T2 | DP | 0.2 | 0.5 | 0.5 |
| T3 | CO | 50 | 3 | 0.5 |
| T3 | THC | 10 | 1 | 0.1 |
| T3 | $NO_x$ | 3 | 10 | 40 |
| T3 | DP | 0.1 | 0.1 | 0.1 |

[a] Minor combustor redesign is assumed to reduce the smoke to invisible or "smokeless" levels for all engine classes.
[b] CO: Carbon monoxide; THC: total hydrocarbon; $NO_x$; oxides of nitrogen.

simple modification of the combustor and fuel nozzles (Table X). It is predicted that all engines in each class can be modified to achieve the "best rates" of Table XI. These "best rates" are not the lowest rates indicated for each engine class, but are rates near the low end of those emission rates that appear to be realistically attainable. The use of the "best rate" basis is necessary to allow effectiveness estimates for each engine class. Because of the wide variations in actual emission rates of turbine engines, an effectiveness analysis based on average rates would be less significant. Some of these estimates are based upon demonstrated performance. Most, however, are not based on direct experience with these control methods on aircraft engines but instead on theoretical analyses of engine performance under the operating conditions associated with the control methods.

No specific estimates have been made for control of reactive hydrocarbons, odor, or aldehydes because control methods applicable to these emissions have not yet been identified. Reductions in these emissions are expected to parallel reductions in total hydrocarbon emissions. Any of the modifications defined for existing turbine engines (t1–t6) can be com-

**Table XI Effectiveness of Engine Modification in Control of Emissions from Turbine Engines, by Operating Mode[a]**

| *Control method*[b] | *Engine class* | *Pollutant*[e] | *Emmision rates during* | | |
|---|---|---|---|---|---|
| | | | *Idle/taxi* | *Approach* | *Takeoff* |
| t2 | T1 | DP | 0.5 | 0.5 | 0.5 |
| t2 | T1 | $NO_x$ | NC[c] | NC | 0.5 |
| t2 | T2 | DP | 0.5 | 0.5 | 0.5 |
| t2 | T3 | $NO_x$ | NC | NC | 0.5 |
| t3 | T1 | THC | NC | NC | 0[d] |
| t3 | T2 | THC | NC | NC | 0[d] |
| t3 | T3 | THC | NC | NC | 0[d] |
| t4 | T1 | CO | 0.25 | NC | NC |
| t4 | T1 | THC | 0.25 | NC | NC |
| t4 | T2 | CO | 0.25 | NC | NC |
| t4 | T2 | THC | 0.25 | NC | NC |
| t4 | T3 | CO | 0.25 | NC | NC |
| t4 | T3 | THC | 0.25 | NC | NC |
| t5 | T1 | $NO_x$ | NC | NC | 0.1 |
| t5 | T2 | $NO_x$ | NC | NC | 0.1 |
| t5 | T3 | $NO_x$ | NC | NC | 0.1 |
| t6 | T1 | CO | 0.5 | NC | NC |
| t6 | T1 | THC | 0.5 | NC | NC |
| t6 | T2 | CO | 0.5 | NC | NC |
| t6 | T2 | THC | 0.5 | NC | NC |
| t6 | T3 | CO | 0.5 | NC | NC |
| t6 | T3 | THC | 0.5 | NC | NC |
| t7 or t8 | T1 | CO | 0.1 | NC | NC |
| t7 or t8 | T1 | THC | 0.1 | NC | NC |
| t7 or t8 | T1 | $NO_x$ | NC | NC | 0.75 |
| t7 or t8 | T1 | DP | 0.5 | 0.5 | 0.5 |
| t7 or t8 | T2 | CO | 0.1 | NC | NC |
| t7 or t8 | T2 | THC | 0.1 | NC | NC |
| t7 or t8 | T2 | $NO_x$ | NC | NC | 0.75 |
| t7 or t8 | T2 | DP | 0.5 | 0.5 | 0.5 |
| t7 or t8 | T3 | CO | 0.1 | NC | NC |
| t7 or t8 | T3 | THC | 0.1 | NC | NC |
| t7 or t8 | T3 | $NO_x$ | NC | NC | 0.75 |
| t7 or t8 | T3 | DP | 0.5 | 0.5 | 0.5 |

[a] Emission rate is fraction of best current rate assumed to be attainable through minor combustion-chamber redesign and with control method cited.

[b] t2 = Major combustion chamber redesign
t3 = Fuel drainage control
t4 = Divided fuel supply system
t5 = Water injection
t6 = Modify compressor air bleed rate
t7 = Variable-geometry combustion chamber
t8 = Staged injection combustor

[c] NC indicates no change.

[d] Refers to raw fuel drainage only.

[e] CO: Carbon monoxide; THC: total hydrocarbon; $No_x$: oxides of nitrogen.

**Table XII Engine Modifications for Emission Control for Existing and Future Piston Engines**

| *Control method* | *Modification* |
|---|---|
| Existing engines | |
| p1—Fuel–air ratio control | Limiting rich fuel–air ratios to only those necessary for operational reliability |
| p2—Simple air injection | Air injected at controlled rate into each engine exhaust port |
| p3—Thermal reactors | Air injection thermal reactor installed in place of, or downstream of, exhaust manifold |
| p4—Catalytic reactors for HC and CO control | Air injection catalytic reactor installed in exhaust system. Operation with lead-free or low-lead fuel required |
| p5—Direct-flame afterburner | Thermal reactor with injection of air and additional fuel installed in exhaust system |
| p6—Water injection | Water injected into intake manifold with simultaneous reduction in fuel rate to provide for cooler engine operation at leaner fuel–air ratios |
| p7—Positive crankcase ventilation (PCV) | Current PCV system used with automotive engines applied to aircraft engines. Effective only in combination with one of preceding control methods |
| p8—Evaporative emission controls | A group of control methods used singly or in combination to reduce evaporative losses from the fuel system. Control methods commonly used include charcoal absorbers and vapor traps in combination with relatively complex valving and fuel flow systems |
| Future engines | |
| p9—Engine redesign | Coordinated redesign of combustion chamber geometry, compression ratio, fuel distribution system, spark and valve timing, fuel–air ratio, and cylinder wall temperature to minimize emissions while maintaining operational reliability |

bined to achieve increased emission control·effectiveness except that modifications t4 and t6 are mutually exclusive.

## 2. Piston Engines

Although aircraft piston engines and automobile engines are fundamentally similar, their applications are significantly different, with different requirements. Reliability is of primary importance in aircraft-piston-engine applications and therefore is given paramount consideration in identifying applicable control methods.

The control methods considered feasible for aircraft piston engines (Table XII) include most of the approaches that have been developed for automotive engines for control of carbon monoxide and total hydrocarbon. Methods for controlling oxides of nitrogen emissions are not included because the fuel-rich operating conditions of aircraft piston engines result in low oxides of nitrogen emission rates.

### *C. Emission Control by Modification of Ground Operations*

Several methods offer some degree of control of carbon monoxide and total hydrocarbon emissions at air carrier airports by modification of turbine-aircraft ground-operation procedures.

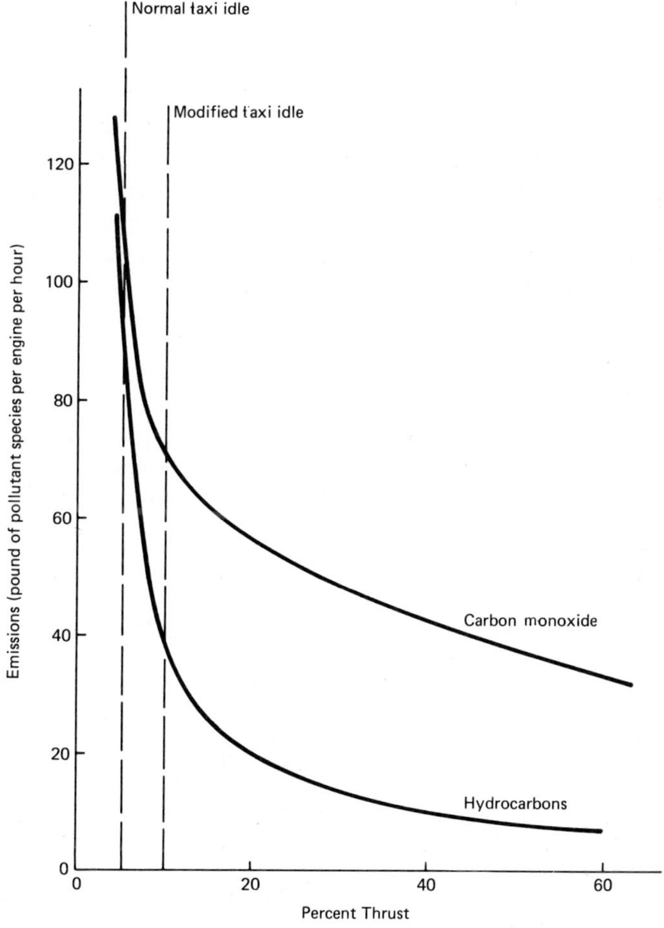


Figure 28. Hydrocarbon and carbon monoxide emissions from a typical aircraft turbine engine (JT3D).

(a) Increase engine speed and reduce number of engines operating during idle and taxi

(b) Reduce idle operating time by controlling departure times from gates

(c) Reduce taxi operating time by transporting passengers to aircraft and by towing aircraft between runway and gate

(d) Reduce operating time of aircraft auxiliary power supply by providing ground-based power supply

The first method reduces emissions by requiring that engines be operated at more efficient power settings than those in current practice (Fig. 28); the next three methods reduce emissions by reducing operating time of either main or auxiliary engines. The effectiveness of these methods in reducing emissions varies considerably.

The control methods listed, with the possible exception of (b), are not applicable to small piston-engine aircraft, and, therefore, do not seem to offer means for controlling emissions at general aviation airports. Periods of delay at takeoff are significant at some general aviation airports; however, aircraft ground traffic at general aviation airports may not be sufficiently controlled to allow an effective system of controlled gate departures or engine start-ups to reduce periods of delay.

## REFERENCES

1. W. Martens and R. C. Stempel, "Fuel Economy Trends and Catalytic Devices," No. 740594. Society of Automotive Engineers, Detroit, Michigan, 1974.
2. C. LaPointe, "Factors Affecting Vehicle Fuel Economy," No. 730791. Society of Automotive Engineers, Detroit, Michigan, 1973.
3. T. C. Austin and C. H. Hellman, "Passenger Car Fuel Economy—Trends and Influencing Factors," No. 730790. Society of Automotive Engineers, Detroit, Michigan, 1973.
4. E. F. Obert, "Internal Combustion Engines Analyses and Practice," 2nd ed. International Textbook, Scranton, Pennsylvania, 1950.
5. B. Lewis and G. von Elbe, "Combustion, Flames and Explosions of Gases," 2nd ed., Chapter IV. Academic Press, New York, New York, 1961.
6. R. Friedman and W. Johnson, *J. App. Phys.* **21**(8), 791 (1950).
7. D. J. Patterson and N. A. Henein, "Emissions from Combustion Engines and Their Control." Ann Arbor Sci. Publ., Ann Arbor, Michigan, 1972.
8. J. D. Ingamells and R. H. Lindquist, "Methanol as a Motor Fuel or a Gasoline Blending Component," No. 750123. Society of Automotive Engineers, Detroit, Michigan, 1975.
9. R. D. Fleming and T. W. Chamberlain, "Methanol as Automotive Fuel, Part 1—Straight Methanol," No. 750121. Society of Automotive Engineers, Detroit, Michigan, 1975.
10. J. F. Stocky, M. W. Dowdy, and T. G. Vanderburg, "An Examination of the

Performance of Spark Ignition Engines Using Hydrogen-Supplemented Fuels," No. 750027. Society of Automotive Engineers, Detroit, Michigan, 1975.

11. J. Houseman and D. J. Gerini, "On-Board Hydrogen Generator for a Partial Hydrogen Injection Internal Combustion Engine," No. 740600. Society of Automotive Engineers, Detroit, Michigan, 1974.
12. J. Meurer, *SAE* (*Soc. Automot. Eng.*) *J.* **64**, 250 (1956); **72**, 712 (1962).
13. V. R. Grundman, Jr., "Submission for Environmental Protection Agency 1977 Suspension Request Hearings." Dresser Industries, Inc., Irvine, California, 1975.
14. "Evaluation of the Dresserator Emission Control System." U.S. Environmental Protection Agency, Ann Arbor, Michigan, 1974.
15. J. G. Rivard, "Closed-Loop Electronic Fuel Injection Control of the Internal-Combustion Engine," No. 730005. Society of Automotive Engineers, Detroit, Michigan, 1973.

15a. E. S. Starkman, *Environ. Sci. Technol.* **9**(9), 820–824 (1975).

16. R. W. Johnson, J. G. Neuman, and P. D. Agarwal, "Programable Energy Ignition System for Engine Optimization," No. 750348. Society of Automotive Engineers, Detroit, Michigan, 1975.
17. G. O. Huntzinger and G. E. Rigsby, "HEI—a New Ignition System through New Technology," No. 750346. Society of Automotive Engineers, Detroit, Michigan, 1975.
18. R. E. Canup, "The Texaco Ignition System—a New Concept for Automotive Engines," No. 750347. Society of Automotive Engineers, Detroit, Michigan, 1975.

18a. T. K. Garrett, *Environ. Sci. Technol.* **9**(9), 826–830 (1975).

19. D. A. Hurter and W. D. Lee, "A Study of Technological Improvements in Automobile Fuel Consumption," No. 750005. Society of Automotive Engineers, Detroit, Michigan, 1975.
20. M. Beltzer, R. J. Campion, J. Harlan, and A. M. Hochhauser, "The Conversion of $SO_2$ Over Automotive Oxidation Catalysts," No. 750095. Society of Automotive Engineers, Detroit, Michigan, 1975.
21. R. L. Bradow and J. B. Moran, "Sulfate Emissions from Catalyst Cars—a Review," No. 750090. Society of Automotive Engineers, Detroit, Michigan, 1975.
22. R. H. Hammerle and M. Mikkor, "Some Phenomena Which Control Sulfuric Acid Emission from Automotive Catalysts," No. 750097. Society of Automotive Engineers, Detriot, Michigan, 1975.
23. J. Mooney, Statement before California Air Research Board Sulfate Hearing, 1975.
24. J. L. Bascunana and M. J. Webb, "Effectiveness and Costs of Retrofit Emission Control Systems for Used Motor Vehicles," No. 720938. Society of Automotive Engineers, Detroit, Michigan, 1972.
25. T. Date, S. Yagi, A. Ishizuya, and I. Fujii, "Research and Development of the Honda CVCC Engine," No. 740605. Society of Automotive Engineers, Detroit, Michigan, 1974.
26. E. Mitchell, M. Alperstein, J. M. Cobb, and C. H. Faist, "Stratified-Charge Multifuel Military Engine (Texaco)—Progress Report," No. 720051. Society of Automotive Engineers, Detroit, Michigan, 1972.
27. M. Alperstein, G. H. Schafer, and F. J. Villforth, III, "Texaco's Stratified Charge Engine—Multifuel, Efficient, Clean, and Practical," No. 740563. Society of Automotive Engineers, Detroit, Michigan, 1974.
28. K. J. Springer and R. C. Stahman, "Emissions and Economy of Four Diesel Cars," No. 750332. Society of Automotive Engineers, Detroit, Michigan, 1975.

29. K. J. Springer and R. C. Stahman, "Diesel Emission Control through Retrofits," No. 750205. Society of Automotive Engineers, Detroit, Michigan, 1975.
30. Committee on Motor Vehicle Emissions, "An Evaluation of Alternative Power Sources for Low Emission Automobiles" (Report of the Panel on Alternate Power Sources). National Academy of Sciences, Washington, D.C., 1973.
31. F. A. Wyczalek, J. L. Harned, S. Maksymiuk, and J. R. Blevins, "EFI Prechamber Torch Ignition of Lean Mixtures," No. 750351. Society of Automotive Engineers, Detroit, Michigan, 1975.
32. T. F. Nagey, P. Mykolenko, M. E. Naylor, and F. J. Verkemp, "Low Emission Gas Turbine Passenger Car—What Does the Future Hold?" Paper No. 73-G7-49. Amer. Soc. Mech. Eng., New York, New York, 1975.
32a. C. A. Amann, "Introduction to the Vehicular Gas Turbine Engine," No. 730618. Society of Automotive Engineers, Detroit, Michigan, 1973.
32b. Anonymous, "Gas Turbines: A Bright Future on Tomorrows Motorways?" pp. 45–48. *Design Engineering,* London, 1973.
33. R. A. Renner and M. Wenstrom, "Experience with Steam Cars in California," No. 750069. Society of Automotive Engineers, Detriot, Michigan, 1975.
34. S. Luchter and W. Mirsky, "A Survey of Automotive Rankine Cycle Combustion Technology," No. 750067. Society of Automotive Engineers, Detroit, Michigan, 1975.
35. C. T. Hare, K. J. Springer, and T. A. Huls, "Locomotive Exhaust Emissions and Their Impact." Amer. Soc. Mech. Eng., New York, New York, 1974.
36. J. O. Storment, K. J. Springer, and K. M. Hergenrother, "$NO_x$ Studies with EMD 2-567 Diesel Engine." Amer. Soc. Mech. Eng., New York, New York, 1974.
37. U.S. Environmental Protection Agency, "Aircraft Emissions: Impact on Air Quality and Feasibility of Control" (An EPA Position Paper). U.S. Environmental Protection Agency, Research Triangle Park, North Carolina, 1972.
38. U.S. Environmental Protection Agency, "Assessment of Aircraft Emission Control Technology," Final Report by Northern Research and Engineering Corporation, Contract No. 68-04-0011. U.S. Environmental Protection Agency, Research Triangle Park, North Carolina, 1971.
39. Aerospace Industries Association, "A Study of Aircraft Gas Turbine Engine Exhaust Emissions." Aerospace Industries Association, Washington, D.C., 1971

# 15

## Agriculture and Agricultural-Products Processing

W. L. Faith



### I. Agriculture

Agriculture, as defined in most air pollution ordinances, refers to those operations involved in the growing of crops or raising of animals. Traditionally, agricultural operations have been exempt from air pollution rules and regulations, but with the rise of the environmental crisis, it has become more difficult to defend the agricultural exemption. Consequently, increasingly stricter regulations have been enacted in the past few years.

Dust and odors are the primary air pollutants arising from agricultural operations. On the basis of soil conservation studies, Vandegrift (*1*) estimated that 65,000,000 tons of "natural" dust is blown into the air over the United States annually. Not all of this is of agricultural origin, but a goodly portion may be. When one adds to this, 40,000,000 tons per year of particulate matter from wild and controlled forest fires, and 2,500,000 tons per year from other agricultural burning, the contribution of agriculture to atmospheric particulate matter is a major one. Further the annual release of pollens and toxins has been estimated to account for 10,500,000 lost work days annually in the United States.

Odor problems are related chiefly to decomposing wastes. The major odorous wastes are of animal origin. There are 2.0 billion tons ($\frac{1}{3}$ liquid) of animal wastes in the United States per year, which is equivalent to the waste production of a human population of 2 billion (*2*).

The major sources of atmospheric pollutants from the agricultural sector of the economy are discussed in detail in the following paragraphs.

### A. Soil Preparation

Whenever natural soil is disturbed so as to produce a dry powder, a latent dust problem is created. The problem becomes particularly bothersome in low rainfall areas during periods of high winds. The conversion of range land in western Kansas and Oklahoma to wheat land in the 1920's led to the devastating dust storms of the early 1930's. The dust was carried by the wind as far as the eastern seaboard of the United States. In the source area, farm homes and outbuildings were inundated with waves of sand and only a hardy few withstood the siege. The "dust bowl" was eventually reclaimed by adoption of contour plowing practices (only short streches of furrows in the direction of the wind) and establishment of windbreaks.

Even in areas with lower wind speeds, soil tilling on marginal land releases great quantities of dust during plowing, harrowing, and planting. Dust raised by high winds in some agricultural areas virtually paralyzes automobile traffic at times and leaves a fine powder in the air for days after a blow. To alleviate this condition some agencies have passed ordinances restricting the time when plowing and cultivation may be done, and the type of cultivation equipment that may be used. Other agencies include agricultural operations in their general fugitive dust control regulation. Thus the state of Kentucky requires "Conduct of agricultural practices such as tilling of land, application of fertilizers, etc. in such manner as to not create a nuisance to others residing in the area." No specific means of control are designated, but some are obvious. Thus in the application of fertilizers and soil conditioners, dust problems may be alleviated by substituting granular for powdered materials. Some highly odorous fertilizers may have to be avoided near populous areas.

### B. Pesticides and Herbicides

Agriculture has always been dependent to a great extent on pest control, but extensive use of pesticides has developed only in the past 30 years. On large farms it is common practice to apply agricultural dusts and sprays by means of aircraft. If the aerosol is toxic and drifts away before settling on the plants, a hazard to man, animals, and other crops may ensue. However, for ecological reasons, use of the more persistent insecticides (and herbicides) has been drastically regulated in the United States and some European countries. Refinements in application equipment and use of reduced-drift formulations have also reduced general atmospheric concentrations. Air monitoring studies at nine localities in the United States in 1967 and 1968 (*3*) indicated that atmospheric concentrations of pesticides ranged from 0.1 ng of pesticide/$m^3$ of air to as high as 2520 ng/$m^3$. The high concentrations were found in agricultural areas. Nearly all levels found in the ambient air were far below levels that might add to the total human intake of pesticides. Herbicide drift may also be reduced by the methods applicable to pesticides.

### C. Open Burning

A major source of atmospheric particulate matter, carbon monoxide, and hydrocarbons is smoke from agricultural burning, still a widespread practice although subject to stricter controls each year.

Agricultural burning includes the burning of a variety of materials including natural ground cover, grasses, cereal crop stubble, weeds, orchard and vine prunings, range brush, and slash timber. The amount of pollutants evolved from open burning (Table I) (*4*) varies widely depending upon the nature of the material burned, particularly, moisture and texture; burning practices, e.g., stacking, spreading, etc.; and meteorological conditions.

The major effect of most open-burning practices is the nuisance caused by the smoke, although occasionally a safety or health hazard may result.

Table I Emission Factors for Agricultural Burning (*4*)

| *Pollutant* | *lb/ton* | *kg/MT*[a] |
|---|---|---|
| Particulate matter | 17 | 7 |
| Carbon monoxide | 60–100 | 25–40 |
| Hydrocarbons ($CH_4$) | 20 | 8 |
| Nitrogen oxides (as $NO_2$) | 2 | 0.8 |

[a] MT = Metric ton.

For example, heavy smoke may interfere with aircraft landings or highway traffic. The burning of some noxious weeds, such as poison ivy, produces volatile irritant oils that may provoke serious illness to sensitive individuals on momentary contact.

Agricultural burning is practiced because it is thought to be necessary to prepare seed beds or planting sites, to clear land for agriculture or other uses, to control plant disease, to make forest fires less severe, to improve the habitat for wildlife, or other similar social or economic reasons.

For some of these practices, alternatives of a less polluting nature are available; for others no suitable alternative exists. For instance, weed control, especially along highways and rights-of-way, may be controlled by use of herbicides. Stubble and some grasses may be plowed under and incorporated with the soil, except where plant disease control is required. Wastes that can be collected readily may be cut into suitable lengths, if necessary, and hauled to sanitary land-fill sites where they may be buried. Such collection and hauling is generally quite expensive.

A mobile field incinerator has been developed at Oregon State University for destroying residual grass after seed has been harvested. The incinerator can process $1\frac{1}{4}$ to 2 acres/hour, and decreases air pollution by 90%.

For many years attempts have been made to utilize farm wastes industrially, but few attempts have been successful commercially, largely because of collection and hauling costs. Oat hulls, cottonseed hulls, rice hulls, and corn cobs have been used as raw material for making furfural (*5*), but only where very large quantities are available at elevators or processing plants. Straw and sugarcane bagasse have been processed to wall board, but again only near huge sources of the waste.

A study at the University of California (*6*) compared four harvesting and handling systems for utilization of rice straw. Field baling, preparatory for shipment to, for instance, a paper mill, cost $7/ton or $28/acre. Field cubing, with a binder and additives to make the straw suitable for livestock feed, cost \$18/ton or \$72/acre. Stationary cubing from long straw cost \$40/ton (\$160/acre). Single-operation harvesting of both rice and straw, with the threshing, drying, and straw processing to be done at a central plant, was found to be feasible, but to present no economic advantage. The same appears to be true of nearly all agricultural wastes. Therefore, unless burning is completely banned, it will continue to be practiced widely. Recently, however, a rice-straw field baler of promise was demonstrated by the California Department of Food and Agriculture (*6a*).

Nevertheless, open burning of agricultural wastes is being increasingly controlled. In many jurisdictions, the practice is prohibited except under permits granted only for when meteorological conditions are favorable for good atmospheric dispersion. For instance, one California air pollution control agency allows burning of fruit and nut tree prunings only, and then only when the atmospheric inversion layer is higher than 2500 ft, wind speed exceeds 10 mph, and the wind is blowing away from populated areas. In Ohio, open burning by farmers is permitted on the farm premises if they are at least 1000 ft from the corporation limits of municipalities of 1000 to 10,000 population, and greater than 1 mile from those of 10,000 or more people. The Oregon law establishes quotas of maximum acreage of grass that can be burned in different localities on "marginal" days (*7*).

Many jurisdictions specify the time of day when open burning is permitted, e.g., from 3 hours after sunrise to 3 hours before sunset. Generally, there is difference of opinion on this matter between air pollution authorities and fire department officials. The former want open burning carried out at times of maximum atmospheric dispersion, i.e., during periods of high winds; the latter prefer calm conditions. Nevertheless, in the long run it appears that, with perhaps the exception of forest management, open burning will be abandoned in favor of alternative methods.

### D. Orchard Heating

In many areas, fruit trees must be protected from frost at least a few nights of the year. One of the methods often used is called "smudging," the production of black smoke by burning heavy oil in pots. Smoke from smudge pots was a serious problem in the citrus belt of southern California in the early twentieth century. Orange and lemon growers believed that smoke was desirable to prevent the sun's rays from striking the fruit after a frosty night. On cold nights oil was burned in open pans, and discarded rubber tires were piled up and fired. The resulting smoke pall had to be seen to be believed.

Eventually it was learned that heat rather than smoke was protecting the fruit, so less smoky heaters were developed. In 1947, Los Angeles County banned the use of all orchard heaters, except certain approved types "of such design that not more than one gram per minute of unconsumed solid carbonaceous matter is emitted" (*8*). In adjacent San Bernardino County, emission of solid carbonaceous matter is restricted to $\frac{1}{2}$ gm/minute, except for specified heaters in which maximum burning rate is limited so as to prevent solid emissions in excess of 1 gm/minute (*9*).

A 1963 census of orchard heaters in San Bernardino County showed over 500,000 heaters for which permits had been issued. In many groves orchard heaters are being replaced by wind machines, in which a rotating horizontal propeller mounted on a vertical pole draws warm air from upper levels to mix with cold air near the ground, thus preventing frost formation.

The use of smudge pots and smoking heaters is not restricted to citrus groves. Stockman and Hildebrandt (*10*) measured the particulate matter in the air of Yakima Valley (Washington) during periods when smudging was being used to protect apple, peach, and similar fruit trees. Suspended particulate matter reached values of 240 $\mu g/m^3/24$ hour in contrast to normal values of 40 to 50. Tape sampler results exceeded 10.4 Cohs/1000 feet on several occasions, in sharp contrast with a normal values of 0.5 Cohs/1000 feet, where the Coh is an arbitrary unit of darkness.

### *E. Fruit and Vegetable Growing*

A common problem in fruit and vegetable districts is odor from plant waste allowed to remain on the ground. Old cabbage leaves remaining on the ground in warm, wet weather after harvest are particularly prone to develop obnoxious odors. Dropped fruit causes a similar problem. Even when rotten fruit is gathered and hauled away, dumping merely transfers the problem to another location. The material should be covered with earth. Obviously, good housekeeping applies to agriculture as well as to the home.

At present, greater attention is being paid to immediate incorporation of waste material into the soil after harvest. In a few cases, treatment in waste ponds or lagoons has proved satisfactory. Pilot-plant composting studies are also said to be promising.

### *F. Alfalfa Dehydration and Silage*

The dehydration and grinding of alfalfa to produce alfalfa meal is an especially dusty operation commonly carried out in rural communities.

Wet, chopped alfalfa is fed into a direct-fired rotary dryer where it meets hot combustion gases at 1800°–2000°C (Fig. 1) *(11)*. The dried alfalfa particles are carried with the combustion gases into an air cooler, then to a collecting cyclone from which hot moist air, carrying some odorous dust escapes. The collected particles are then ground to meal and bagged. Even with cyclones on the grinder and bagger, considerable dust is lost (the overall dust loss may be as high as 7%). The use of cloth collectors after the grinder and bagger cyclones can reduce total losses to 1.0–1.5% without further control of the primary cooler cyclone. Addition of 2–4% fat to dried alfalfa prior to grinding is also said to reduce losses *(12)*.

Green fodder may be preserved either by wet storage (silage) or by drying. In both cases odorous gases are emitted. Those emitted from silos (by fermentation) have been lethal to persons who enter unventilated

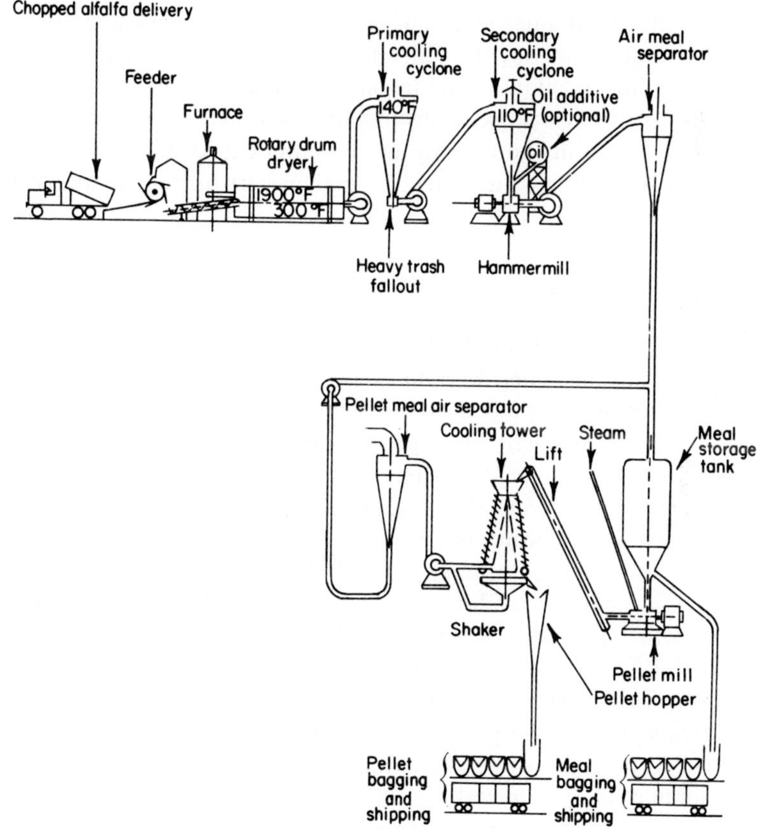


Figure 1. Alfalfa dehydrating process flow diagram. [From "Air Pollution from Alfalfa Dehydrating Mills" *(11)*.]

silos, but are dispersed rapidly outside the silo. Odorous gases from fodder dryers carry a short distance before dissipation. In West Germany, a minimum distance of 250–300 m between the drying plant and other buildings has been suggested.

### G. *Animal Production*

The feeding of cattle, sheep, and hogs for weight gain often produces an obnoxious odor. Considerable feeding is done on the farm, but commercial feeding, particularly of cattle, is a rapidly growing industry. In some localities, the raising of livestock and poultry on the premises of a farm are exempt from odor regulations, but it appears to be only a matter of time until such agricultural exemptions are terminated. Further, it is difficult to determine when an agricultural operation becomes a commercial one, and thus subject to odor control regulations.

In either case, however, the objectionable odor arises from animal excreta, both solid and liquid. For instance, 10,000 head of cattle will produce 280 tons of excreta each day; smaller animals somewhat less. The odorous substances in the excreta are a complex mixture of organic alcohols, aldehydes, ketones, and acids together with nitrogen- and sulfur-containing compounds.

#### 1. Cattle Feed Yards

The cattle population of the United States in 1970 was approximately 50 million, many of which were in cattle feedlots at one time or another. Farm lots may have as few as 100 cattle or less per lot at any one time; commercial feed lots will have populations varying from 2500 to over 80,000 head during the peak season.

Cattle are brought from the range to a feedlot where each animal eats 25 lb of a balanced ration every day for 150 days. During the stay in the feedlot the average animal gains 1 lb in weight for each 8–10 lbs of feed. In a commercial lot, the average 1000-lb animal will produce 26 lb in total excreta each day, of which 15 lb is urine. Accordingly, the potential for an atmospheric odor problem is ever present.

Control of the feedlot odor problem depends primarily on sanitation and housekeeping. If the pens are paved with either concrete or asphalt, a continuous program of pen cleaning and manure removal is necessary. In the summer, daily removal of manure should be practiced. Paved areas should also have good drainage to prevent pooling. Lots following these practices can accommodate one animal for each 80 $ft^2$ of lot.

In arid and semiarid regions, and on farms, feedlots often have earth floors. Here manure need not be removed so frequently inasmuch as moisture does not build up in the pen. Generally the manure dries and does not emit a putrefactive odor if it is not allowed to remain wet or to form a crust. Pooling or wet areas must not be allowed to develop particularly around mangers and water troughs.

As the dry manure builds up on the pen or corral surface, it may be kept relatively odorless by scarifying with a spring-tooth cultivator to allow penetration by sun and air. When the manure becomes too deep for adequate scarifying it should be removed (at least three times a year). It also has been found that odor levels following manure removal and scarifying can be kept to a minimum by spraying the lot with a 1% solution of potassium permanganate so that each treatment amounts to 20 lb $KMnO_4$ per acre (*13*). Earth-floor lots should allow 300-$ft^2$ floor area for each animal.

Stockyard pens should be subject to the same sanitation and space requirements as cattle feed yards. It is usually recommended that they be paved.

Another problem of unpaved corrals in dry areas is dusting if too much manure is allowed to build up. Cattle begin to move around ("play") in the early evening and stir up very fine manure dust. Areas downwind receive the brunt of the dust. Careful spraying of a limited amount of water into the lot by means of fog nozzles can correct the situation.

Nearly all feedlot operations have tried the use of odor counteractant sprays at one time or another. If the yard practices good sanitation methods, odor counteractants are useless and an unnecessary expense.

The same requirements for waste management apply to enclosed feeding facilities and barns, in general. Here it is particularly important that air from manure pits and other collection facilities be handled separately from barn air. In all cases, manure disposal presents problems (see Section I,G,4).

Zoning regulations are also an effective means of controlling odors. Swedish guidelines for animal farms of certain types require a minimum distance of 500 m from any settlement.

## 2. Hog and Sheep Pens and Small-Animal Farms

The keeping of hogs and sheep in an urban community is difficult to justify and is usually prohibited. Still, with housing tracts invading rural areas, hogs often become neighbors to suburbanites. Swine pens range in size from a few head to several thousand. Sheep yards contain as many as 18,000 head.

Usually sheep and hog odors are more offensive than cattle odors. Pens should be located indoors and be well-ventilated and lighted. Floors should be of concrete and have good drainage. The pens should be cleaned and manure removed on a daily basis. Only in this way can sheep and hog pens be acceptable.

Small-animal farms consist primarily of those where animals are raised for pelts, e.g., mink farms. Here, again, good waste management is necessary to control odors. At one mink farm, the sand floor is soaked with a 0.8% solution of potassium permanganate in water (1 liter of solution per square meter of sand). Mink feces falling on the soaked bed are deodorized.

## 3. Poultry and Egg Production

Generally, the raising of poultry and production of eggs are not important contributors to air pollution. Nevertheless two rather specialized problems have arisen, one in connection with poultry ranches, the other with egg production.

The poultry ranch problem is well illustrated by the troublesome problem of dust clouds around turkey ranches. Several hundred large flocks of turkeys (over 50,000 birds) are kept in the San Joaquin and Sacramento Valleys of California on relatively small acreages of land. These ranches soon become barren of vegetation, so at feeding time when thousands of turkeys run across the ground flapping their wings, great clouds of dust, feathers, and feces rise into the air. Under stable atmospheric conditions these clouds may persist for more than a day. Public health authorities are concerned not only with the dust nuisance, but the risk of allergic reactions and transmission of bacterial and virus infections. In some locations it has been claimed that dust from poultry ranches has settled on nearby vegetation and reduced yields.

At egg ranches, the problem is odor. The birds are kept inside houses in small cages. Manure droppings fall to the ground under the houses. A farm with 270,000 laying hens produces 35 to 40 tons of manure per day, and thus presents a major waste management problem. Even in much smaller houses, manure must be removed frequently (even continuously) unless it dries rapidly. Otherwise, flies are attracted and odors develop.

## 4. Manure Utilization

On the farm, manure is never considered as waste. Rather it is a valuable resource which should be recycled for beneficial use. Even feed-yard manure should be recycled to the land for its highest and best use. Because of the large quantities involved, stockpiling must be resorted to.

If adequate steps are taken to avoid water pollution, odor problems may be managed by keeping the piles dry and disturbing them only when the wind is not in the direction of nearby populated areas. In some communities, manure may be sold to commercial fertilizer manufacturers, who dry and market it for home use.

Some attention has been given to composting manure, and successful operations have been reported (*14, 15*). As a final resort, manure may be buried, incinerated, or degraded in lagoons.

## II. Processing Agricultural Products

The processing of agricultural products usually involves the treating of a crop or a farm-raised animal to produce food, feed, or fiber. The major potential air pollution problems are dust and odor emissions. Wherever a dry powder is produced or handled (e.g., in grain milling, various drying operations, and during mixing, conveying, and packaging) dust problems arise. In wet processing, odors from unavoidable decomposition of the materials handled are potential air pollutants. Such processes usually involve meat, fish, fruit, and vegetable processing.

### A. Grain Milling and Handling

All grain-milling operations involve disintegration of the dried grain followed by a series of sizing, separating, and mixing processes. Inasmuch as all these processes involve handling fine dust particles in an air stream, a potential problem always exists. Even before the milling operation begins, the grain must be delivered, unloaded, cleaned, stored, and brought from storage to the processing site.

Of all these operations, the dustiest is the loading and unloading of boxcars, ships, trucks, and barges. Proper control involves loading and unloading in enclosed or semienclosed areas operated under negative pressure to prevent escape of dusty air. Baffles to contain dust and to direct airflow are useful in lessening dust losses during unloading. Finally, any exhausted airstream must be sent through cyclone collectors or baghouses.

The use of special bulk feed and grain cars is highly recommended. These cars have discharge hopper bottoms, sectional compartments, and top spout openings, all of which reduce dust emissions.

Prior to processing, the grain is always cleaned by passing it over a series of scalpers (to remove tramp iron and other large objects), screens (to separate weed seeds, small stones, soil, etc.), and aspirators (where an airstream blows out chaff, grain hairs, pollen, mold spores, parts of

**Table II Dust Control Regulations for United States Grain-Processing Plants**[a]

| *Air pollution control agency* | *Allowable dust emission*[b] |
|---|---|
| Iowa | 0.1 grain/scf effluent air |
| Wisconsin | 0.4 lb/1000 lb of gas |
| Philadelphia, Pennsylvania | 0.02 grain/dry scf air or process-weight regulation (whichever is greater) |

[a] For jurisdictions which have adopted regulations specifically for grain-processing plants. In most States, grain-processing plants must meet process-weight regulations for industry in general.
[b] scf: Standard cubic foot.

insects, and other lightweight contaminants). Uncleaned grains usually contain 3% dockage, of which 90% is removed during the cleaning operation.

Cleaning leads to another major problem, disposal of the collected dust. Regardless of final disposition by the salvager, it must be loaded into dust cars, a difficult and extremely dusty operation. Again, cyclones and baghouses are required if dust is to be controlled. Baghouse air–cloth ratios for this application range from 3 to 5 $ft^3$/minute/$ft^2$ of cloth.

Dust emissions from uncontrolled grain elevators have been estimated to be 27 lb per ton of grain (*16*). Emissions from controlled terminal elevators have been estimated as: shipping and receiving—1 lb per ton; transferring and conveying—2 lb per ton; screening and cleaning—5 lb per ton; drying—6 lb per ton. Shipping and receiving losses at country elevators may be five times greater; emissions from other operations at country elevators are 15 to 60% greater than at terminal elevators.

Usually grain-processing operations must meet general process-weight regulations for stationary sources. In a few instances, grain-processing plants are subject to specific regulations (Table II).

### 1. Milling of Wheat

The milling of wheat flour entails the following steps after the grain is cleaned (*17*): (a) blending of various grades of wheat in a mixer; (b) scalping, screening, and aspirating the grain for a final cleaning; (c) washing grain to toughen the outside bran layer so it will separate easily; (d) rolling, screening, and sifting grain in a series of operations to yield flour; (e) additional processing as desired, e.g., bleaching and enrichment; and (f) packaging.

The only air pollution problem of any consequence is the movement of the raw grain, as previously discussed. All emissions from the milling operation are well contained within the building itself by use of a closed cleaning system and the recycle of air from cloth-type filters (air–cloth ratios 3:1 to 4:1). The chief dust problem specific to the flour-milling operation is the loading of chaff and middlings into rail cars.

## 2. Milling of Other Grains

Secondary to the milling of wheat is the milling of a variety of other grains, e.g., corn, rice, oats, barley, buckwheat, and rye. As in the case with wheat, the major dusty operations are loading and unloading cars and cleaning the grain. Buckwheat and rye are milled in a manner quite similar to wheat. The other grains are processed by modified methods directed to separating the hulls, germ, and bran from the endosperm. In the case of rice, an endosperm of maximum grain size is desired. Inasmuch as the processing of all these grains involves grinding and separation of fine or lightweight material, cyclone separators are attached to nearly every piece of separation equipment. If the dust loss becomes a nuisance, the cyclones are backed up with cloth collectors.

Barley malting is a specialized process in which cleaned barley grain is germinated to increase its content of the enzyme, diastase. When the desired amount of diastase has formed, the germination is stopped by drying the malt in a rotary kiln. The dried malt is separated from sprouts and other waste material and shipped to breweries or food manufacturers. Even though the grain is cleaned prior to malting, continual handling of the malting grain releases dust all through the process. It is captured in cyclone collectors and sold as animal feed.

The wet milling of corn yields cornstarch and its derivatives as end products. It is called wet milling because the corn is soaked (steeped) in a water solution of sulfur dioxide prior to the subsequent separations, and because water is used in large quantities in nearly every step of the process. Grain handling and finished product drying and blending are the only dust problems of note. At one time, sulfur dioxide emissions were a problem, but modern separation processes have reduced losses to almost nil.

In some of the older corn wet-milling plants, odors from the gluten settlers were a problem. Gluten was concentrated by settling in open tanks prior to filtration and drying. When the gluten was allowed to stand too long in the wet state or if spillage was not controlled, putrid odors

**Table III Emission Factors for Grain-Processing Operations (*4*)**

| *Operation* | *lb/ton* | *kg/MT*[a] |
|---|---|---|
| Corn meal manufacturing | 5 | 2 |
| Soybean processing | 7 | 3 |
| Barley or wheat cleaning | 0.2 | 0.1 |
| Milo cleaning | 0.4 | 0.2 |
| Barley flour milling | 3 | 1.25 |
| Barley feed manufacturing | 3 | 1.25 |

[a] MT = Metric ton.

developed. Rapid handling of gluten liquors and good sanitation practice has solved the problem in modern plants.

Particulate emission factors for selected grain-processing operations are shown in Table III.

## 3. Drying and Mixing Operations

Whenever a powder or dry crystalline product is desired as the result of a wet process, the final operation is drying. In the corn wet-milling process, for instance, the final wet products are a starch cake and a gluten cake. Similar cakes result from the wet processing of grain sorghum or various roots, e.g., potatoes, cassava (tapioca), and sago.

In any event, washed starch (or gluten) is fed to a rotary or belt dryer where it meets warm air usually in a countercurrent fashion. Some fine dried particles are picked up by the airstream and blown from the dryer. Usually the product is sufficiently valuable to warrant capture in a cyclone or bag collector. Similar equipment is used in the drying of sucrose (cane sugar) and dextrose (corn sugar).

In the past 15 years, spray drying operations have become widespread in the food industry. Here a water slurry or solution is sprayed under considerable pressure into a chamber where dry air evaporates the moisture from the spray droplets before they meet the side or floor of the chamber. The dryer is so designed and operated that the resulting dried powder is uniform is size, denatured only slightly, if at all, and readily soluble. Among the more common products are milk powder, coffee, tea, corn syrup solids, starches, potatoes, eggs, cheese, fruit and vegetable powders.

All dryers use cyclone collectors to collect the fine particles that do not settle in the dryer chamber, which itself is usually conical in shape. Dust losses are normally less than 0.1% of product. In some jurisdictions, process-weight regulations require even greater efficiency, in which case the cyclone must be backed up with a cloth collector, unless the material is hygroscopic. If this is the case a wet scrubber may be required.

Highly restrictive emission requirements have created difficult economic problems in those operations where large excesses of drying air are required to prevent deposition of moisture in fabric filters, or to prevent dust explosions. In either case, the size of the required fabric filter or scrubber may well be greatly out of proportion to the size of the basic dryer.

After a product is dried, it is often blended with other dry products to form a dry food mix. The manufacture of dry food mixes which require a minimum of preparation before serving is a rapidly expanding business. Among these convenience foods are cake and cookie mixes, puddings, gelatin desserts, pie crusts, hot and cold cereals, soft drink powder, soups, dehydrated vegetables, etc. Processing entails drying, grinding, and pulverization, mixing, conveying and packaging, all operations that produce dust (usually 10–100-$\mu$m particle size). For economic reasons dust recovery is almost universal. Both cyclones and cloth collectors are used. Recommended air–cloth ratios for various food products are shown in Table IV.

## 4. Feed Manufacture

a. Mixed Feeds. Nearly all grain-handling and grain-processing operations involve by-product material which eventually ends up in animal feed, e.g., cracked grain from elevators and cleaning operations; bran middlings and shorts from flour milling; bran and fiber from the milling of minor grains; gluten and steepwater solids from corn wet milling; spent grain and distillers' solubles from the fermentation industry; beet pulp from the beet-sugar industry. In addition, by-products of the meat packing and milk industries also end up in animal feeds. Alfalfa, molasses, calcium salts, and phosphates are further constituents.

Unloading the bulk grains and grain products from box cars and transporting them to bins are the chief dust-producing operations of the industry (Fig. 2). Alfalfa is particularly dusty and must be handled carefully. After the raw materials are unloaded, the grains and other granular materials are ground in a hammer mill, then mixed—usually in continuous operations. When molasses is added, a batch mixing process is normally used. In either case the mixtures are later moistened with steam and forced through dies to produce pellets. Both the grinding and mixing operations are dusty, but the dust may be adequately controlled with cyclones or cloth collectors.

**Table IV Recommended Air–Cloth Ratios for Food Products**

| *Material* | *Ratio of cfm of air to ft² of cloth*[a] *Recommended* | *Range* |
|---|---|---|
| Alfalfa dust (cold) | 6:1 | — |
| Beef scrap dust (dried) | 5:1 | 4:1 to 6:1 |
| Blood (animal, dried) | 5:1 | 4:1 to 6:1 |
| Coffee (ground) | 4:1 | 3:1 to 5:1 |
| Coffee (spray-dried) | 3.5:1 | 2:1 to 4:1 |
| Cornstarch | 2.5:1 | 2:1 to 3:1 |
| Cottonseed meal | 4:1 | 3:1 to 5:1 |
| Egg albumen (dried) | 3:1 | 3:1 to 4:1 |
| Feed | 4:1 | 4:1 |
| Fiber | 4:1 | 4:1 to 5:1 |
| Flour (roll suction) | 3:1 | 3:1 |
| (other) | 4:1 | 4:1 |
| Grain dust | 4:1 | 3:1 to 5:1 |
| (elevator) | 5:1 | 4:1 to 6:1 |
| Milk (powdered, whole) | 2.5:1 | 2.5:1 |
| (powdered, nonfat) | 3:1 | 3:1 |
| Soybean flour | 3:1 | 3:1 |
| Sugar | 2:1 | 2:1 |
| Yeast (dried) | 3:1 | 3:1 |

[a] cfm: Cubic feet/minute.

The principal problem during pelleting is emission of odors. The odors are particularly strong and unpleasant when feeds contain large quantities of tankage and bone meal. The only solution is treating the exhausted air in scrubbers or incinerators, or passing it through a bed of activated charcoal.

b. By-Product Feedstuffs. Air pollution problems accompanying the recovery of by-product feedstuffs are those inherent in the manufacture of the main food product. Whenever the by-product is high in protein value, it is dried, ground, and sold as a high-protein feed, or mixed with low-protein by-products (chaff, hulls, bran, etc.) to increase the food value of the latter. Among the more important high-protein feeds are corn gluten meal, oil-seed meals, dried distillers' solubles, fish meal, and various animal residues (blood, tankage, etc.).

Corn gluten meal is a by-product of the wet milling of corn, a process in which whole grain corn is steeped in a dilute sulfurous acid solution and subsequently separated into starch, fiber, gluten, and corn germ. The separated gluten is concentrated and dried to produce a corn gluten meal.

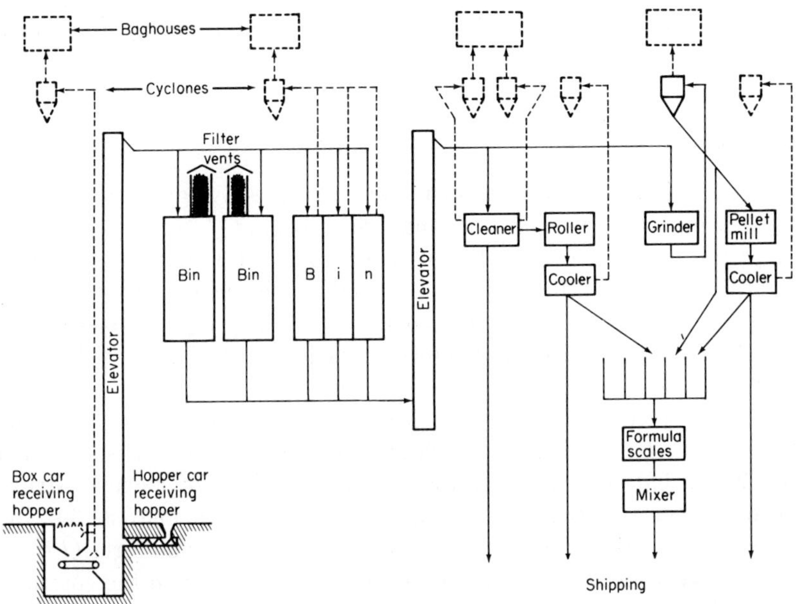


Figure 2. Flow diagram of a simplified feed mill. Basic equipment in solid lines; dust control equipment in dotted lines (*26*).

Some of the gluten meal is mixed with coarse and fine fiber (hulls and bran) and other low-protein by-products to make a lower grade feed than the meal itself. Drying and handling of the final feed produces dust which may be controlled by cyclone collectors, or cloth collectors when necessary.

Oil-seed meals, e.g., cottonseed meal, flaxseed meal, safflower-seed meal, are by-products of the recovery of vegetable oils by pressing or solvent extraction. Excessive loss of solvent vapors during extraction or subsequent steaming of the meal may produce a localized odor problem. Usually the solvent is valuable enough to warrant recovery by condensation or adsorption on activated charcoal. Subsequent processing of the oil may release odorous compounds.

Brewers' grains, distillers' dried grains, and distillers' solubles are all by-products of the brewing and distilling industry. Air pollution may arise from (1) raw material preparation, (2) the fermentation process, and (3) solid and liquid waste disposal. Fermentation odors are usually not strong enough to be objectionable. If they are, vent gases or air from the vicinity of the source may be passed through scrubbers or activated carbon adsorbers. Dust and odors from feed dryers may be controlled by means previously described. However, the economics of feed drying are even more formidable than those of wet-starch drying (see Section II,A,2).

### B. Cotton Processing

The processing of cotton fiber to cloth involves five distinct steps: harvesting, ginning, spinning, weaving, and finishing. Although at one time harvesting was done entirely by hand, today it is done by machines. Mechanical harvesting, either by spindle-type pickers or by strippers, removes not only the seed-cotton (fibers still adhering to the seed) from the plant but also varying amounts of trash (leaves, burrs, sticks, stems, dust, and dirt).

Machine-picked cotton usually contains about 80 lb of foreign matter (29 lb of hulls, 43 lb of leaf trash and dirt, and 9 lb of sticks and stems) per 500-lb bale of ginned cotton. An average bale of machine-stripped cotton is much dirtier; it contains 525 lb of foreign matter (397 lb of hulls, 50 lb of sticks and stems, and 78 lb of leaf trash and dirt) (*18*).

The picked or stripped cotton is delivered by truck to the cotton gin. Originally the purpose of the cotton gin was to remove the cotton fiber from the seed. Now an equally important purpose is also to separate the trash. Thus the products of the modern gin are trash, cottonseed, and cotton fiber.

The cotton gin entails not only the gin stand itself, but also various trash removal units and a dryer (to combat moisture). All streams are handled pneumatically and can produce considerable organic particulate matter to which some people are violently allergic. Adhering pesticide residues also present a problem. In fact arsenic emissions from cotton gins have been estimated at from 27,000 to 6,360,000 $\mu$g/bale (*18*).

The heavier trash may be collected in cyclones and dropped into closed screw conveyors leading to hoppers. Disposal of the trash in the past has caused a secondary air pollution problem when trash was incinerated. Because of growing restrictions on incineration, more and more trash is being returned to the land or to compost piles.

Finely divided fly lint, leaf trash, and dust may be collected in small-diameter cyclones backed up by screen filters, in which the fly lint serves as a filter medium, or cloth filters. Estimated emission factors for cotton-ginning operations are shown in Table V. With adequate collection, emissions may be 90 + % less than shown in the table.

In most United States jurisdictions, gins are subjected to general process-weight regulations and to equivalent opacity requirements. Alabama has a special process-weight regulation for cotton gins, which allows emissions 50 to 100% greater than the general process-weight regulation limit.

**Table V Emission Factors for Particulate Matter from Cotton Ginning Operations without Controls. One Bale Weighs 500 lb (*4*)**

| *Process* | *lb/bale* | *kg/bale* |
|---|---|---|
| Unloading fan | 5 | 2.2 |
| Cleaner | 1 | 0.45 |
| Stick and burr machine | 3 | 1.35 |
| Miscellaneous | 3 | 1.35 |
| | 12 | 5.35 |

In the textile mill, cotton processing consists primarily of spinning, weaving, and finishing. Spinning involves various procedures for opening and cleaning the baled cotton received from the gin and spinning it into yarn. The lint and dust released in the operations associated with these processes are picked up by an exhaust system and conveyed to a centralized filter system, where the lint is removed by a cotton condenser and the dust by one of several types of dust collectors. One type, which uses a traveling paper filter medium mounted on a stainless-steel screen, has an operating air resistance for single-ply paper media of 1-in. water gauge (WG) at 300 to 600 ft/minute (fpm). Another type uses a rotating drum filter with a woven filter medium. About 5 to 8 lb of salable waste lint per 100 lb of cotton processed may be recovered (*19*). Excessive cotton dust in textile mill air produces the occupational disease called "byssinosis."

Dyeing and finishing operations may, at times, cause odor complaints in the vicinity of the mill from evolved formaldehyde, acetic acid, and other organics. If the odor becomes a nuisance it may be alleviated by the usual odor control methods.

## C. Meat and Meat Products

### 1. Meat Packing

Meat packing plants or abattoirs usually consist of ten sections: (a) holding pens and yards; (b) killing floor; (c) hide room; (d) casing room; (e) paunch and intestinal content removal area; (f) coolers; (g) trimming and boning room; (h) pickle room; (i) by-product processing area; and (j) smoke-oven operations.

The control of odors from holding pens and yards are similar to those discussed in Section I,G,1. Some odors arise from the hide room, casing room, and paunch removal area. Odors from the casing area are from manure and the sour, partly digested foodstuffs of the intestine. Further decomposition must be prevented by good sanitation practices. The air from the room may also be treated before it is exhausted to the atmosphere. Usually a water wash is adequate. Scrubbing the air with a potassium permanganate solution is effective but expensive.

Waste water treatment facilities may also produce obnoxious odors if good sanitation methods are not practiced. Common causes of odors are decomposition of screened solids (if not disposed of daily), low flow rates in trickling filters, and spring thawing of stabilization ponds when the ice cover first disappears.

## 2. Inedible Fat Rendering

By-product processing presents the most difficult of all the odor problems of the meat industry, and the paramount process of all is inedible fat rendering. The fat rendering industry is for good reason widely noted for its odor problems. The process involves the application of heat to the raw material (meat scrap, intestines, bones, etc.) to remove water, disintegrate the bone tissue, and release fat or tallow.

In a typical rendering system (Fig. 3) (*20*), meat scrap and other animal residues are ground to a hamburgerlike consistency and charged to horizontal, steam-jacketed kettles. During the cooking process, fat cells are broken down to water, grease, and proteinaceous solids. The solids are then separated from the grease, which may subsequently be purified to produce tallow. Odors are emitted through the entire process, beginning with the raw material storage bins.

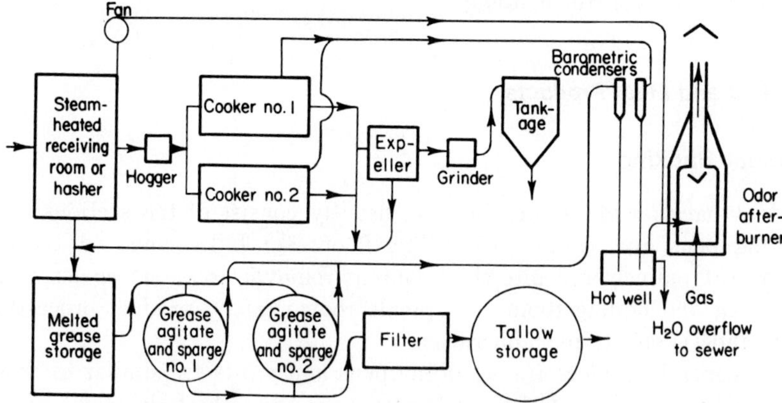


Figure 3. Flow sheet of typical fat rendering plant (*20*).

Chromatographic analysis of the gases evolved in a typical rendering plant show that the odorous materials include various sulfides, disulfides, mercaptans, $C_4$ to $C_7$ aldehydes, trimethylamine and various $C_4$ amines, quinoline, dimethyl and other pyrazines, $C_3$ to $C_6$ acids, and less odorous compounds (*21*).

The greatest odor sources in a conventional batch-cooker rendering plant are the cooker noncondensables and screw-press-vent gases which range from 5000 up to 100,000 odor units (ASTM method) depending on the raw material processed. Percolation-pan odors range from 500 to 5000 odor units (ASTM method) in the ambient air immediately above the percolation pans. Raw material receiving and grinding operations are not significant sources of odors if the material is processed within 8 to 12 hours of receipt, and adequate housekeeping practices are adhered to.

In modern practice, cooker exhaust is cooled by air condensers instead of barometric condensers (to reduce water pollution control costs). The air condensers are designed to reduce condensate temperatures to 100–120°F during the summer months. The noncondensables may then be destroyed by incineration or chemical scrubbing. Screw-press-vent gases are also incinerated or scrubbed. Although incineration is required by some local ordinances, it is being replaced by scrubbing because of the shortage and higher cost of natural gas and of sulfur-free fuel oil.

Incineration is most effective in handling relatively small volumes (up to 1000 or 2000 cfm) of concentrated odors, unless heat can be recovered and used advantageously. In a few plants, cooker noncondensables are incinerated in the plant steam boiler, if it satisfies the necessary combustion design criteria. Wet scrubbers with circulating chemical solutions have been shown to involve lower costs than incineration and to be equally effective. Recent studies (*22*) have shown sodium hypochlorite to be the most effective solution overall (96 to 99% efficient in reducing odors). In some cases, potassium permanganate gives better results. At least one plant uses soda ash effectively.

Even with efficient collection of cooker and screw-press-vent gases, odors from the percolation pan, spills, and minor vents go directly into in-plant air. Odor-masking agents have been widely used in the past but stricter odor regulations have resulted in all exhausted ventilation air being treated by scrubbing or adsorption on activated charcoal. To be of maximum effectiveness, the plant should be in a totally enclosed building. Relative cost data for odor controls for rendering plants are shown in Table VI (*23*).

The most recent trend in the industry is toward continuous rendering systems. The cooker is fed continuously, lower temperatures are used, and the tallow is separated from ground bone and fiber by centrifugation.

**Table VI Relative Costs of Odor Control Methods** (*23*)

| *Method* | *\$/1000 cfm/hour*[a] |
|---|---|
| Fume incineration[b] | 0.83–1.55 |
| Catalytic incineration[b] | 0.83–1.46 |
| Packed-tower scrubber | 0.19–2.50 |
| Activated carbon (ventilation air only) | 0.11 |

[a] cfm: Cubic feet/minute.
[b] Without heat recovery which may reduce costs up to 50%.

Pressing may be obviated by recycling centrifuge solids. In any case, noncondensable gases may be deodorized by either incineration or scrubbing.

## 3. Smoke Ovens

Smoke from ovens and smoke houses used for curing meat is sometimes objectionable because of both smoke and odor. In Los Angeles, California, regulations are extremely strict, and smokehouses cannot comply without the addition of smoke and odor control equipment. One installation required a water scrubber, a low-voltage electrostatic precipitator, and an afterburner. In this case, the air pollution control equipment cost \$42,000 whereas the basic oven cost only \$18,000. In many operations a water scrubber followed by an afterburner is satisfactory. Emission factors for meat smoking are shown in Table VII.

The high cost of smokehouse control may well lead to abandonment of present practices in the domestic smoking of meat. Imposition of 20% plume opacity regulations, as threatened in some quarters may hasten the day. The usual smoking operation may be superseded by liquid smoking which appears to be suitable for some, but not all, meat products.

**Table VII Emission Factors for Meat Smoking (Uncontrolled). Based on 110 lbs of Meat Smoked per lb of Wood Burned** (*4*)

| *Pollutant* | *lb/ton of meat* | *kg/metric ton of meat* |
|---|---|---|
| Particulate matter | 3.0 | 0.15 |
| Carbon monoxide | 0.6 | 0.3 |
| Hydrocarbons ($CH_4$) | 0.07 | 0.035 |
| Aldehydes (HCHO) | 0.08 | 0.04 |
| Organic acids (acetic) | 0.2 | 0.10 |

### D. Fish Processing

The canning, dehydration, and smoking of fish, and the manufacture of fish oil and fish meal are important segments of the fish industry. Odors emanate from the fishing boats and raw storage, and through all processing steps, e.g., cooking, pressing, screening, centrifuging, drying, and waste disposal. The principal malodorants are acrolein, oil decomposition products, hydrogen sulfide, ammonia, butyric and valeric acids, and trimethylamine (*24*).

Odor control entails passing gaseous process effluents and exhaust air from processing areas through control equipment which will either destroy the odorous gas, absorb, or adsorb it. Some of the methods used are activated charcoal adsorbers, scrubbing with chlorinated water or some other oxidizing solution, catalytic combustion, incineration, and the use of masking and counteraction agents.

In Europe, water scrubbing is the principal control method in fish meal plants. If the scrubbed gases are low in volume they are sent through a boiler; if high in volume, chlorine is added. Odor counteractants are also used. Failure to use incineration in European practice probably reflects the high cost of fuel. Generally, the use of chlorine is not favored for fish meal plants in the United States, but chlorine dioxide at 100–150 ppm in scrubber water is said to be effective. Nevertheless, one tuna processing plant in California is effectively using chlorinated seawater to treat cooker effluent gases. Emission factors for fish meal processing are shown in Table VIII. Fish smoking is subject to the same problems as is the smoking of meat (Section II,C,3).

**Table VIII Emission Factors for Fish-Meal Processing (*4*)**

| | *Particulates* | | *Trimethylamine* $(CH_3)_3N$ | | *Hydrogen sulfide* $(H_2S)$ | |
|---|---|---|---|---|---|---|
| *Emission source* | *lb/ton* | *kg/MT*[a] | *lb/ton* | *kg/MT*[a] | *lb/ton* | *kg/MT*[a] |
| Cookers, lb/ton (kg/MT) of fish produced | | | | | | |
| Fresh fish | — | — | 0.3 | 0.15 | 0.01 | 0.005 |
| Stale fish | — | — | 3.5 | 1.75 | 0.2 | 0.10 |
| Driers, lb/ton (kg/MT) of fish scrap | 0.1 | 0.05 | — | — | — | — |

[a] MT = Metric ton.

In many areas fish waste is still dumped into the harbor nearby. Often current flow is sufficient to carry the waste out to sea for dispersal, but wherever stagnant or estuarine conditions exist, solids settle to the bottom, break down anaerobically, and emit highly odorous gases. Fortunately, more and more municipalities are requiring plants to treat their liquid wastes before disposal.

### E. Fruit and Vegetable Processing

The most important processes for the preservation of fruits and vegetables are canning, dehydration, and quick-freezing. The only important air pollution problem is related to the disposal of hulls, leaves, rinds, pods, cuttings, etc. If held too long these materials decay and produce revolting odors.

Water pollution regulations prohibit sending waste to streams unless it has been processed to reduce biological oxygen demand (BOD). One method commonly used to reduce BOD is digestion of the cannery waste in either anaerobic or aerobic lagoons. Even the best-operated lagoons have upsets during which odorous gases (hydrogen sulfide, mercaptans, fatty acids, amines and other nitrogenous compounds, or masking agents) may be released on the downwind side of the lagoon. If the condition persists, addition of a nutrient, precipitation of excess sulfide, or reestablishment of the active organism may be required. Some food processors have replaced lagoons with spray irrigation plants to advantage. The National Canners Association is studying methods of composting (*25*). Leaves, stalks, and cuttings can sometimes be disposed of odorlessly in adequately designed incinerators. Sanitary landfill is an alternate method.

Special situations arise when processing certain fruits and vegetables by specific methods. Thus corn driers often emit "beeswing" chaff and dust (from shelling the corn). The drying of peaches involves the use of sulfur dioxide, some of which escapes to the atmosphere. Both of these are comparatively minor emissions and may be handled by proper zoning. In plants extracting castor beans, castor bean dust must be minimized. The ricinin in the dust may provoke bronchial asthma.

### F. Cooking Processes

The cooking of meat and vegetables by baking, boiling, frying, etc., may give rise to either smoke or odors or both. But it is only in large-scale operations that emissions are significant. Usually the preparation of food, both indoors and outdoors, by the consumer or small groups of consumers is exempt from regulatory requirements. In practice this exemption has usually been extended to restaurants, but control over the more flagrant emitters may well be expected in the future.

Large commercial kitchens and manufacturing establishments, e.g., potato chip producers, fish fryers, meat canners, and soup canners, may produce odors (from burnt fat and protein) in sufficient volume that they will carry a long way when the atmosphere is stable. The usual methods of odor control are adaptable. One large Los Angeles deep-fat-fryer controls smoke with an electrostatic precipitator.

One plant where fish is fried in oil uses a masking agent successfully. A meat-cooking plant passes odorous gases into a scrubbing tower counter current to a solution of 1–2% potassium permanganate buffered with borax to pH 8.5. The solution is recycled. A 90% reduction in total organics is claimed. A potato chip company captures odor-laden air by means of a hood over the fryer, reclaims oil mist in a heat exchanger and passes the oil-free odor-laden air into the furnace of the fryer for odor destruction.

The evolution of smoke from burning fat is usually unnecessary. Either the fire is too hot or the grease trap is poorly designed. Where wood fires are used for cooking, opacity regulations can ordinarily be met by proper firing procedures.

## G. *Coffee Roasting*

Control of coffee roasters (*27*) is required not only because of their characteristic odor emissions but also because of their emission of smoke and particulate matter (dust and chaff). In a typical coffee roasting operation (Fig. 4), green coffee beans are freed of dust and trash by dropping the beans in a current of air. The dust and chaff removed may be recovered in a cyclone collector. The cleaned beans are air blended and

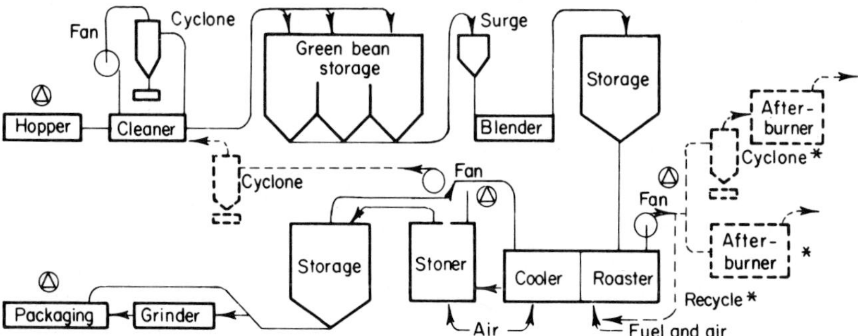


Figure 4. Flow diagram of coffee roasting plant. Ⓐ, points of emissions; dashed line, usual or possible control; *, alternate control possibilities.

**Table IX Particulate Matter Emissions from Coffee Roasting Processes (No Controls) (27)**

| *Processes* | *Particulate matter* | |
|---|---|---|
| | *lb/ton* | *kg/metric ton* |
| Roaster | | |
| Direct fired | 7.6 | 3.8 |
| Indirect fired | 4.2 | 2.1 |
| Stoner and cooler | 1.4 | 0.7 |
| Instant coffee spray dryer[a] | 1.4 | 0.7 |

[a] Assumes use of cyclone and scrubber; cyclones will reduce emissions from other processes by 70%.

sent to a batch or continuous roaster where they contact hot combustion gases. Time and temperature are important variables. During the roasting, moisture is driven off, the beans swell, and certain chemical changes occur that give the roasted beans their typical color and aroma. When the roasting has reached a certain color the beans are quenched, cooled, and stoned (heavy particles removed in an airstream). Emissions from the roasting process are shown in Table IX. As indicated in the table, cyclone collectors are adequate to control particulate emissions.

Smoke and odors pass through the cyclones. In one "smokeless roaster," a damper system recirculates the combustion gases through the gas flame of the roaster to eliminate smoke and odor. Additional fuel consumption of 40% is required. Smoke and odor control may also be accomplished by an afterburner in the roaster stack, but fuel requirements are increased 100–150% over that of a conventional roaster.

Similar dust and odor problems may be expected in the manufacture of instant (spray-dried) coffee, and in the incineration of spent coffee grounds. Cyclones, wet scrubbers, and afterburners are adequate for control.

### H. Spices and Flavoring Materials

The milling of a variety of barks, bulbs, leaves, buds, fruits, roots, and seeds to produce spices and condiments can lead to strong and pungent odors. Subsequent mixing and packaging operations also contribute their share. These operations are carried out in an isolated portion of the food processing plant to prevent odors from contaminating other food products. Exhaust air is commonly run through activated carbon filters to abate the odor present.

Dust control in a modern spice plant may consist of independent high-volume air-handling systems to pick up dust in the plant's processing and packaging rooms. At each grinder, there are two nozzles at points where dust might escape. The dust-laden air is carried through stainless-steel risers to galvanized horizontal ducts, suspended from the ceiling at an incline (to allow flushing with water) and thence to a dust collector.

A wide variety of flavoring materials, usually oils, fats and their extracts, have noticeable odor. These include such materials as peppermint oil, lemon oil, vanilla, chocolate and many others. During their preparation, which may involve distillation, extraction, maceration, or expression, some odorous material may be released to the atmosphere. However, unless the operation is quite large, the odors dissipate within a few yards of the source. In very large operations, e.g., chocolate factories and some candy factories, the odorous air must be treated prior to emission to the atmosphere.

### *I. Miscellaneous and Allied Industries*

In addition to food and feed, a variety of agricultural products and food industry by-products are converted to other articles of commerce. Wherever spoilage (protein and carbohydrate decomposition) takes place there is a potential for an odor problem. Wherever the material is ground, or where dust or other fine material is handled or processed, there is potential for a dust problem. The problems may be abated by methods similar to those described for food and feed products. The more important agricultural process operations where a potential for air pollution exists include glue and gelatin manufacturing; leather tanning, particularly liming and bating; curing of hides and pelts; wool scouring and combing; degreasing; hair recovery and cleaning; feather utilization; tobacco curing and the manufacturing of tobacco products.

Leather tanning is one of the oldest industries in the world, and is still characterized by its many small producers. Cattle-hide tanning dominates the industry, although there is some sheepskin tanning.

The chief air pollutants arising from the several operations in the tannery are reduced sulfides, protein derivatives, organic sulfides, and organic acids, principally the highly repelling compound, caproic acid. All are highly odorous.

Sulfides may be removed from emitted air streams by treatment with activated charcoal; caproic acid by incineration. Considerable dust arises from shaving and buffing leather. It may be controlled by a wet dust collector.

Animal and fish glue manufacture entail the hydrolysis of low-purity waste products of meat packers and tanners and the waste from fish canneries. Inasmuch as these wastes are proteinaceous in character, odorous decomposition products are evolved during cooking and drying operations. Scrubbing effluent gases with a solution of sodium hypochlorite is said to be effective in reducing odors.

## III. Episode Control Planning for Agricultural and Agricultural-Products Processing Operations

Agricultural operations are usually not required to draw up emergency plans for use during air pollution episodes. Nevertheless all air pollution episode regulations call for cessation of all open burning, including land clearing, even at the first stage (usually called the "alert" stage) of the emergency. Unnecessary operation of motor vehicles is also requested. No good citizen will ignore the request.

Agricultural-product processing operations have a low priority in overall episode control plans, simply because they are not important factors in the buildup of pollution during periods of air stagnation. The only portion of the agricultural processing industry usually required to draw up emission reduction plans is the grain industry. However, even where the grain industry and other agricultural-product processing operations are not specifically mentioned, they are required to conform to general regulations which usually require the following actions:

First or "alert" stage (voluntary action only): no open burning, no incinerator operation except from noon to 4 p.m., no boiler lancing or soot blowing except from noon to 4 p.m., and no unnecessary operation of motor vehicles.

Second or "warning" stage: all first stage actions become mandatory; incinerators may not be used at any time; all process steam generating facilities must switch to fuel of lowest available ash and sulfur content.

Third or "emergency" stage: all processes that have not been required previously to submit air pollution emergency plans to authorities must cease operations; heat and steam demands must be reduced to absolute necessity consistent with preventing equipment damage.

Industries specifically mentioned in the regulations and which require only short lead times for shutdown, e.g., the grain industry, must follow the specific plans for emergencies as previously approved by the air pollution control authorities. Usually these plans specify how the shutdown of a few operations in an industry must be carried out to prevent equipment damage or a buildup of hazardous conditions.

In Florida, the grain feed industry is specifically mentioned. At the first stage, or alert level, of the episode, "maximum reduction of fugitive dust by curtailing, postponing, or deferring bulk handling operations" is required. At subsequent episode levels, fugitive dust from this source must be eliminated.

## REFERENCES


1. A. E. Vandegrift and L. J. Shannon, "Particulate Pollutant Emissions Study," Vol. I, Air Pollution Control Office Contract CPA-22-69-104. Midwest Research Institute, Kansas City, Missouri, 1971.
2. Agriculture Posts Waste Problems, *Environ. Sci. Technol.* **4,** 1098–1100 (1970).
3. C. W. Stanley, J. E. Barney II, M. R. Helton, and A. R. Yobs, *Environ. Sci. Technol.* **5,** 430–435 (1971).
4. Office of Air Programs, "Compilation of Air Pollutant Emission Factors" (revised), Publ. No. AP-42. United States Environmental Protection Agency, Research Triangle Park, North Carolina, 1972.
5. W. L. Faith, D. B. Keyes, and R. L. Clark, "Industrial Chemicals," 2nd ed., pp. 404–407. Wiley, New York, New York, 1957.
6. J. B. Dobie, P. S. Parsons, and R. G. Curley, Pap. No. 72-115, Annu. Meet. Amer. Soc. Agr. Eng., 1972.

6a. Bulletin 6, No. 2, pp. 2 and 4. California Air Resources Board, Sacramento, California, 1975.

7. Oregon Administrative Rules, Chapter 340, Department of Environmental Quality, Air Pollution Control, Division 2, Subdivision 6, Salem, Oregon, 1971.
8. "Rules and Regulations," Rule 130. Los Angeles County Air Pollution Control District, Los Angeles, California, 1972.
9. "Rules and Regulations," Rule 131. San Bernardino County Air Pollution Control District, San Bernardino, California, 1970.
10. R. L. Stockman and P. W. Hildebrandt, "The Air Pollution Aspects of Orchard Heating in the Yakima, Washington Area." Washington State Dept. of Health, Seattle, Washington, 1961.
11. Public Health Service, U.S. Dept. of Health, Education, and Welfare, Cincinnati, Ohio, *Robert A. Taft Sanit. Eng. Cent., Tech. Rep.* **A60-4** (1960).
12. J. C. Annis, V. E. Headley, and S. L. Lima, *J. Air Pollut. Contr. Ass.* **20,** 23–30 (1970).
13. W. L. Faith, *J. Air Pollut. Contr. Ass.* **14,** 459–460 (1964).
14. J. S. Wiley, *Compost Sci.* **5,** No. 2, 15–16 (1964).
15. R. C. Hartman, *Compost Sci.* **4,** No. 1, 26–28 (1963).
16. A. E. Vandegrift, L. J. Shannon, E. E. Sallee, P. G. Gorman, and W. R. Park, *J. Air Pollut. Contr. Ass.* **21,** 321–328 (1971).
17. M. E. McLouth and H. J. Paulus, *J. Air Pollut. Contr. Ass.* **11,** 313–317 (1961).
18. V. P. Moore and O. L. McCaskill, Publ. No. 999-AP-31. U.S. Pub. Health Serv., U.S. Dept. of Health, Education, and Welfare, Washington, D.C., 1967.
19. H. A. Schlesinger, E. F. Dul, and T. A. Fridy, Jr, *in* "Industrial Pollution Control Handbook" (H. F. Lund, ed.), Chapter 15. McGraw-Hill, New York, New York, 1971.

20. "Control of Stationary Sources," Vol. 1, Tech Progr. Rep. Los Angeles Air Pollution Control District, Los Angeles, California, 1960.
21. "Investigation of Odor Control in the Rendering Industry," Table 30, Rep. No. IITRI-C8210-15. Fats and Proteins Research Foundation, Inc., Des Plaines, Illinois, 1972.
22. "Investigation of Odor Control in the Rendering Industry," Rep. No. IITRI-C8210-15. Fats and Proteins Research Foundation, Inc., Des Plaines, Illinois, 1972.
23. "Investigation of Odor Control in the Rendering Industry," Appendix B. Fats and Proteins Research Foundation, Inc., Des Plaines, Illinois, 1972.
24. L. C. Mandell, Paper, Annu. Meet. Air Pollution Control Association, Pittsburgh, Pennsylvania, 1961.
25. W. A. Mercer and W. W. Rose, *Compost Sci.* **9**, No. 3 (1968).
26. J. A. Danielson, ed., Publ. No. 999-AP-40, p. 353. Public Health Service, U.S. Dept. of Health, Education and Welfare, Washington, D.C., 1967.
27. F. Partee, Publ. No. 999-AP-9. Public Health Service, U.S. Dept. of Health, Education, and Welfare, Cincinnati, Ohio (1966).

# 16

## The Forest Products Industry

E. R. Hendrickson

## I. The Forest Products Industry

The forest products industry is comprised of a broad spectrum of operations based mainly on utilization of harvested timber and wood residuals. The operations which are involved range from cutting and removing the timber from the forest to productive utilization of wood wastes. In this chapter, the air quality aspects of a number of operations will be considered which span several Standard Industrial Classifications. Forest products operations may include timber harvesting and transportation, saw mills and planing mills, veneer and plywood mills, hardboard and insulation board mills, particleboard mills, pulp and paper mills, and whole wood product mills. Not all of these operations have air quality aspects.

Although originally the industry was noted for a multiplicity of small, specialty operations which burned all of its residuals, it is changing into large integrated operations in which the residuals from one operation become the raw materials for another. For example, a modern forest products installation may include a veneer and plywood operation, a stud mill, and a particleboard mill with the power boiler serving to produce steam for the operations and as an incinerator for all of the solid, liquid, and gaseous materials which cannot otherwise be used. Saw and planing mills, and pulp and paper mills often are found in the same complex.

The latest trend in the industry is toward "total tree harvesting," in which the timber is cut, trimmed, and even barked in the forest. Round wood (whole logs or logs cut to specific lengths) are sent to saw mills, veneer and plywood mills, or pulp mills. The remaining useful wood is chipped in the forest and the fractionated wood sent to pulp mills, fiberboard mills, particleboard mills, and for miscellaneous uses. Bark may be classified for ornamental uses or left in the forest with the leaves and remaining wood wastes.

### *Air Quality Considerations*

For years, the smoky "teepee burner" was the ubiquitous hallmark of the air pollution problems of a large portion of the forest products industry. These burners now have been banned in nearly all major timber producing states. As previously indicated, much of the wood waste which formerly was burned now is used as a raw material for another operation in the industry. Wastes which cannot be utilized in any other way usually are burned in a power boiler or disposed of in land fills.

With the exception of the pulp and paper segment of the industry, air pollution problems cannot readily be classified by the unit operations of the various industry segments. Thus the unit operations which make up a sawmill operation, for example, will not be described in detail because a power boiler and drying kiln may constitute the only points of emission. Pulp mill unit operations will be considered in more detail since many of them are points of emission.

## II. Power Boilers

In the industry, steam and electrical energy are used for a variety of purposes. Sometimes these are purchased but frequently they are generated on-site in a waste boiler or combination boiler burning waste wood and fossil fuel. Power boilers may be found in practically any kind of forest products operation. The power boiler may also serve as an incinerator for liquid wastes such as in a plywood mill where glue wastes which cannot be used for makeup or other purposes may be incinerated.

### *A. Nature of Emissions*

The predominant emissions from these furnaces are in the form of particulate matter whose characteristics depend on the fuel or combination of fuels being burned. The wood waste portion generally consists of charred, partly charred, and unburned particles of a broad size range. The char generally is sliver shaped, is fragile, and has a low specific gravity. Particles from the fossil fuels which may be burned in combination with the bark or wood wastes have the characteristics of those fuels.

Little is known of the nature of the gaseous emissions except for the oxides of nitrogen and sulfur. Most of the sulfur oxides result from any auxiliary fossil fuel which is burned. Little also is known of the nature of gaseous emissions which might result from the incineration of other waste materials in the power boiler.

### *B. Control Devices*

Control of particulate emissions from power boilers is accomplished almost entirely with mechanical collectors. Because of the nature of the particles, differences of opinion exist about the relative merits of small-diameter versus large-diameter collectors, straight versus tapered sides, and reinjection. One combination boiler which burns coal and bark is reported (*1*) to use an electrostatic precipitator; and one hogged fuel boiler is reported (*2*) to use an impingement-type scrubber with some success.

## III. Dryers

In the sawmill and planing mill and veneer and plywood mill segments of the industry, a variety of dryers are used to adjust the natural moisture of the wood to the condition most suitable for its intended use. Dryers (referred to as kilns in sawmills) usually are gas or steam coil heated. Conditions in the dryer depend upon initial moisture, final moisture desired, the kind of wood being dried, and other factors. Kiln drying normally takes 2–5 days, while veneer drying requires several hours.

### A. Nature of Emissions

Studies on 13 veneer dryers (*3*), demonstrated that wood particles at concentrations of less than 0.002 grains per standard dry cubic foot (dscf) were the only significant particulate matter found at stack temperatures. A blue-haze plume consisting of hydrocarbon materials was noted to condense after the plume cooled below stack temperature. About two-thirds of the hydrocarbons were condensable wood resins, resin acids, and wood sugars. The remainder were volatile, and were terpenes in steam-heated dryers, and terpenes plus natural gas components in gas-fired dryers. The nature of the emissions varied depending on wood species, heat source, dryer type, and operational parameters. Plume opacity of the blue-haze emission ranged from 0 to 100%, averaging 20%. Some dryers had visible water plumes (*4*).

### B. Control Devices

Because the concentration of solid particulate matter in dryer exhausts is very low, control is aimed at reduction (with or without recovery) of the hydrocarbons. If the hydrocarbons have an economic value they may be recovered by adsorption, scrubbing, electrostatic precipitation, or filtration. If the concentration or value of recoverable portions is low they may simply be destroyed by incineration at temperatures of 1200° to 1500°F.

## IV. Kraft Pulping

The kraft process produces a strong, dark-colored fiber. Therefore, the market for the unbleached pulp usually is limited to its use in board, bag, and wrapping papers. If kraft pulp is to be used in the manufacture of fine white papers, its fibers must be treated additionally in a bleach plant. The process uses alkaline cooking chemicals, and the presence of caustic soda in the cooking liquor allows the use of practically all wood species.

### A. Description of Processes and Subprocesses

A simplified flow diagram for the kraft pulp process is shown in Figure 1. It must be kept in mind, however, that a variety of combinations of the various unit processes might be used in any given pulp mill depending on the product to be produced, the raw material, yield expected, air quality considerations, and the preferences of the manufacturer's technical staff. Table I shows a number of combinations of unit processes for which flow diagrams were developed and emission controls analyzed by Hendrickson *et al.* (*5*).

The kraft process involves the cooking of wood chips in either a batch or continuous digester, under pressure, in the presence of a cooking liquor. The kraft cooking liquor contains sodium hydroxide and sodium sulfide, the hydroxide being the reagent that dissolves the lignin in the wood chips. During the cooking reaction, the hydroxide is consumed and the sodium sulfide serves to buffer and sustain the cooking reaction. At the same time, small amounts of sulfide react with the lignin giving rise to the odors characteristic of kraft mills.

Upon the completion of the cooking reaction, the residual pressure within the digester is used to discharge the pulp into a blow tank. Gases and flash steam released in the tank are vented through a condenser,

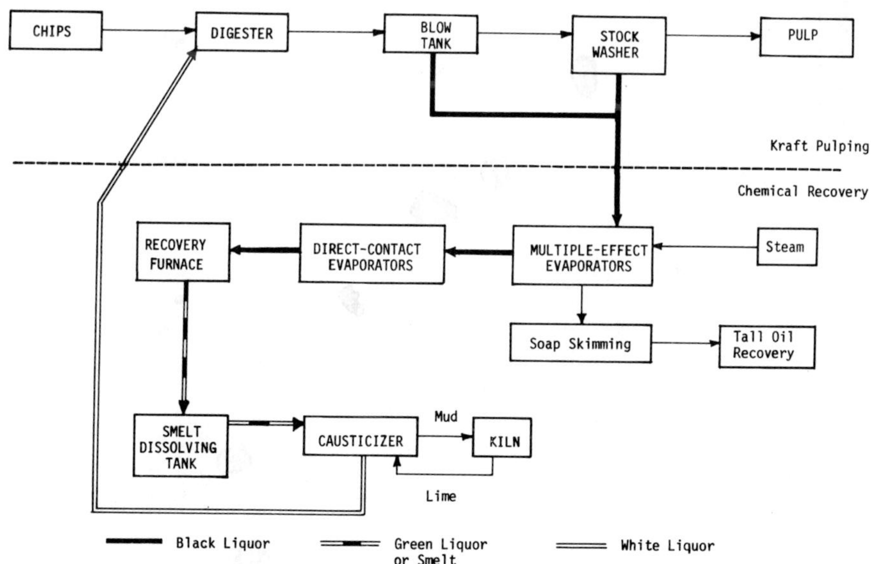


Figure 1. Typical flow diagram of the kraft process.

**Table I Typical Combinations of Unit Processes in Kraft Pulping**

| Pulping operation | Combination type 1 | 2 | 3 | 4 | 5 | 6 | 7 | 8 | 9 | 10 |
|---|---|---|---|---|---|---|---|---|---|---|
| Batch digester | × | | × | | × | | × | | | |
| Continuous digester | | × | | × | | × | | × | × | × |
| Concentrated black liquor oxidation | | | | × | | | | | | |
| Weak black liquor oxidation | | × | × | | | | | | | |
| No oxidation | × | | | | × | × | × | × | × | × |
| Direct-contact evaporator recovery | × | × | × | × | × | × | | × | × | |
| Venturi recovery | | | | | | × | | | | |
| Bleach plant | | × | | | | × | × | × | × | |
| No bleaching | × | | × | × | × | | | | | × |
| Lime kiln—moderate collection efficiency | × | | | | × | × | × | | | |
| Lime kiln—high collection efficiency | | | × | × | | | | × | × | × |
| Fluidized bed calcining—high collection efficiency | | × | | | | | | | | |
| High solids evaporator (63% solids) | | × | | | | | | | | × |
| Long tube vertical evaporators | × | × | × | × | × | × | × | × | × | |
| Hardwood (H) or softwood (S) | S | S | S | S | S | S | S | H | S | S |
| Percent yield | 53 | 45 | 47 | 47 | 47 | 45 | 45 | 46 | 45 | 47 |
| Product | Base liner-board | Bleach-able grades | Paper | Top liner or paper | Top liner | Bleach-able grades | Bleach-able grades | Bleach-able grades | Bleach-able grades | Paper |

where heat is recovered and the condensible vapors removed. The noncondensible gases, which are a source of malodors, are either confined and treated, or released to the atmosphere. At the same time the pulp in the blow tank is being diluted and pumped to washers where the spent chemicals and the organics from the wood are separated from the fibers.

The spent chemicals and the organics, called black liquor, are then concentrated in multiple-effect evaporators and/or direct-contact evaporators for subsequent burning. The evaporators concentrate the liquor to a solids content of 60–70% which is a requirement for combustion in the recovery furnace. During evaporation of the black liquor in the multiple-effect evaporators, volatile malodorous gases are released. These gases escape when entrained gases and vapors are drawn off by the vacuum system. In order to eliminate the venting of these gases to the atmosphere, they can be confined and destroyed.

In most United States mills, the black liquor is concentrated further in a direct-contact evaporator using hot flue gases from the recovery furnace. These hot gases, containing carbon dioxide, react with sulfur compounds in the black liquor leading to the release of malodorous gases such as hydrogen sulfide. Prior oxidation of the black liquor will reduce the sulfide content of the liquor and, hence, the amount of hydrogen sulfide released. Some mills which have gone on stream since about 1969 utilize a new type recovery furnace which eliminates any direct contact between the flue gases and the black liquor, in which case this unit no longer is a source of malodorous emissions.

The concentrated black liquor is then sprayed into the recovery furnace, where its organic content supports combustion. The inorganic compounds, consisting of the cooking chemicals, fall to the bottom of the furnace where chemical reactions occur in a reducing atmosphere. The chemicals are withdrawn from the furnace as a molten smelt, containing mostly sodium sulfide and sodium carbonate, which is dissolved in water in a smelt dissolving tank to form a solution called "green liquor." The green liquor is then pumped from the smelt dissolving tank, treated with slaked lime (calcium hydroxide) in the causticizer, and then clarified. The resulting liquor, referred to as "white liquor," is the cooking liquor used in the digesters.

Most kraft mills recover the sludge resulting from causticizing and burn it to lime in a kiln.

### *B. Sources and Nature of Particulate Emissions*

Particulate emissions from the kraft process occur primarily from the recovery furnace, the lime kiln, and the smelt dissolving tank. They are

**Table II Summary of Particulate Emissions—Kraft Process**

| *Potential source* | *Key to Figure 1* | *Particulate emissions (lb/air-dried ton of pulp)*[a] |
|---|---|---|
| Recovery furnace with direct contact evaporator | E | 75–125 |
| Lime kiln | H | 20–65 |
| Smelt dissolving tank | G | 1–4 |
| Slake tank | — | 4–6 |

[a] Values represent emissions from mills with no control devices.

caused mainly by the carry-over of solids plus the sublimation and condensation of inorganic chemicals. The sublimation and condensation normally produces a plume. Little information is available on the actual range of particle sizes from these sources, especially in the recovery furnace since agglomeration tends to occur readily. In addition, particulate emissions occur from combination and power boilers. Ranges of particulate emissions from uncontrolled sources are summarized in Table II.

## 1. Recovery Furnace/Direct Contact Evaporator System

The kraft recovery furnace potentially is the largest source of particulate emissions in the pulp system. Particulate emissions from the recovery furnace and the direct-contact evaporator system consist primarily of sodium sulfate and sodium carbonate, carried up by the furnace draft or formed in a vaporization-condensation process. There are no particles contributed by the direct-contact evaporator. In fact, the evaporator may serve as a particulate reduction device, the extent of reduction depending upon the type of evaporator used.

## 2. Lime Kiln

Particulate emissions from the lime kiln consist of sodium salts, calcium carbonate, calcium sulfate, calcium oxide, and insoluble ash. The presence of the sodium salts may be accounted for by a sublimation-condensation process and by dust entrainment within the kiln. The presence of the calcium salts must be explained by the entrainment of the calcium carbonate and calcium oxide and subsequent reaction with the flue gases.

## 3. Smelt Dissolving Tank and Slake Tank

The particulate emissions from the smelt dissolving and slake tanks are of a low magnitude and primarily are caused by the entrainment of large particles in the vent gases. Because of the violent reactions taking place in each of these tanks, water droplets containing both dissolved and undissolved inorganic salts are splashed above the surface.

### C. Sources and Nature of Gaseous Emissions

Emissions of total reduced sulfur (TRS) compounds are probably the foremost problem of air quality in the kraft process because of their odors.

The characteristic kraft mill odor is due principally to the presence of a variable mixture of hydrogen sulfide, methyl mercaptan, dimethyl sulfide, and dimethyl disulfide. All of these sulfur-containing gases are referred to as total reduced sulfur (TRS) components, and the latter three gases are usually described as organosulfur compounds. Actually, TRS is defined by the method of analysis.

The principal potential contributors of TRS in the kraft process are the recovery furnace in combination with the direct-contact evaporators, digester relief and blow, multiple-effect evaporators, smelt tank, brown stock washers, and lime kiln. The range of emissions of TRS compounds which might be encountered from uncontrolled kraft mill sources is summarized in Table III.

**Table III Summary of Total Reduced Sulfur (TRS) Emissions—Kraft Process**

| *Potential source* | *Key to Figure 1* | *Total reduced sulfur (TRS) emissions (lb $H_2S$/air-dried ton of pulp)*[a] |
|---|---|---|
| Digester relief and blow (batch) | A and B | 0.63–4.52 |
| Brown stock washer | C | 0.13–0.41 |
| Multiple-effect evaporators | | |
| Oxidized liquor | D | 0.21–0.62 |
| Unoxidized liquor | D | 0.26–4.6 |
| Recovery furnace (after direct-contact evaporator) | | |
| Oxidized liquor | E | 0.17–2.55 |
| Unoxidized liquor | E | 5.7–33.2 |
| Recovery furnace (before direct-contact evaporator) | F | 1.03–5.14 |
| Smelt dissolving tank | G | 0.05–0.13 |
| Lime kiln | H | 0.01–0.83 |

[a] Values represent emissions from mills with no control devices.

## 1. Recovery Furnace and Direct-Contact Evaporators

The theory of TRS generation in the recovery furnace itself is multidimensional and complicated. Sulfur, which exists in organic substances undergoing pyrolysis in a reducing atmosphere, may form volatile reduced sulfur compounds. These gases flow into the upper regions of the furnace.

The TRS gases and the carbon monoxide which are drawn into the oxidizing zone of the furnace should undergo oxidation to sulfur dioxide, carbon dioxide, and water. If proper conditions of temperature, excess oxygen, residence time, and turbulence do not exist, complete oxidation will not occur and the reduced sulfur compounds will escape from the furnace.

Turning to the direct-contact evaporator, the major cause of TRS emissions from this device is the reaction of furnace flue gases, particularly carbon dioxide and sulfur dioxide, with sodium sulfide residual concentrations in black liquor. High sodium sulfide concentrations in black liquor result in the direct contact evaporator becoming a major source of malodorous emissions.

It is difficult to separate emissions from the recovery furnace per se and the direct-contact evaporator since both discharge through the same flue.

## 2. Digester Relief and Blow Tank

The digester relief and blow tank may be the largest source of organosulfur emissions within the kraft mill. This is because these compounds are primarily formed in the digesters and the first opportunity for stripping from the liquor is at this point. These are relatively minor sources of hydrogen sulfide.

## 3. Lime Kiln

There is little published information concerning gaseous emissions from the lime kiln. It has been considered to be a minor source of sulfur gases and thus few studies have been made. Possible sources of sulfur compounds into the kiln system include the fuel used to fire the unit, residual concentrations of reduced sulfur compounds in the lime mud, noncondensible gases burned in the kiln, and scrubbing liquor used in the kiln scrubbers.

## 4. Multiple-Effect Evaporator

The multiple-effect evaporator is the second largest source of organosulfur compounds in the kraft process. The emissions from the multiple-effect evaporator are noncondensible reduced sulfur gases which are vaporized or stripped during the boiling.

### 5. Brown Stock Washer

TRS emissions from the brown stock washer arise primarily from the vaporization of the volatile reduced sulfur compounds. This is considered to be a minor source of gaseous emissions. The malodorous sulfur compounds from this source will be predominantly methyl mercaptan, depending on the liquid used for washing.

### 6. Smelt Tank

The gaseous emissions from the smelt dissolving tank are reduced sulfur compounds. Because the organosulfur compounds could not exist at the smelt temperature, their presence in the vent gases must be accounted for by their introduction from outside sources such as a draft from the furnace or the liquor used for smelt dissolving.

## *D. Control Devices Used*

### 1. Control of Particulate Emissions from the Kraft Process

The add-on devices shown in the following tabulation are commonly used in the kraft industry for control of particulate matter.

| | |
|---|---|
| Recovery furnace | Venturi evaporator/scrubber |
| | Electrostatic precipitator |
| | Electrostatic precipitator plus secondary scrubber |
| Lime kiln | Cyclonic scrubber |
| | Impingement baffle scrubber |
| | Venturi scrubber |
| Smelt dissolving tank | Mesh pads |
| | Cyclonic scrubber |
| | Packed-tower scrubber |
| | Orifice scrubber |
| Lime slaker | Mesh pads |
| | Cyclonic scrubber |

Where scrubbers are indicated, emissions of TRS compounds may be increased or decreased depending on the scrubbing medium that is selected. Generally, the control devices are listed in order of increasing efficiency, although too many factors influence their efficiency to draw specific conclusions.

## 2. Control of TRS Emissions from the Kraft Process

The add-on hardware or process changes shown in the following tabulation are used in the kraft industry for control of TRS emissions from the sources indicated.

| Source | Control |
|---|---|
| Recovery furnace | Strong black liquor oxidation |
| | Weak black liquor oxidation |
| | Elimination of direct contact between flue gases and black liquor |
| Lime kiln | Caustic scrubber |
| Smelt dissolving tank | Caustic scrubber |
| | Incineration |
| Digester relief and blow plus multiple effect evaporators | Chlorination |
| | Incineration |
| Brown stock washers | Continuous diffusion washing |
| | Incineration |

Although oxidation with an oxidizing solution in wet scrubbers has been used extensively, a problem arises in the difficulty of absorbing effectively all of the odorous compounds with a single scrubbing medium. The trend in the industry is toward incineration of both low volume–high concentration sources and high volume–low concentration sources. Usually, incineration control of any source other than digester and multiple-effect-evaporator gases is done only in new mills.

# V. Sulfite Pulping

The sulfite process produces easily bleached pulps from nonresinous woods. Thus sulfite pulps are used in a variety of products including high grade book and bond papers, tissues, dissolving pulp, and others. The acid cooking liquor is made up of sulfurous acid and usually the bisulfite salt of either calcium, sodium, magnesium, or ammonia. Originally the process utilized calcium bisulfite plus free sulfurous acid as the cooking liquor. The trend toward soluble bases has improved the versatility of the sulfite process in terms of wood species which can be pulped, higher yields, and the development of recovery processes.

## *A. Description of Processes and Subprocesses*

A simplified flow diagram for the sulfite pulp process is shown in Figure 2. As in the case of the kraft process a variety of combinations of the

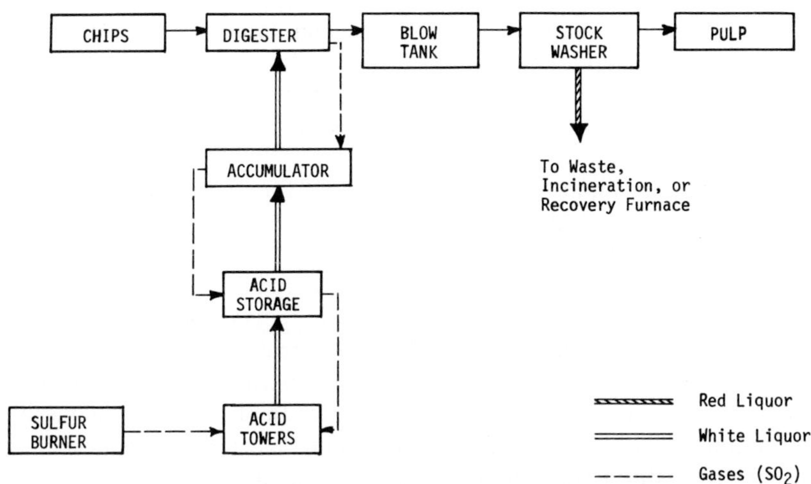


Figure 2. Simplified flow diagram of the sulfite process.

various unit processes and cooking bases might be used in any given pulp mill depending on the product to be produced, the raw material, yield expected, air quality considerations, recovery possibilities, and the preferences of the manufacturer's technical staff. The cations of the cooking liquor include calcium, sodium, magnesium, and ammonia. The cooking salts may be in the form of the acid sulfite or the bisulfite. Sulfurous acid is always used, but various combinations of combined and free sulfur dioxide may be involved. The reader is referred to the previously cited analyses by Hendrickson *et al.* (*5*) for flow diagrams and emission controls on the various modifications of the process.

The sulfite process involves the cooking of wood chips usually in a batch digester under pressure in the presence of an acid cooking liquor. The heat required for cooking is produced by the direct addition of steam to the digester or by steam heating the recirculated acid in an external heat exchanger. The makeup of the cooking liquor has previously been described. The sulfurous acid usually is produced by burning sulfur or pyrites and absorbing the resulting sulfur dioxide in liquor. Normally part of the sulfurous acid is converted to the base bisulfite to buffer the cooking action. During the cooking action it is necessary to vent the digester occasionally as pressure rises within the digester. These vent gases contain large quantities of sulfur dioxide and therefore are recovered into the cooking acid for reuse.

Upon completion of the cooking cycle, the contents of the digester consisting of cooked chips and spent liquor are discharged into a blow tank.

During this operation some water vapor and fumes escape to the atmosphere from the tank vent.

The pulp then goes through a washing stage where the spent liquor is separated from the fibers. The washed pulp is either shipped or kept within the plant for further processing.

Where calcium liquors are used, the spent liquor that is washed out of the pulp or that drains out of the pulp in the blow tank frequently is discarded or undergoes some by-product recovery. If the spent liquor is from a soluble base such as sodium, magnesium, or ammonia the next step may be a recovery cycle. If recovery is to take place the liquor is concentrated by evaporation and the concentrated liquor sprayed into a furnace where the organic compounds are burned. In the case of an ammonium base only the heat is recoverable. In the case of magnesium and sodium bases, residual inorganic compounds may be collected for reuse in the manufacturer of additional cooking acid.

### *B. Sources and Nature of Particulate Emissions*

Particulate emissions from the sulfite process consist mainly of steam and inorganic dust from the recovery furnace if one is used. Data are not available at this time on the range of particulate emissions.

### *C. Sources and Nature of Gaseous Emissions*

Sulfur dioxide is the principal atmospheric emission from the sulfite processes (*6*). The main causes of sulfur dioxide release are stripping by gas streams, inefficient absorption, and volatilization during periods of high liquor temperature. Hydrogen sulfide emissions are possible during recovery of the spent liquors if the recovery system is not maintained under proper oxidizing conditions.

The method of attack by sulfite liquors on lignin is different than the kraft liquor chemical attack. The sulfite processes involve lignin sulfonation, acid hydrolysis, and acid condensation. In sulfite cooking the product of lignin–sulfite reactions do not produce volatile reduced sulfur compounds such as methyl mercaptan and dimethyl sulfide.

#### 1. Absorption Towers and Acid Fortification Towers

Absorption towers for sulfite processes usually are packed towers or venturi absorbers. Sulfur dioxide gas is introduced in the bottom of the tower and a carbonate solution of the desired base is introduced at the top. The quantity of sulfur dioxide gas delivered to the absorption tower, the strength of inorganic chemicals in the cooking liquor, and the absorption conditions determine the emissions. Some processes use acid fortification towers which are merely absorption towers. In this instance weak cooking liquors are passed through the tower for the purpose of absorbing additional sulfur dioxide from various sources.

Currently available data (*5, 7*) show that the sulfur dioxide emissions from absorption towers are in the range of 15 to 20 lb of sulfur dioxide per ton of air-dried pulp before secondary scrubbing of the overhead gases.

## 2. Digester Relief and Blow

Gaseous sulfur dioxide emissions from digester relief and blow gases arise from the temperature increase of the cooking liquor. As the temperature rises the sulfur dioxide becomes less soluble. When relief lines are open and the pressure within the digester is relieved, large quantities of the gas will be emitted with the escaping steam.

Three methods of discharging the digester are used in the sulfite processes—hot blowing, cold blowing, and flushing. The emission of sulfur dioxide from the blow pit or dump tank depends on the system of blowing (*5, 7*). In the hot-blow style, the digester gases are used to discharge the contents of the digester to the blow pit. If no scrubbing is utilized emissions may range from 100 to 150 lb sulfur dioxide per ton of air dried pulp. In the cold blow and flushing style of discharge, the pressure is almost fully relieved from the digester and the relief gases are routed to accumulators for fortifying cooking acid. The contents of the digester are then discharged into a dump tank. Emissions from this source, if no secondary scrubbing is utilized, may range from 10 to 25 lb sulfur dioxide per ton of air dried pulp.

## 3. Recovery System

Sulfure dioxide emissions from the recovery system may occur from evaporators and the recovery furnace. Emissions from these sources usually are routed back to accumulators and eventually utilized for making additional acid. Because of the variety of practices utilized from mill to mill, it is impossible to present meaningful figures on emissions from these sources.

Ammonia-base spent cooking liquors usually are incinerated to recover the heating value. When no attempt at secondary recovery is made, emissions from these incinerators are from 250 to 500 lb of sulfur dioxide per ton of air-dried pulp.

### D. Control Devices Used

Little information is available on control of emissions from the various sulfite processes. Where particulate emissions are involved, the usual control techniques can be used. Gases (commonly sulfur dioxide) are the emissions most in need of control. Control techniques commonly applied are listed in the following tabulation.

| | |
|---|---|
| Acid tower | Vent to additional absorption tower |
| Blow pit | Packed tower scrubber |
| Spent liquors | Incinerate with or without recovery of heat or chemicals (not all recovery combinations are possible with all cooking liquor combinations) |

## VI. Neutral Sulfite Semichemical (NSSC) Pulping

The process is used mainly for the production of high yield pulp having high strength. It utilizes hardwood species that are not adaptable to other processes. The pulp is used mainly in the production of corrugating medium and liner board. If the pulp is cooked to a lower yield, NSSC pulps can be bleached and blended with bleached sulfite for high grade papers. The NSSC process uses a neutral sulfite chemical treatment followed by mechanical defibering.

### A. Description of Processes and Subprocesses

A simplified flow diagram for the NSSC process is shown in Figure 3. Again it must be kept in mind that a variety of combinations of the various unit processes may be used in any given pulp mill depending on the product to be produced, the raw material, the yield expected, air quality considerations, and the preferences of the manufacturer's technical staff. Several combinations of unit processes were investigated by Hendrickson *et al.* (*5*)

The cooking chemicals in this process consist of sodium sulfite and sodium carbonate in a liquor maintained at about a pH of 7. This provides a mild chemical treatment of the wood chips which does not completely remove all of the cementing material. Thus, the chemical stage is followed by some mechanical disintegration to separate the fibers further.

The cooking process is carried out either in batch or continuous digesters. Steam maintains the temperature and the pressure of the cooking reaction within certain limits depending on the end use of the pulp. During the cooking stage, odorous gases are created within the digester. At

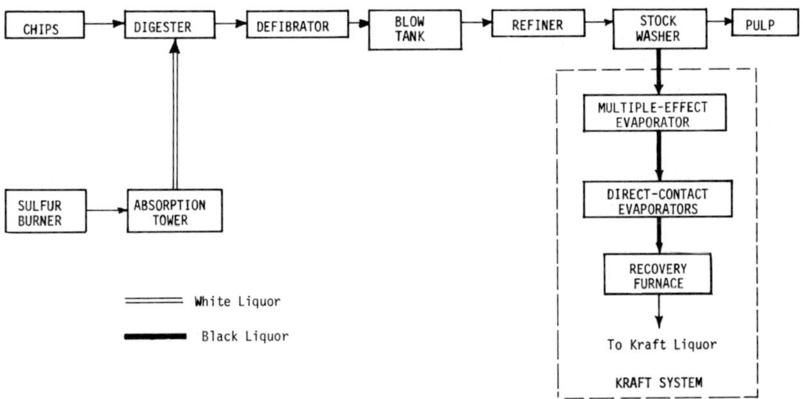


Figure 3. Simplified flow diagram of the Neutral Sulfite Semichemical (NSSC) process (with kraft cross-recovery).

the completion of the cooking cycle, residual pressure in the digester is used to discharge the entire contents of a batch digester into a blow tank. Waste gases usually are vented to the atmosphere. In some cases the sodium sulfite is prepared by burning sulfur and absorbing it in a soda ash solution in a manner similar to that used in the sulfite process.

Following the mechanical treatment, the pulp is sent to multistage drum filters where it is washed. The liquor resulting from the blow tank and the pulp washing may be discharged to the sewer or may be oxidized in a fluid bed reactor to sodium carbonate and sodium sulfate. If a kraft recovery system is adjoining, the NSSC spent liquor can be mixed with the spent kraft liquor up to a limiting percentage and burned in the recovery furnace. The recovered chemicals are used entirely in the kraft system. Emissions of both sulfur dioxide and hydrogen sulfide from the kraft recovery furnace may be increased when NSSC liquor is added.

### B. Sources and Nature of Particulate Emissions

Particulate emissions from the NSSC process occur primarily in the form of steam and dust from the fluidized bed reactor if one is used. In the fluidized bed recovery unit some of the partially reacted liquor may become entrained in the upward flow of gases and escape the reactor. Since this is a valuable makeup chemical, every effort is made to recover these particles and return them to process. Data on the actual emissions are not currently available.

### C. Sources and Nature of Gaseous Emissions

Sulfur dioxide is the principal gaseous emission from the NSSC process itself. Because of the difference in the chemical attack on the lignin using neutral sulfite liquor, compounds such as methyl mercaptan and dimethyl sulfide are not formed. In addition, the absence of sulfide ions from the cooking liquor virtually will eliminate hydrogen sulfide as a possible emission.

Emissions of sulfur dioxide may occur from the sulfur dioxide absorption tower if one is used, from the blow pit, and very small amounts from the fluidized bed reactor, if one is used. Very few studies have been made of sulfur dioxide emissions from the NSSC process. Thus ranges of emissions are not available. Individual mill observations have been reported by Galeano (*8*).

### D. Control Devices Used

Because of the variety of unit process considerations and lack of emission data from NSSC installations, little applicable information on control techniques is available. Control of emissions from unit processes similar to those found in the kraft and sulfite processes have been described previously.

## VII. Effects of Applying the Best Control Technology Available

There is little in the way of new technology available to control particulate emissions from power boilers. The electrostatic precipitator mentioned earlier was applied in a unique situation (coal burning) that is not often repeated in the industry. The wet scrubber mentioned previously is installed at only one mill but may find application elsewhere. The scrubber was an impingement type in series with a bank of multiclones. A performance test was conducted at pressure drops of 7.5 and 8.5 inches watergauge. Collection efficiencies of 96.7 and 98.3%, respectively, were obtained. These efficiencies resulted in emissions to the atmosphere of 0.032 and 0.016 grains/dry standard cubic foot.

Roberson *et al.* (*9*) did a complete analysis of emissions and costs for five hypothetical kraft mills of 500 tons per day capacity. The existing mill was assumed to have no emission control except an electrostatic precipitator on the recovery furnace and a medium energy scrubber on the lime kiln. The new mill was assumed to have a recovery furnace that eliminates direct contact between the black liquor and the flue gases; high efficiency electrostatic precipitator; high efficiency scrubbers on the lime kiln, smelt tank, and lime slaker; plus incineration of all noncondensibles from all other sources mentioned. The comparison is shown in Table IV. It should be noted that existing mills may not be able to achieve the values given here.

**Table IV Comparison of Emissions from an Existing Kraft Mill with No Control Systems and a New Mill Utilizing Best Technology Currently Available**

| | Total reduced sulfur[a] (TRS) | | Particulate matter[a] | |
|---|---|---|---|---|
| *Unit* | *Existing* | *New* | *Existing* | *New* |
| Recovery furnace | 19.5 | <0.5 | 5 | <1 |
| Batch digesters | 3.7 | 0 | 0 | 0 |
| Multiple-effect evaporators | 0.8 | 0 | 0 | 0 |
| Lime kiln | 0.8 | Trace | 10 | 0.3 |
| Smelt tank | 0.05 | Trace | 1 | <0.2 |
| Pulp washers | 0.25 | 0 | 0 | 0 |
| Lime slakers | 0 | 0 | ? | <0.3 |
| | 25.1 | <0.5 | 16 | <1.8 |

[a] Units are in pounds per ton of air dried pulp.

Data are not available to perform equivalent evaluations for sulfite and NSSC pulping.

## REFERENCES

1. R. Nachbar, "Electrostatic Precipitation of Ash from a Coal and Bark Fired Boiler." Air Pollution Control Association, Pittsburgh, Pennsylvania, 1970.
2. H. Effenberger, D. Gradle, and I. Tomany, "Hogged Fuel Boiler Emissions Control, A Case History." Technical Association of Pulp and Paper Industry, Atlanta, Georgia, 1972.
3. F. L. Monroe, R. A. Rasmussen, W. L. Bamesberger, and D. F. Adams, "Investigation of Emissions from Plywood Veneer Dryers." Plywood Research Foundation, Tacoma, Washington, 1972.
4. D. F. Adams, F. L. Monroe, R. A. Rasmussen, and W. L. Bamesberger, *in* "Proceedings of the Fifth Washington State University Symposium on Particleboard," p. 171. Washington State University, Pullman, Washington, 1971.
5. E. R. Hendrickson, J. E. Roberson, and J. B. Koogler, "Control of Atmospheric Emissions in the Wood Pulping Industry," 3 vols., Contract 22-69-18. National Air Pollution Control Administration, Washington, D.C., 1970. (Available from National Technical Information Service, U.S. Department of Commerce, Springfield, Virginia.)
6. H. F. J. Wenzl, "Sulfite Pulping Technology." Lockwood Trade Journals, New York, New York, 1965.
7. Anonymous, "Atmospheric Emissions from the Pulp and Paper Manufacturing Industry," EPA-450/1-73-002. United States Environmental Protection Agency, Research Triangle Park, North Carolina, 1973.
8. S. F. Galeano and B. M. Dillard, *J. Air Pollut. Contr. Ass.* **22,** 195–199 (1972).
9. J. E. Roberson, E. R. Hendrickson, and W. G. Tucker, *Tappi* **54,** 239–244 (1971).

# 17

## Mineral Product Industries

Victor H. Sussman



### I. Introduction

The conversion of naturally occuring minerals into salable products involves various operations and processes. Operations are primarily concerned with physical alteration, and processes are primarily concerned with chemical alteration. Certain unit operations (e.g., material handling, sizing, mixing) are common to practically all mineral production procedures. These are discussed in Section II. Table I lists air pollution control techniques applicable to these unit operations. Section III is concerned with the aspects of specific manufacturing processes related to unique emissions and control technology.

## II. General Operations

### *A. Mining*

#### 1. Deep Mining

Air pollution problems may be created by the discharge of deep mine ventilation air. Such problems are infrequent and easily controlled, most often by relocating the discharge vent.

#### 2. Open-Pit Mining

Particulate matter produced during open-pit mining operations is usually discharged directly to the atmosphere, rather than being captured by a local exhaust ventilation system, and is thus difficult to control. Many open-pit mining operations have an indefinite operating life which may make the installation of a fixed dust collecting system economically impractical. Therefore, dust control should be based upon prevention, at the point at which the contaminants are generated, by wet methods and portable equipment for dust suppression. Drilling, blasting, ore handling operations, and wind erosion are responsible for most of the particulate matter emissions produced during open-pit mining.

### *B. Transportation and Storage*

Mineral producers in the United States handle approximately 3 billion tons of ore and waste a year. Scientific materials handling techniques and automation have reduced air pollution from these operations. Reducing the number of transfers and the period of time required for each should be the first consideration in a materials handling air pollution abatement program.

Loading and unloading, conveyor belt discharge and transfer, and stockpiling operations can produce particulate matter emissions. Local

**Table I Mineral Production Operations Check List**

| *Operation* | *Control techniques* |
|---|---|
| A. Mining | |
| Open pit and quarry roads | Paving; periodic oiling, watering, $CaCl_2$ cover, and/or cleaning; covering trucks to prevent spillage; spray washing truck wheels |
| Blasting | Controlling size of blast, using water sprays immediately after blasting and blasting only when wind direction and other meteorological conditions are such that "neighborhood dusting" will not occur; using "blasting mats" |
| Drilling | Wet drilling or local exhaust ventilation |
| B. Transportation and storage | |
| Conveyor belts | Enclosure and local exhaust ventilation or wet spray—with special attention given to control at transfer points |
| Elevators | Enclosure and local exhaust ventilation |
| Discharge chutes | Telescoping chutes to permit discharge point to be close to surface of pile; spray or local exhaust ventilation at discharge point |
| Storage piles | Enclosure (silos, bins, etc.); covers (plastic coating, tarpaulin, clay, vegetation); wind breaks (trees, barricades) |
| C. Size reduction | |
| Crushing and grinding | Enclosure and local exhaust ventilation; wet sprays and/or exhaust hoods at mill inlet and outlet; where possible employ wet operations |
| D. Concentration, classification, and mixing | Enclosures and local exhaust ventilation; where possible employ wet operations |
| E. Drying | When possible use indirect dryers; exhaust dryer to suitable collector; classification prior to drying to remove fines |

exhaust ventilation and wet sprays can effectively control these emissions.

Lack of attention to wind erosion of stockpiles has often created an "air pollution control paradox." If the dust collected from air pollution control devices is promiscuously piled in open areas, wind erosion of these piles can cause as serious air pollution as that which would have been caused by emission from the operations had they originally not been controlled.

Gaseous emissions can be produced by evaporation of materials from storage piles. Also, there is the possibility of reaction between materials in storage piles. An example of this is the emission of hydrogen sulfide from aggregate produced from blast furnace slag. Hydrolysis occurs when the hot slag is quenched with water during production and when rain falls on a slag storage pile. The hydrolysis of the calcium sulfide in the slag results in the formation of hydrogen sulfide:

$$CaS + 2\ H_2O \rightarrow H_2S + Ca(OH)_2 \tag{1}$$

### C. Size Reduction

Size reduction operations are commonly classified as either crushing or grinding, according to the size range of feeds and the size reduction ratio achieved. Crushing operations usually involve feed sizes of from 20 to 60 in. (50.8 to 152.4 cm) and size reduction ratios of from 3:1 to 10:1. Grinding, pulverizing, and disintegrating operations usually involve feed sizes of from 0.05 to 0.5 in. (1.3 to 12.7 mm) and reduction ratios of from 10:1 to 50:1. Jaw, gyratory, cone, pan, roll, and rotary crushers and ball, pebble, rod, tube, ring-roller, hammer, and disk mill grinders are commonly used. Various types of crushers and grinders are usually operated in series (often close-circuited with a classifier) to obtain desired size reduction.

Dust is discharged from crusher and grinder inlet and outlet ports. For most effective dust control, crushers and grinders should be enclosed and provided with exhaust ventilation discharging to a suitable collector. Hoods at inlet and outlet ports are inefficient since they require the use of large quantities of air to "reach out" and capture dust; and ambient air currents can easily interfere with the flow of dust from the source to the hood. Information on enclosure design and ventilation requirements is available (*1*).

Although generally not as effective as local exhaust ventilation, wet sprays can be used to control dust emissions from crushing, grinding, and other mineral production operations. Wet grinding should be substituted for dry grinding operations, when there is no objection to obtaining a product in slurry form.

### D. Classification, Concentration, and Mixing

Following size reduction operations, ores are generally size classified, concentrated (i.e., separated from gangue by wet or dry processes), and/or mixed to obtain a desired size consist or composition. Local exhaust ventilation systems have been specifically designed to control dust emissions from these operations. Dry concentration operations, e.g., mechanical separation by magnetic, inertial, or electrostatic forces and screening and inertial classification equipment, all generate significant amounts of dust. Unenclosed screening operations at quarries can be a major source of dust concentrations. Screens should be enclosed and provided with exhaust ventilation. The area of openings in the enclosures should be kept to a minimum. Inlet face velocities through enclosure openings should be sufficient to prevent the escape of dust. Velocities of 100–200 ft/minute (30–61 m/minute) have been found to be sufficient, in most cases. Exhaust from the ventilation system should be discharged through an efficient collector.

Dry mixing operations should be enclosed and ventilated in a manner similar to screens. Loading and discharge ports should be enclosed and connected to an exhaust ventilation system.

### *E. Drying*

Wet mixes, slurries, and filter cakes are dried in either direct or indirect dryers In direct dryers, the wet material is brought into direct contact with the hot gases. In indirect dryers, heat transfer takes place through the dryer wall. Direct drying operations are used most often in mineral production. Direct drying usually creates more air pollution than indirect drying since the hot gases remove dust from the material being dried.

## III. Specific Processes*

### *A. Crushed Stone*

With the exception of fuel production, the crushed stone industry is the largest nonmetallic mineral industry in the United States with respect to total value of production. It is second only to sand and gravel in terms of volume produced. There are approximately 5000 crushed stone installations in operation in the United States, with plant capacities ranging from less than 25,000 to more than 14 million tons per year. Most of the rock processed is limestone. The crushed stone industry employs all of the typical mineral production operations discussed in Section II.

#### 1. Process Description and Emissions

After the stone is removed from the quarry, it is processed in a crushing plant (Fig. 1) for size reduction and classification. Although most stone plants are equiped with the types of crushers and inclined screens shown in Figure 1, there is considerable variation in the use of special size classification equipment. Tertiary grinding and air classification may be employed if fine aggregate is required.

* See Reference (*2*).

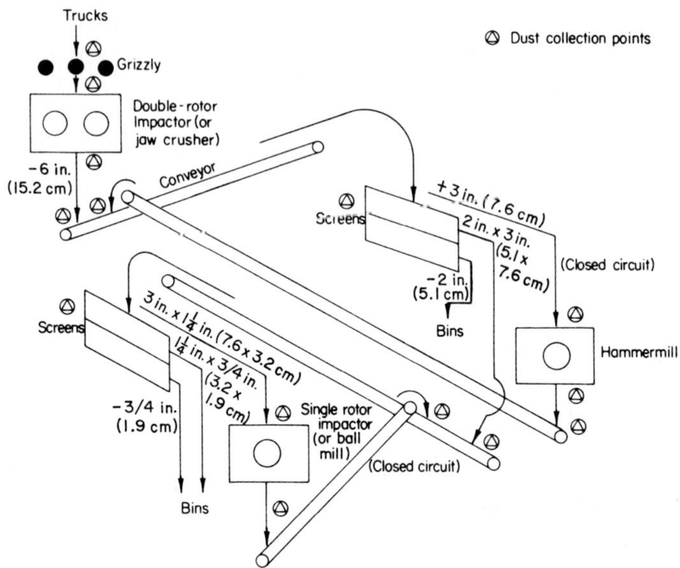


Figure 1. Typical stone crushing plant.

Particulate matter is the only significant emission produced by these operations. Because of the wide variability of the raw material and the characteristics of operations, there are no acceptable emission factors that can be applied to a crushed stone plant. Visual (and thus somewhat subjective) observation indicates that a large percentage of the emissions from a crushed stone plant are fugitive—i.e., from belt transfer points (and associated surge piles), roads, storage piles, and open bins. Some estimates have been made that process sources alone emit about 15 to 17 lb (6.8 to 7.7 kg) of particulate matter per ton of crushed stone produced. Because of difficulties involved in sampling for emissions from process equipment such as open crushers, screenhouses, and elevators, emission estimates based upon plant throughput (i.e., "process weight") are usually of questionable accuracy.

## 2. Dust Controls and Compliance with Regulatory Standards

Because of the difficulty of performing source tests on process equipment and the relatively significant amounts of fugitive emissions from other plant sources, most control standards are based upon opacity. Such subjective standards complicate "efficiency guarantees" for control equipment and the establishment of emission control disciplines within a plant.

Process sources should be enclosed and vented to mechanical or fabric collectors. Fugitive dust sources should be controlled by

(a) Elimination of sources—surge piles at belt transfer points should be eliminated or enclosed; good housekeeping practiced in the plant yard

(b) Dust suppression systems—water, oil, or chemical spray systems on belts, storage piles, open crushing operations and truckloads; storage piles that are not regularly worked should be coated ("stabilized) with plastic spray or covered

Effective control of particulate matter emissions, as indicated by compliance with governmental regulations, should result in no "significant" visible emissions beyond the plant property line.

### *B. Lime Production*

In terms of total tonnages processed, the rotary kiln is the most commonly employed unit process for the conversion of nonfuel minerals into salable products. Most of the lime manufactured in the United States is produced in rotary kilns.

#### 1. Process Description and Emissions

Limestone (calcium carbonate) or dolomite (calcium magnesium carbonate) is sized and charged to a rotary kiln (Fig. 2). As the charge moves down the kiln, countercurrent to the direction of the heated gases,

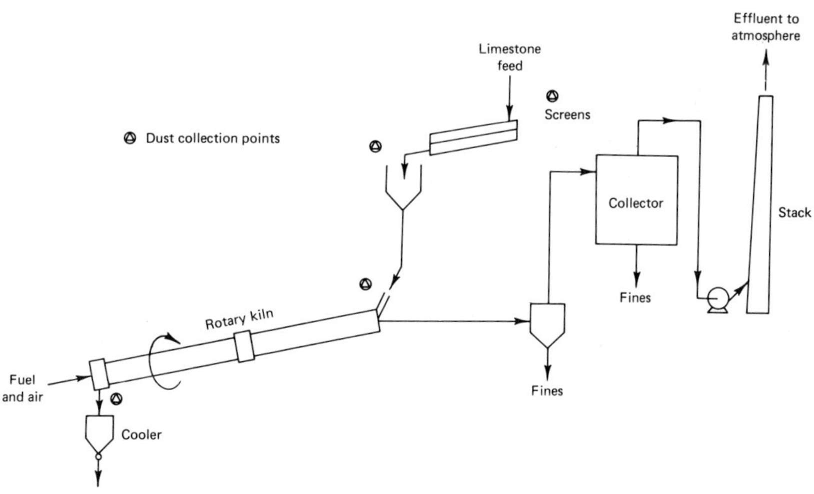


Figure 2. Typical lime kiln.

the carbonate in the limestone is decomposed, releasing carbon dioxide ($CO_2$). The product, calcium oxide, is discharged from the kiln to a cooler prior to screening and packaging. Less than 20% of the lime produced is converted to hydrate [$Ca(OH)_2$] by addition of water in a pugmill.

The hot gases in the kiln carry fine lime dust to the charging end. If coal is used as a fuel, exit gases will also contain flyash.

### 2. Dust Control

In order of decreasing use, kiln emissions are controlled by fabric collectors, water scrubbers, and electrical precipitators. Most collectors are capable of controlling dust emissions to 0.02 grains/scf (46 mg/std $m^3$).* The choice of the type of collector depends upon factors such as:

(a) The type of fuel used to fire the kiln. It has been contended that the tars produced by coal-fired kilns make fabric collectors unsuitable. There are, though, examples of successfully operating baghouses on coal-fired kilns

(b) The use of scrubber water in processes associated with the lime production operation, e.g., kraft paper mills

(c) The expected life of the plant. Because of the lower capital investment required, scrubbers are often the collector of choice for plants with a limited production life (because of age of plant or associated limestone supply facilities)

Regulatory emission standards (both dust loading and opacity) have been based upon performance tests of fabric collectors. Although water scrubbers can comply with these standards, fabric collectors are often considered more desirable because visible plumes are eliminated and the results of "slippages" in required maintenance or operating conditions (e.g., pressure drop) are more obvious. Thus, it is more likely that fabric collectors will be properly maintained. Regulatory agencies prefer fabric collectors over scrubbers because there are less interferences with compliance inspections, i.e., opacity readings.

## *C. Coal Preparation*

The equipment and processes involved in coal preparation are similar to those used in the beneficiation of most mineral ores. Wet processes do not, in themselves, cause air pollution problems. But when the wet product must be dried to facilitate transportation, significant air pollution problems can occur. A by-product of the operations, coal refuse, presents

* scf = Standard cubic feet, std = standard.

a major problem in storing so as to prevent air and water pollution. The construction of large coal-burning power stations near "mine mouths" has resulted in a close physical and operational association between coal preparation plants and power stations. In the absence of a practical system to remove sulfur oxides in flue gases, coal preparation is the prime method for reducing these emissions. For most eastern United States coals, cleaning (which removes only the inorganic sulfur in coal) will not reduce sulfur oxide flue-gas concentrations to the extent necessary for compliance with most United States standards. Particulate emissions from coal-burning sources are affected by preparation techniques. For stoker-fired units, the size consist of the coal is a significant factor (often as significant as the ash content) in reducing particulate emissions. Size consist (and % fines) is one of the specifications to which coal preparation plants operate. In order to fully evaluate the air pollution problems associated with the combustion of coal, consideration must be given to available coal preparation techniques.

The major air pollution problems associated with coal preparation procedures are gaseous emissions from ignited coal refuse disposal areas, dust from refuse and coal storage piles, and particulate matter from coal dryers and dedusting operations.

## 1. Cleaning and Drying

Coal preparation (coal cleaning) operations involve the processing of run-of-mine coal to produce a salable product. Run-of-mine coal contains impurities in varying amounts, and ranges in size from fines to pieces having equivalent diameters of over 10 ft (3 m). The primary purpose of a preparation plant is to crush the coal, remove impurities, and classify the product into standard sizes.

There are four principal coal cleaning techniques (*3*).

a. Cleaning at the Mine Face. This technique is being used less because of the increased use of mechanical mining methods. The removal of unwanted material, usually manually, at the mine face has the advantages of reducing the amount of material which must be transported to the preparation plant, reducing the amount of refuse produced in the plant, and thus reducing the size of coal refuse piles.

b. Picking by Hand or by Mechanical Means. Hand picking, the earliest method used, is usually employed in combination with mechanical methods, e.g., shaking tables.

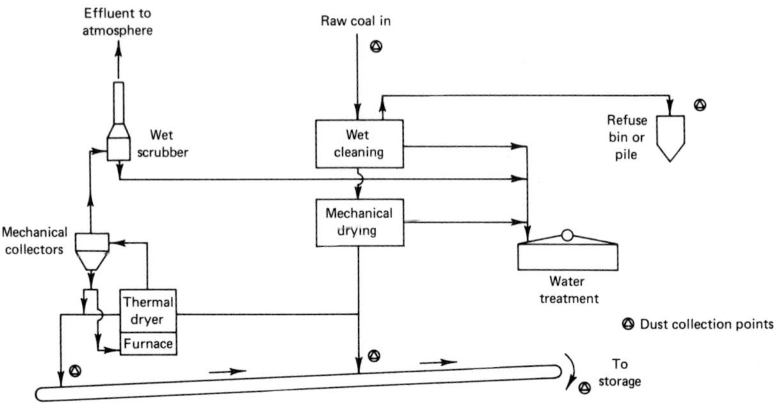


Figure 3. Wet coal cleaning operation (equipped with thermal dryer).

c. Froth Flotation. This is a common ore preparation method in which fine coal particles attach themselves to foam bubbles, while heavier particles of slate do not.

d. Gravity Concentration. This mechanical cleaning method is based upon the difference in the specific gravities of coal and slate or shale. An appropriate separating medium (either liquid, gas, or fluidized solid—or a combination of these) is used to segregate the heavy and light particles.

After wet process cleaning operations (Fig. 3) the coal is usually subjected to thermal drying. Mechanical coal drying usually does not create significant air pollution. Thermal drying usually involves the combustion of coal and the passage of the combustion gases through a bed of wet coal. The size consist of the coal being dried and the velocity of the gases through the bed are major factors in determining the air pollution potential of the plant. Emissions include products of combustion and entrained coal fines.

## 2. Emission Control

The use of inertial collectors and wet scrubbers, or packed towers, in series, can provide efficient control of emissions. Fabric collectors are not usually used (except for control of particulate emissions from air tables) because of fire hazard. Fine particulates are the significant emissions from coal preparation plants. If thermal drying is imployed, oxides of nitrogen and sulfur, and carbon monoxide are also emitted. Surveys have indicated that most thermal dryer capacity is for coals of less than 1% S. Thus, low sulfur coals are most often fired in thermal dryers. Most regulations limit only particulate emissions from air tables and dryer stacks.

Stack tests (using United States Environmental Protection Agency standard techniques) of wet cleaning systems, including fluid bed driers, have led to the following proposed standards of performance for new sources (*4*):

(a) Thermal dryer gases—particulate matter not in excess of 0.070 gram/dry standard cubic meter (gm/dscm) (0.031 grains/dry standard cubic foot (gr/dscf)) and 30% opacity

(b) Pneumatic coal-cleaning equipment (e.g., air table) gases—particulate matter not in excess of 0.040 gm/dscm (0.018 gr/dscf) and 20% opacity

Control costs for wet-scrubbing equipment for thermal driers is estimated at from $0.066–$0.064 per ton of cleaned coal for plant sizes of 3 to 5 million tons per year, respectively. This is comparable to estimates for particulate control scrubbers operating at 20–30 in. (50.8–76.2 cm) water pressure drop on other mineral operations.

Controlling operating procedures and providing for proper maintenance of equipment are the most important factors in preventing air pollution from thermal driers. Even with the best air pollution control equipment, consideration must be continually given to

(a) Not overloading the plant. Increase in the rate of moisture content of coal feed will require an increase in the velocity of drying gases through the bed which will increase particle entrainment from the bed

(b) The prevention of clogging of ducts, cyclone, and other equipment by caked coal fines. This can usually be prevented by maintaining proper gas velocities through equipment, and insulating (or heating, by using a portion of the combustion gases) surfaces on which condensation may take place

## 3. Coal Refuse Disposal

The United States Environmental Protection Agency (USEPA) has stated that

> Coal preparation plants are inherently one of the Nation's major sources of solid waste pollution. This refuse consists of dirt and other contaminants that are mined with the coal and separated from it by the preparation plant. The wet preparation process uses copious quantities of water to accomplish the separation. The additional water requirements and solid waste pollution attributable to control of air pollution are minor when compared to the waste by-products of the basic process (*4*)

Thus, it is surprising that the Federal Government and all but a few states have not attempted to regulate this significant source of various pollutants. With the potential growth of the coal industry, especially in areas (e.g., western United States) where refuse content is high, control efforts must be increased.

From 20 to 50% of the raw coal processed in a cleaning plant is rejected as refuse. Refuse is piled into banks, either near the mine or near the preparation plant, having a base area of from an acre or less to hundreds of acres. The piles vary in height from 20 to 300 ft (6 to 91 m). Many piles contain millions of tons of refuse.

Coal refuse piles ignite either through spontaneous combustion, carelessness, or deliberate action. The oxidation of the carbonaceous and pyritic material in coal refuse is an exothermic reaction. The temperature of a coal refuse pile, or portions of the pile, will increase when the amount of air circulating in the pile is sufficient to cause oxidation, but insufficient to allow for the dissipation of heat. The temperature of the refuse increases until its ignition temperature is reached. Mine timbers, vegetation, or household refuse in the pile may act as kindling. Ten years ago there were approximately 150 burning refuse piles in Pennsylvania. Many have been extinguished through state and federally funded projects. Since the adoption of specification type regulations for the disposal of coal refuse in Pennsylvania, not one refuse pile has ignited and caused an air pollution problem.

Significant concentrations of hydrogen sulfide and sulfur dioxide have been measured in communities adjacent to burning refuse piles (*5*). While passing through the pile, the products of combustion react with each other and with materials in the pile (partially combusted pyritic and carbonaceous material) to form a number of noxious gases including carbon monoxide, ammonia, hydrogen sulfide, oxides of sulfur, and carbon disulfide. The extent and nature of air pollution from coal refuse piles and control techniques have been described by Sussman and Mulhern (*6*).

Proper construction, including compaction and sealing to prevent the circulation of air within refuse piles, will prevent ignition. Harrington and East (*7*) list five methods for controlling a burning coal refuse dump:

(a) Digging out the fire or isolating the fire area by trenching

(b) Pumping water onto the fire area and immediate vicinity

(c) Applying a blanket or cover of incombustible material, such as limestone dust, shale dust, or slag dust over the fire area

(d) Injecting a slurry of rock dust or other incombustible material into the fire area through drill holes; grouting with cement is also practiced

(e) Spraying water over the fire area

Additional techniques employed include:

(a) Removing burning refuse from the pile and extinguishing it by compaction and the use of water sprays

(b) Sealing the surface of burning piles with plastic foam

(c) Using hydraulic mining "cannons" to remove and extinguish burning sections

(d) Using explosives to loosen the pile, and then extinguishing it with water sprays

### D. Cement Production

To produce one barrel of cement, weighing 376 lb (171 kg), approximately 600 lb (272 kg) of raw materials (not including fuel) are required. The raw materials required to make cement may be divided into those supplying the lime component (calcareous), the silica (siliceous), the alumina (argillaceous), and iron component (ferriferous).

The major steps in the production of portland cement are quarrying, crushing, grinding, blending, clinker production, finish grinding, and packaging (Fig. 4). The operations prior to, and subsequent to, clinker production are discussed in Section II. Detailed descriptions of both wet and dry production operations are contained in Kreichelt *et al.* (*8*). The major difference between the wet and dry processes is that in the wet process, the raw materials, i.e., the charge to the kiln, are ground to form

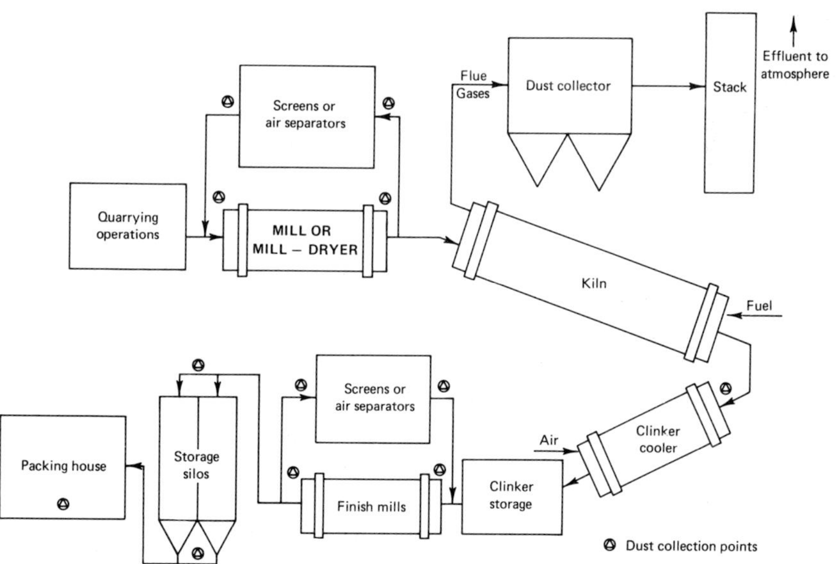


Figure 4. Flow diagram of cement plant operations.

a slurry, whereas in the dry process the free moisture content is reduced to less than 1% prior to or during raw grinding. In the dry process, the kiln feed may be dried in a rotary dryer or in combined drying and grinding units in which drying is accomplished in a separate compartment within the grinding unit.

The kiln is the major source of emissions from a cement plant. The rotary kiln used in most United States plants is a steel cylinder with a refractory lining. Kilns may be as small as 6 ft (1.8 m) in diameter by 60 ft (18.3 m) in length or as large as 25 ft (7.6 m) in diameter by 760 ft (232 m) in length. The kiln is erected horizontally with a gentle slope of $\frac{3}{8}$ to $\frac{3}{4}$ in. per foot (3.1 to 6.2 cm/m) of length and rotates on its longitudinal axis.

The kiln feed, commonly referred to as "slurry" for wet process kilns or "raw meal" for dry process kilns, is fed into the upper end of the revolving sloped kiln. As the feed flows slowly toward the lower end, it is exposed to increasing temperatures. During the passage through the kiln (1–4 hours), the raw materials are heated, dried, calcined, and finally heated to a point of incipient fusion [about 2900°F (1593°C)], a temperature at which a new mineralogical substance, called clinker, is produced. In the lower portion of the kiln, coal, fuel oil, or gas is burned to produce a process temperature of 2600° to 3000°F (1427° to 1649°C). The combustion gases pass through the kiln counterflow to the material and leave the kiln, along with carbon dioxide driven off during calcination, at a temperature of from 300° to 1800°F (149° to 982°C), depending on the kiln length and the process used.

Table II indicates typical dust loading characteristics of gases leaving kilns. It has been estimated that uncontrolled particulate emissions are 122 kg/metric ton (MT) for dry process kilns and 113 kg/MT for wet process kilns (*9*). Between 50–60% of kiln dust is finer than 10–20 μm.

## 1. Emission Control Techniques

a. Kilns (Clinker Production). A typical dry process kiln operating at a burning rate of 3000 barrels of clinker per day will use 150 tons of coal and approximately 900 tons of solid raw material to produce 540 tons of clinker and 4320 tons of kiln gas. It can readily be seen that the kiln gases far outweigh the other materials handled and that this is one reason dust collection equipment and auxiliaries represent such a large capital expenditure in a cement plant (*10*).

Some degree of control of dust from kiln operation may be obtained by adjusting operating conditions and kiln design to keep dust within

**Table II Ranges of Dust Emissions from Control Systems Serving Dry- and Wet-Type Cement Kilns** (*8*)

| Source | Type of dust collector | *Range of dust emissions from collector* grain/scf (gm/scm) | lb/ton of cement (kg/ton) | lb/ton of cement (kg/MT)[a] |
|---|---|---|---|---|
| Kiln—dry type | Multicyclones | 1.55–3.06 (3.55–7.00) | 26.2–68.6 (11.9–31.1) | 26.2–68.6 (13.1–34.3) |
| | Electrical precipitators | 0.04–0.15 (0.09–0.34) | 1.7–5.7 (0.77–2.6) | 1.7–5.7 (0.85–2.8) |
| | Multicyclones and electrical precipitators | 0.03–1.3 (0.07–3.0) | 0.6–29.4 (0.27–13.3) | 0.6–29.4 (0.30–14.7) |
| | Multicyclones and cloth filters | 0.039 (0.089) | 0.7 (0.32) | 0.7 (0.35) |
| Kiln—wet type | Electrical precipitators | 0.03–0.73 (0.07–1.67) | 0.52–9.9 (0.24–4.5) | 0.52–9.9 (0.26–4.9) |
| | Multicyclones and electrical precipitators | 0.04–0.06 (0.09–0.14) | 4.3–24.2 (2.0–11.0) | 4.3–24.2 (2.1–12.1) |
| | Cloth filters | 0.015 (0.034) | 0.35 (0.16) | 0.35 (0.17) |

[a] MT = Metric ton.

the kiln. Reducing gas velocities within the kiln (some of the newer designed kilns have larger diameters at the feed end), modification of the rate and location of feed introduction, and hanging a dense curtain of lightweight chain at the discharge end of the kiln appear to have some effect on reducing the amount of dust discharged.

The weight-rate of dust emission from kilns is of such a magnitude that efficient collection is required, irrespective of the location or size of the plant. Since most of the kiln dust is usually smaller than 10 μm in size, inertial collectors alone are not adequate for efficient control. Electrostatic precipitators or fabric filters (often in series with inertial collectors) can effectively collect fine kiln dust. Siliconized glass fabric is most often used in baghouses to remove dust from hot kiln gases. Fabric filters have been used successfully on both wet and dry processes.

The weight of kiln dust per unit weight of clinker produced varies, depending upon the type of process (wet or dry), operating conditions, and whether or not collected dust is reintroduced to the kiln. A 4000 barrels (bbl)/day wet process plant can produce 60 tons/day of kiln dust. A 90% efficient collection system would only reduce emissions to 6 tons/day. This would create an air pollution problem, even if the plant were located in a rural area. At present, manufacturers will guarantee 99.5%+ efficiency, and this is the range usually required to prevent air pollution problems.

The United States federal performance standard for new kilns is no more than 0.30 lb (0.136 kg) particulate per ton of feed to the kiln (maximum 2-hour average). It is clear from Table II that fabric filters or high efficiency electrical precipitators will be required to meet this standard. The use of scrubbers for the control of kiln dust is decreasing because of increasingly stringent water pollution control regulations and odor problems resulting from waste water treatment. In the United States, state and local regulations for existing kilns are usually of the process-weight type and permit emissions of from 0.3 to 0.9 lb (0.136 to 0.408 kg) per ton. But, these state and local agencies are increasingly using opacity standards (usually 10 to 20%) to require upgrading of collectors and the use of fabric filters. Federal opacity standards for kilns has been the subject of litigation (*11*). The originally adopted standard of 10% opacity was revised to 20% (*12*). The interference of steam plumes (e.g., from scrubbers) in determining opacity was recognized and readings were permitted at points downwind of stack discharge, where water vapor is no longer present. In this instance, it is clear that a uniform national new source performance standard does not necessarily represent "best available control technology." Opacity is often affected by local weather conditions, and long-term observations have indicated that well-controlled and maintained kilns in some areas of the country are capable of meeting the 10% opacity standard.

b. Storage Silos. Storage silos are under slight pressure as a result of material displacing air during the filling operation. In modern installations displaced dust-laden air is vented to a bag-type dust collector. This is especially true for silos with pneumatic loading and circulating systems.

c. Clinker Cooler. The clinker coming from the kiln is normally cooled in rotary drum, shaking, inclined, horizontal, or traveling-grate coolers. When a rotary drum cooler is used, cooling air induced through it may be utilized in the kiln as secondary combustion air, whereas the air drawn through a grate cooler is in excess of the quantity that can be used as secondary combustion air in the kiln. However, this excess can be used for drying purposes or can be discharged to the atmosphere.

The United States federal new source performance standard for clinker coolers is no more than 0.10 lb (0.045 kg) of particulate per ton of feed to the kiln (maximum 2-hour average) and 10% opacity. In the United States, state and local standards permit from 0.2 to 0.6 lb (0.091 to 0.272 kg) of particulate per ton for existing sources.

Clinker cooler emissions have been controlled with electrical precipitators and baghouses. Because clinker cooler dust is not as fine as kiln dust (10–15% below 10 $\mu$m), mechanical collectors have been used at older plants, without causing opacity violations.

d. ROTARY DRIERS. Rotary driers are a major source of dust generation in a cement plant, especially if kiln exit gases are utilized (waste-heat driers). Uncontrolled dust concentrations in excess of 10 grains/ft$^3$ (22.9 gm/m$^3$) can be expected. This operation is discussed in Section III,E.

## 2. Handling Collected Dust

Considerable progress has been made in developing techniques to reintroduce collected kiln dust into the kiln. Reintroduction methods vary from plant to plant and may consist of one or a combination of the following:

(a) Mixing with the raw feed either prior to or at the kiln charging end

(b) Dust return by scoop feeders located in front of the chain system

(c) Leaching dust with large volumes of water, mechanical dewatering, and introducing the resultant slurry into the kiln feed, onto the chain system, or into the kiln as raw feed. Leaching is one method employed to control the alkalinity of feed dust (see below)

(d) Introducing dust into the burning zone (insufflation), often through the fuel pipe. The high temperature at the burning zone causes the dust to sinter

The characteristics of collected dust, such as degree of calcination, alkalinity, sulfur content, and fineness, are important considerations in determining if reintroduced kiln dust will adversely affect clinker quality, kiln operation, or dust collector efficiency. The percentage of alkali permissible in finished cement is limited by the American Society for Testing and Materials (ASTM) specifications. Very fine collected dust usually has a high alkali content and often cannot be introduced into the kiln without pretreatment. The coarser dust captured by primary inertial collectors is usually low in alkali content and can be directly introduced. The introduction of very fine partially calcined dust can increase the undesirable mud ring-forming qualities of the raw feed. The introduction of very fine high alkali dust with the raw feed increases the tendency of the kiln dust to blind fabric collectors.

Collected kiln dust which cannot be reintroduced into the kiln may be used as a substitute for agricultural limestone, fertilizer, or mineral filler. If collected dust cannot be reused, it is often disposed of in abandoned quarries or storage piles. When stored in this manner, the dust piles should be covered, enclosed, or sprayed with water to form a surface crust.

### *E. Asphaltic Concrete*

A flow diagram of an asphaltic concrete plant is shown in Figure 5. Aggregate (gravel, rock, and sand) is proportioned and charged to a rotary dryer and heated to about 300°–350°F (148.9°–176.7°C). The dried aggregate is screened, transferred to storage bins, and introduced into a mill where it is mixed with hot asphaltic oil.

#### Emissions and Compliance with Regulatory Standards

Air pollution sources include the rotary dryer, transfer points, bucket elevators, screens, weighing and mixing operations, storage piles, plant roads, and sometimes the boiler facility. The dryer is the major source of dust emissions. A direct oil-fired dryer [24 ft (7.3 m) long, 60 in. (152 cm) diameter; capacity: 50 to 60 tons/hour; approximate moisture reduction: from about 12% to 0.5%] requires 11,830 $ft^3$/minute (cfm) (335 $m^3$/minute) exhaust ventilation capacity. Associated with this dryer, the hot elevator requires 1000 cfm (0.47 $m^3$/sec), vibration screens 2000 cfm (0.94 $m^3$/sec), bins 1500 cfm (0.71 $m^3$/cm), and scales 1500 cfm (10.71 $m^3$/cm). Tests for dust loading from this type of equipment indicate a wide variation depending primarily upon the aggregate size consist. Table III (*8*) indicates emission factors for plants equipped with various type of collectors.

Test of scrubber stack particulate emissions show a linear increase with the increase in the amount of minus 200-mesh aggregate charged to the dryer. Electrical precipitators have not been successfully used in controlling particulate emissions.

Because of the many plant variables that affect emissions and control equipment costs, the establishment of a United States federal "new source performance" standard for asphalt concrete plants has been a long and controversial process. The promulgated standards [0.04 grain/dscf (0.092

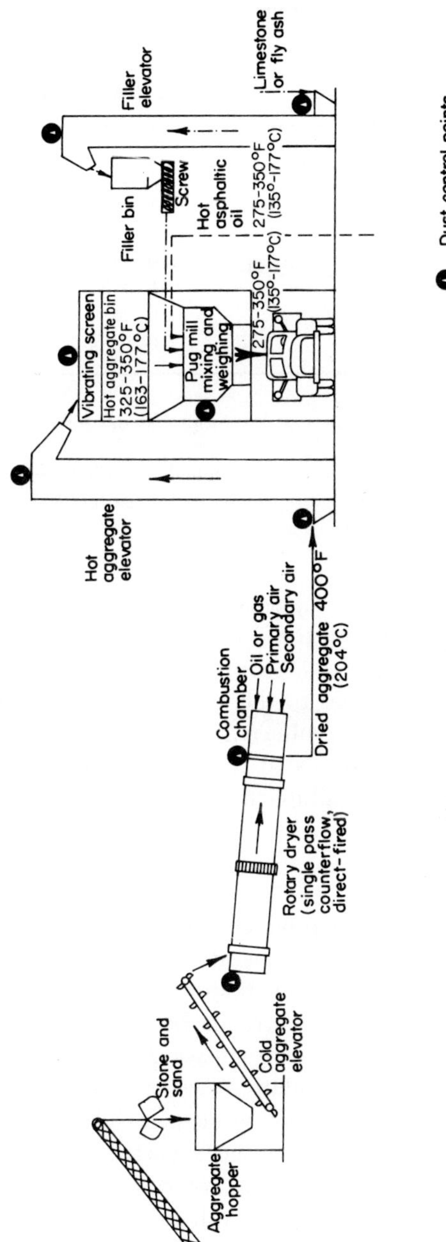


Figure 5. Flow diagram of asphalt (paving) plant operations.

**Table III Particulate Emission Factors for Asphaltic Concrete Plants** (*8*)

| Type of control | Emissions lb/ton | kg/MT[a] |
|---|---|---|
| Uncontrolled[b] | 45.0 | 22.5 |
| Precleaner | 15.0 | 7.5 |
| High-efficiency cyclone | 1.7 | 0.85 |
| Spray tower | 0.4 | 0.20 |
| Multiple centrifugal scrubber | 0.3 | 0.15 |
| Baffle spray tower | 0.3 | 0.15 |
| Orifice-type scrubber | 0.04 | 0.02 |
| Baghouse[c] | 0.1 | 0.05 |

[a] MT = Metric ton.
[b] Almost all plants have at least a precleaner following the rotary dryer.
[c] Emissions from a properly designed, installed, operated, and maintained collector can be as low as 0.005 to 0.020 lb/ton (0.0025 to 0.010 kg/MT).

gm/dscm) and less than 20% opacity] have been justified, and the costs of achievement detailed, by the USEPA (*13*). Compliance with this standard (which is comparable to many state and local governmental standards for existing sources in the United States) requires the use of fabric filters or high energy scrubbers. Some state and local agencies have revised their emission standards to require the use of baghouses, on the basis that this is the "best available demonstrated" control technique.

### F. Glass Manufacturing

The glass industry makes products from common minerals, silicates and alkalies, employing high temperature processes. This industry has usually not been considered as a major source of air pollutants. The primary reason for requiring control is because of plume opacity (the average particle size of particulate emissions is in the range of 0.1 μm) or because mass rates of particulate emissions marginally exceed regulatory emission standards (such as "process weight") that are unrelated to glass manufacturing. Arrandale has reported that "the rate of stack emission is actually dependent upon furnace operating temperatures, instead of upon process weight fed per hour" (*14*). Recent studies have confirmed that reduction in melter temperatures significantly affects particulate generation (*15*). Uncontrolled particulate matter concentrations in glass furnace stacks are normally less than 0.1 gr/scf (0.23 gm/scm) (*14*) and therefore would comply with most "grain loading" regulations. Many glass furnaces operate on natural draft, which has complicated evaluation to define "best available control technology," (which usually implies collector type) for regulatory purposes.

## 1. Melting

Glass furnaces are of various sizes and configurations, depending upon the type and quantity ("pull rate") of glass desired. Pot furnaces, for production of specialty glasses, are batch operations which usually do not exceed pull rates of 50 tons per day. Heat is provided by electricity, oil, or gas. Additives such as lead and metallic oxides or salts are used to color the final product or to facilitate molding or etching for decorative purposes. Large (i.e., pull rates up to 500 tons per day or greater) direct-fired (oil or gas) continuous melting furnaces are used in the production of container and flat glass. Approximately 90% of all glass production in the United States is for the container industry. These large refractory lined furnaces are divided into two sections—the melter area, where temperatures of approximately 2700°F (1482°C) are maintained, and the fining area [at aproximately 2200°F (1204°C)] where working viscosity is obtained and impurities and bubbles are released from the melt prior to forming.

Regenerative firing systems using brickwork chambers (similar to open-hearth steelmaking furnaces) are employed to recover heat from furnace gases. The thermal requirement for large furnaces is approximately $5.5 \times 10^6$ Btu ($11.4 \times 10^6$ kcal) per ton of glass.

## 2. Emissions

a. Batch Preparation and Furnace Charging. The batch usually consists of silica ($SiO_2$), alkalies (e.g., limestone), soda ash ($Na_2CO_3$), salt cake ($Na_2SO_4$), and cullet (broken glass). Dust emissions during storage, transfer, mixing, and weighing are controlled by techiques discussed in Section II. During charging of the furnace, dust may be released at the charging port, or part of the batch may be swept through the melting area by the furnace draft. Such emissions are usually of minor significance. Fabric collectors have been used to capture dust released from automatic batch feeding equipment. Adjusting furnace operating conditions (e.g., draft, charging configurations) can reduce "swept through" dust. Pelletizing the charge eliminates this source.

**Table IV Emissions from the Manufacture of Glass (*16*)**

| *Emissions* | *Flint glass furnace* | *Amber glass furnace* |
|---|---|---|
| Solids, gr/scf (gm/scm) | 0.029 (0.066) | 0.041 (0.094) |
| Solids, lb/hr (kg/hr) | 2.1 (0.95) | 5.4 (2.45) |
| Sulfur trioxide, ppm[a] | 17 | 15 |
| Sulfur dioxide, ppm | 250 | 315 |
| Fluorine, ppm | 2.2 | 1.9 |
| Chlorine, ppm | 4.9 | 4.1 |
| Nitrogen oxides, ppm | 340 | 640 |
| Carbon monoxide, ppm | 375 | 40 |

[a] Parts per million by volume for all gaseous pollutants.

b. Furnace Stack Gases. Stockham (*16*) has reported on emissions from two types of glass furnaces, Table IV.

Particle sizing of the captured particulate indicated that "the geometric median particle diameter of the flint glass particles was 0.13 $\mu$m and the geometric standard deviation was 1.5. Corresponding values for the amber glass furnace effluent particles were 0.11 $\mu$m and 1.7, respectively."

Furnace particulate matter emissions consist primarily of $Na_2SO_4$ that is formed and condenses (to a solid particle) in the regenerator checkerwork. Therefore, modifying batch composition and furnace operating conditions can reduce particulate emissions.

Various "emission factors" for particulate matter have been reported—3 lb/ton (1.36 kg/ton) (*17*), 5.3 lb/ton (2.40 kg/ton) (*14*), 2 lb/ton (0.91 kg/ton) (*9*). Caution must be used in applying these factors since, as pointed out earlier, the rate of particulate emission is not direcly related to production rate. Also, the presence of sulfur oxides in the stack gases can result in the formation of "pseudo-particulate" (condensibles) in a stack sampling train and lead to erroneously high results.

The source of sulfur oxides in stack gases are primarily sulfur in the furnace fuel and sulfates in the batch (e.g., salt cake). Recent stack tests on large gas-fired furnaces have shown sulfur oxide concentrations to be within regulatory requirements.

The amount of fluorides (and the ratio of gaseous to particulate fluoride) present in stack gases varies greatly with the type of glass being produced and the geographical source of batch materials. It is believed that gaseous fluorides are generally produced by the following reaction:

$$CaF_2 + SiO_2 + H_2O \rightarrow CaSiO_3 + 2\,HF \qquad (2)$$

and that any $SiF_4$ formed is likely to be hydrolyzed (*18*).

### 3. Particulate Control Equipment

The submicron particulate, produced at low concentrations, complicates selection of control equipment—especially since requirements for such equipment are based primarily on opacity problems. On a cost-effectiveness basis, incremental reductions in particulate emissions through process modification (e.g., furnace temperature or batch composition) is the most desirable approach. Collectors for large furnaces may be required to handle 200,000 cfm (94.4 $m^3$/second). Stack gas temperatures range from 700° to 1000°F (371° to 538°C). Although high collection efficiencies are not required to meet mass emission standards, reduction in plume opacity is difficult to predict.

Fabric filters have been employed to collect particulate from both batch charging and furnace emissions. Electrical precipitators have been installed on a number of container glass furnaces. High energy scrubbers have also been used in limited applications.

*Note on brick and ceramic ovens.* Smoke discharged at low levels from brick and ceramic curing ovens were, at one time, a major source of local nuisances. Improved combustion controls and fuel switching have significantly reduced these smoke emissions. It is often necessary to use soft coal to impart certain desired color characteristics to brick. Heavy smoke usually occurs during oven start-up. By using gas or oil during oven start-up periods and changing to soft coal later in the heating cycle, it is possible to obtain the desired brick color with a significant reduction in air pollution.

During curing operations smoke-producing fuels or smoke-producing firing techniques are sometimes employed to create a reducing atmosphere in ovens. Secondary combustion chambers in ovens or afterburners in stacks are used to control the resultant smoke emissions. Consideration should be given to modifying raw materials so that desired product characteristics may be obtained by curing in a smokeless atmosphere.

## *G. Concrete Batching*

Concrete batching plants are generally simple arrangements of steel hoppers, elevators, and batching scales for proportioning rock, gravel, and sand aggregates with cement for delivery, usually in transit mixer trucks. Aggregates are usually crushed and sized in separate plants and are delivered by truck or belt conveyors to ground or other storage from which they can be reclaimed and placed in the batch plant bunkers. Dust control procedures for these operations are described in Section II.

By careful use of sprays, felt, or other filter material over breathers in the cement silos, and canvas curtains drawn around the cement dump trucks while dumping, dust losses can be controlled. Aggregate stocks in bunkers should be wet down with sprays to prevent dusting. With careful operation, losses in these plants can be held to about 0.02 lb (0.91 gm) of dust/yd$^3$ of concrete. Uncontrolled plants have emissions of about 0.2 lb (9.1 gm) of dust/yd$^3$ of concrete handled. The air pollution problems and methods of control associated with wet and dry batching plants have been extensively reviewed by the Los Angeles County, California, Air Pollution Control District (*19*). The extent of emissions and methods of control can vary significantly depending upon type of plant (wet, dry, portable, permanent) and the characteristics of the aggregate (natural, expanded slag, size consist) used (*9*).

### H. Asbestos Processing

The special biological effects resulting from occupational exposure to asbestos dust have been recognized since the late 1920's. More recent studies have been conducted with respect to nonoccupational community exposures to asbestos dust (*20, 21*). Considerable controversy exists regarding the nature and extent of community air pollution resulting from asbestos emissions. Asbestos is unique as an air pollutant. No ambient air quality standards or standard sampling methods have been established. Standard source sampling methods also have not been developed. Therefore, it has not been possible to establish emission standards in terms of performance (i.e., kg/hour, gm/scm, etc.).

Regulatory specification type standards have been promulgated (*22*) and are being proposed (*23*) to control asbestos emissions. The standards apply to asbestos mills and the manufacture of cement products, fireproofing and insulating materials, friction products, paper, millboard, felt, floor tile, paints, coatings, caulks, adhesives, sealants, plastics and rubber materials, and chlorine, and have been proposed for the manufacture of shotgun shells and asphalt concrete. The standards require that either there be no visible emissions to the outside air from manufacturing operations, or specific air pollution control devices (primarily fabric filters) be used to control emissions. Operating and design characteristics of the collectors are specified in these unique standards. Operating type standards have been established for the surfacing of roadways and demolition of buildings (*22*).

The major sources of asbestos emissions were identified in a USEPA contracted study (Table V) (*24*). This "emission inventory" (which was not developed from emission tests) and a review of health effects information, provided the basis for the types of sources to be controlled and a rationale for requiring control. Recent regulatory standards are addressed primarily toward the control of asbestos emissions from milling and manufacturing operations.

**Table V Asbestos Emissions—1968** *(24)*

| *Category* | *Source* | *Short tons* |
|---|---|---|
| Mining and other basic processing | Mining and milling | 5610 |
| Reprocessing | Friction materials | 312 |
| | Asbestos cement products | 205 |
| | Textiles | 18 |
| | Paper | 15 |
| | Floor tile | 100 |
| | Miscellaneous | 28 |
| Consumptive uses | Construction | 61 |
| | Brake linings | 190 |
| | Steel fireproofing | 15 |
| | Insulating cement | 25 |
| | | 6579 |

## 1. Milling

Fiber production from ore presents unique problems in air pollution control because of the large amounts of dust produced by milling equipment and the large amounts of air used in these operations. Milling operations vary, depending upon the type of rock to be processed. The basic operations usually are crushing and drying the ore, fiber separation, and grading.

Primary and secondary crushing, close-circuited with vibrating screens (Fig. 1), is used to prepare mine rock for drying and milling. The crushed rock is dried in vertical or rotary dryers to remove surface moisture. Dryer gas temperatures vary between 200° (93°C) and 350°F (177°C). Emissions are usually controlled by a wet-type collector. Because of possible plugging, fabric collectors are not often employed. Regulatory specification standards for wet collectors employed to control asbestos dusts provide that they be designed to operate with a unit contacting energy of at least 40 in. (102 cm) water gauge (wg) pressure (*22*).

After drying, the ore may be subjected to additional crushing before being processed to separate fibers. During milling the fiber is released from the ore, fluffed up, and separated by aspiration (*25*). Shaking screens release the fiber and the lighter fibers are aspirated through an air exhaust hood and duct system to cyclone-type collectors, while coarse rock particles and unopened bundles of fiber cascade over the end of the screens, and fines drop through the screens. As indicated above, emissions from all collectors (including inertial types) are required to meet zero opacity standards. Inertial collectors are usually considered as part of the processing equipment and are followed, in series, by fabric filters.

This extraction process is repeated several times over a series of successively finer mesh screens, each treating the fines from the previous screen. The unaspirated overs from these screens go through fiberizers to release and open up the rest of the fiber and these products are also fed to aspirating screens. An asbestos mill may use several hundred screens. A variety of crushing machines are used, such as cone, gyratory, hammermill, or disk type. Fiberizers are often hammermills, but are generally run at slower speeds than the usual type hammermill.

The fibers recovered from the various milling stages are cleaned and graded to market specifications. This involves another series of screening operations with fiber aspiration, using shaking-type screens or special forms of revolving trommel screens and various types of air separators. Fractions of the various sizes obtained from these operations may be blended to produce a wide range of fiber grades, before being conveyed to bins from which they are drawn off to mechanical packing or bagging machines.

Large quantities of dust are generated during dry asbestos milling. However, pneumatic transport is an inherent process component so that only a fraction of the total air moving capacity in a mill is provided exclusively for dust control. About 75 tons of process air may be handled for each ton of asbestos fiber produced, supplemented by an additional 25 tons of air exhausted for dust removal at nonaspirated screens, belt conveyors, bucket elevators, rock and fiber bins, graders, hoppers, and pressure-packing machines. Central air handling and filtering systems usually provide for combined process and dust control air (*26*).

## 2. Product Manufacturing

Chrysotile, a hydrous magnesium silicate having the theoretical formula $3\ MgO \cdot 2\ SiO_2 \cdot 2\ H_2O$, is the most often used type of asbestos. Asbestos fiber is used to manufacture textiles (safety clothing, curtains, lagging cloth, brake linings, clutch facings), cement products (pipes, sheets, shingles), plastics (as filler), and many other products. The asbestos cement industry is the principal consumer of chrysotile asbestos—on a worldwide basis, approximately two-thirds of the asbestos produced is used to manufacture cement products.

Because of the awareness of the industry to the occupational health aspects of asbestos, processing and handling equipment is routinely equipped with local exhaust ventilation systems. These systems usually discharge to a fabric filter collector—or a wet scrubber, if moisture in the exhaust gases may cause filter plugging.

## IV. Emission Reduction during Air Pollution Episodes

Governmental requirements, for procedures to be followed during air pollution episodes, treat the mineral product industries in the same manner as other manufacturing industries. Reduction in emissions is required through curtailing, postponing, or deferring production. At "warning" or "emergency" levels, elimination of emissions by ceasing or curtailing production, to the extent possible without causing injury to employees or equipment, is required.

Except for glass manufacturing, where emission rates are primarily related to furnace operating temperature, mineral process emission rates can be reduced by reducing production rates. Damage to equipment and refractory furnace linings from thermal shock caused by abrupt shutdown is an obvious concern. Consideration should also be given to preventing excessive emissions while increasing production to normal levels, following an episode.

## V. Conclusion

The control of particulate matter emissions, the primary pollutant emitted during mineral processing, is a major activity of governmental air pollution control agencies. Particulate matter is the most obvious pollutant and (with the possible exception of odors) the most common reason for air pollution complaints by the public. Particulate matter emission standards have become increasingly severe. Technology is generally available to control this pollutant.

Experience has shown that many air pollution problems associated with mineral processing are caused by overloading or not operating process equipment at design capacity. Also, the importance of good plant "housekeeping" cannot be overstressed. Overloading kilns, dryers, and furnaces results in higher gas velocities which tend to pick up more dust and overload collectors. Many air pollution problems can be abated or substantially reduced by modifying operations—changing processing techniques and material handling procedures; providing for proper equipment maintenance—hopper clean-out, enclosure and duct-work repair; and good "housekeeping"—clean roadways, enclosed storage piles, and vacuum sweeping systems. In selecting dust collectors, it is important to consider the size consist and other characteristics of the collected dust which will permit recovery in a usable form. The size consist is usually the most important factor in determining if recovered mineral dust can be reintroduced into the process or directly become a salable product. The proper selection and arrangement of collectors (often placing two or three types in series to fractionate the dust stream) will often permit recovery of some of the dust in a usable form.

## REFERENCES


1. American Conference of Governmental Industrial Hygienists, "Industrial Ventilation, A Manual of Recommended Practice," 13th ed. Committee on Industrial Ventilation, Lansing, Michigan, 1974.
2. Much of the information for Sections III, A, B, C and D related to pending EPA technical reports on new source control technology was obtained from S. T. Cuffe [personal communications, U.S. Environmental Protection Agency (EPA). Research Triangle Park, North Carolina].
3. D. R. Mitchell, "Coal Preparation," 2nd ed., p. 820. Amer. Inst. Mining Met. Eng., New York, New York, 1950.
4. United States Federal Register, *Fed. Regist.* **39** (207), 37922-37924 (1974).
5. Community Air Sampling Files. Pennsylvania Dept. of Health, Harrisburg, Pennsylvania, 1959-1966.
6. V. H. Sussman and J. J. Mulhern, *J. Air Pollut. Contr. Ass.* **14,** 279–284 (1964).
7. D. Harrington and J. H. East, Jr., *U.S., Bur. of Mines, Inform. Circ.* **1C 7439** (1948).
8. T. E. Kreichelt, D. A. Kemnitz, and S. T. Cuffe, "Atmospheric Emissions From the Manufacture of Portland Cement," U.S. Pub. Health Serv. Publ. No. 999-AP-17. U.S. Dept. of Health, Education, and Welfare, Cincinnati, Ohio, 1962.
9. "Compilation of Air Pollutant Emission Factors," 2nd ed., Publ. AP-42. U.S. Environmental Protection Agency, Research Triangle Park, North Carolina, 1973.
10. R. J. Plass, *in* "Proceedings of the Electrostatic Precipitation Seminar." Pennsylvania State University, State College, Pennsylvania, 1960.
11. United States Federal Register, *Fed. Regist.* **39** (219), 39872-39877 (1974).
12. United States Federal Register, *Fed. Regist.* **36** (247), 24880 (1971).
13. "Background Information for New Source Performance Standards," EPA Publ. 450/2-74-003 (APTD-1352C). United States Environmental Protection Agency, Washington, D.C., 1974.
14. R. S. Arrandale, *in* "Symposium sur fusion verre," pp. 619–644. Brussels, Belgium, 1958 (in English).
15. Engineering Study Program prepared for the Glass Container Manufacturers' Institute, Washington, D.C., by TRW Systems Group, 1971.
16. J. D. Stockham, *J. Air Pollut. Contr. Ass.* **21,** 713-715 (1971).
17. G. P. Larson, *Ind. Eng. Chem.* **45,** 1070–1074 (1953).
18. A. P. Konopka and N. W. Frisch, 73rd Annu. Meet. Amer. Ceram. Soc., Columbus, Ohio, 1971.
19. J. A. Danielson, ed., "Air Pollution Engineering Manual," 2nd ed., Publ. No. AP-40, pp. 334–346. U.S. Environmental Protection Agency, Research Triangle Park, North Carolina, 1973.

20. "Background Information—Proposed National Emission Standards for Hazardous Air Pollutants: Asbestos, Beryllium, Mercury," Publ. No. APTD-0753. U.S. Environmental Protection Agency, Research Triangle Park, North Carolina, 1971.
21. "Background Information on Development of National Emission Standards for Hazardous Air Pollutants: Asbestos, Beryllium, and Mercury," Publ. No. APTD-1503. U.S. Environmental Protection Agency, Research Triangle Park, North Carolina, 1973.
22. United States Federal Register, *Fed. Regist.* **38** (66), 8820-8830 (1973).
23. United States Federal Register, *Fed. Regist.* **39** (208), 38064-38073 (1974).
24. "National Inventory of Sources and Emissions—Cadmium, Nickel, and Asbestos." Report by W. E. Davis & Associates under Contract No. CPA 22-69-131 to Department of Health, Education, and Welfare, Washington, D.C., 1970.
25. J. C. Kelleher, *Asbestos* **27**(3), 2-10; **27**(4), 3-10; **27**(5), 6-12 (1945).
26. L. Rispler and C. R. Foss, "Flow Sheet for Milling of Asbestos Ores," ACGIH Process Flow Sheets. American Conference of Governmental Industrial Hygienists, Cincinnati, Ohio, 1961.

# 18

## Chemical Industries

Stanley T. Cuffe, Robert T. Walsh, and Leslie B. Evans

## A. INORGANIC CHEMICAL PROCESSES

### I. Introduction

This section discusses the air pollution control aspects of the manufacture of eight principal inorganic acids and alkalies, phosphate fertilizers, ammonium nitrate, and the halogens bromine and chlorine (Table I) (*1*). Of these processes, phosphate fertilizers and sulfuric acid have undergone

**Table I Production of Major Inorganic Chemicals—United States (*1*)**

| *Chemical* | *Number of plants (year)* | *Thousands of short tons per year*[a] *1970* | *1971* |
|---|---|---|---|
| Ammonium nitrate | 76 (1970) | 6,456 (5855) | 6,605 |
| Fertilizer | 63 | 5,502 (4990) | 5,660 |
| Industrial explosives and other uses | 29 | 954 (865) | 945 |
| Bromine | 10 (1971) | 175 (158) | 178 |
| Chlorine | 76 (1971) | 14,192 | 13,824 |
| Hydrochloric acid, total (100%) | 88 (1971) | 2,013 | 2,099 |
| From salt and acid | | 125 | 116 |
| From chlorine | | 95 | 98 |
| By-product and other | | 1,793 | 1,886 |
| Hydrofluoric acid (100%) | 15 (1971) | 325 (295) | 333 |
| Lime, includes quick line, hydrated lime and dead-burned dolomite | 188 (1971) | 19,747 | 20,290 |
| Nitric acid (100%) | 100 (1968) | 6,679 | 6,792 |
| Phosphate rock | 74 (1971) | 38,739 | 38,886 |
| Phosphoric acid (100% $P_2O_5$) | 57 (1971) | 5,682 | 6,240 |
| From phosphorous (thermal) | 27 | 1,041 | 954 |
| From other sources (wet) | 30 | 4,642 | 5,286 |
| Sodium carbonate | 12 (1971) | 7,092 (6432) | 7,153 |
| Synthetic | 7 | 4,414 (4003) | 4,275 |
| Natural | 5 | 2,678 (2429) | 2,878 |
| Sodium hydroxide (100% NaOH) | 70 (1971) | 10,141 | 9,667 |
| Liquid | | 9,577 | 9,123 |
| Solid | | 564 | 544 |
| Sulfuric acid (100% $H_2SO_4$) | 184 (1971) | 29,524 | 29,422 |
| Chamber process | 16 | 320 | 226 |
| Contact process | 167 | 29,204 | 29,196 |

[a] Values in parentheses are for metric tons.

the most significant changes, particularly in marketing, during the past several years. Because of the increased world population and resulting demand for more agricultural products, fertilizer consumption in the United States and other noncommunist countries should double between 1965 and 1980 (*2*).

The advent of many sulfur dioxide recovery plants for making contact sulfuric acid has changed competition in the marketplace considerably. Recovery plants are expected to continue to operate at full capacity, regardless of the market price of acid, to conform to air pollution control regulations. They therefore dispose of acid at any obtainable price and may well sell acid at a zero dollar price as an alternate to other means of disposal (*3*). Conventional sulfuric acid plants can therefore compete with recovery plants only when they are outside of the shipping radius of the recovery plant.

## II. Hydrochloric Acid

Most of the hydrochloric acid manufactured in the United States is a by-product of organic chlorination processes. The remaining 10% is split about equally between the salt process and the direct synthesis of hydrogen and chlorine. In recent years, the by-product process has produced increasingly greater amounts of hydrochloric acid while the salt and synthesis processes have declined in use (*4*).

All of the processes use water or weak hydrochloric acid to absorb hydrogen chloride and produce acid of varying strengths. Hydrogen chloride gas is extremely soluble in water. Losses of hydrogen chloride to the atmosphere can occur by transferring acid and from leaks and spills. However, the greatest single loss is in tail gases from absorbers.

### *A. By-product Hydrogen Chloride*

Hydrogen chloride is released in exhaust gases from many processes in which organic materials are chlorinated. Examples are the production of carbon tetrachloride from methane and chlorine, and chlorobenzene from chlorine and benezene. Much of the product acid is, in turn, used to manufacture organic compounds. In most of these applications, the hydrogen chloride is contaminated with chlorine, air, organic products, and moisture. In a few instances, it may be in relatively pure form. The concentration of hydrogen chloride can vary significantly from one process to another. By-product hydrogen chloride processes for the production of acid differ mainly in the makeup of feedstocks and in the devices or processes used to remove or further process other components of the gas stream.

The flow diagram of Figure 1 shows a sample situation in which hydrogen chloride, chloromethane, and methane are evolved from a pressurized process used to chlorinate methane (*5*). Hydrogen chloride is absorbed in a series of three cocurrent absorbers. In the first tower, rich gases from chlorinators are contacted with medium strength acid to produce strong acid containing some organics. After decanting carbon tetrachloride, the product acid is stripped of organics, cooled, and filtered. Overhead gases from the stripper are contacted with makeup water to produce 5–10% strength acid that is fed to the second absorber. Chloromethane and light organics are present in gases exhausted from the stripper absorber and

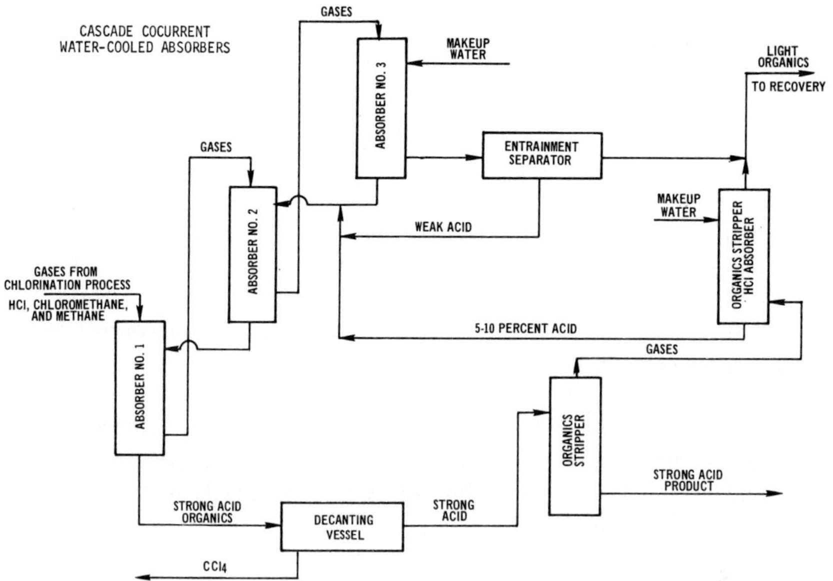


Figure 1. Flow diagram of the by-product process for manufacture of hydrochloric acid, noting potential air pollution sources.

from the third-stage absorber. These potential air pollutants can be controlled by cooling, condensation, or adsorption.

## Air Pollutant Emissions and Their Control

Hydrogen chloride emissions from by-product processes depend on a number of factors including the chlorination process from which the feedstocks are derived and the concentration of hydrogen chloride entering the system. It is possible to remove almost all of the hydrogen chloride from the exhaust gases if so desired. Reported losses range from negligible to 8.5 lb of hydrogen chloride per ton (4.3 kg/metric ton) of 20° to 22° Baumé acid produced (*6*). Where tail gas scrubbers are not used, emissions have been estimated to average about 3.5 lb per ton (1.8 kg/metric ton). In most cases, at least two absorption towers are utilized with weak acid from the last absorber returned to the primary tower. If greater control of hydrogen chloride is required, the absorbing media can be chilled or pressures greater than atmospheric can be utilized.

### *B. Salt Process*

The salt process involves common salt (NaCl) and 60° or 66° Baumé sulfuric acid. The products of the reaction are hydrogen chloride, sodium bisulfate ($NaHSO_4$), and the normal sulfate ($Na_2SO_4$). The equations involved are

$$NaCl + H_2SO_4 \rightarrow HCl + NaHSO_4 \quad (1)$$

$$NaCl + NaHSO_4 \rightarrow HCl + Na_2SO_4 \quad (2)$$

$$2\ NaCl + H_2SO_4 \rightarrow 2\ HCl + Na_2SO_4 \text{ (salt cake)} \quad (3)$$

Salt and sulfuric acid are indirectly heated in a closed furnace (usually Mannheim) to 1400°–1600°F (Fig. 2). Modern furnaces are kept under negative pressure [—0.05 to —0.50 in. of water (0.1 — 1.0 mm Hg)]. Entrained solids in the exhaust gases are controlled by dust control equipment such as settling chambers and cyclones.

The gases evolved, 30–70 vol % hydrogen chloride, are cooled to about 100°F and purified prior to passing to a water absorber. Different types of absorbers have been used, but, in the main, a packed-tower system or a cooled absorption tower followed by a packed tail-gas tower or falling-film-type tower are used.

## Air Pollutant Emissions and Their Control

Losses of hydrogen chloride may occur in the exhaust gases from the furnace, the hot salt cake, or the tail gases (*7*). Furnace losses in older units may occur at the doors, particularly if they are hand operated. Control usually consists of hooding the doors and scrubbing the exit gases with water. In more modern furnaces, control is exercised by maintaining a slight negative pressure in the muffle. The potential loss of hydrogen

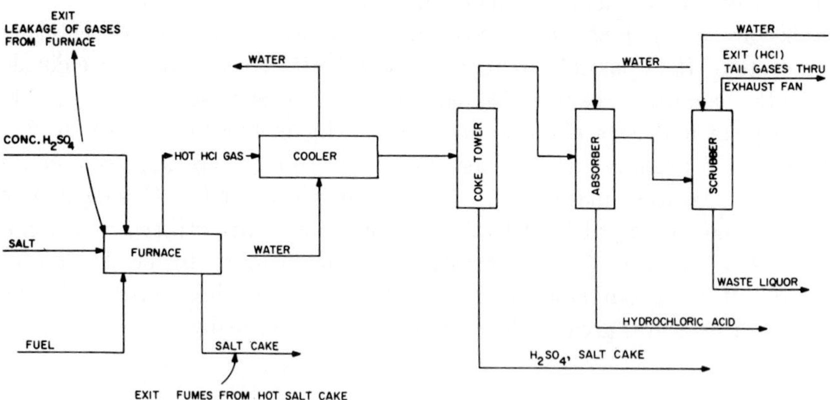


Figure 2. Flow diagram of the salt process for the manufacture of hydrochloric acid, noting potential air pollution sources.

chloride from salt cake (removed from the oven at about 1000°F) can be controlled by cooling the cake in a water-cooled screw conveyor prior to dumping.

Tail-gas emission of hydrogen chloride has been reported to range from 1.3 to 3.9 lb per ton (2.0 kg/ton) of acid produced (0.06 to 0.46% by volume). Much of this emission is attributed to inadequate maintenance (*8*).

Upsets in the absorption system due to improper temperature control or insufficient feed water usually can be corrected quickly with minor adjustments. In the case of cooled-tower absorption systems, the addition of water to the tail tower of the absorption system may be automatically controlled. The temperature of the tail-tower acid exit gas is correlated with the strength of the product acid from the absorber. Control of water addition to the tail tower is usually gauged by product acid strength and tower temperature.

### C. Synthetic Hydrogen Chloride

Production of hydrochloric acid by direct synthesis is used where high purity hydrochloric acid is required. The acid produced by this method in the United States has shown a steady decline: 120,000 tons in 1969 versus 150,000 tons in 1966 and 184,300 tons in 1959.

Hydrogen is burned with chlorine in a reactor to produce a high purity hydrogen chloride gas (98 to 99.7%) (*9*) (Fig. 3). Product gas purity is dependent on the purity of the raw materials. The absorption and recovery of acid is similar to the processes previously described.

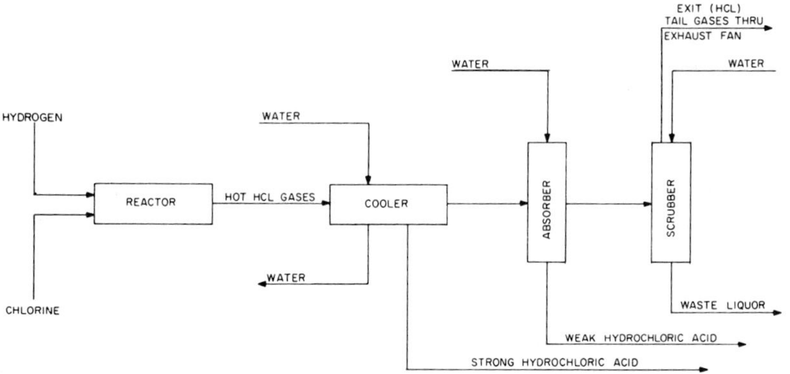


Figure 3. Flow diagram of the synthetic process for the manufacture of hydrochloric acid, noting potential air pollution sources.

### Air Pollutant Emissions and Their Control

Since the high purity acid of the synthesis process commands a higher price than lesser grades, operators generally absorb almost all hydrogen chloride from tail gases. Emissions have been reported at less than 0.01% hydrogen chloride in tail gases (*10*).

### *D. Emission Reduction during Air Pollution Episodes*

During episode conditions, hydrocholic acid plants can be shut down, or hydrogen chloride emissions can be curtailed by reducing production or increasing absorption. At by-product plants, the acid plant, which is in reality an air pollution control device, cannot be shut down independently of the process it serves. Significant hydrogen chloride reductions can be effected by reducing production rate or improving absorption from tail gases. Additional absorption capacity can be employed during episodes, if necessary.

## III. Hydrofluoric Acid

In 1972 approximately 360,000 tons (326,520 metric tons) of hydrofluoric acid (100% HF equivalent) were produced in the United States in 15 plants located in 9 states (*11*).

The aluminum industry used 44% of the total production in the manufacture of aluminum fluoride and synthetic cryolite. Fluorocarbon production, used primarily for propellants in aerosol cans, consumed 42%. Use as alkylation catalyst in the petroleum industry and the separation of uranium by the United States Energy Research and Development Agency together accounted for 5%. All hydrofluoric acid is manufactured by the fluorspar–sulfuric acid process (*12*).

### *A. Fluorspar–Sulfuric Acid Process*

The primary reaction for the production of hydrofluoric acid from fluorspar and sulfuric acid is

$$CaF_2 + H_2SO_4 \xrightarrow{\text{heat}} CaSO_4 + 2\ HF \tag{4}$$

Acid grade fluorspar concentrate contains about 98% calcium fluoride. Any impurities present enter into side reactions which lower yield and may produce gases other than hydrogen fluoride, which make pollution control more difficult. Typically, the reaction takes place in a rotary kiln which is externally fired to about 300°C to provide the heat required to drive the reaction to completion. The solid calcium sulfate residue from the kiln is slurried with lime to neutralize any remaining acid and to prevent further evolution of hydrogen fluoride. The gas stream evolved from the kiln also contains water vapor, silicon tetrafluoride, sulfur dioxide, carbon dioxide, and fluorspar–calcium sulfate dust. There are several different methods for recovering the hydrogen fluoride from this stream. The method used depends upon the purity of the acid required, and whether an anhydrous acid or water solution is most convenient. In the classic plant shown in Figure 4, the solids are removed by a dust separator and by scrubbing with the incoming sulfuric acid. Aqueous hydrofluoric acid solutions are formed by absorbing the gases in a series of countercurrent absorption and cooling towers. The acid obtained can be blended to the strength required or distilled to the anhydrous acid. In some more modern plants, the water from the off-gas in the kiln is removed by scrubbing with sulfuric acid and sufficient sulfur trioxide (oleum) to react with the water

$$H_2O + SO_3 \rightarrow H_2SO_4 \tag{5}$$

The anhydrous hydrogen fluoride is then condensed in refrigerated exchangers and redistilled to 99.9% purity.

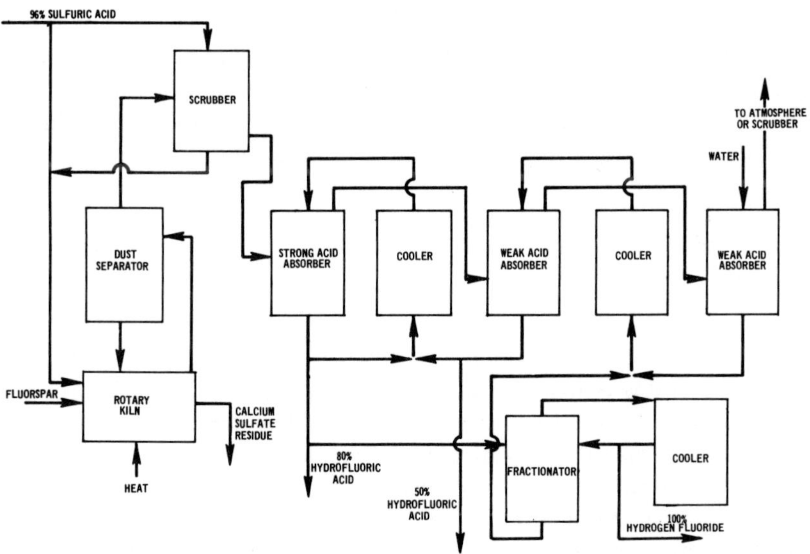


Figure 4. Flow diagram of the fluorspar–sulfuric acid process for the manufacture of hydrofluoric acid, noting potential pollution sources.

### Air Pollutant Emissions and Their Control

The major potential source of air pollutants are the tail gases from the final acid absorber or condenser which contain unrecovered hydrogen fluoride and various amounts of air, water, carbon dioxide, and silicon tetrafluoride ($SiF_4$). The amount of these gases depends on the impurities in the fluorospar and the type and efficiency of the recovery system for hydrogen fluoride.

The silicon tetrafluoride and remaining hydrogen fluoride are usually removed by water scrubbing. The 30–35% fluosilicic acid produced is available for use as a raw material for other products. The vent gas from the scrubber will contain carbon dioxide and air with trace quantities of sulfur dioxide and silicon tetrafluoride. A caustic soda scrubber could be used for more complete clean up.

One source estimates that a solution plant similar to that shown in Figure 4, producing 25 tons of HF per day (50% as anhydrous, 25% each as 50 and 80% HF) would emit 26 lb/hour of hydrogen fluoride and 34 lb/hour of silicon tetrafluoride from the weak acid absorber vent (*13*). It is further estimated that a spray scrubber which could reduce this emission by 90% would have a total installed cost of $11,000 (1971).

### *B. Emission Reduction during Air Pollution Episodes*

Atmospheric emissions from hydrofluoric acid plants can be reduced significantly or stopped completely within a period of 4 hours. Emissions from the process are proportional to production rate, and the production rate may be reduced to less than half of maximum without operating difficulties. There is no increase in emissions during shutdown or start-up.

## IV. Nitric Acid

Almost all of the 6.5 million tons of nitric acid produced annually in the United States is weak acid, i.e., 55 to 70% strength and is produced by the ammonia oxidation process (AOP). About 75% of production is utilized in the manufacture of ammonium nitrate which in turn is used primarily as fertilizer and only secondarily as a commercial explosive. High strength nitric acid also finds use in nitration processes and as a component of rocket fuels. Existing plants are generally of less than 500 tons per da' capacity (100% acid basis). New plants are being built with capacities ranging up to 880 tons per day.

### A. Pressure Process

In the pressure process (Fig. 5) which is used most extensively in the United States, a preheated mixture of 90% air and 10% ammonia by volume is passed through a platinum–rhodium catalyst. The oxidation of ammonia takes place at about 1650°F and 112 psig (5790 mm Hg) attaining a nitric oxide (NO) conversion of about 95%. Resultant gases containing about 10% nitric oxide are cooled to about 100°F and acid mist is removed. Cooled nitric oxide reacts with oxygen to form nitrogen dioxide ($NO_2$) and its dimer, dinitrogen tetroxide ($N_2O_4$). In the absorber, nitrogen dioxide or dinitrogen tetroxide is absorbed in water and dilute acid to produce nitric acid of 50 to 70% strength.

Exit gases from the absorber contain both nitric oxide and nitrogen dioxide. Tower exit temperatures are approximately 85°F, but can vary considerably depending upon the temperature of the cooling water and the makeup water. Tail gases from the pressure process are heated by exchange with hot process gases, and energy is recovered in centrifugal expanders that drive the air compressor. In those cases where catalytic abatement devices are utilized to destroy oxides of nitrogen or decolorize nitrogen dioxide, further heat exchange equipment is used to recover energy before releasing gases through the centrifugal expander.

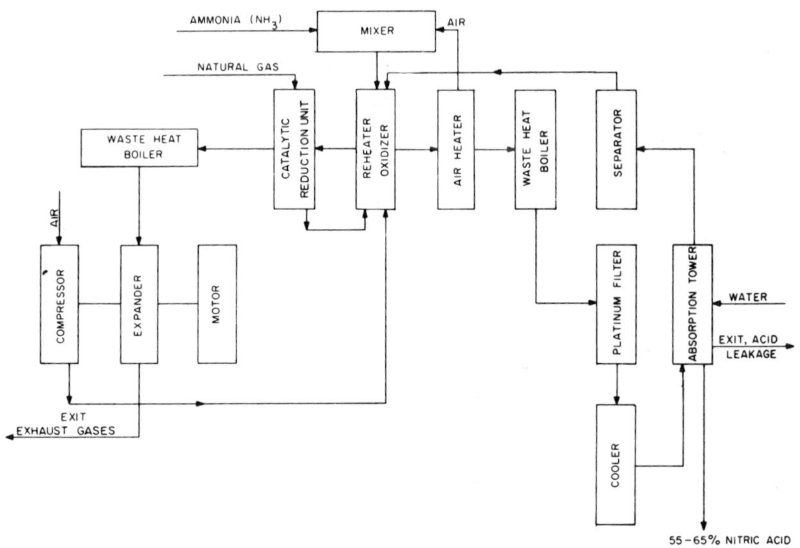


Figure 5. Flow diagram of a nitric acid plant using the pressure process, noting potential air pollution sources.

The original AOP plants were operated at atmospheric pressure and absorption of nitrogen dioxide was relatively slow. Variations of the pressure process are sometimes utilized to obtain more efficient conversion of ammonia to nitric oxide. In one variation, ammonia is oxidized at pressures of zero to 30 psig with absorption being accomplished at 30 to 50 psig (1550 to 2585 mm Hg). In the intermediate pressure process, both oxidation and absorption are conducted at 20 to 60 psig (1000 to 3000 mm Hg).

### B. Direct Strong Acid Process

Nitric acid of 98% strength can be produced by concentrating weak acid or by the direct strong acid process. In the concentration process (Fig. 6), a mixture of weak acid and concentrated sulfuric acid is distilled. Nitric acid is evaporated and condensed in concentrated form while sulfuric acid and water remain. An alternative concentration process uses magnesium nitrate instead of sulfuric acid (*14*).

The direct strong acid process is an ammonia oxidation system in which dinitrogen tetroxide is condensed from the system by refrigeration. The condensed dinitrogen tetroxide is autoclaved at 750 psig (40,000 mm Hg) together with water or weak acid and oxygen to produce an acid of about 98% strength. The direct strong acid process has been utilized in Europe to a greater degree than in the United States (*15*).

## Air Pollutant Emissions and Their Control

The major source of air pollution from nitric acid manufacture is the tail gas stream from weak acid production processes. The principal un-

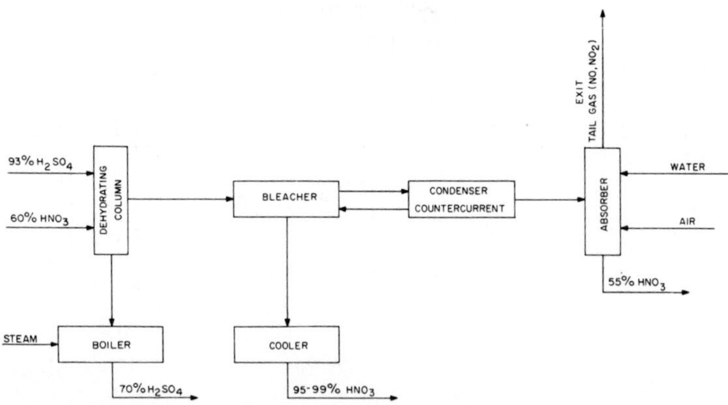


Figure 6. Flow diagram of the process for concentrating nitric acid, noting potential air pollution sources.

absorbed oxides of nitrogen are nitric oxide and nitrogen dioxide. Emissions vary with plant design, acid strength, maintenance practices, production rate, and absorber efficiences. For well-maintained and properly operated plants without air pollution control devices, nitrogen oxide concentrations normally average between 0.2 and 0.4% by volume. However, levels as low as 0.1% have been reported. In the acid concentration process, about 0.7% of the acid processed is lost in the distillation process (*16*).

If there are significant concentrations of nitrogen dioxide, a visible reddish-brown color will be noted in the exit gases. Such color is attributable to nitrogen dioxide and dinitrogen tetroxide. Nitric oxide is colorless. In general, those factors which decrease absorption tend to increase visible emissions.

Nitric acid plants were estimated to release about 120,000 tons of nitrogen oxides to the atmosphere of the United States in 1970. This figure would have been approximately 50% greater had not emission controls been utilized. In 1971, the United States Environmental Protection Agency (USEPA) promulgated nitrogen oxides emission standards which are applicable to all new nitric acid plants installed in the United States (*17*). The standards govern only weak acid plants, not those which produce strong acid directly or which concentrate weak acid into strong acid. Oxides of nitrogen (measured as nitrogen dioxide) are limited to 3 lb per ton of 100% acid produced or about 210 ppm nitrogen oxides. Visible emissions are to be held to less than 10% opacity.

Emissions of nitrogen oxides from weak acid plants can be controlled by several means including catalytic reduction, adsorption, and absorption. To date, the catalytic reduction process (Fig. 7) has been utilized to a much greater degree than any other nitrogen oxides reduction scheme. Exit gases are heated to about 900°F and mixed with fuel, usually natural gas or hydrogen-rich ammonia plant purge gas, before passing over a catalyst bed (*18, 19*). Reduction to nitric oxide and subse-

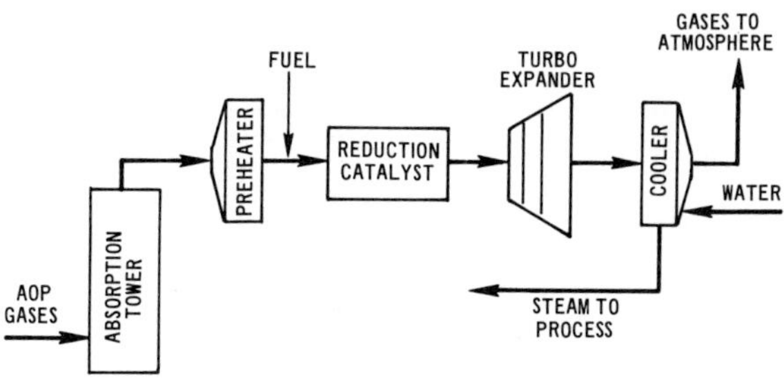


Figure 7. Flow diagram of a catalytic abatement system for a nitric acid plant.

quently to nitrogen, water, and carbon dioxide occurs stepwise. The following reactions, using methane as fuel, illustrate the major reactions:

$$CH_4 + 2\ O_2 \rightarrow CO_2 + 2\ H_2O \quad (6)$$

$$CH_2 + 4\ NO_2 \rightarrow 4\ NO + CO_2 + 2\ H_2O \quad (7)$$

$$CH_4 + NO \rightarrow 2\ N_2 + CO_2 + 2\ H_2O \quad (8)$$

Reactions (6) and (7) proceed rapidly with the evolution of considerable heat. Reaction (7) converts opaque nitrogen dioxide and dinitrogen tetroxide, if present, to colorless nitric oxide; thus, the first two reactions describe "decolorizers" which have been utilized by many acid plants to render tail gases invisible. In Reaction (8), nitric oxide is decomposed to nitrogen, carbon dioxide, and water. The operation of decolorizing catalytic reduction systems is often profitable because of increased energy recovery. Full decomposition to nitrogen usually represents a fuel penalty. The effectiveness of catalytic reduction systems is highly dependent upon the catalyst and upon the fuels employed. Where high reactivity catalyst and hydrogen-rich fuels are utilized, it is usually possible to obtain exit levels of about 100 ppm nitrogen oxides (measured as nitrogen dioxide). Where natural gas is the fuel, it is often not possible to obtain levels much below 200 ppm nitrogen oxides. Catalysts can lose activity in stages. Loss of initial activity greatly inhibits Reaction (8) such that a full abatement system can become a decolorizer as the catalyst ages.

Extended absorption systems which can significantly reduce nitrogen oxides from weak acid plants are offered commercially. Extended absorption provides greater contact time in vessels which are considerably larger than typical AOP absorbers (*20*). The process was developed in Europe and has only recently been introduced in the United States. Designers are planning to achieve 200 ppm exit concentrations of nitrogen oxides by this process.

Molecular sieve absorption systems (Fig. 8) installed at two United

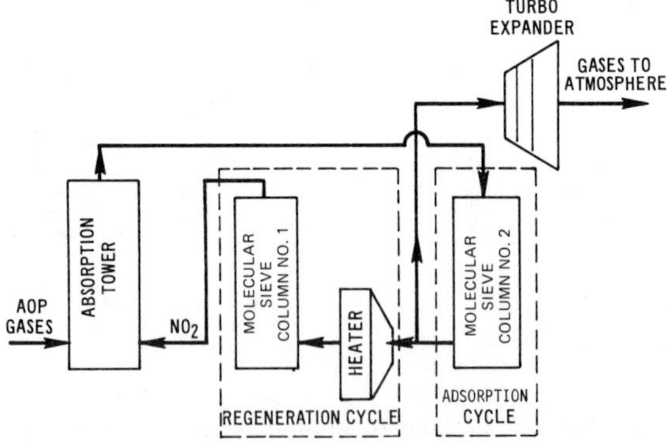


Figure 8. Flow diagram of a molecular sieve control system for a nitric acid plant.

States nitric acid plants became operational in 1974 (*21*). The sieves are synthetic zeolites which adsorb nitrogen dioxide. Desorbed nitrogen dioxide is pumped to the absorption tower to produce more acid. Regeneration is accomplished at 400°–500°F. Molecular sieves offer a possible means of achieving extremely low (less than 50 ppm nitrogen oxides) emissions from nitric acid plants. The viability of this system will depend to a large degree on the usable life of the sieves as well as the economics of regeneration.

Nitrogen oxides from nitric acid plants can be controlled by various alkaline scrubbing systems (*22*) as well as adsorption on activated carbon and silica gel. All of these have been studied in the laboratory but none have been installed in full-scale plant operation. A principal drawback to alkaline scrubbing systems is the extremely slow reaction time between alkali and nitrogen oxides.

### *C. Emission Reduction during Air Pollution Episodes*

AOP plants cannot be shut down abruptly since components are sensitive to thermal shock. Start-up and shutdown will result in above average nitrogen oxides emissions for varying periods of time. At uncontrolled plants, reducing production rate reduces the quantity of nitrogen oxides released per ton of product as well as the product rate itself. The extent of such reduction will vary, but it is reasonable to expect that many uncontrolled weak acid plants could reduce nitrogen oxides emissions below 1000 ppm by cutting back production. Catalytic reduction systems may operate at greater or lesser efficiency under partial load. Conceivably, it could be difficult to maintain the proper temperature profiles and oxygen levels in the system at reduced loads. Plants equipped with molecular sieves as well as extended absorption should provide better than normal control at reduced production rates.

## V. Elemental Phosphorus and Phosphoric Acid

### *A. Elemental Phosphorus*

An elemental phosphorus plant is essentially a solids handling process. Blended phosphate ore of sand-to-gravel size is charged to a rotary kiln where it is heated to incipient fusion and agglomerated to a clinker of 6-in. maximum particle size. The hot clinker from the kiln is cooled by direct contact with air, crushed, and sized. The resulting phosphate nodules, coke that has been dried and screened, and silica are fed by gravity from overhead silos into an electric furnace. A typical 25,000-ton/year furnace with prebaked carbon electrodes consumes about 25,000 kW/hour and operates between 2300° and 2700°F. The furnace and phosphorus recovery system operate at a slight positive pressure to preclude in-leakage of air. The overall furnace reaction is

$$2\ Ca_3(PO_4)_2 + 10\ C + 6\ SiO_2 \rightarrow P_4 + 10\ CO + 6\ CaSiO_3 \quad (9)$$

Elemental phosphorus vapor and carbon monoxide flow overhead from the furnace as a gas and through high temperature electrostatic precipitators for removal of dust (Fig. 9). The exit gases from the precipitators pass through direct-contact water scrubbers to condense the phosphorus. The carbon monoxide stream discharged from the scrubber contains about 0.4% phosphorus and is burned as fuel in the calcining operations and, in some cases, in process steam boilers. The furnace slag is tapped to open pits and cooled with water while the ferrophosphorus is tapped into molds and cooled. The condensed phosphorus is pumped to steam-heated tanks which store the liquid phosphorus under a blanket of water (*24*).

## Air Pollutant Emissions and Their Control

The principal air pollutants are particulate matter and fluorides. The main sources of atmospheric emissions are the phosphate rock calciner and the electric furnace slag-tapping operation. Air pollutant emissions from the calciner include fluorides, nitrogen oxides, sulfur oxides, and particulate matter consisting of phosphorus pentoxide fume and phosphate rock dust. Atmospheric emissions from the furnace slag-tapping

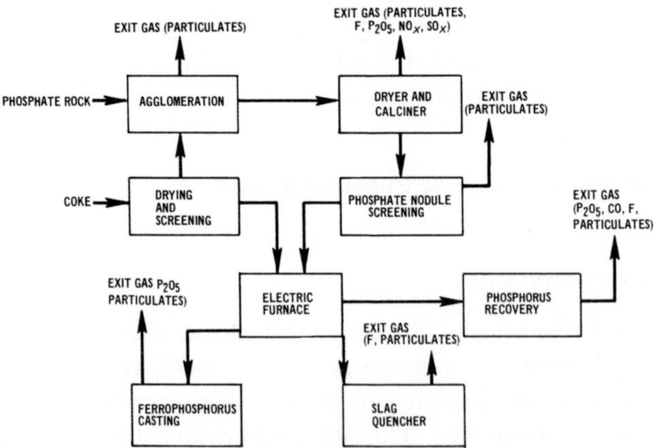


Figure 9. Flow diagram for the manufacture of elemental phosphorus, noting potential air pollution sources.

operation include fluorides, phosphorus pentoxide fume, and sulfur oxides. Phosphorus pentoxide is also emitted from the top of the furnace and from the electrode seals. Other sources of particulate emissions include blending of phosphate rock; clinker cooling; coke drying; screening of coke and phosphate nodules; recovery of phosphorus from sludge; and the casting, sizing, and shipping of ferrophosphorus.

Fluorides and particulate matter from the calciner are normally controlled by centrifugal collectors followed by water scrubbers having a low pressure drop, i.e., 3 in. of water (5.61 mm Hg). Emissions from this type of control system will range from 0.002 to 0.009 grain of total F per dry standard cubic foot (dscf) (0.0046 to 0.021 gm/N $m^3$) (N $m^3$ normal cubic meter) (1 to 2 lb/hour) and 0.003 to 0.004 grain of $P_2O_5$/dscf (0.007 to 0.009 gm/N $m^3$) (3 to 4 lb/hour). Nitrogen oxide emissions range from 260 to 394 ppm (*25*). Atmospheric emissions from furnace slag-tapping operations are collected by hooding over the slag and ferrophosphorus runners and treated in a venturi scrubber having a pressure drop of 30 to 40 in. of water (107.2 to 139.9 mm Hg). Emissions of total fluorides range from 0.03 to 0.05 grain/scf (0.157 to 0.262 gm/N $m^3$) and $P_2O_5$ emissions range from 0.001 to 0.01 grain/scf (0.0023 to 0.023 gm/N $m^3$). About 98% of the fluoride is water soluble. Typically, scrubber water is neutralized with calcium hydroxide to precipitate the fluorides as calcium fluoride. Sulfur oxide emissions range from about 15 to 50 ppm (*26*).

Phosphorus pentoxide emissions between the furnace top and the moving electrodes can be controlled effectively by water seals. Emissions of phosphorus pentoxide from the cooling and condensation of phosphorus can be abated by using closed water handling systems. Most of the particulate from solids handling can be controlled by use of fabric filter baghouses which are capable of emitting no visible emissions.

### *B. Thermal Process*

Liquid phosphorus is pumped to a cylindrical combustion tower at feed rates that range from 1 to 5 gal/minute. The phosphorus is mixed with air at the burner and oxidized at temperatures of 3000° to 5000°F in the combustion tower (*27*) according to the reaction

$$P_4 + 5\ O_2 \rightarrow P_4O_{10} \tag{10}$$

The resulting mixture of phosphorus pentoxide, nitrogen oxides, steam, and excess air passes from the combustion tower to the hydrator (Fig. 10). The phosphorus pentoxide vapors are contacted with weak and prod-

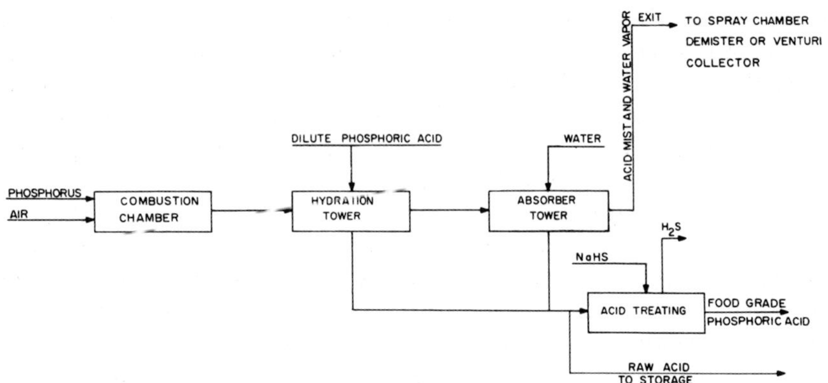


Figure 10. Flow diagram of thermal phosphoric acid process, noting potential air pollution sources.

uct acid to hydrate the oxide to phosphoric acid. The hydrated acid is passed through a water-spray-absorbing tower where 85% phosphoric acid is produced. In some plants, the hydration and absorption are accomplished in one tower. The hydration reaction is

$$P_4O_{10} + 6\ H_2O \rightarrow 4\ H_3PO_4 \tag{11}$$

For that portion of the acid to be used in food-grade products, treatment with sodium hydrosulfide is necessary to remove traces of arsenic and lead ions. These heavy metal ions are precipitated as insoluble sulfides and removed by filtration before the acid is transferred to storage (*28*).

## Air Pollutant Emissions and Their Control

The principal air pollutant emission is phosphoric acid mist in the absorber-tower exit gases. The more efficient glass-fiber mist eliminators and high-pressure-drop wire-mesh mist eliminators are operated at 99.9% efficiency. Emissions from these mist eliminators range from 0.1 to 0.6 mg of $P_2O_5$ per scf (2.29 to 13.8 mg/N $m^3$). About 90% of the acid mist particles at the eliminator outlet are less than 5 $\mu$m in diameter (*29*).

A significant intermittent emission in the exit gas from the acid treater tank is hydrogen sulfide, which may range from 10 to 2500 ppm. These emissions are controlled efficiently by alkaline scrubbing or by venting the hydrogen sulfide through a flashback device to a phosphorus furnace where it is oxidized to sulfur dioxide. Minor amounts of nitrogen oxides, i.e., 10 ppm, are discharged from the absorbing tower (*29*).

### C. Wet Process

In essence, the manufacture of phosphoric acid by the wet process involves the decomposition of phosphate rock with sulfuric acid. Although nitric or hydrochloric acid could be used, they are not used in commercially significant quantities for this purpose at the present time (*28, 30*).

The phosphate rock is ground to a size which allows at least 50% to pass through a 200-mesh screen (Fig. 11). The ground phosphate rock is contacted with concentrated sulfuric acid (93–98%) in multicompartment reaction tanks. The reactors typically consists of a series of tanks with the slurry alternately overflowing and underflowing from one compartment to the next. The basic reaction is

$$Ca_3(PO_4)_2 + 3\ H_2SO_4 + 6\ H_2O \rightarrow 2\ H_3PO_4 + 3\ CaSO_4{\cdot}2\ H_2O \qquad (12)$$

The acidulation reaction essentially goes to completion in less than 1 hour, but the slurry is retained from 5 to 8 hours to grow large gypsum crystals for efficient filtration. The acid slurry is fed from the last reactor compartment to a continuous rotary filter where the gypsum is separated. The gypsum filter cake is countercurrently washed with weak acid and water. After the gypsum is vacuum dried, it is discharged to nearby ponds. The reactor-filter product acid is first pumped to a storage vessel and then to forced-circulation vacuum evaporators. The orthophosphoric acid is then concentrated from 30 to 54% $P_2O_5$ (*31, 32*).

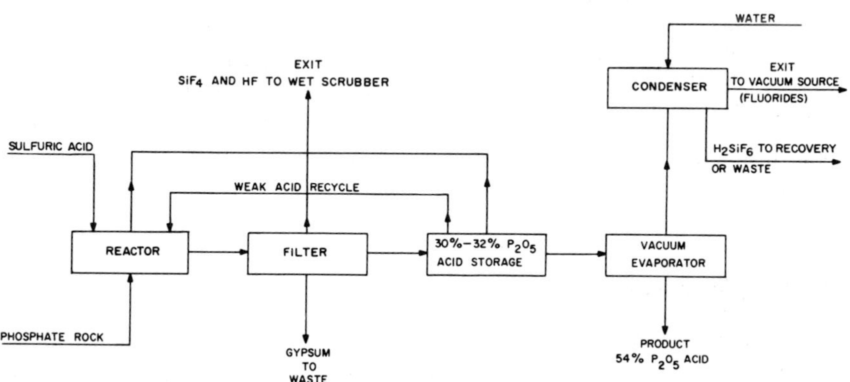


Figure 11. Flow diagram of wet-process phosphoric acid plant, noting potential air pollution sources.

## Air Pollutant Emissions and Their Control

The air pollutants of major concern are gaseous fluorides which occur as silicon tetrafluoride ($SiF_4$) and hydrogen fluoride (HF). The highest concentrations of fluorides are discharged from the reaction system. Other sources of low-concentration fluorides, which in many cases are also collected for treatment, include the filter hood, filter feed box, filtrate seal tanks, flash-cooler seal tank, and evaporator hot wells. Because of the excess silicon dioxide normally present in the acid and available for reaction with hydrogen fluoride, the fluorine evolved in the reaction and filtration steps is presumed to be emitted primarily as silicon tetrafluoride. Uncontrolled emissions of gaseous fluorides will typically range from 0.2 to 0.9 lb F per ton of $P_2O_5$ (0.1 to 0.45 kg/metric ton) (200 to 400 ppm of F) for the reaction system; from 0.005 to 0.02 lb F per ton of $P_2O_5$ (0.0025 to 0.01 kg/metric ton) (10 to 25 ppm of F) for the filtration system; and from 0.05 to 0.07 lb F per ton of $P_2O_5$ (0.025 to 0.035 kg/metric ton) (5 to 10 ppm of F) for the evaporator hot wells (*23*). The fluoride emissions from evaporator hot wells are based on using modern vacuum evaporation. If submerged combustion evaporation is used, considerable acid mist will be emitted along with gaseous fluorides.

The modern wet-process phosphoric acid plant will commonly use one scrubber system to serve all of the previously listed sources of fluorides. A typical packed scrubber using cross flow or concurrent spray patterns will handle 50,000 to 75,000 scfm of gas per ton of $P_2O_5$ [1450 to 2.75 scmm (standard cubic meter per minute)] in a low-pressure-drop scrubber (5 to 6 in. of water). Fluoride emissions from this type of system range from 0.005 to 0.05 lb F per ton of $P_2O_5$ (0.0025 to 0.025 kg/metric ton) (2 to 5 ppm of F). Scrubber efficiencies will range from 1000 99% when using recycled gypsum pond water which contains from 1000 to 9000 ppm of fluorine. The 1972 installed cost for a spray cross flow packed scrubber serving a 500-ton/day acid plant was about $120,000. A total annual cost for this system, which includes capital charges, maintenance, and power, would be $40,000.

### *D. Superphosphoric Acid*

The Tennessee Valley Authority first produced superphosphoric acid, i.e., greater than 70% phosphorus pentoxide, by concentrating thermal phosphoric acid. This process does not produce significant amounts of acid for fertilizer use, however. Superphosphoric acid for fertilizer consumption is currently produced by either the submerged-combustion or vacuum-evaporation processes. One of the main advantages of superphosphoric acid is its sequestering power which permits higher nutrient concentrations in liquid fertilizers and reduced sludge handling costs. In the submerged-combustion process, flue gases from a burner are sparged directly into 54% phosphoric acid. In this process, substantial entrainment losses and resultant emissions of acid mist and fluorides will normally require expensive vent-gas treatment systems, e.g., wet scrubbing plus mist elimination.

The most likely process to be used in the future is vacuum evaporation (*23*). This process incorporates natural circulation, forced circulation, or falling-film evaporators in much the same way that a concentrator is used in a typical wet-process phosphoric acid plant. The temperature required for superphosphoric acid is higher, however. A clarified feed acid is pumped to an evaporator where it is concentrated by a high temperature heating medium such as Dowtherm or high pressure steam (Fig. 12). Vapors from the evaporator are condensed in an overhead barometric condenser, where the evolved gaseous fluorides are absorbed in the condenser water. Concentrated acid flows to a cooling tank where it is cooled by water-cooling coils prior to being pumped to superphosphoric acid storage.

## Air Pollutant Emissions and Their Control

Gaseous fluorides are the major air pollutant emissions from the vacuum-evaporation process. In addition to fluorides, emissions from the submerged-combustion process include acid mist and fuel combustion products, e.g., nitrogen oxides and/or sulfur dioxide. The two sources of emissions from vacuum-evaporation units are the product cooling tank and the evaporator barometric condenser hot well. The main source of emissions from a submerged-combustion process is the flue gas from the evaporator burner.

Low-pressure-drop water scrubbers, e.g., 1 to 10 in. (1.87 to 18.7 mm Hg) of water, are used to absorb emissions of gaseous fluoride from the cooling tank and condenser hot well of the vacuum-evaporation process. Gypsum pond water is normally used as the scrubbing medium. The gaseous fluoride emissions from the scrubber will range from 0.001 to 0.002 lb F per ton of $P_2O_5$ (0.0005 to 0.001 kg/metric ton) (5 to 15 ppm of F). These emission levels are small compared to those from other sources of fluoride emissions from processing phosphate rock.

Based upon 1972 prices, the installed cost for a low-pressure-drop air-induced venturi scrubber treating 2500 scf/minute (70 scmm) of gas

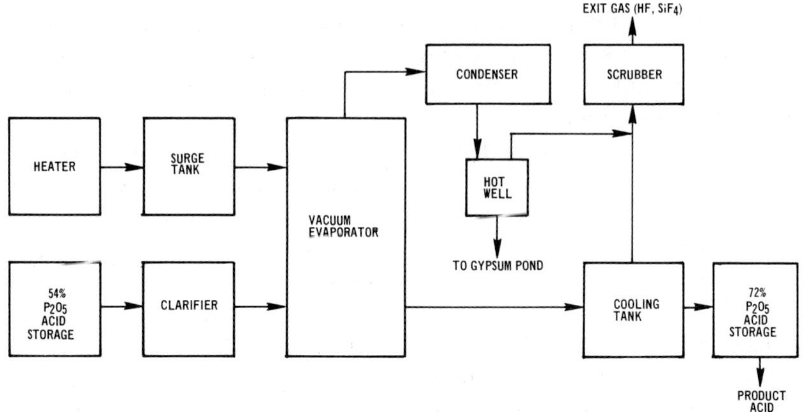


Figure 12. Flow diagram of process for manufacture of superphosphoric acid by the vacuum process, noting potential air pollution sources.

from a 300-ton/day superphosphoric acid plant is about $30,000. The combined capital charges and maintenance and power costs should be about $8500 per year.

### *E. Emission Reduction during Air Pollution Episodes*

In the manufacture of elemental phosphorus, the power to the electric furnace and the fuel to the calciner can be reduced or shut off immediately. Both of these units can operate at 50% of capacity with a significant reduction in emissions of fluorides and particulates. If the furnace is shut down for longer than a day, the slag must be tapped to preclude its solidifying in the furnace. During a shutdown, the furnace is purged with steam to prevent an explosive mixture. The purge gas containing fluorides, carbon monoxide, and phosphorus pentoxide fume is usually vented to the atmosphere. Although the fuel to the calciner can be shut off within 5 minutes, the traveling grate and rotary kiln must be kept operating for several hours to prevent heat damage and warping. The scrubber serving the calciner should continue to operate during this period.

Atmospheric emissions from the manufacture of phosphoric acid can be reduced significantly or stopped within a time period of 5 to 30 minutes. The phosphorus feed to a thermal-process furnace can be reduced or shut off in 5 to 10 minutes. The plant can operate at 50% of capacity with an accompanying reduction in emission of phosphoric acid mist. The feed stock to the reactor in a wet-process acid plant or to the concentrator in a superphosphoric acid plant can be reduced or shut off within 10 to

15 minutes. Both of these plants can operate at 50% of capacity with reductions in emissions of gaseous fluorides. Because the fluoride emissions from a vacuum-process superphosphoric acid plant are so small (less than 1 lb/day), the impact of a shutdown of this type of unit is negligible.

## VI. Sulfuric Acid

Sulfuric acid is manufactured principally by the contact process which accounted for 99% of the 30 million tons produced in the United States in 1971. The older lead chamber process accounts for only a small fraction of current production. In 1971, there were 214 contact process plants and 37 chamber process plants in the United States (*10, 33*). The contact process yields acid of up to and greater than 100% acid (fuming sulfuric acid or oleum), the latter being fortified with sulfur trioxide ($SO_3$). Some contact plants are actually air pollution control devices installed primarily to prevent the release of sulfur dioxide to the atmosphere.

Although there is a trend to greater utilization of waste sulfur gases, elemental sulfur is still the principal raw material. Other common raw materials are spent acid, acid sludge, and hydrogen sulfide from petroleum refineries and chemical plants and sulfur dioxide from nonferrous smelters. Pyrites supply only a small fraction of United States feedstocks. Most by-products or nonsulfur plants were installed as a result of environmental pressures. Spent acid and acid sludge are converted to new acid to avoid water and land pollution while sulfur dioxide and hydrogen sulfide are converted to acid to prevent air pollution. Furthermore, a growing fraction of elemental sulfur feedstocks are produced as by-products of air pollution control measures. Most of the by-product sulfur to date has been generated at refineries, but recovery systems are now being applied to copper smelters and steam-electric power plants. By-product sulfur from the two largest sources—smelters and power plants—could supply all United States acid plant requirements. In 1970, coal-fired power plants in the United States released enough sulfur dioxide to produce 27 million tons of 100% sulfuric acid. Nonferrous smelters (copper, lead, and zinc) in the United States produced 1.8 million tons of acid in 1970 and could have produced another 6.5 million tons if all uncontrolled sulfur dioxide had been converted to acid.

### A. Contact Process

All contact plants utilize a catalyst, usually vanadium pentoxide, to oxidize sulfur dioxide to sulfur trioxide and a strong acid absorber to absorb sulfur trioxide. The method of generating and introducing sulfur dioxide can vary widely. Pretreatment of gases prior to catalysis is often necessary to remove moisture or particulates.

Where elemental sulfur is the raw material (Fig. 13), it is burned directly to produce a stream of about 8% sulfur dioxide and 11–14% oxygen. In a single absorption system, 96–98+% conversion of sulfure dioxide to sulfur trioxide is accomplished in the catalytic converter at about 900°F. Gases are cooled to about 210°F, and sulfur trioxide is absorbed in sulfuric acid of 98.5 to 99% strength circulated in a packed tower. Product acid is bled from the tower. Where oleum or fuming sulfuric acid is produced, sulfur trioxide gases from the converter are passed through an oleum production tower before absorbing the remaining sulfur trioxide in an acid production tower.

Acid towers are normally followed by mist collectors which yield some product, however, the principal purpose of such collectors is to remove acid mist air pollutants.

The contact process can be varied considerably depending on feedstocks. Where spent acid, acid sludge, or similar contaminated feedstocks are utilized, auxiliary fuel is frequently required to burn the material. Resultant gases must be cleaned of dust and mist which would otherwise interfere with catalysis. At many plants, combinations of spent acid, hydrogen sulfide, and elemental sulfur are fed to the process.

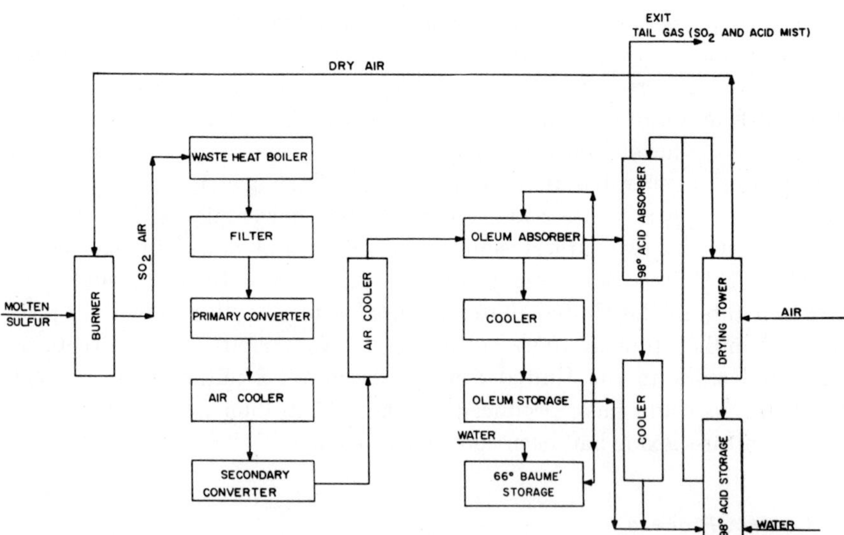


Figure 13. Flow diagram of contact process for the manufacture of sulfuric acid, noting potential air pollution sources.

## Air Pollutant Emissions and Their Control

Exhaust gases from the acid absorption tower are the principal source of sulfur dioxide and acid mist emissions. Where oleum is produced, additional sulfur trioxide/acid mist may be released from storage tanks and tank cars and from transfer operations.

Concentrations of sulfur dioxide in the stack gases from most single absorption plants range from 0.15 to 0.5% by volume. An efficient new single absorption plant can achieve an exit concentration of about 2000 ppm at 98% sulfur conversion (*34*). Somewhat lower levels can be realized with fresh catalyst or during periods of partial load.

Operating parameters which affect sulfur dioxide emissions are concentrations of sulfur dioxide and oxygen in gases to the converter, temperatures in converters, catalyst volume, and plant load factor. Factors which maximize conversion to sulfur trioxide minimize sulfur dioxide emissions.

Acid mist emissions in exhaust gases range from as low as 0.1 mg/scf with high efficiency mist eliminators to over 40 mg/scf (1100 mg/N $m^3$) where no controls are utilized (*34*). Factors which affect acid mist emissions are type of feedstocks, design and operation of acid absorber, presence or absence of a sulfur dioxide control system. Regardless of the degree to which these factors are optimized, there is significant acid mist in exhaust gases unless effective mist eliminators are utilized.

### *B. Bayer Dual-Absorption Process*

Sulfur dioxide emissions can be reduced either through process modifications or tail gas desulfurization. Process changes are more adaptable to new plants whereas tail-gas treatment can be used as either new or existing plants.

There are two commercially offered processes which are modifications of the basic contact system, i.e., the dual-absorption process and the pressurized single-absorption process. Both are reportedly capable of reducing sulfur dioxide emission levels well below 500 ppm. However, there has been considerably more experience with dual absorption than with pressurized plants.

The Bayer dual-absorption (DA) process (Fig. 14) has a second catalytic converter and a second absorber. Gases from the first absorber are reheated to about 900°F and passed through the second converter. Resultant sulfur trioxide is absorbed in the second acid scrubbing tower. The

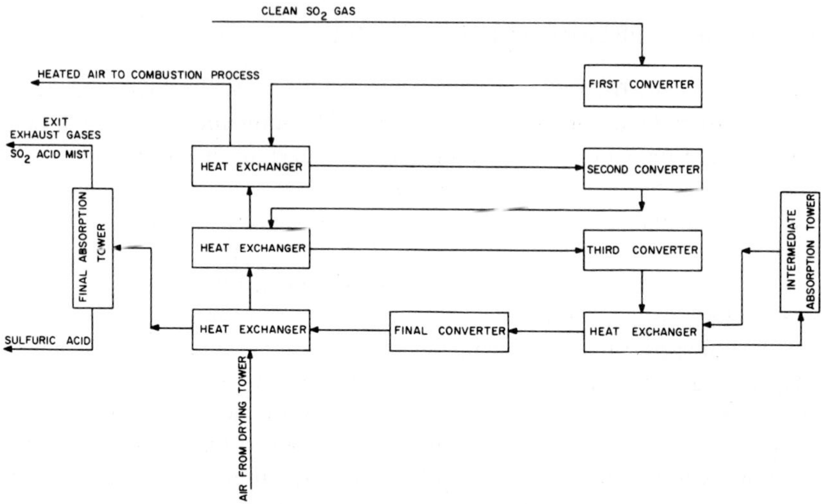


Figure 14. Flow diagram of Bayer dual-absorption process for sulfuric acid production showing gas-phase flow, noting potential air pollution sources.

sulfur conversion efficiency in the combined system is 99.6 to 99.8% or greater for elemental sulfur plants (*35*). Dual-absorption systems were originally used only for elemental sulfur feedstocks but more recently have been applied to spent acid plants and smelters.

The pressurized process is a single-absorption system that takes advantage of the more favorable kinetics and greater solubility of sulfur dioxide in acid at increased pressure. Sulfur is burned in a pressurized furnace at about 150 psig (8000 mm Hg). The higher pressure allows 99+% conversion of sulfur dioxide to sulfur trioxide while using less catalyst than with typical atmospheric plants. Cooling resultant gases condenses a portion of the sulfur trioxide and sulfur dioxide. Remaining gases are passed through an acid scrubbing tower. Here sulfur trioxide is absorbed and much of the sulfur dioxide is also dissolved in the acid. After the acid is depressurized, dissolved sulfur dioxide is air-stripped and fed back to the furnace to make more acid. Pilot studies indicate that sulfur dioxide emissions can be held to well below 500 ppm in exit gases. The first full-scale plant became operational in 1974.

Where essentially pure sulfur dioxide is available, other new processes could be used to yield significantly reduced quantities of sulfur dioxide in the tail gases. Under the Partial Recycle Air Process about two-thirds of the absorber tail gases are recirculated to the acid converter with the result that exit gases to the atmosphere are at only one-third the normal rate. The Total Recycle Oxygen Process (TROP) goes a step further using oxygen instead of air for makeup to the converter. With the TROP system, exhaust gases to the atmosphere are only a small fraction of the volume released from a normal contact plant (*36*).

Tail-gas desulfurization systems which are capable of controlling sulfur dioxide emissions to almost any level are commercially available. They include scrubbing with alkali solutions, dilute sulfuric acid and hydrogen peroxide, and absorption with molecular sieves.

The regenerative Wellman–Lord sodium sulfite system (Fig. 15) has been employed at acid plants as well as in sulfur recovery units and oil-fired power plants. Sodium sulfite reacts with sulfur dioxide and forms bisulfite in the scrubber. Bisulfite is regenerated thermally to produce sulfur dioxide which is fed back to the acid plant. A portion of the sulfite is converted to sodium sulfate which must be purged from the system. Current efforts are aimed at reducing the volume of the purge stream and of finding markets for waste sulfate (*31*).

Ammonia liquor scrubbing was first used to treat smelter gases (*38*) but more recently it has been applied to acid plants. Ammonium sulfate is produced as a by-product. Where no market exists, the ammonium sulfate can be burned in the acid plant to produce sulfur dioxide and more acid. This approach is particularly adaptable to spent acid plants and similar by-product plants which are already equipped with dust and mist collection equipment upstream of the converter.

Ammonia liquor scrubbing can produce appreciable quantities of salts. Some of these sublime at stack conditions and subsequently condense as a very fine particulate. These materials are amenable to collection with the same types of collectors as are used to remove acid mist from stack gases.

Scrubbing with hydrogen peroxide produces weak acid (50% sulfuric

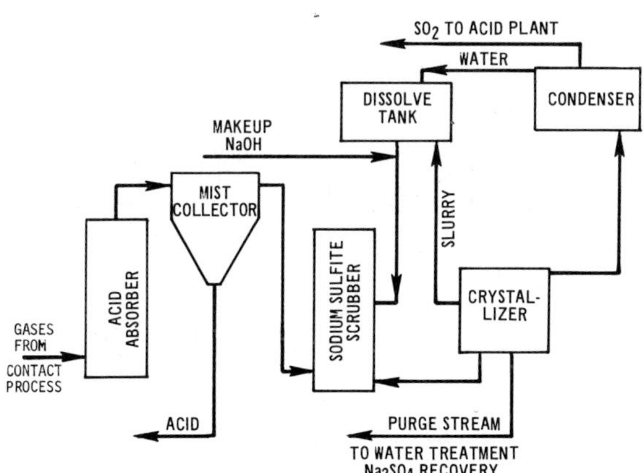


Figure 15. Flow diagram of a Wellman–Lord control system for a sulfuric acid plant.

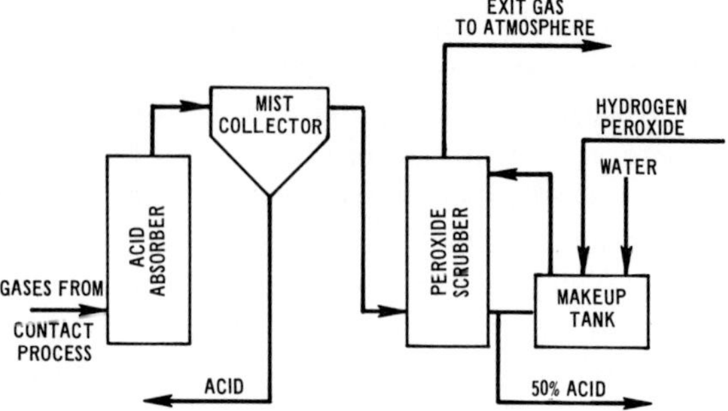


Figure 16. Flow diagram of a peroxide control system for a sulfuric acid plant.

acid by weight) directly (Fig. 16). The weak acid is blended with other streams to increase product yield. As with sodium and ammonia systems, sulfur dioxide levels can be reduced below 100 ppm if necessary.

Sulfur dioxide can be absorbed from tail gases with molecular sieves (Fig. 17). The sieves (synthetic zeolites) are desorbed thermally and sulfur dioxide is fed back to the converter to produce more acid. The process has been used successfully at a spent acid plant. Molecular sieves are capable of achieving sulfur dioxide levels below 10 ppm (*39*).

The United States Environmental Protection Agency has established standards which limit sulfur dioxide emissions from new contact acid plants to 4 lb per ton of acid produced (2 kg/ton) (100% basis). This level is equivalent to an exit concentration of about 380 ppm for an elemental sulfur plant.

Emissions of sulfuric acid mist and sulfur trioxide to the atmosphere can be controlled through the use of fiber demisters and electrostatic precipitators. Proper tower operation is a must regardless of whether tail gas treatment is employed for acid mist reduction. In order to minimize emissions, the acid concentration in the tower must be maintained at close

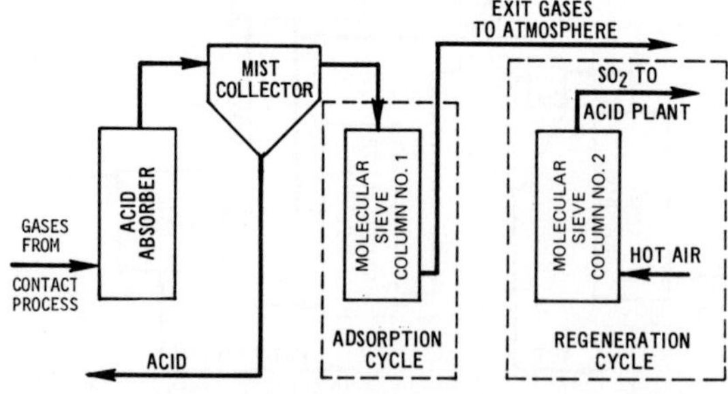


Figure 17. Flow diagram of a molecular sieve control system for a sulfuric acid plant.

to 98.5%, and acid must be circulated at a sufficient rate and evenly distributed across the tower.

In recent years very few precipitators have been installed for mist collection. Fiber mist eliminators are generally as effective as precipitators and require less maintenance. Fiber demisters introduce pressure drops of 8 to 14 in. of water (15 to 26 mm Hg). The most effective devices are dense, high-efficiency tubular demisters that take advantage of Brownian diffusion as well as mist impaction. Selection of mist collection equipment depends on the inlet mist concentration and desired emission levels. New source performance standards established by the United States Environmental Protection Agency limit acid mist emissions to 0.15 lb of acid (100% basis) per ton of acid produced (0.075 kg/ton). This is roughly equivalent to a concentration of 0.8 mg/scf (23 mg/N $m^3$) of exhaust gases. In addition, visible mist emissions are to be held to less than 10% opacity (*35*).

### C. Emission Reduction during Air Pollution Episodes

During periods of severe air pollution, emissions of both sulfur dioxide and acid mist can be curtailed and, if necessary, eliminated at contact sulfuric acid plants. While it is possible to shut down the equipment quickly, resultant rapid temperature changes can damage equipment. Gradual cutback of production can effect significant decreases of both sulfur dioxide and acid mist emissions. An indirect result of complete cessation is the large emission of sulfur dioxide subsequently encountered when a contact acid plant undergoes cold start-up. Reducing the throughput of a typical contact plant will cause sulfur dioxide emissions to be reduced to a somewhat greater degree than acid production. In both single- and dual-absorption processes, reducing production provides greater contact between gases and catalyst so that a higher fraction of the sulfur input is converted to acid. Most tail-gas systems can improve sulfur dioxide control in emergencies either through operational changes in the control system or through production cutback.

## VII. Calcium Oxide (Lime)

In 1972, about 20 million tons of lime in the form of quicklime (calcium oxide) and hydrated lime (calcium hydroxide) were produced in the United States. This production resulted from the operation of 187 plants in 42 different States (*40*). Most lime is produced in the United States in rotary kilns (85%) with the remainder in vertical or shaft kilns and other types such as fluidized bed furnaces. The metallurgy and alkali industries are the largest lime consumers, but several other industries consume significant quantities (*9*). A potentially large market exists in the control of sulfur dioxide from fossil fuel-fired power plants.

### A. Calcining and Hydrating of Limestone

Quicklime is produced by burning (i.e., heating) limestone (calcium carbonate). It is necessary to raise the temperature to at least 2100°F to decompose the carbonate. Lime is often shipped as quicklime and hydrated at the point of usage.

Rotary kilns (Fig. 18) can handle limestone of a variety of sizes and produce quicklime of uniform quality. Capacities range from 50 to 650 tons/day. Kiln sizes range from 6 to 12 ft (2 to 4 m) in diameter and up to 400 ft (120 m) in length. With long kilns, exhaust gas temperatures are usually between 1100° and 1400°F, while with short kilns, gas temperatures range from 1700° to 2100°F (*41*).

Shaft or vertical kilns are essentially batch-type operations in which relatively large limestone lumps are allowed to fall by gravity through hot gases. The product is not normally as uniform as that produced in rotary kilns, however, fuel usage is more efficient and exit temperatures are usually around 600°F.

Lime hydrators vary in shape and operation. A common type is the rotary pan. Quicklime reacts with water to form calcium hydroxide and liberate heat. While no gases are pulled through the system, hydration normally generates steam which will carry some dust out of the system.

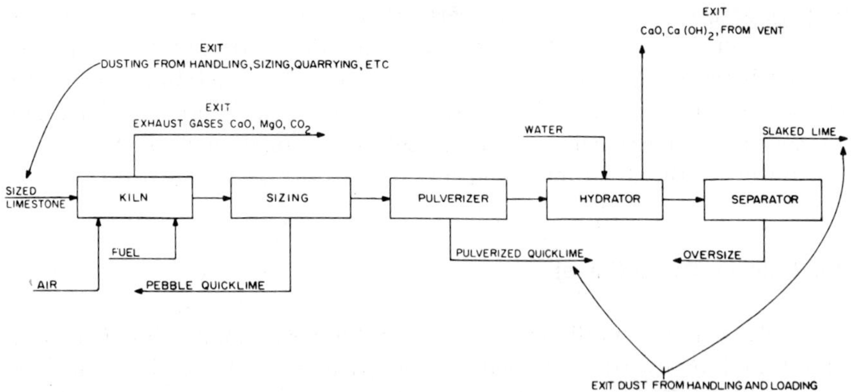


Figure 18. Flow diagram of process for manufacture of calcium oxide (lime), noting potential air pollution sources.

The volume of exhaust gases are small compared with that from calcining kilns.

## Air Pollutant Emissions and Their Control

The principal pollutants from lime manufacturing operations are limestone and quicklime dust from kilns, and limestone dust from primary and secondary limestone crushing operations. Lesser sources of these particulates are quarrying operations, screening, material transfer, hydration, pulverizing, and bagging and bulk loading of products. In addition, trace materials present in limestone feedstocks may be volatilized in the kiln. For instance, mercury compounds and low boiling metal compounds, if present in the ore, will be vaporized in kilns and could be released to the atmosphere.

Rotary kilns generate considerably more dust than vertical kilns because of smaller limestone feed size, higher gas velocities, and other design features. Dust loading in gases from rotary kilns generally range from 5 to 15 grains/scf (11 to 33 gm/N $m^3$) (*42*). Emission factors for rotary kilns have been estimated at 200 lb per ton of feed material (100 kg/ton) for rotary kilns and 8 lb per ton (4 kg/ton) for vertical kilns.

Particulate emissions from lime production in the United States in 1970 were estimated at 1,060,000 tons or about 107 lb per ton (54 kg/ton) of production (*16*). This would indicate that a large number of plants were not equipped with high efficiency dust collectors.

Many lime kilns are equipped with at least primary dust collectors, usually simple cyclones, which probably collect 60–80% of the dust. Several types of secondary collectors have been employed including fabric filters, electrostatic precipitators, scrubbers, and granular bed filters. Based on tests of similar sources (*43*) and on a few tests of lime plants, these appear capable of removing particulates down to the level of 0.02 to 0.03 grains/scf (40 to 70 mg/N $m^3$). Nonetheless, testing using the USEPA particulate sampling train has been sparse. The lowest levels, 0.001 grains/scf (2 mg/N $m^3$) were reported for a baghouse. Other reported levels are 0.02 (45) for high energy scrubbers; 0.22 (500) for electrostatic precipitators; 0.3 to 0.4 for impingement plate scrubbers; and 0.05 (110) for granular bed filters (*44*). These test values are not necessarily indicative of emission levels that can be realized with the best device of each type. High energy scrubbers and baghouses have found the greatest use for control of kilns and may be the most effective means of achieving levels of 0.02 to 0.03 grains/scf (40 to 70 mg/N $m^3$). The resistivity of lime dust makes it difficult to collect in electrostatic precipitators. However, this factor can be offset by injecting water into the gas stream. The first granular bed filter installed in the United States was applied to a lime kiln in Arizona. These filters are new, and more experience with their operation will be necessary before their day-to-day emission levels can be ascertained.

Lime hydrators release relatively small quantities of dust and steam. These can be controlled effectively with dust collectors and simple scrubbers. In fact, many hydrators are so equipped (*45*). The relatively high purity lime is recovered as milk of lime and piped back to the hydrators as part of the slaking water. In some instances, wet cyclones may also be employed.

### B. Emission Reduction during Air Pollution Episodes

During air pollution episodes, blasting, crushing, and conveying operations from the quarry can be halted at almost any time with no danger to the equipment and without increasing air pollution emissions. Kilns can be shut down over short periods. However, it is reasonable to allow operators to empty the charge in a kiln before cessation of operations. For installations with marginally effective dust collection systems, some decrease in air pollution emissions usually can be expected with reduced production.

## VIII. Sodium Carbonate (Soda Ash)

Sodium carbonate (soda ash) is manufactured by the Solvay (ammonia-soda) process or is extracted from naturally occurring brines or ores. In 1972 in the United States, 4,301,000 tons (3,901,000 metric tons) of soda ash were produced by the Solvay process in seven plants located east of the Rocky Mountains. Five plants in Wyoming and California extracted 3,218,000 tons (2,919,000 metric tons) from mined trona and natural lake brines. Although no new Solvay process plants have been built in the United States since 1934, it is still the dominant process. Because trona and brine processes are less significant sources of air pollution nationally, they are not discussed. Of all soda ash produced, 47% is used in the manufacture of glass and 25% is used in the production of other chemicals. Pulp and paper and alkali cleaners account for most of the remaining consumption (*46*).

### A. Solvay Process

Ammonia, limestone, and salt are the basic raw materials for the manufacture of sodium carbonate by the Solvay process. The limestone (calcium carbonate) is first heated in a conventional lime kiln to form lime (calcium oxide) and carbon dioxide. The lime is then converted to milk of lime (calcium hydroxide slurry) by water hydration. These two processes are covered in Section VII of this chapter.

A flow diagram for the Solvay process is shown in Figure 19. The salt brine is first purified of calcium and magnesium ions in a washing tower by off-gas ammonia and carbon dioxide from the absorber and carbonating tower which precipitates the impurities as a mud. The purified salt brine is then ammoniated in an absorber and treated with carbon dioxide in two carbonating towers in series. The overall reaction can be shown simply as:

$$NaCl + NH_3 + CO_2 + H_2O \rightarrow NaHCO_3 + NH_4Cl \qquad (13)$$

The sodium bicarbonate precipitates and is removed from the solution in a rotary drum vacuum filter. The filter cake is calcined in externally heated rotary kilns where carbon dioxide is released and sodium carbonate is formed:

$$2\,NaHCO_3 \xrightarrow{\text{heat}} Na_2CO_3 + CO_2 + H_2O \qquad (14)$$

The carbon dioxide is returned to the process along with makeup carbon dioxide from the lime kilns. The sodium carbonate may be cooled and screened and sold as "light ash." Some product is granulated with water to increase density and reduce dusting. This material is dried and sold as "dense ash."

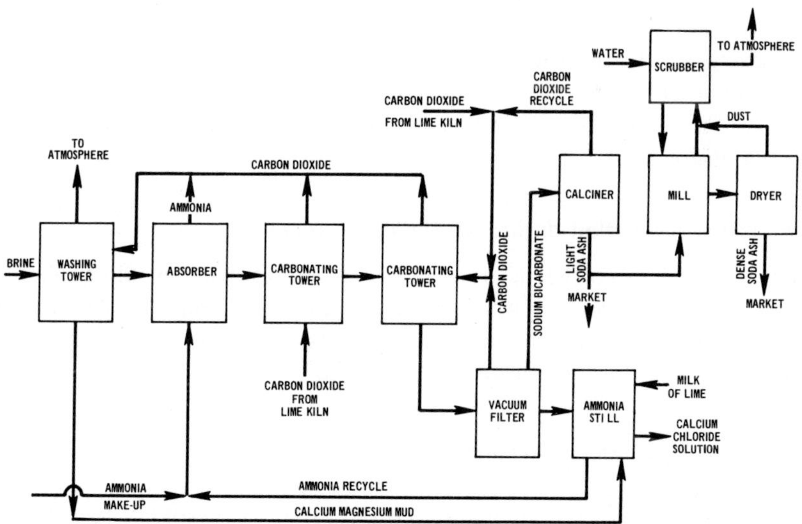


Figure 19. Flow diagram of the Solvay process for the manufacture of sodium carbonate (soda ash), noting potential air pollution sources.

The ammonia in the vacuum filter liquid is removed in the ammonia still by reaction with milk of lime and the application of heat.

$$2\ NH_4Cl + Ca(OH)_2 \xrightarrow{\text{heat}} 2\ NH_3 + CaCl_2 + 2\ H_2O \tag{15}$$

The free ammonia is recycled to the process and the calcium chloride solution is discharged as waste. Some ammonia makeup must be added to compensate for losses from the system (*47–49*).

### Air Pollutant Emissions and Their Control

The major emissions from the manufacture of sodium carbonate are the dust from the handling and transfer of the dry product and the ammonia which is released from the system.

Major sources of dust include the rotary dryers, all conveyor transfer points, air classification equipment, and car loading and packaging systems. Water scrubbers can be used for efficient control of dust from the dryers. Well-maintained hooding systems can be used with small bag collectors to clean up dust emissions from transfer points and loading equipment. It has been estimated that about 6 lb of dust per ton (3 kg/metric ton) of product would be emitted from a plant without control (*50*). Controls of the types mentioned should reduce this emission rate by 95–98%.

Small amounts of ammonia are emitted in the off-gases from the washing tower and as fugitive losses throughout the system. The total ammonia makeup ranges between 0.9 to 3.0 lb per ton (0.45 to 1.5 kg/metric ton) of product, most of which is fugitive loss. However, some ammonia is discharged with the waste calcium chloride solution and therefore ammonia makeup is not exactly equal to the ammonia loss to the air. The amount of this loss depends primarily on how well the plant is operated and maintained.

### *B. Emission Reduction during Air Pollution Episodes*

The particulate emissions from soda ash manufacture would probably not be reduced in proportion to a reduction in production rate. An orderly shutdown would require 8 hours and there would be no increase in emissions during shutdown or subsequent start-up.

## IX. Sodium Hydroxide (Caustic Soda)

Sodium hydroxide is now manufactured almost exclusively by electrolytic processes, all of which produce both chlorine and sodium hydroxide.

The principal alternative, the lime–soda process, is no longer used to produce sodium hydroxide in the United States. Information concerning electrolytic chlorine-caustic soda production will be found elsewhere in Section XII of this chapter.

## X. Phosphate Fertilizers

### *A. Phosphate Rock Preparation*

Phosphate rock consists primarily to fluorapatite [3 $Ca_3(PO_4)_2 \cdot CaF_2$] and hydroxyapatite [$Ca_3(PO_4)_2 \cdot Ca(OH)_2$]. In the United States, economically recoverable deposits are mined in Florida, North Carolina, Tennessee, Arkansas, and the Rocky Mountain region (Idaho, Montana, Utah, and Wyoming). Of the 36.7 million tons of rock processed and shipped in the United States in 1969, 79% was mined in Florida. This phosphate rock is obtained primarily by strip mining. The rock is found as a phosphate rock matrix about 20 ft in depth located beneath an overburden of 10 to 30 ft (3.05 to 9.15 m). After a dragline has removed the overburden, the phosphate rock is mined hydraulically.

The mined rock is beneficiated to remove impurities by washing, sizing, and concentrating. The washed rock may contain 7 to 20% moisture. The washed rock is normally dried to between 1 and 2% moisture before grinding. Rotary, direct-fired kilns, 8 to 10 ft in diameter by 25 to 100 ft (7.62 to 30.5 m) long, are used to dry the phosphate rock (Fig. 20).

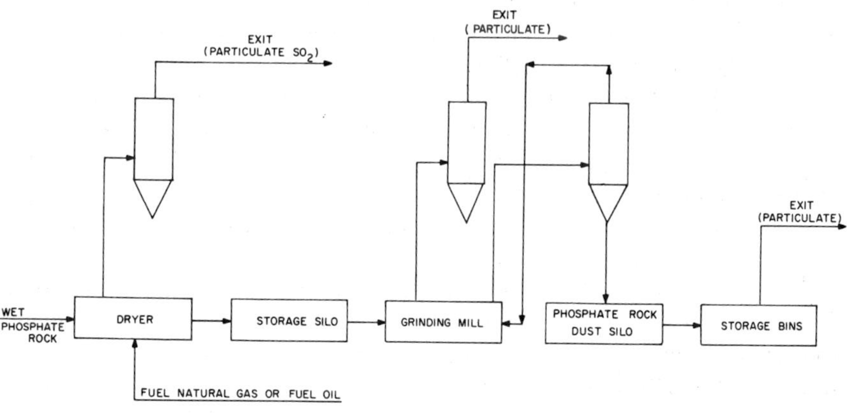


Figure 20. Flow diagram of phosphate rock storage and grinding facilities, noting potential air pollution sources.

These countercurrent or parallel flow dryers, which are fired with natural gas or fuel oil, range in size from 10 tons per hour to over 200 tons per hour. The dried rock is ground in ball or roller mills and finally conveyed to large storage silos (*51*).

## Air Pollutant Emissions and Their Control

Because phosphate rock beneficiating plants typically use wet-processing methods, there are no significant atmospheric emissions. The drying and grinding operations, however, do generate fine rock dust. Some sulfur dioxide may also be present in the dryer exhaust from the combustion of sulfur in the fuel. Phosphate rock dryers are usually equipped with wet scrubber systems for control of rock dust. The wet scrubbers will treat gas streams ranging from 10,000 to 100,000 scfm (283 to 2830 scmm) at pressure drops of 5 to 10 in. of water. The scrubber systems will operate at typical particulate collection efficiencies of 97 to 98%, with emissions ranging from 0.2 to 0.6 lb of particulate per ton (0.1 to 0.3 kg/metric ton) of $P_2O_5$.

The ball mills and roller mills used to grind phosphate rock to very fine 200-mesh particle size are closed systems except for vents to the atmosphere for control of moisture. To preclude visible dust emissions in the vented air, fertilizer plants normally equip the mills with baghouse collectors. These fabric filter dust-control systems will usually treat exit-gas streams of 5000 to 15,000 scfm (140 to 420 scmm) at pressure drops of 2 to 6 in. of water (3.74 to 11.22 mm Hg). Particulate emissions from the baghouses range from 0.01 to 0.06 lb per ton (0.005 to 0.03 kg/metric ton) of $P_2O_5$. The particulate collection efficiency for the fabric filters is normally between 98 and 99.9% (*23*).

Additional sources of emissions in a phosphate rock grinding and preparation plant are transfer points on conveying systems and discharge points at storage hoppers and silos. These sources can be equipped with hoods and an exhaust system to gather the generated dust. The gas streams can be further treated in baghouses with dust-collection efficiencies of over 99% and no visible emissions.

The 1972 installed cost for a scrubber system that treats 70,000 scfm from a phosphate rock dryer was about $140,000. Total annual cost for this system, which includes capital charges, maintenance, and power, would be $45,000. The installed cost during 1972 for a baghouse treating 13,000 scfm (3640 scmm) of dusty gases from a phosphate rock mill system was approximately $40,000. A typical annual cost for this system is $9000 (*23*).

### *B. Normal Superphosphate Production*

The material resulting from the reaction of ground phosphate rock and 55 to 75% sulfuric acid is called normal superphosphate. This fertilizer product contains from 16 to 20% phosphoric anhydride ($P_2O_5$). Because most of these plants are located near the consumer, the phosphate rock is shipped from the mining area to the plant. In 1970, there were 99 normal superphosphate plants throughout the United States. Because of the demand for more concentrated $P_2O_5$ products, the total number of plants has been declining in recent years. By 1980, it is expected that the present number of plants will have been reduced by about 50%. The United States production of normal superphosphate in 1970 was about 670,000 tons (603,000 metric tons) of $P_2O_5$ equivalent (*23*).

The three basic production steps in the manufacture of normal superphosphate which have remained essentially the same over the years are: (a) finely ground phosphate rock is mixed with sulfuric acid, (b) the slurry mix is dropped into a den to allow sufficient time to set in a solid porous form, and (c) the solid material is stored in a curing building to permit the acidulation reaction to go to completion (Fig. 21). The cone mixer developed by the Tennessee Valley Authority is used in many plants because of its low capital cost, low maintenance expense, and simple operation. A normal mixer charge consists of 5 parts of ground phosphate rock and 4 parts of sulfuric acid. The charge is mixed for 1 to 3 minutes and dropped into either a batch or a continuous den. A typical batch den is a fixed large wooden box with one end that is removable. The set material is moved out of the open end where it contacts a set of rotating knives, which reduce the large mass to lumps suitable for conveying to storage for aging. A commonly used continuous den is an apron-type conveyor belt that moves the material continuously to a cut-

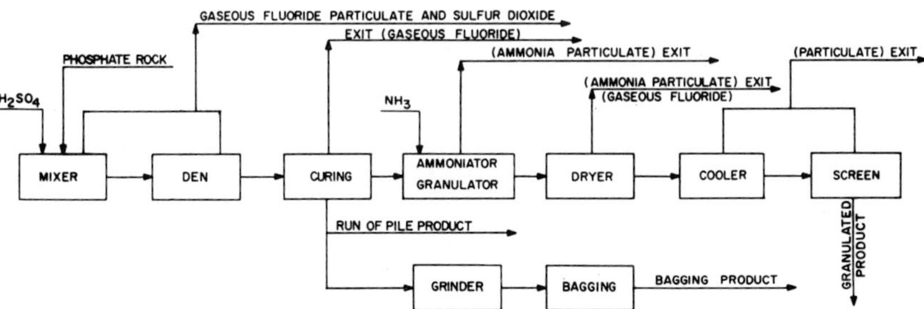


Figure 21. Flow diagram of a normal superphosphate plant, noting potential air pollution sources.

ter. Den capacities may range from 40 to 300 tons (36 to 270 metric tons). Depending upon the type of den, the mixture may be held for 30 minutes up to overnight before being transferred to storage. Normal superphosphate is uncured as it comes from the dens and must be held in a curing building for up to 6 weeks to allow the acidulation reaction to go to completion (*51*). Equations showing the reactions in the acidulation of phosphate rock are

$$3\,Ca_{10}(PO_4)_6 + 21\,H_2SO_4 + 9\,H_2O \rightarrow 21\,CaSO_4 + 9\,CaH_4P_2O_8 \cdot H_2O + 6\,HF \quad (16)$$

$$6\,HF + SiO_2 \rightarrow H_2SiF_6 + 2\,H_2O \quad (17)$$

$$H_2SiF_6 \rightarrow 2\,HF\uparrow + SiF_4\uparrow \quad (18)$$

After the curing period, the normal superphosphate is (a) sold directly as run-of-pile product, (b) ground and bagged for sale, or (c) granulated for sale as granular mixed fertilizer or granulated superphosphate. If normal superphosphate is granulated, it is blended with some or all of the following ingredients: ammonia, sulfuric acid, phosphoric acid, triple superphosphate, and potash. In some cases, water or steam is added to aid the granulation process. To ensure that the granulated product will retain its form, the mixture is then dried by passing through a rotary dryer fired with natural gas or fuel oil. After going through a cooler, the granulated fertilizer is bagged for sale or transferred to a storage bin.

## Air Pollutant Emissions and Their Control

The principal air pollutants from the acidulation of phosphate rock are gaseous emissions of silicon tetrafluoride and hydrogen fluoride. Most of the fluorine is evolved as silicon tetrafluoride. With Western United States phosphate rock containing appreciable amounts of organic matter, some sulfur dioxide is released from the reduction of sulfuric acid by the organic matter (*51*). Between 20 and 40% of the total fluoride in the phosphate rock is evolved during the acidulation and denning operations. In normal superphosphate plants, emissions from the curing building are not usually controlled. The uncontrolled fluoride emissions from a normal superphosphate acidulation plant will average between 2 to 5 gm/scf (*52*). These gases may also contain trace amounts of particulates ranging from 5 to 10 mg/scf (141.5 to 283 mg/N $m^3$) of stack gas (*23*).

The discharge gases from the mixer and den are normally treated in a water scrubber for removal of gaseous fluorides and trace amounts of particulate matter. In older plants the spray towers are designed locally and often made of wood. More efficient control systems utilize venturi, cyclonic, or water-eductor scrubbers. The water-eductor scrubbers have the advantage of providing their own motive force to draw gases from the mixer and den through the scrubber without a separate fan. The scrubbers will treat gas streams ranging from 5000 to 35,000 scfm (141.5 to 990.5 scmm) at pressure drops of 3 to 8 in. of water (5.6 to 15 mm Hg). The total fluoride collection efficiencies for the scrubber systems are from 94 to 99%. Atmospheric emissions from the scrubber range from 0.05 to 0.80 lb of total fluoride per ton (0.025 to 0.040 kg/metric ton) of $P_2O_5$ (*23*).

Until recently, it has been the practice to discharge the fluoride-containing scrubber water to a waste gypsum pond. With the present interest in fluoridation of municipal water supplies, and the need of aluminum and steel makers for a less expensive source of fluoride than fluorspar, some plants are recovering fluorine from water scrubbers as a 25% fluosilicic acid (*53, 54*). One Florida plant can produce 10,000 tons per year (9000 metric tons per year) of fluosilicic acid (100% basis) that could add an annual gross revenue of a one-half million dollars based on an acid price of $50 per ton (*53*).

Vent gases from the granulator-ammoniator may contain gaseous fluorides, ammonia, ammonium chloride, and fertilizer dust. Emissions from the dryer will include gaseous fluorides, fertilizer dust, and ammonia. If the dryer is fired with high-sulfur fuel oil, the exit gases will also contain sulfur dioxide. Fertilizer dust will also be emitted from the cooler.

Atmospheric emissions from ammoniation-granulation plants can be controlled in several ways. In some plants the granulator exit gases pass through a wet scrubber for collection of fluorides and particulates. Scrubbers are normally inefficient, however, in collecting ammonium chloride fume. To preclude the formation of ammonium chloride, the Tennessee Valley Authority has utilized specially designed sulfuric acid spargers which neutralize the ammonia before it can react with potassium chloride (*55, 56*). By changing the formulation to replace sulfuric acid with superphosphoric acid, no ammonium chloride fume will be formed. The use of superphosphoric acid will also improve granulation, eliminate the need for a dryer, and decrease the amount of cooling required (*31*). Because of the appreciable cost of ammonia, its losses are usually kept to a minimum. By utilizing a phosphoric acid scrubber ahead of the water scrubber, efficient collection of ammonia can be achieved.

Emissions from the dryer and cooler are passed through cyclone collectors in some plants for removal of particulate. Frequently, however, secondary wet collectors or baghouses are used which have collection efficiencies ranging from 90 to 99%.

### *C. Diammonium Phosphate Production*

In 1970, there were 41 diammonium phosphate plants in the United States. It is expected that by 1980, the number of these plants will increase by over 50% (*2*). Diammonium phosphate is continuing to retard the growth of triple superphosphate because it has a higher content of plant food and a lower shipping cost per pound of $P_2O_5$.

The manufacturing process most prevalent in the United States today was developed in pilot-plant work at the Tennessee Valley Authority in 1961. The process consists of an acid brick-lined reactor for ammoniation of wet-process phosphoric acid, a rotary drum ammoniator-granulator, and accessory equipment for drying, cooling, and screening the product (Fig. 22). The primary reaction is

$$2\,NH_3 + H_3PO_4 \rightarrow (NH_4)_2HPO_4 \tag{19}$$

Ammonia, fresh phosphoric acid, and acid that recycles from the scrubber serving the reactor are used as feedstock. In many plants, 93% sulfuric acid is used to control the analysis of the final product. The hot liquid diammonium phosphate slurry is pumped from the reactor to the granulator where additional ammonia is added along with recycle product to form a solid material averaging 18% nitrogen and 46% $P_2O_5$. To ensure maximum solubility and fluidity, the reactor is usually operated at a mole ratio ($NH_3/H_3PO_4$) of about 1:4. About two-thirds of the total ammonia charged to the process goes to the reactor while the remaining one-third

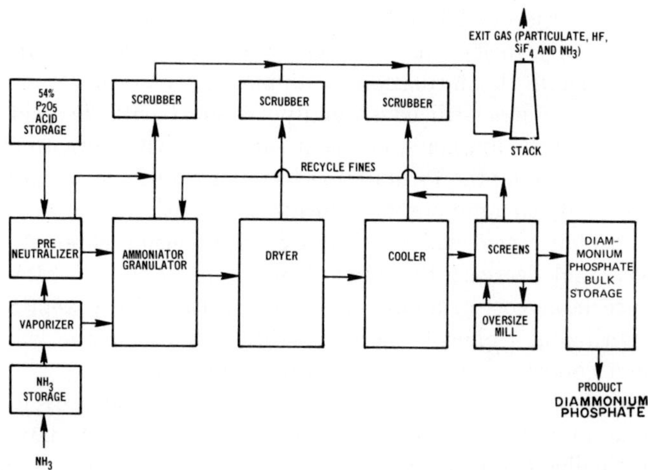


Figure 22. Flow diagram of a diammonium phosphate plant, noting potential air pollution sources.

is fed to the granulator. The diammonium phosphate from the granulator is then passed through a rotary dryer, a screening station, and a rotary cooler before being conveyed to storage.

## Air Pollutant Emissions and Their Control

Gaseous fluorides, ammonia, and particulates are the significant air pollutants from production of diammonium phosphate. The six main sources of emissions are the reactor, granulator, dryer, cooler, screens, and mills. In larger plants, these exit gases are usually combined into three main ducts, each with its own water-scrubbing device. The three systems include the (a) reactor-granulator; (b) dryer; and (c) cooler, screens, and mills. The reactor-granulator exit gases are rich in ammonia while the dryer off-gases contain less ammonia and more gaseous fluorides in addition to particulates entrained by burner flue gases. Although particulates are generated from the cooler, screens, and mills, relatively little fluoride and ammonia are evolved from these units. Primary and secondary scrubbers with packing, or of venturi or cyclonic design, are often used in newer plants to control these emissions. A solution of 28 to 30% phosphoric acid in water is used in the primary scrubber to absorb ammonia while gypsum pond water is employed in the secondary scrubber for collection of fluorides and particulates. The exit gases from the granulator, dryer, screens, coolers, and mills usually pass through a cyclone located ahead of the scrubber to recover particulate for recycle to the process.

Each of the wet scrubber systems may process gas streams ranging from 20,000 to 50,000 scfm (566 to 1415 scmm) at pressure drops of 15 to 20 in. of water (28 to 37.4 mm Hg). The primary scrubbers collect ammonia at efficiencies of greater than 99%. Emissions of ammonia from the scrubber systems serving the reactor-granulator, dryer, and cooler range from 0.02 to 0.20, 0.005 to 0.20, and 0.005 to 0.01 lb per ton (0.01 to 0.10, 0.0025 to 0.10, and 0.0025 to 0.005 kg/metric ton) of ammonia fed, respectively. The secondary scrubbers recover fluorides and particulates at efficiencies ranging from 80 to 95%. Total fluoride emissions from the secondary scrubber serving the reactor-granulator, dryer, and cooler range from 0.01 to 0.10, 0.01 to 0.20, and 0.01 to 0.02 lb per ton (0.005 to 0.05, 0.005 to 0.10, and 0.05 to 0.10 kg/metric ton) of $P_2O_5$, respectively (*23*). Efficient control of ammonia and gaseous fluorides by scrubbing also ensures effective control of particulates. Visible emission from the scrubber discharge (excluding condensed water vapor) is normally less than 10% opacity.

The installed cost (1972 basis) for a scrubber system processing

120,000 scfm (3396 scmm) from a 500-ton/day diammonium phosphate plant was about $175,000. An estimated annual cost for this system, which includes capital charges, maintenance, and power, is $32,000.

### *D. Triple Superphosphate—Run of Pile*

The reaction between ground phosphate rock and phosphoric acid yields a product called triple superphosphate. This product usually contains about 2½ times the $P_2O_5$ contained in normal superphosphate. In 1970, there were 17 triple superphosphate plants in the United States with a production of 500,000 tons of $P_2O_5$ per year (450,000 metric tons per year) (*23*). Only one new plant has been constructed in the United States since 1963.

The production of run-of-pile triple superphosphate is normally a continuous operation in large plants located near deposits of phosphate rock. Measured quantities of ground phosphate rock (32 to 34% $P_2O_5$) and phosphoric acid (54% $P_2O_5$) are combined in a TVA cone-type continuous mixer (Fig. 23). The principal reaction is

$$Ca_3(PO_4)_2 + 4\ H_3PO_4 + 3\ H_2O \rightarrow 3\ CaH_4(PO_4)_2 \cdot H_2O \tag{20}$$

The resultant viscous slurry drops onto a slowly moving belt called a "den" on which the reactions continue until the slurry solidifies. When the porous mass reaches the end of the belt, it is reduced to small chunks by a cutter. This material is then conveyed to a storage pile where the reaction continues slowly. After about 30 days, the run-of-pile triple superphosphate is considered "cured" and is ready for shipment.

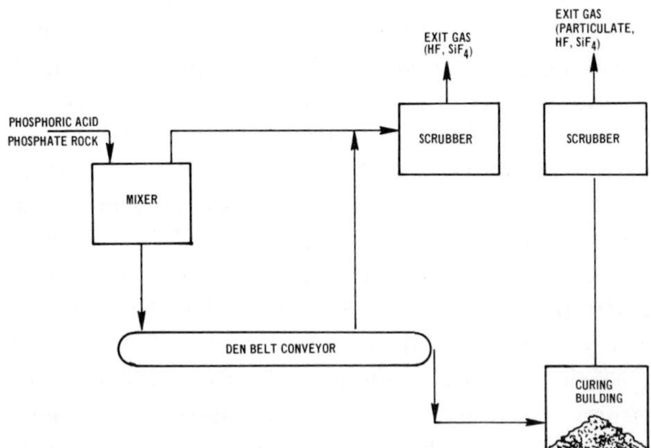


Figure 23. Flow diagram for the manufacture of run-of-pile triple superphosphate, noting potential air pollution sources.

## Air Pollutant Emissions and Their Control

The primary pollutants from run-of-pile triple superphosphate (ROP–TSP) plants are gaseous fluorides consisting of silicon tetrafluoride and hydrogen fluoride. There are also small amounts of dust from processing phosphate rock and ROP–TSP. The two major point sources of the gaseous fluorides are the curing belt and the storage pile. Phosphate rock dust is emitted from the ground-rock storage bin and the phosphoric acid–phosphate rock mixing cone. Some ROP–TSP dust is emitted from crushing and sizing operations. Although many plants retain the triple superphosphate in the curing building for 6 weeks to allow completion of the acidulation reaction, most of the gaseous fluorides are emitted during the first 2 weeks. The rate of fluoride evolution is increased by increasing the temperature of the mix on the belt or on the pile. Fluoride evolution will also depend on the surface area of the pile exposed to air flow. A porous light pile will release more gaseous fluorides than a dense compact pile.

Cyclonic or venturi-type scrubbers are used to collect fluoride emissions from ROP–TSP production and storage. In most plants the off-gases from the mixer, belt, and curing building are combined for treatment in a single scrubber. A typical pond water scrubber system will treat gas streams ranging from 150,000 to 300,000 scfm (4245 to 8490 scmm) at pressure drops of 10 to 15 in. of water (18.7 to 28.1 mm Hg). Emissions of total fluorides from a modern scrubber system range from 0.05 to 0.30 lb per ton (0.025 to 0.15 kg/metric ton) of $P_2O_5$ input. Visible emissions (excluding condensed water vapor) are normally less than 10% opacity (*23*).

The 1972 installed cost for a scrubber system treating 200,000 scfm (5660 scmm) from a 300-ton/day ROP–TSP plant was approximately $350,000. An annual cost for capital charges, maintenance, and power is estimated to be $85,000.

### *E. Triple Superphosphate—Granular*

The predominant process for manufacturing granular triple superphosphate (GTSP) is direct granulation. The chemistry of the manufacture of GTSP is essentially the same as that described for ROP–TSP. Phosphoric acid (54% $P_2O_5$) is mixed with ground phosphate rock to produce a slurry which is pumped to a granulator (Fig. 24). The slurry is mixed with recycled fine material in the granulator which in turn is coated with slurry to build it up to product size. This material is discharged to a rotary gas- or oil-fired dryer where combustion gases reduce the moisture content to about 3% or less. The product is then cooled and screened

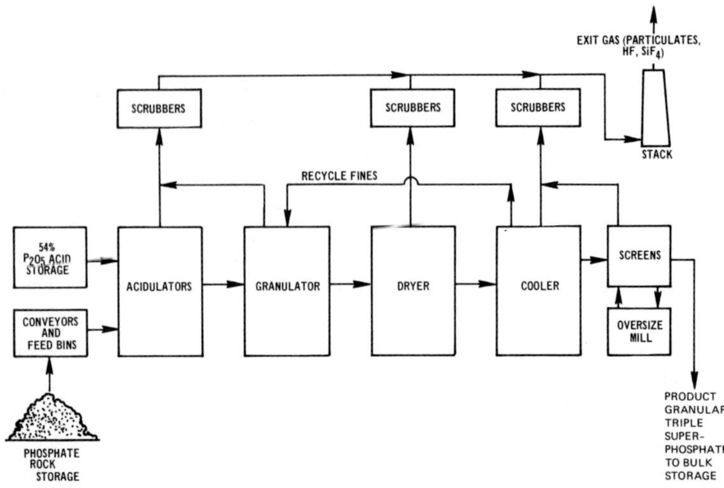


**Figure 24. Flow diagram for the manufacture of granular triple superphosphate, noting potential air pollution sources.**

before being conveyed to storage. Granular triple superphosphate is produced in some plants by granulating ROP–TSP after it has been cured. Except for the mixer, the same facilities would be used for these plants as those described for the direct-granulation process.

## Air Pollutant Emissions and Their Control

Atmospheric emissions of primary concern are gaseous fluorides (silicon tetrafluoride and hydrogen fluoride). Emissions of lesser importance include airborne particulates from the granulation and drying of triple superphosphate. Most of the gaseous fluorides are emitted from the reactor and granulator, while particulates are discharged from the granulator, dryer, cooler, screens, and mill. In large plants, the gaseous fluorides and particulate emissions are combined into several main ducts, each with its own pond water scrubbing system. Essentially the same configuration would be used as that for a diammonium phosphate plant. The cone mixer and granulator in a direct-granulation GTSP plant would generally have fluoride emissions of the same order of magnitude as the cone mixer and den belt of a conventional ROP–TSP plant. Water scrubbers with packing or of venturi design are commonly used in newer plants for emission control.

The pond water scrubbers may each process gas streams ranging from 20,000 to 50,000 scfm (566 to 1400 scmm). Pressure drops of 15 to 20 in. of water (28.1 to 37.4 mm Hg) across the scrubber would be typical.

Emissions of total fluorides from scrubbers serving a modern GTSP plant will range from 0.05 to 0.30 lb per ton (0.025 to 0.15 kg/metric ton) of $P_2O_5$ input. A well-controlled 500-ton/day GTSP plant achieving 0.20 lb fluoride per ton of $P_2O_5$ (0.10 kg fluoride per metric ton of $P_2O_5$) would discharge about 100 lb (45.3 kg) of fluorides each day. Visible emissions from such a plant would be less than 10% opacity during normal operations (*23*).

For a 500-ton/day granulated triple superphosphate plant treating 120,000 scfm (336 scmm) exhaust gases, the installed cost (1972 basis) for an efficient scrubber system was about $180,000. An estimated annual cost for this system is $32,000.

### F. Bulk-Dry Blending and Liquid Mixing

Phosphate fertilizers, particularly the granular types, are typically marketed in the consuming areas through small bulk-blend and liquid mixed-fertilizer plants. During the period of 1959 to 1969, the number of bulk-blend units throughout the United States increased from 200 to over 4100. During the same period, liquid mix units increased from 335 to over 1700 (*57*). This trend is expected to continue.

A mixed fertilizer is an "NPK" product if it contains the three main ingredients, i.e., nitrogen, phosphorous, and potassium. The NPK product may be a liquid, e.g., a water solution of potash and ammonium phosphate; or it may be a dry bulk blend, e.g., urea, triple superphosphate, and potash. A common practice in the central area of the United States is the mixing of dry granular materials such as ammonium sulfate, superphosphate, and potassium chloride in small plants and transporting the product in bulk to the farmer.

The term NPK complex is usually reserved for three-element granular materials produced by simultaneous ammoniation and granulation. A widely practiced NPK process is ammoniation (with anhydrous ammonia or ammonia solutions) of normal or triple superphosphate with the addition of potash and sulfuric acid and with simultaneous granulation. The nitrogen in the product is built up to the desired ratio by addition of sulfuric acid which reacts with the ammonia without increasing the $P_2O_5$ content. The granular product and bulk delivery of the product to the farmer distinguish present-day bulk blending from the dry mixing operation of the past.

### Air Pollutant Emissions and Their Control

Particulates are emitted from dry mixers, granulators, dryers, and from rock unloading and fertilizer loading points. Ammonia and gaseous fluorides are discharged from the granulator and dryer. Particulate emissions from mixers and from loading, unloading, and packaging operations are normally collected by cyclones or fabric filters. Recovery of particulate emissions can be above 99% with properly designed equipment. Phosphoric acid scrubbers are used to recover ammonia from the granulator and dryer discharge gases, while water scrubbers are used to control fluoride emissions from these sources. Scrubbers may process exit gas streams ranging from 500 to 20,000 scfm (140 to 560 scmm) at pressure drops of 5 to 10 in. of water (9.35 to 18.7 mm Hg). Fluoride collection efficiencies may range from 80 to 97%.

### G. Emission Reduction during Air Pollution Episodes

Atmospheric emissions from most phosphate fertilizer processes can be reduced significantly or stopped within a time period of 5 minutes to 1 hour. Gaseous fluoride emissions from dens and curing buildings are not however, amenable to such short-term reduction or elimination.

In phosphate rock preparation, emissions of particulates and sulfur oxides can be significantly reduced from the dryer by operating at half load, or shut off, by stopping feed to the dryer and fuel to the burner within 15 to 30 minutes. Operations for conveying and grinding dried phosphate rock can be shut down in 5 to 10 minutes, thereby eliminating emissions of particulate.

In the production of normal superphosphate, triple superphosphate, and diammonium phosphate, feed to the mixer or reactor and to the ammoniator-granulator can be reduced or shut off within 15 to 30 minutes. It may take 30 minutes to an hour to empty and shut down the mixers, ammoniator-granulators, dryers, and coolers. Product grinding, screening, and packaging can be shut down within 10 to 15 minutes. It may take 2 to 3 hours, however, to complete reaction in the den sufficiently to transfer the batch to the curing building.

In bulk-blending and liquid-mixed plants, the dry blending, liquid mixing, and packaging operations can be shut down within 10 to 15 minutes. For plants employing ammoniation and/or granulation, the previous comments for these operations apply concerning shutdown.

The particulate, ammonia, and fluoride emission control system should operate during the shutdown and emptying of mixers, granulators, dryers, and coolers. The control system should continue to operate throughout an air pollution episode because of the continuous emission of gaseous fluorides from the superphosphate storage pile. These emissions are not amenable to appreciable reduction or elimination.

## XI. Ammonium Nitrate

In 1972, 6,872,000 tons (6,233,000 metric tons) of ammonium nitrate was produced in the United States. Of this amount, about 85% was used in fertilizers, while the remaining 15% was consumed mostly in munitions and industrial explosives (*10*).

Ammonium nitrate is the primary source of nitrogen for the fertilizer industry. Many of the direct application liquid fertilizers are blended from ammonium nitrate solutions. Ammonium nitrate solutions are also used in combination with phosphate and potassium salts for the manufacture of many granulated grades of fertilizer. Most ammonium nitrate, however, is sold as a prilled, solid, pure material which is applied directly to the soil. Some ammonium nitrate is prilled or granulated with limestone to produce a fertilizer with a lower nitrogen content but with improved handling properties.

### *A. Neutralization Process*

Ammonium nitrate is produced by the neutralization of nitric acid with ammonia. A very typical plant for producing both ammonium nitrate solution and high density prills is shown in Figure 25. Essentially there

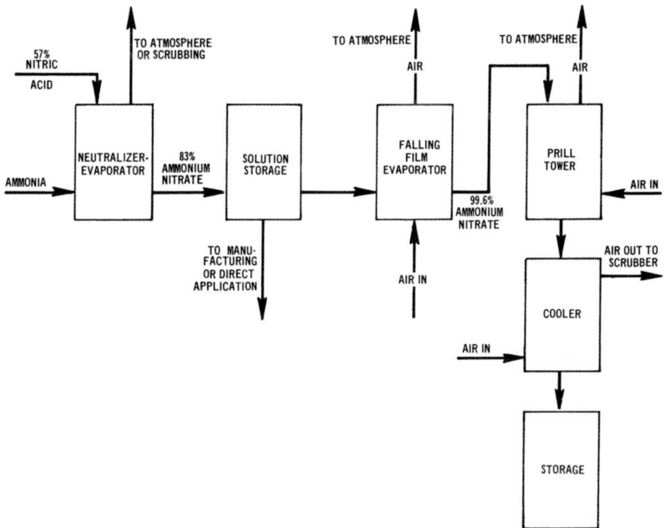


Figure 25. Flow diagram for the manufacture of ammonium nitrate solution and high-density prill, noting potential air pollution sources.

are three steps: neutralization, evaporation of water from the ammonium nitrate solution, and prill or granule formation (*58, 59*).

The neutralizer will usually operate at atmospheric pressure although a few units operate at pressures of up to 50 or 60 lb per $in.^2$ (4.4 to 5.0 atm). The basic reaction is

$$NH_3 + HNO_3 \rightarrow NH_4NO_3 + \text{heat} \tag{21}$$

Because the reaction is highly exothermic, the heat of reaction is used to drive off part of the water present in the nitric acid. In some modern plants, where a slight excess of ammonia is maintained in the neutralizer, the ammonia present in the neutralizer overhead vapor is neutralized with nitric acid and recycled. If the neutralizer is operated with a slight excess of acid there is less ammonia in the evaporator overhead but more danger of equipment corrosion. In either case, final "touch-up" adjustments are made in the exit stream to maintain a near-neutral ammonium nitrate solution product. This solution may be used in the manufacture of direct application solutions or granular "complete" fertilizers. Most, however, is concentrated further to produce prills or granules.

In the process shown, the ammonium nitrate solution is evaporated to a melt containing 0.4% moisture in a conventional falling-film evaporator. The molten ammonium nitrate is then sprayed through nozzles in the top of the prill tower. The drops formed fall downward and are cooled and solidified by an up-flowing stream of cool air that is blown into the tower by large, low pressure fans in the base of the tower. The prills are removed from the bottom of the tower, cooled, and stored.

In some processes the ammonium nitrate solution is concentrated to only about 95% in the evaporators. A tower is required to prill the solution while a drier must be used after the prill tower to reduce moisture to about 0.5%. (Little drying takes place in the prill tower.) This type of processing produces a relatively "low density" prill in contrast to the so-called "high density" prill made from 99.6% melt. The products are interchangeable, except that the low density prill is much more suitable for use as an explosive.

The use of some type of prill tower is almost universal for the manufacture of pure solid ammonium nitrate. Other types of granule formation have their special advantages but are in use in only a few plants.

## Air Pollutant Emissions and Their Control

The primary emission in the manufacture of ammonium nitrate is dust or fume from the prilling tower. Emissions of ammonia, ammonium nitrate, and nitric oxides from the neutralizer and falling-film evaporator are less significant and more easily controlled. The amount of dust emitted from a prilling tower depends primarily on the velocity of air in the tower and on the temperature of the ammonium nitrate at the spray nozzles. When the ambient air is cold, less air is required and dust production is less. Emissions are higher for high density plants because the 99.6% solution must be hotter than the 95.0% solution at the spray nozzle, more ammonium nitrate is vaporized, and more air is required to cool the particle to solidification.

The dust from a high density prill tower consists primarily of submicrometer particulate, formed from condensed ammonium nitrate vapor, which makes a highly visible fume and is difficult to remove.

Emissions of particulate from high density plants range from a minimum of 5 lb per ton (2.5 kg/metric ton) of product to 20 lb per ton (10 kg/metric ton) of product. Particulate emissions are usually less from low density plants, and, because the particle size of the particulate is larger, the emissions are less visible.

Several methods for the control of particulate emissions from prill towers are under development. Although conventional scrubbing appears feasible, large air volumes require treatment for particulate removal. A typical tower making 15 tons (13.6 metric tons) an hour of high density prills has an airflow of about 120,000 scfm (3400 scmm) with less than 0.5 in. (1.3 cm) of water pressure drop. Much larger fans would be required to develop the differential pressure required for an efficient scrubbing system. Several plants are investigating the possibility of scrubbing using a fine spray of water followed by a wire mesh or fiber bed mist eliminator.

A novel method now under development provides manifolding within the tower to separate the tower off-gas into two streams. A small stream, collected from around the spray nozzles, contains most of the particulate. This stream may be treated economically by conventional methods. The larger stream, with up to 80% of the total flow, can be vented without treatment. Another control method would be to recycle all the prill tower cooling air after it had been cooled.

For new plants, a different type of granulation could be used which does not require a large airflow for cooling. Pan granulation is one possibility. Another type of granulating device sprays a molten stream of ammonium nitrate onto a falling or tumbling bed of granules.

### B. Emission Reduction during Air Pollution Episodes

The particulate emissions from an ammonia nitrate prill tower can be reduced by reducing the amount of solution prilled. This reduction might be made by decreasing the amount of solution manufactured and prilled, or by leaving solution manufacture constant, prilling less, and building up solution inventory. A prill tower can be shutdown in a few minutes and there is no increase in emissions on shutdown or start-up.

## XII. Chlorine

Chlorine is used primarily in the manufacture of chlorinated hydrocarbons, the bleaching of pulp and paper, and in the purification of municipal water supplies. In 1971, about 5 million tons of chlorine were produced by 76 plants in the United States. Ninety-seven percent of the installed chlorine production capacity in the United States is by two electrolytic methods—the diaphragm-cell process (70%) and the mercury-cell process (27%). The remaining 3% is by fused-salt and nonelectrolytic processing (*60*).

### *A. Electrolytic Process*

Chlorine and caustic are produced concurrently in electrolytic cells. A chloride salt, which is usually fed to the cell as a water solution, is decomposed by an electric current. Chlorine is produced at the anode in both the diaphragm and the mercury cell. The electrical energy requirements are 2500 to 4000 kW-hours/ton of chlorine produced (*61, 62*). In the diaphragm cell, gaseous hydrogen and sodium hydroxide are liberated at the cathode. A diaphragm is used to prevent contact of the chlorine produced with the hydrogen or the alkali hydroxide. In the mercury cell, the cathode consists of liquid mercury, which forms an amalgam with the alkali metal. After the amalgam is removed from the cell, it reacts with water in a separate chamber called a denuder to form alkali hydroxide and hydrogen. The hydrogen gas, saturated with water vapor, passes overhead from the cathode normally with a purity of greater than 99.9% (Fig. 26). The hydrogen is cooled to condense moisture, compressed, and used as process hydrogen or fuel. The water vapor in the chlorine gas leaving the cell is also removed by condensation. The gas is further dried by direct contact with strong sulfuric acid. The dried chlorine gas is then compressed for in-plant use or is cooled further by refrigeration to liquefy the chlorine. About half of the total chlorine production in the United States is liquid chlorine (*61*).

In mercury-cell plants, high-purity caustic can be produced in any desired strength and requires no concentration. In the diaphragm-cell

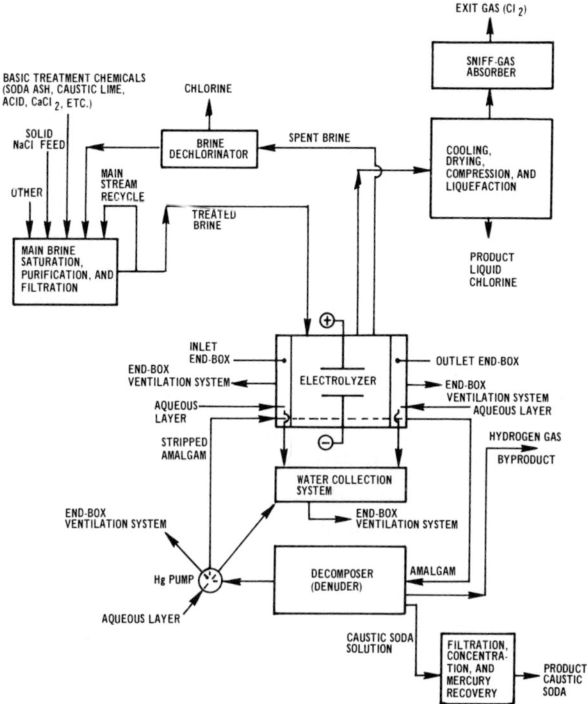


**Figure 26.** Flow diagram for chlor–alkali mercury-cell process, noting potential air pollution sources.

plants, however, the caustic leaves the cell as a dilute solution. The caustic brine solution is then evaporated to increase the concentration to either 50 or 73% solutions. Most of the residual salt is precipitated and removed by filtration.

## Air Pollutant Emissions and Their Control

a. Chlorine. The major source of chlorine from the electrolytic cell is the liquefaction vent gas. Other sources include vents from tank cars, ton containers, and cylinders during loading and unloading and from storage and process transfer tanks. The chlorine content of blow gas streams ranges from 4000 to 16,000 lb per 100 tons of chlorine (2000 to 8000 kg/100 tons) for mercury cells and from 2000 to 10,000 lb per 100 tons (1000 to 5000 kg/100 tons) for diaphragm cells (*61*). These chlorine emissions are normally controlled by (a) recovery of chlorine from vent gas streams, (b) use of dilute concentrations of chlorine in other plant processes, and (c) neutralization in alkaline scrubbers.

Vent gas streams containing more than 10% chlorine are economically attractive for recovery of chlorine by absorption in water or a carbon tetrachloride solution by use of packed or spray towers. Chlorine is subsequently stripped from the absorbing medium in a distillation tower, thus regenerating the absorption medium for recycle. A properly designed and operated water scrubber can operate at chlorine recovery efficiencies of 97% or greater, with exit-stream concentrations of less than 0.5% chlorine. This represents a chlorine loss of less than 100 lb per 100 tons (50 kg/100 tons) of chlorine produced (*61, 63*).

It has not been economical, however, to recover chlorine from vent gas streams containing less than 1% chlorine. Typical practice for these streams includes scrubbing with alkaline solutions in packed, plate, or spray towers. Alkaline scrubbers that react caustic or lime with dilute concentrations of chlorine have absorption efficiencies approaching 99.9% for a well-operated unit. Exit chlorine concentrations of less than 10 ppm can be expected (*61*).

About 450 lb (204 kg) of chlorine per 55-ton tank car is vented from uncontrolled tank cars. The handling and loading of shipping containers generate about 1700 lb per ton (770 kg/ton) of liquefied chlorine, if uncontrolled. However, most of this gas is returned to the liquefaction system or controlled by means of scrubbing systems.

b. Mercury. The major sources of emissions of mercury to the atmosphere include the (a) hydrogen by-product stream, (b) end-box ventilation system, and (c) cell-room ventilation air (*64*). The hydrogen by-product stream is saturated with mercury vapor. If this stream is vented directly to the atmosphere without prior cooling, an estimated 220 lb (100 kg) of mercury would be emitted for each 100 tons of chlorine produced. Mercury and mercury compounds are collected from the end boxes, the mercury sump pumps, and their water collection systems by the end-box ventilation system. Emission measurements by the U.S. Environmental Protection Agency show that mercury emissions from untreated end-box ventilation systems range from 2 to 15 lb (1 to 8 kg) for each 100 tons of chlorine produced (*62*). The cell-room ventilation system cools the cell-room environment but also provides a means of reducing the cell-room concentration of mercury to within the threshold limit value (50 $\mu g/m^3$). The air rate of the cell-room ventilation system varies from 100,000 to 1,000,000 $ft^3$/minute (170,000 to 1,700,000 N $m^3$/hour) for each 100 tons of daily chlorine capacity. The estimated mercury emissions from the cell-room ventilation system, based on data from operating plants, vary from 0.5 to 5.0 lb (0.2 to 2 kg) per day per 100 tons of daily chlorine capacity (*62*).

The various control techniques for mercury that have been applied to the by-product hydrogen stream and the end-box ventilation system include (a) cooling and condensation, (b) mist eliminators, (c) chemical scrubbing techniques, (d) treated activated carbon, and (e) molecular sieves.

After the hydrogen gas stream has been cooled to 90° to 100°F by shell-and-tube heat exchangers, residual mercury mist "remains" in the hydrogen gas stream. Much of the mist can be removed by use of a direct-contact cooler with chilled water or brine. The bleed-off liquor from the direct-contact cooling can be treated with chemicals such as sodium hydrosulfide to remove the mercury.

End-box ventilation air is sometimes cooled by a direct-contact packed tower using water as a coolant. Further cooling is provided by an indirect water cooler. The residual particulate mercury is removed by a mist eliminator (*62*). The mercury emission rate for this system has been estimated at 1 lb (0.5 kg) per day for each 100 tons per day of chlorine capacity. The capital and annual operating costs (1972 basis) for a control system using only primary cooling and partial mist elimination for a 100-ton/day chlorine plant were $49,000 and $15,000, respectively. The addition of a secondary cooler, a knockout drum, and a mist-elimination device would increase the capital cost by $202,000 and the annual operating expense by $60,000.

Depleted-brine scrubbing techniques have been used for removal of mercury from hydrogen and end-box ventilation gas streams at a few installations. The solution is used as a scrubbing medium in a sieve-plate tower or in a packed-bed scrubber. Mercury losses from the treated hydrogen stream are reported to be less than 0.01 lb (0.005 kg) per day on a basis of 100 tons of chlorine per day (*62*). A hypochlorite scrubbing technique has also been developed which employs a dilute solution of sodium hypochlorite with a large excess of sodium chloride. A mercury-collection efficiency of 95 to 99% has been reported for this system (*62*). The range of costs (1972 basis) for a depleted-brine scrubbing system which will serve both the hydrogen and the end-box ventilation stream was $160,000 to $350,000 for a 100-ton/day plant. Annual operating costs are estimated to be $48,000 to $105,000.

Activated carbon impregnated with iodine or sulfur is used by several mercury-cell chlor-alkali plants for reducing the concentration of mercury in the hydrogen stream. In this type of control system, the mercury vapor is adsorbed by the carbon and reacts chemically with the iodine or sulfur to form mercury compounds (*62*). The treated activated carbon can adsorb from 10 to 20% of its weight in mercury before it requires replacement. The capital investment (1972 basis) for this system was estimated to be $279,000 for a 100-ton/day chlorine plant. The operating cost is estimated at $83,000 per year.

A molecular sieve system has recently been installed on a chlor-alkali plant in Maine (*65, 66*). This type of system, which uses a sieve-adsorbent blend, reportedly reduces the mercury concentration from hydrogen gas streams to 0.50 $mg/m^3$ (*67*). This concentration is equivalent to a hydrogen stream emission of 0.04 lb (0.02 kg) of mercury per 100 tons of daily chlorine capacity.

Last, a control method which eliminates all mercury emissions from chlor-alkali plants is the conversion of mercury cells to diaphragm cells. The cost of converting a 300-ton/day chlorine plant is estimated at $3,410,000 (*68*). In Japan, poisonous levels of methyl mercury were found in fish. Japan's Ministry of International Trade and Industry therefore ordered chlorine producers to end all discharges from mercury-cell plants and switch to diaphragm cells (*69*). By December 1975, 60% of the mercury cell plants had been converted to diaphragm cells.

### B. Emission Reduction during Air Pollution Episodes

In the manufacture of chlorine and caustic by the electrolytic process, the power to the diaphragm or mercury cells can be reduced or shut off immediately. The cells can operate at half capacity with an accompanying reduction in emissions of chlorine and mercury. To operate at reduced capacity, however, many plants must be capable of adjusting their refrigeration capacity to condense chlorine. If a plant shutdown were required, the comparatively large drop in power should not present a major problem to the utility if they are notified an hour before the shutdown. In plant start-up, however, the cells may require purging to the atmosphere with inert gas to preclude explosive mixtures of hydrogen and oxygen.

## XIII. Bromine

In 1972 all the bromine produced in the United States was extracted from the naturally occurring brines of Arkansas, Michigan, and California. The principal use for bromine, the production of ethylene dibromide, consumed 74% of United States production in 1972. Ethylene dibromide is used in gasoline antiknock fluids to prevent lead deposition from lead alkyls. Flame-retardant and agricultural chemicals account for most of the remaining 26%. United States production of bromine was 193,400 tons (175,400 metric tons) in 1972 (*46*) from 10 plants.

It is estimated that 70% of the bromine plant capacity in the United States is in Arkansas and 29% in Michigan. The same steam extraction process is used at both locations. There are some variations in the process depending on the bromine content of the brine but all processes require: (a) oxidation of bromide to bromine, (b) removal of bromine from solution by vaporization, (c) condensation of the vapor, and (d) purification. Although steam is usually used to drive out the bromine vapors, air is more economical when the bromine content is low (*70*).

Bromine is no longer extracted from sea water in the United States.

### A. Stream Process for Natural Brines

Raw Arkansas brine, from the well, contains a significant amount of dissolved hydrogen sulfide. The various brine sources have varying amounts of hydrogen sulfide but there may be as much as 0.32 kg of sulfur per kg of bromine. The hydrogen sulfide is removed from the brine in a stripper and sent to a flare or sulfur recovery plant (Fig. 27). The stripped brine is heated and passed to the top of the steaming-out tower, a packed column lined with acid resistant material. The top part of the tower acts as a chlorinator. The downflowing brine contacts and absorbs the rising chlorine gas. The bromides present are converted to the corre-

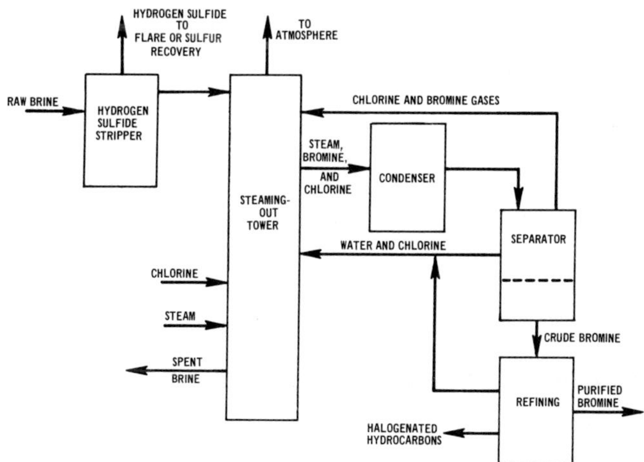


Figure 27. Flow diagram of steam-out process for manufacture of bromine, noting potential air pollution sources.

sponding chlorides and bromine. Makeup chlorine is added at the bottom of the chlorinating section. The chlorinated brine then enters the steam-out section of the tower. In this section the molecular bromine and excess dissolved chlorine are removed by steam stripping. The gaseous steam–bromine–chlorine mixture is cooled to condense the water and bromine which form two layers. The chlorine is recycled to the chlorinating section of the tower. The water is separated and returned to the steam-out section of the tower, and the crude bromine is refined by distillation. The overhead gases from the distillation, chlorine, and water vapor are returned to the steaming-out tower for recovery. In Arkansas the distillation bottoms are halogenated hydrocarbons formed from traces of crude oil in the raw brine. The purified bromine is shipped as a liquid, usually in tank car quantities. The spent brine is pumped back into the ground some distance from the brine well. In Michigan, calcium chloride may be recovered before disposal of the brine.

### Air Pollutant Emissions and Their Control

The major source of emissions from bromine manufacture from Arkansas brine is the hydrogen sulfide stripper. The hydrogen sulfide may be oxidized to sulfur dioxide in a flare or sent to a sulfur recovery plant. A typical sulfur recovery plant of the Claus type would convert 95% of the hydrogen sulfide to sulfur. The remaining 5% would be vented to atmosphere as sulfur dioxide. Newer types of sulfur recovery plants have efficiencies of greater than 99%. In 1973 less than 50% of Arkansas plant capacity had some type of sulfur recovery system. There is no hydrogen sulfide present in Michigan brines.

Very little chlorine is lost from the steaming-out tower. The excess chlorine added in the chlorinator is continually returned to the tower and the amount consumed is essentially that needed to replace the bromine and react with reducing substances.

Some bromine and chlorine air pollution occurs during infrequent emergency upsets which make it necessary to dump the steaming-out tower to a holding pond.

## *B. Emission Reduction during Air Pollution Episodes*

Sulfur dioxide emissions are directly proportional to the amount of bromine produced in most Arkansas plants. However, if a producer had several brine sources with different amounts of hydrogen sulfide, some reduction in emissions could be made by using as much as possible of the low sulfide material. A bromine plant can be shut down in a few hours and there is no increase in emissions on shutdown or start-up.

## B. PETROCHEMICAL PROCESSES

### XIV. Introduction

There are over 200 petrochemicals in production in the United States that have an annual production value of $10 million or more (*71*). Most of the petrochemical processes are essentially enclosed and normally vent a comparatively small amount of fugitive emissions. Petrochemical processes that use air-oxidation-type reactions, however, normally have large, continuous amounts of gas emissions to the atmosphere. The six processes considered in this section (Table II) employ reactions using air oxidation.

**Table II Production of Selected Petrochemicals, United States (*64*)**

| *Chemical* | *Number of plants (year)* | *Thousands of short tons per year*[a] | |
|---|---|---|---|
| | | *1970* | *1971* |
| Acrylonitrile | 7 (1970) | 520 (471) | 489 (443) |
| Carbon black | 37 (1971) | 1466 (1330) | 1509 (1369) |
| Ethylene dichloride | 17 (1971) | 3730 (3383) | 3779 (3428) |
| Ethylene oxide | 20 (1970) | 1933 (1753) | 1799 (1632) |
| Formaldehyde (37%) | 57 (1970) | 2214 (2008) | 2261 (2050) |
| Phthalic anhydride | 17 (1970) | 367 (333) | 397 (360) |

[a] Values in parentheses are for metric tons.

### XV. Acrylonitrile

Although other processes have been important in the past, in 1973 acrylonitrile was produced in the United States only by propylene ammoxidation; the Sohio fluid bed catalytic process was used universally. United States production of acrylonitrile in 1972 was 1.1 billion lb (1 billion kg), 61% of which was used in the manufacture of acrylic and modacrylic fibers. The second largest use is for the production of acrylonitrile resins. Most of the remaining production is used by the nitrile rubber industry (*71, 72*).

### *Ammoxidation Process*

Vaporized propylene and ammonia are mixed with air and introduced into a fluid bed catalytic reactor which operates at 5 to 30 lb/in.$^2$ (1.4 to 3.0 atm) and 750° to 950°F (400° to 510°C). The reaction is

$$2\ CH_2{=}CHCH_3 + 2\ NH_3 + 3\ O_2 \rightarrow 2\ CH_2{=}CHCN + 6\ H_2O \qquad (22)$$

As shown on the flow diagram (Fig. 28) the product gases pass to a water absorber where most of the reaction products are recovered. The remaining gas, mostly nitrogen, carbon dioxide, and water is vented to the atmosphere. The aqueous product solution is purified through various fractionations, and the product acrylonitrile is stored and shipped as a liquid. Acetonitrile and hydrogen cyanide are produced as by-products at a rate of about 0.1 kg of each per kilogram of acrylonitrile. Although some part of the by-products can be marketed, the major portion must be disposed of by incineration. Although acrylonitrile yield depends on many factors, a typical plant requires about 1.28 kg of propylene and 0.53 kg of ammonia for each pound of acrylonitrile produced (*73*).

## Air Pollutant Emissions and Their Control

The major source of emissions from an acrylonitrile plant is the exit-gas stream from the product absorber. For a typical plant producing 200 million lb (90 million kg) a year of acrylonitrile, the gas flow rate would be 45,370 scfm (1284 scmm). The pollutants in pounds per ton (kilograms per metric ton) of acrylonitrile produced are shown in Table

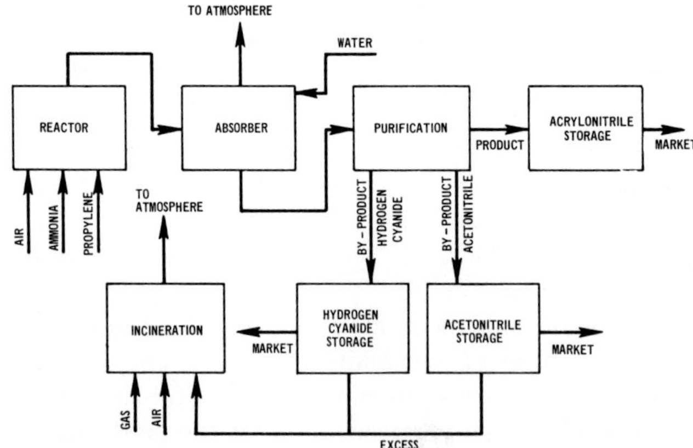


Figure 28. Flow diagram of the Sohio process for the manufacture of acrylonitrile, noting potential air pollution sources.

**Table III Pollutant Production in Acrylonitrile Manufacture by Ammoxidation**

| *Pollutant* | *Pounds per ton of acrylonitrile produced*[a,b] |
|---|---|
| Carbon monoxide | 244 (122) |
| Propylene | 76 (38) |
| Propane | 122 (61) |
| Hydrogen cyanide | 1.0 (0.5) |
| Acrylonitrile | 0.5 (0.25) |
| Acetonitrile | 13.0 (6.5) |

[a] With no control.
[b] Values in parentheses are for kilograms per metric ton.

III. The composition of the stream will depend primarily on the selectivity of the catalyst in the reactor and the efficiency of the absorber. All acrylonitrile plants in the United States vent this gas stream directly to the atmosphere. In Europe, at least one producer combusts the exit gases from the absorber and generates steam with the by-product heat.

A second, less important, source of emissions from this process is the by-product incinerator. About 1 to 3% of the nitrogen content of the excess hydrogen cyanide and acetonitrile is converted into nitrogen oxides in a commercial incinerator operating at 1600°F (871°C). There is presently no demonstrated control device for the oxides of nitrogen. The amount of the emission depends upon the amount of by-product gas burned, the temperature of combustion, and the percent of excess air.

A thermal incinerator will oxidize essentially all of the contaminants from the main process vent stream. However, the gas has such a low heat content (20 to 40 BTU per scf) that supplemental fuel must be added to achieve stable flame control and complete combustion. Various combinations of heat exchange could be used. Because acrylonitrile plants generate more steam than they can utilize, any steam generated by the incinerator has to be exported. The most feasible air pollution control system for new plants would be a thermal incinerator with process vent gas preheat, combustion air preheat, and a waste heat boiler (*73*).

For the typical 200-million-lb (90-million-kg)/year plant, the purchase and installation cost for a thermal incinerator was $760,000 (1973 basis) and the total annual operating cost would be $117,000, if it is assumed that all steam produced can be exported. An efficient air pollution control system for existing plants would be a thermal incinerator with combustion air and process vent gas preheat but without steam generation. The purchase and installation cost would be $350,000 (1973 basis) with an annual operating cost of about $136,000 (*73*).

## XVI. Carbon Black

Of all carbon black produced in the United States, about 84% is manufactured by the furnace process. The thermal process which produces 14% of the carbon black is a minor source of air pollution. The two United States plants using the channel process presently account for less than 2% of production.

United States production of carbon black in 1972 was 3.4 billion pounds (1.5 billion kilograms). Approximately 95% of all the carbon black produced was used by the rubber industry, primarily for the production of motor vehicle tires. The remaining 5% was used in news ink and as a colorant in paint, paper and plastics (*46*).

### *Furnace Process*

In the furnace process, natural gas and a high carbon aromatic oil are preheated and introduced into a furnace with a limited amount of air (Fig. 29). A combination of cracking and combustion occurs.

$$CH_4 + 1\tfrac{1}{2}\,O_2 \longrightarrow CO + 2\,H_2O + \text{heat} \quad (23)$$

$$(\text{—}CH_2\text{—}) \xrightarrow{\text{heat}} C + H_2 \quad (24)$$

$$H_2 + \tfrac{1}{2}\,O_2 \longrightarrow H_2O + \text{heat} \quad (25)$$

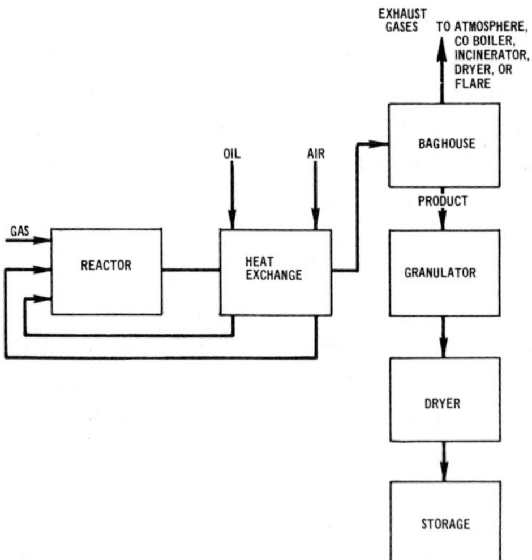


Figure 29. Flow diagram of furnace process for the manufacture of carbon black, noting potential air pollution sources.

The hot exit gas containing finely divided carbon particles is cooled by quenching and heat exchange. The black is then removed from the gas stream, usually by a bag filter. In the past the black was removed by various combinations of cyclones, electrostatic precipitators, and water scrubbers. Today the use of bag filters is almost universal. Typically, two to four reactors will be used to provide the black for a bank of bag filters containing 6 to 18 compartments. Each compartment will contain 300 to 400 bags, each about $5\frac{1}{2}$ in. (14 cm) in diameter and 126 in. (320 cm) long. The exit gas from the filter is usually discharged to the atmosphere.

The black, discharged from the bag collectors, is very light and dusty and is always granulated with water to improve its handling properties. The wet granules are dried and shipped (usually) in bags. Product yield is determined by the grade of black produced, but a typical plant requires about 1.63 kg of aromatic oil and 0.54 kg of natural gas for each kilogram of black produced (*74*).

## Air Pollutant Emissions and Their Control

The major source of emissions from a furnace black plant is the exit-gas stream from the bag collector. For a typical plant producing 90 million lb (41 million kg) of black per year the gas flow rate would be 51,000 scfm (1440 scmm). The pollutants in pounds per ton of black produced are shown in Table IV. The exact composition of the stream will depend upon the grade of black produced and the composition of the aromatic oil. Grades which require a larger proportion of gas feed will have increased emissions of carbon monoxide. The amount of hydrogen sulfide formed is proportional to the sulfur content of the aromatic oil.

Almost all of the carbon black plants in the United States vent this gas stream directly to the atmosphere. In 1974 only one plant incinerated

**Table IV Pollutant Production in the Furnace Black Process** (*69*)

| *Pollutant* | *Pounds per ton of carbon black produced*[a,b] |
|---|---|
| Hydrogen | 233 (116.5) |
| Carbon monoxide | 2540 (1270) |
| Hydrogen sulfide | 55 (27.5) |
| Sulfur dioxide | trace trace |
| Methane and acetylene | 109 (54.5) |
| Particulate matter (black) | 4 (2) |

[a] No control after baghouse.
[b] Values in parentheses are for kilograms per metric ton.

this gas and recovered the heat content by steam generation. Several other plants flare the gas. In Europe, where fuel prices are higher, many plants burn their off-gases and recover heat.

Minor amounts of carbon black dust may be emitted from the black conveying system, from the drier vent, or from the bagging and storage area. Separate dust collecting systems and baghouses are usually used to recover this relatively small amount of material.

Any type of incineration device will oxidize most of the carbon monoxide and hydrogen sulfide in the vent stream. However, a carbon monoxide boiler or thermal incinerator would probably be more efficient than a flare. Because off-gas is low in heat content [about 40 BTU per scf (356,000 cal per scm)], most combustion devices require supplemental fuel (natural gas) to maintain combustion. Various combinations of heat exchange, supplemental fuel addition, and heat recovery can be used. The choice is an economic one determined by the costs of fuel and equipment and the ability to use any steam generated. The most efficient air pollution control system for new plants under most circumstances would be a process vent gas thermal incinerator (with combustion air heat exchange) plus a waste heat boiler and steam-driven process equipment. This should oxidize all the carbon monoxide, hydrogen sulfide, and hydrocarbons. For a typical plant with a capacity of 90 million lb (41 million kg) a year, the cost of the incinerator and all of the associated equipment would be $720,000 (1973 basis). There would be a net annual operating return of $10,600 from recovered energy.

The least expensive air pollution control system for existing plants would be a flare. In this application a flare would not completely oxidize all of the combustible gases. The purchase and installation cost for a flare would be $150,000 and there would be a net annual operating cost of $33,000.

## XVII. Ethylene Dichloride

In the United States, ethylene dichloride is presently produced by the direct chlorination or oxychlorination of ethylene. About 45% of the 6.9 billion lb (3.1 billion kg) produced in the United States in 1972 was produced by oxychlorination (*71*). The atmospheric pollutants from the direct chlorination process are much less than those from the oxychlorination process described here. Almost all ethylene dichloride is used for the manufacture of vinyl chloride monomer. A small part is used for the manufacture of chlorinated solvents and for other miscellaneous usage (*75*).

### *Oxychlorination Process*

Vaporized ethylene, anhydrous hydrogen chloride, and air are fed to a catalytic reactor which operates at 20 to 75 lb/in.$^2$ (2.3 to 6.1 atm) and 400° to 600°F (200° to 315°C). The reaction is

$$2\,CH_2{=}CH_2 + O_2 + 4\,HCl \rightarrow 2\,CH_2Cl{-}CH_2Cl + 2\,H_2O \tag{26}$$

Although there are several variations in the processes used to recover the ethylene dichloride product, the most typical scheme is shown in Figure 30. The reactor effluent gases are cooled by direct water quench and pass to the phase separator where the oily ethylene dichloride is separated and sent to purification. The water phase is returned to the quencher. Some excess water must be purged to waste treatment. The off-gas from the phase separator still contains some ethylene dichloride, most of which is removed by scrubbing with an aromatic solvent. The ethylene dichloride is separated from the solvent by steam stripping and sent to purification. The product of the final distillation is stored and shipped as a liquid. The gaseous effluent from the solvent scrubber is vented to the atmosphere. A typical plant of this type requires about 0.78 kg of hydrogen chloride and 0.32 kg of ethylene for each kilogram of ethylene dichloride produced (*76, 77*).

## Air Pollutant Emissions and Their Control

The major source of emissions from an ethylene dichloride plant is the exit-gas stream from the solvent scrubber. For a typical plant of the

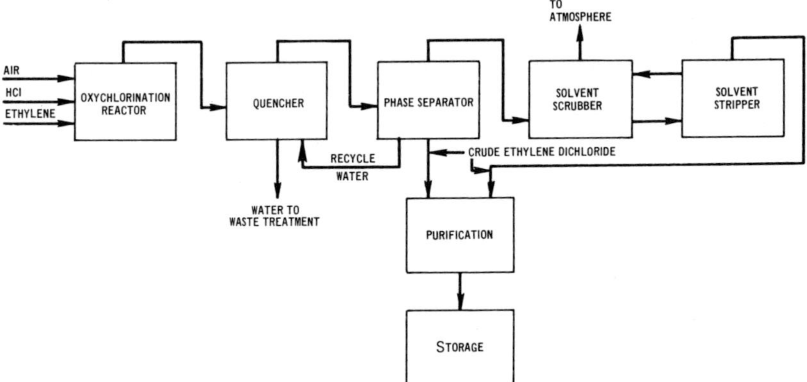


Figure 30. Flow diagram of the oxychlorination of ethylene for the manufacture of ethylene dichloride, noting potential sources of air pollution.

type described producing 700 million lb/year (317 million kg per year), the gas flow rate would be 17,280 scfm (489 scmm).

The pollutants in pounds per ton (kilograms per metric ton) of ethylene dichloride produced are shown in Table V. The values given are average values for several different oxychlorination processes. The composition of the stream will depend on catalyst activity, reactor operating conditions, and the solvent scrubber efficiency. All existing ethylene dichloride plants in the United States vent this stream directly to the atmosphere with no treatment. Any one of several combustion devices could be used to destroy the contaminants present. A water or caustic scrubber could be used after the combuster to remove the hydrogen chloride generated. Several different types of incineration would be feasible. The most efficient method for new plants would be a thermal incinerator followed by a waste heat boiler and a final caustic scrubber. The steam generated could be used in the plant. Because this system has not been demonstrated in the United States, it is possible that there might be significant operating problems. Careful control of temperature would be necessary to prevent condensation of hydrogen chloride and resultant corrosion of metal. Also, because of the low heating value of the gas stream [30 to 55 BTU per scf (267,000 to 445,000 cal per scm)], supplemental fuel would have to be added to achieve complete combustion and satisfactory flame control. For the typical 700-million pounds per year plant, a control system of this type would cost about $1,000,000 (1973 basis) and have an annual operating cost of $390,000 (*76*).

The most feasible air pollution control system for an existing plant would be a thermal incinerator and scrubber on the main process vent, for which the estimated purchase and installation cost would be $750,000 and the net annual operating cost $500,000.

**Table V Pollutant Production in Ethylene Dichloride Manufacture by Oxychlorination** (*71*)

| *Pollutant* | *Pounds per ton of ethylene dichloride produced*[a,b] |
|---|---|
| Carbon monoxide | 1.3 (0.65) |
| Methane | 4.0 (2.0) |
| Ethylene | 9.6 (4.8) |
| Ethane | 12.7 (6.3) |
| Ethylene dichloride | 13.8 (6.9) |
| Ethyl chloride | 11.7 (5.9) |
| Aromatic solvent | 2.2 (1.1) |

[a] No control.
[b] Values in parentheses are for kilograms per metric ton.

## XVIII. Ethylene Oxide

The oxidation of ethylene is the most widely used process for the production of ethylene oxide and is used by all producers in the United States. In 1972, 4.2 billion lb (1.9 billion kg) was produced in the United States in 16 plants (*71*). The production of ethylene glycol, used primarily as automotive antifreeze, consumes more than half the ethylene oxide produced. The second largest use of ethylene oxide is in the manufacture of nonionic surfactants (*78*).

### *Oxidation of Ethylene*

Ethylene oxide is produced by passing ethylene and air (or oxygen) over a silver catalyst and recovering ethylene oxide from the gas stream by water absorption. In 1973, about one-third of all United States production used pure oxygen which was extracted from air in a separate plant. Ethylene oxide is stripped from the water solution and refined. The reaction is

$$H_2C{=}CH_2 \ + \ \tfrac{1}{2}O_2 \ \longrightarrow \ H_2C\underset{O}{\diagdown\!\!\!-\!\!\!\diagup}CH_2 \ + \ \text{heat} \qquad (27)$$

(air or oxygen)

Flow diagrams for the air and oxygen processes are given in Figures 31 and 32. In the air process, air and ethylene are added to a recycle gas stream and fed to the main reactor where the reaction takes place

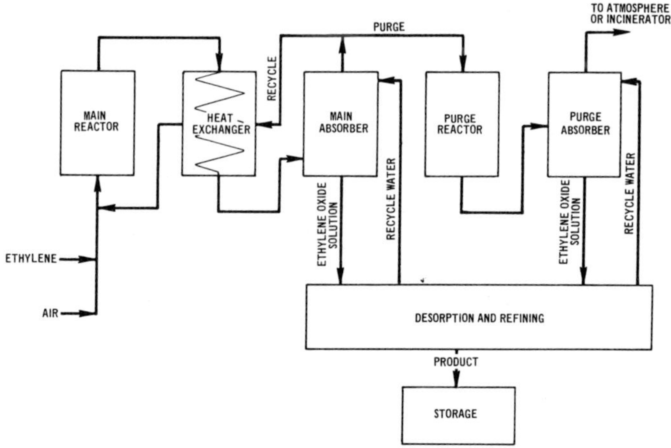


Figure 31. Flow diagram of the direct oxidation process for the manufacture of ethylene oxide using air, noting potential sources of air pollution.

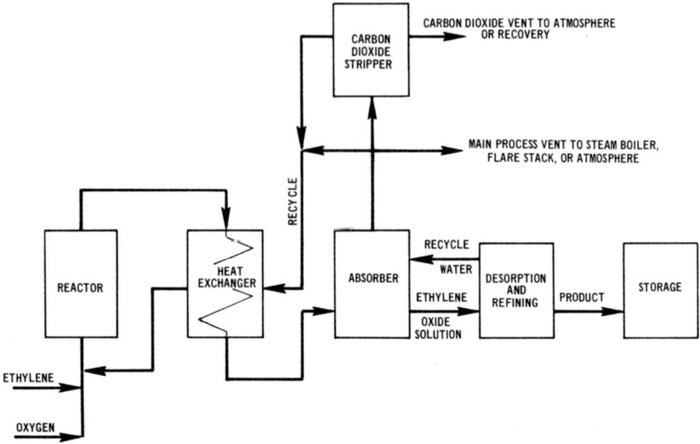


Figure 32. Flow diagram of the direct oxidation process for the manufacture of ethylene oxide using oxygen, noting potential sources of air pollution.

in tubes containing silver catalyst. After heat exchange with recycle gas, the reactor effluent gases pass to the main absorber where the ethylene oxide is absorbed in a water solution. Approximately two-thirds of the absorber overhead gas (which contains nitrogen, carbon dioxide, unreacted ethylene, and air) is returned as recycle to the main reactor. The remaining one-third of the main absorber overhead gas is purged from the system. Most of the ethylene in the purge stream is recovered as ethylene oxide which is formed in the purge reactor and removed in the purge absorber. The vent from the purge absorber, mostly nitrogen and carbon dioxide, contains some unreacted ethylene, a small amount of ethylene oxide, and all the ethane in the ethylene raw material. The ethylene oxide in the main absorber and purge absorber bottoms is removed by desorption, refined, and sent to storage.

Although the oxygen process is similar to the air oxidation process, there is usually no purge reactor or absorber. Also, because the conversion of ethylene per pass is lower, the recycle is larger. Product is recovered by absorption as in the air oxidation process. A carbon dioxide absorber, used on a portion of the recycle stream, controls the buildup of carbon dioxide. Other inert gases are removed by a small absorber vent gas purge. In a variation of the process, methane is added to the recycle gas to act as an inert in the reactor and yield a high calorie vent gas which burns easily in a boiler (*79–81*).

## Air Pollutant Emissions and Their Control

The only important source of emissions from ethylene oxide manufacture are the purge vents. Typical pollutants for the purge absorber vent

**Table VI Pollutant Production in Ethylene Oxide Manufacture by Air Oxidation**

| Pollutant | *Pounds per ton of ethylene oxide produced—purge absorber vent*[a] | |
|---|---|---|
| | *With no control* | *With combustion* |
| Ethane | 12 (6) | Nil |
| Ethylene | 185 (92) | Nil |
| Ethylene oxide | 2 (1) | Nil |

[a] Values in parentheses are for kilograms per metric ton.

of an ethylene oxide plant using air oxidation are shown in Table VI. For a typical 200-million-pound (90-million-kg) per year plant the vent gas rate would be 34,500 scfm. Typical main process vent pollutants for a plant using oxygen oxidation are shown in Table VII. For a 200 million lb (90 million kg) per year plant the main process vent gas rate would be 120 scfm (3.4 scmm). Because the carbon dioxide vent of 5800 scfm (164 scmm) contains an insignificant amount of contaminants, it can be vented or sold as carbon dioxide.

The most efficient method to control emissions from an ethylene oxide plant is to feed the main process vent to a catalytic combustor and use the hot gases from this unit in expanders to drive the process compressors. More than 99% removal of combustible gases would be expected. A new plant built with this control device would cost an additional $300,000 (1973 basis) and have a net operating cost of $42,000 per year.

## XIX. Formaldehyde

In the United States, in 1973, all formaldehyde was produced by the air oxidation of methanol. About 57% was used in combination with

**Table VII Pollutant Production in Ethylene Oxide Manufacture by Oxygen Oxidation**

| Pollutant | *Pounds per ton of ethylene oxide produced—main process vent*[a] | |
|---|---|---|
| | *With no control* | *With combustion* |
| Ethane | 6 (3) | Nil |
| Ethylene | 5 (2.5) | Nil |

[a] Values in parentheses are for kilograms per metric ton.

phenol, urea, or melamine to form resins and adhesives. The largest and fastest growing use of phenol–formaldehyde resins is as an adhesive for plywood. Urea–formaldehyde resin is primarily used as an adhesive for particle board. Formaldehyde is also used to manufacture hexamethylenetetramine, pentaerythritol, and several miscellaneous resins (*82*). United States production of formaldehyde in 1972 was 7.5 billion lb (3.4 billion kg) of 37% water solution (*71*).

### *Air Oxidation of Methanol*

There are two methods of producing formaldehyde by the air oxidation of methanol. The first uses a mixed metal oxide catalyst and a large excess of air to produce formaldehyde by the following oxidation reaction:

$$\underset{}{CH_3OH} + \underset{\text{(air)}}{\tfrac{1}{2}\,O_2} \xrightarrow{\text{catalyst}} CH_2O + H_2O + \text{heat} \tag{28}$$

The methanol content is kept below the lower explosive limit.

The second method employs a combined oxidation-dehydrogenation reaction using a silver catalyst:

$$CH_3OH + \underset{\text{(air)}}{\tfrac{1}{2}\,O_2} \xrightarrow{\text{catalyst}} CH_2O + H_2O + \text{heat} \tag{29}$$

$$CH_3OH \xrightarrow{\text{catalyst}} CH_2O + H_2 \tag{30}$$

Less air is used than with the mixed-oxide process to maintain the methanol content above the upper explosive limit. Although the dehydrogenation reaction is endothermic and there is less net heat of reaction than with the metal oxide catalyst, the reaction is still thermally self-supporting.

It is estimated that in 1973, 22% of formaldehyde produced in the United States was made by the mixed-oxide catalyst processes and the balance by the silver catalyst processes. In both methods the air–methanol mix is fed to a fixed-bed reactor where the reaction takes place (Fig. 33). In the mixed-oxide catalyst process there are usually one or two shell and tube reactors per train with catalyst in the tubes. A heat-transfer fluid in the shell removes the heat of reaction. In the silver catalyst process, there may be as few as three or as many as 40 packed-bed reactors per train. In both processes the exit gases are cooled before entering the absorber tower where the formaldehyde gas is absorbed by water and a 37–53% solution is formed. The formaldehyde solution from the absorber is passed to a purification section where unreacted methanol may be removed and returned to the reactor (*83–85*).

The off-gas from the absorber passes through an entrainment separator

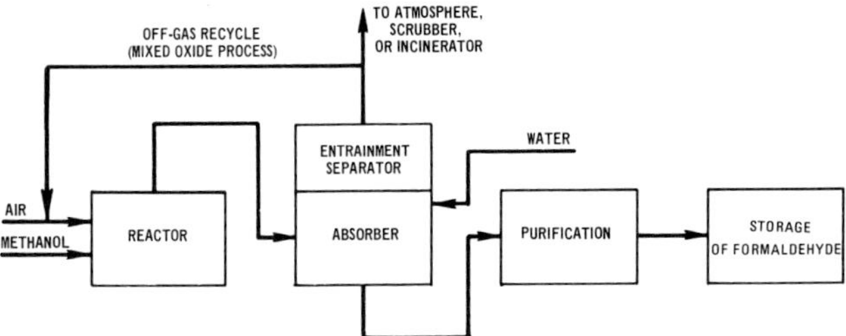


Figure 33. Flow diagram of the methanol process for the manufacture of formaldehyde, noting potential sources of air pollution.

to remove entrained liquids and is then discharged to the atmosphere or treated in a control device. The off-gas contains mostly nitrogen but includes some formaldehyde, methanol, and carbon monoxide. The silver-catalyst-process plants also emit hydrogen, while the mixed-oxide plants vent dimethyl ether. The composition of the gas at the exit of the absorber depends primarily on the number of trays, the degree of cooling of the trays, and the strength of formaldehyde produced.

In the mixed-oxide process, part of the absorber vent gas is usually recycled through the process. The recycling lowers the oxygen concentration in the reactor which allows increased concentration of methanol without danger of explosion. This results in increased capacity and higher yield of product.

## Air Pollutant Emissions and Their Control

The major source of emissions from a formaldehyde plant is the exit-gas stream from the scrubber. The amount and composition of gaseous emissions and the type of control device that can be used depend upon the type of process.

For the mixed metal oxide catalyst process the pollutants in pounds per ton of 37% formaldehyde solution produced are shown in Table VIII. The values given are for a plant operating at maximum recycle; for non-recycle operation these values would be higher. For a typical plant producing 100 million lb (45 million kg) per year of 37% formaldehyde, the exit-gas flow rate at maximum recycle would be 3400 scfm (9.6 scmm). In most plants of this type in the United States, this stream is vented directly to the atmosphere. A water scrubber is used in one plant to remove formaldehyde, the most odorous and obnoxious constituent, but essentially none of the carbon monoxide or dimethyl ether is removed. The contaminants absorbed require water treatment, which must be added to the cost of operating this system. An incinerator could be used

**Table VIII Maximum Absorber Vent Gas Recycle Pollutants in Formaldehyde Production by Mixed-Oxide Catalyst Process[a]** *(76)*

| | *With no control*[b] | *With incineration* | *With water scrubbing*[b] |
|---|---|---|---|
| Formaldehyde | 1.6 (0.8) | Nil | 0.2 (0.1) |
| Methanol | 4 (2) | Nil | 0.5 (0.2) |
| Carbon monoxide | 34 (17) | Nil | 34 (17) |
| Dimethyl ether | 1.6 (0.8) | Nil | 1.6 (0.8) |

[a] Values are given in pounds per ton 37% solution produced.
[b] Values in parentheses are for kilograms per metric ton.

to oxidize all the contaminants from this stream, but because of the very low heat content of the gas, [less than 10 BTU/scf (89,000 cal/scm)] considerable auxiliary fuel would be required.

A water scrubber for this operation would cost about $81,000 (1973 basis), purchased and installed, and would have an annual operating cost of approximately $20,000 not including water treatment costs. A thermal incinerator would cost $44,000 and have an annual operating cost of about $35,000. Heat recovery is not economical for an incinerator of this small size.

For the silver catalyst process the pollutants in pounds per ton of 37% formaldehyde solution produced are shown in Table IX. For a typical plant producing 100 million lb (45 million kg) per year of 37% formaldehyde, the gas flow rate would be 2200 scfm (62 scmm) with a heat content of about 62 BTU/scf (552,000 cal/scm). The composition of the vent gas depends again on design and operating conditions of the absorber, but the absorption is easier because of the lower gas volume. Incineration can be used to oxidize all the contaminants in this stream. In the United States, about one-half of the silver catalyst plants vent the stream di-

**Table IX Pollutant Production in Formaldehyde Manufacture by Silver Catalyst Process** *(75)*

| | *Pounds per ton of 37% formaldehyde solution produced*[a] | |
|---|---|---|
| *Pollutant* | *With no control* | *With incineration* |
| Formaldehyde | 1.1 (0.6) | Nil |
| Methanol | 5.1 (2.5) | Nil |
| Carbon monoxide | 10.5 (5.2) | Nil |
| Hydrogen | 20.1 (10.0) | Nil |

[a] Values in parentheses are for kilograms per metric ton.

rectly to the atmosphere. The rest either flare the gas, burn it in an incinerator, or use it as a boiler house fuel.

If by-product steam can be used, the boiler house vent gas burner would seem to be the most attractive method of control. Although some auxiliary fuel must be used, there should be a net annual return by the use of this procedure of about \$3000 [for a 100-million-lb (45 million kg) per year plant]. Total purchase and installation cost to allow this to be done would be \$44,000 (1973).

## XX. Phthalic Anhydride

In the United States, all phthalic anhydride is produced by the vapor-phase air oxidation of either *o*-xylene or naphthalene. About 56% goes into the production of plasticizers for vinyl resins, and phthalic-based alkyd resins, used primarily for exterior surface coatings, account for about 22% of the material produced. About 17% is used in the manufacture of polyester resins, most of which are glass reinforced (*86*). United States production of phthalic anhydride in 1972 was 900 million lb (408 million kg) (*71*).

The vapor-phase air oxidation of *o*-xylene in the most economical process for the manufacture of phthalic anhydride. Although naphthalene was the raw material for about 46% of United States production in 1971, probably very few new naphthalene plants will be built. Except for the feedstock, the processes are similar and their emissions are comparable.

### *Oxidation of o-Xylene*

Air and *o*-xylene are fed to a catalytic (usually fixed-bed) reactor (Fig. 34). A small amount of sulfur dioxide is usually added to maintain catalyst activity. Typical operating conditions are 10 lb/in.$^2$ (1.7 atm) gauge and 700°F (370°C). The *o*-xylene content of the feed is held below the lower explosive limit, which is around 1.0%. The reactor is cooled by a molten-salt heat transfer fluid which is used to generate steam in a waste-heat boiler. A yield of about 1.27 kg of phthalic anhydride per kilogram of *o*-xylene is achieved when using a vanadium pentoxide catalyst.

The basic reaction is

$$C_6H_4(CH_3)_2 + \underset{\text{(air)}}{3O_2} \longrightarrow C_6H_4(CO)_2O + 3H_2O + \text{heat} \qquad (31)$$

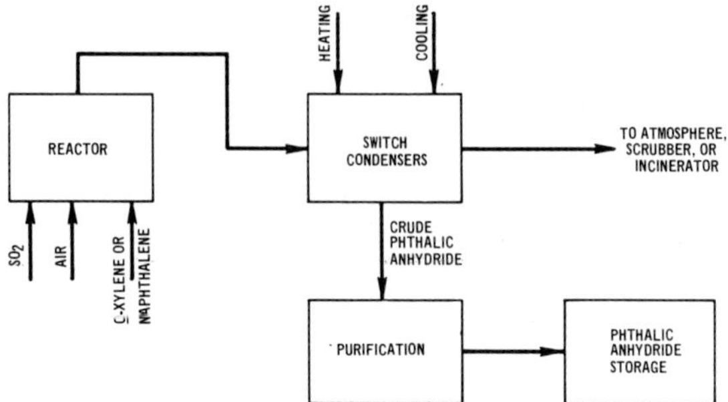


Figure 34. Flow diagram of the oxidation of *o*-xylene or naphthalene for the manufacture of phthalic anhydride, noting potential sources of air pollution.

The reactor off-gas, after heat exchange, passes through a bank of switch condensers where the phthalic anhydride is solidified. When the amount of solid crystalline phthalic on the tubes begins to obstruct the flow of gas through one condenser the reactor off-gas is "switched" to an empty cold condenser. The product is removed from the full condenser by heating and melting. Part of the bank is therefore on freeze-out while another part is on melt. The crude phthalic anhydride from the switch condensers is purified by fractionation and stored. Most phthalic anhydride is shipped in tank car quantities as a liquid. Some small part is flaked and shipped in bags or drums.

The off-gas from the switch condensers contains mostly nitrogen, oxygen, water, and carbon dioxide. Smaller amounts of carbon monoxide, maleic anhydride, phthalic anhydride, and sulfur oxides are also present (*87*, *88*).

## Air Pollutant Emissions and Their Control

The major source of emissions from phthalic anhydride manufacture is the off-gas from the switch condensers. For a typical 130,000,000-lb per year (58 million kg per year) plant this stream has a flow rate of 130,000 scfm (3680 scmm). In a few old plants off-gas is discharged to the atmosphere without control. The composition of this stream is shown in Table X for plants using *o*-xylene feedstock. The organic acids and anhydrides are malodorous lachrymators.

The most common type of emission control is water scrubbing. A typical combination for this application is a venturi scrubber followed by a cyclone separator and a packed countercurrent scrubber. The absorption efficiencies of the various scrubbers will vary considerably, but typical emissions are given in Table X. The scrubbing liquid will usually be incinerated in a small natural gas incinerator which will itself have

**Table X Pollutant Production in Phthalic Anhydride Manufacture** (*78*)

| Pollutant | Pounds per ton phthalic anhydride produced[a] | | |
|---|---|---|---|
| | *With no control* | *With incineration* | *With water scrubbing* |
| Organic acids and anhydrides | 130 (65) | 5 (2.5) | 6 (3) |
| Sulfur dioxide | 10 (5) | 10 (5) | 10 (5) |
| Carbon monoxide | 280 (140) | 0.2 (0.1) | 280 (140) |
| Particulate matter | Trace | — | Trace |

[a] Values in parentheses are for kilograms per metric ton.

some emissions. The values given in Table X include the emissions from the scrubber and the scrubber water incinerator. The combination water scrubber and associated incinerator would have an installed capital cost of about $1,500,000 (1973 basis) and an approximate annual operating cost of $420,000 (*87*).

A few *o*-xylene phthalic anhydride plants incinerate the entire off-gas stream from the switch condensers. The typical unit uses natural gas as fuel and has a retention time of 0.9 seconds at 600°C. Steam generation recovers the fuel value of the natural gas. A typical unit on a 130,000,000-lb a year (58 million kg per year) plant might cost $1,300,000 installed, and have an annual net operating cost of about $250,000. As shown in Table X this typical incinerator would perform more efficiently than the scrubber in removing the obnoxious organic acids and anhydrides. Minor additional quantities of phthalic anhydride vapors are released from storage tanks, tank car loading stations, and from the flaking and packaging equipment.

## XXI. Emission Reduction during Air Pollution Episodes

Atmospheric emissions from the petrochemical processes described in this section can be reduced significantly or stopped completely within a period of 1–3 hours. Petrochemical plants usually have several independent trains of processing equipment, and these trains may be shutdown as required to reduce emissions. Most operators also have the option of reducing production by leaving all trains on line and reducing the throughout per train. In either case the emissions from the process would be approximately proportional to operating level.

All of the processes described may be shutdown without any increase in emissions, and all but acrylonitrile may be started up without any increase in emissions. During an acrylonitrile reactor start-up the reactors are vented directly to the atmosphere for 1 hour, and the average hydrocarbon emission during this period is three to seven times the normal on-stream value. Therefore, for this process, a partial reduction of air emissions during an air episode might be best accomplished by leaving all trains on line and reducing the throughput per train.

It is difficult to predict the economic impact of reducing production in a petrochemical plant. Petrochemical processes are typically integrated with other processes, often with little storage between plants. If the production rate of one process is reduced, other upstream and downstream processes may also be forced to reduce because of lack of storage capacity or raw material.

## REFERENCES

1. Chemical Information Services, "Chemical Economics Handbook and 1971 Directory of Chemical Producers." Stanford Research Institute, Menlo Park, California, 1971.
2. *Chem. Week* **111,** 11–12 (1972); **112,** 12 (1973); **113,** 6 (1973).
3. Annual Report of the Administrator of the Environmental Protection Agency to the Congress of the United States, "The Economics of Clean Air," p. 162. U.S. Environmental Protection Agency, Research Triangle Park, North Carolina, 1972.
4. Q. R. Stahl, "Preliminary Air Pollution Survey of Hydrochloric Acid," APTD 69-36, Contract No. PH-22-68-25, p. 63. U.S. Dept. of Health, Education, and Welfare, Public Health Service, National Air Pollution Control Administration, Raleigh, North Carolina, 1969.
5. R. E. Kirk and D. F. Othmer, "Encyclopedia of Chemical Technology," 2nd ed., Vol. 11, p. 318. Wiley, New York, New York, 1967.
6. U.S. Dept. of Health, Education, and Welfare, "Atmospheric Emissions from Hydrochloric Acid Manufacturing Processes," Air Pollution Series, A.P. 54, p. 11. Public Health Service, National Air Pollution Control Administration, Durham, North Carolina, 1969.
7. R. W. Gerstle and T. W. Devitt, Annu. Meet. Air Pollution Control Association, Pittsburgh, Pennsylvania, 1971.
8. U.S. Dept. of Health, Education, and Welfare, "Atmospheric Emissions from Hydrochloric Acid Manufacturing Processes," Air Pollution Series, A.P. 54, p. 20. Public Health Service, National Air Pollution Control Administration, Durham, North Carolina, 1969.
9. R. N. Shreve, "Chemical Process Industries," 3rd ed. McGraw-Hill, New York, New York, 1967.
10. U.S. Dept. of Health, Education, and Welfare, "Atmospheric Emissions from Hydrochloric Acid Manufacturing Processes." Air Pollution Series, A.P. 54, p. 21.

Public Health Service, National Air Pollution Control Administration, Durham, North Carolina, 1969.

11. Bureau of Census, "Current Industrial Report Series," M 28 A (72)-14, Inorganic Chemicals, 1972. U.S. Dept. of Commerce, Washington, D.C., 1973.
12. *Oil Paint Drug Rep.* **197**(11), (1970).
13. Office of Air Programs, "Engineering and Cost Effectiveness Study of Fluoride Emission Control," PB-209-647, p. 3–295. U.S. Environmental Protection Agency, Research Triangle Park, North Carolina, 1972.
14. T. L. Chilton, "Strong Waters," pp. 127–131. MIT Press, Cambridge, Massachusetts, 1968.
15. L. Hellmer, *Hydrocarbon Process.* **45,** No. 11, 183–188 (1966).
16. Manufacturing Chemists Association and the U.S. Department of Health, Education, and Welfare, "Atmospheric Emissions from Nitric Acid Manufacturing Processes," Coop. Study Proj., Environ. Health Ser. 999-AP-27, p. 27. Pub. Health Serv., U.S. Dept. of Health, Education, and Welfare, Cincinnati, Ohio, 1965.
17. "Background Information for Proposed New Source Performance Standards," APTD-0711, p. 37. U.S. Environmental Protection Agency, Research Triangle Park, North Carolina, 1971.
18. O. J. Adlhart, S. G. Hindin, and R. E. Kenson, *Chem. Eng. Progr.* **67**(2), 73–78 (1971).
19. D. J. Newman, *Chem. Eng. Progr.* **67**(2), 79–84 (1971).
20. G. Nonhebel, "Gas Purification Processes," pp. 188–215. J. W. Arrowsmith, Ltd., Bristol, England, 1964.
21. L. L. Fornoff, Annu. Meet. Amer. Inst. Chem. Eng., New York, New York, 1971.
22. G. A. Chappell, "Development of Aqueous Processes for Removing $NO_x$ from Flue Gases-Addendum," Contract 68-02-0220. U.S. Environmental Protection Agency, Research Triangle Park, North Carolina, 1973.
23. U.S. Environmental Protection Agency, "Background Information for Standards of Performance: Phosphate Fertilizer Industry," Vol. 1, EPA-450/2-74-01971. U.S. Environmental Protection Agency, Research Triangle Park, North Carolina, 1974.
24. H. S. Bryant, N. G. Holloway, and A. D. Selber, *Ind. Eng. Chem.* **62,** 8–23 (1970).
25. Pedco Environmental Specialist, Test Numbers 76-MM-26 and 27, Contract No. 68-02-0237. U.S. Environmental Protection Agency, Cincinnati, Ohio, 1973.
26. Pedco Environmental Specialist, Test Numbers 76-MM-04 and 05, Contract No. 68-02-0237. U.S. Environmental Protection Agency, Cincinnati, Ohio, 1972.
27. H. J. Allgood, F. E. Lancaster, and J. A. McCollum, Nat. Meet. Amer. Chem. Soc., Washington, D.C., 1966.
28. A. V. Slack, "Phosphoric Acid," 1st, ed., p. 927. Dekker, New York, New York, 1968.
29. "Atmospheric Emissions from Thermal-Process Phosphoric Acid Manufacture," pp. 16–22. U.S. Dept. of Health, Education, and Welfare, National Air Pollution Control Administration, Durham, North Carolina, 1968.
30. K. Karbe, *Chem. Eng.* **46,** 268 (1968).
31. J. G. Kronseder, *Chem. Eng. Progr.* **44,** 97–102 (1968).
32. *Chem. Week* **112,** 33–34 (1972).
33. Chemical Construction Corporation, "Engineering Analysis of Emissions Control Technology for Sulfuric Acid Manufacturing Processes," Final Report of Contract 22-69-81, Publ. No. PB-190-393. National Air Pollution Control Administration, U.S. Dept. of Health, Education, and Welfare, Public Health Service, Washington, D.C., 1970.

34. U.S. Dept. of Health, Education, and Welfare, "Atmospheric Emissions from Sulfuric Acid Manufacturing Processes," Environ. Health Ser. 999-AP-13, pp. 51–55. Public Health Service, U.S. Dept. of Health, Education, and Welfare, Cincinnati, Ohio, 1965.
35. "Background Information for Proposed New Source Performance Standards," APTD-0711, p. 44. United States Environmental Protection Agency, Research Triangle Park, North Carolina, 1971.
36. W. G. Tucker and J. R. Burleigh, *Chem. Eng. Progr.* **67**, No. 5, 57–63 (1971).
37. B. H. Potter and T. L. Craig, *Chem. Eng. Progr.* **68**, 53–54 (1972).
38. J. N. Robinson, *Int. J. Sulfur Chem., Part B* **7**, No. 1, 51–56 (1972).
39. J. J. Collins, L. L. Fornoff, Manchanda, W. C. Miller, and D. C. Lovell, Annu. Meet. Amer. Inst. Chem. Eng., New York, New York, 1973.
40. U.S. Dept. of Interior, "Minerals Industry Surveys." U.S. Bureau of Mines, Washington, D.C., 1971.
41. Industrial Gas Cleaning Institute, "Air Pollution Control Technology and Costs for Seven Selected Areas," Contract No. 68-02-0289. U.S. Environmental Protection Agency, Research Triangle Park, North Carolina, 1973.
42. C. J. Lewis and B. B. Crocker, *J. Air Pollut. Contr. Ass.* **19**, 1 (1969).
43. "Background Information for Proposed New Source Performance Standards," Vol. 2, APTD-1352(b). U.S. Environmental Protection Agency, Research Triangle Park, North Carolina, 1973.
44. J. A. Schueler, "Gravel Bed Filters for Lime Kilns." National Lime Association, Hershey, Pennsylvania, 1972.
45. L. J. Minnick, *J. Air Pollut. Contr. Ass.* **21**, No. 4, 195–200 (1971).
46. U.S. Department of Interior, "Minerals Yearbook 1972." U.S. Bureau of Mines, Washington, D.C., 1973.
47. R. N. Shreve, "Chemical Process Industries," 3rd ed., pp. 226–228. McGraw-Hill, New York, New York, 1967.
48. W. L. Faith, D. B. Keyes, and R. L. Clark, "Industrial Chemicals," pp. 664–669. Wiley, New York, New York, 1966.
49. R. E. Kirk and D. F. Othmer, "Encyclopedia of Chemical Technology," 2nd ed., Vol. 1, pp. 707–740. Wiley, New York, New York, 1963.
50. Office of Air Programs, "Air Pollutant Emission Factors," AP-42. U.S. Environmental Protection Agency, Research Triangle Park, North Carolina, 1973.
51. W. H. Waggaman, "Phosphoric Acid, Phosphates, and Phosphatic Fertilizers," 2nd ed., p. 50. Hafner, New York, New York, 1969.
52. A. J. Teller, *Chem. Eng. Progr.* **63**, 3 (1967).
53. *Environ. Sci. Technol.* **6**, 5 (1972).
54. F. Molyneux, *Aust. Chem. Process. & Eng.* **23**, 2 (1970).
55. F. P. Achorn and J. S. Lewis, Jr., *Agr. Chem. & Commer. Fert.* **27**, 10–14 (1972).
56. F. E. Gartrell and J. C. Barber, *Proc. Amer. Soc. Civil Eng.* pp. 1321–1334 (1970).
57. E. A. Harre, "Fertilizer Trends—1969," p. 72. Tennessee Valley Authority, Muscle Shoals, Alabama, 1969.
58. W. L. Faith, D. B. Keyes, and R. L. Clark, "Industrial Chemicals," pp. 89–94. Wiley, New York, New York, 1966.
59. R. E. Kirk and D. F. Othmer, "Encyclopedia of Chemical Technology," 2nd ed., Vol. 2, pp. 320–329. Wiley, New York, New York, 1963.
60. North American Chlor-Alkai Industry Plants and Production Data Book, p. 15. The Chlorine Institute, New York, New York, 1972.
61. "Atmospheric Emissions from Chlor-Alkali Manufacture," No. AP 80, pp. 1–4.

U.S. Environmental Protection Agency, Research Triangle Park, North Carolina, 1971.
62. "Control Techniques for Mercury Emissions from Extraction and Chlor-Alkali Plants," No. AP-118, pp. 24–33. U.S. Environmental Protection Agency, Research Triangle Park, North Carolina, 1973.
63. F. J. Flewelling, *Chem. Can.* **23**, (1971).
64. U.S. Environmental Protection Agency, Contract CPA 70-132, Task Orders No. 2 and 3. Roy F. Weston, Inc., Westchester, Pennsylvania, 1971.
65. *Chem. &Eng. News* pp. 14–15 (1972).
66. *Chem. Week* **111,** 45 (1972).
67. J. J. Collins, W. C. Miller, and J. E. Philcox, Annu. Meet. Air Pollution Control Association, Pittsburgh, Pennsylvania, 1972.
68. D. H. Porter and J. D. Watts, Nat. Meet. p. 14. Amer. Inst. Chem. Eng., New York, New York, 1971.
69. *Chem. Week* **113,** 31–32 (1973).
70. R. E. Kirk and D. F. Othmer, "Encyclopedia of Chemical Technology," 2nd ed., Vol. 3, pp. 750–766. Wiley, New York, New York, 1964.
71. "Synthetic Organic Chemicals, U.S. Production and Sales, 1972." U.S. Tariff Commission, Washington, D.C., 1973.
72. *Oil Paint Drug Rep* **200,** No. 6 (1971).
73. Office of Air Quality Planning and Standards, "Engineering and Cost Study of Air Pollution Control for the Petrochemical Industry, Acrylonitrile," EPA-450/3-73-006-b. U.S. Environmental Protection Agency, Research Triangle Park, North Carolina, 1974.
74. Office of Air Quality Planning and Standards, "Engineering and Cost Study of Air Pollution Control for the Petrochemical Industry, Carbon Black," EPA-450/3-73-006-a. U.S. Environmental Protection Agency, Research Triangle Park, North Carolina, 1974.
75. *Oil Paint Drug Rep.* **200,** No. 12 (1971).
76. Office of Air Quality Planning and Standards, "Engineering and Cost Study of Air Pollution Control for the Petrochemical Industry, Ethylene Dichloride," EPA-450/3-73-006-c. U.S. Environmental Protection Agency, Research Triangle Park, North Carolina, 1974.
77. A. M. Brownstein, "U.S. Petrochemicals," pp. 250–253 and 286–290. Petroleum Publ., Tulsa, Oklahoma, 1972.
78. *Oil Paint Drug Rep.* **196,** No. 14 (1969).
79. Office of Air Quality Planning and Standards, "Engineering and Cost Study of Air Pollution Control for the Petrochemical Industry, Ethylene Oxide," EPA-450/3-73-006-f. U.S. Environmental Protection Agency, Research Triangle Park, North Carolina, 1974.
80. A. M. Brownstein, "U.S. Petrochemicals," pp. 280–285. Petroleum Publ., Tulsa, Oklahoma, 1972.
81. A. V. G. Hahn, R. Williams, Jr., and H. W. Zabel, "The Petrochemical Industry," pp. 274–277. McGraw-Hill, New York, New York, 1970.
82. *Chem. Mkt. Rep.* **201,** No. 18 (1972).
83. Office of Air Quality Planning and Standards, "Engineering and Cost Study of Air Pollution Control for the Petrochemical Industry, Formaldehyde via the Silver Catalyst Process," EPA-450/3-73-006-d. U.S. Environmental Protection Agency, Research Triangle Park, North Carolina, 1974.
84. Office of Air Quality Planning and Standards, "Engineering and Cost Study

of Air Pollution Control for the Petrochemical Industry, Formaldehyde via the Mixed Oxide Catalyst Process," EPA-450/3-73-006-e. U.S. Environmental Protection Agency, Research Triangle Park, North Carolina, 1974.
85. A. V. G. Hahn, R. Williams, and H. W. Zabel, "The Petrochemical Industry," pp. 75–88. McGraw-Hill, New York, New York, 1970.
86. *Oil Paint Drug Rep.* **200,** No. 7 (1971).
87. Office of Air Quality Planning and Standards, "Engineering and Cost Study of Air Pollution Control for the Petrochemical Industry, Phthalic Anhydride," EPA-450/3-73-006-g. U.S. Environmental Protection Agency, Research Triangle Park, North Carolina, 1974.
88. A. M. Brownstein, "U.S. Petrochemicals," pp. 220–230. Petroleum Publ., Tulsa, Oklahoma, 1972.

# 19

# Petroleum Refining

Harold F. Elkin

## I. Introduction

As of January 1, 1974, there were 247 petroleum refineries in the United States capable of processing 14.2 million barrels of crude oil daily (*1*). This reflects a 6% growth in refining capacity during 1973 compared to a 2.2% gain during 1972 and a 3.2% gain during 1971 (*2*). This relatively slow growth in domestic refining capacity was attributed to various factors such as dwindling domestic crude reserves; a tight money market; price controls; uncertainty of future environmental requirements relative to product quality; uncertainty of stationary source emission controls; and, the difficulty of site selection for new grass root facilities.

By comparison, the refining capacity as of January 1, 1974 in noncommunist countries outside the United States and Canada was 38.3 million barrels per calendar day, a gain of 10.1% over a period of 1 year (*3*). The outlook for 1974 and 1975 shows a slower rate of growth for the noncommunist countries, at about 2% per year, with a corresponding upswing in domestic and Canadian capacity of just under 6% per year (*3*). Table I shows a state-by-state tabulation of refineries operating in the United States (*1*).

United States refineries represent a gross investment of almost $18 billion (*4*), and according to the American Petroleum Institute, the industry will need to invest over $350 billion (*4*) worldwide between 1971 and 1980 to explore for more oil, build new refineries, and provide marketing and transportation facilities to meet the needs of consumers in the United States and worldwide.

In 1972, oil and natural gas supplied 77% (*4*) of the total fuel requirements of the United States compared with about 73% in 1966 and 63% in 1952. Taking into consideration the October 1973 Arab oil embargo, the United States refined petroleum product demand for 1974 was projected to be 18.7 million barrels per day (*5*). Of this volume, about 40% was expected to be directly imported or manufactured in the United States from imported crude oil. It is difficult to say precisely what impact conservation practices and emerging domestic and foreign energy policies will have on petroleum consumption in the United States. Estimates prior to the 1973 Arab embargo predicted oil consumption in the United States

**Table I Survey of Operating Refineries in the United States (State Capacities as of Jan. 1, 1974)[a,b]**

| State | No. plants | Crude capacity b/cd | Crude capacity b/sd | Vacuum distillation | Thermal operations | Cat. cracking[c] Fresh feed | Cat. cracking[c] Recycle | Cat. reforming | Cat. hydrocracking | Cat hydrorefining | Cat. hydrotreating | Alkylation | Aromatics/isomerization | Lubes | Asphalt | Coke (t/d) |
|---|---|---|---|---|---|---|---|---|---|---|---|---|---|---|---|---|
| | | | | Charge capacity—b/sd | | | | | | | | Production capacity—b/sd | | | | |
| Alabama | 4 | 35,421 | 37,010 | 8,660 | | | | 1,400 | | | 5,000 | | | | 6,926 | |
| Alaska | 4 | 66,050 | 69,020 | | | | | | | | | | | | 300 | |
| Arizona | 1 | 9,000 | 10,000 | 4,000 | | | | | | | | | | | | |
| Arkansas | 4 | 56,125 | 58,000 | 23,800 | 5,000 | 15,000 | 5,200 | 5,500 | | 5,000 | 8,000 | 4,000 | | 2,100 | 9,650 | |
| California | 34 | 1,809,295 | 1,894,800 | 958,200 | 452,820 | 481,610 | 122,730 | 476,540 | 317,629 | 138,250 | 662,020 | 83,390 | 12,490 | 23,770 | 98,410 | 15,165 |
| Colorado | 3 | 55,800 | 57,920 | 10,500 | 14,500 | 20,500 | 2,450 | 11,900 | | | 17,000 | | | | 3,300 | 400 |
| Delaware | 1 | 140,000 | 150,000 | 90,700 | 44,000 | 62,000 | 15,000 | 42,000 | 17,000 | | 110,000 | 8,000 | | | | 1,800 |
| Florida | 1 | 5,000 | 5,500 | 2,400 | | | | | | | | | | | 2,500 | |
| Georgia | 2 | 14,375 | 15,130 | | | | | | | 1,200 | | | | | 9,300 | |
| Hawaii | 2 | 70,000 | 73,689 | 15,000 | | 14,100 | 8,900 | | | | 1,400 | 4,130 | 1,350 | | 1,300 | |
| Illinois | 11 | 1,151,500 | 1,206,390 | 415,730 | 143,300 | 390,270 | 81,240 | 312,280 | 66,500 | 107,000 | 435,280 | 89,020 | 10,200 | 5,600 | 42,800 | 4,380 |
| Indiana | 7 | 551,000 | 588,050 | 243,000 | 22,500 | 206,500 | 36,780 | 98,000 | | 29,500 | 213,000 | 35,700 | 6,500 | 8,670 | 47,000 | 730 |
| Kansas | 11 | 401,823 | 418,050 | 93,800 | 43,200 | 155,900 | 46,700 | 81,900 | 2,950 | 7,500 | 100,400 | 38,070 | 3,400 | 4,000 | 16,400 | 1,310 |
| Kentucky | 3 | 159,800 | 167,000 | 65,000 | 4,000 | 63,000 | 6,700 | 25,500 | | | 42,500 | 6,000 | 21,500 | | 13,500 | |
| Louisiana | 18 | 1,666,600 | 1,733,180 | 461,090 | 147,830 | 598,280 | 130,830 | 362,930 | 78,500 | 99,000 | 403,410 | 138,290 | 23,500 | 48,450 | 35,880 | 5,250 |
| Maryland | 2 | 23,500 | 24,740 | 13,800 | | | | | | | | | | | 19,000 | |
| Michigan | 6 | 137,080 | 147,230 | 44,800 | | 38,000 | 9,950 | 32,750 | | 17,700 | 31,100 | 5,450 | 2,000 | | 11,650 | |
| Minnesota | 3 | 190,420 | 198,000 | 78,000 | 23,000 | 65,400 | 8,200 | 30,100 | | 27,000 | 65,100 | 11,500 | | | 42,000 | 1,300 |
| Mississippi | 5 | 289,700 | 298,390 | 152,563 | 6,700 | 70,500 | 6,350 | 70,700 | 62,000 | | 41,450 | 9,200 | 6,000 | | 7,270 | 320 |
| Missouri | 1 | 105,000 | 110,530 | 40,000 | 11,000 | 38,000 | 12,000 | 14,000 | | | 45,000 | 5,000 | | | 6,500 | 450 |
| Montana | 8 | 158,481 | 166,200 | 48,350 | 8,600 | 49,800 | 26,200 | 36,850 | 4,900 | | 104,400 | 10,400 | 4,700 | | 17,625 | 250 |
| Nebraska | 1 | 5,000 | 5,500 | 2,000 | | 2,000 | 600 | 1,000 | | | | | | | | |
| New Jersey | 5 | 619,000 | 646,131 | 318,380 | 40,340 | 249,440 | 63,830 | 100,440 | | 50,000 | 236,450 | 21,610 | | 6,400 | 71,000 | 975 |
| New Mexico | 6 | 55,359 | 57,130 | 12,500 | 1,250 | 12,200 | 5,060 | 11,700 | | | 11,450 | 3,100 | | | 2,240 | |
| New York | 2 | 107,600 | 111,000 | 43,000 | | 39,000 | 12,000 | 22,700 | | 20,000 | 26,100 | 4,800 | 2,500 | | 17,500 | |
| North Dakota | 2 | 52,658 | 55,530 | | 1,100 | 23,000 | 11,000 | 10,200 | | | 12,600 | 2,900 | | | | |
| Ohio | 7 | 572,400 | 606,500 | 173,500 | 27,800 | 200,000 | 52,100 | 170,700 | 82,000 | 42,000 | 159,000 | 35,300 | | 2,100 | 27,900 | 1,260 |
| Oklahoma | 12 | 481,000 | 497,695 | 156,110 | 69,070 | 184,100 | 42,230 | 112,490 | 4,500 | | 136,590 | 41,280 | 16,610 | 10,600 | 21,010 | 1,445 |
| Oregon | 1 | 14,000 | 14,740 | 15,000 | | | | | | | | | | | 8,600 | |
| Pennsylvania | 11 | 689,620 | 729,215 | 247,778 | 26,750 | 196,900 | 20,700 | 197,208 | 32,700 | 133,000 | 227,750 | 38,100 | 57,900 | 29,628 | 35,500 | |
| Tennessee | 1 | 29,000 | 30,000 | | | 12,000 | | | | | | 3,500 | | | 3,000 | |
| Texas | 40 | 3,732,875 | 3,896,560 | 1,277,560 | 344,750 | 1,199,370 | 268,535 | 980,510 | 151,170 | 289,500 | 1,427,650 | 211,160 | 172,410 | 86,020 | 64,720 | 6,025 |
| Utah | 6 | 133,150 | 140,620 | 34,800 | | 53,400 | 17,000 | 18,500 | 1,000 | 7,000 | 13,600 | 8,850 | 2,550 | | 4,700 | |
| Virginia | 1 | 53,000 | 55,790 | 28,000 | 14,000 | 26,500 | 4,000 | 8,000 | | | 22,000 | | | | | 710 |
| Washington | 7 | 346,500 | 361,100 | 132,700 | 36,000 | 89,280 | 27,330 | 76,890 | 35,000 | 20,500 | 133,780 | 21,330 | 2,900 | 1,900 | 9,600 | 1,500 |
| West Virginia | 3 | 19,550 | 20,500 | 8,675 | | | | 6,160 | | 4,440 | 7,510 | | | 6,600 | | |
| Wisconsin | 1 | 37,000 | 38,000 | 15,500 | | 9,700 | 1,000 | 8,200 | | 5,800 | 8,200 | 1,200 | | | 12,000 | |
| Wyoming | 10 | 172,605 | 181,210 | 65,110 | 4,440 | 52,830 | 16,420 | 30,990 | | 15,440 | 52,540 | 7,800 | 1,500 | 1,280 | 14,520 | 140 |
| | 247 | 14,216,287 | 14,876,050 | 5,300,006 | 1,491,950 | 4,618,580 | 1,061,035 | 3,358,038 | 855,840 | 1,019,830 | 4,760,280 | 849,080 | 348,010 | 237.118 | 683,601 | 43,410 |

[a] J. A. Bauer and C. W. Peters, *Oil Gas J.* **72** (13) (1974).
[b] State totals include figures converted to calendar day or stream day basis (barrels per calendar day, b/cd; barrels per stream day, b/sd).
[c] Cat. = catalyst.

would rise by 50% by 1985. With a slowdown, or long-term cutback in foreign crude imports and a continued decline in domestic production, the United States refining industry will be forced to expand capacity to process synthetic petroleum feedstocks originating from shale, tar sands, and coal. This development will require expanded efforts in processing "heavier" materials and will necessitate the use of new and more costly technology to produce "clean fuels" from environmentally acceptable refining facilities. Even with a slowdown in demand for petroleum products, it appears that new refining capacity would have to increase at the rate of two 200,000 barrel per day (BPD) plants each year for the next 10 years. The total capital costs required for this amount of conventional crude refining capacity would be about $8 billion; of this total capital, 5–10% would be for air pollution control out of a total of 15–20% for overall environmental control. An additional $2.7 billion would be required for desulfurization by assuming that two-thirds of the new capacity is for high-sulfur crude, thus bringing the total to $10.7 billion.

## II. Oil Refining Technology

Because of the technological intricacy of petroleum refining processes, an understanding of air pollution control techniques is necessarily dependent on a general knowledge of oil industry manufacturing operations. Descriptions and flow diagrams for commercial refining processes as well as capabilities of the various operations for each refinery are readily available in the literature (*6*). However, this information is difficult to keep up to date because of the changing nature of the industry.

Although each modern refinery is unique in design and contains a composite of many processes employing a multitude of towers, vessels, piping, valves, tubes, exchangers, and storage tanks, the operations can be classified into four basic procedures—separation, conversion, treating, and blending (*7*).

Crude oil is initially separated into its various components or fractions, e.g., gas, gasoline, kerosine, middle distillates such as diesel fuel and fuel oil, and heavy bottoms. Since these initial fractions seldom conform to either the relative demand for each product or to its qualitative requirements, the less desirable fractions are subsequently converted to more salable products by splitting, uniting, or rearranging the original molecular structure. Separation and conversion products are subsequently treated for removal or inhibition of undesirable components. The refined base stocks may then be blended with each other and with various additives to develop the most useful products.

Individual refineries differ widely, not only as to crude oil capacity, but also as to the degree of processing sophistication employed. Simple refineries may be confined to crude separation and limited treating (Fig. 1). Intermediate refineries may add catalytic or thermal cracking, catalytic reforming, additional treating, and manufacture of such heavier products as lube oils and asphalt (Fig. 2). Complete refineries, generally large in capacity, include crude distillation, cracking, treating, gas processing, manufacture of lube oils, asphalts, waxes, as well as gasoline upgrading processes such as catalytic reforming, alkylation, or isomerization (Fig. 3).

### *A. Separation*

Crude oil is a mixture of many different hydrocarbons combined with small quantities of sulfur, oxygen, nitrogen, and traces of other elements. Distillation is employed to separate crude oil into its various components with different boiling ranges such as wet gas, gasoline, kerosine, fuel oil, middle distillate, lube distillate, and heavy bottoms. The relative quantity of these "straight-run" products is determined by the composition of the original crude oil. Depending on the completeness of the particular refinery, these initial products may be treated and sold as feedstock for subsequent processing. The crude oil is distilled by heating in furnaces followed by vaporization and separation in fractionating towers. Heavy fractions of the original crude may be subsequently vaporized by steam or vacuum distillation.

### *B. Conversion*

Conversion processes are employed to increase the yield of the more desirable products and to provide the refiner with the necessary flexibility to meet seasonal changes in product demand. In addition to shifting the product distribution, conversion processes also upgrade component quality. The most familiar technique is cracking, whereby relatively heavy hydrocarbon molecules are split under heat and pressure to produce smaller, lower-boiling hydrocarbons. Cracking by catalytic action is still the most significant conversion process employed. United States refineries have catalytic cracking facilities for approximately 50% of the total crude capacity. Catalytic cracking yields additional "synthetic" gaseous hydrocarbons, gasoline, and reduced quantities of heavy fuel oil. Coke

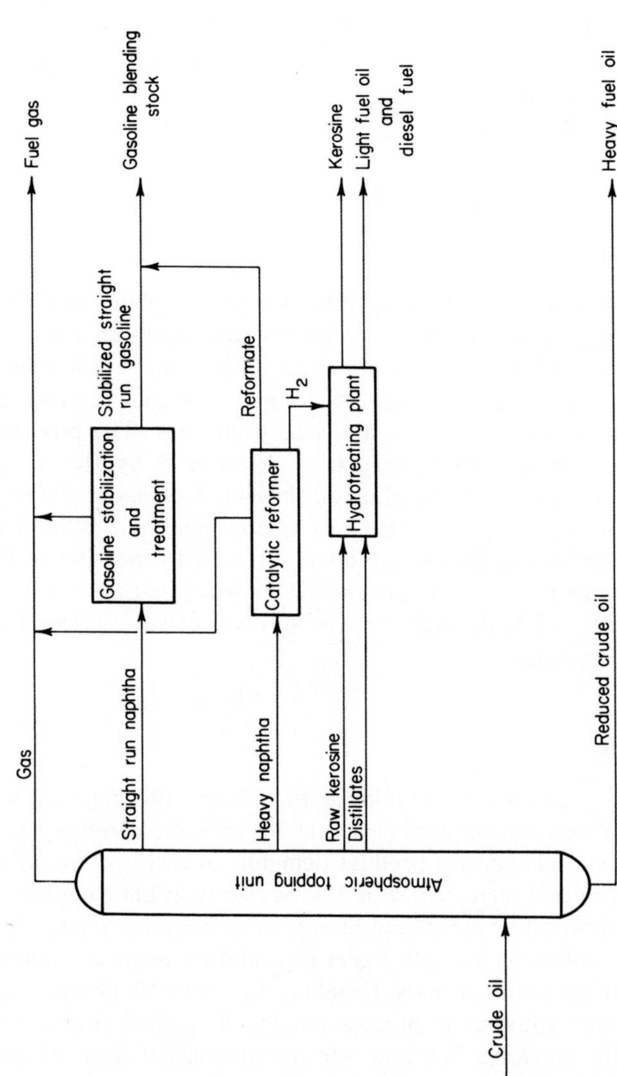


**Figure 1. Processing plan for typical minimum refinery.**

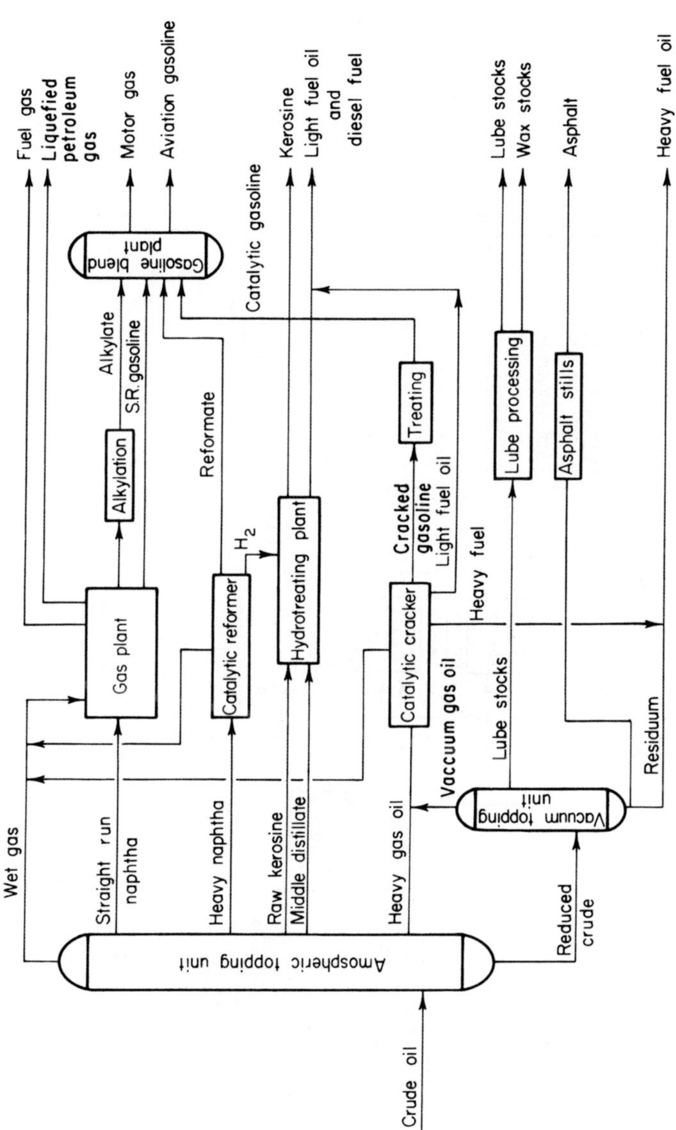


Figure 2. Processing plan for typical intermediate refinery. Lube: Lubricating oil; S.R. straight run.

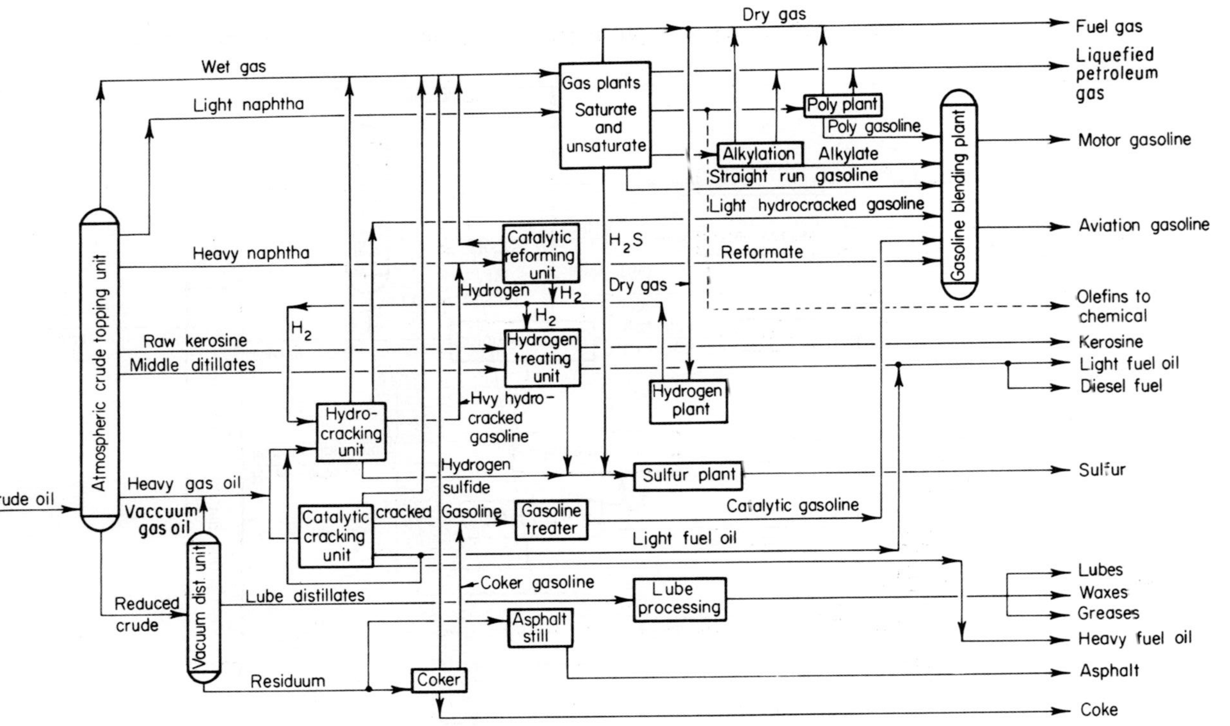


Figure 3. Processing plan for typical complete refinery. Lube: Lubricating oil; poly: polymerization.

deposits which form on the catalyst, usually an alumina-silicate, are burned off in separate regeneration vessels. Earlier catalytic processes employed fixed-bed or once-through catalysts. Moving-bed and fluid-bed systems are now commonly used.

Hydrocracking, cracking in the presence of a high hydrogen partial pressure, is experiencing rapid growth. Hydrocracking complements catalytic cracking and adds flexibility in meeting seasonal product-volume demand fluctuations. This process is characterized by high liquid yields of saturated isomerized products.

Virtually supplanted by catalytic cracking and hydrocracking, some thermal cracking units are still operating. Coking (severe cracking) and visbreaking (mild cracking) are still useful forms of thermal cracking.

Catalytic reforming is a rearrangement of molecular structure to produce higher quality gasoline and large quantities of hydrogen. Polymerization combines two or more gaseous olefins (unsaturated hydrocarbons), while alkylation joins an olefin and an isoparaffin (branched-chain hydrocarbon) to form liquid hydrocarbons in the gasoline range. Isomerization alters the arrangement of the atoms in a molecule to form branched-chain hydrocarbons for higher octane fuel. These processes utilize catalysts such as phosphoric, sulfuric, and hydrofluoric acid; platinum; and aluminum chloride.

### C. Treating

Crude oil may contain small quantities of impurities, such as sulfur and certain trace amounts of metals. Sulfur in oil generally occurs as sulfides, mercaptans, polysulfides, and thiophenes. A substantial portion of the domestic crude in the United States contains over 0.5% sulfur, and perhaps 10% of the crude contains about 2% sulfur. Imported crudes from North Africa are typically low sulfur ($<0.5\%$) while Persian Gulf and Mid-East crudes range up to 2% sulfur. Environmental restraints have made the lower sulfur crudes more desirable and hence more costly. Up through late 1973 the United States was planning to rely on the more abundant high-sulfur foreign crudes to meet its surging demand for petroleum products. The extent of this United States reliance on high-sulfur foreign crudes will no doubt be altered by the effects of the 1973–1974 Arab oil embargo. While sulfur removal from basic crude is neither technically nor economically feasible at this time, desulfurization of products and intermediate stocks has come into wide use. Desulfurization of products is practiced because of the effect on product quality, catalyst sensitivity, odor, and corrosivity.

Both physical and chemical procedures are available for treating products and feedstocks. Physical methods include electrical coalescence, filtration, absorption, and air blowing. Chemical methods include acid treatment of hydrocarbon streams to remove sulfur and nitrogen compounds, sweetening processes to oxidize mercaptans to disulfides, solvent extraction such as removing sulfur compounds with strong caustic, and the use of additives to oxidize mercaptans to disulfides.

Solvent extraction for removal of aromatics is also widely practiced. The lighter aromatics are extracted from gasoline boiling-range material for sale as petrochemicals, and heavier aromatics are extracted from fuel oil and lube oil fractions for quality improvement.

The use of hydrogen for desulfurization of petroleum products has become widespread. The hydrogenation converts organic sulfur compounds into hydrogen sulfide for subsequent disposal or recovery. The extracted hydrogen sulfide is generally converted and recovered as elemental sulfur or burned to sulfur dioxide in plant boilers when the quantities do not justify recovery. This process also converts gum-forming hydrocarbons and diolefins into stable compounds. Hydrogen required for the process is commonly furnished by catalytic reforming units and is frequently supplemented by generated hydrogen from processes such as steam-methane reforming.

### *D. Blending*

The relatively few base and intermediate stocks are blended in innumerable combinations to produce over 2000 finished products including liquefied gases, motor and aviation fuels, lubricating oils, greases, waxes, and heating fuels. Rigorous specifications as to vapor pressure, viscosity, specific gravity, sulfur content, octane number, initial boiling point, etc., must be satisfied, depending on the product.

### *E. Desulfurization*

Air pollution considerations have focused attention on the technology available and progress to date in desulfurizing petroleum products. Since 1950 there has been a significant reduction in the total fuel sulfur content of petroleum products which could have contributed to atmospheric pollution even though the average sulfur content of crude charged has increased slightly as shown in the tabulation on page 823.

It has been estimated that capital costs for desulfurization facilities

| Fuel | Average sulfur content wt % | | |
|---|---|---|---|
| | *1950* | *1965* | *1974* |
| Motor gasoline | 0.084 | 0.042 | 0.03–0.04 |
| Kerosine | 0.20 | 0.084 | 0.05–0.07 |
| Heating oils | 0.37 | 0.27 | 0.10–0.30 |
| Diesel fuels | 0.65 | 0.36 | 0.20–0.40 |
| Residual fuel oil | 1.19 | 1.47 | 0.50–2.0 |
| Crude oil | 0.72 | 0.76 | 0.80–1.0 |

will be as much as $1000 per barrel if high-sulfur crude processed (*8*). In terms of capital investment, this means $1 billion for every million barrels of high-sulfur crude processed.

As refiners converted more heavy residual materials into lighter products, the volume of residual fuel decreased from 20 to 7% of crude charged between 1950 and 1973. This reduction in liquid output resulted in concentrating high molecular weight sulfur-bearing compounds in the heavy residual oil fractions, and thus increased the sulfur content of certain residual fuels. Residual fuel sulfur levels vary considerably from refiner to refiner and are determined by the type of crude processed, desulfurization facilities available, and blending capabilities of the particular refinery.

Between 1970 and 1973, demand for residual fuel rose tremendously. During this period many utilities switched from coal to cleaner burning residual oil to comply with environmental regulations. Nearly all of this new demand was satisfied by imported residual fuels. Efforts initiated, as a result of residual fuel shortages during 1974, resulted in a renewed emphasis on the use of coal as the primary utility fuel. No longer a marginal product, residual fuel oil has received considerable study as the principal remaining petroleum product which generally contains a substantial sulfur content. Specifications are generally concerned with flash point, viscosity, and a maximum limit on sediment and water. Residual fuel sulfur content depends not only on the properties of the crude but also on the degree of processing that each of the blend components has received.

Hydrogen treating at high pressure in the presence of a catalyst is thus far the most feasible means of effecting a significant degree of sulfur reduction for most refinery stocks. Hydrodesulfurization processes for treating gasolines, middle distillates, and lubricating oils are in extensive use at this time. Extension of this technology toward residual oils has proved costly, and has found only limited application at this writing although certain commercial processes are available. Processing a high sulfur residual oil through a hydrodesulfurization unit converts this feedstock into a "synthetic crude" with a full range of materials ranging from gas and distillates down to a true residual fraction. The various fractions can be utilized for sale, as feedstocks for other units, and in part for blending with the reduced volume residual from the desulfurization step.

A 1964 study (*9*) suggested an approximate cost of $0.40–0.65 per barrel for desulfurizing typical West Coast United States residual fuels from a typical initial sulfur content of 1.5% down to 0.5% sulfur content.

A 1967 study (*10*) determined the approximate costs for desulfurizing residual fuel oil produced in Caribbean refineries which have supplied 85% of the residual fuel imported into the United States. The principal crude charged to these refineries originates in Venezuela. Based on a "typical" 281,500 barrel per day Caribbean refinery producing 161,600 barrels of residual oil, or 57% of crude charge, the study considered reductions from an initial 2.6% sulfur in the residual fuel. Reduction to 2% sulfur would require an additional facilities investment of approximately $45 million and would result in an incremental cost of the order of $0.30 per barrel. Reduction down to 1% sulfur would require double this investment and cost approximately $0.72 per barrel on an equivalent Btu basis. Desulfurization to a 0.5% sulfur level could require an investment in the order of $100 to $150 million and would result in incremental costs per barrel of $0.90–$1.00 on an equivalent Btu basis.

Since there is very limited experience with residual desulfurization of the high metals content materials produced from refining Caribbean crude, there are still areas of uncertainty as to the costs involved and the ultimate sulfur levels attainable. Metals content, which varies with crude type, has an advese effect on catalyst life and thereby influences processing costs.

Cost estimates (*1*) indicate a range of $1.00 to $1.30 for direct desulfurization of 4% sulfur Kuwait reduced crude to the 1 and 0.3% levels; and, a range of $0.84 to $1.30 for direct desulfurization of 2.5% sulfur Iranian reduced crude to the 1 and 0.3% level. Additional cost estimates (*12–15*) have been presented for various process schemes. The costs range from $0.60 to $1.30/barrel depending on the processing scheme selected and degree of desulfurization required.

Atmospheric residual material is being desulfurized, but not to the extremely low levels (0.3%) called for by certain East Coast United States areas. Desulfurization of residual vacuum-tower bottoms (VTB) has yet to be commercially demonstrated although one major United States West Coast refiner has started construction of a 25,000 BPD unit. The most serious problem with desulfurizing VTB is that the high metals (vanadium and nickel) content of the material tends to quickly deactivate the catalysts used, and accumulates in the catalyst, causing plugging of the bed. The asphaltene content of the residual material is another important variable involved in direct desulfurization. In order to remove the asphaltene sulfur that is bound in multi-ring sulfur-containing structures, a high severity operation is required. The asphaltenes are thermally unstable and are easily converted to coke on the catalyst surface, which contributes to deactivation and plugging of the system. If a deeper cut into a barrel of oil is made, it tends to concentrate the metals content and the asphaltene content, thus making it extremely difficult to desulfurize these "heavier" oil cuts before the catalyst becomes plugged and deactivated.

The most widely practiced desulfurization scheme for heavy material is that employed by the Caribbean refineries. These plants are currently desulfurizing vacuum gas oil to the 0.3% level then back blending this with VTB, desulfurized atmospheric bottoms, and/or naturally occurring low-sulfur atmospheric bottoms, to satisfy the 0.5 and 1% sulfur fuel demand. Desulfurized vacuum gas oil is sold to meet the 0.3% residual oil market, and in some cases naturally occurring low-sulfur crudes or low-sulfur atmospheric bottoms are used for this purpose.

Factors influencing residual fuel sulfur content include the sulfur content of the crude oil prior to processing, availability of low sulfur crude and blending stocks, and whether all residual fuel produced at a particular refinery has to meet a single sulfur specification or whether two or more sulfur levels are produced.

### F. Lead Antiknock Additives

Inquiries on the feasibility and economics of producing motor gasoline in the United States without lead antiknock additives prompted a study of this subject in 1967 (*9*). This study concluded that, to maintain gasoline volume without lead with similar antiknock and performance qualities as leaded fuels would require an investment of some $4.24 billion by United States refiners. Added manufacturing costs would range from $0.018 to $0.047 per gallon, depending on refinery size, existing processing and, to a lesser extent, location, and would average somewhat over $0.02 per gallon. Processes such as high severity reforming, aromatic extraction, alkylation, olefin disproportionation, isomerization, naphtha steam cracking, and hydrocracking were included in the technology available for application by the study. Total crude oil processed would increase approximately 5.5%, largely due to the extra energy consumed in the additional refining processing. The unleaded gasolines of equivalent octane value would be considerably higher in aromatic content and somewhat lower in olefins than conventional leaded fuels. More recent estimates (*8*), based on the United States Environmental Protection Agency's unleaded gasoline requirement and their lead phase-out schedule, indicate a $2.5 billion cost for unleaded fuel manufacture. The primary cost to the petroleum industry will be for revision of the gasoline distribution system to handle the unleaded fuel and to keep it segregated from leaded grades. The cost could be higher than $2.5 billion if 91 octane unleaded fuel is not sufficient to satisfy customers' driving needs. If unleaded fuel of higher than 91 octane is required, additional refinery processing will be necessary.

## III. Type of Emissions

The estimation and evaluation of refinery atmospheric emissions is an immensely complicated project. Individual refineries vary greatly as to the character and quantity of emissions Controlling factors include crude oil capacity, type of crude processed, type and complexity of the processing employed, air pollution control measures in use, and the degree of maintenance and good housekeeping procedures in force. Refinery emissions may be classified as smoke and particulate matter, hydrocarbons, and other gaseous compounds, principally oxides of sulfur and nitrogen, but also including malodorous vapors. The technology of emission evaluation and abatement has been established through the cooperative efforts of the industry and regulatory authorities, particularly in the Los Angeles, California area (*7, 17–19*) (Table II).

While the quantitative aspects of refinery emissions are generally considered by type of equipment, the major refining operations offer a guide as to the type of releases that are encountered. In crude separation the use of barometric condensers in vacuum distillation can release noncondensable hydrocarbons to the atmosphere. Regeneration of catalyst in cracking by controlled combustion can release unburned hydrocarbons, carbon monoxide, ammonia, and sulfur oxides although these gases are commonly rendered less objectionable through the use of CO boilers. Catalyst handling systems can also be the source of discharge of catalyst fines, but conventional mechanical or electrical separation equipment has proved successful for control of this material.

The molecular rearrangement processes—alkylation, reforming, polymerization, and isomerization are significant in that they handle volatile hydrocarbons; and tight equipment such as valves and pumps becomes

**Table II Potential Sources of Specific Emissions from Oil Refineries (*8*)**

| *Emission* | *Potential sources* |
|---|---|
| Oxides of sulfur | Boilers, process heaters, catalytic cracking unit regenerators, treating units, $H_2S$ flares, decoking operations |
| Hydrocarbons | Loading facilities, turnarounds, sampling, storage tanks, waste water separators, blowdown systems, catalyst regenerators, pumps, valves, blind changing, cooling towers, vacuum jets, barometric condensers, air blowing, high pressure equipment handling volatile hydrocarbons, process heaters, boilers, compressor engines |
| Oxides of nitrogen | Process heaters, boilers, compressor engines, catalyst regenerators, flares |
| Particulate matter | Catalyst regenerators, boilers, process heaters, decoking operations, incinerators |
| Aldehydes | Catalyst regenerators |
| Ammonia | Catalyst regenerators |
| Odors | Treating units (air blowing, steam blowing) drains, tank vents, barometric condenser sumps, waste water separators |
| Carbon monoxide | Catalyst regeneration, decoking, compressor engines, incinerators |

of greater emission control importance than in the heavier oil processes yielding fuel oils, lubes, or asphalts.

Pollution aspects of treating operations are concentrated on methods of disposing of the spent chemicals and extracted impurities such as spent acids, spent caustics, and hydrogen sulfide. Air agitation and blowing, vessel vents, drains, valves, and pump seals are possible sources of loss or leakage of sulfur oxides and hydrocarbons in any of the above processes or handling procedures.

Before considering individual emission sources, an overall estimation of refinery releases will be of interest. The following data are interpolated from the exhaustive studies in Los Angeles where control procedures have achieved a degree of completeness that is as unique as the local climatic conditions that motivated their adoption (*18, 20, 21*). Estimated refinery contaminants, before the rigorous controls of the 1950's, are compared with the relatively complete containment of the Los Angeles refineries that process over 700,000 barrels per day of crude oil (Table III).

## Relative Significance of Emissions

As would be expected, refineries and refining areas vary considerably in the degree of control of the above emissions. The extent and need for control depend on local meteorological and topographical conditions, size and type of plant, quantity and height of release, and proximity to com-

**Table III Estimated Daily Refinery Emissions**[a,b]

| *Substance* | *Without controls* | *After rigorous controls* |
|---|---|---|
| Hydrocarbons | 100 | 7 |
| Nitrogen oxides | 4 | 4 |
| Sulfur dioxide | 145 | 5 |
| Sulfur trioxide | >3 | *c* |
| Carbon monoxide | 220 | 13 |
| Particulate matter | 14 | <2 |
| Ammonia | <1 | <1[c] |
| Aldehydes | <1 | <1[c] |
| Organic acids | <1 | <1[c] |
| Aerosols | <1 | <1 |

[a] Los Angeles California refinery emissions, tons per 100,000 barrels crude capacity.
[b] Note: Emissions from combustion of fuels not included in above totals.
[c] Not reported or assumed value.

munity activity. Most refineries have facilities to reduce hydrocarbon and sulfur emissions to an economic level, and many so-called air pollution control measures are customarily included by refiners as accepted good conservation practices. Few, if any United States refineries still discharge emissions to the extent of the "Without controls" column in Table III.

Conversely, the high degree of containment accomplished in the Los Angeles refineries is probably unmatched in other refining areas. The acute photochemical smog problem in southern California, which had its origin in the unique local meteorology, actuated an intensive control program in the late 1950's which effectuated numerous rules and regulations, many of which apply specifically to oil industry operations.

Although all petroleum vapors have been subject to control in Los Angeles refineries, a relatively low percentage of these hydrocarbons are believed to participate significantly in the chemical reactions associated with air pollution manifestations (*7*). Most paraffins, naphthenes, benzene, and acetylene (*22*) are considered essentially as slow or nonreactive photochemically (see also Chapter 6, Vol. I). Hydrocarbon emissions in amounts normally released by refinery operations are invisible, have no offensive odor, are nontoxic, are relatively inert, and not all of them enter into photochemical smog reactions.

Only the olefinic or unsaturated hydrocarbons and certain aromatics (*22*) have been demonstrated to react rapidly enough in the atmosphere to participate in the complex photochemical reaction mechanism involving nitrogen oxides and resulting ozone in the presence of ultraviolet light from the sun. A survey of Los Angeles refineries indicated that olefins represented only 6.5% of the total hydrocarbon evaporation loss (*23, 24*), the remainder consisting principally of the relatively inert paraffins. Los Angeles refinery survey data for 1971 (*21*) classify only 10–12% of the remaining uncontrolled hydrocarbons emitted to the atmosphere as "high reactivity."

Emissions of gaseous contaminants other than hydrocarbons, principally oxides of sulfur and nitrogen, originate mainly from combustion sources. Emissions of sulfur oxides result from the sulfur content of the crude oil being charged to the plant and the type of processing employed. Most of the original crude sulfur content is driven into the hydrocarbon gas and heavy fuel fractions during distillation, cracking, and other processing. Hydrogen sulfide in light hydrocarbon fractions and fuel gas can be extracted prior to burning. Process waste waters can likewise be stripped of hydrogen sulfide. The extracted hydrogen sulfide can be recovered as elemental sulfur.

Along with hydrocarbons and sulfur oxides, refineries discharge relatively small quantities of carbon monoxide, nitrogen oxides, aldehydes, organic acids, ammonia, and particulate matter. The technology of dust control by mechanical means has been well established and is in extensive use. The other miscellaneous discharges originate from combustion reactions in boilers, process heaters, and catalyst regeneration units.

## IV. Source and Control

Refinery atmospheric emissions and control procedures are customarily considered by types of equipment employed rather than by refinery process operation (*25*).

### *A. Storage Tanks*

Hydrocarbon vapors may be released through a number of mechanisms in storage tanks, including tank breathing due to temperature changes, direct evaporation, and displacement during filling. The principal source of potential loss is from crude oil and light distillate products. The hydrocarbon content of crude oil is substantially saturated and, as explained above, is not believed to be involved in the photochemical smog complex. Light distillates have considerable value and are normally controlled to a practical economic level. Vapor conservation storage may involve tanks with floating roof covers, pressurized tanks, and connections to vapor recovery systems. The United States Environmental Protection Agency (USEPA) has established performance standards for new and modified petroleum storage vessels that have capacities greater than 40,000 gal (151,412 liters). These standards specify the type of equipment required depending on the volatility or true vapor pressure of the stored material. For materials exhibiting a true vapor pressure (TVP) at storage conditions of below 1.52 psia (78 mm of Hg), no controls are required; for materials with a TVP between 1.52 and 11.1 psia (78 and 570 mm of Hg), the storage vessel must be equipped with a floating roof or equivalent; for materials with a TVP exceeding 11.1 psia, the vessel must be equipped with a vapor recovery system or equivalent. In general, these requirements follow accepted industry practice aimed at preventing the escape of, and thus conserving, volatile hydrocarbons. Gasoline vapor pressure has historically varied depending on the climate and geographical area in which the gasoline is to be used. During colder months refiners add more butane to increase the vapor pressure of gasoline in order to ensure proper automotive engine performance. This has presented some problem for United States Gulf Coast refiners who produce winter grade (high vapor pressure) gasoline for use in northern markets. At times the true vapor pressure of this winter grade gasoline stored in United States Gulf Coast terminals equipped with floating roof tanks could approach or exceed 11.1 psia.

### B. Catalyst Regeneration Units

Coke formed on the surface of catalysts during catalytic cracking, reforming, and hydrogenation is burned off in regenerating vessels by controlled combustion. Flue gases from regenerators may contain catalyst dust, carbon monoxide, hydrocarbon (principally methane), and sulfur and nitrogen oxides. The catalyst dust may be controlled by mechanical or electrical collecting equipment. The carbon monoxide and unburned hydrocarbons are generally dispersed in the atmosphere, but may be eliminated by burning in a waste heat boiler which also generates additional steam. New United States federal performance standards limit the amount of catalyst emitted from fluid catalytic cracking regenerators to 1 lb of catalyst per 1000 lb of coke make, and carbon monoxide to 500 ppm by volume. *In situ* type operation is finding wider use in fluid catalytic cracking units. This process directionally reduces particulates, carbon monoxide, and even sulfur oxide emissions from catalytic cracking regenerators.

### C. Waste Water Separators

Waste water gravity separators are commonly used to trap and recover oil discharged to the sewer system from equipment leaks and spills, shutdowns, sampling, process condensate, pump seals, etc. Depending on the quantity and type of oil in the sewers, some hydrocarbon vapors may evaporate from the drainage and separator system. If this vaporization is sizable and control is indicated, the front end of the separators may be covered. Catch basin liquid seals, manhole covers, and good housekeeping practices will likewise control drainage system vapor losses.

### D. Loading Facilities

While most petroleum products leave the refinery though pipelines with no emission to the atmosphere, loading into tank trucks, tank cars, and drums can result in hydrocarbon vapor loss by displacement or evaporation. Careful operation to minimize spillage, and vapor collection and recovery equipment will control vapor loss from this operation.

### E. Pipeline Valves

The typical refinery contains a maze of piping, mostly above grade. The effects of heat, pressure, vibration, and corrosion may cause leaks in valved connections. Depending on the product carried and the temperature, the leaks may be liquid, vapor, or both. Regular inspection and prompt maintenance will correct vapor loss from this source.

### F. Pumps and Compressors

Hydrocarbons can leak at the contact between the moving shaft and stationary casing in pumps and compressors. Asbestos or other fibers are packed around the shaft to retard leakage from shaft motions. Mechanical seals, consisting of two plates, perpendicular to the shaft, forced tightly together, are also used. Wear can cause both packed and mechanical seals to leak product. Inspection and maintenance, sealing glands under pressure, and use of mechanical seals in light hydrocarbon service are useful control measures.

### G. Blowdown Systems, Flares, and Shutdowns

Refinery process units and equipment are periodically shut down for maintenance and repair. Since these turnarounds generally occur about once a year, losses from this source are sporadic. Hydrocarbons purged during shutdowns and start-ups may be manifolded to blowdown systems for recovery, safe venting, or flaring. Vapors can be recovered in a gas holder or compressor and discharged to the refinery fuel gas system. Flares should be of the smokeless type, utilizing either steam or air injection. Design data for smokeless flares are readily available from combustion equipment manufacturers and in industry technical manuals (*26*). For esthetic purposes, ground flares are becoming more popular.

### H. Boilers and Process Heaters

Refineries depend on boilers and heaters to supply high pressure steam at elevated temperatures. Fuels may include refinery or natural gas, heavy fuel oil, and coke, often in various changing combinations. Sulfur oxides in the flue gas are, of course, a result of the sulfur in the fuel feed. Nitrogen oxides and small quantities of hydrocarbons, organic acids, and particulate matter are also present. Sulfide stripping of fuel gas prior to burning and selective blending of fuels may be employed to control sulfur emissions. Normally, good combustion practices will control smoke and particulate matter.

Stacks on boilers and heaters are elevated to improve atmospheric dispersion and further diminish resulting ground-level concentrations of gases such as sulfur dioxide and nitrogen oxides.

### I. Sulfur Recovery Units

There are about 170 sulfur recovery units presently in operation in United States refineries and gas processing plants (*27*). Nearly all of these are of the Claus type and can attain 90–95% recovery under ideal operating conditions. Three- and four-stage units claim up to 97% recovery. In some areas, emission limits require 99.5% recovery. To accomplish this high degree of recovery, industry has had to install add-on tail-gas treating units to existing Claus units. At the time of this writing, there are about 10 tail-gas units on stream in the United States and Canada, none of which have had sufficiently long enough operating experience to fully evaluate their control efficiency.

The potential for $H_2S$ loss to the atmosphere occurs at two points in the Claus process: (a) the interface between the $H_2S$ recovery process and the feed to the unit; (b) the tail gas from the plant. If the Claus plant breaks down and cannot take the feed, $H_2S$ has to be flared. Some refineries utilize spare Claus capacity so that flaring is minimized. The presence of $H_2S$ in the tail gas indicates that the conversion of $H_2S$ to S is not complete. This tail gas is either incinerated at 1000° to 1200°F temperature thus releasing all sulfur compounds as sulfur oxides, or further treated in a tail-gas recovery system. The tail-gas treating systems generally convert all sulfur compounds to $H_2S$, after which the stream is scrubbed to remove the $H_2S$.

### J. Incinerators

There are a number of incineration processes available for the incineration of refinery sludges, solids, and spent caustics. Two fluid bed incinerators have been installed at midwestern United States refineries. These fluid bed incinerators operate at temperatures of 1300°F. As the sludge is burned, the solids from the sludge remain in the bed while gaseous products of combustion, water vapor and fine particulate matter pass overhead through a cyclone separator and a water scrubber before venting to the atmosphere.

### K. Light and Noise

Light and noise are being controlled to an increasing extent at refineries. During normal refinery operations, the amount of gas burned in flares is, for conservation reasons, very small. However, at certain times, sudden increases have been the cause of complaints. It is, therefore, considered good practice to situate flares as far away as practicable from residential areas or other locations where they are exposed to the public, or to enclose the flares to reduce light emission.

The major noise sources in oil refineries can be classified into four general categories: (a) noise produced by moving fluids, (b) noise produced by high-velocity jets of gas or vapor discharging to atmosphere, (c) noise produced by mechanical equipment, and (d) noise produced by combustion processes.

In new plants, noise abatement begins with the selection of quiet equipment plus a plot plan layout which allows higher noise sources to be located away from the community or shielded by other facilities.

### L. Liquid Natural Gas (LNG), Synthetic Natural Gas (SNG), and Liquefied Petroleum Gas (LPG)

#### 1. LNG

Liquified natural gas (LNG) is a relatively new commercial operation which is expected to assume a significant role in supplying the expanding natural gas market. Stronger demands for clean air, coupled with growing energy needs, will mean intense activity in developing technology for manufacturing, transporting and storing vast quantities of LNG. LNG plants require the following processing steps: feed gas purification ($CO_2$ removed by amine washing, or adsorption on solid beds and dehydration); cooling; liquefaction via refrigeration; liquefaction via compressors; compression via compressor drives (steam or gas turbines); and storage (above-ground metal or prestressed concrete, excavated or, rock reservoirs). For the most part, these processes require the use of vapor control principles available in the hydrocarbon–cryogenic processing industry.

### 2. SNG

Since basically no new processing technology is involved, the extent to which plants will be built and controlled depends entirely on the type of feedstocks employed.

### 3. LPG

The extensive use of cracking and reforming processes have resulted in the production of large quantities of LPG. Prior to the start of the tremendous growth of the use of LPG in ethylene production, the major use was for household and industrial fuel. The use of ethyl mercaptan for odor detection of LPG has on occasion resulted in release of mercaptans to the atmosphere.

## *M. Miscellaneous*

Various other miscellaneous emission sources, usually of lesser significance, will be found in refinery operations. Pressure relief valves may be manifolded into vapor recovery or flare systems to control leakage and relief discharge. Steam-driven vacuum jets, employed to induce negative pressure in process equipment, may discharge light hydrocarbons with the exhaust steam. These gases may be vented and burned in an adjacent boiler or heater firebox.

Fumes from air blowing operations may be consumed by incineration or absorbed by scrubbing. Gases from spent caustic and mercaptan disposal may be burned in fireboxes.

Hydrogen sulfide and mercaptans are the principal potential pollutants that may cause odors. These gases can be released from process steam condensates, drain liquids, barometric condenser sumps, sour volatile product tankage, and spent caustic solutions from treating operations.

Odorous compounds in steam condensates can be removed by stripping with air, flue gas, or steam and offensive gases can be burned in furnaces or boilers. Drain liquids can be collected in closed storage systems and recycled to the process. Barometric condensers are being replaced by more modern surface condensers, and the noncondensables may be burned in process heaters or in a separate incinerator.

Spent caustic can be degasified, neutralized with flue gas, and/or stripped before disposal. Sulfides can also be removed from sour process water and spent caustic solutions by air oxidation to thiosulfates and sulfates.

Refinery waste gases that contain hydrogen sulfide are generally scrubbed with appropriate solutions for extraction of the sulfide by nonregenerative or heat-regenerative procedures. In the former method, the waste gases are scrubbed with a caustic solution, producing a solution of sodium sulfide and acid sulfide. As described above, the spent caustic may be oxidized by air blowing or sold. Vent gas from the blowing operation should be burned.

Heat-regenerative methods involve scrubbing sour gases with various types of amine, phenolate, or phosphate solutions which absorb hydrogen sulfide at moderately low temperatures and release it at higher temperatures. These methods are cyclic and consist of an absorption step, in which the hydrogen sulfide is scrubbed from the absorbing solution at approximately 100°F, followed by a regeneration step in which the solution is reactivated for use by heating it to its boiling point to drive off the hydrogen sulfide. The released hydrogen sulfide is then burned or oxidized to form sulfur.

Refinery process changes may have the net effect of reducing overall emissions to the atmosphere. Examples of such process changes include substitution of hydrogen treating for chemical treatment of distillates, use of harder catalysts to reduce attrition losses, and regeneration of spent chemicals for reuse.

## V. Estimation of Quantities

When an accurate estimate of refinery atmospheric emissions is required, there is no substitute for in-plant field testing, preferably by plant personnel familiar with the equipment and operations involved. Although on-site evaluation programs can be complicated, time-consuming, and costly, they are required for reliable determination of the highly variable emission rates from such contributing elements as tank evaporation, catalytic cracking regeneration, vacuum jet exhausts, and air blowing operation.

The purpose of the survey should be clearly defined in order to establish the extent and depth of the evaluation. If a rough order-of-magnitude appraisal is indicated, available data from neighboring or similar refining operations may be extrapolated to the operations of the plant under consideration. However, emission data on which to base the need for control should always be developed by actual in-plant appraisal.

Before any effort is made to apply survey results from one refinery study to another, the many variables of type of crude run, processes employed, existing status of control facilities, and procedures must be carefully considered. Despite apparent similarities between refineries, experienced judgment is required to establish the necessary basis of comparison and avoid misleading correlations.

When it is determined to employ comparative refinery emission data in a particular study, the information developed by the Los Angeles Joint Project study will be useful (*7, 19*). Comparisons of refinery emissions on the basis of crude capacity or fuel consumption can be grossly misleading and are not recommended. Very rough estimates of total hydrocarbon emissions by type of equipment and capacity can be developed from the above study. However, hydrocarbon losses from refineries may range from 0.1 to 0.6% by weight of crude throughput, depending on the complexity of refining procedures and the employment of abatement facilities.

The Joint Project Study has developed average emission factor data (*7, 19*) for various discharges including hydrocarbons, particulate matter, and nitrogen oxides which, when judiciously applied to specific refinery sources, will provide an estimate of the magnitude of emissions. Actual in-plant surveys should accompany such estimates to develop reliable data. The reader is referred to Tables IV and V (*7, 28*) for information on typical petroleum refinery emission and loss factors.

## VI. Economics of Control

The economics of air pollution control are a function of specific sources and methods of control in individual refineries. It has become industry-wide practice to include pollution control equipment and costs as an integral part of new process and equipment design. A well-designed new plant may have a satisfactory pollution control posture incorporated in original construction that will obviate additional expenditures for many years. An older refining operation, particularly with encroaching community development, may require relatively high expenditures to correct an objectionable condition.

Partial economic return may be realized when pollution control facilities recover such components as hydrocarbons, sulfur, and catalyst. The amount of this return will, of course, be a function of the value of the material recovered, capital investment for equipment, amortization period, interest rates, maintenance and operating expenses. In general, rates

**Table IV Emission Factors Developed from Los Angeles, California Survey[a]**

| Combustion sources | Units of emission factor: Pounds per | Emission factors: Hydrocarbons | Aldehydes as HCHO | Carbon monoxide | $NO_x$ as $NO_2$ | Ammonia as $NH_3$ | Particulate matter |
|---|---|---|---|---|---|---|---|
| Boilers and process heaters | Bbl of fuel oil burned | 0.14 | 0.025 | Neg. | 2.9 | Neg. | 0.8 |
| Boilers and process heaters | 1000 $ft^3$ of gas burned | 0.026 | 0.0031 | Neg. | 0.23 | Neg. | 0.02 |
| Compressor internal combustion engine | 1000 $ft^3$ of gas burned | 1.2 | 0.11 | Neg. | 0.86 | 0.2 | — |
| Fluid bed catalytic cracking unit[b] | 1000 bbl of fresh feed | 220. | 19. | 13,700 | 63. | 54. | c |
| Moving bed catalytic cracking unit[b] | 1000 bbl of fresh feed | 87. | 12. | 3,800 | 5.0 | 5.0 | d |

[a] Data adapted from Elkin and McArthur *(8)*.
[b] Before CO waste heat boiler.
[c] With electrical precipitators: 0.0009% of catalyst circulated; without electrical precipitators: 0.005% of catalyst circulated.
[d] With centrifugal separators: 0.002% of catalyst circulated.

**Table V Hydrocarbon Loss Factors Developed from Los Angeles, California Survey**[a]

| *Evaporation sources* | *Units of emission factor* — *Pounds of hydrocarbon per* | *Emission factor* |
|---|---|---|
| Pipeline valves | Valve per day | 0.15 |
| Vessel relief valves | Valve per day | 2.4 |
| Pumps seals | Seal per day | 4.2 |
| Compressor seals | Seal per day | 8.5 |
| Cooling towers | Million gal of cooling water circulated | 6.0 |
| Blowdown systems | 1000 bbl of refining capacity | 5–300[b] |
| Vacuum jets | 1000 bbl of vacuum distillation capacity | 0–130[b] |
| Process drains | 1000 bbl of waste water processed | 8–210[b] |
| Storage tanks | Tank per day | *c* |
| Other sources | 1000 bbl of refinery capacity | 10.0 |

[a] Data adapted from Public Health Service (*7*).
[b] Range of values.
[c] For computation method, see Luche and Deckert (*28*).

of return on pollution control expenditures do not approach those for normal processing installations. In some cases, processing revisions have been combined with plant modernization steps, resulting in modest economic return. More frequently, little or no payout results from pollution control installations. In extreme cases, an existing operation may even be discontinued or replaced with newer equipment at great cost. Odor and particulate matter control are particularly unattractive categories for economic justification. However, the amount expended is not always an accurate measure of the effectiveness of air pollution controls. Relatively small expenditures for process alterations and for supporting good operating and housekeeping procedures may result in significant improvements, whereas excessively large sums may be required to comply with unduly restrictive limits established without full consideration of economic justification.

### A. *Impact of Pollution Control Costs on Overall Refinery Expenditures*

An American Petroleum Institute survey indicates that the petroleum industry spent more than $4.4 billion for environmental protection between 1966 and 1972 (*29*). Nearly 50% of the total was spent on preventing air pollution. The petroleum industry environmental expenditures per barrel of crude processed rose to $0.34 per barrel in 1972. The refining segment of the oil industry spends anywhere from 5 to 10% of its capital investment in new refining facilities toward the control of air pollution.

### B. Examples of Cost-Effectiveness to Achieve Incremental Levels of Control

A useful tool in determining a reasonable level of pollution control is the method of cost-effectiveness analysis. Traditionally, the cost of abatement of any pollutant rises exponentially with percentage removal required. Both industry and government have the obligation to determine that regulations are developed in such a way that cost-effectiveness levels of control are selected wherever possible. The following examples illustrate the cost-effectiveness method of analysis.

#### 1. Hydrocarbon Emissions

Hydrocarbon vapors are emitted to the atmosphere during storage and transfer of crude oil and refined petroleum products. The three main emission control techniques are (a) floating roof tankage; (b) submerged or bottom loading during shipping and transfer of products; and (c) vapor recovery using refrigeration, condensation, or absorption. Table VI shows the incremental costs for a 6000-barrel/day gasoline loading facility, for incremental levels of control. Recovery of valuable gasoline vapor more than pays for the first two steps of control, but the added cost for vapor recovery equipment appears expensive.

#### 2. Particulate Emissions

The major sources of particulate emissions in petroleum refining are fluid catalytic cracking units. These emissions are normally controlled by the use of internal cyclones with the addition of either external cy-

**Table VI Incremental Costs for Incremental Levels of Control for a 6000-Barrel per Day Gasoline Loading Facility**

| | *Incremental installed cost ($1000)* | *Cumulative percent removal* | *Cost/percent incremental removal ($1000/ton)* |
|---|---|---|---|
| Floating roofs | 48 | 34 | Saving |
| Submerged fill | 32 | 81 | Saving |
| Vapor recovery | 130 | 90 | 240 |

clones or electrostatic precipitators. Table VII illustrates the cost and efficiency of these devices for a typical 50,000 BPD fluid catalytic cracker. The cyclone is cheaper on a dollar percent removal basis than either high or low efficiency electrostatic precipitators. However, once efficiencies of removal of greater than 78% are required, the high efficiency electrostatic precipitator has a lower unit cost than the low efficiency electrostatic precipitator showing that it is sometimes more cost-effective to select new equipment with a high degree of removal efficiency.

**Table VII Cost and Efficiency of Particulate Matter Control Devices for a Typical 50,000-Barrel per Day Fluid Catalytic Cracker**

| *Control* | *Installed cost (1972) ($1000)* | *Cumulative percent removal* | *Incremental cost ($1000)* | *Incremental percent removal* | *Cost/percent incremental removal ($1000)* |
|---|---|---|---|---|---|
| External cyclone | 930 | 78 | 930 | 78 | 12 |
| Low efficiency electrostatic precipitator | 1260 | 85 | 330 | 7 | 47 |
| High efficiency electrostatic precipitator | 1520 | 94 | 260 | 9 | 29 |

A slightly different situation exists in controlling particulates from an existing cracker. Many existing catalytic crackers are partly controlled with either external cyclones or low to medium efficiency electrostatic precipitators. In such cases, the incremental cost of moving to a slightly higher degree of control becomes exhorbitant. Figure 4 illustrates this concept. The lowest curve shows the incremental cost for a new unit while the two steep curves represent the added costs to go from existing 80 and 90% control to a final level of 95%.

### 3. Sulfur Plant Emissions

Assuming a typical Claus sulfur plant design will effect 95% removal, then the incremental cost for a tail-gas treating system would be as shown in Table VIII for a 250-ton/day plant.

## *C. Economic Incentives and Disincentives*

Economic incentives and disincentives are being proposed by government as a method of accelerating pollution abatement or as a substitute

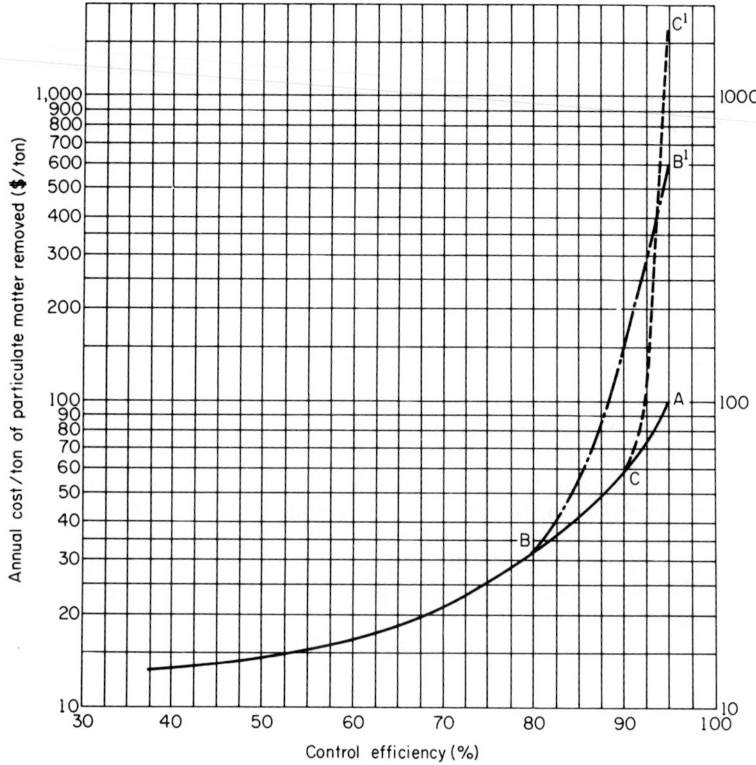


Figure 4. Fluid catalytic cracker particulate control costs on an annual basis vs control efficiency. (A) Solid line: can be applied to a previously uncontrolled or new unit (provided no serious retrofit problems are involved); (B–B′) dash-dot line: added costs for a unit presently operating at 80% control efficiency; (C–C′) dashed line: added costs for a unit presently operating at 90% control efficiency.

**Table VIII Incremental Cost for a Tail-Gas Treating System for a 250-Ton/Day Claus Sulfur Plant**

| *Process* | *Installed cost ($1000)* | *Percent removal* | *Incremental cost ($1000)* | *Incremental removal (%)* | *Cost per incremental percent removal ($1000)* |
|---|---|---|---|---|---|
| Claus sulfur plant | 2600 | 95 | 2600 | 95 | 27 |
| Claus plant with tail-gas system designed into the original plant | 3800 | 99.5 | 1200 | 4.5 | 267 |
| Claus plant with tail-gas system "added on" | 4800 | 99.5 | 2200 | 4.5 | 489 |

for pollution abatement. These incentives can be regarded as opportunities or obstacles to profit maximization. Suggested incentives have been in the form of: tax exempt bonds, rapid amortization, investment tax credit, state tax credit, and government grants. Generally speaking, the majority of businesses, government, and the public appear to favor incentives. There is considerably more controversy, however, about the desirability of utilizing disincentives such as taxes and effluent charges to bring about a reduction in emissions.

## REFERENCES


1. F. J. Gardner, *Oil Gas J.* **71** (53), 87 (1973).
2. L. R. Aalund, *Oil Gas J.* **71** (14), 92 (1973).
3. F. J. Gardner, *Oil Gas J.* **71** (53) (1973).
4. American Petroleum Institute, Committee on Public Affairs, "Oilman's Fact Finder," Brochure No. 1-73-50. American Petroleum Institute, Washington, D.C., 1973.
5. "Emergency Preparedness for Interruption of Petroleum Imports into the United States." A Supplemental Interim Report of the National Petroleum Council, Washington, D.C., 1973.
6. "Process Handbook," *Hydrocarbon Process, Petrol. Refiner* **45** (9) (1966).
7. Public Health Service, "Atmospheric Emmissions from Petroleum Refineries; a Guide for Measurement and Control," Publ. No. 763. Pub. Health Serv., U.S. Dept. of Health, Education, and Welfare, Washington, D.C., 1960.
8. H. F. Elkin and J. A. McArthur, "Impact of Federal/State Environmental Regulations on the Petroleum Industry." Paper, 72nd Annu. Meet. National Petroleum Refiners Association, Washington, D.C., 1974.
9. Bechtel Corporation, "The Economics of Fuel Oil Desulfurization," Pub. Health Serv., U.S. Dept. of Health, Education, and Welfare, Washington, D.C., 1964.
10. "Desulfurization of Caribbean Residual Fuel Oils." Report by the Bechtel Corporation to the American Petroleum Institute, 1967.
11. R. E. Conser, "Management of Sulfur Emissions," Paper, 72nd Annu. Meet. National Petroleum Refiners Association, Washington, D.C., 1974.
12. W. L. Nelson, Ed., *Oil Gas J.* **71** (47) (1973).
13. W. L. Nelson, Ed., *Oil Gas J.* **71** (48) (1973).
14. W. L. Nelson, Ed., *Oil Gas J.* **71** (49) (1973).
15. W. L. Nelson, Ed., *Oil Gas J.* **71** (50) (1973).
16. "The Economics of Manufacturing Unleaded Motor Gasoline." Report by Bonner and Moore Associates to the American Petroleum Institute, Washington, D.C., 1967.
17. Technical Progress Report, "Control of Stationary Sources," Vol. I, Chapter VI. Petrol. Ind., Los Angeles County Air Pollution Control District, Los Angeles, California.
18. "Emissions to the Atmosphere from Petroleum Refineries in Los Angeles County," Joint District, Federal, and State Project, Final Rep. No. 9. Los Angeles County Air Pollution Control District, Los Angeles, California, 1958.

19. M. Mayer, "A Compilation of Air Pollution Emission Factors for Combustion Processes, Gasoline Evaporation, and Selected Industrial Processes." Pub. Health Serv., U.S. Dept. of Health, Education, and Welfare, Washington, D.C., 1965.
20. S. S. Griswold, G. Fisher, and C. V. Kanter, *Proc. Nat. Conf. Air Pollut., 1958,* pp. 140–146. Pub. Health Serv., U.S. Dept. of Health, Education, and Welfare, 1959.
21. "Profile of Air Pollution Control—1971." Los Angeles County Air Pollution Control District, Los Angeles, California, 1971.
22. E. R. Stephens, E. F. Darley, and F. R. Burleson, *Proc. Amer. Petrol. Inst., SECT. 3* p. 47 (1967).
23. "Air Pollution and Smog," Air Pollut. Found., Los Angeles, California, 1960.
24. W. L. Faith, *Ind. Wastes* **5,** No. 4 (1960).
25. J. A. Danielson, ed., "Air Pollution Engineering Manual, Los Angeles," Pub. Health Serv. Publ. 999-AP-40, Chapters 10 and 11. U.S. Dept. of Health, Education, and Welfare, Cincinnati, Ohio, 1967.
26. D. H. Stormont, *Oil Gas J.* **63,** No. 48 (1965).
27. Process Research, Inc., "Characterization of Claus Plant Emissions." U.S. Environmental Protection Agency, Research Triangle Park, North Carolina, 1973.
28. R. G. Lunche and I. S. Deckert, "Hydrocarbon Losses from Petroleum Storage Tanks," Air Pollut. Eng. Rep. Los Angeles County Air Pollution Control District, Los Angeles, California, 1956.
29. Automated Services, Inc., "Environmental Expenditures of the U.S. Petroleum Industry 1966–72," API Publ. No. 4176. American Petroleum Institute, Washington, D.C.

# 20

## Nonferrous Metallurgical Operations

Kenneth W. Nelson, Michael O. Varner, and Thomas J. Smith

## I. Introduction

Nonferrous metals production has been very important in the development of the science of air pollution control. Smelting processes are carried on at high temperatures, and large quantities of dusts and metal oxide fumes are generated to contaminate the huge volumes of gas which flow through process equipment. Some idea of the gas cleaning problem may be gained from the fact that the mass of air handled daily in a smelter is far greater than the mass of solid materials. The latter may be several thousand tons per day.

In addition to the potential for substantial emissions of particulate matter, some of which could be toxic, there is a concomitant high production and emission of sulfur dioxide. Most of the copper, lead, and zinc in the earth's crust occurs as complex sulfide minerals. After mining, milling, and concentration, the minerals shipped to smelters contain as much as 30% sulfur, all of which is oxidized during the various smelting steps. And nearly all of it appears as $SO_2$. It has been estimated that smelters contribute about 13% of the total $SO_2$ emitted annually in the United States.

The sulfur is a mixed blessing. Its oxidation provides valuable process heat and conserves energy that would have to be supplied from other sources. Also, sulfur dioxide can be a useful raw material for sulfuric acid production in areas where acid can be used and where smelter acid can compete in the market with acid derived from burning elemental sulfur. But in the main, sulfur dioxide has been an onerous problem.

In the early days of United States smelting, sulfide ores were simply piled on open ground and ignited. Heap-roasting, as it was called, created high $SO_2$ concentrations at ground level and caused extensive vegetation damage nearby. Later $SO_2$ was vented with combustion gases into flues and low stacks which provided draft for furnaces, and smelters were established in isolated areas to prevent operations from creating a smoke nuisance. As the scale of operations increased, and as lands near smelters were inhabited and cultivated by farmers, smelter smoke* created serious problems.

In 1950 Haywood of the United States Department of Agriculture examined plants grown near smelters and showed that foliage injured by effluent gases had a higher sulfate content than intact samples (*1*). In 1915 the Selby Smelter Commission reported upon its study of the effects of smoke from the Selby smelter location on the northeast shore of San Francisco Bay (*2*). The study group measured the effects of different concentrations of $SO_2$ on vegetation and clearly demonstrated that some growing plants are more sensitive to the gas than are animals and humans. The Commission's report is a classic in air pollution literature and contains a good bibliography of that era.

In 1914 the American Smelting and Refining Company established a Department of Agricultural Investigation for systematic research into the effects of smelter effluents and means of their control. So far as is known the department laboratory was the first private facility in the United States founded specifically for air pollution research. Numerous experiments were done to determine the precise conditions required to produce vegetation injury with $SO_2$ and to classify the differing sensitivities of various plant species.

Better means of measuring low concentrations of $SO_2$ in air were sought to replace laborious manual procedures. Thus, in 1928, Dr. M. D. Thomas, a department scientist, invented an automatic measuring and recording instrument, in which $SO_2$ is collected in dilute, acidified hydrogen peroxide solution and oxidized to sulfuric acid. The resultant change in conductivity is proportional to the concentration of $SO_2$ in sampled air. The Thomas instrument has been modified over the years and has proved to be highly reliable and accurate in smelter areas where interfering air contaminants are likely to be absent or minimal.

In addition to the instrumentation of Thomas which made $SO_2$

* Smelter smoke means the gaseous effluent including gases and fumes as well as "smoke."

measurements relatively simple and popular, the earlier research contributions of O'Gara (*3*), the Department's first director, and of Thomas's contemporaries and colleagues have also been noteworthy (*4–6*).

Injury to crops and trees by $SO_2$ from smelters has been the basis for much protracted and costly litigation. Court-ordered shutdowns of California smelters prompted the appointment of the Selby Smelter Commission. Since then there have been numerous lawsuits, especially by farmers. In one instance the United States was the complainant. It was claimed that smoke from the Consolidated Mining and Smelting Company at Trail, British Columbia, was crossing the United States–Canadian border and causing damage to forests. An International Joint Commission began hearings on the matter in 1928. Not until 1941 did the Trail Smelter Arbitral Tribunal come to a final decision and set up a meteological regime under which the smelter was required to operate.

Some smelter owners have felt that farmers sometimes claimed compensation for crop damage by smoke when crop losses were actually caused by poor farming practices or unfavorable weather. Habitual claimants were referred to as "smoke farmers" and their actions forced smelting companies to keep detailed records of $SO_2$ concentrations in potentially affected farm areas and to inspect crop conditions regularly during the growing season.* A typical program designed to protect the interests of the farmers as well as those of the company has been described by Davis (*7*).

Damage from smelter particulate emissions has also been the subject of litigation. Farmers near lead smelters have claimed that lead fallout has accumulated in soil and pasture grasses in sufficient concentrations

* An amusing bit of doggerel printed in a Tacoma newspaper 40 years ago tells the story (by "W.C.H." from the files of the late Paul H. Ray, Salt Lake City, Utah):

*The Joke Is on the Smelter*

If the horses have the glanders,
If the turkeys have the roup,
If the deadly hawk is flying
Into his chicken coop.

The farmer has his inning,
The matter is no joke
For he traces down his losses
Direct to smelter smoke.

The frost may blight the melons,
The crows may get his corn,
And the pigs may have the cholera,
His cow a crumpled horn.

The farmer grabs his pencil
He charges all to smoke,
He swiftly sends his little (?) bill,
And thinks it is a joke.

The water in the stream dries up,
The south wind blasts his fields,
His daughter has the whooping cough,
His wheat it fails to yield.

But the farmer's never troubled,
He banks his wealth in town,
He never feels the want of cash,
Till the smelter closes down.

to poison farm animals, usually horses. In some cases there has been no doubt of the validity of the claims; others have been disputed. Fluoride fallout and gases from aluminum smelters have also been responsible for harm to grazing animals. Agate *et al.* (*8*) have reported on episodes in Scotland.

## II. Copper

### *A. Mining, Milling, and Concentrating*

Both open-pit and underground mining are practiced (see also Chapter 17, this volume); however, the principal method employed is open pit. During 1969 about 88% of the ore and 84% of the recoverable copper was produced by the open-pit method (*9*).

Open-pit operations are conducted successfully with ores containing as little as 0.5% copper. Costs of operating underground mines are higher, and such mines usually must have a somewhat better grade of ore. Ore, incidentally, may be defined as rock from which metals may be recovered at a profit. Otherwise the rock is waste to a miner.

A major development in open-pit mining has been the leaching of waste dumps. Large tonnage of wastes known as overburden are stockpiled or dumped into gulches as the mine develops. When these low-grade copper minerals are exposed to slight acid water, the soluble salts dissolve and are carried to precipitation plants. The dissolved copper precipitates on scrap iron, is washed into settling tanks, and is reclaimed as slurry or pulp. Precipitated copper, known as "cement copper," usually assays 70–90% copper and is suitable smelter feed.

Most of the primary domestic copper is recovered from low-grade sulfide ores using pyrometallurgical procedures, to be described later in this section. Leaching (hydrometallurgy) can sometimes be used for these ores, although it is normally employed only for oxide and mixed ores in which oxide minerals predominate. The four principal hydrometallurgical methods practiced are in-place leaching, heap leaching, dump leaching, and vat leaching. Copper-containing solids are leached with a dilute solution of sulfuric acid in all four of the methods.

Copper ores are handled in tremendous quantities. Mills processing 500 tons or more per hour are not unusual. Enormous jaw, gyratory, and cone crushers are used to reduce large boulders to pebble-sized pieces.

Separation of the traces of copper-bearing minerals from the mass of waste is essential to economic recovery of the metal. Crushed ore is ground wet in ball or rod mills to produce a thin slurry. The slurry plus

**Table I Typical Analysis of a High-Quality Copper Concentrate**

| *Constituent* | *Percent (dry basis)* |
|---|---|
| Copper (Cu) | 27.5 |
| Iron (Fe) | 24.5 |
| Sulfur (S) | 31.5 |
| Silica ($SiO_2$) | 11.0 |
| Other | 5.5 |

a variety of flotation reagents is piped to flotation cells where air is beaten into the mixture. A copper-rich froth is formed and skimmed off. Solids in the froth are dewatered by settling and filtration, and are shipped to a smelter. The concentrate contains 15–30% copper and overall copper recovery averages 80–85%. Table I shows a typical analysis of a high-quality copper concentrate (*10*).

### B. Smelting

A typical copper smelter in the United States uses roasters, reverberatory furnaces, and converters in the sequence shown in Figure 1 (*11*).

Copper sulfide concentrates received at the smelter are normally roasted in multiple hearth roasters to remove moisture, to oxidize part of the contained sulfur, and to preheat the material before smelting. Some plants use fluid bed roasters, while many smelters have eliminated the roasting step altogether.

In most United States smelters, roaster calcines or wet concentrates are charged into large reverberatory furnaces where copper and iron in the form of oxides, sulfides, and sulfate are melted to form copper matte, a mixture of cuprous and ferrous sulfide. Some iron, calcium, magnesium, and aluminum—all as silicates—are removed in the form of a viscous slag.

The copper matte is tapped off into large ladles and transferred to converters. Siliceous fluxes are added to the matte in the converter and air is blown into the hot, molten mass through tuyeres. Iron and other impurities form a silicate slag on top of the denser cuprous sulfide. The slag is poured off at intervals and returned to the reverberatory furnaces for recovery of entrained copper. Further blowing of the converter oxidizes the sulfur—leaving free copper according to the overall equation:

$$Cu_2S + O_2 = 2\ Cu + SO_2 \tag{1}$$

Copper is then poured off and transferred to a holding furnace for deoxi-

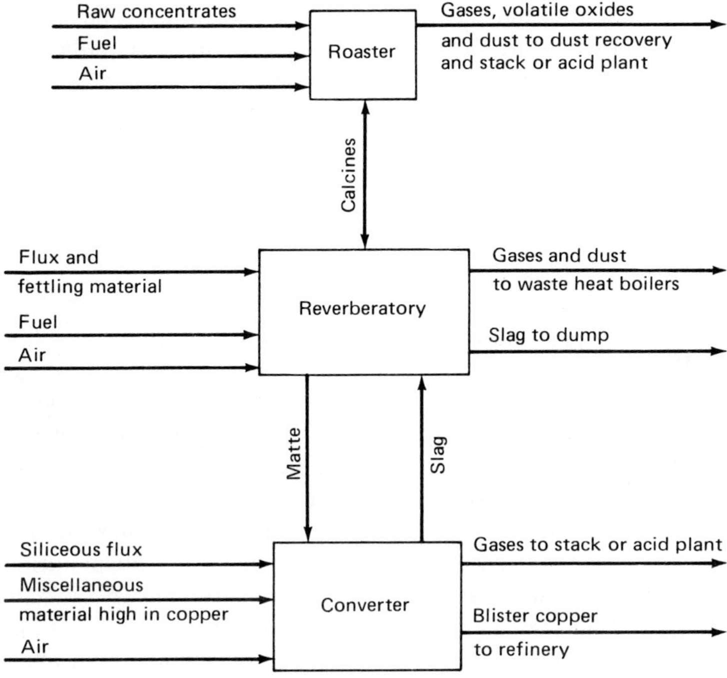


Figure 1. Production of blister copper sulfide concentrates (*11*).

dizing and casting as slabs of blister copper—about 98.5% pure—or as anodes for shipment to an electrolytic refinery.

### *C. Refining*

A small percentage of virgin copper is sufficiently refined by special furnace treatment at the smelter to be used directly in certain applications. However, the bulk of the copper is intended for electrical uses, and must be highly purified. Very small amounts of certain impurities greatly reduce copper's electrical conductivity, and adversely affect its annealability.

Anodes of blister copper—so named because of its rough surface—are arranged alternately, face to face with thin sheets of pure copper in large tanks. The sheets are the cathodes of the multiple electrolytic cells to be formed in each tank. The tanks are nearly filled with a dilute solution of copper sulfate and sulfuric acid. A low voltage current passes between anodes and cathodes, causing dissolution of copper at the anodes and deposition at the cathodes. Trace metals present in the impure anode either dissolve in the circulating electrolyte or settle to the bottoms of the tanks as a black sludge called anode mud, or slimes. The slimes may contain selenium, tellurium, gold, silver, platinum, and palladium. All these elements are frequently associated with copper in its ores.

### D. Emissions and Controls

#### 1. Mining Operations

Dust can be a problem in underground mines in spite of wet collaring and drilling. Free silica may be present in appreciable percentages and pose a health risk to miners. Ventilation must be provided to keep dust concentrations down to safe levels. In most instances power ventilation is also necessary to remove nitrogen oxides and carbon monoxide from blasting and to control heat and humidity.

Although dust and blasting gases are eventually discharged from the mine, there is no significant contribution to community air pollution. The concentrations of dust particles are low—of the order of 5 million particles per cubic foot ($<1.76 \times 10^8/m^3$) of air or less. Blasting gases are absent except during the short period of blasting and then are highly diluted by the large volume of ventilating air constantly being moved through the mine.

Open-pit mining may create localized dustiness near operating drills, power shovels, and other equipment. But it is standard practice to use water while drilling and to wet down ore and waste piles, when necessary, before loading the material into trucks or railway cars for transport from the mine. Roadways are sprinkled to reduce dusting.

Blasting, properly done, disperses surprisingly little dust in open-pit mining. Nitrogen dioxide may be produced from explosions of ammonium nitrate–fuel oil (ANFO) or other ammonium nitrate combinations. When the air in the pit is cool in relation to air at the brim of the pit, as on a summer morning, blasting at times creates a visible cloud of $NO_2$–$N_2O_4$. Viewed from a distance of several hundred feet, the cloud looks formidable. But tests show the nitrogen oxide concentration to be less than 30 ppm. And the gases dissipate in minutes. Improved formulations and practices have virtually eliminated production of nitrogen oxides from exploding ANFO.

Ore-crushing generates considerable dust. Water is used to wet the ores but it must be applied judiciously to avoid difficulties such as overwetting, buildups of mud on conveyor belts and pulleys, and clogging of chutes. Furthermore, as ore passes through successive crushing stages, new, dry surfaces are exposed and dust is readily abraded from them. Wetting, therefore, must usually be supplemented by appropriate exhaust ventilation for good dust control. Because dust burdens in ventilating air may be heavy, effluents may be neighborhood nuisances unless effective collectors are incorporated in the exhaust systems. In most modern ore mills dust concentrations are kept below 5 million particles per cubic foot ($<1.76 \times 10^8/m^3$).

The flotation process itself creates no dust problems but waste material, or tailings, issuing as a sandy slurry from the concentrators may do so. The slurry is channeled into ponds from which water, free of settleable solids, may be recovered. As the solids accumulate, a tailings dump covering many acres is formed. Surface drying of the dump and high winds may result in localized dust storms. Keeping the dump wet is an effective means of control. In recent years cohesion compounds have been used to suppress dust emissions from tailings dams. Planting and cultivation of ground cover and shrubs to act as windbreaks are also successful. They also improve appearance.

## 2. Smelters

a. Emissions. The high temperatures necessary in roasting, smelting, and converting cause volatilization of a number of the trace elements, such as As, Pb, Cd, etc., which may be present in copper ores and concentrates. The waste gases from these processes thus may contain not only fumes of these elements, but also concentrate and calcine dust and $SO_2$ (Tables II and III) (*9, 12*). Roasting drives off a portion of any arsenic, antimony, and lead as the oxides. More of these elements plus bismuth and some selenium and tellurium may be eliminated as fume in the reverberatory furnace. They also to some extent become incorporated into furnace slag.

The more highly oxidizing conditions of the copper converting operation lead to further removal of the remaining volatile elements except selenium and tellurium. Nickel, cobalt, and the precious metals are also not volatilized significantly and remain dissolved in the crude copper.

b. Control Methods. The value of the volatilized elements, as well as air pollution considerations, dictates efficient collection of fumes and dusts from process off-gases. Balloon flues in which gases move at low velocities serve as gravity collectors of the larger particles and provide low resistance ducting for the large volumes of combustion gases and ventilating air that must be moved. Cyclones also may be used. For collection of the finer particulates, electrostatic precipitators, in which col-

**Table II Example Analysis of Collected Dust and Fume in Effluent Gases from Copper Roasters, Reverberatory Furnaces, and Converters (9)**

| *Element* | *From roasters* | *From reverberatory furnaces* | *From converters* |
|---|---|---|---|
| Copper % | 5.2 | 2.9 | 1.12 |
| Iron % | 6.6 | 1.6 | 1.2 |
| Lead % | 7.6 | 30.5 | 47.1 |
| Zinc % | 1.7 | 8.3 | 3.2 |
| Arsenic % | 43.0[a] | 25.7[a] | 9.6[a] |
| Antimony % | 5.3 | 3.0 | 1.6 |
| Bismuth % | 0.4 | 1.11 | 1.64 |
| Cadmium % | — | 0.71 | 1.15 |
| Silver oz/ton | 7.6 | 18.6 | 10.9 |
| Gold oz/ton | 0.02 | 0.04 | 0.02 |

[a] Present to this extent only in smelting of concentrates high in arsenic.

lection efficiencies up to 99.7% for copper dust and fume are attained by careful conditioning of the flue gases, are most often used.

Cleaned, hot flue gases are vented to the atmosphere via tall stacks for maximum dispersion and dilution of contained sulfur dioxide. A major proportion of the gas may be used to produce sulfuric acid. Not all of it is used for this purpose because the $SO_2$ concentration in gases available from some sources within the smelter may be too low for efficient catalytic oxidation to $SO_3$ by available commercial processes. To make sulfuric acid efficiently, $SO_2$ concentrations in the gas must be between 4 and 8%.

**Table III $SO_2$ Emission Rates from Copper Smelting (12)**

| | *Emission rate* | |
|---|---|---|
| | *lb $SO_2$/ ton of metal* | *gm $SO_2$/ kg of metal* |
| Primary copper smelting | | |
| If roasting is practiced | | |
| Roasting | 650–1350 | 325–675 |
| Reverberatory furnaces | 300–950 | 150–475 |
| Converters | 1950–2150 | 975–1075 |
| If roasting is not practiced | | |
| Reverberatory furnaces | 550–1600 | 275–800 |
| Converters | 1700–3600 | 850–1800 |

Two United States copper smelters have installed dimethylaniline (DMA) scrubbing plants to recover and concentrate the sulfur dioxide contained in the converter off-gases, in the form of liquid $SO_2$.

Utilization of $SO_2$ reduces the potential air pollution problem associated with copper smelting; hence acid production would seem to be an obvious and simple solution. The facts are that smelter acid plants entail heavy capital investments and substantial operating costs. For break-even or profitable operation of acid plants, costs must be returned by sale of the product.

There may not be a local market for acid. Distances to demand areas and consequent shipping charges may be prohibitive for smelter acid to compete with plants using Frasch sulfur as a source of $SO_2$. Each copper smelter is unique in some ways, and the addition of an acid plant may or may not be economical. However, the passage of the United States Clean Air Act of 1970 forced many smelters to install acid plants regardless of economic consequences.

The idea of recovering sulfur dioxide from smelter gases is, of course, an attractive one, and a large number of processes have been described in the technical and patent literature. A summary of them is given in the United States Environmental Protection Agency (USEPA) Document, "New Source Performance Standards," August, 1973 (*13*). So far as major copper smelters in the United States are concerned, however, none of the processes, other than direct conversion to sulfuric acid and the DMA process, had been proved practicable as of 1974.

On occasion, cold tail-gas discharges from by-product acid plants have caused higher than expected ambient $SO_2$ concentrations in nearby neighborhoods. Tail-gas $SO_2$ stack concentrations will typically range from 200 to 700 ppm (532–1862 $mg/m^3$ at 20°C) for a double contact acid plant and run from 1000 to 3000 ppm (2662–7986 $mg/m^3$ at 20°C) in the tail gases of single pass plants.

Tall stacks, up to 1000 ft (305 m) as at Hayden, Arizona, are depended upon for dispersion and dilution of $SO_2$ to keep ground-level concentrations to a minimum. The effectiveness of high smelter stacks was shown by Hill *et al.*, in 1944 (*14*) and has been reemphasized by Smith (*15*), Ross *et al.* (*16*), Frankenberg (*17*), Pooler and Neimeyer (*18*), Lee and Stern (*19*), and Engdahl (*20*).

$SO_2$ monitoring stations at selected points around a smelter are helpful in carrying out a "sea captain" method of controlling $SO_2$ emissions, known as "closed loop control" or "intermittent control" (*21*). This technique makes use of measurement of meterological conditions followed by their expert evaluation. If conditions are unfavorable for adequate dispersion of gas, or are predicted to be unfavorable, $SO_2$ emitting operations are curtailed. However, forecasting is not perfect and the detection of significant $SO_2$ at ground level by an automatic instrument may be the first indication that dispersion and dilution are not adequate. Telemetering of detector information to the smelter control center computer system permits immediate curtailment action to be taken. If there are multiple sources of $SO_2$ in a given area in the same direction from a detecting station, smelter control by monitoring is hampered. This type of emissions control has proved effective in reducing ambient $SO_2$ concentrations at El Paso, Texas, and Tacoma, Washington (*22*).

Although limestone scrubbing has been proposed as an acceptable method of removing $SO_2$ from smelter stack gases, it has proved to be neither feasible nor economical, according to pilot-plant experiments performed by the Smelter Control Research Association (*23*). The solid waste produced by a typical copper smelting scrubber operation creates a substantial disposal problem. The capture of 113 tons per day (TPD) of sulfur (226 TPD of $SO_2$) yields 1000 TPD of wet sludge which settles and becomes 565 TPD of dried $CaSO_3 \cdot 2\ H_2O$. The wet sludge would require a 350,000 tons per year disposal effort (*12*).

## 3. Refinery Emissions and Controls

Electrolytic copper refining operations, by their nature, do not create significant air pollution. In the tank house, traces of electrolyte mist are generated by splashing of liquid as it is circulated by gravity and pumps among the tanks. Repeated tests by the authors have shown acid mists in work areas to be within the United States Occupational Safety and Health Act (OSHA) 8-hour time-weighted average standard.

In one part of the copper refining process called electrolyte purification, there are possibilities of excessive acid mist generation and of arsine evolution. The copper content of the electrolyte becomes very low during final electrolysis, and hydrogen is produced at the cathode. If traces of arsenic are carried in the solution, arsine, $AsH_3$, will be evolved in sufficient amounts to be dangerous to men working in the vicinity. Hence, exhaust ventilation of purification tanks is essential. Purification tanks are few in number compared to the total number of tanks in a refinery and are easily isolated and ventilated. Based on stack sampling tests conducted by the authors, typical stack concentrations average about 20 ppm $AsH_3$.

Treatment of slimes for recovery for silver, gold, selenium, tellurium, and traces of other elements usually entails fusion and oxidation in a furnace of appropriate size. Some selenium is volatilized during the process and is captured as the furnace gases pass successively through

a wet scrubber and an electrostatic precipitator. The latter removes mists that escape the scrubber.

### E. Impact of Regulations on Emission Control Practices

The impact of the 1970 United States Clean Air Act regulations on copper operations has been severe. The United States Environmental Protection Agency has estimated the investment required for copper smelters to achieve compliance with Federal air pollution standards at \$313 million (*24*). This figure has been criticized by industry as being too low. Industry figures put the capital investment at \$540 million. Installations at copper smelters seem to bear this out. Installation of a sulfuric acid plant at Hayden, Arizona, designed to remove approximately 55% of the sulfur emissions, cost \$17.5 million; installation of a similar but lower capacity and more efficient plant at El Paso, Texas, cost \$14.3 million; installation of a DMA process liquid $SO_2$ plant at Tacoma, Washington, cost \$17 million. Much higher expenditures would be required for higher percentages of $SO_2$ abatement.

Alternative methods to reduce sulfur oxide emissions include process changes such as the use of electric smelting and flash smelting furnaces. Both types of furnaces have the advantage of producing a gas stream of higher $SO_2$ concentration than streams from reverberatory furnaces. An experimental plant has been put in operation in El Paso, Texas, to study the feasibility of converting $SO_2$ in smelter strong gas streams to elemental sulfur (*25*).

Although some of these newer approaches look promising, pilot-plant studies are showing that there are still many difficult problems to be ironed out before they can be applied to complex smelter operations. In addition, some of the contemplated $SO_2$ removal processes have the potential of causing severe adverse effects on water and land. There are also the problems of an increasing supply of collected $SO_2$ by-products on markets traditionally supplied by elemental sulfur, and of high energy consumption by $SO_2$ recovery facilities. A sulfuric acid plant at one Arizona smelter requires more electrical power than the remainder of the smelter complex.

## III. Lead

### A. Mining, Milling, and Concentrating

Most lead ores are mined underground. Crushing, grinding, and concentrating follow the same general pattern as the processing of copper ores.

Normally, however, lead ores are of higher metal content. The run of the mill contains between 6 and 10% lead. The lead mineral of greatest importance is galena, PbS. Traces of silver, cadmium, iron, zinc, and other metals usually are present in lead ores, and they accompany the lead in the ore concentrate.

### B. Smelting

The sulfur content of lead concentrates is reduced by sintering them on Dwight–Lloyd sintering machines (Fig. 2), most commonly on updraft machines. Moistened concentrate–flux–fuel mix is fed to an endless belt of cast iron grate sections. The charge is ignited, burns under forced draft, and is finally discharged from the machine as grates flip over at the head pulley of the belt. The charge is now a fused mass of material called sinter. It may be crushed, mixed with other materials, such as flue dusts, and sintered a second time, or it may be fed directly to the blast

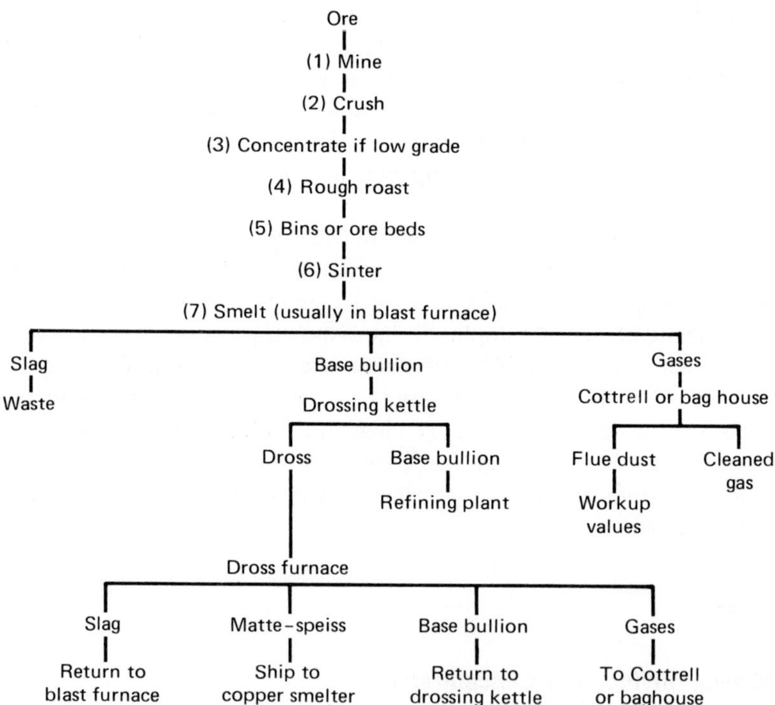


Figure 2. Usual treatment of a sulfide lead ore (*11*).

furnace. In addition to eliminating most of the sulfur from PbS concentrates, the sintering procedure prevents dust losses which would occur if concentrates were smelted directly. Also, it creates a more porous raw material to facilitate smelting. Up to 85% of the sulfur in the smelter feed is released during the sintering process.

A mixture of sinter, iron, and coke is charged into blast furnaces. Contact with carbon monoxide and possibly free carbon at high temperatures reduces lead compounds to metallic lead. A mixture of molten lead and siliceous slag accumulates in the hearth of the furnace and is tapped off, either continuously or intermittently. Gravity separation of the lead from the slag takes place in heavily insulated settlers. Slag is allowed to overflow from the top into slag pots for transport to the dump or to a fuming furnace for zinc recovery.

As it is tapped periodically from the settlers, the crude lead is at red heat and contains considerable amounts of dissolved impurities which become insoluble as the metal cools. Hence the hot lead is transferred to holding kettles for cooling and subsequent skimming of impurities from the surface. At this stage the metal is about 95 to 99% pure and is further refined on the premises or cast into blocks for shipment to a refinery.

### C. Refining

Lead bullion is purified by a number of different processes. The Parkes process consists, in broad outline, of heating under oxidizing conditions in a reverberatory furnace for removal of arsenic, antimony, and tin (softening), dissolving zinc in softened lead in kettles, cooling, skimming of a silver-rich crust (desilverizing), and removing dissolved zinc by vacuum distillation (dezincing). If bismuth is present, it is removed by treatment with calcium and magnesium (debismuthizing).

The Harris process employs treatment with molten sodium hydroxide and sodium nitrate as a substitute for furnace softening. Arsenic, antimony, and tin, if present, are separated as sodium salts.

The Betts process is electrolytic and produces pure lead cathodes plus a slime containing impurities derived from the crude lead bullion.

An important part of lead refining is the recovery of silver and gold. Zinc crust collects both metals. The crust, containing a considerable percentage of entrained lead, is heated in graphite or clay retorts to distill off the zinc. The residual bullion is transferred to cupel furnaces for separation of lead. This is done by means of an air blast directed on the molten bullion. The litharge (PbO) produced is molten and is carefully decanted as it accumulates on the surface. Doré metal (Ag–Au) is the final product of cupellation and is tapped off for further treatment.

### D. Emissions and Controls

#### 1. Mining Operations

Dust problems in the mining, milling, and concentrating of lead are the same as those outlined for copper.

#### 2. Smelter Operations

a. EMISSIONS. Hot gases from the lead concentrate sintering process carry $SO_2$, dust, and the oxide fume of volatile metals such as antimony, cadmium, lead, and zinc (Tables IV and V). Also, some lead is volatilized and oxidized. Blast furnaces emit similar particulates plus low concentrations of $SO_2$ and carbon monoxide.

b. CONTROL METHODS. Dust and fume are recovered from the gas stream by settling in large flues and by precipitation in electrostatic precipitators or filtration in large baghouses, more commonly the latter.

Collection efficiencies are up to 96% for precipitators and 99.5% for baghouses. A well-controlled lead smelter will discharge on the order of 5.0 lb/hour (2.3 kg/hour) of lead from the sinter plant baghouse stack and approximately 1.0 lb/hour (0.45 kg/hour) from the blast furnace baghouse stack.

Sulfur dioxide derived from sintering may not be concentrated enough

**Table IV Emissions from Lead Sintering, Blast Furnaces, and Reverberatory Furnaces**[a]

| | | Raw gas | |
|---|---|---|---|
| | *Waste gas ($m^3$)* | *Dust content ($gm/m^3$)* | *$SO_2$ content (%)* |
| Sintering | 3,000[b] | 2–15 | 1.2–5.0 |
| Blast furnaces | 15,000–50,000[c] | 5–15 | — |
| Reverberatory furnaces | 100–500[d] | 3–20 | — |

[a] Clean Air Guide 2285, Kommission Reinhaltung der Luft, Verein Deutscher Ingenieure, VDI—Verlag GmbH, Dusseldorf, West Germany, September, 1961.
[b] Per ton of sinter.
[c] Per ton of coke.
[d] Per ton of charge.

**Table V $SO_2$ Emission Rates from Lead Smelting** (*12*)

| | *Emission rate* | |
|---|---|---|
| | *lb $SO_2$/ ton of metal* | *gm $SO_2$/ kg of metal* |
| Primary lead smelting | | |
| Sintering | 1150–2150 | 575–1075 |
| Blast furnaces | 5–10 | 2.5–5 |
| Dross reverberatory furnace | 5–10 | 2.5–5 |

to be used directly by available commercial processes for sulfuric acid production. It is possible, however, by recirculation of gases, to build up the $SO_2$ concentration sufficiently to permit production of $H_2SO_4$, or liquid sulfur dioxide, from a portion of the gases. A practicable process for liquid $SO_2$ production at a lead smelter has been described by Fleming and Fitt (*26*).

Because the sulfur proportion in galena, PbS, is only 13.4%, the total amount of by-product $SO_2$ theoretically available from lead smelting is considerably less than copper or zinc smelting for equivalent rates of metal production. A 5000-ton (4.54-million kilogram)/month lead plant thus would have, in theory, only 670 tons (0.61 million kilograms) of sulfur available which would yield about 2000 tons (1.81 million kilograms) of $H_2SO_4$ per month. Acid plants would therefore be relatively small and in certain circumstances uneconomic. Low $SO_2$ emission potential, on the other hand, diminishes the dispersion problem and the necessity for $SO_2$ recovery.

Even though the use of an acid plant appears superficially to be a feasible solution for $SO_2$ control at a lead smelter, acid production, because of technical and disposal problems, may be a less than satisfactory and reliable control method. A sinter machine operates intermittently; therefore, an acid plant attached to a sinter machine is subject to corrosive effects of the condensation of moisture during alternating heating and cooling cycles. Cold, concentrated $H_2SO_4$ does not attack steel, but dilute acid does. Also, organic compounds released in the sintering operation will be captured in the acid yielding a low-grade "black" acid which is marginally marketable at best. Because of severe corrosion and other operating problems, an acid plant built at a new lead smelter in the late 1960's has had a maximum monthly operation of 75%.

Lead blast furnace gases, after cooling, are amenable to treatment in large baghouses having several chambers, each of which will contain hundreds of bags. A common size of bag used is 18 in. (46 cm) diameter by 30 ft (9.15 m) long. Wool has been traditionally used for the bags, and service lives in many cases have been remarkably good. Synthetic materials, including glass fibers, are competing successfully with wool, however, because of superior resistance to high temperatures.

Filtered blast furnace gases contain traces of $SO_2$ which cannot be removed by any feasible method. Dilution and dispersion by discharge from stacks of appropriate height are the only practicable means of preventing excessive concentrations in ambient air.

Fugitive emissions from sinter plant operations are in many instances collected by appropriate exhaust hoods and are passed into baghouses or high efficiency scrubbers. Wetting with water spray nozzles is effective in abating fugitive dust in the sinter plant.

#### 3. Refinery Emissions and Controls

Lead refineries have baghouses for recovery of fume from softening furnaces and cupeling furnaces. Zinc oxide fume release during distillation of zinc from zinc–silver skims may be captured by a local exhaust ventilation system and passed into the flue-serving cupel furnaces, with ultimate discharge through the baghouse. The baghouses do an excellent job when properly maintained.

### *E. Impact of Regulations on Emission Control Practices*

The impact of air pollution regulations on lead operations is similar to the impact on copper operations, previously discussed.

## IV. Zinc

### *A. Mining, Milling, and Concentrating*

The processes used for zinc are essentially the same as for copper and lead. The bulk of zinc ore contains zinc as sphalerite, ZnS, which is separated as a concentrate from accompanying minerals by selective flotation. Concentrates contain about 60% zinc.

Another important source of raw material for zinc metal is zinc oxide from fuming furnaces. Zinc as an impurity in lead smelting is recovered from lead blast furnace slag by heating the slag to high temperatures and blowing pulverized coal and air through it. Zinc is reduced, volatilized, reoxidized, and is collected as ZnO in bag filter units. The baghouse product is passed through a rotary kiln to reduce the lead and/or cadmium content by volatilization and to increase the density of the material for easier handling and shipping.

### *B. Roasting and Retorting*

For efficient recovery of zinc, sulfur must be removed from concentrates to less than 2%. This is done by roasting. Multiple hearth, flash, or fluid-bed roasting may be followed by sintering; or double-pass sintering may be used alone (Fig. 3).

The liberation of zinc from roasted concentrates involves simple heating of a mixture of roast and coke breeze to about 1100°C. Simultaneous reduction of zinc from the oxide to the metal and distillation of the metal takes place. Zinc vapor passes from the heated vessel into a condenser where it condenses to a liquid which is drained off at intervals into molds.

Reduction and distillation of zinc may be done as a batch process in banks of cylindrical retorts—the Belgian retort process—or in continuously operating vertical retorts. Gas is the preferred fuel, hence most smelters are located in natural gas fields. Electric distillation furnaces are used to a small extent.

A very pure grade of zinc is produced by a continuous fractional distillation process developed by the New Jersey Zinc Co.

A process for simultaneous smelting of roasted lead and zinc concentrates has been developed by the Imperial Smelting Corp. The process makes use of carbon monoxide generated from coke to reduce lead and

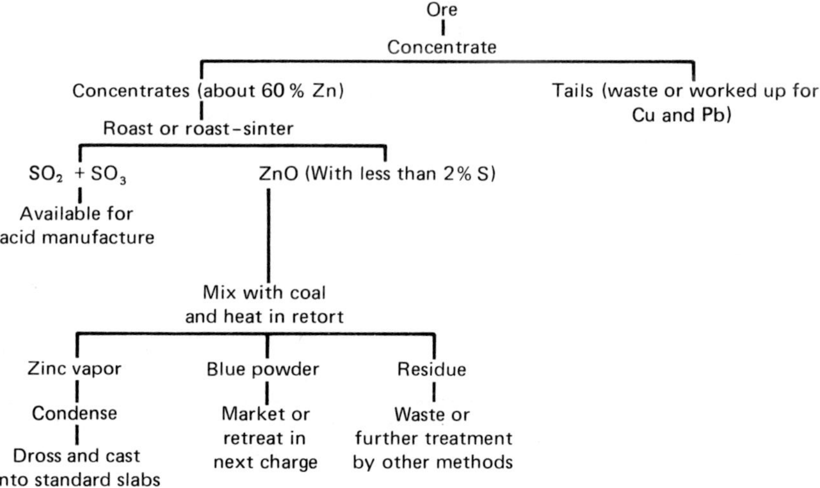


Figure 3. Outline of zinc production from sulfide ore (*11*).

zinc oxides in a sealed shaft furnace. Lead bullion accumulates in the furnace bottom and acts as a collector for copper, silver, and gold. Zinc passes as a vapor out the top of the furnace into condensers in which a shower of molten lead is continuously maintained. Zinc vapor is condensed quickly to a liquid which dissolves in the molten lead. Outside the condenser the lead–zinc solution is cooled and a 98% pure zinc floats to the surface and overflows into containers. The cooled lead is pumped back to the condensers (*27*).

### C. Leaching and Electrolysis

Zinc of high purity may be produced from roasted concentrates, from densified zinc oxide from fuming furnaces, or from impure metallic zinc by solution in sulfuric acid, removal of impurities from the solution by appropriate chemical treatment, and, finally, electrolysis of the purified electrolyte.

Electrolysis is done in tanks containing alternating anodes of lead and cathodes of aluminum. Essentially, pure zinc is deposited on the cathodes and later stripped from them by hand. The zinc is then melted in a small reverberatory or electric furnace and cast into slabs or other forms for shipment.

### D. Emissions and Controls

Dust, fume, and $SO_2$ are evolved from zinc concentrate roasting or sintering (Tables VI and VII). During 1969 emissions to the atmosphere

**Table VI Emissions from Zinc Sintering and Horizontal Retorts**[a]

| | | *Raw gas* | | |
|---|---|---|---|---|
| | *Waste gas* ($m^3$) | *Dust content* ($gm/m^3$) | *% Particles* <10 *μm* | $SO_2$ *content* (%) |
| Sintering | 4,000[b] | 10 | 100 | 4.5–7 |
| Horizontal retorts | 12,000–18,000[c]<br>96,000–144,000[d] | 0.1–0.3 | 100 | |

[a] Clean Air Guide 2284, Kommission Reinhaltung der Luft, Verein Deutscher Ingenieure, VDI—Verlag GmbH, Dusseldorf, West Germany, September, 1961.
[b] Per ton of ZnS.
[c] Waste flue gas per ton of Zn.
[d] Condenser waste gas (including extraneous air) per ton of Zn.

**Table VII $SO_2$ Emission Rates from Zinc Smelting** *(12)*

| | *Emission rate* | |
|---|---|---|
| | *lb $SO_2$/ ton of metal* | *gm $SO_2$/ kg of metal* |
| Primary zinc smelting | | |
| Roasting | 1650–2400 | 825–1200 |

in the United States resulting from metallurgical processing of zinc-bearing ores and concentrates was estimated to be 50,000 tons (*28*).

Particulates are caught in conventional baghouses or electrostatic precipitators. Sulfur dioxide attains concentrations of 6–7% in roaster gases which, after being freed of particulates, may be passed directly into a sulfuric acid plant or vented from tall stacks. An interesting and very successful scheme for recovery of zinc roaster $SO_2$ as well as of the more dilute $SO_2$ from lead roasting has been developed at Trail, British Columbia, by the Consolidated Mining and Smelting Co. (COMINCO) (Fig. 4).

Sulfur dioxide in the more concentrated gas streams is utilized for direct production of sulfuric acid by the contact process, while more dilute gas streams are scrubbed with an aqueous ammonia solution. Ammonia is fortunately economically available because of an abundance of natural gas and hydroelectric power. An ammonium silfite–bisulfite mixture is produced according to the following reactions:

$$2\ NH_4OH(\text{solution}) + SO_2 = (NH_4)_2SO_3(\text{solution}) + H_2O \quad (2)$$
$$(NH_4)_2SO_3(\text{solution}) + SO_2 + H_2O = 2\ NH_4HSO_3(\text{solution}) \quad (3)$$

In the second step of the COMINCO process the ammonium sulfite–bisulfite solution from the absorbers is acidulated with sulfuric acid to free $SO_2$ as a concentrated gas, which is used for further production of acid and to form a solution of ammonium sulfate as follows:

$$(NH_4)_2SO_3(\text{solution}) + H_2SO_4 = SO_2 + (NH_4)_2SO_4(\text{solution}) + H_2O \quad (4)$$
$$2\ NH_4HSO_3(\text{solution}) + H_2SO_4 = 2\ SO_2 + (NH_4)_2SO_4(\text{solution}) + 2\ H_2O \quad (5)$$

The ammonium sulfate solution is evaporated to form crystalline ammonium sulfate for marketing as a fertilizer (*29*).

Here again, at a specific location and under a specific set of circumstances, $SO_2$ has been and still is being recovered commercially to produce

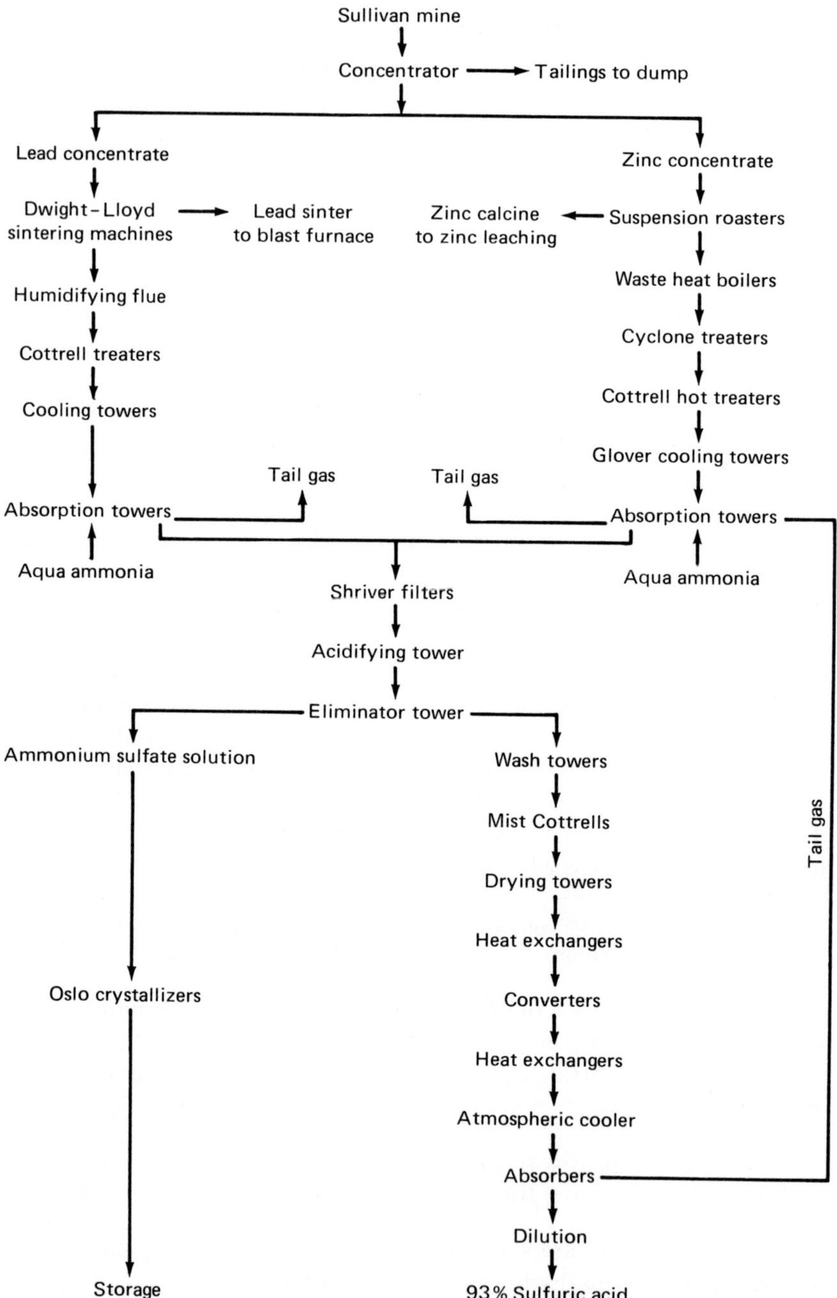


**Figure 4.** Flow sheet of ammonia process at Trail, British Columbia (*29*).

sulfuric acid and ammonium sulfate. However, the market for ammonium sulfate is limited. The salt is highly water soluble and, obviously, because of rainfall cannot be easily stored in unlimited quantities as a waste product. Production must be geared to market needs. In much of the United States ammonium sulfate is not an acceptable fertilizer.

Other variations of ammonia scrubbing have been proposed or have been employed on a limited basis in either pilot or prototype plants. Most such process variations produce either a concentrated $SO_2$ stream or ammonium sulfate and, therefore, like the DMA process and the Wellman–Lord process,* practical applications of ammonia scrubbing techniques are restricted.

In zinc distillation by the retort process small holes are left in vapor condensers to vent gases from charged retorts as they are heated. When the temperature becomes high enough for zinc vapor to distill over, some vapor escapes from the hole and ignites spontaneously. The total zinc oxide fume so produced is appreciable—as much as 8% of the zinc throughput. The ZnO is carried by convection currents up the fronts of the condenser banks and out to the atmosphere with combustion gases through ridge ventilators of the furnace buildings. A distinctive characteristic of a zinc retort plant is a flag of highly visible white ZnO fume. The actual fume concentrations are low and studies have shown the cost of its collection cannot be recovered from the value of the zinc in the collected oxide fume.

Leaching and electrolysis do not emit significant amounts of particulates or gases. Tanks in which leaching and electrolyte purifications are done are covered and ventilated to prevent worker exposure to possible toxic gases or mists. The electrolytic process itself does disperse some electrolyte mist because of gas evolution, but this is a potential problem only within the confines of the tank houses. Stack sampling surveys and continuous ambient monitoring have verified that the outplant emissions from zinc leaching and electrolysis operations are insignificant.

### E. Impact of Regulations on Emission Control Practices

In recent years the United States zinc smelting industry has undergone tremendous change due to imposition of environmental control measures, sharply rising costs, and the lack of an attractive domestic climate for the investment of capital in production facilities. Between 1970 and 1973,

* Absorption of $SO_2$ in sodium sulfite solution with subsequent steam stripping of concentrated $SO_2$, which may be converted to sulfuric acid, and reforming of sodium sulfite for reuse.

7 of the 14 United States zinc plants, constituting more than 40% of domestic capacity, shut down. Only one horizontal retort zinc smelter remains.

## V. Aluminum

### A. Mining and Ore Treatment

Bauxite is the base ore for aluminum production. It is a hydrated oxide of aluminum associated with silicon, titanium, and iron, and contains 30–70% $Al_2O_3$.

Most bauxite ore is purified by the Bayer process. The ore is dried, ground in ball mills, mixed with sodium hydroxide solution, and autoclaved for several hours to dissolve the $Al_2O_3$ as sodium aluminate, $NaAlO_2$. By settling, dilution, and filtration, iron oxide, silica, and other insoluble impurities are removed. Aluminum hydroxide is precipitated from the diluted, cooled solution—the reaction being initiated or "seeded" by introduction of a small amount of freshly precipitated $Al(OH)_3$. The precipitate is filtered, washed, and calcined to produce pure alumina, $Al_2O_3$.

### B. Electrolysis

Commercial recovery of aluminum from the oxide is accomplished by a unique electrolytic process discovered simultaneously in 1886 by Hall in the United States and Heroult in France. Alumina is dissolved in a fused mixture of fluoride salts and dissociated electrically into metallic aluminum and oxygen. Fused natural or synthetic cryolite ($3NaF \cdot AlF_3$) with about 10% fluorspar and 5% dissolved alumina is contained in carbon-lined cells or pots. Heavy carbon anodes are immersed in the mixture to within about 2 in. (5 cm) of the cathode, a heel of molten aluminum covering the carbon lining A heavy electric current between anode and cathode reduces $Al_2O_3$ to Al, which accumulates and is drawn off at intervals into crucibles. Alumina is steadily fed onto the top of the molten electrolyte to replace that which has been decomposed. Crucible aluminum is skimmed of dross, has alloying metals added to it, and is charged into holding furnaces before being cast out as salable ingot.

### C. Emissions and Controls

Calcining of aluminum hydroxides for the production of alumina entails mechanical dust dispension. Approximately 300 to 400 gm of alumina dust is generated per cubic meter of waste gas, and 3000 $m^3$ of waste gas is produced per ton of alumina. The valuable dust is recovered from kiln effluents by a combination of cyclone collectors and electrostatic precipitators.

Iverson has recently reviewed primary aluminum industry emissions and controls as a part of a United States Environmental Protection Agency study of American industry (*30*). His findings are summarized below.

The type and character of emissions depend on the type of electrolytic cell used to electrolyze the alumina feed material. Figure 5 shows the basic configuration and the primary emission controls of the various aluminum reduction cells. The major pollutants for all types are gaseous and particulate fluorides. Hydrogen fluoride accounts for approximately 90% of the gaseous fluoride emitted. The only other important gaseous pollutant is sulfur dioxide resulting from the oxidation of sulfur present in the coke and pitch used in the electrodes. Minor gaseous emissions are carbon dioxide, carbon monoxide, nitrogen dioxide, hydrogen sulfide, carbonyl sulfide (COS), carbon disulfide, sulfur hexafluoride, and gaseous fluorocarbons. Nonfluoride dust emissions are principally mechanically generated carbon dust from the electrodes and mechanically dispersed alumina dust from the charge materials (*31*).

Other smelter operations normally make only slight contributions to overall emissions. However, the anode plant for a prebake electrode operation can be the source of significant sulfur dioxide and hydrocarbon emissions, particularly tarry and distillate hydrocarbons from the pitch in the anodes.

Given the same operating conditions and metal production rates, the vertical and horizontal stud Sodenberg cells generate about the same quantities of total emissions, the total being higher than emissions from prebaked electrode cells. Volatile constituents of pitch binder used in the electrodes are removed during baking, which is done in facilities separate from the pot rooms. In Soderberg cells baking occurs in the reduction process, using the heat of the process to bake or "coke" the carbon electrodes in place.

Prebake and horizontal stud cells are usually completely closed by exhaust hoods in order to capture emissions. Vertical stud cells cannot be completely enclosed because of the physical arrangements of the cell components and because of the working requirements. Emissions capture is not complete in any of the cell types, however. Access to the cells is necessary to add reagents, punch gas holes in the frozen alumina crust, break the crust, and perform maintenance (*32*). For such tasks hooding usually must be partially opened, allowing some emissions to escape into the pot-

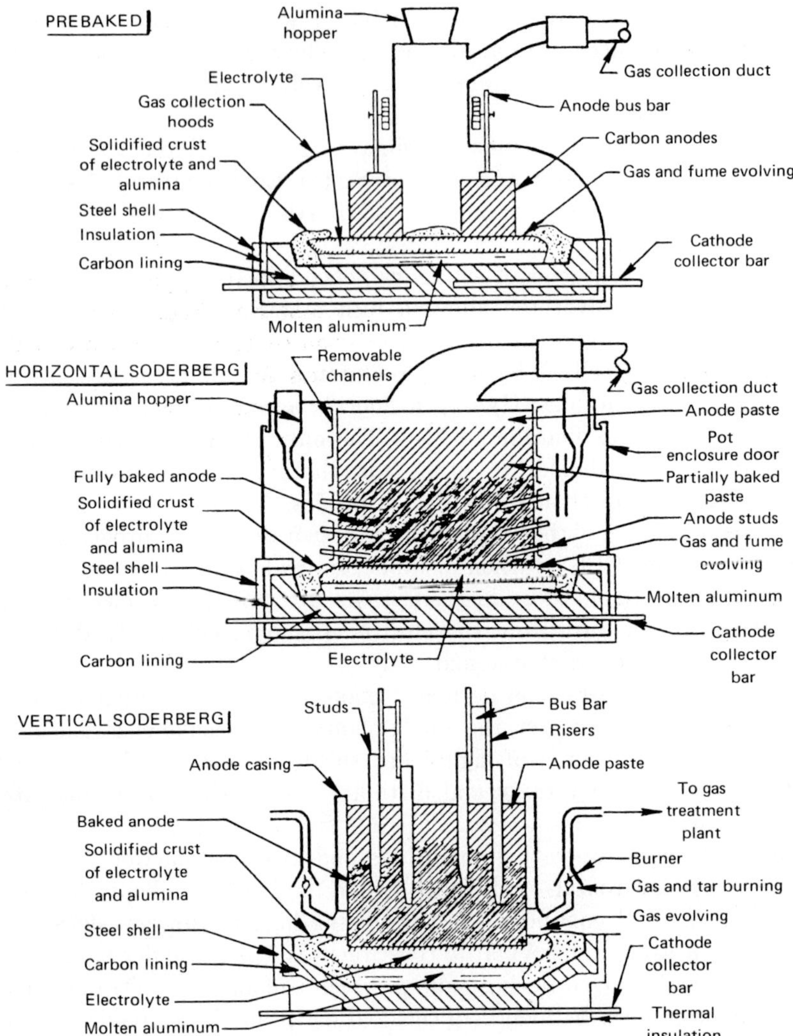


**Figure 5. Details of prebaked and Soderberg aluminum reduction cells (*30*).**

room. The prebake-type cell having crust-breaking equipment between the two rows of anodes is considered to offer the highest possible hooding effectiveness because hoods need not be removed or disturbed in order to break the crust. Further, metal-tapping can be done through the end of the pot with minimal disturbance of hooding.

Dust and gases collected by the primary systems are removed from the exhausted air by a variety of means. Gaseous fluorine and fluoride compounds are commonly recovered by wet contact processes such as floating beds, venturi scrubbers, impingement baffles, and towers with sprays, packed beds, cyclonic scrubbers, and bubblers.

Recently, several dry systems have been developed which use chemisorption of hydrogen fluoride on a dry packed or fluidized bed. The Aluminum Company of America's (ALCOA) 398 process uses a fluidized bed of alumina to capture hydrogen fluoride with greater than 99% efficiency (*33*). The fluidized bed is also efficient in capturing particulate matter with efficiencies ranging from 91 to 98%. A series of filter bags is used to remove any dust entrained in the gases leaving the fluidized bed. Spent alumina from the bed is fed into the reduction pots and the fluoride content used to meet cryolite makeup requirements. Table VIII shows the performance of this process at three ALCOA plants. Table IX shows the cost of using this process on prebaked electrode cell emissions.

A second type of dry collection process uses alumina as a coating on bags in a baghouse collector. Other units inject alumina into the waste gas stream and recapture it in a baghouse collector. These units have costs and efficiencies similar to the ALCOA 398 process.

Some dust collection occurs in all of the systems designed for gaseous fluoride control. Multiclones, venturi scrubbers, and wet and dry electrostatic precipitators are all used.

Primary collection systems on the electrolytic cells are not 100% efficient because not all emissions are captured by the hoods and because some emissions are released during maintenance and operating procedures

**Table VIII Performance of ALCOA 398 Process—Smelting Division—Aluminum Company of America (*33*)**

| | *Total F (mg/scm)*[a] | | *F Efficiencies (%)* | | | *Particulate loss* | |
|---|---|---|---|---|---|---|---|
| | *Inlet* | *Emission* | *Total* | *Gas* | *Solids* | lb/*ton Al* | *gm/kg Al* |
| Plant 1 | | | | | | | |
| Average | 195 | 1.6 | 99.2 | 99.5 | 98.3 | 0.14 | 0.07 |
| Range | 135–215 | 0.4–4.5 | 97.9–99.7 | 97.6–99.9 | 91.8–99.8 | | |
| Plant 2 | | | | | | | |
| Average | 125 | 1.9 | 98.5 | 99.4 | 96.3 | 1.41 | 0.71 |
| Range | 100–180 | 0.1–13.3 | 92.5–99.9 | 98.5–100.0 | 66.7–99.9 | | |
| Plant 3 | | | | | | | |
| Average | 100 | 3.0 | 96.7 | 99.2 | 91.8 | 4.54 | 2.27 |
| Range | 90–200 | 0.2–15.9 | 92.0–99.9 | 97.5–99.9 | 73.6–99.3 | | |

[a] scm: Standard cubic meter.

**Table IX ALCOA 398 Process Costs (Prebake Cells)** *(33)*

| | *Alcoa 398 process (8 installations 1967–1970)* | | *Electrostatic precipitators plus water scrubbers* | |
|---|---|---|---|---|
| | *6 New* | *2 Conversion* | *8 Installations—1950–1956* | *Escalated to 1970* |
| Investment cost, \$/annual ton aluminum (\$/annual 1000 kg aluminum) | | | | |
| Fans | \$ 3.30 (\$ 3.64) | | \$ 3.10 (\$ 3.42) | \$ 5.40 (\$ 5.95) |
| Processing equipment | 23.40 ( 25.80) | | 24.20 ( 26.68) | 42.35 ( 46.69) |
| Alumina handling | 4.00 ( 4.41) | | | |
| Total | \$30.70 (\$33.85) | | \$27.30 (\$30.10) | \$47.75 (\$52.65) |
| Range | \$28–37 (\$31–41) | \$45–\$51 (\$50–56) | \$23–44 (\$25–49) | |
| Operating cost, \$/ton aluminum, (\$/1000 kg aluminum) | | | | |
| Operating Labor | \$ — | | \$ — | |
| Operating Supplies | 0.97 (1.07) | | 0.72 (0.79) | |
| R and M Labor[a] | 1.00 (1.10) | | 0.35 (0.39) | |
| R and M Supplies[a] | 0.45 (0.50) | | 0.34 (0.37) | |
| Power | 1.73 (1.91) | | 0.72 (0.79) | |
| Total operating cost | \$ 4.15 (4.58) | | \$ 2.13 (2.35) | \$ 2.13 (2.35) |
| Range | \$2.90–\$4.70 (3.20–5.18) | | \$2.01–\$2.73 (2.22–3.01) | |
| Amortization (15 year) | \$ 2.05 (2.26) | | \$ 1.80 (1.98) | \$ 3.18 (3.51) |
| F Recovery credit @ \$15/lb $AlF_3$ (\$33.08/kg $AlF_3$) | [\$8.00] [(17.63)] | | — | — |
| Net cost or [credit] | [\$1.80] [(1.98)] | | \$ 3.93 (4.33) | \$ 5.31 (5.85) |

[a] R and M: Repair and maintenance.

when hoods are opened for access to the cells. The quantity of these releases depends upon the amount of effort given to preventing emissions during operations on the cells, and on the type of cells. These secondary emissions are released through the roof ventilators and constitute 5–40% of the total smelter emissions. Secondary collection systems mounted in the roof monitors just below the ceiling fans are used to collect these emissions. Spray screens and packed beds are the most frequently used collection devices. The collection efficiencies of selected primary and secondary systems have been reported by Iverson (Table X).

Aluminum chloride may also be released by aluminum smelters because either it or chlorine gas is used to treat aluminum in holding furnaces to flux and degas the molten metal. Aluminum chloride is volatile, subliming at 180°C. It reacts with hot, moist air to form a highly visible "smoke" cloud. Electrostatic precipitation, scrubbing, and condensation of the volatized salt are effective means of control for the aluminum chloride. Chlorine gas is too reactive with aluminum to be released from this operation.

### *D. Impact of Regulations on Emission Control Practices*

Historically emission regulations for aluminum production facilities have been designed to prevent damage to vegetation and livestock in the surrounding areas. These effects have been largely avoided by dependence

**Table X Fluoride Removal Efficiencies of Selected Primary and Secondary Control Systems** (*30*)

| | *Fluoride removal efficiencies, %* | | |
|---|---|---|---|
| *Control system* | *HF* | *Particulate* | *Total F* |
| Coated filter dry scrubber | 90 | 98 | 94 |
| Fluid bed dry scrubber | 99 | 98 | 99 |
| Injected alumina dry scrubber | 98 | 98 | 98 |
| Wet scrubber + wet ESP[a] | 99 | 99 | 99+ |
| Dry ESP + wet scrubber | 98 | 90–95 | 94–96 |
| Floating bed | 98 | 87 | 95 |
| Spray screen | 93–95 | 45–85 | 62–77 |
| Venturi | 99 | 96 | 98 |
| Bubbler scrubber + wet ESP | 99 | 98 | 99 |

[a] Electrostatic precipitator.

on a combination of emission controls and environmental dispersion of emissions through tall stacks. Regulations were written around acceptable environmental fluoride concentrations which would not lead to damage to vegetation and animals.

In recent times there has been a change in the United States to an emphasis on "best available technology" as a result of the Clean Air Amendment of 1970. Work has been done by the United States Environmental Protection Agency and its consultants for the purpose of defining the best available emission control technology for the primary aluminum industry (Table XI).

**Table XI Best Available Control Technology Emissions for Three Electrolytic Cell Types** *(30)*

| | *Emissions with primary and secondary control* | | | |
|---|---|---|---|---|
| | *Fluorides* | | *Particulates* | |
| | *lb/ton Al* | *gm/kg Al* | *lb/ton Al* | *gm/kg Al* |
| Cell type | | | | |
| Prebaked | 1.60 | 0.8 | 6.0 | 3.0 |
| Vertical stud Soderberg | 2.8 | 1.4 | 9.1 | 4.6 |
| Horizontal stud Soderberg | 3.6 | 1.8 | 12.3 | 6.2 |

The pressure for low fluoride emissions and the associated cost of control will most likely lead to a clear preference for prebaked electrode facilities when new primary aluminum plants are being developed. However, economic and operational factors may force the improvement of controls for Soderberg cells and/or may lead to entirely new primary production processes.

## VI. Beryllium

### *A. Production and Fabrication*

Beryllium is widely distributed in the crust of the earth, constituting approximately 0.001%. However, it seldom exists in a concentrated form economically suitable for mining. The only beryllium-containing ores presently mined for their beryllium content are beryl and bertrandite. The majority of beryl ore processed in the United States is imported, and the only large-scale domestic mine produces bertrandite ore (*34*).

There are three basic processes commercially used for extracting beryllium from ores, all of which extract beryllium in the form of beryllium hydroxide (*34*). The hydroxide is subsequently converted into the desired product—metal, oxide, or alloy.

### B. Emissions and Controls

The emissions generated by beryllium extraction processes include beryllium salts, acids, beryllium oxides, and other beryllium compounds in the form of dust, fume, or mist. Emissions from machine shops and fabrication facilities are usually limited to beryllium dust and small amounts of beryllium oxide dust. Beryllium oxide dust is generated during ceramic production; and beryllium containing fumes and dust are produced by beryllium copper foundry operations.

Emissions generated by beryllium operations are successfully controlled by the following classes of gas cleaning equipment:

(a) Prefilters of the viscous impingement and dry extended-medium types
(b) Dry mechanical collectors
(c) Wet collectors
(d) Fabric filters (baghouses)
(e) High efficiency particulate air filters (HEPA filters)

Dry cyclones and baghouses in series are commonly used to control beryllium emissions from ore handling operations such as crushing and milling. Wet chemical extraction processes normally employ wet collectors, such as venturi and packed-tower scrubbers. Beryllium foundries and machine shops utilize dry cyclones and, more commonly, baghouses. Limited small-scale operations sometimes employ HEPA filters. Beryllium ceramic plants and propellant plants usually operate series arrangements of prefilters and HEPA filters.

### C. Impact of Regulations on Emission Control Practices

In 1973 the United States Environmental Protection Agency classified beryllium as a "hazardous air pollutant." As such, beryllium emissions to the atmosphere are limited to a maximum of 10 gm over a 24-hour period, or as an alternative, ambient concentration in the vicinity of the source shall not exceed 0.01 $\mu g/m^3$, averaged over a 30-day period.

Because of beryllium's extreme toxicity and the well-known "neighborhood cases," beryllium operations, for the most part, have been closely scrutinized and well controlled. Therefore, the above regulation has not caused a significant impact on beryllium control operations. The greatest impact has been the added emphasis on stack testing and ambient monitoring.

## VII. Mercury

### A. Mining

Mercury is produced commercially by processing mercury sulfide ore, cinnabar. Some native mercury is always present in shales that occur with cinnabar. This native mercury may be a health hazard to the miners working in the confined spaces of a mine if the mine is not well-ventilated. However, only a slight elevation in the ambient concentration of mercury vapor has been noted in the vicinity of mining operations. Historically, ambient levels of mercury vapor have been used to detect zinc ore bodies that lie close to the surface (*35*). There is a natural background of mercury present in many areas due to the off-gassing of the earth's crust (*36*).

### B. Production

Mercury sulfide is thermally decomposed by a variety of means such as a retort, or a Wedge or Herreshoff roaster, to produce elemental mercury and sulfur dioxide. The volatilized mercury is captured in condensers after the gases are cleaned of entrained dust by cyclone separators. Retorts are arranged in banks connected to a series of air-cooled condensers. The operating cycle consists of charging, raking, and discharging the retorts. A similar operating cycle is used on a continuous basis in other production facilities. An undesirable condensation product is "quicksilver flour," sometimes formed when finely divided mercury droplets fail to coalesce. Two products are recovered from the condensers—fluid mercury and soot. The soot is composed of finely divided mercury mixed with mercurial salts, ore dust, ash from fuel, tar, etc.

### C. Emissions and Controls

The majority of ore processing facilities in the United States are located in California near a major cinnabar deposit which has been mined for decades.

There are three major sources of mercury vapor emissions from a mercury ore processing facility: (a) mercury vapor not captured by the condensers; (b) leaks from broken retorts or poorly sealed retorts, roasters, or condensers; and (c) poor housekeeping practices. Data on the quality of these emissions are very sparse.

Mercury capture by the condensers is at least 95%. Stack losses of mercury from ore processing facilities range approximately from 2 to 3% of total production capacity, but higher losses have been recorded (*37*).

A major concern with mercury emissions lies with emissions from other sources where mercury is a trace contaminant, e.g., coal burning electric power plants, and chlor–alkali plants. These sources are dealt with elsewhere in this book. Other primary smelting operations may also emit mercury as a trace contaminant.

Mercury vapor is very hard to collect on a large scale. One of the most effective and practical means is condensation of the vapor in a refrigerated unit followed by filtration or electrostatic precipitation. Other methods use scrubbing with refrigerated chlorinated brine or basic solutions of sodium hypochlorite and sodium chloride. Two methods are commercially available which use solid adsorbents; one uses iodine-impregnated activated charcoal, and the other uses a proprietary adsorbent developed by Monsanto Environ-Chem (*38*). Where minor mercury emissions occur from other nonferrous smelting activities such as zinc smelting, it has been found that the sulfuric acid production facilities used to process the waste gases effectively capture mercury vapor (*39*).

### D. Impact of Regulations on Emission Control Practices

It is likely that United States regulations will have considerable direct and indirect impact on mercury smelting activities. Strict regulations have been promulgated by the United States Environmental Protection Agency for all operations emitting mercury. No more than 2300 gm of mercury may be emitted per 24 hours. These regulations require very stringent controls, and will doubtlessly force the closure of less profitable operations. It will also have the effect of returning some mercury to the market place that is now lost; or, alternatively, will cause the generation of difficult-to-dispose-of wastes, because it is no longer acceptable to trade air pollution for water pollution, via scrubber wastes, or solid waste problems, via spent adsorbents. A likely effect is a substantial reduction in mercury smelting as more mercury is recycled and less mercury is used as other substances or processes are substituted for those requiring mercury.

## VIII. Nonferrous Metals of Minor Significance

### *A. Arsenic*

Although arsenic occurs naturally in some minerals it is not commercially mined. It is produced only as a by-product of various primary smelting operations. Copper and lead ores from certain ore deposits are high in arsenic content, which generally must be removed before they can be processed for copper and lead recovery. Dust recovered from emission control devices on copper and lead smelters, is in many cases, high in arsenic content. Materials from these two sources may have sufficient arsenic content, 6–45%, to make processing economical.

Arsenic is produced commercially by heating arsenic-containing fumes and dust in Godfrey roasters or in reverberatory furnaces to volatilize arsenic trioxide. The roaster or reverberatory gases are gradually cooled in large brick condensers, called kitchens, where the arsenic trioxide condenses into a fine powder or dust of varying qualities. The least pure arsenic trioxide condenses from the gas stream in the first part of the kitchen and the highest grade product is deposited in the last part of the kitchen. High-quality dust is screened to 6 mesh; particles larger than this are reduced to size in a hammermill, and stored or packaged for shipment. Low-grade dust is reprocessed.

The majority of arsenic is used commercially in herbicides, rodenticides, insecticides, and in paints as a fungicide. It is also used in glass manufacture as a bronzing or decolorizing agent, and in the production of opal glass and enamels. Small amounts of arsenic are added to alloys, 0.3–0.5%, to increase hardening and heat resistance. Arsenic is also used to manufacture organic arsenicals for therapeutic use. There is a small but increasing demand for high-purity arsenic for use in semiconductors.

Emissions from arsenic production facilities are primarily uncondensed and unsettled arsenic trioxide in the gases leaving the kitchens. These releases are controlled by passing gases through electrostatic precipitators and/or baghouses. Recovered dusts are reprocessed. Because arsenic trioxide sublimes at a relatively low temperature, 600°F (315°C), the Godfrey roasters produce little else as potential air pollutants. Only a small percentage of arsenic trioxide remains in suspension after passing through the kitchens because the conditions during condensation are such that there is little formation of particles less than 1 $\mu$m. Hence, collection efficiencies are generally very good, 95–99%.

It is anticipated that there will be little direct impact of present pollution control regulations on arsenic production because it is considered that "best available" control technology is being used. The low dollar value of arsenic, its presence in certain copper and lead ores, and pollution control requirements have the effect of restricting the sale of those ores to specialized smelters. It should be noted that much of the high arsenic copper concentrates come from mines outside the United States.

### *B. Cadmium*

As with arsenic, most cadmium production is the result of processing primary smelter by-products which are rich in cadmium, particularly those resulting from zinc smelting. Cadmium is a minor component of most zinc ores. A small amount of cadmium is produced by processing Greenockite, the cadmium sulfide mineral.

Cadmium is largely volatilized during the roasting of zinc or lead concentrates and is collected as an oxide dust. Recycling of the dust through roasters will increase the cadmium content. High CdO dust is then treated with sulfuric acid to make acid-soluble cadmium sulfate. Zinc dust may be added to the purified solution and cadmium precipitated as an impure "sponge" which can be further refined by distillation. High-purity cadmium metal is produced from acid cadmium sulfate solution by electrolysis in a tank-house plating operation.

Most of the cadmium metal produced in the United States is sold for use in electroplating and alloying applications. A portion of the metal is converted to cadmium sulfide and cadmium oxide for use as paint pigments.

Baghouses are considered to be the emission control devices of choice and the "best available technology" for preventing cadmium releases to the atmosphere. Because part of the production process involves volatilization of cadmium as cadmium oxide fume, the amount of cadmium lost through emissions is kept as low as possible. The cadmium is, after all, the product, and maximum control efficiency is maintained to minimize these losses. Future air pollution regulations are unlikely to change present practices significantly.

## IX. Secondary Copper, Lead, Zinc, and Aluminum

### *A. Sources*

Scrap provides an important source of metals for the market. Copper in substantial tonnage is recovered from electrical cable and automobile radiators. About 85% of the lead used in automobile batteries is collected by scrap dealers and eventually sold to secondary lead smelters. Secondary zinc comes from galvanizing baths and die-casting scrap. Secondary aluminum is recovered mostly from industry-generated scrap, but aircraft assemblies and even pots, pans, and beverage cans may be used. Economics governs the flow of material in the secondary market. Secondary operations are receiving added emphasis because of the trend to push recycling.

### B. Recovery Processes

Scrap may be converted directly to usable metal by simple melting and casting into salable ingots. More often, scraps of similar composition are melted together and the composition of the molten metal is adjusted by removal or addition of some constituent elements. When the right proportions of each are present, the charge is cast. Brass, type metal, babbitts, and solders are produced in this way.

Insulation is stripped or burned from electrical cable and the copper wire is melted and cast into anodes for direct conversion by electrolysis into refined copper.

Lead battery plates are charged into reverberatory furnaces with coke and fluxes for smelting. The product is an antimonial lead bullion which can be refined to pure lead or adjusted in composition for sale as an alloy.

Zinc scrap is commonly refined by distillation from retorts, the vapor being condensed to a liquid and cast into slabs, or condensed directly from the vapor state to a zinc powder of controlled particle size.

Aluminum alloys are melted in reverberatory furnaces and adjusted in composition by drossing, chlorination, and addition of alloying metals.

### C. Emissions and Controls

Emissions from secondary metal processes are similar to those from primary metallurgical operations except that little or no sulfur dioxide is evolved, and, in general, smaller quantities of metal oxide dusts and fumes are produced. There are no sulfides in secondary metal furnace charges as there are in primary smelting furnaces, and only small amounts of sulfur are used to kettle-refine lead-base alloys.

Lead oxide is volatilized from secondary lead smelting. Zinc oxide fume is a by-product of zinc-alloy distillation and of brass furnace operation. Baghouses, more commonly, and electrostatic precipitators are successfully used for collection.

Visible emissions from small kettles and pots have been a problem at some secondary metal plants. This problem has been solved usually by hooding the kettles and pots and venting them through the baghouses used for dust and fume control on the reverberatory, sweating, and cupola furnaces.

Chlorination of molten aluminum in secondary refining furnaces produces aluminum chloride fume which creates a high-opacity white cloud when it comes in contact with moist air. Properly designed scrubbers or condensers are effective control devices. The Derham process, ALCOA "Fumeless Demagging Unit," and other emission abatement devices have been used successfully (*40–42*).

### *D. Impact of Regulations on Emission Control Practices*

Recent United States air pollution regulations have caused a significant impact on secondary nonferrous operations—particularly the smaller ones. The impact on the larger manufacturers has been minimal because most plants of this nature were already effectively and successfully controlling emissions with baghouses and/or electrostatic precipitators. In some instances it has been necessary to hood sources of fugitive emissions and vent them through baghouses in order to comply with opacity regulations. Small plants, usually employing less than 20 people, have closed, partially because of more stringent opacity and particulate emission standards.

## X. Nonferrous Foundries

### *A. Alloys and Operations*

#### 1. Copper-Based Alloys

There are two groups of these alloys; the brasses containing 60–65% copper and a major percentage of zinc, and the bronzes which usually contain 85–90% copper and a major percentage of tin. Some of these alloys may also contain small amounts of lead tin (in the case of brasses), zinc (in the case of bronzes), manganese, silicon, or phosphorus where special properties are desired.

There are four basic types of furnaces used to melt and prepare these alloys; pit, tilting crucible induction, and cylindrical rotary or stationary reverberatory furnaces. In pit furnaces, a silicon carbide crucible is indirectly heated in a ceramic-lined pit using an operating cycle of melting the charge and flux, removing the crucible from the pit, and pouring the metal into molds. Tilting crucible furnaces have an operating cycle similar to pit furnaces but differ in that they are directly heated, have a larger capacity, and are floor mounted. Electric induction furnaces are the most rapid means of melting metal and are capable of the best temperature control on the melt. In these furnaces granulated charcoal is used as a flux instead of glass or borax because it does not attack the furnace walls. Cylindrical rotary and stationary reverberatory furnaces are the largest alloy production units. In these furnaces the flame impinges directly upon the surface of the metal. The cylindrical rotary furnace rotates on its horizontal axis and tilts for pouring and charging.

A foundry will also have facilities for core making, mechanical sand mullers, sand handling, grinding, buffing and polishing, and plating. Which of these operations are present depends upon the foundry's size and end products.

### 2. Zinc-Based Alloys

Zinc die-casting metal accounts for most of the zinc used in alloys, excluding brass. The most widely used die-casting metal alloys contain high-purity zinc (99.99% pure), mixed with up to 4% aluminum and small amounts of magnesium (0.04%). Another proprietary die-casting metal called Mazak uses a portion of copper. Small amounts of other metals may also be present.

Pit and tilting crucible, pot, or induction furnaces are all used, usually without fluxes, to melt and prepare zinc alloys. These are charged with ingots and rejects, before the addition of scrap and thin-walled materials, in order to build up a bath and avoid burning. Dross resulting from dirty charge materials and metal oxidation must be skimmed off the molten metal surface before pouring. These alloys are poured or injected into the molds at temperatures of 775°–825°F (412°–441°C).

## *B. Emissions and Controls*

Metal fume will be evolved during the melting and casting of alloys, if the temperatures are high enough for volatile constituent elements to have appreciable vapor pressure, and if no intermetallic compounds are formed.

Zinc is the major source of fume in brass and bronze foundry operations. Pure zinc boils at 1663°F (906°C) and the usual pouring temperature of high zinc copper alloys is 1900°–2100°F (1038°–1148°C). In general, copper alloys with more than 5% zinc are likely to fume. For a given percentage of zinc, an increase of 100°F (37.7°C) will increase the fuming rate by a factor of 3. Fuming does not occur for die-casting metals because they are not heated to sufficiently high temperatures. Under some conditions high lead alloys will produce lead fumes.

Another source of emissions during charging and melting is smoke production from the heating of the grease, oil and mold release compounds used in casting, dirty scrap, and miscellaneous scrap materials.

The type and degree of emission controls required depends upon the furnace and the alloys. Crucible and pot furnaces control fuming during melting by maintaining a layer of molten flux on top of the metal. Crucible heating and pouring stations must be hooded if excessive fuming occurs. Cylindrical rotary and reverberatory furnaces must invariably be hooded because the flame impinges directly on the metal so that zinc and lead fumes will be produced if these metals are present. Electric induction furnaces have the lowest emissions as a rule because they allow the closest temperature control and do not have large volumes of hot combustion gases to aid metal vaporization and oxidation. These emission characteristics can be seen in the data shown in Table XII.

Operating procedures can be as important as the type of furnace in determining emissions. Improper combustion, overheating the charge, addition of zinc at the maximum furnace temperature, heating the metal too rapidly, and insufficient flux cover will all significantly affect the emissions.

The design of control systems for these collection problems is difficult because it normally involves high temperature gases from a source that must be accessible to workmen, and, in the case of ladles and crucibles, may be moveable. Once the emissions are collected, baghouses and electrostatic precipitators are the methods of choice to remove dust and fumes from the gases (*43, 44*).

Shakeout of castings from sand molds is a dusty operation and may be conducted under hoods or over downdraft gratings. Dust loadings in ventilating air range from 0.25 to 1.0 grains/ft$^3$ (0.57–2.23 gm/m$^3$). Collection of up to 97% of such material is effected with wet or dry centrifugal collectors: up to 99 + % with dry fabric-type collectors (*45*).

### C. Impact of Regulations on Emission Control Practices

Emission control technology is well-developed for nonferrous foundries. It is unlikely that there will be any future improvements which would lead to changes in basic foundry practices. It is possible that small opera-

**Table XII Dust and Fume Discharge from Brass Furnaces (*43*)[a]**

| Type of furnace | Composition of alloy, % | | | | | Type of control | Fuel | Pouring temp. °F | Process wt | | Fume emission | |
|---|---|---|---|---|---|---|---|---|---|---|---|---|
| | *Cu* | *Zn* | *Pb* | *Sn* | *Other* | | | | *lb/hr* | *kg/hr* | *lb/hr* | *kg/hr* |
| Rotary | 85 | 5 | 5 | 5 | — | None | Oil | No data | 1104 | 501 | 22.5 | 10.2 |
| Rotary | 76 | 14.7 | 4.7 | 3.4 | 0.67 Fe | None | Oil | No data | 3607 | 1636 | 25 | 11.3 |
| Rotary | 85 | 5 | 5 | 5 | — | Slag cover | Oil | No data | 1165 | 528 | 2.73 | 1.24 |
| Elec. ind.[b] | 60 | 38 | 2 | — | — | None | Elect. | No data | 1530 | 694 | 3.47 | 1.57 |
| Elec. ind.[b] | 71 | 28 | — | 1 | — | None | Elect. | No data | 1600 | 726 | 0.77 | 0.35 |
| Elec. ind.[b] | 71 | 28 | — | 1 | — | None | Elect. | No data | 1500 | 680 | 0.64 | 0.24 |
| Cyl. reverb.[c] | 87 | 4 | 0 | 8.4 | 0.6 | None | Oil | No data | 273 | 124 | 2.42 | 1.10 |
| Cyl. reverb.[c] | 77 | — | 18 | 5 | — | None | Oil | 2100 | 1267 | 575 | 26.1 | 11.84 |
| Cyl. reverb.[c] | 80 | — | 13 | 7 | — | Slag cover | Oil | 2100 | 1500 | 680 | 22.2 | 10.07 |
| Cyl. reverb.[c] | 80 | 2 | 10 | 8 | — | None | Oil | 1900–2100 | 1250 | 567 | 10.9 | 4.94 |
| Crucible | 65 | 35 | — | — | — | None | Gas | 2100 | 470 | 213 | 8.67 | 3.93 |
| Crucible | 60 | 37 | 1.5 | 0.5 | 1 | None | Gas | 1800 | 108 | 49 | 0.05 | 0.02 |
| Crucible | 77 | 12 | 6 | 3 | 2 | Slag cover | Gas | No data | 500 | 227 | 0.822 | 0.37 |

[a] Source: Los Angeles County Air Pollution Control District.
[b] Elec. ind.: Electric induction.
[c] Cyl. reverb.: Cylindrical reverberatory.

tions, which cannot comply with existing regulations and a rigorous enforcement program, may have to close.

## ACKNOWLEDGMENT

The authors gratefully acknowledge the advice and assistance of J. M. Henderson, Superintendent—Chemical and Engineering, Central Research Laboratories, American Smelting and Refining Company.

## GENERAL REFERENCE

W. H. Davis, "Metallurgy of the Non-Ferrous Metals." Pitman, New York, New York, 1961.

## REFERENCES

1. J. K. Haywood, *U.S., Dept. Agr., Bull.* **89** (1905); *U.S., Bur. Mines, Bull.* **537** (1954) (abstr. no. 1231).
2. J. A. Holmes. C. C. Franklin, and R. A. Gould, *U.S., Bur. Mines, Bull.* **98** (1915).
3. P. J. O'Gara, *Met. Chem. Eng.* **17** (12) (1917).
4. G. R. Hill and M. D. Thomas, *Plant Physiol.* **8,** 223 (1933).
5. M. D. Thomas, J. O. Ivie, J. N. Abersold, and R. H. Hendricks, *Ind. Eng. Chem., Anal. Ed.* **15,** 28 (1943).
6. M. D. Thomas, R. H. Hendricks, and G. R. Hill, *Proc. Nat. Air Pollut. Symp., 1st, 1949* p. 142 (1950).
7. C. R. Davis, "Pay Dirt," No. 326, p. 5. C. F. Willis, Phoenix, Arizona, 1966.
8. J. N. Agate *et al., Med. Res. Counc. Memo.* **22** (1949); quoted by P. Drinker, *J. Roy. Inst. Pub. Health* **20,** 307 (1957).
9. W. E. Davis, "National Inventory of Sources and Emissions," Sect. III, COPPER, PB-210-678. U.S. Environmental Protection Agency, Research Triangle Park, North Carolina, 1972. (Distributed by National Technical Information Service, U.S. Dept. of Commerce, Springfield, Virginia.)
10. M. I. Weisburd, "Field Operations and Enforcement Manual for Air Pollution Control," Vol. 3, PB-213-010. Pacific Environmental Services, Inc. U.S. Environmental Protection Agency, Research Triangle Park, North Carolina, 1972. (Distributed by National Technical Information Service, U.S. Dept. of Commerce, Springfield, Virginia.)
11. C. R. Hayward, "An Outline of Metallurgical Practice," 3rd ed. Van Nostrand-Reinhold, Princeton, New Jersey, 1952.
12. Personal communication with J. M. Henderson, American Smelting and Refining Company, South Plainfield, New Jersey.
13. "Background Information—Proposed New Source Performance Standards for Pri-

mary Copper, Zinc, and Lead Smelters." U.S. Environmental Protection Agency, Research Triangle Park, North Carolina, 1973.
14. G. R. Hill, M. D. Thomas, and J. M. Abersold. *Proc. 9th Annu. Meet. Ind. Hyg. Found, 1944* p. 11. Industrial Hygiene Foundation, Pittsburgh, Pennsylvania, 1944.
15. M. E. Smith, *Combustion,* pp. **23–27** (1967); *Proc. Nat. Conf. Air Pollu., 3rd, 1966,* Publ. No. 1669, p. 151. Pub. Health Serv., U.S. Dept. of Health, Education, and Welfare, Washington, D.C., 1967.
16. F. F. Ross, A. J. Clarke, and D. H. Lucas, *Proc. Int. Clean Air Congr., 2nd, 1970* p. 1041. (1971).
17. T. T. Frankenberg, *J. Amer. Ind. Hyg. Ass.* **29,** 181–185 (1968).
18. F. Pooler, Jr. and L. E. Neimeyer, *Proc. Int. Clean Air Congr., 2nd, 1970* pp. 1049–1056 (1971).
19. W. L. Lee and A. C. Stern, *Air Pollut. Contr. Ass., J.* **23,** No. 6, 503–513 (1973).
20. R. B. Engdahl, *Air Pollut. Contr. Ass., J.* **23,** No. 5, 364–375 (1973).
21. K. W. Nelson, M. A. Yeager, and C. K. Guptill, "Closed-Loop Control System for $SO_2$ Emissions from Non-Ferrous Smelters," ECE Seminar on the Control of Emissions from the Non-Ferrous Metallurgical Industries, Dubrovnik, Yugoslavia, Nov. 19–24, 1973. Economic Commission for Europe, Geneva, Switzerland.
22. K. W. Nelson, "Statement Regarding Alternatives to an Emission Standard," Att. 18. Letter from J. F. Boland, attorney for American Smelting and Refining Company to Cassandra Jencks, Legal Counsel, EPA Region IX, October 16, 1972.
23. I. E. Campbell, "Abatement of Pollutants from Primary Smelting of Copper, Lead, Zinc, Nickel, and Other Metals," ECE Seminar on the Control of Emissions from the Non-Ferrous Metallurgical Industries, Dubrovnik, Yugoslavia, Nov. 19–24, 1973. Economic Commission for Europe, Geneva, Switzerland.
24. W. D. Ruckelshaus, "The Economics of Clean Air," Annual Report of the Administrator of the United States Environmental Protection Administration to the Congress of the United States, 92nd Congress, Senate Document 67 (1972).
25. J. M. Henderson, "Reduction of $SO_2$ to Sulfur," ECE Seminar on the Control of Emissions from the Non-Ferrous Metallurgical Industries, Dubrovnik, Yugoslavia, Nov. 19–24, 1973. Economic Commission for Europe, Geneva, Switzerland.
26. E. P. Fleming and T. C. Fitt, *Ind. Eng. Chem.* **42,** 2249 (1960).
27. P. J. Callahan and T. D. Parker, *Chem. Eng.* **74,** 159 (1967).
28. W. E. Davis, National Inventory of Sources and Emissions, Sect. V, ZINC, PB-210-680. U.S. Environmental Protection Agency, Research Triangle Park, North Carolina, 1972. (Distributed by National Technical Information Service, United States Department of Commerce, Springfield, Virginia.)
29. B. Johnson, "The Removal of Sulphur Gases from Smelter Fumes." Ontario Research Foundation, Toronto, Ontario, 1949.
30. R. E. Iverson, *J. Metals* **25,** 19–23 (1973).
31. "Air Pollution by Fluorine Compounds from Primary Aluminum Smelting." Organization for Economic Cooperation and Development, Paris, France, 1973.
32. I. Nestaas, "Abatement of Pollutants from Smelting of Primary Aluminum—Introductory Report on Topic A," ECE Seminar on the Control of Emissions from the Non-Ferrous Metallurgical Industries, Dubrovnik, Yugoslavia, Nov. 19–24, 1973. Economic Commission for Europe, Geneva, Switzerland.
33. C. C. Cook, G. R. Swany, and J. W. Colpitts, *Air. Pollut. Contr. Ass., J.* **21** (8), 479–483 (1971).

34. "Control Techniques for Beryllium Air Pollutants," EPA Publ. No. AP-116. U.S. Environmental Protection Agency, Research Triangle Park, North Carolina, 1973.
35. Dr. L. James, American Smelting and Refining Company, Department of Exploration Services, Salt Lake City, Utah (personal communication).
36. H. V. Weiss, M. Koide, and E. D. Goldberg, *Science* **174,** 629–694 (1971).
37. C. L. Nobbs, "Mercury Use and Social Choice." Organization for Economic Cooperation and Development, Paris, France, 1972.
38. *Chem. & Eng. News* **50** (7), 14–15 (1972).
39. J. Kangas, E. Nyholm, and J. Rastas, *Chem. Eng. News* **20,** 55–57 (1972).
40. C. C. Andrews, "Supplementary Note on Abatement of Pollutants from Secondary Metal Recovery—Emission Control in the Secondary Aluminum Industry," ECE Seminar on the Control of Emissions from the Non-Ferrous Metallurgical Industries, Dubrovnik, Yugoslavia, Nov. 19–24, 1973. Economic Commission for Europe, Geneva, Switzerland.
41. "Alcoa Fumeless Demagging Unit." International Alloys, Ltd., Aylesbury, Bucks, London, England. (ECE Seminar on the Control of Emissions from the Non-Ferrous Metallurgical Industries, Dubrovnik, Yugoslavia, Nov. 19–24, 1973. Economic Commission for Europe, Geneva, Switzerland.)
42. W. R. King, "Abatement of Pollutants from Secondary Metal Recovery—Emission Control in the Secondary Aluminum Industry," ECE Seminar on the Control of Emissions from the Non-Ferrous Metallurgical Industries, Dubrovnik, Yugoslavia, Nov. 19–24, 1973. Economic Commission for Europe, Geneva, Switzerland.
43. "Air Pollution Engineering Manual," 2nd ed., Publ. No. AP-40. U.S. Environmental Protection Agency, Research Triangle Park, North Carolina, 1973.
44. K. D. Green, *in* "Gas Purification Processes" (G. Nonhebel, ed.), Chapter **13,** pp. 559–572. Butterworth, London, England, 1964.
45. J. M. Kane, *Amer. Foundryman* **19,** 34 (1951).

# 21

## Ferrous Metallurgical Operations

Bruce A. Steiner

## I. Introduction

Iron and steel industry production facilities can be grouped generally into steel mills and ferrous foundries. Steel mills are concerned with the production of steel shapes for subsequent use in other manufacturing industries, whereas ferrous foundries produce iron and steel castings. Worldwide, the greatest percentage of steel manufacture is undertaken at large integrated steel mills which encompass the facilities necessary to convert the basic raw materials of steelmaking—iron ore, coke, and limestone—into finished steel shapes. Nonintegrated steel mills, sometimes referred to as mini-plants, lack the necessary ironmaking capability, but are capable of producing steel shapes from scrap steel or other direct sources of iron and steel.

A completely integrated steel mill includes facilities for cokemaking, sinter production, ironmaking, steelmaking, rolling, processing, and finishing. Not all integrated mills have facilities to produce coke or sinter, but these facilities are normally identified with the steel industry. Some steel mills include facilities such as utility boilers or lime kilns which are not normally identified with the industry. These are discussed elsewhere (Chapters 12, 16, and 17, this volume.)

## II. Coke Production

### *A. General Process Description*

Carbon has been found to be the reducing agent best suited for the reduction of iron ore to metallic iron. Although early blast furnaces utilized wood charcoal for this purpose, coke is the material used in modern blast furnaces. Coke is made by heating blended coals to a temperature of 900°–1100°C over a period of 10–20 hours to drive off the volatile matter, while retaining certain physical and chemical properties. The manufacture of metallurgical coke is accomplished by the nonrecovery beehive process or the by-product process (*1*), the latter accounting for the vast majority of world coke production.

In the beehive process, the coke oven consists of a dome-shaped refractory lined structure resembling a beehive. Other nonrecovery type ovens have also been employed (*2*). Coal is charged through a hole in the top of a hot oven; the charge is leveled; and then air is introduced into the oven in controlled amounts. This air burns in the oven and the volatile matter is distilled from the coal. The heat of its combustion heats the oven and its contents thereby completing the coking process. When essentially all the volatile matter has been driven from the coal, the material remaining in the oven, which is now coke, is quenched with water and removed from the oven. Beehive ovens are normally arranged in banks or rows and are located near a mine supplying coal suitable for this process. All the products of combustion of the volatile matter and the unburned volatile matter produced during coking are normally discharged directly to the atmosphere.

The by-product coking process is conducted in a series of slot ovens alternated with heating chambers or flues. Each oven may be up to 18 m long, 6.5 m high, and 50 cm wide. A battery of ovens may contain as many as 100 chambers. The by-product process differs from the beehive process in that air is excluded from the coking chamber, the heat necessary for distillation being supplied from external combustion of coke oven gas. The flues between the ovens are heated with hot combustion products. The major advantage of the by-product process is recovery of certain constituents in the volatile matter driven from the coal, the value of which partially offsets the cost of producing coke.

In the conventional coke plant (Fig. 1), after coal is received, blended, and weighed, a hopper-equipped vehicle operating on top of the ovens is loaded with a charge of coal. This vehicle, called a larry car, moves to the oven to be charged, and discharges the coal from these coal hoppers through three to five charging ports in the top of the oven. The charging port lids are closed, the coal is leveled by the leveling bar, and the destructive distillation process is initiated by heat conducted into the coal from the gas being burned in the two adjacent heating chambers. At the completion of the coking cycle, doors on each end of the oven are opened, and the coke is pushed from the oven by a pusher ram through the coke

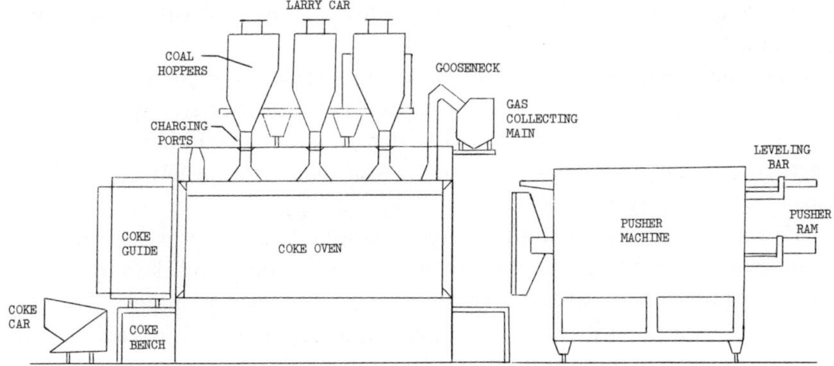


Figure 1. Section of coke oven and associated equipment.

guide into the coke car. The car is then transferred to a quenching station where water is sprayed onto the hot coke to cool it for further transport, use, or preparation. Gases evolved during the coking process are withdrawn through the gooseneck and the gas collecting main to a by-product recovery system for extraction of tars, light oil, naphthalene, and other compounds. After such extraction, the cleaned coke oven gas is used as fuel for heating the ovens and for other fuel requirements in the plant (*3*).

### B. Emissions and Their Control

The generation of air pollutants in the cokemaking process is associated with (a) coal and coke handling, (b) coke oven charging, (c) coke oven discharging, (d) coke quenching, (e) leaking oven doors, and (f) by-product processing. Coal and coke handling problems are similar to those found in other bulk material handling industries (Chapter 17, this volume). By-product processing employs numerous processes similar to those in petroleum refineries (Chapter 19, this volume) and chemical plants (Chapter 18, this volume).

#### 1. Coke Oven Charging

As coal is charged into the hot oven, the gases in the oven are displaced, and the coal immediately begins to volatilize. Pollutants are emitted through the charging ports at this time. These pollutants consist of vaporized hydrocarbons, graphite, tars, char, and coal dust. Quantities of emissions reported show a wide range, with European coke plants reporting particulate matter emissions of from 0.04 to 1.25 kg/ton of coal charged and hydrocarbon emissions of from 0.06 to 0.80 kg/ton of coal charged (*4*). North American ovens are reported to have higher emissions due to higher volatile coals, hotter ovens, lower coal moisture, and higher charging rates.

Charging emissions represent 60% of overall coke plant particulate emissions. Efforts to control them can be grouped into five general categories: (a) aspiration systems, (b) larry-mounted scrubbers, (c) fixed-duct secondary collectors, (d) sequential charging practices, and (e) closed charging systems (*5*).

Aspiration systems include a range of equipment and practices intended to increase the draft on the coke oven during charging and thereby extract the charging fumes into the coke oven gas collecting mains. This can be accomplished with steam or liquor spray ejection using various configurations and pressures. The effectiveness of this technique varies from facility to facility depending upon the configuration of the gas mains and their connecting legs (goosenecks), the presence of other control devices, and the dimensions of the oven.

Larry-mounted scrubbers consist of complete self-contained gas cleaning systems which are mounted on the traveling larry car (Fig. 2). The charging ports are hooded, and the gases are ignited and/or scrubbed before being exhausted from the stacks on the larry car. While difficult to reliably operate and maintain, such systems are widely used, especially in conjunction with aspiration systems (*6*). Preferred design includes provision of auxiliary ignition, independent scrubber systems for each charging port, and the use of high energy venturi scrubbers.

Fixed-duct secondary collectors, found principally in Japan, are used in conjunction with both aspiration systems and larry-mounted scrubbers, and have developed from the shortcomings of performance of the latter. Such systems consist of a stationary duct and scrubber system into which the discharge of a larry-mounted scrubber is connected through multiple doors corresponding to larry car positions. This technology is relatively undeveloped and is not needed if the larry-mounted scrubber system operates as intended.

Sequential or staged charging systems encompass devices or operating practices which control the charging operation in such a manner as to render aspiration alone adequate as a primary control (*7*). In this practice, emissions are prevented by limiting the number of oven openings open at any one time and assuring adequate in-draft at those openings. To accomplish this, it is necessary to control individual charging port lids and coal hoppers and to provide adequate aspiration and simultaneous drafting from both ends of the oven. Because this control technique is a practice rather than a system of equipment, variations are numerous and the technique must be tailored to each plant. For example, simultaneous drafting from both ends of the oven may be accomplished with a separate collecting main at plants employing double mains, whereas, a separate duct, or "jumper pipe," connected to a nearby oven is required for drafting at plants having a single collecting main. In general, how-

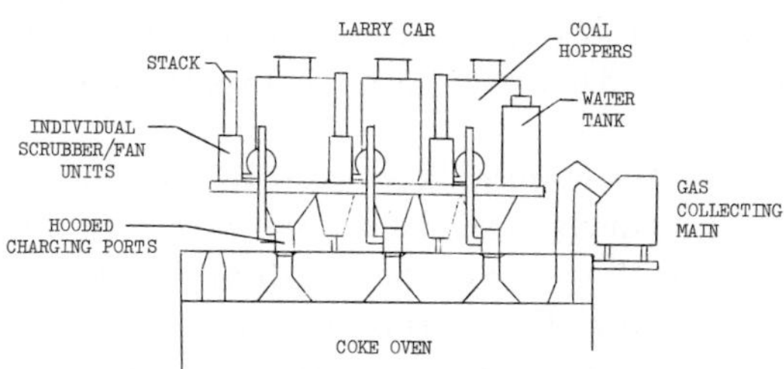


Figure 2. Typical larry-mounted scrubber system.

ever, this control measure is considered by most sources to be the superior method for controlling charging emissions.

In sealed or closed charging systems, aspiration removes only the charging gases, i.e., they exclude induced air. Such systems are of two types: (a) truly sealed systems that charge coal by unconventional techniques, and (b) mechanically sealed systems utilizing conventional larry cars. The former type was developed primarily for increasing coke oven throughout by charging preheated coal, and cannot, by itself, be considered an air pollution control technique (*8, 9*). The latter type system employs mechanical drop sleeves on the coal hoppers that prevent fumes from escaping through the charging ports and thereby force the gases to be extracted through the collector main or mains.

## 2. Coke Oven Discharging

When the hot coke is discharged or pushed from the oven into an open coke car, thermal drafts in the vicinity of the operation give rise to the emission of abraded coke dust or, in the event of incomplete coking, dust and hydrocarbons from partial combustion of the uncoked coal. The most important factor in minimizing pushing emissions is the prevention of operations giving rise to uncoked coals or "green" coke. Even with adequate operation, some green coke is often found in the corners or ends of the oven. The quantity of emissions associated with coke oven pushing, in plants relatively free of green coke pushes, is reported to be in the vicinity of 30% of the total coke plant particulate emissions. European sources report dust emission levels ranging from 0.12 to 0.40 kg/ton of coal (*4*). Factors that affect the level of emissions and control techniques employed include the degree and frequency of incompletely coked discharges, exposure of the coke side of the ovens to the elements, oven and machinery design, hooding clearance, power requirements, and interference with conventional coke bench operations.

Methods of controlling emissions from discharging ovens have been nearly as varied as for charging. These methods can be grouped into five general categories: (a) bench-mounted self-contained hoods, (b) coke-car-mounted hoods, (c) fixed-duct hoods, (d) spray systems, and (e) coke side enclosures (*5*).

Bench-mounted self-contained hood systems include designs incorporating mobile hoods, ducts, scrubbers, and fans that are mounted on a separate vehicle which traverses the length of the battery on the bench with the coke guide. Utilization of this approach is confined to West Germany. An advanced version of this concept utilizes a steam ejector venturi scrubber rather than an induced draft fan.

Coke-car-mounted hood systems include a family of designs employing hoods, ductwork, scrubbers, and fans which are mounted on and travel with the coke car. Two of the three designs included in this grouping operate in conjunction with continuous underground quenching systems, while the third relies upon a scrubber trailer which travels the full route of the coke through the quenching operation *(10)*. Of primary importance with systems of this type are proper mating connections with the coke guide, hoods, cars, and quenching apparatus. Also, as with bench-mounted hoods, a steam ejector scrubber may be used rather than an induced draft fan to minimize rail-mounted rotating equipment.

Fixed-duct hood systems provide a stationary duct, fan, and scrubbing system with duct ports for connecting to a mobile hood arrangement over the pushing operation (Fig. 3). Such systems are prevalent in Japan. Adequate capture and emission control is dependent upon proper mating of the fixed duct and mobile hood, adequate gas volume, and proper gas cleaning capability, exemplified as high energy venturi scrubbing or low energy scrubbing followed by use of electrostatic precipitators.

Water spray or fogging systems can be employed to minimize pushing emissions. Such sprays can be located at the coke guide or the coke car. Whereas sprays by themselves can be partially effective, they are more often used to enhance the effectiveness of hood and scrubber systems. Electrical safety, ice formation, and water removal are necessary concerns with this type of system.

Coke side enclosures or sheds entail nearly complete enclosure of the coke side of the battery *in lieu* of local hooding (Fig. 4). Numerous design advantages are cited, not the least of which are simplicity of design and operation, ease of retrofitting to existing batteries, and the ability to collect emissions from leaking oven doors *(11)*. The major objection to sheds centers around the necessity of maintaining a safe, clean working environment on the coke side.

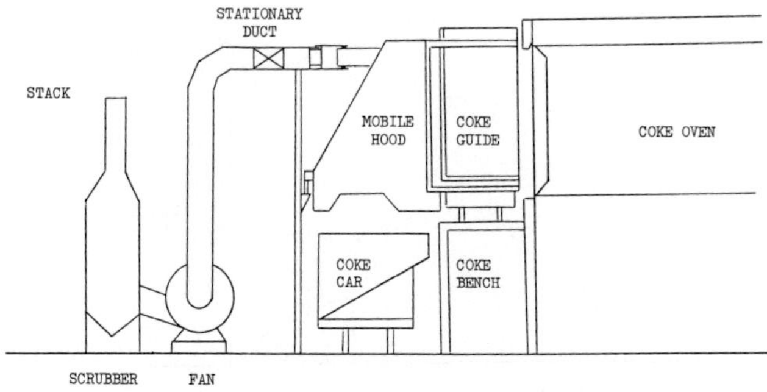


**Figure 3. Example of fixed-duct hood system for coke pushing.**

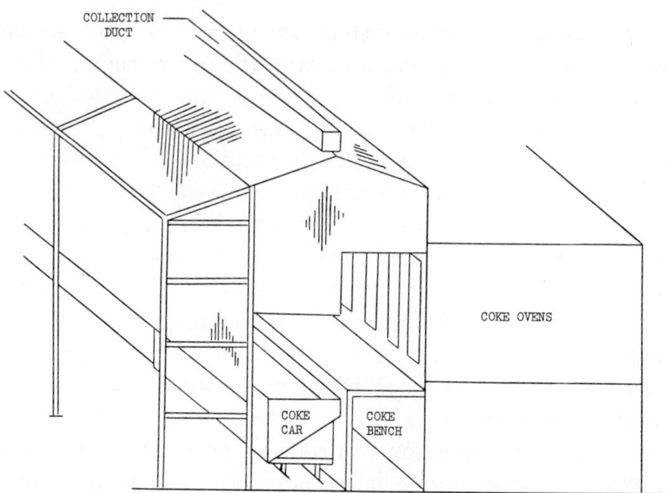


Figure 4. Typical coke side enclosure.

## 3. Coke Quenching

In conventional coke quenching practice, the hot coke is transported to a quenching station that consists of a large brick tower erected over a large sump (Fig. 5). The coke is deluged with water sprays. The rising cloud of heated air and water vapor lifts particles of fine coke into the size to 1 mm. Uncontrolled particulate emissions from quenching have been reported in a range of 0.06 to 0.24 kg/ton of coal. However, these emissions are routinely reduced by approximately 80% by simple baffles

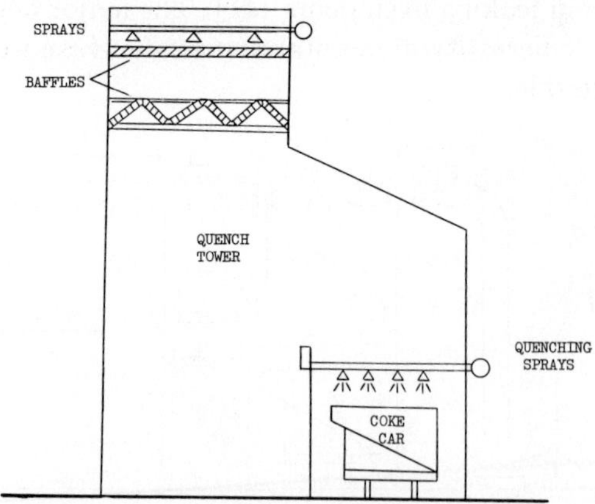


Figure 5. Coke quenching tower and controls.

(of various configuration) mounted in the quench tower. Water sprays are provided for periodically cleaning the baffles (*6*).

In underground and continuous quenching systems, particulate emissions are suppressed, or are contained and removed in systems integrated with the coke pushing emission control system.

Dry quenching of coke, accomplished by circulating unreactive gases through the coke and a cooling system, is widely used in the U.S.S.R. The primary incentive for dry quenching is energy conservation through heat transfer and recovery, but minimization of quenching emissions is an advantage cited. Decisions for the widespread application of dry quenching will be made on factors other than pollution control (*12*).

### 4. Leaking Oven Doors

Shortly after coke ovens are charged, sufficient internal positive pressure is developed to result in leakage at all available ports in the ovens, including the end doors. Most coke plants built since 1945 have been equipped with self-sealing doors which are designed to force a strip of steel against a flat-machined jamb, forming a close fit with narrow gaps. As pressure increases, a deposition of carbon and tar occurs at the gaps as the oven leaks until the door seals. By design, then, doors will leak for the first 15 to 60 minutes after a charge until this carbon and tar seal is established.

Factors that affect the extent of door leakage include the type of seal (diaphragm versus strip type), adjustment practice, door and jamb cleaning practice and equipment, height of the ovens, type of latching mechanism, coking practice (especially the use of preheated coal), and other less well-documented factors. No quantitative determinations have been made of door leaking emissions, but the absence of visible emissions 15 minutes following the charge is considered indicative of a well-sealed oven (*5*). With the exception of improved maintenance and operating practice, the principal effort made to control door leaking emissions is the use of ventilated coke side sheds discussed in Section II,B,2.

### 5. Coke Oven Gas Desulfurization

Sulfur occurs in raw coke oven gas primarily as hydrogen sulfide from the reaction of ferrous sulfide and hydrogen. Its normal range is from tower vent. This particulate matter varies widely in size from micron 5.5 to 11.0 gm hydrogen sulfide per standard cubic meter (scm) of coke oven gas, depending upon the sulfur content of the coal, which is normally low for metallurgical coking coals. Some steel companies remove hydrogen sulfide from coke oven gas before its use in metallurgical operations sensitive to sulfur and to minimize corrosion problems in gas distribution systems. Some plants convert stripped hydrogen sulfide to sulfuric acid for use in by-product operations (production of ammonium sulfate), but others vent the stripped hydrogen sulfide directly to the atmosphere or burn it, thus failing to reduce overall emissions to the atmosphere (*13*).

Worldwide, there are eight commercial processes for coke oven gas desulfurization. They can be subdivided into three categories: (a) dry oxidative, (b) wet oxidative, and (c) liquid absorption. The dry oxidative process, employing iron oxide boxes, was historically the first technique used. In this process hydrogen sulfide is absorbed by solid iron oxide and disposed of as a sulfide or oxidized to sulfur. This process has generally been replaced by more modern processes (*14*).

Wet oxidative processes utilize an absorption step followed by direct oxidation to elemental sulfur. The four wet processes include the Stretford (*15*), Takahax, Fumaks (*16*), and Giammarco Vetrocoke, each differing primarily in its oxidizing agent.

The liquid absorption processes include the vacuum carbonate, Sulfiban (*17*), and Firma Carl Still or ammonia processes (*18*). These systems employ absorption of the acid gases in solution, desorption of the acid gases, and conversion of the hydrogen sulfide in the acid gases to elemental sulfur or sulfuric acid. Each varies in the composition of the absorbing solution and the method of desorption of the acid gases.

The selection of a coke oven gas desulfurization process at any single plant requires an exhaustive analysis of the degree of hydrogen sulfide removal sought, the desired sulfur end product (as influenced by local market conditions), the need for treating discharge water contamination, the presence of substances that might foul the process, the desirability of simultaneous recovery or destruction of other by-products, and operating and capital costs. The presence of cyanides in coke oven gas has been a major deterrent in the development of suitable desulfurization processes and has prevented transferral of much of the technology of the petroleum coke industry. Cyanides contaminate the feed stream to sulfur processing equipment in the case of absorption processes and result in undesirable waste discharges in the case of wet oxidative processes (*14*).

## III. Sintering

### A. General Process Description

Ideal blast furnace feed material should contain at least 60% iron and a minimum of undesirable elements, be relatively uniformly sized, and have the physical and chemical characteristics to withstand heat and degradation, while at the same time being reasonably reducible (*1*). These requirements have given rise to two major agglomerating processes: sintering and pelletizing. Pelletizing is normally conducted in conjunction with a mining operation rather than at a steelmaking complex and will not be discussed here.

Sintering was developed to beneficiate low grade ores and to effectively utilize ore fines which, unless agglomerated, are detrimental to efficient blast furnace operation. The sintering process (Fig. 6) consists of mixing moist iron ore fines and other iron-bearing materials with a solid fuel, usually coke fines, and feeding the mixture onto a permeable traveling grate or strand. The upper surface of the bed of material is fired in an ignition zone and air is drawn through the bed into a series of distribution chambers called windboxes. After the mixture is ignited, the combustion zone moves downward through the bed as the grate travels, progressively igniting, drying, heating, and fusing the mixture into a sinter. At the end of the grate, the sinter is tipped into a crusher and is screened, cooled, and screened again for subsequent transport to the blast furnace. All screened fines are returned to the sintering process. Because sinter degrades and weathers relatively easily, stockpiles are small and plants are most often located very near blast furnaces.

Sintering machines process a wide variety of materials and produce substantially different sinter depending upon blast furnace requirements. Some plants process only low grade ore fines whereas others process mostly iron-bearing fines generated within a steelmaking complex. Also, some plants will elect to produce a sinter high in lime which can result in substantially different waste gas characteristics.

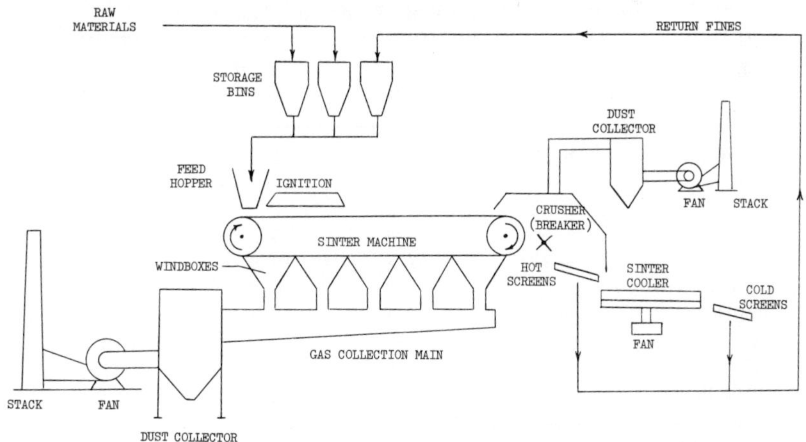


Figure 6. Schematic diagram of a sinter plant.

### *B. Emissions and Their Control*

The discharge of pollutants from sinter plants can occur principally at four locations within the plant: (a) the windbox waste gases or main stack gases, (b) the discharge end of the sinter machine, (c) materials handling points throughout the plant, and (d) the sinter cooler.

#### 1. Windbox Waste Gases

As combustion air is pulled through the sinter bed into the windboxes below, dust is extracted from the fine material and entrained in the gas stream. In addition to dust, the gases will contain many products of partial combustion as well as volatilized contaminants driven from the mixture. Contaminants reported include particulate matter, carbon monoxide, sulfur oxides, chlorides, fluorides, ammonia, hydrocarbons, and arsenic (*19*).

Particulate matter is the predominant contaminant in windbox waste gases. Material balances indicate dust concentrations in uncleaned gas as high as 25 kg/ton of sinter produced. The dust is generally relatively coarse. Therefore, most sinter plants are equipped with a series of gravity separators or dropout chambers, followed by mechanical collectors such as cyclones, to remove the coarse, abrasive dust to protect the induced draft fan from severe wear. The chemical analysis of the dust is highly dependent upon the sintering practice. Typical analyses are shown in Table I (*19–21*). Particle-size distribution of typical sinter plant dusts are shown in Table II (*19–22*).

**Table I Composition of Sinter Plant Particulate Emissions**[a]

| *Component* | *Range reported* (*19*) | *Plant A* (*20*) | *Plant B* (*20*) | *Typical range* (*21*) |
|---|---|---|---|---|
| $Fe_2O_3$ | 36–78 | 33.9 | 11.7 | — |
| Fe | — | — | — | 50 |
| $SiO_2$ | 6–12 | 4.8 | 2.4 | 9–15 |
| $Al_2O_3$ | 1–5 | 2.6 | 4.3 | 2–8 |
| CaO | 4–13 | 7.1 | 10.9 | 7–12 |
| MgO | 1–6 | 5.3 | 0.4 | 1–2 |
| ZnO | 0–1.3 | 0.4 | 0.1 | — |
| C | — | — | — | 0.5–5 |
| S | — | — | — | 0–2.5 |
| Alkalies | — | — | — | 0–2 |

[a] Weight %.

**Table II Particle-Size Distribution of Sinter Plant Emissions**

| Weight percent less than (μm) | Overall size distribution | | | After mechanical collection | | |
|---|---|---|---|---|---|---|
| | Reported range (19) | Plant B (20) | Range median (21) | Plant A (20) | Range median (21) | Typical values (22) |
| 100 | 40–89 | 40 | 55 | 100 | 97 | — |
| 40 | 14–50 | 30 | 30 | 95 | 90 | 98 |
| 20 | 6–33 | 20 | 16 | 90 | 72 | 87 |
| 10 | 2–19 | 12 | 8 | 80 | 50 | 73 |
| 5 | 2–7.5 | 8 | 3 | 55 | 25 | 25 |
| 1 | — | 2 | 0.5 | 2 | 2 | — |

The gas characteristics of the windbox waste gases are dependent upon the material mix, the condition of the plant machinery, and the product. About 25% of the carbon in the coke is reported to be present as carbon monoxide. Sulfur is derived primarily from the feed materials and coke, with 70% reportedly driven off to the waste gases as oxides (*19*). The presence of free lime dust for high lime sinter production can substantially reduce sulfur oxide concentrations by absorption. Moisture content in sinter plant waste gases will range from 8–15%. Hydrocarbons can occur in widely varying quantities in windbox waste gases. These hydrocarbons are volatilized from sintering mix materials such as blast furnace filter cake, mill scale, coke breeze, turnings, borings, and even certain ores. Fluorides can occur in plants processing certain high fluoride ores, but lime percentages will influence these concentrations (*20*).

Efforts to clean sinter plant waste gases have been directed toward removing the residual dust concentrations, normally in the range of 450 to 700 mg/Nm$^3$ (Nm$^3$, normal cubic meters), which remain following the mechanical collectors. Electrostatic precipitators have been the most widely used; efficiencies are reported to exceed 99% (*21*). However, some investigators have reported substantially reduced performance associated with electrical resistivity problems when sinter practice was altered to produce a highly basic sinter. Such occurrences may be associated with reduced sulfur oxide concentrations. Others have reported a concern over inadequate removal of hydrocarbons with precipitation. Although precipitators continue to be widely used in Europe, the trend in North America and Japan is away from precipitators due to the widespread production of high lime sinters.

A fabric filter has been installed on the windbox discharge of one United States plant with good success (*23*). However, oily feed materials must be restricted to circumvent the problem of the condensation of hydrocarbons. Investigations by others of the application of fabric filters for windbox gases are known to have resulted in unfavorable evaluations.

Wet scrubbers have been installed on sinter plants in the United States and the U.S.S.R. (*24, 25*). Although reasonable performance has been obtained with low pressure drops, high efficiencies required to achieve discharge concentrations of 50 mg/Nm$^3$ or lower require pressure drops of 85 mm Hg or greater, especially if condensed hydrocarbons are present in substantial quantity (*24, 26*). Wet scrubbers have a substantial advantage in that gaseous sulfur oxides, fluorides, or chlorides can be removed if they are present in excessive quantities. However, to accomplish this, the water chemistry of the scrubbing medium must be controlled closely for best scrubber operation.

## 2. Sinter Machine Discharge

Most modern sinter plants are equipped with crushers or breakers at the discharge end of the machine. As the fused sinter falls from the strand, it is broken up and screened prior to being cooled. The breaker and hot screen area is a source of dust and is normally hooded and exhausted to the atmosphere. Dust generation may be as high as 11 kg/ton of sinter produced (*27*). Other material transfer points in the vicinity of the discharge end may be exhausted by the same system.

Mechanical collectors, fabric filters, and low energy wet scrubbers have all been used successfully on this application, although mechanical collectors seldom are efficient enough by themselves. Another technique used is to direct the exhaust discharge to the hood located over the sinter machine and use it as a portion of the sinter machine air. When this is done, the dust is filtered by the sinter or passed to the windbox waste gas system.

## 3. Materials Handling

The vast quantity and variety of materials handled in sinter plants invariably require that certain materials handling points be hooded and exhausted, and the gases cleaned prior to discharge. Such materials will include dry, fine steelmaking dusts, certain ore fines, screened sinter fines, or a multitude of other materials which, by their physical characteristics or handling methods, represent potential dust generation.

Fabric filters are the common means for controlling such dust problems, but mechanical collectors may be adequate in some instances. Water sprays and chemical dust suppressants have also been used successfully (*28*).

### 4. Sinter Cooler

After hot screening, the sinter is cooled prior to final sizing. There are numerous cooler configurations, but many are forced draft rotary coolers discharging through a stack. Cooler exhausts can contain entrained sinter dust, but most cooler discharge stacks contain no visible emissions if the sinter is properly screened ahead of the cooler. One investigator has reported cooler emissions of 1.5 kg/ton of sinter (*19*) and at least one installation is known to have a fabric filter on a cooler discharge stack vent.

## IV. Ironmaking

### *A. General Process Description*

Most iron is produced in blast furnaces that are large refractory-lined structures producing up to 10,000 tons per day of molten iron (Fig. 7). Iron-bearing material (ore, pellets, or sinter), limestone, and coke are charged at the top of the furnace by the charging skip through a series of seals. Preheated blast air is introduced through a bank of tuyeres near the bottom of the furnace and rises through the charge, reacting with the coke to form a reducing gas which reduces the oxides in the ore to metallic iron. Steam, oil, gas, or coal may be injected with the blast air. As the burden moves downward, the molten iron collects in the hearth

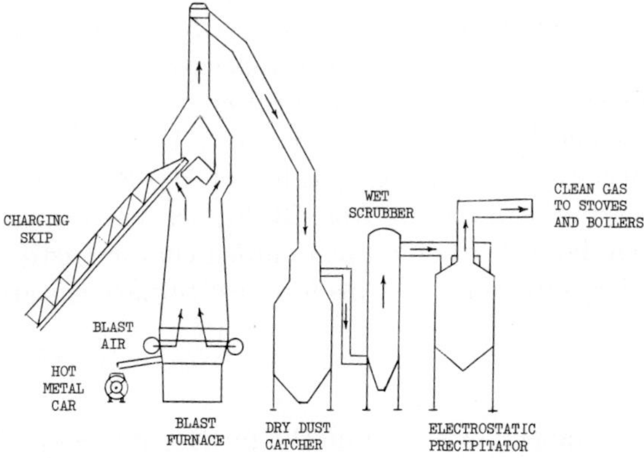


Figure 7. Blast furnace and associated gas-cleaning system (typical).

and is periodically tapped from the furnace into hot metal cars. The limestone in the furnace reacts with impurities and forms a layer of molten slag above the iron. Slag is tapped from the furnace into steel pots or a slag pit for air cooling, or is granulated by water sprays.

Blast furnace gas is extracted from the top of the furnace, and successively cooled, cleaned of dust, and then used to fire the regenerative stoves for blast air heating. A portion of the gas is also used in boilers to generate steam for driving the turbocompressors that furnish the blast air.

### B. Emissions and Their Control

The operations associated with potential blast furnace emissions consist of (a) charging, (b) blast furnace gas handling, (c) casting, and (d) slag handling. Raw materials handling prior to charging is relatively free of emissions because of the preparation which takes place at the limestone quarry, the pellet or sinter plant, and the coke plant. Charging through a sealed system creates no serious emission problem if the seals are adequately maintained.

#### 1. Blast Furnace Gas Handling

Dust generated and discharged with raw blast furnace gas can range from 14 to 150 kg/ton of molten metal depending upon the type and amount of burden preparation and screening (*29*). Dust concentrations may be as high as 30 gm/$Nm^3$ of gas.

Most furnaces are equipped with multiple-stage dust collection systems consisting of either a dry cyclone, wet scrubber, and wet electrostatic precipitator in series, or a two-stage wet scrubber. These devices have been installed for the purpose of cleaning the blast furnace gas to a level suitable for use in the stoves and boiler house. Multiple-stage gas cleaning is generally capable of reducing dust concentrations to approximately 10 mg/$Nm^3$ of gas (*30, 31*). The chemical composition and size analysis of blast furnace flue dust are shown in Tables III and IV, respectively.

Blast furnace gas has a heating value of 700–850 kcal/$Nm^3$ of gas. The gas composition will contain from 23–40% carbon monoxide and from 2–6% hydrogen, both depending upon burden characteristics and operating practices. Pressure relief systems for blast furnace gas are flared (*22*).

#### 2. Casting

Molten iron is tapped or cast from the furnace into a system of refractory lined troughs called runners which transport the iron to hot metal

**Table III Composition of Blast Furnace Flue Dust** *(21)*[a]

| *Component* | *Range* | |
|---|---|---|
| | *United States plants* | *European plants* |
| Fe | 36.5–50.3 | 5.0–40.0 |
| $SiO_2$ | 8.9–13.4 | 9.0–30.0 |
| $Al_2O_3$ | 2.2–5.3 | 4.0–15.0 |
| CaO | 3.8–4.5 | 7.0–28.0 |
| MgO | 0.9–1.6 | 1.0–5.0 |
| Mn | 0.5–0.9 | 6.3–1.5 |
| Pb | — | 0–15.0 |
| Zn | — | 0–35.0 |
| P | 0.5–0.2 | 0.3–1.2 |
| C | 3.7–13.9 | 5.0–10.0 |
| S | 0.2–0.4 | −0.1 |
| Alkalies | — | 0–20.0 |

[a] Weight %.

cars. The emerging iron is saturated with carbon, and as it cools, graphite flakes are emitted from the surface of the molten metal. This graphite, called "kish," as well as other minor metallic oxide emissions, escape during the casting operation. These emissions can be minimized by using shorter runs to the cars. However, residual emissions which emerge from the cast house, the enclosure surrounding this operation, can be visible to the extent that a separate control system is required.

Two North American companies exhaust large volumes of air from the space above the casting operation and direct it to a fabric filter. Japanese

**Table IV Particle-Size Distribution of Blast Furnace Flue Dust** *(22)*

| *United States series screen mesh* | *μm* | *Range (wt %)* |
|---|---|---|
| 20 | 833 | 2.5–20.2 |
| 30 | 589 | 3.9–10.6 |
| 40 | 414 | 7.0–11.7 |
| 50 | 295 | 10.7–12.4 |
| 70 | 208 | 10.0–15.0 |
| 100 | 147 | 10.2–16.8 |
| 140 | 104 | 7.7–12.5 |
| 200 | 74 | 5.3–8.8 |
| −200 | −74 | 15.4–22.6 |

efforts to minimize cast house emissions have centered around suppression of fumes by covering the runner system and hooding the tap spout.

### 3. Slag Handling

Except for small percentages of sulfur retained in the iron, all remaining sulfur charged as part of the blast furnace burden is removed in the slag, which will range from 1.2–1.8% sulfur. Slag is removed from the furnace through a runner system at the rate of 200–350 kg/ton of iron. As the sulfur in the slag is exposed to air, some sulfur dioxide is formed. The presence of moisture can result in formation of hydrogen sulfide.

Most sulfur emissions are associated with the subsequent slag handling and result from quenching with water in various ways. Depending upon the method of processing slag, sulfur emissions may range from nil for air cooling (rain contact only) to 1.4 kg/ton of iron for the expanded slag process. Most slag is processed either by air cooling (55%) or by the hard slag process (30%), the latter creating an estimated sulfur emission of 0.15 kg/ton of iron. Except for changes in slag practice, no successful control efforts have been widely used (*32*).

## *C. Direct Reduction Processes*

In the late 1960's and early 1970's, a number of commercially available processes emerged for converting iron oxides to metallic iron without melting as in the conventional blast furnace. These direct reduction processes hold promise for ore-rich areas seeking a steel industry but lacking capital or product demand to justify a blast furnace/basic oxygen steelmaking complex. A direct reduction/electric furnace steelmaking complex can be economical for 3–5% of the size of a conventional complex. The process also can minimize the impact of scrap quality and prices, and the availability and cost of metallurgical coal (*33*).

The competitive processes are of sufficient complexity and diversity to prohibit detailed description in this text, but, in general, ores are reduced by natural gas, liquid hydrocarbons, or standard-grade coals. The available process types are (a) fluid beds, including Exxon's FIOR process and United States Steel's HIB process; (b) vertical shaft moving beds, including the Armco process, Midrex process, and Thyssen-Purofer process; (c) the vertical fixed-bed process by Hylsa; and (d) rotary kilns, including the SL/RN process, the Krupp process, and the Allis-Chalmers process.

Because of the proprietary nature of these processes, the magnitude and extent of control of air pollution is not well documented. In general, however, because most processes deal with closely controlled recirculated reducing gas systems, emissions would be expected to be minimal. Material handling emissions associated with storage and transport to and from the furnaces may require conventional controls. Emission levels would be expected to be much less than blast furnace operation due to the elimination of casting and slag handling requirements.

## V. Open Hearth Steelmaking

### A. General Process Description

The open hearth steelmaking process was developed in the 19th century and maintained its dominance in steelmaking capacity (80–90%) into the 1960's when the basic oxygen process prevailed. By 1980 United States open hearth production is expected to be only 20% of total raw steel production. In spite of reduced levels, worldwide production continues to be significant.

The open hearth furnace (Fig. 8) consists of a shallow rectangular hearth with a brick arch roof and refractory ducts at each end of the furnace for removal of hot gases to regenerative checker chambers prior to their discharge to the atmosphere in order to preheat combustion air. The flame for heating is swept across the length of the open bath. Fuel may consist of oil, tar, coke oven gas, or natural gas.

Material consisting of solid pig iron, scrap steel, limestone, and ore is charged through the side doors of the furnace by the charging machine and the furnace is fired. Following initial melting, additional scrap

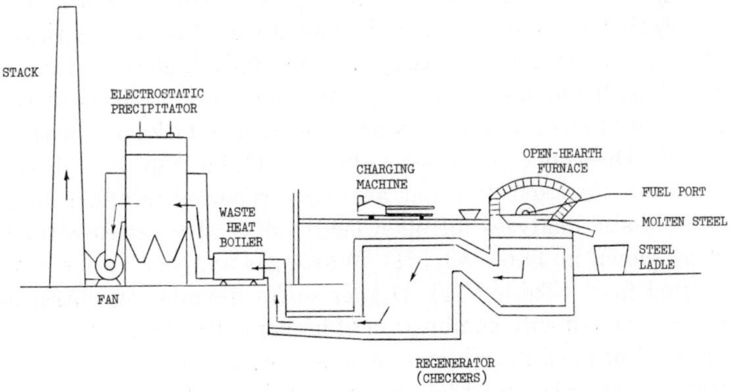


Figure 8. Open hearth furnace and associated gas-cleaning system (typical).

and/or molten iron may be subsequently charged and melted. Carbon in the iron reacts with oxygen in the ore creating an ore boil. This is followed by the lime boil during which the limestone is calcined and a slag is formed. The batch of molten metal in the hearth, which is called a heat, is then refined to obtain the desired metallurgical characteristics. Oxygen may be injected to hasten melting or to aid in refining. Following refining the heat is tapped into a steel ladle through a tap hole in the back wall of the furnace. Alloying elements may be added to the ladle during tapping. Ingot molds are filled with molten steel from the ladle. The entire heat cycle may last 6–12 hours, depending upon whether or not molten iron is charged.

Gases leave the open hearth furnace at a temperature in excess of 1650°C and are reduced to 650°–850°C in the regenerators (checkers). The installation of waste heat boilers or double-pass checkers on some furnaces further reduces the gas temperature to 300°–400°C.

### B. Emissions and Their Control

Minor emissions of dust are associated with charging and tapping the open hearth and teeming the molten steel into ingots. The magnitude and characteristics of these emissions are highly dependent upon the quality of scrap, the extent to which molten iron is employed, and the nature of ladle additives. Control of these emissions in open hearth shops has not been defined or controlled to the extent of those in electric furnace or basic oxygen shops. However, emissions are similar, and the reader is referred to those portions of the text dealing with this subject (Sections VI,B,2 and VII,B).

The waste gas emissions from open hearths consists primarily of iron oxides. Chemical compositions of some open hearth dusts are shown in Table V. Both the composition and quantity of dust varies throughout the heat cycle. Dust concentrations are normally highest during oxygen lancing and with hot metal charges. Dust generation has been reported to be 6 kg/ton of steel without oxygen lancing and 11 kg/ton with oxygen lancing (*27*). Dust concentrations as high as 11.5 g/Nm$^3$ have been noted during lancing. Fuel selection has also been reported to affect dust loadings. Particle-size analyses for open hearth dust also varies with furnace activity as shown in Table VI (*34*). Waste gas analysis varies somewhat with the fuel fired (Table VII). Higher sulfur dioxide concentrations are naturally associated with combustion of higher sulfur fuels.

The control of open hearth emissions is accomplished with electrostatic precipitators (*21*, *35*) and high energy scrubbers (*34*, *36*). On furnaces with waste heat boilers, precipitators represent a clear choice because

**Table V Composition of Open Hearth Particulate Emissions**

| *Component* | *Ranges reported (wt %)* |
|---|---|
| $Fe_2O_3$ | 61.3–96.5 |
| Fe | 55.9–68.0 |
| $SiO_2$ | 0.3–3.8 |
| $Al_2O_3$ | 0.1–1.9 |
| CaO | 0.3–6.5 |
| MgO | 0.3–1.9 |
| MnO | 0.3–0.7 |
| ZnO | 0.3–2.0 |
| P | 0.1–1.2 |
| C | 0.1–0.6 |
| S | 0.3–2.8 |
| Pb | 0.1–1.0 |
| Alkalies | 0.2–2.9 |

the gas characteristics are ideal for precipitator application. For furnaces without waste heat boilers, scrubbers will offer an economic advantage since the additional cost of installing waste heat boilers to accommodate precipitators can be prohibitive. Most electrostatic precipitators installed on open hearths since 1960 have exceeded 98% overall efficiency (*21*). The installation of venturi scrubbers requires pressure drops in excess of 85 mm Hg to reduce dust concentrations to 70 mg/Nm$^3$ or less. If a particular open hearth is confronted with a gaseous emission problem, scrubbers can reduce these pollutants also. However, problems with corrosion can be severe for wet scrubbers, and have been reported for precipitators and even for one fabric filter installation.

**Table VI Particle-Size Distribution of Open Hearth Emissions (*34*)**

| *Weight percent less than (μm)* | *Composite sample* | *Lime boil sample* |
|---|---|---|
| 40 | 94 | — |
| 20 | 85 | 98 |
| 10 | 70 | 92 |
| 5 | 48 | 75 |
| 1 | — | 22 |

**Table VII Composition of Open Hearth Waste Gases (21)**

| | *Flue-gas composition (%) with 40% excess air for fuel specified* | | | | | |
|---|---|---|---|---|---|---|
| *Waste constituent* | *Producer gas* | *Coke oven gas* | *Fuel oil* | *Coal tar* | *Pitch creosote* | *Natural gas* |
| $CO_2$ | 14.2 | 6.05 | 9.6 | 11.2 | 11.5 | 7.26 |
| $H_2O$ | 6.9 | 16.4 | 13.4 | 11.3 | 8.9 | 13.45 |
| $N_2$ | 74.8 | 72.0 | 71.5 | 72.0 | 73.95 | 73.75 |
| $O_2$ | 4.1 | 5.4 | 5.4 | 5.4 | 5.6 | 5.54 |
| $SO_2$ | 0.07 | 0.15 | 0.06 | 0.05 | 0.04 | Nil |

## VI. Basic Oxygen Steelmaking

### A. General Process Description

Basic oxygen steelmaking was developed in Linz-Donawitz, Austria, in the 1950's and is a variation of the older Bessemer process. The basic oxygen process now accounts for the majority of steelmaking capacity worldwide and is expected to comprise 55% of United States steel production by 1980. Over 80% of Japan's steel is produced in basic oxygen furnaces. There are several types of vessels and variations of waste gas handling which distinguish between processes generally categorized as basic oxygen steelmaking.

Most basic oxygen furnaces consist of large refractory-lined, pear-shaped vessels mounted in a trunnion to permit full rotation for various operations (Fig. 9). Furnace capacities range from 15 to 400 tons. The

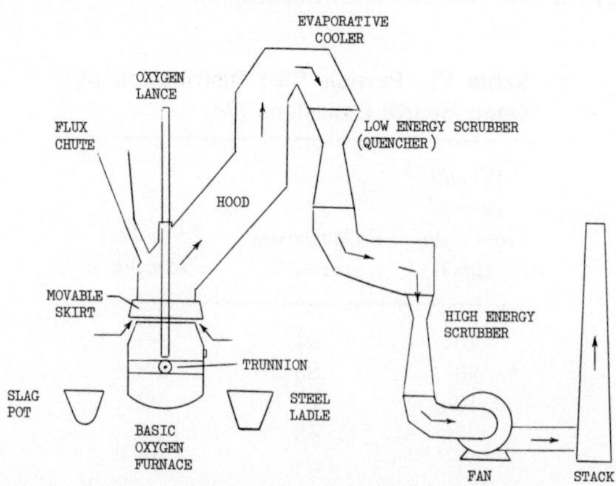


Figure 9. Basic oxygen furnace and associated gas-cleaning system (typical).

vessel is tilted and charged with scrap and molten iron in proportions of approximately 30 and 70%, respectively. Burnt lime is added for fluxing through the flux chute. A water-cooled oxygen lance is lowered through the top of the uprighted vessel, and high purity oxygen is blown into the molten metal bath at rates up to 60,000 $Nm^3$/hour. Violent agitation and intimate mixing of the bath materials cause rapid oxidation of carbon and other impurities, as well as of some iron. The reaction is exothermic and requires no fuel. A slag is formed on the surface of the bath. When the entire charge is melted and the metallurgical requirements are met, the vessel is tilted to tap the molten steel through a tap hole in the vessel into a steel ladle for subsequent teeming or continuous casting. Slag is removed by tilting the furnace in the opposite direction so that it pours into the slag pot. An entire heat cycle may require only 30–50 minutes.

Two other processes, the Stora–Kaldo process developed in Sweden and the Rotor process developed in Germany, basically employ cylindrical, horizontally oriented vessels with different configurations of hooding, oxygen lancing, and tapping. However, vessel reactions are similar and pollutants generated would be expected to be similar.

A bottom blown oxygen process (referred to as Q–BOP) has been developed in the United States by the United States Steel Corporation (*37*). This technique differs from conventional practice by injecting oxygen below the bath through tuyeres. The major advantage of the Q–BOP process is the elimination of high buildings required for conventional oxygen lance clearances, thus making it possible to retrofit open hearth shops with an oxygen steelmaking process. Higher metallic yields and the ability to charge greater percentages of scrap are also claimed advantages.

Other process variations involve the different techniques for hooding and combustion of the gases evolved from the basic oxygen process. Hoods can be open combustion hoods, in which excess air is introduced in quantities of from 10 to 300%, or they can be closed hoods, into which only 5 to 70% of the air theoretically required for combustion is allowed to infiltrate. With open hoods, there is a gap between the hood and the furnace top into which air can be induced. With closed hoods, a movable skirt seals as tightly as feasible to the furnace top to discourage air inflow. Closed hoods have the advantages of reduced gas volume and reduced heat in the hood, reduced fume generation, and possible savings through increased metallic yield or off-setting credits for recovered carbon monoxide. Several partial (suppressed) combustion systems have been developed independently and differ only slightly in the method of controlling infiltration and combustion levels at the interface of the vessel mouth and hood. These systems include the Oxygen Converter Gas Recovery System (OG system) developed in Japan (*38*), the

IRSID–CAFL system from France (*39*), and the Baumco partial combustion system and Krupp system, both developed in Germany. The selection of a closed- versus open-hood system is based upon a very complex set of conditions and operating requirements which do not make the choice clear. However, there appears to be a trend worldwide to install or convert to suppressed combustion systems on basic oxygen furnaces.

### *B. Emissions and Their Control*

The potential sources of emissions in a conventional basic oxygen shop are associated with the following activities: (a) hot metal transfer, slag skimming, and/or desulfurization, (b) charging and tapping, and (c) furnace waste gases. Minor emissions may also occur during teeming or casting, but these have been neither defined nor controlled.

#### 1. Hot Metal Transfer

In all basic oxygen shops, molten iron arrives from the blast furnace in hot metal cars and must be transferred to a ladle for charging to the furnace. This transfer of hot metal evolves graphite, or "kish," from the cooling, carbon-saturated molten metal, and iron oxide as some of the metal is oxidized on contact with air. Particulate concentrations have been reported in the range of 0.9–1.8 $gm/Nm^3$ in the exhaust from hoods mounted over the ladle at the transfer station during the short duration of the transfer operation. Much of the material is coarse kish with 16% reported to be less than 10 $\mu m$ and 3% less than 1 $\mu m$ (*40*).

Early control efforts consisted of the installation of multiple-tube cyclones which removed most of the kish particles but passed the finer, more visible iron oxide particles. Fabric filters are now widely used to reduce total particulate discharges from hot metal transfer operations.

Other hot metal conditioning practices can occur in basic oxygen shops. Some shops rake or skim excess slag from the ladle of molten iron and some employ various techniques for desulfurization of the hot metal. These activities generate emissions similar to those at hot metal transfer stations, and similar control efforts, i.e., local hooding and fabric filtration, are being applied.

#### 2. Charging and Tapping

When scrap and molten iron are charged to the vessel, pollutants may be evolved. The emissions consist of kish and iron oxide from the hot metal, and dirt and volatile materials which may be burned or driven from the scrap. The quality of scrap is the single most important factor affecting the magnitude of charging emissions.

The control of charging emissions can be accomplished by changes in practice or with control systems. Most emissions occur during the charging of molten metal on top of the scrap. If the scrap is preheated, either externally or in the vessel with the main exhaust hood in place, the potential emissions are driven off and collected prior to the hot metal charge. However, preheating of scrap is dictated by production requirements and is not practical in many plants. If the molten metal can be charged slowly with the vessel in a nearly vertical position, emissions can often be extracted by the main exhaust hood. However, production requirements, shop configurations, and the type of hood and gas-cleaning equipment will often not permit this practice.

The installation of auxiliary control systems for charging emissions has been employed in the United States, Canada, and Japan. The most common practice is to install auxiliary hoods on the charging side above the vessel and direct the exhaust fumes to a dust collector. Fabric filters and wet scrubbers have been used. If scrubbers are employed, pressure drops in the 75–95 mm Hg range are required to satisfactorily remove the fine iron oxide particles. Auxiliary canopy hoods may also be installed in the roof trusses of the building to extract fumes that the local hoods do not capture. For existing shops which lack space to install close hooding, this approach may be the only choice. Canopy hoods are exhausted to fabric filters and may be combined with local hooding to dilute high temperature gases from the local hoods. Auxiliary hoods have also been exhausted to electrostatic precipitators serving the main waste gas system, but this practice can be detrimental to optimum precipitator operation.

Tapping emissions are generally of less consequence than charging emissions, but can be substantial depending upon ladle additions. Control efforts have been few but have been similar to charging emission controls to the extent that the same system often serves both purposes.

### 3. Basic Oxygen Furnace Waste Gases

The violent reaction occurring in a basic oxygen furnace evolves substantial quantities of particulate emissions and gases. Dust and fume generation has been reported in a range of 7–30 kg/ton of steel produced. Dust concentrations at the mouth of the vessel may be as high as 45 gm/$Nm^3$. Examples of chemical composition of the dust generated are shown in Table VIII (*21, 22, 41*). Particle-size distributions reported are shown in Table IX.

**Table VIII Composition of Basic Oxygen Process Particulate Emissions**[a]

| Component | Noncombusted gas system (21) | Combusted gas systems (21) | (22) | (21) |
|---|---|---|---|---|
| $Fe_2O_3$ | 4.0 | 90.0 | 80 | — |
| FeO | 21.4 | 1.5 | — | 11.5–16.4 |
| Fe | 66.7 | — | 56.0–57.7 | 65.1–68.8 |
| $SiO_2$ | 1.3 | 1.3 | 1.3–2.0 | 1.7–2.6 |
| $Al_2O_3$ | — | 0.2 | 0.1–0.4 | — |
| CaO | 3.7 | 0.4 | 3.1–5.1 | 1.8–12.3 |
| MgO | Trace | — | 0.6–1.1 | 0–0.6 |
| P | — | — | 0.1–0.2 | — |
| $P_2O_5$ | 0.4 | 0.3 | — | 0.1–0.2 |
| Mn | — | — | 0.4–1.5 | — |
| MnO | — | — | — | 1.6–2.2 |
| $Mn_3O_4$ | — | 4.4 | — | — |
| Zn | — | — | 1.9–4.8 | — |

[a] Weight %.

The reaction of carbon and oxygen in the furnace generates substantial quantities of carbon monoxide. In an open-hood system, sufficient air is introduced to assure combustion of the carbon monoxide to carbon dioxide in the hood above the vessel. In closed-hood systems, the carbon monoxide is handled in a partially burned condition and then flared or recovered for fuel or a chemical feedstock.

The Kaldo process is reported to generate less carbon monoxide and

**Table IX Particle-Size Distribution of Basic Oxygen Process Emissions**

| Weight percent less than (μm) | Noncombusted gas system (21) | Combusted gas systems (21) | (41) |
|---|---|---|---|
| 60 | — | — | 75 |
| 40 | — | — | 66 |
| 30 | 86.6 | — | — |
| 20 | 57.8 | — | 56 |
| 15 | — | 100 | — |
| 10 | 18.0 | — | 50 |
| 5 | 8.9 | — | 6 |
| 1 | — | 85 | — |

less particulate matter, about 5 kg/ton of steel. Particle size of the dust is also larger (*22*). The closed-hood systems generally are claimed to produce less dust due to the restriction of oxidizing turbulence-creating air at the mouth of the vessel. The Q–BOP also is reported to generate less dust by reducing turbulence in the furnace through submerged oxygen injection.

Control of basic oxygen emissions is accomplished effectively with both electrostatic precipitators and high energy scrubbers (*42*). Precipitators are normally associated with open-hood systems with excess air in the range of 200–300% to minimize the possibility of exposing a combustible gas to sparking in the precipitator. High volumes of air are therefore required, and efficiencies must be in excess of 99.5% to reduce emissions to levels acceptable for regulatory standards. To attain such efficiencies, care must be taken to assure optimum design conditions for precipitator performance. Moisture conditioning and gas distribution to and through the collector have been reported as primary factors for high efficiency. Precipitators are often preceded by gravity separators to remove the larger particles for improved performance.

Wet scrubbers are employed with all closed-hood systems and with open-hood systems having 10–70% excess air. Processing of smaller gas volumes is essential due to the requirement for high scrubber pressure drops (85–100 mm Hg) which lead to high power requirements and attendant operating costs. Gas temperature and volume can be reduced by the use of an evaporative cooler after the hood. Wet scrubbers properly designed are capable of reducing dust concentrations to the level of 45 mg/Nm$^3$. Performance of the high energy scrubber may be enhanced by removing larger particles with a low energy scrubber (quencher) preceding it. Because of the particle size of the dust removed, water treatment facilities must be designed to remove a more finely divided material.

## VII. Electric Furnace Steelmaking

### *A. General Process Description*

The electric arc furnace process as it is commonly known today, employing a cylindrical vessel and three-phase power, was first installed in 1909. However, significant percentages of steelmaking capacity were not attained until World War II. The process has grown rapidly in importance since that time with the increased availability of low cost electrical power, the phasing out of open hearth steelmaking, and the availability of low cost scrap. By 1980 electric arc furnaces are expected to account for 25% of United States steelmaking capacity. The growth of the electric furnace is due to its extremely versatile capabilities. Virtually every known grade of steel can be made in an electric furnace.

Electric arc furnace steelmaking (Fig. 10) is a batch process with heats generally ranging in duration from 1½ to 5 hours for carbon steel production and 5 to 10 hours for alloy steel production. The cyclic operation consists of one or more charges of scrap steel from the scrap charging bucket, followed by melting, refining, and tapping through the tap spout into the steel ladle. The furnace is a cylindrical, refractory-lined shell into which the steel scrap is dropped. An electric arc is imparted to the charge through a carbon electrode system, and the electrical energy melts the scrap. Although the vast majority of electric furnaces operate with 100% cold scrap, molten iron, preheated scrap, and/or prereduced iron pellets have been charged. Oxygen may or may not be injected into the furnace for refining purposes. A few furnaces are charged through side doors, but most are charged from the top with the furnace roof swung aside. The furnaces are tilted for tapping or removal of slag. Furnaces range from 1- to 400-ton capacity and from 1 to 9.75 m in diameter.

Some electric furnace shops producing stainless steels also utilize a process known as argon–oxygen refining (AOR) which employs an open top vessel resembling a basic oxygen furnace, for refining the molten metal following melting of the heat in an electric furnace. Argon and oxygen are introduced in controlled quantities through tuyeres below the charge to attain the proper metallurgical requirements (*43*).

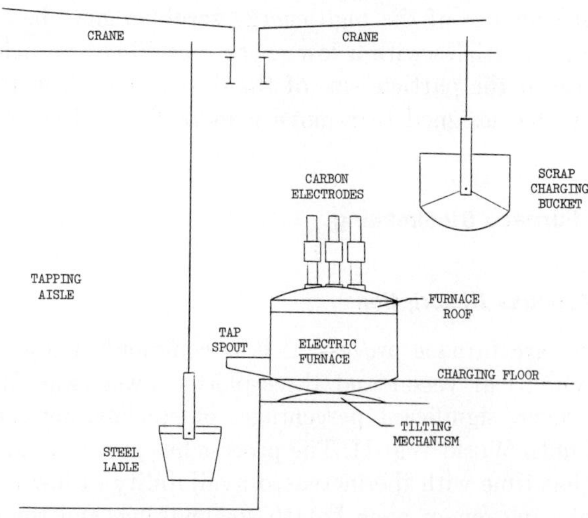


Figure 10. Electric arc furnace facilities.

### *B. Emissions and Their Control*

Pollutants generated during operation of electric furnaces are primarily limited to particulate matter and carbon monoxide. The majority of particulate emissions occur during the melting and oxygen blowing cycles. Lesser amounts also occur during charging and tapping. Reported dust evolution figures range from 5 to 25 kg/ton of steel produced, depending upon such variables as charging methods, scrap quality, oxygen blowing rates, furnace operating practices, and the type of steel produced (*44*). The ranges of chemical composition of electric furnace dusts are shown in Table X. The wide variations are attributed to the effect of the type of steel produced which is extremely diverse in electric furnaces. Ranges of particle-size distributions of electric furnace dusts are given in Table XI.

Carbon monoxide is generated upon reaction of the carbon in the electrodes or steel with the directly applied oxygen, or from air induced into the furnaces. Much of the carbon monoxide is combusted as it leaves the furnace and mixes with air.

The control of emissions from electric furnaces has revolved around variations in the method of fume capture and in the method of gas cleaning (*45*). There are basically three methods of fume capture which are

**Table X Composition of Electric Furnace Dusts**

| *Component* | *Range of reported values (wt %)* |
|---|---|
| $Fe_2O_3$ | 19–60 |
| FeO | 4–11 |
| Fe | 5–36 |
| $SiO_2$ | 1–9 |
| $Al_2O_3$ | 1–13 |
| CaO | 2–22 |
| MgO | 2–15 |
| MnO | 3–12 |
| $Cr_2O_3$ | 0–12 |
| NiO | 0–3 |
| PbO | 0–4 |
| ZnO | 0–44 |
| P | 0–1 |
| S | 0–1 |
| C | 1–4 |
| Alkalies | 1–11 |

**Table XI Particle-Size Distribution of Electric Furnace Emissions**

| *Weight percent less than, μm* | *Ranges of commonly reported percentages* |
|---|---|
| 40 | 82–100 |
| 20 | 67–98 |
| 10 | 61–95 |
| 5 | 43–72 |

employed for electric furnace fume control systems: (a) direct shell evacuation, (b) roof-mounted hoods, and (c) canopy hoods.

Direct shell evacuation (Fig. 11) employs an induced draft which withdraws emissions from the furnace through a special hole in the furnace roof. Generally, a water-cooled duct system, suitably flanged to permit roof swing and furnace tilting, is used for heat protection to transport gases to a point at which further cooling is accomplished either by evaporative cooling, dilution air additions, or radiant-convective gas cooling. Air is induced through the gaps in the ductwork to assure complete combustion of the carbon monoxide. Direct shell evacuation is commonly applied to the larger furnaces (those greater than 4.6 m in diameter).

Roof-mounted hoods of various designs have been employed, but the most common is the side-draft or lateral exhaust type. This arrangement

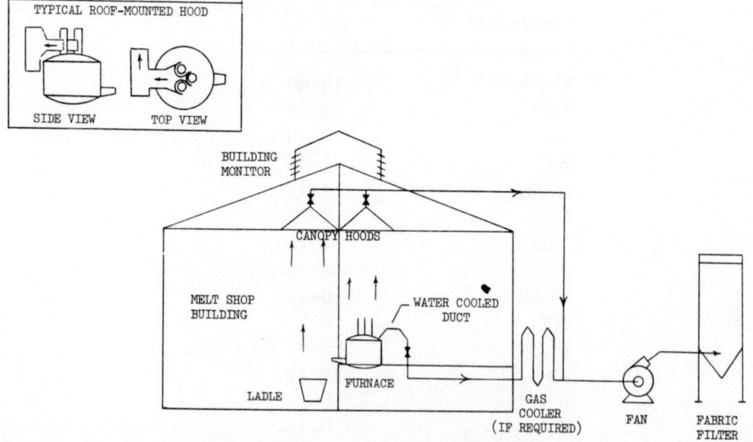


Figure 11. Electric arc furnace shop with combination canopy hood/direct evacuation control system. Insert shows typical roof-mounted hoods.

consists of a hood on the furnace roof which draws the fumes laterally as they escape the electrode ports of the furnace. The fume is then transported via ductwork to a fabric filter. Gas cooling is of minor consequence because sufficient dilution air is pulled into the hood along with the furnace fumes; however, water-cooled hoods are sometimes required depending on design heat conditions. Side-draft hoods are widely used on furnaces of 4.5 m diameter and smaller, but systems have been installed on furnaces up to 6.7 m.

Canopy hoods consist of large hoods located well above the furnace roof in the melt shop roof trusses. Fumes are permitted to escape the furnace uncontrolled and are withdrawn through the canopy hood to a fabric filter system. Cooling of gases is inherently by dilution with melt shop air. A variation of the canopy-hood approach is to sectionalize portions of the furnace building and withdraw fumes from the top of the building. This approach is commonly referred to as building evacuation and can result in undesirable working conditions in the shop. Canopy hoods have been used on all furnace sizes, with the decision being based more on melt shop configuration.

The vast majority of electric furnace emission control systems have employed fabric filters as the gas-cleaning device of choice. Baghouses offer advantages in overall cost, simplicity, reliability, performance, and ease of expansion. Depending on the relative significance of other factors, however, high energy wet scrubbers can be competitive and have been used for several installations, primarily on large furnaces and in multiple furnace shops. Electrostatic precipitators have also been used, primarily in Europe. As with other steelmaking processes, high efficiencies are required to contend with the high rates of dust evolution, and scrubbers require high pressure drops for satisfactory performance.

Although charging and tapping emissions represent only 2–5 wt% of the total emissions from electric furnaces, the concentrated discharge of these fumes over relatively short periods of time, coupled with the regulatory requirements for diminished visible emissions, has led to extensive efforts to control these emissions, particularly in the United States (*46–53*). Because of the configuration of most melt shops, control of these emissions is largely confined to the use of canopy hoods. The use of these canopy hoods has strengthened the position of fabric filters for two reasons: (a) the large volumes of air associated with canopy hoods cannot be economically handled through high energy scrubbers, and (b) the large volumes of relatively cool air has tended to encourage combination direct evacuation–canopy hood systems (Fig. 11) owing to the gas cooling effect of this air, thereby resulting in mixed gas temperatures suitable for baghouse operation. Also, since charging/tapping emissions are generally associated with those periods when the direct evacuation system is not connected, it is desirable to divert that capacity to the control of charging and tapping emissions, and this is readily possible with a combination system.

Emissions from an AOR are similar to those during oxygen blowing in the electric furnace, and consist of a high concentration of iron oxide particles. AOR vessels are closely hooded and exhausted to wet scrubbers or are controlled with canopy hoods exhausted to fabric filters. Combination of their fume control system with an electric furnace system is usually possible.

## VIII. Rolling, Processing, and Finishing

### A. General Process Description

After steel is tapped in molten form from an open hearth, basic oxygen, or electric furnace, it is either poured into ingots or continuously cast into slabs or billets. Whether the steel is made into ingots or continuously cast, it must be further heated to proper temperature before being further reduced to the desired shape and gauge in a rolling mill. Ingots require an extra heating step, which takes place in furnaces called soaking pits, prior to being rolled into slabs and heated again in a slab furnace. The various steps of reheat furnaces may be fired with natural gas, oil, and/or coke oven gas.

Slabs or blooms made from ingots often require surface preparation to remove defects which would otherwise be rolled into the steel. The process of removing a thin layer of the steel surface by directing an oxygen–gas flame at the surface is known as scarfing. This process can be done manually or automatically by machine in a continuous rolling mill. Continuously cast shapes are normally free of surface defects and do not require scarfing.

Reduction of the gauge of steel strip is accomplished initially on a hot rolling mill. A cold rolling mill is used for further gauge reduction. After hot rolling, pickling of the steel is required to remove scale and oxides on the strip surface prior to cold reduction. Pickling of most other steel shapes is usually required at some point in the processing and finishing steps. Pickling is accomplished by passing steel through a bath of sulfuric, hydrochloric, or nitric–hydrofluoric acid. The process can consist of batch or continuous strip pickling.

Numerous other metallurgical or chemical coating processes for finishing steel may be conducted at any given steel mill. These steps may consist of normalizing, annealing, heat treating, galvanizing, tin plating, aluminizing, or painting.

### *B. Emissions and Their Control*

Significant emissions from rolling and processing of steel are limited to only a few operations. The potential sources of emissions that will be discussed here are (a) reheat furnaces, (b) scarfing, (c) pickling, and (d) galvanizing. Particulate emissions from high speed hot rolling mills have been reported but are minor. Oil mist can be generated from the intense heat and pressure on rolling oils used on high speed cold reduction mills, but simple exhaust hooding and wetted baffles serve to minimize these emissions.

#### 1. Reheat Furnaces

Emissions from reheat furnaces are limited to products of combustion. Most furnaces are very efficient and are equipped with elaborate controls to assure adequate combustion. Depending upon the sulfur content of the fuel used, sulfur oxides may be emitted in significant quantities. Desulfurization of the fuel (coke oven gas, for example) or switching to alternative fuels may be the only practical solution.

#### 2. Scarfing

Hand scarfing emissions are localized and, in general, are minor in comparison to machine scarfing. The scarfing process volatilizes the steel at the surface of the slab or bloom, creating a fine iron oxide fume in concentrations up to 1150 mg/$Nm^3$. The fume may consist of a particle-size distribution of 70 wt% less than 1 $\mu$m and 30 wt% less than 0.1 $\mu$m. Because of water sprays used to blast the scarfed material from the steel surface, the exhaust gas from a scarfing machine is saturated at a temperature of 50°–60°C. The quantity of fume is dependent upon the amount of material removed which may range from 2% for slabs up to 8% for blooms.

Successful control installations for scarfers have been made with wet scrubbers and wet electrostatic precipitators (*54*). Because of the particle size, pressure drops as high as 110 mm Hg are required for scrubbers to attain levels of 70 mg/$Nm^3$, a level which still may be unacceptable from the standpoint of opacity. Wet precipitators at higher initial cost appear to be capable of higher efficiencies for much less energy input. Both systems require water treatment that is normally readily available at rolling mill complexes. Severe corrosion, believed to be due to oxygen carryover from the scarfer, has been reported on earlier attempts to use conventional dry precipitators.

## 3. Pickling

In both batch and continuous strip pickling, acid mists are evolved from the hot acid baths. Historically most acid tanks are hooded and/or exhausted to prevent corrosion in buildings and to ventilate the working environment. Packed towers and wet scrubbers of various designs have been used with success for removing acid mists from the exhausted air. Removal efficiencies in excess of 95% are common for hydrochloric and sulfuric acids, the most common pickling agents. Slightly lower efficiencies are achieved with nitric–hydrofluoric acid used in stainless-steel pickling. Emissions of nitrogen dioxide may be excessive from this type of operation. In cases such as this, special scrubbing solutions may be required.

## 4. Galvanizing

Hot dip galvanizing, as distinguished from electrolytic galvanizing, is the process of applying a zinc coating to a steel shape by submerging the steel in a bath of molten zinc. A fluxing agent, which serves as a final surface preparation and consists of ammonium chloride or zinc ammonium chloride is normally floated on the surface of the zinc bath. The presence of flux is the cause of the release of air contaminants in galvanizing, since emissions are released each time the flux is disturbed. Emissions consist chemically of about 70% ammonium chloride, 15% zinc oxide, and lesser amounts of zinc, ammonia, oil, and water. The particle size of these emissions is generally in the range of 0.1 to 1.0 $\mu$m. Exhaust air for a well-controlled source can contain from 20 to 200 mg/Nm$^3$.

Control of galvanizing emissions requires the installation of local hoods where possible or canopy hoods located above the zinc kettle when freedom of movement to and from the kettle prevents close hooding. Electrostatic precipitators, wet scrubbers, and fabric filters have been used for control devices. Precipitators are normally uneconomical for small galvanizing operations. Wet scrubbers are more common but require pressure drops in the vicinity of 110 mm Hg to achieve acceptable performance on such small particles. Fabric filters are widely used, but care must be taken to consider the deliquescent characteristics of galvanizing fumes which can cause the fabric to blind. Precautions which have been taken include provisions for heating the gas ahead of the baghouse and injection of a dry additive to the gas stream for the purpose of absorbing moisture picked up by the ammonium chloride.

## IX. Ferrous Foundry Operations

### A. General Process Description

Ferrous foundries produce castings of gray iron, malleable iron, ductile iron, and/or steel. Scrap metal is melted with the proper alloys to attain the desired metallurgy, and the molten metal is cast directly, usually into sand molds. When the casting is solidified, it is stripped from the mold, cleaned, heat treated, and finished. The mold sand is reclaimed, conditioned, and reformed into cores and molds.

Molten steel used to produce castings in steel foundries is produced in electric melting furnaces or open hearth furnaces. The processes in foundries and their associated pollution problems and controls are smaller in scale but similar to those already discussed in Sections V and VII. Gray iron, as well as ductile and malleable irons, can also be produced in electric furnaces.

The predominant means of melting iron, however, is the cupola (Fig. 12), a refractory-lined cylindrical vessel equipped with combustion air tuyeres at the base and a charging door or opening near the top, through which coke, limestone, and scrap metal are charged. A charge of coke is made initially, and the cupola is heated by burning some of the coke. The

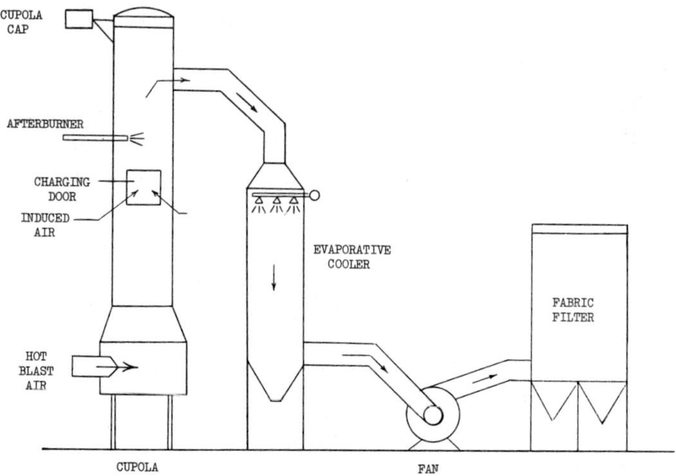


Figure 12. Hot blast cupola and associated gas-cleaning system (typical).

cupola is then charged with scrap metal, limestone, and more coke to a level just below the charging door. The heat from the burning coke melts the iron. As the charge melts and descends through the shaft, additional charge materials are added at the top, and molten iron and slag are extracted from tapholes near the base. The combustion air may be heated to improve fuel economy and hasten melting. Waste gases are emitted from the top of the cupola above the charging doors. A cupola cap is installed on furnaces with control systems. The tapped molten iron is poured into casting molds.

The cooled castings are separated from the sand molds either manually or on vibrating screens or shake out stations. The sand is reclaimed and reconditioned by screening, mixing, moisture control, and binder additions and is made into new molds. Cores are specially formulated from sand and organic or inorganic binders and are cured in core ovens. Many different coremaking processes are used (*56*).

### *B. Emissions and Their Control*

Potential emissions from ferrous foundry operations are primarily associated with (a) the melting facility, (b) the shake out and sand conditioning system, and (c) coremaking. Discussion of melting emissions and controls is limited to cupola operations.

#### 1. Cupolas

Particulate emissions emitted from a cupola consist of metallic oxides, coke ash, and volatilized materials driven from the scrap. Emissions may range from 5 to 22 kg/ton of iron with emissions tending to be higher for hot blast cupolas. Blast volume per unit of furnace area, coke/scrap ratio, melting rate, and scrap quality are among the factors which cause dust generation to vary. Typical chemical compositions of cupola dust are given in Table XII. Particle-size distributions reported are shown in Table XIII.

Gaseous emissions from cupolas may include carbon monoxide and sulfur dioxide. Carbon monoxide is created by the reaction of carbon in the coke with oxygen in the blast air. Sufficient air is induced through the charging door as the gases leave the furnace to burn the carbon monoxide to carbon dioxide. An afterburner may be installed to assure combustion. Sulfur dioxide is evolved from the sulfur in the coke and in the scrap, and may range from 25 to 250 ppm.

Control of cupola emissions has been primarily confined to the use of fabric filters and wet scrubbers, although several installations have in-

**Table XII Composition of Cupola Particulate Emissions**

| *Component* | *Range of reported values (wt %)* |
|---|---|
| $Fe_2O_3$, $FeO$, Fe | 5–26 |
| $SiO_2$ | 10–45 |
| $Al_2O_3$ | 1–25 |
| CaO | 2–18 |
| MgO | 1–5 |
| MnO | 1–9 |
| Loss on ignition | 10–64 |

cluded electrostatic precipitators. Fabric filters can attain very high efficiencies, but it is necessary to assure proper combustion of carbon monoxide, followed by gas cooling to assure protection of the baghouse from high gas temperatures. Cooling can be accomplished by radiant-convective or evaporative methods. Condensation and corrosion problems can occur with improper cooling. To maintain temperatures above the acid dewpoint and to minimize cooling requirements, high temperature fiberglass filters are normally used.

Wet scrubbing is common, but designs vary widely depending upon the desired gas-cleaning efficiency, which is a function of pressure drop. As with other metallurgical dusts, pressure drops in excess of 75 mm Hg are normally required to achieve outlet dust concentrations of 70 mg/Nm$^3$ or less. Scrubbers have the advantage of requiring no precooling of gases,

**Table XIII Particle-Size Distribution of Cupola Emissions**

| *Weight percent less than μm* | *Range of reported values* | |
|---|---|---|
| | *Hot blast cupolas* | *Cold blast cupolas* |
| 200 | 80–100 | 60–90 |
| 100 | 45–100 | 30–80 |
| 50 | 25–100 | 15–60 |
| 20 | 10–95 | 5–55 |
| 10 | 10–90 | 5–50 |
| 5 | 10–80 | 5–45 |
| 1 | 10–70 | 5–40 |

but costs for water treatment requirements and corrosion protection may be excessive (*57*).

Electrostatic precipitators are not well-suited for cupola operation because of the highly variable characteristics of emissions, high dust resistivity, and the requirements for positive assurance of carbon monoxide combustion ahead of the unit. Also some cooling is required as with a fabric filter. In spite of these difficulties, precipitators have been widely applied to cupolas in Europe.

## 2. Sand Handling

Control of emissions in foundries from sand conditioning, shake out, and mold preparation areas is well developed due to the health hazard from occupational exposure to silica sand dust. Local hooding and ventilation systems are installed and dust-laden gases are directed to wet scrubbers, fabric filters, or, in some cases, mechanical collectors. Medium energy wet collectors are best suited due to the moisture in the exhausted gases and the fairly coarse dust particle sizes. However, fabric filters can attain much higher efficiencies, and the collected dust, which may contain valuable binders as well as sand, may be more easily recovered.

## 3. Coremaking

Gases and odorous compounds derived from binders in the coremaking process are emitted from core baking ovens. Thermal or catalytic oxidation has been employed to burn these gases. The reader is referred to Chapter 9, this volume, for characterization and control of these types of emissions.

## REFERENCES

1. H. E. McGannon, ed., "The Making, Shaping and Treating of Steel," 9th ed. United States Steel Corp., Pittsburgh, Pennsylvania, 1971.
2. J. W. Miller, *J. Metals* **19**, 74 (1967).
3. T. M. Barnes, A. O. Hoffman, and H. W. Lownie, "Evaluation of Process Alternatives to Improve Control of Air Pollution from Production of Coke," Doc. No. PB 189–266. Nat. Tech. Inform. Serv., U.S. Dept. of Commerce, Springfield, Virginia, 1970.
4. "Air Pollution by Coking Plants," UN Publ. ST/ECE/COAL/26. United Nations, New York, New York, 1968.
5. T. M. Barnes, H. W. Lownie, and J. Varga, "Summary Report on Control of Coke-Oven Emissions to the American Iron and Steel Institute." Battelle Columbus Laboratories, Columbus, Ohio, 1973.

6. W. D. Edgar, *Iron Steel Eng.* **49,** 86 (1972).
7. J. G. Munson, R. E. Lewis, G. T. Weber, and W. E. Brayton, *Air. Pollut. Contr. Ass., J.* **24,** 1059 (1974).
8. G. E. Balch, *Air Pollut. Contr. Ass., J.* **22,** 187 (1972).
9. G. E. Balch, "Experience with Pipeline Charging of Coke Ovens," Paper No. 74–180, Presented at Annual Meeting of the Air Pollution Control Association, Denver, Colorado, June 9–13, 1974. Air Pollut. Contr. Ass., Pittsburgh, Pennsylvania, 1974.
10. R. S. Patton, "Hooded Coke Quenching System for Air Quality Control," Paper No. 74-182, Presented at Annual Meeting of the Air Pollution Control Association, Denver, Colorado, June 9–13, 1974. Air Pollut. Contr. Ass., Pittsburgh, Pennsylvania, 1974.
11. E. H. Roe and J. D. Patton, "Coke Oven Pushing Emission Control System," Paper No. 74-183, Presented at Annual Meeting of the Air Pollution Control Association, Denver, Colorado, June 9–13, 1974. Air Pollut. Contr. Ass., Pittsburgh, Pennsylvania, 1974.
12. R. Kemmetmueller, *Iron Steel Eng.* **50,** 71 (1973).
13. R. W. Dunlap, W. L. Gorr, and M. J. Massey, "The Desulfurization of Coke Oven Gas: Technology, Economics and Regulatory Activity," Presented at the C. C. Furnas Memorial Conference, November 8–9, 1971. State University of New York at Buffalo, Buffalo, New York, 1971.
14. M. L. Massey and R. W. Dunlap, "Economics and Alternatives of Sulfur Removal from Coke Oven Gas," Paper No. 74-184, Presented at Annual Meeting of the Air Pollution Control Association, Denver, Colorado, June 9–13, 1974. Air Pollut. Contr. Ass., Pittsburgh, Pennsylvania, 1974.
15. J. E. Ludberg, "Removal of Hydrogen Sulfide from Coke Oven Gas by the Stretford Process," Paper No. 74-185, Presented at Annual Meeting of the Air Pollution Control Association, Denver, Colorado, June 9–13, 1974. Air Pollut. Contr. Ass., Pittsburgh, Pennsylvania, 1974.
16. S. Kambara, "Desulfurization of Coke Oven Gas," Symposium on Environmental Control in the Steel Industry, Tokyo, Japan, February 18–21, 1974. SEC/3/E/115. Int. Iron Steel Inst., Brussels, Belgium, 1974.
17. F. M. Temmel, "Desulfurization of Coke Oven Gas by the Sulfiban Process," Symposium on Environmental Control in the Steel Industry, Tokyo, Japan, February 18–21, 1974, SEC/3/E/117. Int. Iron Steel Inst., Brussels, Belgium, 1974.
18. B. Bussman, "Description and Operational Results of a Plant for Desulphurisation of Coke Oven Gas by Ammonia and Production of Sulphuric Acid with Simultaneous Destruction of the Ammonia," Symposium on Environmental Control in the Steel Industry, Tokyo, Japan, February 18–21, 1974, SEC/3/E/118. Int. Iron Steel Inst., Brussels, Belgium, 1974.
19. D. F. Ball, A. F. Bradley, and A. Grieve, "Environmental Control in Iron Ore Sintering," Paper No. 7, Presented at Minerals and the Environment Symposium, June 4–7, 1974. Inst. Mining Met., London, England, 1974.
20. G. E. Manning and F. E. Rower, *Ironmaking Proc.* **30,** 452 (1971).
21. S. Oglesby and G. B. Nichols, "A Manual of Electrostatic Precipitator Technology, Part II—Application Areas," Doc. No. PB 196-381. Nat. Tech. Inform. Serv., U.S. Dept. of Commerce, Springfield, Virginia, 1970.
22. J. Varga and H. W. Lownie, "A Systems Analysis Study of the Integrated Iron and Steel Industry," Doc. No. PB 184–577. Nat. Tech. Inform. Serv., U.S. Dept. of Commerce, Springfield, Virginia, 1969.

23. T. T. Nowak, *Ironmaking Proc.* **31,** 75 (1972).
24. B. A. Steiner and F. E. Rower, *Ironmaking Proc.* **31,** 59 (1972).
25. R. S. Suprunenko, *Metallurgiya* **9,** 539 (1964).
26. R. B. Bayr and R. J. Wachowiack, *Ironmaking Proc.* **31,** 55 (1972).
27. R. L. Duprey, "Compilation of Air Pollutant Emission Factors," Doc. No. EP 4.9.42. U.S. Govt. Printing Office, Washington, D.C., 1968.
28. T. A. Young, *Blast Furn. Steel Plant* **56,** 1057 (1968).
29. J. M. Uys and J. W. Kirkpatrick, "The Beneficiation of Raw Materials in the Steel Industry and its Effect upon Air Pollution Control," Paper No. 62-47, Presented at Annual Meeting of the Air Pollution Control Association, Chicago, Illinois, May 20–24, 1962. Air Pollut. Contr. Ass., Pittsburgh, Pennsylvania, 1962.
30. N. E. Hipp and J. R. Westerholm, *Iron Steel. Eng.* **44,** 101 (1967).
31. O. Bohus, "Blast Furnace Gas Cleaning by Means of Venturi Washers," Presented at the Seminar on Air and Water Pollution Arising in the Iron and Steel Industry, Leningrad, U.S.S.R., August 23–28, 1971. Economic Commission for Europe, Geneva, Switzerland, 1971.
32. R. S. Kaplan and G. W. P. Rengstorff, "Emission of Sulfurous Gases from Blast-Furnace Slags," Paper No. 71-174, Presented at Annual Meeting of the Air Pollution Control Association, Atlantic City, New Jersey, June 27–July 1, 1971. Air Pollut. Contr. Ass., Pittsburgh, Pennsylvania, 1971.
33. N. R. Iammartino, *Chem. Eng.* **81,** 56 (1974).
34. C. A. Bishop, W. W. Campbell, D. L. Hunter, and M. W. Lightner, *Air Pollut. Contr. Ass., J.* **11,** 83 (1961).
35. A. C. Elliott and A. J. Lafreniere, "The Collection of Metallurgical Fumes from an Oxygen Lanced Open Hearth Furnace," Paper No. 64-63, Presented at Annual Meeting of the Air Pollution Control Association, Houston, Texas, June 21–25, 1964. Air Pollut. Contr. Ass., Pittsburgh, Pennsylvania, 1964.
36. J. E. Johnson, *Iron Steel Eng.* **44,** 96 (1967).
37. H. N. Hubbard and W. T. Lankford, *Iron Steel Eng.* **50,** 37 (1973).
38. K. Tagiri, "Performance of Gas Cleaning Systems-OG Equipment," Symposium on Environmental Control in the Steel Industry, Tokyo, Japan, February 18–21, 1974, SEC/5/E/124. Int. Iron Steel Inst., Brussels, Belgium, 1974.
39. A. Maubon, *Iron Steel. Eng.* **50,** 87 (1973).
40. L. A. Thaxton, F. A. Lindahl, and K. J. Aken, "Kish and Fume Control and Collection-Basic Oxygen Plant," Paper No. 69-214, Presented at Annual Meeting of the Air Pollution Control Association, New York, New York, June 22–26, 1969. Air Pollut. Contr. Ass., Pittsburgh, Pennsylvania, 1969.
41. E. G. Apukhtina, F. E. Dubinskaya, and V. N. Uzhov, "Dust-Extraction from Steelworks Fumes," Presented at the Seminar on Air and Water Pollution Arising in the Iron and Steel Industry, Leningrad, U.S.S.R., August 23–28, 1971. Economic Commission for Europe, Geneva, Switzerland, 1971.
42. D. H. Wheeler, "Fume Control in L-D Plants," Paper No. 67-94, Presented at Annual Meeting of the Air Pollution Control Association, Cleveland, Ohio, June 11–16, 1967. Air Pollut. Contr. Ass., Pittsburgh, Pennsylvania, 1967.
43. "Argon in Steelmaking," *33 Mag.* **7,** 72 (1969).
44. J. H. Flux, *Iron Steel Int.* **47,** 185 (1974).
45. R. J. Wright, *Air Pollut. Contr. Ass., J.* **18,** 175 (1968).
46. J. L. Venturini, *Air Pollut. Contr. Ass., J.* **20,** 808 (1970).
47. M. S. Wilcox and R. T. Lewis, *Iron Steel Eng.* **45,** 113 (1968).
48. J. R. Brough and W. A. Carter, *Air Pollut. Contr. Ass., J.* **22,** 167 (1972).

49. K. A. Brown, "USS Duquesne Works Electric Furnace Shop Gas Cleaning System," Symposium on Environmental Control in the Steel Industry, Tokyo, Japan, February 18–21, 1974, SEC/5/E/126. Int. Iron Steel Inst., Brussels, Belgium, 1974.
50. L. T. Kaercher and J. D. Sensenbaugh, *Iron Steel Eng.* **51**, 47 (1974).
51. J. H. Flux, "The Containment of Melting Shop Roof Emissions in Electric Arc Furnace Practice," Symposium on Environmental Control in the Steel Industry, Tokyo, Japan, February 18–21, 1974, SEC/5/E/125. Int. Iron Steel Inst., Brussels, Belgium, 1974.
52. Y. Yamashita, "A Counter Measure for Dusts at the Electric Furnace Plants," Symposium on Environmental Control in the Steel Industry, Tokyo, Japan, February 18–21, 1974, SEC/5/E/128. Int. Iron Steel Inst., Brussels, Belgium, 1974.
53. S. Iida, *Southeast Asian Iron Steel Inst. Quart.* p. 17–22 (1973).
54. A. C. Elliott and A. J. Lafreniere, "The Design and Operation of a Wet Electrostatic Precipitator to Control Billet Scarfing Emissions," Paper No. 71-159, Presented at Annual Meeting of the Air Pollution Control Association, Atlantic City, New Jersey, June 27–July 1, 1971. Air Pollut. Contr. Ass., Pittsburgh, Pennsylvania, 1971.
55. T. S. Miller, "The State of the Art in Controlling and Filtering Hot Dip Galvanizing Kettle Emissions," Presented at Annual Meeting of the American Hot Dip Galvanizers Association, Houston, Texas, March 15, 1973. Amer. Hot Dip Galvanizers Ass., Washington, D.C., 1973.
56. A. T. Kearney & Co., "Systems Analysis of Emissions Control in the Iron Foundry Industry," Doc. No. PB 198–348. Nat. Tech. Inform. Ser., U.S. Dept. of Commerce, Springfield, Virginia, 1971.
57. B. A. Callen, *Clean Air J.* **5**, 3 (1971).

# Subject Index

## A

## E



## F

## G

## H

## L



## M



## N

## O



## P

## Q

## R



## S

## T



## U



## V



## W

## X



## Z

# ENVIRONMENTAL SCIENCES

An Interdisciplinary Monograph Series



ARTHUR C. STERN, editor, AIR POLLUTION, Second Edition, Volumes I–III, 1968; Third Edition, Volumes I, III, 1976; Volume IV, 1977; Volumes II, V, in preparation

L. FISHBEIN, W. G. FLAMM, and H. L. FALK, CHEMICAL MUTAGENS: Environmental Effects on Biological Systems, 1970

DOUGLAS H. K. LEE and DAVID MINARD, editors, PHYSIOLOGY, ENVIRONMENT, AND MAN, 1970

KARL D. KRYTER, THE EFFECTS OF NOISE ON MAN, 1970

R. E. MUNN, BIOMETEOROLOGICAL METHODS, 1970

M. M. KEY, L. E. KERR, and M. BUNDY, PULMONARY REACTIONS TO COAL DUST: "A Review of U. S. Experience," 1971

DOUGLAS H. K. LEE, editor, METALLIC CONTAMINANTS AND HUMAN HEALTH, 1972

DOUGLAS H. K. LEE, editor, ENVIRONMENTAL FACTORS IN RESPIRATORY DISEASE, 1972

H. ELDON SUTTON and MAUREEN I. HARRIS, editors, MUTAGENIC EFFECTS OF ENVIRONMENTAL CONTAMINANTS, 1972

RAY T. OGLESBY, CLARENCE A. CARLSON, and JAMES A. MCCANN, editors, RIVER ECOLOGY AND MAN, 1972

LESTER V. CRALLEY, LEWIS T. CRALLEY, GEORGE D. CLAYTON, and JOHN A. JURGIEL, editors, INDUSTRIAL ENVIRONMENTAL HEALTH: The Worker and the Community, 1972

MOHAMMED K. YOUSEF, STEVEN M. HORVATH, and ROBERT W. BULLARD, PHYSIOLOGICAL ADAPTATIONS: Desert and Mountain, 1972

DOUGLAS H. K. LEE and PAUL KOTIN, editors, MULTIPLE FACTORS IN THE CAUSATION OF ENVIRONMENTALLY INDUCED DISEASE, 1972

Merril Eisenbud, ENVIRONMENTAL RADIOACTIVITY, Second Edition, 1973

James G. Wilson, ENVIRONMENT AND BIRTH DEFECTS, 1973

Raymond C. Loehr, AGRICULTURAL WASTE MANAGEMENT: Problems, Processes, and Approaches, 1974

Lester V. Cralley, Patrick R. Atkins, Lewis J. Cralley, and George D. Clayton, editors, INDUSTRIAL ENVIRONMENTAL HEALTH: The Worker and the Community, Second Edition, 1975

A 7
B 8
C 9
D 0
E 1
F 2
G 3
H 4
I 5
J 6